American Mathematical Society
Colloquium Publications
Volume 65

Graphs and Geometry

László Lovász

AMERICAN MATHEMATICAL SOCIETY

2010 *Mathematics Subject Classification.* Primary 05C99.

For additional information and updates on this book, visit
www.ams.org/bookpages/coll-65

Library of Congress Cataloging-in-Publication Data

Names: Lovász, László, 1948– author.
Title: Graphs and geometry / Laszlo Lovasz.
Description: Providence, Rhode Island: American Mathematical Society, [2019] | Series: American Mathematical Society colloquium publications; volume 65 | Includes bibliographical references and indexes.
Identifiers: LCCN 2019011532 | ISBN 9781470450878 (alk. paper)
Subjects: LCSH: Mathematics–Graphic methods. | Geometry–Graphic methods. | Graph theory. | Representations of graphs. | Geometry. | AMS: Combinatorics – Graph theory – None of the above, but in this section. msc
Classification: LCC QA90 .L5885 2019 | DDC 511/.5–dc23
LC record available at https://lccn.loc.gov/2019011532

∞ The paper used in this book is acid-free and falls within the guidelines established to ensure permanence and durability.
Visit the AMS home page at `https://www.ams.org/`

10 9 8 7 6 5 4 3 2 1 24 23 22 21 20 19

To Kati

Contents

Preface

Graphs are usually represented as geometric objects drawn in the plane, consisting of nodes and curves connecting them. The main message of this book is that *such a representation is not merely a way to visualize the graph, but an important mathematical tool.* It is obvious that this geometry is crucial in engineering if you want to understand rigidity of frameworks and mobility of mechanisms. But even if there is no geometry directly connected to the graph-theoretic problem, a well-chosen geometric embedding has mathematical meaning and applications in proofs and algorithms. This thought emerged in the 1970s, and I found it quite fruitful: Among its first applications, it led to a classification result in the theory of node-coverings, to the theory of orthogonal representations, and through this, to several combinatorial applications of semidefinite optimization.

I have given quite a few courses and lectures about the fast expanding theory of geometric representations of graphs. In 2014, I gave an advanced course on this topic at the Eidgenossische Technische Hochschule (ETH) in Zurich, and the interest and active participation of the audience, faculty, postdocs, and graduate students provided great inspiration. I am most grateful for the hospitality of ETH.

By that time, I had decided that it was worthwhile to merge my handouts and lecture notes about this general topic into a book. This took quite some time; partly because other duties limited the time I could spend on this project, but also because I kept recognizing common concepts that extended through several seemingly unrelated topics, and holding, I hope, the material better together (for example, randomization, non-degeneracy and duality). Many new results and new applications of the topic have also been emerging, even outside mathematics, like in statistical and quantum physics and computer science (learning theory). At some point I had to decide to round things up and publish this book.

Acknowledgments. I am grateful to a number of colleagues for sharing with me their insight and expertise on one or more topics. Tibor Jordán updated me on new developments in rigidity theory, in particular about global rigidity, and gave very useful comments on the chapter on that subject. András Patkós advised me on quantum physics, saving me from serious blunders in a field not my own. Robert Weismantel pointed out that the procedure of creating random orthogonal representations can be viewed as a generalization of the Gram–Schmidt orthogonalization. Boaz Klartag called my attention to the connection between canonical stress matrices and the Alexandrov–Fenchel Inequality, and advised me on other issues of volume and convexity. Tomasz Łuczak told me about the applications of orthogonal representations in quantum physics.

Several colleagues have read parts or all of the manuscript and made valuable comments. I must express my thanks to Lilla Tóthmérész and above all to Kati Vesztergombi, who read the manuscript at various stages of its preparation, and suggested many improvements.

CHAPTER 1

Why are Geometric Representations Interesting?

To represent a graph geometrically is a natural goal in itself, since it provides visual access to the abstract structure of the graph. But, in addition, it is an important tool in the study of various graph properties, including their algorithmic aspects. To illustrate what this means, let us describe a few examples of increasing complexity.

1.1. Edge coloring

We start with a very simple application of a well-chosen geometric image. How many colors are needed to color the edges of a complete graph so that no two edges incident with the same node get the same color? The answer depends on the parity of the number of nodes n. If n is odd, then at most $(n-1)/2$ edges can use the same color, so we need at least

$$\binom{n}{2}\Big/\frac{n-1}{2}=n$$

colors. To show that this is sufficient, we consider the nodes as the vertices of a regular polygon. An edge and all diagonals parallel to it can be colored with the same color. Rotating this set about the center of the polygon, we get n such sets of edges, which together cover all edges of the graph. So they form an edge-coloring with n colors (Figure 1.1, left).

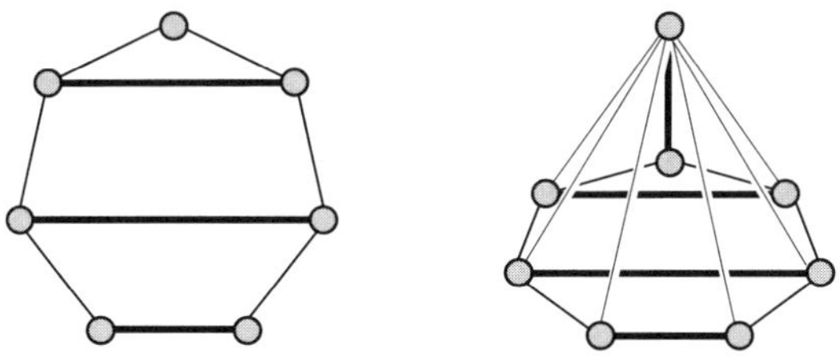

FIGURE 1.1. Optimal edge coloring of a complete graph with an odd and even number of nodes.

If n is odd, then the analogous computation gives a weaker lower bound of

$$\binom{n}{2}\Big/\frac{n}{2}=n-1$$

for the number of colors. Is this sufficient? To improve the construction by a similar use of rotations as for odd n, we step up to dimension 3. We represent the nodes as the vertices of a pyramid with an $(n-1)$-sided regular polygon as its base. For any edge e of the base, we can color the diagonals parallel to e as well as the edge connecting its apex to the base vertex opposite to e with the same color. Rotating this set again about the axis of the pyramid, we get a coloring of the edges of K_n with $n-1$ colors (Figure 1.1, right).

There are of course many other ways to solve this simple problem; but the above construction using geometry is perhaps the nicest.

1.2. Disjoint paths

Suppose that we have a 2-connected graph G and two specified nodes s and t. Two "requests" come in for two nodes x and y, and we have to find two disjoint paths connecting s and t to x and y (it does not matter which of x and y will be connected to s). This can be computed by one of zillions of flow or connectivity algorithms in reasonable time.

Now suppose that we have to compute such paths for many requests $\{x, y\}$. Do we have to repeat the computation each time? We can do much better if we use the following theorem: *Given a 2-connected graph and two specified nodes s and t, we can order all nodes so that s is first, t is last, and every other node v has a neighbor that comes earlier as well as a neighbor that comes later.* Such an ordering is called an *s-t numbering*. It is not hard to construct such a numbering (Exercise 1.2).

Once we know an s-t numbering, and a request $\{x, y\}$ comes in, it is trivial to find two disjoint paths: Let (say) x precede y in the ordering, then we can move from x to an earlier neighbor x', then to an even earlier neighbor x'' of x' etc. until we reach s. Similarly, we can move from y to a later neighbor y', then to an even later neighbor y'' of y' etc. until we reach t. This way we trace out two paths as requested.

The ordering can be thought of as representing the nodes of G by points on the line, and the easy procedure to find the two paths uses this geometric representation.

1.3. Hitting times

In this book, we will discuss some properties of random walks on a graph (as far as they are related to geometric representations). We'll also use the physical representation of a graph with the edges replaced by rubber bands. As a sampler, let us describe a very simple but nice connection.

Let a be a node of a connected graph G. We start a walk at $v^0 = a$. We select one of the neighbors v^1 of v^0 at random (every neighbor has the same probability of being selected), and we move to v^1. Then we move from v^1 to v^2 in the same way, etc. This way we get an infinite sequence of random nodes $(v^0, v^1, v^2, \dots)$, which we call a *random walk* on G.

Many important questions can be asked about a random walk (we are going to talk about some of them later). Perhaps the following is the simplest. Let b be another node of G. We define the *hitting time* $H(a, b)$ as the expected number of steps before a random walk, started at a, will reach b. (This number is known to be finite for a finite connected graph.)

There are many interesting questions you can ask about hitting times. To begin with, what are the hitting times on particular graphs like paths, cycles, or trees?

The following construction is very useful answering quite a few of these basic questions (of course, there are questions about hitting times whose answer requires a much more sophisticated approach). Consider the edges of the graph as rubber bands; these are ideal (or really high tech) rubber bands, contracting to zero length when not stretched, and not getting tangled. Attach a weight of $\deg(v)$ to each node v. Nail the node b to the wall and let the graph find its equilibrium (Figure 1.2).

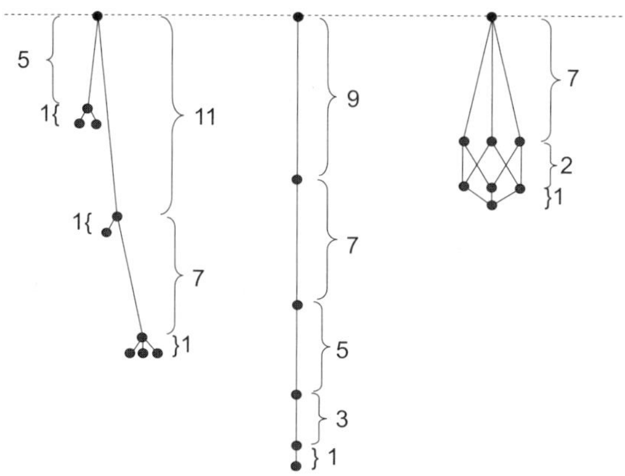

FIGURE 1.2. Hanging a tree, a path, and a 3-cube from a node. The graphs are horizontally distorted to show their structure. The stretching of the edges is easily computed using Hooke's Law.

The simple but useful fact about this construction is that *each node v will be at a distance of $H(v,b)$ below b.* This is not hard to prove noticing that the hitting times to b and the distances from b satisfy the same equations (see Exercise 1.4).

Using this geometric/physical picture, we can derive several interesting facts about hitting times. Applying the construction to a path with k nodes, we see that the k-th edge from the bottom is stretched to a length of $2k-1$ (this is the total weight of the nodes below it). Hence the total length of the path will be $1+3+\cdots+(2n-3) = (n-1)^2$; this is the hitting time from one endpoint of the path to the other. Or, using the symmetries of the skeleton of the 3-dimensional cube, we can see that the lowest edges are stretched to length 1 (3 edges carry a weight of 3), the next layer, to length 2, and the edges incident to b, to length 7. So the hitting time from a node to the opposite node is 10. Exercises 1.5–1.7 offer some further applications of this geometric construction.

1.4. Shannon capacity

The last two examples used just a single dimension; of course, one-dimensional geometry is not "really" geometry, and we better give a higher-dimensional example, which is substantially more involved. The following problem in information theory was raised by Claude Shannon, and it motivated the introduction of orthogonal representations [**Lovász 1979b**] and several of the results to be discussed in this book.

Consider a noisy channel through which we are sending messages composed of a finite alphabet V. There is an output alphabet U, and each $v \in V$, when transmitted through the channel, can come out as any element in a set $U_v \subseteq U$. Usually there is a probability distribution specified on each set U_v, telling us the probability with which v produces a given $u \in U_v$, but for the problem we want to discuss, these probabilities do not matter. As a matter of fact, the output alphabet will play no role either, except to tell us which pairs of input characters can be confused: those pairs (v, v') for which $U_v \cap U_{v'} \neq \emptyset$.

One way to model the problem is as follows: We consider V as the set of nodes of a graph, and connect two of them by an edge if they can be confused. This way we obtain a graph G, which we call the *confusion graph* of the alphabet. The maximum number of non-confusable messages of length 1 is the maximum number of nonadjacent nodes (the maximum size of a stable set) in the graph G, which we denote by $\alpha(G)$.

Now we consider longer messages, say of length k. We want to select as many of them as possible so that no two of them can possibly be confused. This means that for any two of these selected messages, there should be a position, where the two characters are not confusable. As we shall see, the number of words we can select grows as Θ^k for some $\Theta \geq 1$, which is called the *Shannon zero-error capacity* of the channel.

A simple and natural way to create such a set of words is to pick a non-confusable subset of the alphabet, and use only those words composed from this set. So if we have α non-confusable characters in our alphabet, then we can create α^k non-confusable messages of length k. But, as we shall see, making use of other characters in the alphabet we can create more! How much more, is the issue in this discussion.

Let us look at two simple examples (Figure 1.3).

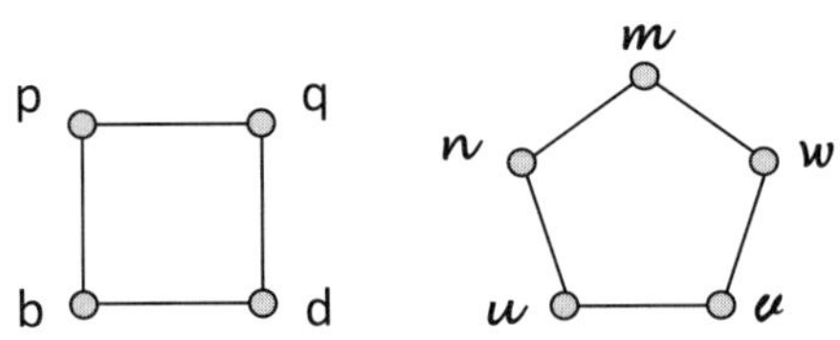

FIGURE 1.3. Two confusion graphs. In the alphabet $\{p, q, b, d\}$ two letters that are related by a reflection in a horizontal or vertical line are confusable, but not if they are related by two such reflection. The confusability graph of the alphabet $\{m, n, u, v, w\}$ is only convincing a little in handwriting, but this graph plays an important role in this book.

EXAMPLE 1.1. Consider the simple alphabet (p, q, d, b), where the pairs $\{p, q\}$, $\{q, d\}$, $\{d, b\}$ and $\{b, p\}$ are confusable (Figure 1.3, left). We can just keep p and d (which are not confusable), which allows us to create 2^k non-confusable messages of length k. On the other hand, if we use a word, then all the 2^k words obtained from it by replacing some occurrences of p and q by the other, as well as some occurrences of b and d by the other, are excluded. Hence the number of messages we can use is at most $4^k/2^k = 2^k$. ♦

EXAMPLE 1.2 (5-cycle). If we switch to alphabets with 5 characters, then we get a much more difficult problem. Let $V = \{m, n, u, v, w\}$ be our alphabet, with confusable pairs $\{m, n\}$, $\{n, u\}$, $\{u, v\}$, $\{v, w\}$ and $\{w, m\}$ (Figure 1.3, right; we refer to this example as the "pentagon"). Among any three characters there are two that can be confused, so we have only two non-confusable characters. Restricting the alphabet to two such characters (say, m and v), we get 2^k non-confusable messages of length k.

But we can do better: the following 5 messages of length two are non-confusable: mm, nu, uw, vn and wv. This takes some checking: for example, mm and nu cannot be confused, because their second characters, m and u, cannot be confused. If k is even, then we can construct $5^{k/2}$ non-confusable messages, by concatenating any $k/2$ of the above 5. This number grows like $(\sqrt{5})^k \approx 2.236^k$ instead of 2^k, a substantial gain! ♦

Can we do better by looking at longer messages (say, messages of length 10), and by some *ad hoc* method finding among them more that 5^5 non-confusable messages? We are going to show that we cannot, which means that the set of words composed of the above 5 messages of length 2 is optimal.

The trick is to represent the alphabet in a different way. Let us assign to each character $i \in V$ a vector $\mathbf{u}_i$ in some Euclidean space $\mathbb{R}^d$. If two characters are not confusable, then we represent them by orthogonal vectors.

If a subset of characters S is non-confusable, then the vectors $\mathbf{u}_i$ $(i \in S)$ are mutually orthogonal unit vectors, and hence for every unit vector $\mathbf{c}$,

$$\sum_{i\in S}(\mathbf{c}^\mathsf{T}\mathbf{u}_i)^2 \le 1.$$

Hence $|S| \min_{i\in S}(\mathbf{c}^\mathsf{T}\mathbf{u}_i)^2 \le 1$, or

$$|S| \le \max_{i\in S} \frac{1}{(\mathbf{c}^\mathsf{T}\mathbf{u}_i)^2} \le \max_{i\in V} \frac{1}{(\mathbf{c}^\mathsf{T}\mathbf{u}_i)^2}.$$

So if we find a representation $\mathbf{u}$ and a unit vector $\mathbf{c}$ for which the squared products $(\mathbf{c}^\mathsf{T}\mathbf{u}_i)^2$ are all large (which means that the angels $\measuredangle(\mathbf{c}, \mathbf{u}_i)$ are all small), then we get a good upper bound on $|S|$.

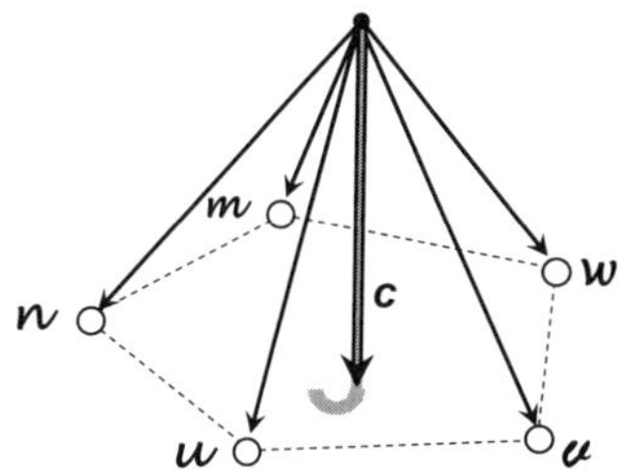

FIGURE 1.4. An umbrella representing the pentagon.

For the alphabet in Example 1.2 (the pentagon), we use the 3-dimensional vectors in Figure 1.4. To describe these, consider an umbrella with 5 ribs of unit length. Open it up to the point when nonconsecutive ribs are orthogonal. This way we get 5 unit vectors $\mathbf{u}_m, \mathbf{u}_n, \mathbf{u}_u, \mathbf{u}_v, \mathbf{u}_w$, assigned to the nodes of the pentagon so that each $\mathbf{u}_i$ forms the same angle with the "handle" $\mathbf{c}$ and any two nonadjacent nodes are labeled with orthogonal vectors. With some effort, one can compute that $(\mathbf{c}^\mathsf{T}\mathbf{u}_i)^2 = 1/\sqrt{5}$ for every i, and so we get that $|S| \le \sqrt{5}$ for every non-confusable set S. Since $|S|$ is an integer, this implies that $|S| \le 2$.

This is ridiculously much work to conclude that the 5-cycle does not contain 3 nonadjacent nodes! But the vector representation is very useful for handling longer messages. We define the *tensor product* of two vectors $\mathbf{u} = (u_1, \dots, u_n) \in \mathbb{R}^n$ and $\mathbf{v} = (v_1, \dots, v_m) \in \mathbb{R}^m$ as the vector

$$\mathbf{u} \circ \mathbf{v} = (u_1 v_1, \dots, u_1 v_m, u_2 v_1, \dots, u_2 v_m, \dots, u_n v_1, \dots, u_n v_m)^{\mathsf{T}} \in \mathbb{R}^{nm}.$$

It is easy to see that $|\mathbf{u} \circ \mathbf{v}| = |\mathbf{u}|\,|\mathbf{v}|$, and (more generally) if $\mathbf{u}, \mathbf{x} \in \mathbb{R}^n$ and $\mathbf{v}, \mathbf{y} \in \mathbb{R}^m$, then $(\mathbf{u} \circ \mathbf{v})^{\mathsf{T}}(\mathbf{x} \circ \mathbf{y}) = (\mathbf{u}^{\mathsf{T}}\mathbf{x})(\mathbf{v}^{\mathsf{T}}\mathbf{y})$. For a $k \ge 1$, if we represent a message $i_1 \dots i_k$ by the vector $\mathbf{u}_{i_1} \circ \dots \circ \mathbf{u}_{i_k}$, then non-confusable messages will be represented by orthogonal vectors. Indeed, if $i_1 \dots i_k$ and $j_1 \dots j_k$ are not confusable, then there is at least one subscript r for which i_r and j_r are not confusable, hence $\mathbf{u}_{i_r}^{\mathsf{T}}\mathbf{u}_{j_r} = 0$, which implies that

$$(\mathbf{u}_{i_1} \circ \dots \circ \mathbf{u}_{i_k})^{\mathsf{T}}(\mathbf{u}_{j_1} \circ \dots \circ \mathbf{u}_{j_k}) = (\mathbf{u}_{i_1}^{\mathsf{T}}\mathbf{u}_{j_1}) \dots (\mathbf{u}_{i_k}^{\mathsf{T}}\mathbf{u}_{j_k}) = 0.$$

Using $\mathbf{c} \circ \dots \circ \mathbf{c}$ (k factors) as the "handle", we get that for any set S of non-confusable messages of length k,

$$|S| \le \max_{i_1,\dots,i_k} \frac{1}{\big((\mathbf{c} \circ \dots \circ \mathbf{c})^{\mathsf{T}}(\mathbf{u}_{i_1} \circ \dots \circ \mathbf{u}_{i_k})\big)^2} = \max_{i_1,\dots,i_k} \frac{1}{(\mathbf{c}^{\mathsf{T}}\mathbf{u}_{i_1})^2 \dots (\mathbf{c}^{\mathsf{T}}\mathbf{u}_{i_k})^2} = (\sqrt{5})^k.$$

So every set of non-confusable messages of length k has at most $(\sqrt{5})^k$ elements. We have seen that this bound can be attained, at least for even k. Thus we have established that *the Shannon zero-error capacity of the pentagon is* $\sqrt{5}$.

We will return to this topic in Sections 11.5.1 and 12.2, where the zero-error capacity problem will be discussed for general confusability graphs, in classical and quantum information theory.

1.5. Vector-labeled graphs or frameworks?

One of the most basic objects we study in this book is a graph whose nodes are labeled by vectors from a euclidean space $\mathbb{R}^d$. We can also think of these vectors as the positions of the nodes in $\mathbb{R}^d$. A mapping $\mathbf{u} : V \to \mathbb{R}^d$ can be thought of as a "drawing", or "embedding", or "geometric representation" of the set V in a Euclidean space. Most of the time, V will be the node set of a graph, and we should think of $\mathbf{u}_i$ as the position of node i; in this case, we think of the edges as straight line segments connecting the points in the corresponding position. The main point in this book is to relate geometric and graph-theoretic properties, so this way of visualizing is often very useful. On the other hand, all three phrases above are ambiguous, and we are going to use "vector labeling" in most of the formal statements. This is the computer science view: we have a graph and store additional information for each node (see Figure 1.5). A vector-labeled graph $(G, \mathbf{u})$ will also be called a *framework*, motivated by the important topic of rigidity.

The difference between "vector-labeled graphs" and "geometric representations" is a bit more than just different usage of words. In computer science (indeed, in any area that uses and analyzes data), one considers large tables where each row corresponds to an item of some sort, and the numbers in the row represent different data (for example, age, weight, height, income of a person, or frequencies of a word in various types of documents). Often the set of rows has a network structure (say, people who know each other or words with similar meaning can be considered "adjacent" in these networks).

Surprisingly, the geometry of the row vectors often contains important information about the data, and geometric manipulation can lead to a better handle on the data, even though these data had nothing to do with geometry. While obviously

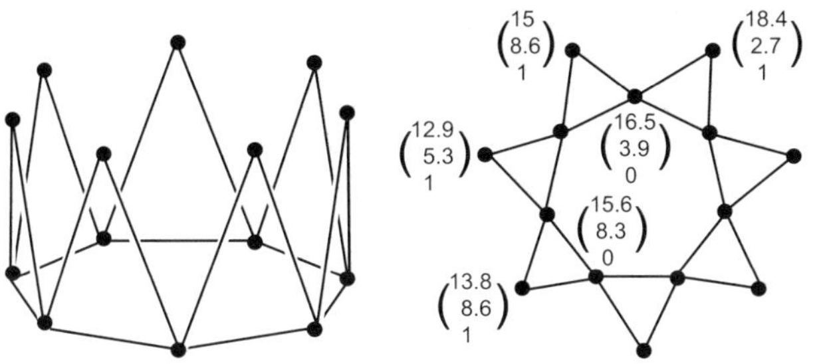

FIGURE 1.5. Two ways of looking at a graph with vector labeling (not all vector labels are shown).

closely related to our more mathematical material, we cannot treat in detail this emerging line of research.

While this is not strictly true, most of the time we can distinguish two kinds of vectors: those which are "true" vectors (best to think of them as geometric objects, like point positions or velocities), and those which are assignments of numbers, like weighting the nodes or edges of a graph. "True" vectors will be denoted by boldface characters. Mathematically, there is no difference of course.

EXERCISE 1.1. Prove that s-t numberings can be characterized as orderings of the nodes in which the nodes preceding any node induce a connected subgraph, and the same holds for the nodes following a given node.

EXERCISE 1.2. Design an algorithm that constructs an s-t numbering of a graph in polynomial time.

EXERCISE 1.3. Let us assign a real weight $f(e)$ to every edge of a transitive tournament on n nodes.
(a) For $v \in V$, let P_v be the set of all pairs (a,b) such that there is a directed path of length a leaving v along which the weights are monotone increasing, as well as a directed path of length b leaving v along which the weights are monotone non-increasing, starting with the same edge. Prove that the sets P_v are different.
(b) Prove that if $n \le \binom{p+q}{p}$, then either there is a directed path of length $p+1$ along which the weights are monotone increasing, or there is a directed path of length $q+1$ along which the weights are monotone non-increasing.[**Chvátal–Komlós 1971**]:

EXERCISE 1.4. (a) Prove that the hitting times to a given node of a connected graph G satisfy the equations

$$H(u,b) = 1 + \frac{1}{\deg(u)} \sum_{v \in N(u)} H(v,u), \qquad (u \neq b).$$

(b) Prove that the distances to b in the construction in Section 1.3 satisfy the same equations. (c) Prove that this implies that the hitting times to b are equal to the distances from b.

EXERCISE 1.5. Compute the hitting time (a) between two nodes of a cycle of length n; (b) between two nodes of the skeleton of the dodecahedron; (c) from one vertex of the d-dimensional cube to the opposite one.

EXERCISE 1.6. Let G be a tree, and let $ij \in E$. Let k be the number of nodes of $G \setminus ij$ in the component containing i. Then the hitting time from i to j is the $2k-1$.

EXERCISE 1.7. Let G be a simple connected graph, and let $ij \in E$. Then $|H(i,j) - H(j,i)| \leq n^2 - 3n + 3$, and this can be attained.

EXERCISE 1.8. Prove that for every graph with at most 5 nodes, with the exception of the pentagon, its Shannon capacity equals to the maximum cardinality of a stable set (mutually nonadjacent nodes) of the graph.

CHAPTER 2

Planar Graphs

In the first half of this book, we consider mostly planar graphs and their geometric representations, mostly in the plane. We start with a survey of basic results on planar graphs. This chapter is of an introductory nature; we give very few proofs, and this material might go as well to the Appendix as well-known background. Still, planar graphs play a very important role, and some of the considerations here provide a good introduction into the methods in this book, so this chapter deserves to be part of the main material. We assume familiarity with basic graph theory; notation and terminology are summarized in Appendix B.

2.1. Graphs and maps

2.1.1. Planarity and duality. A multigraph G is *planar*, if it can be drawn in the plane so that its edges are Jordan curves and they are disjoint except for their endnodes. A *planar map* is a planar multigraph with a fixed embedding. We also use this phrase to denote the image of this embedding, i.e., the subset of the plane which is the union of the set of points representing the nodes and the Jordan curves representing the edges. It is trivial that every subgraph of a planar graph is planar. It is slightly less trivial that contracting an edge to a single node preserves planarity.

The complement of a map G decomposes into a finite number of arcwise connected pieces, which we call the *countries* of the planar map. (Often the countries are called "faces", but we reserve the word "face" for the faces of polyhedra.) The set of its countries will be denoted V^*, and often we use the notation $f = |V^*|$. There is one special country, which is unbounded; the other countries are bounded. The unbounded country most often does not play any special role; we can make any other country p unbounded by applying an inversion with respect to a circle contained in p.

A planar map is called a *triangulation* if every country is bounded by a triangle. Note that a triangulation may have parallel edges, but two parallel edges cannot bound a country. In every simple planar map G we can introduce new edges to turn all countries into triangles while keeping the graph simple.

If the planar map is connected (which we are going to assume most of the time), then every country is homeomorphic to an open disk, except for the unbounded country, which is homeomorphic to an open ring. Each country has a *boundary* consisting of a cyclic sequence of edges. An edge can occur twice in the boundary sequence; this happens if and only if it is a cut-edge (isthmus) of the multigraph. If the multigraph is 2-edge-connected, then no boundary sequence contains a repeated edge, and if it is 2-node-connected, then every boundary is a (simple) cycle. The country also defines a cyclic sequence of nodes; a node may occur many times in this sequence. Each occurrence of a node in this sequence is called a *corner*. In other words, a corner is defined by two edges of the boundary of the country, called the *edges of the corner*, which are consecutive in the cyclic order along the boundary, but also in the cyclic order of the edges around the node. These two edges are different except if the corner is a node of degree 1.

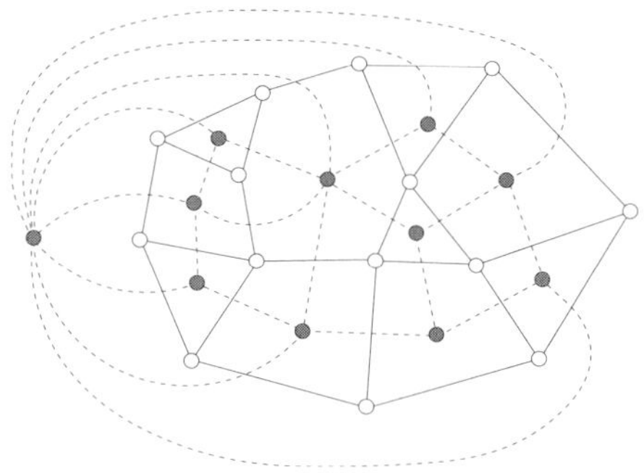

FIGURE 2.1. A planar map and its dual.

Every connected planar map G has a *dual map* $G^* = (V^*, E^*)$ (Figure 2.1). As an abstract graph, this can be defined as the multigraph whose nodes are the countries of G, and if two countries share k edges, then we connect them in G^* by k edges. So each edge $e \in E$ will correspond to a unique edge e^* of G^*, and vice versa; hence $|E^*| = |E| = m$. If the same country is incident with e from both sides, then e^* is a loop. For a country p, we denote by $\deg(p)$ its degree in the dual map. This is the number of edges on its boundary, with the addition that if the boundary passes an edge twice, then this edge is counted twice in $\deg(p)$.

This dual map has a natural drawing in the plane: in the interior of each country F of G we select a point v_F (which can be called its *capital*), and on each edge $e \in E$ we select a point u_e (this will not be a node of G^*, just an auxiliary point). We connect v_F to the points u_e for each edge on the boundary of F by nonintersecting Jordan curves inside F. If the boundary of F goes through e twice (i.e., both sides of e belong to F), then we connect v_F to u_e by two curves, entering e from two sides. The two curves entering u_e form a single Jordan curve representing the edge e^*. It is not hard to see that each country of G^* will contain a unique node of G, and so $(G^*)^* = G$. Contracting an edge that is not a loop in a planar map corresponds to deleting it from the dual, and vice versa.

Often we will need an orientation of G. Then each (oriented) edge $e \in E$ has a *tail* $\mathsf{tl}(e) \in V$, a *head* $\mathsf{hd}(e) \in V$, a *right shore* $\mathsf{rs}(e) \in V^*$, and a *left shore* $\mathsf{ls}(e) \in V^*$. There is a natural way to define an orientation of the dual map $G^* = (V^*, E^*)$, so that the dual edge $e^* \in E^*$ crosses the corresponding edge $e \in E$ from left to right. The dual of the dual of an oriented map is not the original map, but the original map with its orientation reversed.

From a planar map, we can create further planar maps that are often useful. The *lozenge map* $G^\diamondsuit = (V^\diamondsuit, E^\diamondsuit)$ of a planar map G is defined by $V^\diamondsuit = V \cup V^*$, where we connect $p \in V$ to $i \in V^*$ if p is a node of country i. So $G^\diamondsuit$ has an edge for every corner of G. We do not connect two nodes in V nor in V^*, so $G^\diamondsuit$ is a bipartite graph. It is easy to see that $G^\diamondsuit$ is planar.

The dual map of the lozenge map is also interesting. It is called the *medial map* of G, denoted by $G^{\bowtie}$, and it can be described as follows: we subdivide each edge by a new node; these will be the nodes of the medial map. For every corner of every country, we connect the two nodes on the two edges bordering this corner by a Jordan curve inside the country (Figure 2.2). The medial graph has two kinds of countries: those corresponding to nodes of G and those corresponding to countries of G. Thus the countries of $G^{\bowtie}$ can be two-colored so that every edge separates countries of different colors. This of course is equivalent to saying that $G^{\diamond}$ is bipartite.

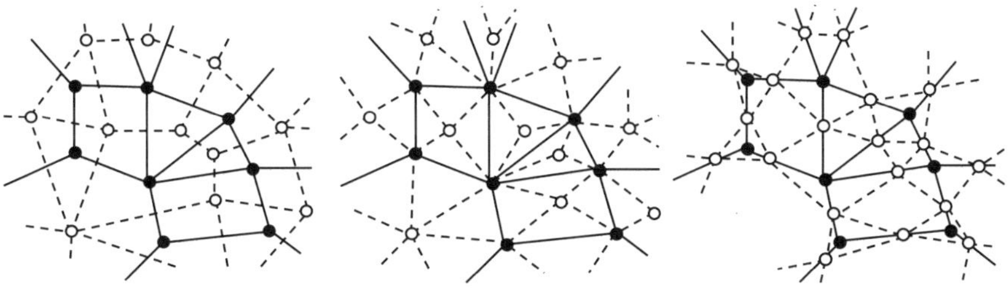

FIGURE 2.2. The dual map, the lozenge map and the medial map of a part of a planar map.

We can avoid the complications of the exceptional unbounded country if we consider maps on the sphere. Every planar map can be projected onto the sphere by inverse stereographic projection (Figure 2.3, see also Appendix A.4.2), and vice versa. Often this leads to simpler statements, since one does not need to distinguish an unbounded country. (On the other hand, it is easier to follow arguments in the plane.)

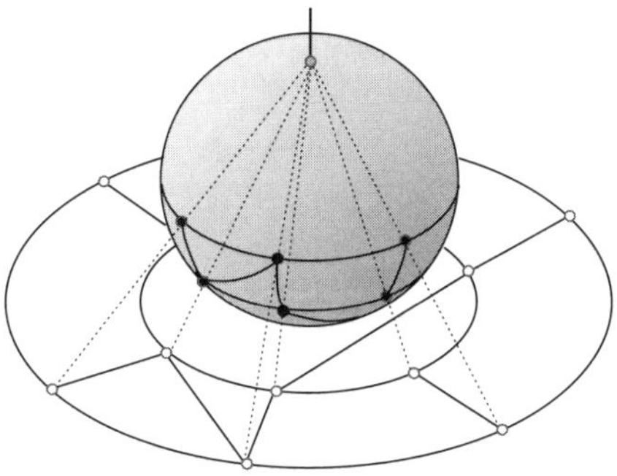

FIGURE 2.3. Projecting a planar map onto the sphere by inverse stereographic projection.

2.1.2. Euler's Formula. We often need the following basic identity for connected planar maps, called Euler's Formula (where n denotes the number of nodes, m denotes the number of edges, and f denotes the number of countries):

$$n-m+f=2. \tag{2.1}$$

This follows easily from the observation that if we take a spanning tree of a planar map, then those edges of the dual graph which correspond to edges outside this tree form a spanning tree of the dual graph (Figure 2.4). The spanning tree has $n-1$ edges, the spanning tree of the dual has $f-1$ edges, whence $(n-1)+(f-1)=m$.

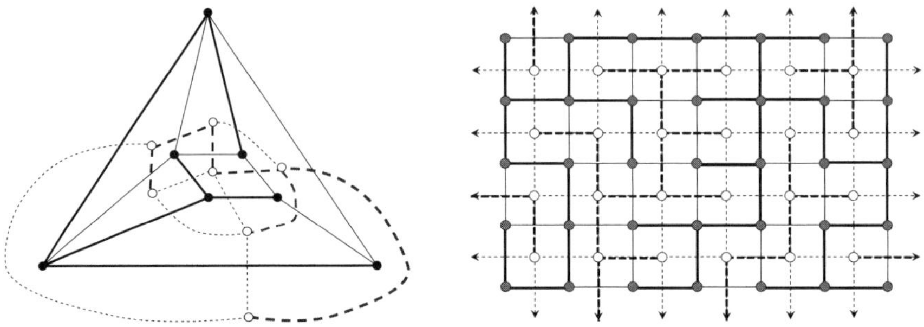

FIGURE 2.4. Spanning trees in two planar graphs, and the corresponding spanning trees in their duals. For the grid on the right, the broken edges emanating through the boundary go to the capital of the unbounded country, which is not shown.

Let us mention some important consequences of Euler's Formula. For a simple planar map, each country has at least 3 edges, and adding these up, we count every edges twice. So we have $2m \geq 3f = 3(m+2-n)$, and rearranging, we get

$$m \leq 3n-6. \tag{2.2}$$

If, in addition, the graph has no triangles, then a similar simple counting argument gives that

$$m \leq 2n-4. \tag{2.3}$$

From these inequalities it is easy to derive the following fact.

PROPOSITION 2.1. *Every simple planar graph has a node with degree at most* 5, *and every triangle-free simple planar graph has a node with degree at most* 3. □

From (2.2) and (2.3) it follows immediately that the "Kuratowski graphs" K_5 and $K_{3,3}$ (see Figure 2.5) are not planar. This observation leads to the following characterization of planar graphs.

THEOREM 2.2 (Kuratowski's Theorem). *A graph G is embeddable in the plane if and only if it does not contain a subgraph homeomorphic to the complete graph K_5 or the complete bipartite graph $K_{3,3}$.* □

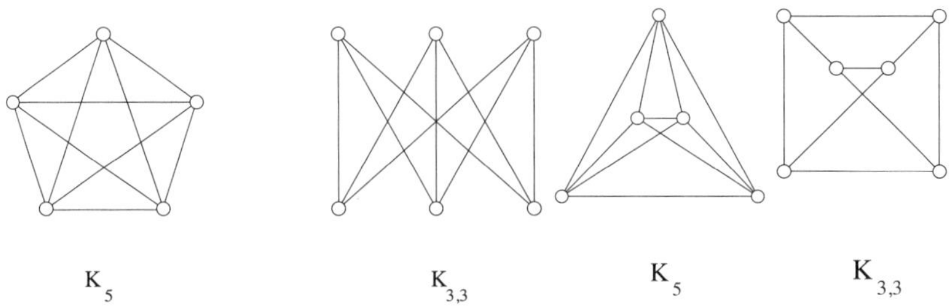

FIGURE 2.5. The two Kuratowski graphs. The two drawings on the right show that both graphs can be drawn in the plane with a single crossing.

To show a further typical application of Euler's Formula, we prove a lemma, which will be useful in several arguments later on. In a planar map whose edges are 2-colored, we call a node *quiet* if the edges of each color emanating from it are consecutive in the cyclic order of the embedding.

LEMMA 2.3. *In every simple planar map whose edges are 2-colored there are at least two quiet nodes.*

PROOF. We may assume that the graph is 3-connected: we can add edges to make it 3-connected, and color them arbitrarily; a quiet node in the extended graph is quiet in the original graph as well. We call a corner *bicolored*, if its two edges are differently colored. We are going to estimate the number N of bicolored corners and the number h of quiet nodes in two different ways.

If a node v is not quiet, then the color of edges incident with it changes at least four times in their cyclic order according to the embedding. (In particular, every such node must have degree at least four.) This means that v is incident with at least 4 bicolored corners, nd hence

$$N \geq 4(n-h). \tag{2.4}$$

On the other hand, let f_r be the number of countries with r edges on their boundary. The maximum number of bicolored corners of a country with r edges is

$$\begin{cases} r-1, & \text{if } r \text{ is odd,} \\ r, & \text{if } r \text{ is even.} \end{cases}$$

Thus the number of bicolored corners is at most

$$N \leq 2f_3+4f_4+4f_5+6f_6+6f_7+\dots \tag{2.5}$$

Using the trivial identities

$$f = f_3+f_4+f_5+\dots \qquad \text{and} \qquad 2m = 3f_3+4f_4+5f_5+\dots$$

we get by Euler's formula

$$\begin{aligned} N &\leq 2f_3+4f_4+4f_5+6f_6+6f_7+\dots \leq 2f_3+4f_4+6f_5+8f_6+\dots \\ &= 4m-4f = 4n-8. \end{aligned}$$

By (2.4), it follows that $h \geq 2$. □

2.1.3. 3-connected planar graphs.

Among planar graphs, 3-connected planar graphs are especially important. We start with a simple but useful fact. A cycle C in a graph G is called *separating*, if either it has a chord, or $G\setminus V(C)$ has at least two connected components. (We can think of considering every chord of C as a connected component of the point set $G\setminus C$, where G and C are considered as subsets of the plane.)

PROPOSITION 2.4. *In a 3-connected planar graph a cycle bounds a country if and only if it is not separating.* □

Since to be separating is an intrinsic property of a cycle in the graph (i.e., it does not depend on the way the graph is embedded), this proposition implies:

COROLLARY 2.5. *Every simple 3-connected planar graph has an essentially unique embedding in the plane in the sense that the set of cycles that bound countries is uniquely determined.* □

Once we have identified these cycles, gluing a topological disk on each of them, we get a (topological) sphere with the graph embedded in it.

The following related characterization of 3-connected planar graphs was proved in [**Tutte 1963**].

THEOREM 2.6. *Let G be a 3-connected graph. Then every edge of G is contained in at least two nonseparating cycles. G is planar if and only if every edge is contained in exactly two nonseparating cycles.* □

2.1.4. Planar separation. The following important theorem is the basis of many recursive algorithms for planar graphs [**Lipton–Tarjan 1979**].

THEOREM 2.7 (Planar Separator Theorem). *Every planar graph G on n nodes contains a set $S \subseteq V$ such that $|S| \leq \sqrt{8n}$, and every connected component of $G \setminus S$ has at most $2n/3$ nodes.* □

One can play with the two constants in the bounds on the size of S and the size of the connected components of $G \setminus S$, and many variations and improvements have been found, which we don't discuss here. We will describe a proof based on circle packing, with a weaker bound of $3n/4$ on the sizes of the components, but stronger bound $2\sqrt{n}$ on the size of any component of $G \setminus S$, in Section 5.4. The simple example of a $k \times k$ grid shows that (up to the constants) the theorem is best possible (Exercise 2.10).

2.2. Straight line representation

Planarity is defined in terms of embedding the graph in the plane in the topological sense. How "nice" can we make this drawing? We'll talk a lot about this question; but as a warm-up, let us answer the first question that comes to mind: do we need curves to represent the edges in the planar embedding of a graph, or can we get by using straight line segments only? It is clear that to make two parallel edges disjoint (up to their endpoints), at least one of them must bend. The Wagner–Fáry Theorem [**Wagner 1936, Fáry 1948**] shows that this is the only situation that forces bending.

THEOREM 2.8. *Every simple planar map can be drawn in the plane with straight edges.* □

Figure 2.6 illustrates the theorem. Since this theorem will follow from several of our later constructions, we do not give the simple induction proof here (which is described in many textbooks and internet sites).

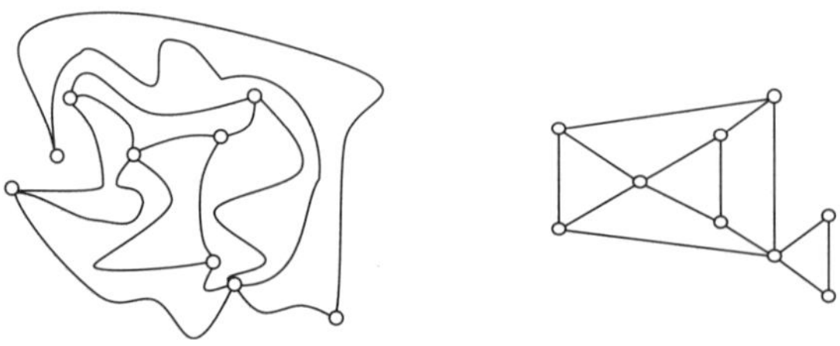

FIGURE 2.6. Drawing a simple planar map with curvy and straight edges.

2.2.1. Skeletons of polytopes. Let P be a 3-polytope (see Section C.1 for basic terms and facts about polytopes).

PROPOSITION 2.9. *The skeleton of every 3-polytope is a 3-connected planar graph.*

We describe the simple proof, because this is our first example of how a geometric representation implies a purely graph-theoretic property, namely 3-connectivity.

PROOF. Planarity of G_P can be proved by constructing an embedding called the *Schlegel diagram* of the polytope. Let F be any facet of P, and let x be a point that is outside P but very close to an interior point of F; more precisely, assume that the plane Σ of F separates x from P, but for every other facet F', x is on the same side of the plane of F' as P. Let us project the skeleton of P from x to the plane Σ. Then we get an embedding of G_P in the plane (Figure 2.7).

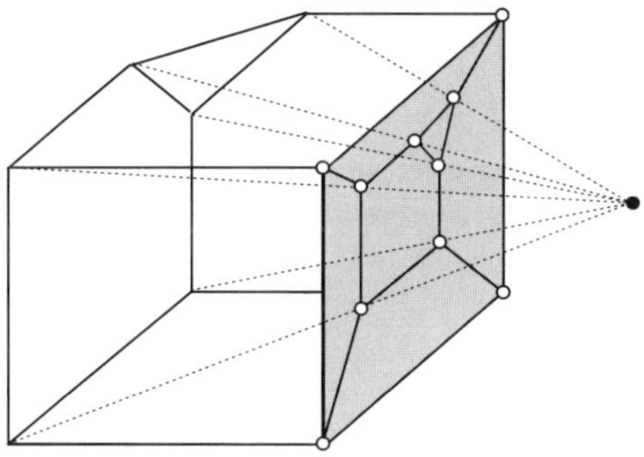

FIGURE 2.7. Projecting the skeleton of a polytope into one of the facets.

To see that G_P is 3-connected, it suffices to show that for any four distinct nodes a, b, c, d there is a path from a to b which avoids c and d.

If a, b, c, d are not coplanar, then let Π be a plane that separates $\{a, b\}$ from $\{c, d\}$. We can connect a and b by a polygon consisting of edges of P that stays on the same side of Π as a and b, and so avoids c and d.

If a, b, c, d are coplanar, let Π be a plane that contains them. One of the open halfspaces bounded by Π contains at least one vertex of P. We can then connect a and b by a polygon consisting of edges of P that stays on this side of Π (except for its endpoints a and b), and so avoids c and d. $\square$

The converse of this last proposition is an important and much more difficult theorem [**Steinitz 1922**]:

THEOREM 2.10 (Steinitz's Theorem). *A simple graph is isomorphic with the skeleton of a 3-polytope if and only if it is 3-connected and planar.*

Figure 2.8 illustrates the theorem. A bijection between the nodes of a simple graph G and the vertices of a convex polytope P in $\mathbb{R}^3$ that gives an isomorphism between G and the skeleton of P is called a *Steinitz representation* of the graph G. We do not prove Steinitz's Theorem right now, but constructions of representations by polytopes (in fact with special properties) will follow from the material in sections 3.2, 5.2.2 and 16.2.

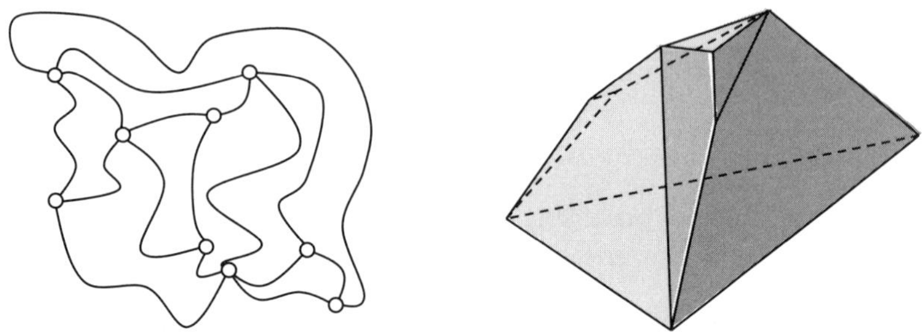

FIGURE 2.8. Representing a 3-connected planar map by a polytope.

There may be many representations of a graph by a 3-polytope, but the set of these representations, up to reflection, has nice properties (among others, it is connected in the topological sense). In higher dimensions graphs that can be represented as skeletons of polytopes are not characterized, and the set of representations of one and the same graph can be very complicated. We refer to [**Richter-Gebert 1996**] for an in-depth presentation of this problem.

The construction of the planar embedding of G_P in the proof of Proposition 2.9 gives an embedding with straight edges. Therefore Steinitz's Theorem also proves the Fáry–Wagner theorem, at least for 3-connected graphs. It is easy to see that the general case can be reduced to this by adding new edges so as to make the graph 3-connected (see Exercise 2.1).

Finally, we note that the Steinitz representation is also related to planar duality.

PROPOSITION 2.11. *Let P be a convex polytope in $\mathbb{R}^3$ with the origin in its interior, and let P^* be its polar. Then the skeletons G_P are G_{P^*} are dual planar graphs.* □

2.2.2. Voronoi and Delaunay diagrams. We define two related important constructions. Both of them can be generalized to any dimension, but we start in the plane, since this is the version leading to straight line embedded planar graphs.

Let $S \subseteq \mathbb{R}^2$ be a finite set. For every $\mathbf{s} \in S$ we define the *Voronoi cell* $V_{\mathbf{s}}$ consisting of those points $\mathbf{v} \in V$ for which $\mathbf{s} \in S$ is a point of S closest to $\mathbf{v}$. It is easy to see that Voronoi cells are convex polygons, whose interiors are disjoint, and which cover the whole plane; some cells will be unbounded (Figure 2.9, left). The collection of Voronoi cells is the *Voronoi diagram*. Each cell contains a single point of S, which we call its *capital*.

The intersection of two Voronoi cells is a convex subset of the perpendicular bisector of the segment connecting their capitals, and hence this intersection may be a semiline, a segment, a point, or the empty set. If it is a semiline or a segment, we call it an *edge* of the diagram. The intersection of three cells can be a single point or empty; if it is a single point, we call it a *vertex*. For every vertex $\mathbf{x}$, the capitals of those cells containing it are at the same distance from $\mathbf{x}$, and so they form a convex polygon, and the vertex is the center of the circumscribed circle of this polygon.

This way, the Voronoi cells define a planar graph, together with a straight line embedding in the plane, with the caveat that we need to create a "vertex at infinity", and think of each semiline edge as ending at this new vertex. We call this graph G the *Voronoi graph* defined by S.

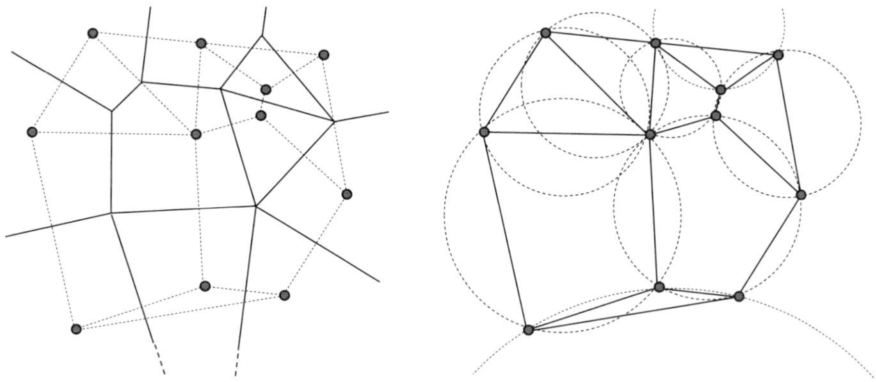

FIGURE 2.9. Left: the Voronoi and Delaunay diagrams of a set of points in the plane, as duals of each other. (Two vertices of the Voronoi graph are not shown: a vertex way down, and the "vertex at infinity". Right: the Delaunay graph and the circumscribed circles of its countries.

The dual graph G^* of the Voronoi graph G, called the *Delaunay graph* on S, is also very interesting. Its nodes are the points in the starting set S, and two of these nodes $\mathbf{u}$ and $\mathbf{v}$ are connected by an edge if the two Voronoi cells share an edge. The circle about any internal point $\mathbf{y}$ of this edge, going through the two nodes, contains no further point of S in its interior or boundary: indeed, the cell defined by such a point would contain $\mathbf{y}$ in its interior or on its boundary, so $\mathbf{y}$ could not be in the cell of $\mathbf{u}$, or there would be a third cell containing $\mathbf{y}$.

This leads to an alternative definition of the Delaunay graph: we connect two points of S by an edge if and only if there is a circle through both of them that contains not third point of S in its interior or on its boundary. In particular, every edge of the convex hull of S is subdivided by the points of S it contains, to give a path in G^*.

Circles around a vertex $\mathbf{x}$ of the Voronoi graph that go through the nearest points of S contain three or more points of S. These points form a convex polygon, which is the country of the Delaunay graph corresponding to $\mathbf{x}$ under duality. So every bounded country of the Delaunay graph has a circumscribed circle.

REMARK 2.12. We can generalize these constructions to Euclidean spaces of any dimension. For every finite set $S \subseteq \mathbb{R}^d$, and every subset $T \subseteq S$, we say that T is a *Delaunay cell* if there is a point $\mathbf{x} \in \mathbb{R}^d$ such that the set of points of S closest to $\mathbf{x}$ is exactly T. The *Voronoi cell* corresponding to this Delaunay cell is the set of all such points $\mathbf{x}$. The dimensions of a Voronoi cell and of the corresponding Delaunay cell are inversely related: they sum to $d+1$. These cells have many nice geometric properties (for example, they are convex polyhedra; see Exercise 2.13).

The definitions above can be generalized to any metric space (V, d), but we only need a simplified form in this book. We define the *Voronoi partition* induced by a finite set $S \subseteq V$ as the partition that has a partition class V_s for each $s \in S$, and every point $v \in V$ is put in a class V_s for which $s \in S$ is a point of S closest to v. Ties can be broken arbitrarily. The partition classes in the Voronoi partitions will not be so nice in general: see Figure 2.10 for the Euclidean and ℓ_1 ("Manhattan") distance in the plane. The construction is useful nevertheless.

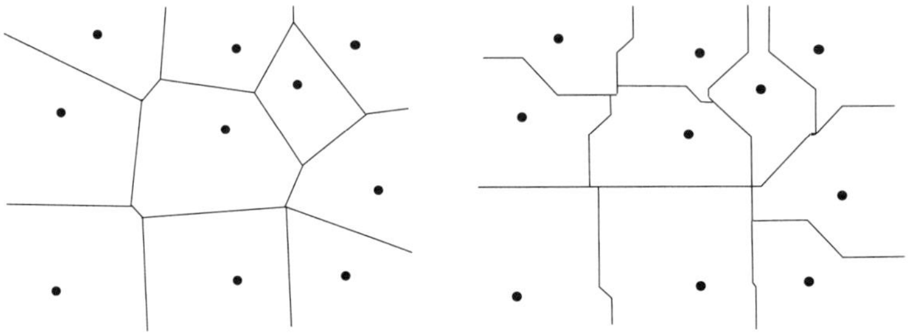

FIGURE 2.10. Voronoi cells of a finite point set in the plane in the Euclidean and in the Manhattan distance

EXERCISE 2.1. Let G be a simple planar graph. Prove that you can add edges to G so that you make it 3-connected while keeping it simple and planar.

EXERCISE 2.2. (a) Prove Euler's Formula. (b) Find a relationship between the numbers of nodes, edges and countries of a map in the projective plane and on the torus.

EXERCISE 2.3. Let G be a planar map and H, a spanning subgraph of G. Let H^* denote the spanning subgraph of the dual map G^* consisting of those edges in $E(G^*)$ that are not crossed by any edge of H. Prove that the number of components in H and H^* are related by $c(H^*) = c(H) + |E(H)| - |V(G)| + 1$.

EXERCISE 2.4. Let G be a simple 3-connected planar map. Prove that we can assign each corner either to its country or to its node so that each country is assigned to 2 corners, two nodes are not assigned to any corner, and each of the remaining nodes is assigned to 2 corners.

EXERCISE 2.5. Construct a 2-connected simple planar graph on n nodes which has an exponential (in n) number of different embeddings in the plane. (Recall: two embeddings are considered different if there is a cycle that is the boundary cycle of a country in one embedding but not in the other.)

EXERCISE 2.6. Prove that in every simple graph embedded in the projective plane whose edges are 2-colored, there is at least one quiet node. No quiet node can be guaranteed on the torus.

EXERCISE 2.7. (a) Let L be a finite set of lines in the plane, no two of which are parallel, and not all going through the same point. Prove that there is a point where exactly two lines in L go through.

(b) Let S be a finite set of points in the plane, not all on a line. Prove that there is a line which goes through exactly two points in S.

EXERCISE 2.8. Let G be a triangulation of the plane. (a) Prove that G has an edge which is contained in two triangles only. (b) Prove the Wagner–Fáry Theorem 2.8 based on (a).

EXERCISE 2.9. (a) Let L be a finite set of lines in the plane, not going through the same point. Color these lines red and blue. Prove that there is a point where at least two lines in L intersect and all the lines through this point have the same color.

(b) Let S be a finite set of points in the plane, not all on a line. Color these points red and blue. Prove that there is a line which goes through at least two points in S and all whose points have the same color.

EXERCISE 2.10. If G is a $k\times k$ grid, then every set S of nodes separating G into parts smaller that $(1-c)k^2$ has at least $\sqrt{2c}k$ nodes.

EXERCISE 2.11. The skeleton of every d-polytope is d-connected.

EXERCISE 2.12. The skeleton of every d-polytope has a K_{d+1} minor [**Grünbaum–Motzkin 1963**]).

EXERCISE 2.13. Prove that Voronoi cells are convex polyhedra in every dimension, and every face of a Voronoi cell is itself a Voronoi cell.

EXERCISE 2.14. Let S be a finite set in the x-y plane. Introduce a third coordinate z, consider the set S' of points (x, y, x^2+y^2) in 3-space, where $(x, y) \in S$, and the convex hull $P = \text{conv}(S')$. Prove that projecting those facets of P to the x-y plane whose outer normal points "down", you get exactly the bounded countries of the Delaunay graph on S.

CHAPTER 3

Rubber Bands

3.1. Energy, forces, and centers of gravity

Let us start with an informal description. Let G be a connected graph. Replace the edges by ideal rubber bands (satisfying Hooke's Law). Think of the nodes in a nonempty subset $S \subseteq V$ as nailed to given positions in d-space (thinking of the plane, $d = 2$, will be enough for a while), but let the other nodes settle in equilibrium (Figure 3.1). (We are going to see that this equilibrium position is uniquely determined.) The nodes in S will be called *nailed*, and the other nodes, *free*. We call this equilibrium position of the nodes the *rubber band representation* of G in $\mathbb{R}^d$ extending the representation of the nailed nodes. We represent the edges by straight line segments (the rubber bands don't get tangled).

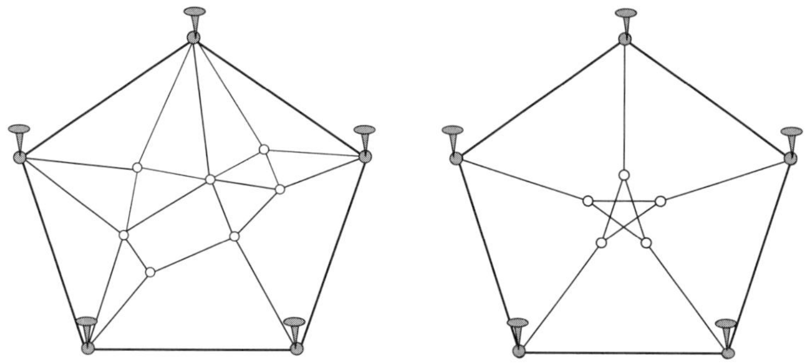

FIGURE 3.1. Rubber band representation of a planar graph and of the Petersen graph.

To make this precise, let $\mathbf{u}_i = (u_{i1}, \dots, u_{id})^\mathsf{T} \in \mathbb{R}^d$ be the position of node $i \in V$. By definition, $\mathbf{u}_i = \overline{\mathbf{u}}_i$ is prescribed for $i \in S$, but arbitrary for the remaining nodes. The *energy* of this representation is defined as

$$\mathcal{E}(\mathbf{u}) = \sum_{ij \in E} |\mathbf{u}_i - \mathbf{u}_j|^2 = \sum_{ij \in E} \sum_{k=1}^{d} (u_{ik} - u_{jk})^2. \tag{3.1}$$

We want to find the representation with minimum energy, subject to the boundary conditions:

$$\begin{aligned} &\text{minimize } \mathcal{E}(\mathbf{u}) \\ &\text{subject to } \mathbf{u}_i = \overline{\mathbf{u}}_i \text{ for all } i \in S. \end{aligned} \tag{3.2}$$

Note that while we use phrases like energy, the representation is defined in exact mathematical terms.

LEMMA 3.1. *The function* $\mathcal{E} : \mathbb{R}^{d\times(V\setminus S)} \to \mathbb{R}$ *is strictly convex.*

PROOF. In (3.1), every function $(u_{ik} - u_{jk})^2$ is convex, so $\mathcal{E}$ is convex. Furthermore, moving along an (affine) line in $\mathbb{R}^{d\times(V\setminus S)}$, this function is strictly convex unless $u_{ik} - u_{jk}$ remains constant along this line. If this applies to each coordinate of each edge, then moving along the line means parallel translation of the vectors $\mathbf{u}_i$, which is impossible if at least one node is nailed. □

It is trivial that if any of the vectors $\mathbf{u}_i$ tends to infinity, then $\mathcal{E}(\mathbf{u})$ tends to infinity (still assuming the boundary conditions 3.2 hold, where S is nonempty). Together with Lemma 3.1, this implies that the representation with minimum energy is uniquely determined. If $i \in V\setminus S$, then for the representation minimizing the energy, the partial derivative of $\mathcal{E}(\mathbf{u})$ with respect to any coordinate of $\mathbf{u}_i$ must be 0:

$$\sum_{j\in N(i)} (\mathbf{u}_i - \mathbf{u}_j) = 0 \qquad (i \in V\setminus S). \tag{3.3}$$

We can rewrite this as

$$\mathbf{u}_i = \frac{1}{\deg(i)} \sum_{j\in N(i)} \mathbf{u}_j. \tag{3.4}$$

This equation means that *every free node is placed in the center of gravity of its neighbors.* Equation (3.3) has a nice physical meaning: the rubber band connecting i and j pulls i with force $\mathbf{u}_j - \mathbf{u}_i$, so (3.3) states that the forces acting on i sum to 0 (as they should at the equilibrium). It is easy to see that (3.3) characterizes the equilibrium position.

We will return to the 1-dimensional case of rubber band representations in Chapter 4, studying harmonic functions on a graph. In those terms, equation (3.4) asserts that *every coordinate function is harmonic at every free node.*

It will be useful to extend the rubber band construction to the case when the edges of G have arbitrary positive weights (or "strengths"). Let $c_{ij} > 0$ denote the strength of the edge ij. We define the energy function of a representation $\mathbf{u}$ by

$$\mathcal{E}_c(\mathbf{u}) = \sum_{ij\in E} c_{ij} |\mathbf{u}_i - \mathbf{u}_j|^2. \tag{3.5}$$

The simple arguments above remain valid: $\mathcal{E}_c$ is strictly convex if at least one node is nailed, there is a unique optimum, and for the optimal representation every $i \in V\setminus S$ satisfies

$$\sum_{j\in N(i)} c_{ij} (\mathbf{u}_i - \mathbf{u}_j) = 0. \tag{3.6}$$

This can be rewritten as

$$\mathbf{u}_i = \frac{1}{\sum_{j\in N(i)} c_{ij}} \sum_{j\in N(i)} c_{ij} \mathbf{u}_j. \tag{3.7}$$

Thus $\mathbf{u}_i$ is no longer at the center of gravity of its neighbors, but it is still a convex combination of them with positive coefficients. In particular, it is in the relative interior of the convex hull of its neighbors.

3.2. Rubber bands, planarity and polytopes

3.2.1. How to draw a graph? The rubber band method was first analyzed in [**Tutte 1963**]. In this classical paper he describes how to use "rubber bands" to draw a 3-connected planar graph with straight edges and convex countries.

Let G be a 3-connected planar graph, and let p_0 be any country of it. Let C_0 be the cycle bounding p_0. Let us nail the nodes of C_0 to the vertices of a convex polygon P_0 in the plane, in the appropriate cyclic order, and let the rest find its equilibrium. We draw the edges of G as straight line segments connecting the appropriate endpoints. Figure 3.2 shows the rubber band representation of the skeletons of the five platonic bodies.

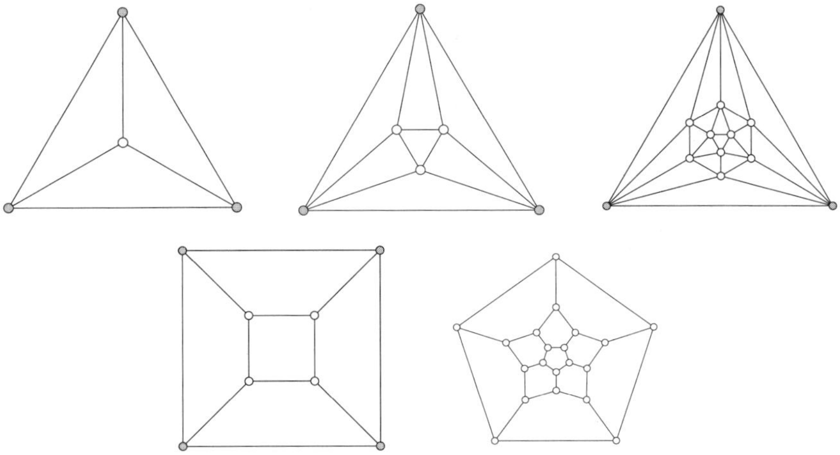

FIGURE 3.2. Rubber band representations of the skeletons of platonic bodies

By the above, we know that each node not on C_0 is positioned at the center of gravity of its neighbors. Tutte's main result about this representation is the following:

THEOREM 3.2. *If G is a simple 3-connected planar graph, then every rubber band representation of G (with the nodes of a particular country p_0 nailed to a convex polygon) gives a straight-line embedding of G in the plane. In addition, each country is a convex polygon.*

PROOF. Let $\mathbf{u} : V \to \mathbb{R}^2$ be this rubber band representation of G. Let ℓ be a line intersecting the interior of the polygon P_0, and let U_0, U_1 and U_2 denote the sets of nodes of G mapped on ℓ and on the two (open) sides of ℓ, respectively. The key to the proof is the following claim.

Claim 1. *The sets U_1 and U_2 induce connected subgraphs of G.*

Let us prove this for U_1. Clearly the nodes of the outer cycle p_0 in U_1 form a (nonempty) path P_1. We may assume that ℓ does not go through any node (by shifting it very little in the direction of U_1) and that it is not parallel to any line connecting two distinct positions (by rotating it with a small angle). Let $a \in U_1 \setminus V(C_0)$, we show that it is connected to P_1 by a path in U_1. Since $\mathbf{u}_a$ is a convex combination of the positions of its neighbors, it must have a neighbor a_1

such that $\mathbf{u}_{a_1}$ is in U_1 and at least as far away from ℓ as $\mathbf{u}_a$. By our assumption that ℓ is not parallel to any edge, either $\mathbf{u}_{a_1}$ is strictly farther from ℓ than $\mathbf{u}_a$, or $\mathbf{u}_{a_1} = \mathbf{u}_a$.

At this point, we have to deal with an annoying degeneracy. There may be several nodes represented by the same vector $\mathbf{u}_a$ (later it will be shown that this does not occur). Consider all nodes represented by $\mathbf{u}_a$, and the connected component H containing a of the subgraph of G induced by these nodes. If H contains a nailed node, then it contains a path from a to P_1, all in U_1. Else, there must be an edge connecting a node $a' \in V(H)$ to a node outside H (since G is connected). Since the system is in equilibrium, a' must have a neighbor a_1 such that $\mathbf{u}_{a_1}$ is farther away from ℓ than $\mathbf{u}_a = \mathbf{u}_{a'}$ (here we use again that no edge is parallel to ℓ). Hence $a_1 \in U_1$, and thus a is connected to a_1 by a path in U_1.

Either a_1 is nailed (and we are done), or we can find a node $a_2 \in U_1$ such that a_1 is connected to a_2 by a path in U_1, and $\mathbf{u}_{a_2}$ is farther from ℓ than $\mathbf{u}_{a_1}$, etc. This way we get a path Q in G that starts at a, stays in U_1, and eventually must hit P_1. This proves the claim (Figure 3.3).

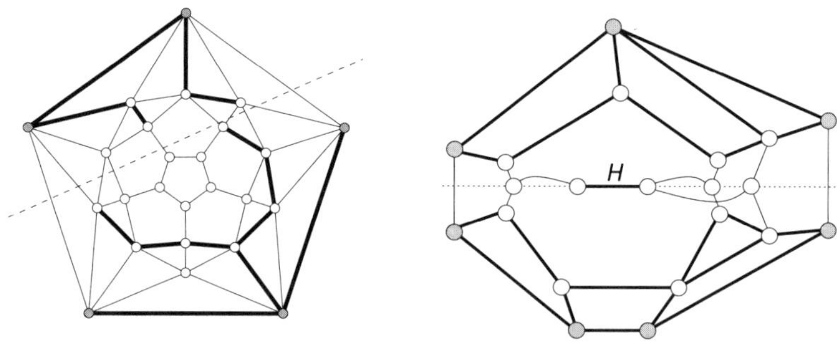

FIGURE 3.3. Left: every line cuts a rubber band representation into connected parts. Right: Each node on a line must have neighbors on both sides of the line.

Next, we exclude some possible degeneracies. (We are not assuming any more that no edge is parallel to ℓ: this assumption could be made for the proof of Claim 1 only.)

Claim 2. *Every node $u \in U_0$ has neighbors both in U_1 and U_2.*

This is trivial if $u \in V(C_0)$, so suppose that u is a free node. If u has a neighbor in U_1, then it must also have a neighbor in U_2; this follows from the fact that $\mathbf{u}_u$ is the center of gravity of the points representing its neighbors. So it suffices to prove that not all neighbors of u are contained in U_0.

Let T be the set of nodes $u \in U_0$ with $N(u) \subseteq U_0$, and suppose that this set is nonempty. Consider a connected component H of $G[T]$ (H may be a single node), and let S be the set of neighbors of H outside H. Since $V(H) \cup S \subseteq U_0$, the set $V(H) \cup S$ cannot contain all nodes, and hence S is a cutset. Thus $|S| \geq 3$ by 3-connectivity.

If $a \in S$, then $a \in U_0$ by the definition of S, but a has a neighbor not in U_0, and so it has neighbors in both U_1 and U_2 by the argument above (see Figure 3.3). The set $V(H)$ induces a connected graph by definition, and U_1 and U_2 induce connected subgraphs by Claim 1. So we can contract these sets to single nodes. These three nodes will be adjacent to all nodes in S. So we have contracted G to $K_{3,3}$, which is a contradiction, since G is planar. This proves Claim 2.

Claim 3. *Every country has at most two nodes in U_0.*

Suppose that $a, b, c \in U_0$ are nodes of a country p. Clearly $p \neq p_0$. Let us create a new node d and connect it to a, b and c; the resulting graph G' is still planar. On the other hand, the same argument as in the proof of Claim 1 (with $V(H) = d$ and $S = \{a, b, c\}$) shows that G' has a $K_{3,3}$ minor, which is a contradiction.

Claim 4. *Let p and q be the two countries sharing an edge ab, where $a, b \in U_0$. Then $V(p_1) \setminus \{a, b\} \subseteq U_1$ and $V(p_2) \setminus \{a, b\} \subseteq U_2$ (or the other way around).*

Suppose not, then p has a node $c \neq a, b$ and q has a node $d \neq a, b$ such that (say) $c, d \in U_1$. (Note that $c, d \notin U_0$ by Claim 3.) By Claim 1, there is a path P in U_1 connecting c and d (Figure 3.4). Claim 2 implies that both a and b have neighbors in U_2, and again Claim 1, these can be connected by a path in U_2. This yields a path P' connecting a and b whose inner nodes are in U_2. By their definition, P and P' are node-disjoint. But look at any planar embedding of G: the edge ab, together with the path P', forms a Jordan curve that does not go through c and d, but separates them, so P cannot exist.

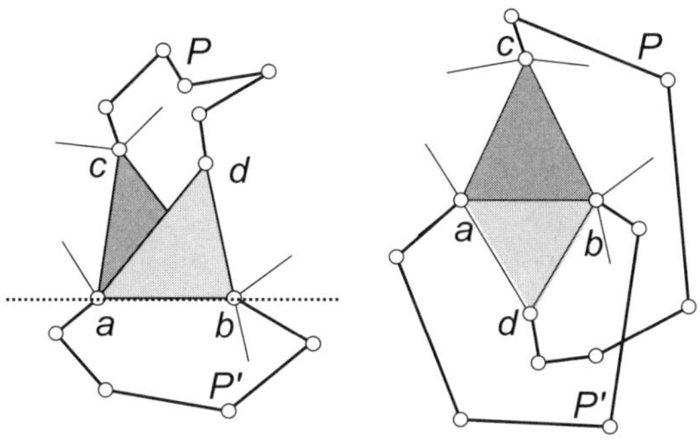

FIGURE 3.4. Two adjacent countries having nodes on the same side of ℓ in the rubber band representation (left), and the supposedly disjoint paths in the planar embedding (right).

Claim 5. *The boundary of every country q is mapped onto a convex polygon P_q.*

This is immediate from Claim 4, since no edge of a country, extended to a line, can intersect its interior.

Claim 6. *The interiors of the polygons P_q ($q \neq p_0$) are disjoint.*

Let $\mathbf{x}$ be a point inside P_{p_0}, we want to show that it is covered by one P_q only. Clearly we may assume that $\mathbf{x}$ is not on the image of any edge. Draw a line through $\mathbf{x}$ that does not go through the position of any node, and see how many times its points are covered by interiors of such polygons. As we enter P_{p_0}, this number is clearly 1. Claim 4 says that as the line crosses an edge, this number does not change. So $\mathbf{x}$ is covered exactly once.

Now the proof is essentially finished. Suppose that the images of two edges have a common point (other than their common endpoints). Then two of the countries incident with them would have a common interior point, which is a contradiction except if these countries are the same, and the two edges are consecutive edges of this country. □

Before going on, let's make a couple of remarks about this drawing method. The key step, namely Claim 1, is very similar to a basic fact concerning convex polytopes, mentioned in Appendix C.1, that every hyperplane intersecting the interior of the polytope cuts the skeleton into connected parts. Note that the proof of Claim 1 did not make use of the planarity of G (see Exercise 3.4, and also Section 16.5).

Tutte's method, as described above, is a very efficient procedure to find straight-line embeddings of 3-connected planar graphs in the plane. These embeddings look nice for many graphs (as our figures show), but they may have bad parts, like points getting too close. Figure 3.5 shows a simple situation in which positions of nodes get exponentially close to each other and to the midpoint of an edge. You may play with the edge weights, but finding a good weighting adds substantially to the algorithmic cost.

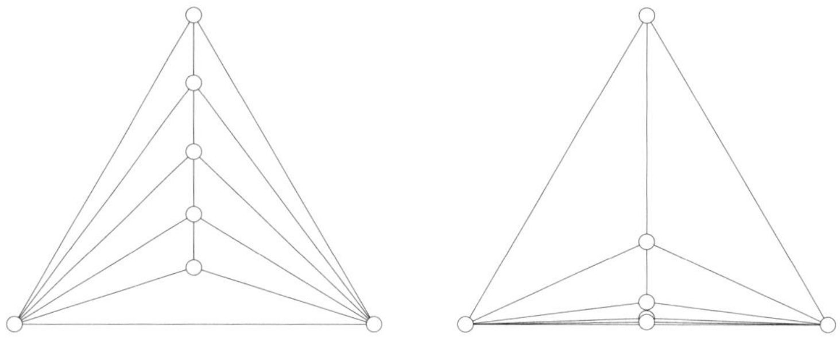

FIGURE 3.5. The rubber band representation can lead to crowding of the nodes. Each node on the middle line will be placed at or below the center of gravity of the triangle formed by the lower edge and the node immediately above it. So the distance between the k-th node from the top and the lower edge decreases faster than 3^{-k}.

A reasonable way to exclude nodes being positioned too close is to require that their coordinates are integers. Then the question is, of course, how to minimize these coordinates. In which rectangles $[0, a] \times [0, b]$ can every planar graph on n nodes be squeezed in so that we still get a straight line embedding? It turns out that this can be achieved with $a, b = O(n)$ [**Fraysseix–Pach–Pollack 1990, Schnyder 1990**], with several good bounds on the appropriate pairs (a, b) [**Chrobak–Nakano 1998**].

Representing edges by straight lines is not always the most important format of useful drawings. In some very important applications of planar embedding, most notably the design of integrated circuits (chips), the goal is to embed the graph in a grid, so that the nodes are drawn on gridpoints, and the edges are drawn as zig-zagging grid paths. (Of course, we must assume that no degree is larger than 4.) This way all nonzero distances between nodes and/or edges are automatically at least 1 (the edge-length of the grid). Besides trying to minimize the size or area of the grid in which the embedding lies, it is very natural (and practically important) to minimize the number of bends. The good news is that this can be achieved within a very reasonable area ($O(n^2)$) and with $O(n)$ bends [**Tamassia 1987**, **Tamassia–Tollis 2014**]. See the Handbook of Graph Drawing and Visualization [**Tamassia 2014**] for details.

3.2.2. How to lift a graph? An old construction of Cremona and Maxwell can be used to "lift" Tutte's rubber band representation to a Steinitz representation. We begin with analyzing the reverse procedure: projecting a convex polytope onto a face. This procedure is similar to the construction of the Schlegel diagram from Section 2.2, but we consider orthogonal projection instead of projection from a point.

Let P be a convex 3-polytope, let F be one of its faces, and let Σ be the plane containing H. Suppose that for every vertex $\mathbf{v}$ of P, its orthogonal projection onto H is an interior point of F; we say that the polytope P is *straight over the face* F.

THEOREM 3.3. (a) *Let P be a 3-polytope that is straight over its face F, and let G be the orthogonal projection of the skeleton of P onto the plane of F. Then we can assign positive strengths to the edges of G so that G will be the rubber band representation of the skeleton with the vertices of F nailed.*

(b) *Let G be a 3-connected planar graph, and let T be a triangular country of G, and let Δ be a triangle in a plane Σ. Then there is a convex polytope P in 3-space such that T is a face of P, and the orthogonal projection of P onto the plane Σ gives the rubber band representation of G obtained by nailing T to Δ.*

In other words (b) says that we can assign a number $\eta_i \in \mathbb{R}$ to each node $i \in V$ such that $\eta_i = 0$ for $i \in V(T)$, $\eta_i > 0$ for $i \in V \setminus V(T)$, and the mapping

$$i \mapsto \mathbf{v}_i = \begin{pmatrix} \mathbf{u}_i \\ \eta_i \end{pmatrix} = \begin{pmatrix} u_{i1} \\ u_{i2} \\ \eta_i \end{pmatrix} \in \mathbb{R}^3$$

is a Steinitz representation of G.

EXAMPLE 3.4 (Triangular Prism). Consider the rubber band representation of a triangular prism (or of a triangular frustum, a truncated tetrahedron) in Figure 3.6, left. If this is an orthogonal projection of a convex polyhedron, then the lines of the three edges pass through one point: the point of intersection of the planes of the three quadrangular faces. It is easy to see that this condition is necessary and sufficient for the picture to be a projection of a triangular frustum. To see that it is satisfied by a rubber band representation, it suffices to note that the inner triangle is in equilibrium, and this implies that the lines of action of the forces acting on it must pass through one point. ♦

Now we are ready to prove theorem 3.3.

PROOF. (a) Let's call the plane of the face F "horizontal", spanned by the first two coordinate axes, and the third coordinate direction "vertical", so that the polytope is "above" the plane of F. For each face p, let $\mathbf{g}_p$ be a normal vector. Since no face is vertical, no normal vector $\mathbf{g}_p$ is horizontal, and hence we can normalize $\mathbf{g}_p$ so that its third coordinate is 1. Clearly for each face p, $\mathbf{g}_p$ will be an outer normal, except for $p = F$, when $\mathbf{g}_p$ is an inner normal (Figure 3.6).

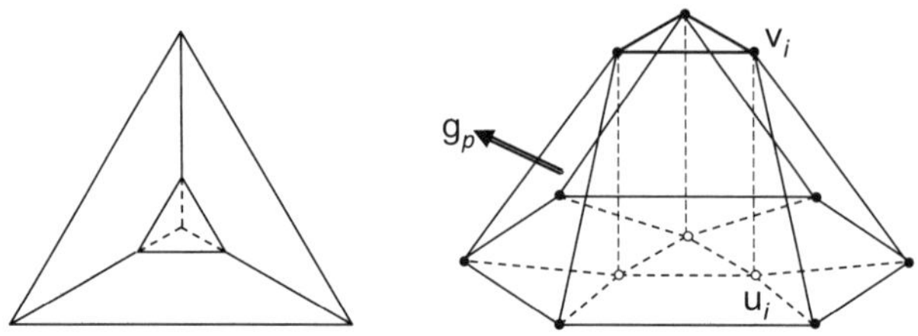

FIGURE 3.6. Left: The rubber band representation of a triangular prism is the projection of a polytope. Right: Vertical projection of a polytope into one of its faces.

Write $\mathbf{g}_p = \binom{\mathbf{h}_p}{1}$. Let ij be any edge of G, and let p and q be the two countries on the left and right of ij. Then

$$(\mathbf{h}_p - \mathbf{h}_q)^\mathsf{T}(\mathbf{u}_i - \mathbf{u}_j) = 0. \tag{3.8}$$

Indeed, both $\mathbf{g}_p$ and $\mathbf{g}_q$ are orthogonal to the edge $\mathbf{v}_i\mathbf{v}_j$ of the polytope, and therefore so is their difference, and

$$(\mathbf{h}_p - \mathbf{h}_q)^\mathsf{T}(\mathbf{u}_i - \mathbf{u}_j) = \begin{pmatrix}\mathbf{h}_p - \mathbf{h}_q\\ 0\end{pmatrix}^\mathsf{T}\begin{pmatrix}\mathbf{u}_i - \mathbf{u}_j\\ \eta_i - \eta_j\end{pmatrix} = (\mathbf{g}_p - \mathbf{g}_q)^\mathsf{T}(\mathbf{v}_i - \mathbf{v}_j) = 0.$$

We have $\mathbf{h}_T = 0$, since the face T is horizontal.

Let R denote the counterclockwise rotation in the plane by $90°$, then it follows that $\mathbf{h}_p - \mathbf{h}_q$ is parallel to $R(\mathbf{u}_j - \mathbf{u}_i)$, and so there are real numbers c_{ij} such that

$$\mathbf{h}_p - \mathbf{h}_q = c_{ij}R(\mathbf{u}_j - \mathbf{u}_i). \tag{3.9}$$

We claim that $c_{ij} > 0$. Let k be any node on the boundary of p different from i and j. Then $\mathbf{u}_k$ is to the left from the edge ij, and hence

$$(\mathbf{u}_k - \mathbf{u}_i)^\mathsf{T}R(\mathbf{u}_j - \mathbf{u}_i) > 0. \tag{3.10}$$

Going up to the space, convexity implies that the point $\mathbf{v}_k$ is below the plane of the face q, and hence $\mathbf{g}_q^\mathsf{T}\mathbf{v}_k < \mathbf{g}_q^\mathsf{T}\mathbf{v}_i$. Since $\mathbf{g}_p^\mathsf{T}\mathbf{v}_k = \mathbf{g}_p^\mathsf{T}\mathbf{v}_i$, this implies that

$$c_{ij}\big(R(\mathbf{u}_j - \mathbf{u}_i)\big)^\mathsf{T}(\mathbf{u}_k - \mathbf{u}_i) = (\mathbf{h}_p - \mathbf{h}_q)^\mathsf{T}(\mathbf{u}_k - \mathbf{u}_i) = (\mathbf{g}_p - \mathbf{g}_q)^\mathsf{T}(\mathbf{v}_k - \mathbf{v}_i) > 0. \tag{3.11}$$

Comparing with (3.10), we see that $c_{ij} > 0$.

To complete the proof of (a), we argue that the projection of the skeleton is indeed a rubber band embedding with strengths c_{ij}, with F nailed. We want to prove that every free node i is in equilibrium, i.e.,

$$\sum_{j\in N(i)} c_{ij}(\mathbf{u}_j - \mathbf{u}_i) = 0. \tag{3.12}$$

Using the definition of c_{ij}, it suffices to prove that

$$\sum_{j\in N(i)} c_{ij}R(\mathbf{u}_j - \mathbf{u}_i) = \sum_{j\in N(i)} (\mathbf{h}_{p_j} - \mathbf{h}_{q_j}) = 0,$$

where p_j is the face to the left and q_j is the face to the right of the edge ij. But this is clear, since every term occurs once with positive and once with negative sign.

Let us make a remark here that will be needed later. Using that not only $\mathbf{g}_p - \mathbf{g}_q$, but also $\mathbf{g}_p$ is orthogonal to $\mathbf{v}_j - \mathbf{v}_i$, we get that

$$0 = \mathbf{g}_p^\mathsf{T}(\mathbf{v}_j - \mathbf{v}_i) = \mathbf{h}_p^\mathsf{T}(\mathbf{u}_j - \mathbf{u}_i) + \eta_j - \eta_i,$$

and hence

$$\eta_j - \eta_i = -\mathbf{h}_p^{\mathsf{T}}(\mathbf{u}_j - \mathbf{u}_i). \tag{3.13}$$

(b) The proof consists of going through the steps of the proof of part (a) in reverse order: given the Tutte representation, we first reconstruct the vectors $\mathbf{h}_p$ so that all equations (3.8) are satisfied, then using these, we reconstruct the numbers η_i so that equations (3.13) are satisfied. It will not be hard to verify then that we get a Steinitz representation.

We need a little preparation to deal with edges on the boundary triangle. Recall that we can think of $\mathbf{F}_{ij} = c_{ij}(\mathbf{u}_j - \mathbf{u}_i)$ as the force with which the edge ij pulls its endpoint i. Equilibrium means that for every free node i,

$$\sum_{j \in N(i)} \mathbf{F}_{ij} = 0. \tag{3.14}$$

This does not hold for the nailed nodes, but we can modify the definition of $\mathbf{F}_{ij}$ along the three boundary edges so that $\mathbf{F}_{ij}$ remains parallel to the edge $\mathbf{u}_j - \mathbf{u}_i$ and (3.14) will hold for *all* nodes (this is the only point where we use that the outer country is a triangle). The only complication is that if we write $\mathbf{F}_{ij} = c_{ij}(\mathbf{u}_j - \mathbf{u}_i)$ for the boundary edges, then we have to allow negative strengths c_{ij}; this will cause not trouble in the computations. The existence of such forces $\mathbf{F}_{ij}$ is natural by a physical argument: let us replace the outer edges by rigid bars, and remove the nails. The whole framework will remain in equilibrium, so appropriate forces must act in the edges ab, bc and ac to keep balance. To translate this to mathematics, one has to work a little; this is left to the reader as Exercise 3.5.

We claim that we can choose vectors $\mathbf{h}_p$ for all countries p so that

$$\mathbf{h}_p - \mathbf{h}_q = R\mathbf{F}_{ij} \tag{3.15}$$

if ij is any edge and p and q are the two countries on its left and right. This follows by a "potential argument", which will be used several times in our book. Starting with $\mathbf{h}_T = 0$, and moving from country to adjacent country, equation 3.15 will determine the value of $\mathbf{h}_p$ for every country p. What we have to show is that we do not run into contradiction, i.e., if we get to the same country p in two different ways, then we get the same vector $\mathbf{h}_p$. This is equivalent to saying that if we walk around a closed cycle of countries, then the total change in the vector $\mathbf{h}_p$ is zero. It suffices to verify this when we move around countries incident with a single node. In this case, the condition boils down to

$$\sum_{i \in N(j)} R\mathbf{F}_{ij} = 0,$$

which follows by (3.14). This proves that the vectors $\mathbf{h}_p$ are well defined.

Second, we construct numbers η_i satisfying (3.13) by a similar argument (just working on the dual graph). We set $\eta_i = 0$ if i is a node of the unbounded country. Equation (3.13) tells us what the value at one endpoint of an edge must be, if we have it for the other endpoint.

One complication is that (3.13) gives two conditions for each difference $\eta_i - \eta_j$, depending on which country incident with it we choose. But if p and q are the two countries incident with the edge ij, then

$$\mathbf{h}_p^{\mathsf{T}}(\mathbf{u}_j - \mathbf{u}_i) - \mathbf{h}_q^{\mathsf{T}}(\mathbf{u}_j - \mathbf{u}_i) = (\mathbf{h}_p - \mathbf{h}_q)^{\mathsf{T}}(\mathbf{u}_j - \mathbf{u}_i) = (R\mathbf{F}_{ij})^{\mathsf{T}}(\mathbf{u}_j - \mathbf{u}_i) = 0,$$

since $\mathbf{F}_{ij}$ is parallel to $\mathbf{u}_i - \mathbf{u}_j$ and so $R\mathbf{F}_{ij}$ is orthogonal to it. Thus the two conditions on the difference $\eta_i - \eta_j$ are the same.

As before, equation (3.13) determines the values η_i, starting with $\eta_a = 0$. To prove that it does not lead to a contradiction, it suffices to prove that the sum of changes is 0 if we walk around a country p. In other words, if C is the cycle bounding a country p (oriented, say, clockwise), then

$$\sum_{\overrightarrow{ij} \in E(C)} \mathbf{h}_p^{\mathsf{T}}(\mathbf{u}_j - \mathbf{u}_i) = 0,$$

which is clear. It is also clear that $\eta_b = \eta_c = 0$.

Now define $\mathbf{v}_i = \binom{\mathbf{u}_i}{\eta_i}$ for every node i, and $\mathbf{g}_p = \binom{\mathbf{h}_p}{1}$ for every country p. It remains to prove that $i \mapsto \mathbf{v}_i$ maps the nodes of G onto the vertices of a convex polytope, so that edges go to edges and countries go to facets. We start with observing that if p is a country and ij is an edge of p, then

$$\mathbf{g}_p^{\mathsf{T}}\mathbf{v}_i - \mathbf{g}_p^{\mathsf{T}}\mathbf{v}_j = \mathbf{h}_p^{\mathsf{T}}(\mathbf{u}_i - \mathbf{u}_j) + (\eta_i - \eta_j) = 0,$$

and hence there is a scalar α_p so that all nodes of p are mapped onto the hyperplane $\mathbf{g}_p^{\mathsf{T}}\mathbf{x} = \alpha_p$. We know that the image of p under $i \mapsto \mathbf{u}_i$ is a convex polygon, and so the same follows for the map $i \mapsto \mathbf{v}_i$.

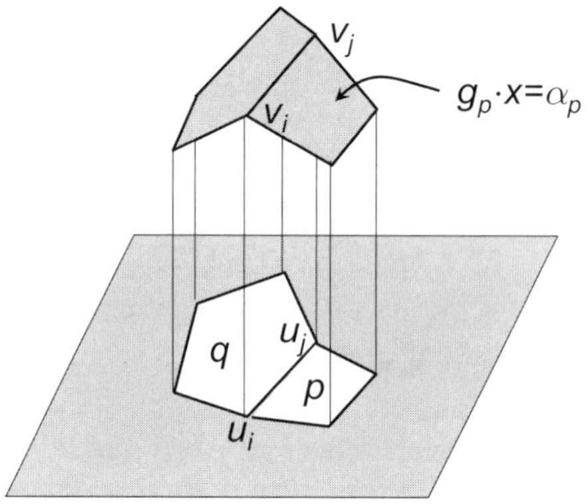

FIGURE 3.7. Lifting a rubber band representation to a polytope.

To conclude, it suffices to prove that if ij is any edge, then the two convex polygons obtained as images of countries incident with ij "bend" in the right way; more exactly, let p and q be the two countries incident with ij, and let Q_p and Q_q be two corresponding convex polygons (see Figure 3.7). We claim that Q_p lies on the same side of the plane $\mathbf{g}_p^{\mathsf{T}}\mathbf{x} = \alpha_p$ as the bottom face. Let $\mathbf{v}_k$ be any vertex of the polygon Q_q different from $\mathbf{v}_i$ and $\mathbf{v}_j$. We want to show that $\mathbf{g}_p^{\mathsf{T}}\mathbf{v}_k < \alpha_p$. Indeed,

$$\mathbf{g}_p^{\mathsf{T}}\mathbf{v}_k - \alpha_p = \mathbf{g}_p^{\mathsf{T}}\mathbf{v}_k - \mathbf{g}_p^{\mathsf{T}}\mathbf{v}_i = \mathbf{g}_p^{\mathsf{T}}(\mathbf{v}_k - \mathbf{v}_i) = (\mathbf{g}_p - \mathbf{g}_q)^{\mathsf{T}}(\mathbf{v}_k - \mathbf{v}_i)$$

(since both $\mathbf{v}_k$ and $\mathbf{v}_i$ lie on the plane $\mathbf{g}_q^{\mathsf{T}}\mathbf{x} = \alpha_q$),

$$= \begin{pmatrix} \mathbf{h}_p - \mathbf{h}_q \\ 0 \end{pmatrix}^{\mathsf{T}} (\mathbf{v}_k - \mathbf{v}_i) = (\mathbf{h}_p - \mathbf{h}_q)^{\mathsf{T}}(\mathbf{u}_k - \mathbf{u}_i) = (R\mathbf{F}_{ij})^{\mathsf{T}}(\mathbf{u}_k - \mathbf{u}_i) < 0$$

(since $\mathbf{u}_k$ lies to the right from the edge $\mathbf{u}_i\mathbf{u}_j$). $\square$

Theorem 3.3 proves Steinitz's theorem in the case when the graph has a triangular country. If this is not the case, then Proposition 2.1 implies that the graph has a node of degree 3, which means that the dual graph has a triangular country. So we can represent the dual graph as the skeleton of a 3-polytope. Choosing the origin in the interior of this polytope, and considering its polar, we obtain a representation of the original graph.

3.3. Rubber bands and connectivity

Rubber band representations can be related to graph connectivity, and this relation can be used to give a test for k-connectivity of a graph [**Linial–Lovász–Wigderson 1988**]. It turns out that connectivity of the underlying graph is related to the degeneracy of the representation—a notion various versions of which will be crucial in many parts of this book.

3.3.1. Degeneracy: essential and nonessential. We start with a discussion of what causes degeneracy in rubber band embeddings. Consider the two graphs in Figure 3.8. It is clear that if we nail the nodes on the convex hull, and then let the rest find its equilibrium, then there will be a degeneracy: the grey nodes will all move to the same position. However, the reasons for this degeneracy are different: In the first case, it is due to symmetry; in the second, it is due to the node that separates the grey nodes from the rest, and thereby pulls them onto itself.

One can distinguish the two kinds of degeneracy in another way: In the first graph, the strengths of the rubber bands must be strictly equal; varying these strengths it is easy to break the symmetry and thereby get rid of the degeneracy. However, in the second graph, no matter how we change the strengths of the rubber bands (as long as they remain positive), the grey nodes will always be pulled together into one point.

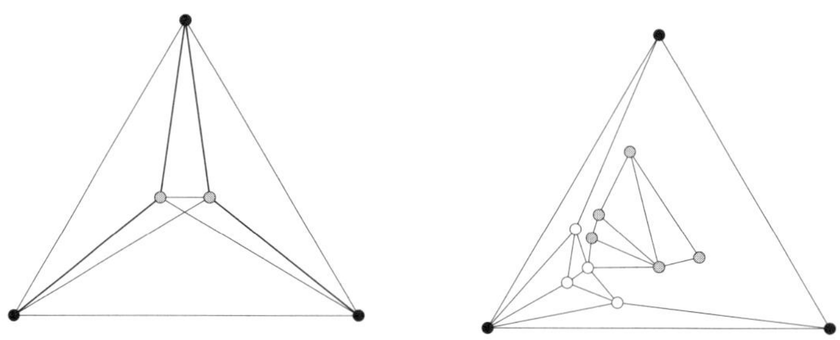

FIGURE 3.8. Two reasons why two or more nodes end up on top of each other: symmetry, or a separating node

Figure 3.9 illustrates a bit more delicate degeneracy. In all three pictures, the grey points end up collinear in the rubber band embedding. In the first graph, the reason is symmetry again. In the second, there is a lot of symmetry, but it does not explain why the three grey nodes are collinear in the equilibrium. (It is not hard to argue though that they are collinear: a good exercise!) In the third graph (which is not drawn in its equilibrium position, but before it) there are two nodes separating the grey nodes from the nailed nodes, and the grey nodes will end up on the segment connecting these two nodes. In the first two cases, changing the strength of the rubber bands will pull the grey nodes off the line; in the third, this does not happen.

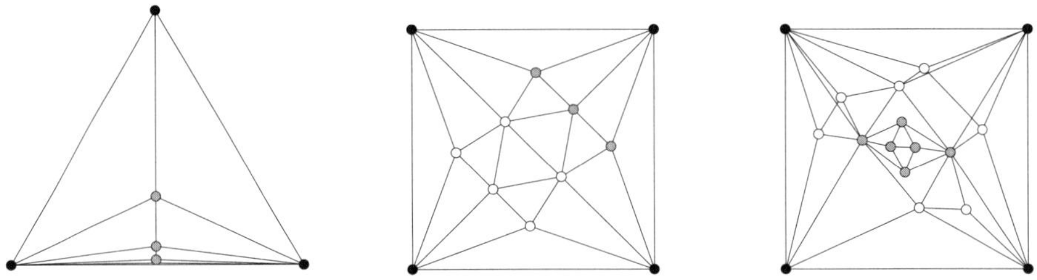

FIGURE 3.9. Three reasons why three nodes can end up collinear: symmetry, just accident, or a separating pair of nodes

3.3.2. Connectivity and degeneracy. Our goal is to prove that essential degeneracy in a rubber band embedding is always due to low connectivity. We start with the easy direction of the connection, relating connectivity and affine rank, formalizing the examples from the previous section.

LEMMA 3.5. *Let G be a graph and $S,T \subseteq V$. Then for every rubber band representation $\mathbf{u}$ of G with S nailed,* $\mathrm{rk}_{\mathsf{aff}}(\mathbf{u}(T)) \le \kappa(S,T)$.

PROOF. There is a subset $U \subseteq V$ with $|U| = \kappa(S,T)$ such that $V \setminus U$ contains no (S,T)-paths. Let W be the union of connected components of $G \setminus U$ containing a vertex from T. Then $\mathbf{u}$, restricted to W, gives a rubber band representation of $G[W]$ with boundary U. Clearly $\mathbf{u}(W) \subseteq \mathrm{conv}(\mathbf{u}(U))$, and so

$$\mathrm{rk}_{\mathsf{aff}}(\mathbf{u}(T)) \le \mathrm{rk}_{\mathsf{aff}}(\mathbf{u}(W)) = \mathrm{rk}_{\mathsf{aff}}(\mathbf{u}(U)) \le |U| = \kappa(S,T). \qquad \square$$

The Lemma gives a lower bound on the connectivity between two sets S and T. The following theorem asserts that if we take the best convex representation, this lower bound is tight:

THEOREM 3.6. *Let G be a graph and $S,T \subseteq V$ with $\kappa(S,T) = d+1$. Then G has a rubber band representation in $\mathbb{R}^d$, with suitable rubber band strengths, with S nailed such that* $\mathrm{rk}_{\mathsf{aff}}(\mathbf{u}(T)) = d+1$.

This theorem has a couple of consequences about connectivity not between sets, but between a set and any node, and between any two nodes.

COROLLARY 3.7. *Let G be a graph, $d \ge 1$ and $S \subseteq V$. Then G has a rubber band representation in general position in $\mathbb{R}^d$ with S nailed if and only if no node of G can be separated from S by fewer than $d+1$ nodes.*

COROLLARY 3.8. *A graph G is k-connected if and only if for every $S \subseteq V$ with $|S| = k$, G has a rubber band representation in $\mathbb{R}^{k-1}$ in general position with S nailed.*

To prove Theorem 3.6, we choose generic edge weights.

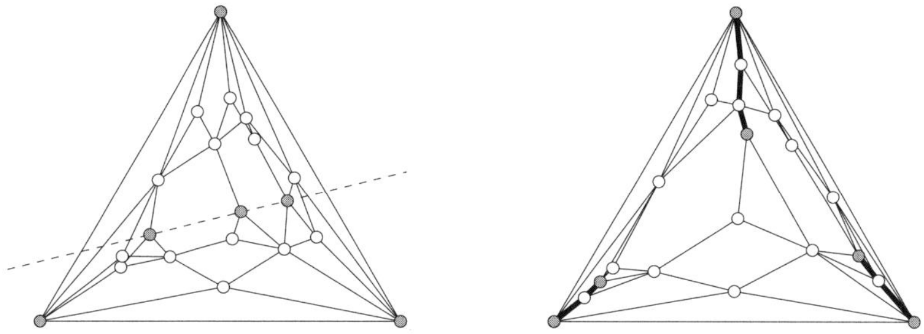

FIGURE 3.10. Three nodes accidentally on a line. Strengthening the edges along the three paths pulls them apart.

THEOREM 3.9. *Let G be a graph and $S,T \subseteq V$ with $\kappa(S,T) \ge d+1$. Choose algebraically independent edgeweights $c_{ij} > 0$. Map the nodes of S into $\mathbb{R}^d$ in general position. Then the rubber band extension of this map satisfies $\mathrm{rk}_{\mathrm{aff}}(\mathbf{u}(T)) = d+1$.*

If the edgeweights are chosen randomly, independently and uniformly from $[0,1]$, then the algebraic independence condition is satisfied with probability 1.

PROOF. The proof will consist of two steps: first, we show that there is *some* choice of the edgeweights for which the conclusion holds; then we use this to prove that the conclusion holds for *every* generic choice of edge-weights.

For the first step, we use that by Menger's Theorem B.2, there are $d+1$ disjoint paths P_i connecting a node $i \in S$ with a node $i' \in T$. The idea is to make the rubber bands on these paths very strong (while keeping the strengths of the other edges fixed). Then these paths will pull each node i' very close to i. Since the positions of the nodes $i \in S$ are affine independent, so will be the positions of the nodes i' (Figure 3.10).

To make this precise, let D be the diameter of the set $\{\mathbf{u}(S)\}$. Fix any $R > 0$, and define strengths c_{ij} of the rubber bands by

$$c_{ij} = \begin{cases} R, & \text{if } ij \text{ is an edge of one of the paths } P_r, \\ 1, & \text{otherwise.} \end{cases}$$

Let $\mathbf{u}$ be the rubber band extension of the given mapping of the nodes of S with these strengths.

Recall that f minimizes the potential $\mathcal{E}_c$ (defined by (3.5)) over all representations of G with the given nodes nailed. Let $\mathbf{v}$ be the representation with $\mathbf{v}_j = \mathbf{u}_i$ if $j \in P_i$ (in particular, $\mathbf{v}_i = \mathbf{u}_i$ if $i \in S$); for any node j not on any P_i, let $\mathbf{v}_j$ be any point in $\mathrm{conv}\{\mathbf{u}(S)\}$. In the representation $\mathbf{v}$ the edges with strength R have 0 length, and so

$$\mathcal{E}_S(\mathbf{u}) \le \mathcal{E}_S(\mathbf{v}) \le D^2 m.$$

On the other hand, for every edge uv on one of the paths P_i,

$$\mathcal{E}_S(\mathbf{u}) \ge R|\mathbf{u}_u - \mathbf{u}_v|^2,$$

and hence

$$|\mathbf{u}_u - \mathbf{u}_v| \le \frac{D\sqrt{m}}{\sqrt{R}}.$$

So these edges can be made arbitrarily short if we choose R large enough, and hence every node i' will be arbitrarily close to the corresponding nailed node i. Since $\{\mathbf{u}_i : i \in S\}$ are affine independent, this implies that the nodes in $\{\mathbf{u}_{i'} : i \in T\} = \{\mathbf{u}_i : i \in S\}$ are affine independent for a large enough value of R.

This completes the first step. Now we argue that this holds for all possible choices of generic edge-weights. To prove this, we only need some general considerations. The embedding minimizing the energy is unique, and so it can be computed from the equations (3.6) (say, by Cramer's Rule). What is important from this is that the vectors $\mathbf{u}_i$ can be expressed as rational functions of the edgeweights. Furthermore, the value $\det\big((1+\mathbf{u}_{t_i}^\mathsf{T}\mathbf{u}_{t_j})_{i,j=0}^d\big)$ is a polynomial in the coordinates of the $\mathbf{u}_i$, and so it is also a rational function of the edgeweights. We know that this rational function is not identically 0; hence it follows that it is nonzero for every algebraically independent substitution. □

REMARK 3.10. Instead of rubber band embeddings, it would suffice to assume that each free node is placed in the convex hull of its neighbors [**Cheriyan–Reif 1992**].

3.3.3. Rubber bands and connectivity testing.

Rubber band representations yield a (randomized) graph connectivity algorithm with good running time [**Linial–Lovász–Wigderson 1988**]; however, a number of algorithmic ideas are needed to make it work, which we describe in their simplest form. We refer to the paper for a more thorough analysis.

Connectivity between given sets. We start with describing a test checking whether or not two given k-tuples S and T of nodes are connected by k node-disjoint paths. For this, we assign random weights c_{ij} to the edges, and map the nodes in S into $\mathbb{R}^{k-1}$ in general position. Then we compute the rubber band representation extending this map, and check whether the points representing T are in general position.

This procedure requires solving a system of linear equations, which is easy, but the system is quite large, and it depends on the choice of S and T. With a little work, we can save quite a lot.

Let G be a connected graph on $V = [n]$ and $S \subseteq V$, say $S = [k]$. Given a map $\mathbf{u} : S \to \mathbb{R}^{k-1}$, we can compute its rubber band extension by solving the system of linear equations

$$\sum_{j\in N(i)} c_{ij}(\mathbf{u}_i - \mathbf{u}_j) = 0 \qquad (i \in V \setminus S). \tag{3.16}$$

This system has $(n-k)(k-1)$ unknowns (the coordinates of nodes $i \in S$ are considered as constants) and the same number of equations, and we know that it has a unique solution, since this is where the gradient of a strictly convex function (which tends to ∞ at ∞) vanishes.

At the first sight, solving (3.16) takes inverting an $(n-k)(k-1)\times(n-k)(k-1)$ matrix, since there are $(n-k)(k-1)$ unknowns and $(n-k)(k-1)$ equations. However, we can immediately see that the coordinates can be computed independently, and since they satisfy the same equations except for the right hand side, it suffices to invert the matrix of the system once.

Below, we will face the task of computing several rubber band representations of the *same* graph, changing only the nailed set S. Can we make use of some of the computation done for one of these representations when computing the others?

The answer is yes. First, we do something which seems to make things worse: We create new "equilibrium" equations for the nailed nodes, introducing new variables for the forces that act on the nails. Let $\mathbf{f}_i = \sum_{j\in N(i)} c_{ij}(\mathbf{u}_j - \mathbf{u}_i)$ be the force acting on node i in the equilibrium position. Note that $Y\mathbf{e}_i = \mathbf{f}_i = 0$ for $i \notin S$. Let X and Y be the matrices of the vector labelings $\mathbf{u}$ and $\mathbf{f}$. Let L_c be the symmetric $V\times V$ matrix

$$(L_c)_{ij} = \begin{cases} -c_{ij} & \text{if } ij \in E, \\ \sum_{k\in N(i)} c_{ik}, & \text{if } i = j, \\ 0, & \text{otherwise} \end{cases}$$

(the weighted Laplacian of the graph). Then we can write (3.16) as $XL_c = Y$. To simplify matters a little, let us assume that the center of gravity of the node positions is 0, so that $XJ = 0$. We also have $YJ = XL_cJ = 0$. It is easy to check that

$$X = Y(L_c + J)^{-1} \tag{3.17}$$

satisfies these equations (for more about this, see the "inverting the Laplacian" trick in Section 4.3.2).

It is not clear that we are nearer our goal, since how do we know the matrix Y (i.e., the forces acting on the nails)? But the trick is this: we can prescribe these forces arbitrarily, as long as the conditions $\sum_{i\in S}\mathbf{f}_i = 0$ and $\mathbf{f}_j = 0$ for $j \notin S$ are satisfied. Then the rows of X will give some position for each node, for which $XL_c = Y$. So in particular the nodes in $V\setminus S$ will be in equilibrium, so if we nail the nodes of S, the rest will be in equilibrium. If Y has rank d, then so does X, which implies that the vectors representing S will span the space $\mathbb{R}^d$. Since their center of gravity is 0, it follows that they are not on one hyperplane. We can apply an affine transformation to move the points $\mathbf{u}_i$ ($i \in S$) to any other affine independent position if we wish, but this is irrelevant: what we want is to check whether the nodes of T are in general position, and this is not altered by any affine transformation.

To sum up, to check whether the graph is k-connected between S and T (where $S, T \subseteq V$, $|S| = |T| = k$), we select random weights c_{ij} for the edges, select a convenient matrix Y, compute the matrix $X = Y(L_c+J)^{-1}$, and check whether the rows with indices from T are affine independent. The matrix L_c+J is positive definite, and it has to be inverted only once, independently of which sets are we testing for connectivity. The matrix Y has dimensions $(k-1)\times n$, but most of its elements are zero; a convenient choice for Y is

$$Y = \begin{pmatrix} 1 & 0 & \dots & 0 & -1 & 0\dots \\ 0 & 1 & \dots & 0 & -1 & 0\dots \\ \vdots & & & \ddots & & \\ 0 & 0 & \dots & 1 & -1 & 0\dots \end{pmatrix}.$$

and then the matrix multiplication $Y(L_c+J)^{-1}$ is trivial.

Connectivity between all pairs. If we want to apply the previous algorithm for connectivity testing, it seems that we have to apply it for all pairs of k-sets. Even though we can use the same edgeweights and we only need to invert L_c+J once, we have to compute $X = Y(L_c+J)^{-1}$ for potentially exponentially many different sets S, and then we have to test whether the nodes of T are represented in general position for exponentially many sets T. The following lemma shows how to get around this.

LEMMA 3.11. *For every vertex $v \in V$ we select an arbitrary k-subset $S(v) \subseteq N(v)$. Then G is k-connected if and only if $S(u)$ and $S(v)$ are connected by k node-disjoint paths for every u and v.*

PROOF. The "only if" part follows from the well-known property of k-connected graphs that any two k-subsets are connected by k node-disjoint paths. The "if" part follows from the observation that if $S(u)$ and $S(v)$ are connected by k node-disjoint paths, then u and v are connected by k openly disjoint paths. □

Thus the subroutine in the first part needs be called only $O(n^2)$ times. In fact, we do not even have to check this for every pair (u, v), and further savings can be achieved through randomization, using Exercises 3.11 and 3.12.

Numerical issues. The computation of the rubber band representation requires solving a system of linear equations. We have seen (Figure 3.5) that for a graph with n nodes, the positions of the nodes can get exponentially close in a rubber band representation, which means that we might have to compute with exponentially small numbers (in n), which means that we have to compute with linearly many digits, which gives an extra factor of n in the running time. One way out is to solve the system in a finite field rather than in $\mathbb{R}$. Of course, this "modular" embedding has no physical or geometrical meaning any more, but the algebraic structure remains!

For the analysis, we need the Schwartz–Zippel Lemma [**Schwartz 1980**, **Zippel 1979**]:

LEMMA 3.12. *Given a polynomial over any field of degree k in m variables, the probability that it vanishes for a random substitution, where each variable is chosen uniformly from N possible values, is bounded by k/N.* □

Let G be a graph and let $S \subseteq V$, $|S| = k$, and $d = k-1$. Let p be a prime and $c_e \in \mathbb{F}_p$ for $e \in E$. A *modular rubber band representation* of G (with respect to S, p and c) is defined as an assignment $i \mapsto \mathbf{u}_i \in \mathbb{F}_p^d$ satisfying

$$\sum_{j\in N(i)} c_{ij}(\mathbf{u}_i - \mathbf{u}_j) = 0 \qquad (\forall i \in V \setminus S). \tag{3.18}$$

This is formally the same equation as for real rubber bands, but we work over $\mathbb{F}_p$, so no notion of convexity can be used. In particular, we cannot be sure that the system has a solution. But things work if the prime is chosen at random.

LEMMA 3.13. *Let $N > 0$ be an integer. Choose uniformly a random prime $p < N$ and random weights $c_e \in \mathbb{F}_p$ $(e \in E)$.*

(a) *With probability at least $1 - n^2/N$, there is a modular rubber band representation of G (with respect to S, p and c), such that the vectors $\mathbf{u}_i$ $(i \in S)$ are affine independent. This representation is unique up to affine transformations of $\mathbb{F}_p^d$.*

(b) *Let $T \subseteq V$, $|T| = k$. Then with probability at least $1 - n^2/N$, the representation $\mathbf{u}_c$ in* (a) *satisfies $\mathrm{rk}_{\mathrm{aff}}(\{\mathbf{u}_i : i \in T\}) = \kappa(S, T)$.*

PROOF. The determinant of the system (3.18) is a polynomial of degree at most n^2 of the weights c_{ij}. The Schwartz–Zippel Lemma gives (a). The proof of (b) is similar. □

We sum up the complexity of this algorithm without going into fine details. In particular, we restrict this short analysis to the case when $k < n/2$. We have to fix an appropriate N; $N = n^5$ will do. Then we pick a random prime $p < N$. Since N has $O(\log n)$ digits, the selection of p and every arithmetic operation in $\mathbb{F}_p$ can be done in polylogarithmic time, and we are going to ignore these factors of $\log n$. We have to generate $O(n^2)$ random weights for the edges. We have to invert (over $\mathbb{F}_p$) the matrix $L_c + J$, this takes $O\big(M(n)\big)$ operations, where $M(n)$ is the cost of multiplying two $n \times n$ matrices (currently known to be $O(n^{2.3727})$ [**Vassilevska Williams 2012**]). Then we have to compute $Y(L_c + J)^{-1}$ for a polylogarithmic number of different matrices Y (using Exercise 3.12(b) below). For each Y, we have to check affine independence for one k-set in the neighborhood of every node, in $O\big(nM(k)\big)$ time. Up to polylogarithmic factors, this takes $O\big(M(n) + nM(k)\big)$ time.

Using much more involved numerical methods, solving linear equations involving the Laplacian and similar matrices can be done more efficiently. For details, we refer to [**Spielman 2011**] and [**Spielman–Teng 2014**].

EXERCISE 3.1. Prove that $\min_{\mathbf{u}} \mathcal{E}_c(\mathbf{u})$ (where the minimum is taken over all positions $\mathbf{u}$ with some nodes nailed) is a concave function of the rubber band strengths $c \in \mathbb{R}^E$.

EXERCISE 3.2. Let $\mathbf{x}$ be the equilibrium position of a rubber band framework with the nodes in S nailed, and let $\mathbf{y}$ be any other position of the nodes (but with the same nodes nailed at the same positions). Then

$$\mathcal{E}_c(\mathbf{y}) - \mathcal{E}_c(\mathbf{x}) = \mathcal{E}_c(\mathbf{y} - \mathbf{x}).$$

EXERCISE 3.3. Let G be a connected graph, $\emptyset \neq S \subseteq V$, and $\overline{\mathbf{u}} : S \to \mathbb{R}^d$. Extend $\overline{\mathbf{u}}$ to $\mathbf{u} : V \setminus S \to \mathbb{R}^d$ as follows: starting a random walk at j, let i be the (random) node where S is first hit, and let $\mathbf{u}_j$ denote the expectation of the vector $\overline{\mathbf{u}}_i$. Prove that $\mathbf{u}$ is the same as the rubber band extension of $\overline{\mathbf{u}}$.

EXERCISE 3.4. Let G be a connected graph, and let $\mathbf{u}$ be a vector-labeling of an induced subgraph H of G (in any dimension). If $(H, \mathbf{u})$ is section-connected, then its rubber-band extension to G is section-connected as well.

EXERCISE 3.5. Let $\mathbf{u}$ be a rubber band representation of a planar map G in the plane with the nodes of a country T nailed to a convex polygon. Define $\mathbf{F}_{ij} = \mathbf{u}_i - \mathbf{u}_j$ for all edges in $E \setminus E(T)$. (a) If T is a triangle, then we can define $\mathbf{F}_{ij} \in \mathbb{R}^2$ for $ij \in E(T)$ so that $\mathbf{F}_{ij} = -\mathbf{F}_{ji}$, $\mathbf{F}_{ij}$ is parallel to $\mathbf{u}_j - \mathbf{u}_i$, and $\sum_{i \in N(j)} \mathbf{F}_{ij} = 0$ for every node i. (b) Show by an example that (a) does not remain true if we drop the condition that T is a triangle.

EXERCISE 3.6. Prove that every Schlegel diagram with respect to a face F can be obtained as a rubber band representation of the skeleton with the vertices of F nailed (the strengths of the rubber bands must be chosen appropriately).

EXERCISE 3.7. Let G be a 3-connected simple planar graph with a triangular country $p = \overline{abc}$. Let q, r, s be the countries adjacent to p. Let G^* be the dual graph. Consider a rubber band representation $\mathbf{u} : V \to \mathbb{R}^2$ of G with a, b, c nailed down (both with unit rubber band strengths). Prove that the segments representing the edges can be translated so that they form a rubber band representation of $G^* - p$ with q, r, s nailed down (Figure 3.11).

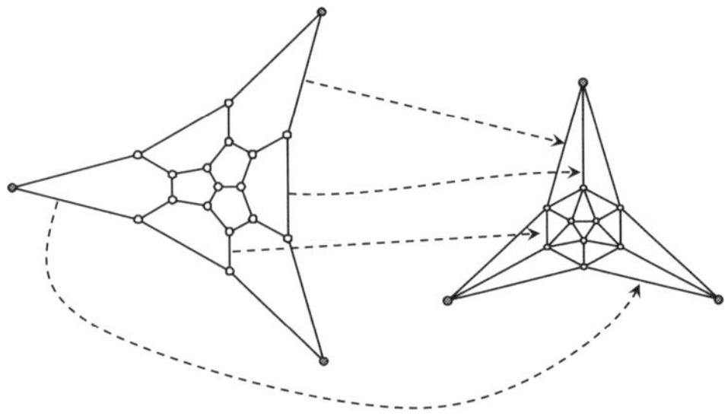

FIGURE 3.11. Rubber band representation of a dodecahedron with one node deleted, and of an icosahedron with the edges of a triangle deleted. Corresponding edges are parallel and have the same length.

EXERCISE 3.8. A *convex representation* of a graph G (in dimension d, with boundary $S \subseteq V$) is a mapping of $V \to \mathbb{R}^d$ such that every node in $V \setminus S$ is in the relative interior of the convex hull of its neighbors. (a) The rubber band representation extending any map from $S \subseteq V$ to $\mathbb{R}^d$ is convex with boundary S. (b) Not every convex representation is constructible this way.

EXERCISE 3.9. In a rubber band representation, increase the strength of an edge between two unnailed nodes (while leaving the other edges invariant). Prove that the length of this edge decreases.

EXERCISE 3.10. Prove that a 1-dimensional rubber band representation of a 2-connected graph, with boundary nodes s and t, nondegenerate in the sense that the nodes are all different, is an s-t-numbering (as defined in the Introduction). Show that instead of the 2-connectivity of G, it suffices to assume that deleting any node, the rest is either connected or has two components, one containing s and one containing t.

EXERCISE 3.11. Let G be any graph, and let H be a k-connected graph with $V(H) = V$. Then G is k-connected if and only if u and v are connected by k openly disjoint paths in G for every edge $uv \in E(H)$.

EXERCISE 3.12. Let G be any graph and $k < n$. (a) For t pairs of nodes chosen randomly and independently, we test whether they are connected by k openly disjoint paths. Prove that if G is not k-connected, then this test discovers this with probability at least $1-\exp(-2(n-k)t/n^2)$. (b) For r nodes chosen randomly and independently, we test whether they are connected by k openly disjoint paths to every other node of the graph. Prove that if G is not k-connected, then this test discovers this with probability at least $1-\big((k-1)/n\big)^r$.

EXERCISE 3.13. Let G be a graph, and let Z be a matrix obtained from its Laplacian L_G by replacing its nonzero entries by algebraically independent transcendentals. For $S,T \subseteq V$, $|S| = |T| = k$, let $Z_{S,T}$ denote the matrix obtained from Z by deleting the rows corresponding to S and the columns corresponding to T. Then $\det(Z_{S,T}) \neq 0$ if and only if $\kappa(S,T) = k$.

CHAPTER 4

Discrete Harmonic Functions

Having discussed the nice geometry of rubber band embeddings, let us have a closer look at the one-dimensional case (or, if you prefer, at one of the coordinate functions). This will allow us to make some arguments from the previous chapter ("potential argument" and "inverting the Laplacian") more general and precisely stated.

Discrete harmonic functions are discrete analogues of harmonic functions in analysis, where they play a fundamental role. A "true" harmonic function can be defined as a twice differentiable function $f: \mathbb{R}^d \to \mathbb{R}$ (perhaps defined only in some domain) satisfying the differential equation $\Delta f = 0$, where $\Delta = \sum_{i=1}^{d} \partial^2/\partial x_i^2$ is the Laplace operator. It is a basic fact that such functions can be characterized by the "mean value property": their value at any point equals to their average on any ball around this point. We take this second characterization as our starting point for introducing an analogous notion of harmonic functions defined on the nodes of an arbitrary graph.

Discrete harmonic functions play an important role in the study of random walks (after all, the averaging in the definition can be interpreted as expectation after one move). They also come up in the theory of electrical networks, and also in statics. This provides a connection between these fields, which can be exploited. In particular, various methods and results from the theory of electricity and statics, often motivated by physics, can be applied to provide results about random walks, and vice versa. From our point of view, their main applications will be in the theory of discrete analytic functions in later chapters.

4.1. Basics

4.1.1. Definition, uniqueness and existence. Let G be a connected simple graph. A function $h: V \to \mathbb{R}$ is called *harmonic* at a node $v \in V$ if

$$\frac{1}{\deg(v)} \sum_{u \in N(v)} h(u) = h(v) \tag{4.1}$$

(the value of h at a node v is the average of its values at the neighbors of v.) We can also write this condition as

$$\sum_{u \in N(v)} \big(h(v) - h(u)\big) = 0. \tag{4.2}$$

A node where a function is not harmonic is called a *pole* of the function. Figure 4.1 shows a couple of very simple functions harmonic at many nodes, with black nodes denoting their poles.

If we consider h as a vector indexed by the nodes, and Lh is the vector obtained by applying the Laplacian $L = L_G$ to h, then we can express (4.1) as $(Lh)_v = 0$.

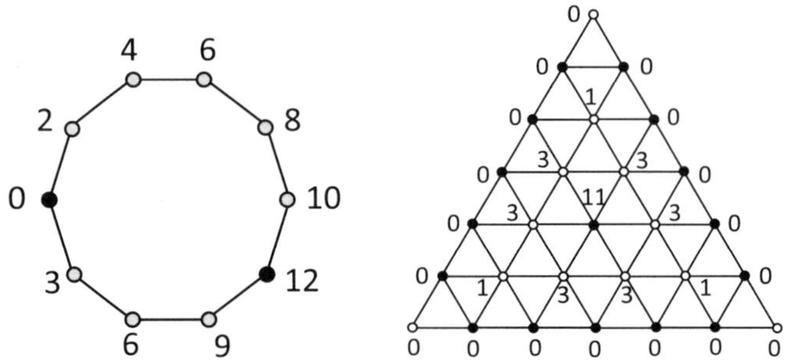

(a) (b)

FIGURE 4.1. (a) A harmonic function on a cycle, with two poles (black nodes). (b) A harmonic function on a triangular grid, with poles at the center and on the boundary.

It is useful to note that, since each column sum of L is zero,

$$\sum_{v\in V}(Lh)_v = \mathbb{1}^{\mathsf{T}}Lh = 0. \tag{4.3}$$

So every function is harmonic "on the average".

We can extend this notion to multigraphs by replacing (4.1) by

$$\frac{1}{\deg(v)}\sum_{uv=e\in E} h(u) = h(v). \tag{4.4}$$

Another way of writing this is

$$\sum_{u\in V} a_{uv}\big(h(v)-h(u)\big) = 0, \tag{4.5}$$

where a_{uv} is the multiplicity of the edge uv. We could further generalize this by assigning arbitrary positive "strengths" β_{uv} to the edges of G, and say that h is harmonic at v (on the weighted graph (G,β)) if

$$\sum_{u\in V} \beta_{uv}\big(h(v)-h(u)\big) = 0. \tag{4.6}$$

In what follows, we will restrict our arguments to simple graphs (to keep things simple), but all of the results below could be extended to multigraphs and weighted graphs in a straightforward way.

Every constant function is harmonic at each node. On the other hand, it was observed very early [**Katona–Korvin 1968**] that this is all:

PROPOSITION 4.1. *Every nonconstant function on the nodes of a connected graph has at least two poles.*

This implies that if $h : V \to \mathbb{R}$ is a nonconstant function, then $Lh \neq 0$, so the nullspace of L is one-dimensional: it consists of the constant functions only.

PROOF. Let S be the set where the function assumes its maximum, and let S' be the set of those nodes in S that are connected to any node outside S. Then every node in S' must be a pole, since in (4.1), every value $h(u)$ on the left had side is at most $h(v)$, and at least one is less, so the average is less than $h(v)$. Since the function is nonconstant, S is a nonempty proper subset of V, and since the graph is connected, S' is nonempty. So there is a pole where the function attains its maximum. Similarly, there is another pole where it attains its minimum. $\square$

The proof above shows that every non-constant function on the nodes of a graph assumes its maximum as well as its minimum on one of its poles. This can be viewed as a (very simple) version of the *Maximum Principle* familiar from the theories of harmonic and analytic functions.

For any two nodes there is a nonconstant harmonic function that is harmonic everywhere else. More generally, we have the following theorem.

THEOREM 4.2. *For a connected simple graph G, nonempty set $S \subseteq V$ and function $h_0 : S \to \mathbb{R}$, there is a unique function $h : V \to \mathbb{R}$ extending h_0 that is harmonic at each node of $V \setminus S$.*

We call this function h the *harmonic extension* of h_0. Note that if $|S| = 1$, then the harmonic extension is a constant function (and so it is also harmonic at S, and it does not contradict Proposition 4.1). If $S = \{a, b\}$, then a function that is harmonic outside S is uniquely determined by two of its values, $h(a)$ and $h(b)$. Scaling the function by a real number and adding a constant preserves harmonicity at each node, so if we know how to construct the harmonic extension h with (say) $h(a) = 0$ and $h(b) = 1$, then $g(v) = g_0(a) + \big(g_0(b) - g_0(a)\big)h(v)$ describes the harmonic extension of the function $g_0 : \{a, b\} \to \mathbb{R}$. In this case equation (4.5) is equivalent to saying that $F_{ij} = h(j) - h(i)$ defines a flow from a to b.

PROOF. The uniqueness of the harmonic extension is easy by the argument in the proof of Proposition 4.1. Suppose that h and h' are two harmonic extensions of h_0. Then $g = h - h'$ is harmonic on $V \setminus S$, and satisfies $g(v) = 0$ at each $v \in S$. If g is the identically 0 function, then $h = h'$ as claimed. Else, either its minimum or its maximum is different from 0. But we have seen that both the minimizers and the maximizers contain at least one pole, which is a contradiction.

The existence is just the 1-dimensional case of the construction of the rubber band extension in Section 3.1. $\square$

There are several other ways of proving the existence; a few of these could be based on the models below and on the linear algebra considerations in the next sections.

4.1.2. Random walks. Random walks were considered in the Introduction, but we need a few more definitions. Recall that a *random walk* on a graph G is a random sequence $(v^0, v^1, \dots)$ of nodes constructed as follows: We pick a starting point v^0 from a specified initial distribution σ (often a specific node), we select a neighbor v^1 of it at random (each neighbor is selected with the same probability $1/\deg(v^0)$), then we select a neighbor v^2 of this node v^1 at random, etc. We denote by σ^k the distribution of v^k.

For any event A expressible in terms of the random walk $(v^0, v^1, \dots)$, started from the distribution σ, we denote by $\mathsf{P}_\sigma(A)$ its probability. For example, $\mathsf{P}_u(v^t = u)$ is the probability that after t steps, the random walk started at u returns to the starting node. Expectations $\mathsf{E}_\sigma(X)$ are defined similarly for random variables X depending on the walk.

In the language of probability theory, a random walk is a finite time-reversible Markov chain. There is not much difference between the theory of random walks on directed graphs and the theory of finite Markov chains; every Markov chain can be viewed as random walk on a directed graph, if we allow weighted edges, and every time-reversible Markov chain can be viewed as a random walk on an edge-weighted undirected graph. The theory of random walks (finite Markov chains) has a large literature, and we will not be able to provide here even an introduction. The following discussion, based on well known and intuitive facts, can be followed even by those not familiar with the subject. For example, we need the fact that, with probability 1, a random walk on a connected graph hits every node infinitely often.

Let π denote the probability distribution in which the probability of a node is proportional to its degree: $\pi_v = \deg(v)/(2m)$. This distribution is called the *stationary distribution* of the random walk. It is easy to check that if v^0 is selected from π, then after any number of steps, v^k will have the same distribution π. (This explains the name "stationary distribution").

Using the adjacency matrix $A = A_G$ and the diagonal matrix $D = D_G$ composed of the degrees, we define the *transition matrix* of the random walk on G as $P = D^{-1}A = I - D^{-1}L$. If we are at node i, then the probability of moving to node j is P_{ij}. The eigenvalues of P are the same as the eigenvalues of the symmetric matrix $I - \widehat{L} = D^{1/2}PD^{-1/2}$, so in particular they are real. (This shows why the spectrum of the normalized Laplacian plays an important role in the study of random walks.) The largest eigenvalue of the transition matrix P is 1, and it has multiplicity 1 for connected graphs. It is straightforward to check that the right eigenvector belonging to it is $\mathbb{1}$, and the left eigenvector is given by π.

Let us return to harmonic functions. Given a nonempty subset $S \subseteq V$ and a function $h_0 : S \to \mathbb{R}$, we define

$$h(v) = \mathsf{E}_v\big(h_0(a_v)\big), \tag{4.7}$$

where a_v is the (random) node where a random walk first hits S (this node exists with probability 1). Clearly $h(v) = h_0(v)$ for $v \in S$. Furthermore, for a node $v \in V \setminus S$ we have

$$h(v) = \mathsf{E}_v\big(h_0(a_v)\big) = \sum_{u \in N(v)} \mathsf{P}_v(w^1 = u)\mathsf{E}_u\big(h_0(a_u)\big) = \frac{1}{\deg(v)} \sum_{u \in N(v)} h(u).$$

This shows that h is harmonic on $V \setminus S$.

As a special case, if $S = \{s,t\}$, $h_0(s) = 0$ and $h_0(t) = 1$, then $h(v)$ is the probability that a random walk starting at v visits t before s.

REMARK 4.3. Random walks on infinite graphs have quite different problems. Consider an infinite connected graph G with finite degrees, and specify a "root" node r. For each $v \in V$, let $h(v)$ denote the probability that starting a random walk at v, we hit r at some time. Clearly $h(r) = 1$, and h is harmonic at every node other than r.

If the function h is identically 1, then the random walk is *recurrent*. In this case, it almost surely visits every node infinitely often. If it is not recurrent, then it is called *transient*; it almost surely visits every node only a finite number of times. (These properties seem to depend on the root, but in fact if the graph is recurrent for one root, then it is recurrent for every other root.)

For a recurrent walk, the function h is a trivial harmonic function. For a transient walk, we get an interesting function with a single pole. We can of course restrict the function to a finite induced subgraph, but then it will have more poles, at all neighbors of the complementary node set.

Basic results in probability tell us that the grid graph on $\mathbb{Z}^d$ (in which two nodes are connected whenever their distance is 1) is recurrent for dimensions $d \le 2$, but transient for $d > 2$ [**Pólya 1921**]. Random walks on infinite graphs is a very important topic, connected to our theme in many ways, but this topic does not fit in this book.

4.1.3. Electrical networks. In this third construction for harmonic functions, we use facts from the physics of electrical networks like Ohm's Law and Kirchhoff's two laws. This looks like stepping out of the realm of mathematics; however, these results are all purely mathematical.

To be more precise, we consider these laws as axioms. Given a connected network whose edges are conductors, and an assignment of potentials (voltages) to a subset S of its nodes, we want to find an assignment of voltage differences U_{ij} and currents I_{ij} to the edges ij so that $U_{ij} = -U_{ji}$, $I_{ij} = -I_{ji}$ (so the values are defined if we give the edge an orientation, and the values are antisymmetric), and the following laws are satisfied:

Ohm's Law: $U_{ij} = R_{ij} I_{ij}$ (where $R_{ij} > 0$ is the resistance of edge e; this does not depend on the orientation).

Kirchhoff's Current Law: $\sum_{u\in N(v)} I_{uv} = 0$ for every node $v \notin S$;

Kirchhoff's Voltage Law: $\sum_{ij\in E(C)} U_{ij} = 0$ for every cycle $\vec{C}$ (the order of i and j is given by fixing a direction of traversing C).

Note that every cycle has two orientations, but because of antisymmetry, these two oriented cycles give the same condition. Therefore, in a bit sloppy way, we can talk about Kirchhoff's Voltage Law for an undirected cycle. As a degenerate case, we could require the Voltage Law for oriented cycles of length 2, consisting of an edge and its reverse, in which case the law says that $U_{ij} + U_{ji} = 0$, which is just another way of saying that U is antisymmetric.

The following reformulation of Kirchhoff's Voltage Law will be useful for us: *there exist potentials $p(i) \in \mathbb{R}$ $(i \in V)$ so that $U_{ij} = p(j) - p(i)$.* This sounds like a statement in physics, but in fact it is a rather simple fact about graphs:

PROPOSITION 4.4. *Given a connected graph G and an antisymmetric assignment U_{ij} of real numbers to the oriented edges, there exist real numbers $p(i)$ $(i \in V)$ so that $U_{ij} = p(j) - p(i)$ if and only if $\sum_{ij\in E(C)} U_{ij} = 0$ for every cycle C.*

PROOF. It is trivial that if a "potential" exists for a given assignment of voltages U_{ij}, then Kirchhoff's Voltage Law is satisfied. Conversely, suppose that Kirchhoff's Voltage Law is satisfied. We derive a couple of consequences.

Claim 1. *If W is any closed walk on the graph, then $\sum_{ij\in E(W)} U_{ij} = 0$.*

This follows by induction on the length of the walk. Consider the first time when we encounter a node already visited. If this is the end of the walk, then the walk is a single oriented cycle, and the assertion follows by Kirchhoff's Voltage Law. Otherwise, this node splits the walk into two closed walks, and since the sum of voltages is zero for each of them by induction, it is zero for the whole walk.

Claim 2. *If W_1 and W_2 are open walks with the same starting nodes and the same ending nodes, then $\sum_{ij\in E(W_1)} U_{ij} = \sum_{ij\in E(W_2)} U_{ij}$.*

Indeed, concatenating W_1 with the reversal of W_2 we get a closed walk whose total voltage is $\sum_{ij\in E(W_1)} U_{ij} - \sum_{ij\in E(W_2)} U_{ij}$, and this is 0 by Claim 1.

Now let us turn to the construction of the potentials. Fix any node v as a root, and assign to it an arbitrary potential $p(v)$. For every other node u, consider any walk W from v to u, and define $p(u) = \sum_{ij\in E(W)} U_{ij}$. This value is independent of the choice of W by Claim 2. Furthermore, for any neighbor u' of u, considering the walk $W + uu'$, we see that $p(u') = p(u) + U_{uu'}$, which shows that p is indeed a potential for the voltages U. □

To check Kirchhoff's Voltage Law for every cycle in the graph is a lot of work; but these equations are not independent, and it is enough to check it for a basis of these equations. There are several useful ways to select a basis; here we state one that is particularly useful, and another one as Exercise 4.1.

COROLLARY 4.5. *Let G be a 2-connected planar graph and let U_{ij} be an antisymmetric assignment of real numbers to the oriented edges. There exist real numbers $p(i)$ $(i \in V)$ so that $U_{ij} = p(j) - p(i)$ if and only if Kirchhoff's Voltage Law is satisfied by the boundary cycle of every bounded country.*

PROOF. We want to show that Kirchhoff's Voltage Law is satisfied by every cycle C. Fix any embedding of the graph in the plane, and (say) the counterclockwise orientation of C. Let us sum Kirchhoff's Voltage Law corresponding to countries contained inside C, with the counterclockwise orientation of each; the terms corresponding to internal edges cancel (by antisymmetry), and we get Kirchhoff's Voltage Law for C. □

We have given these simple arguments in detail, because essentially the same argument is used repeatedly in the book (already used in Section 3.2), and we are not going to describe it every time; we will just refer to it as the "potential argument". We could formulate and prove Proposition 4.4 and Corollary 4.5 for an arbitrary abelian group instead of the real numbers. The group of complex numbers, and the group of vectors in a linear space, will be used several times.

The main result from the theory of electrical networks we are going to use is that *given a connected graph with voltages assigned to a nonempty set of nodes, there is always a unique assignment of potentials to the remaining nodes and currents to the edges satisfying Ohm's and Kirchhoff's Laws.*

Returning to the proof of Theorem 4.2, consider the graph G as an electrical network, where each edge represents a conductor with unit resistance. Given a set $S \subseteq V$ and a function $h_0 : S \to \mathbb{R}$, let a current flow through the network while keeping each $v \in S$ at potential $h_0(v)$. Then the potential $h(v)$ defined for all nodes v is an extension of h_0, and it is harmonic at every node $v \in V \setminus S$. Indeed, the current through an edge uv is $h(u) - h(v)$ by Ohm's Law, and hence by Kirchhoff's Current Law, $\sum_{j\in N(i)} \big(h(j) - h(i)\big) = 0$ for every node i.

As a special case, if we send unit current from node s to node t, then the potentials h of the nodes satisfy the equation $Lh = \mathbb{1}_t - \mathbb{1}_s$.

4.1.4. Rubber bands.

We have seen that harmonic functions arise as 1-dimensional special cases of rubber band extensions. Let us discuss a further property for this model.

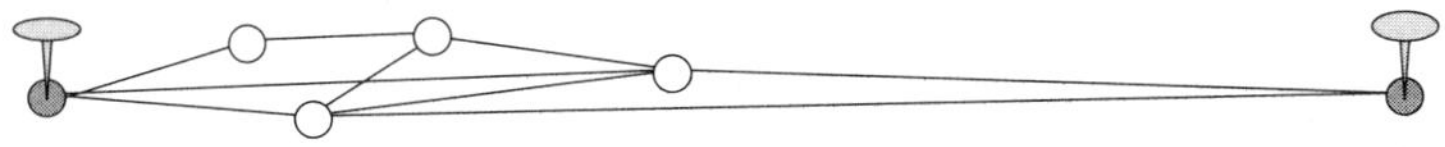

FIGURE 4.2. Stretching two nodes of a graph with edges replaced by rubber bands. The figure should be one-dimensional, but it is vertically distorted so that the graph can be seen.

Consider the case of stretching just two nodes s and t to a unit distance apart (Figure 4.2). We see that they pull our hands with a force of

$$F(s,t) = \sum_{u\in N(s)} (h(u)-h(s)) = \sum_{u\in N(t)} (h(t)-h(u)) \tag{4.8}$$

(by basic physics, both of our hands feel the same force, in opposite directions). Considering the energy balance gives further important information. Imagine that we slowly stretch the graph until nodes s and t will be at distance 1. When they are at distance x, the force pulling our hands is $xF(s,t)$ (everything scales by a factor of x), and hence the energy we have to spend is

$$\int_0^1 xF(s,t)\,dx = \frac{1}{2}F(s,t).$$

This energy accumulates in the rubber bands. By a similar argument, the energy stored in the rubber band uv is $\big(h(v)-h(u)\big)^2/2$. By conservation of energy, we get the equation

$$\sum_{uv\in E} \big(h(v)-h(u)\big)^2 = F(s,t). \tag{4.9}$$

(It is easy to prove identities (4.8) and (4.9) for harmonic functions without any reference to physics.)

4.1.5. Relating different models. A consequence of the uniqueness property in Theorem 4.2 is that the harmonic functions constructed in sections 4.1.2, 4.1.3 and 4.1.4 are the same. As an application of this idea, we show an interesting identity [**Nash-Williams 1959**, **Chandra et al. 1989**].

Considering the graph G as an electrical network, where every edge has resistance 1, we define the *effective resistance* $R(s,t)$ between nodes s and t by sending a unit current from s to t, and measuring the voltage between s and t. Considering the graph G as a rubber band structure in equilibrium, with two nodes s and t nailed down at 1 and 0, let $F(s,t)$ denote the force pulling the nails (cf. (4.10)). Doing a random walk on the graph, let $\mathsf{comm}(s,t)$ denote the commute time between nodes s and t (i.e., the expected time it takes to start at s, walk until you first hit t, and then walk until you first hit s again).

THEOREM 4.6. *Let G be a connected graph and $s,t\in V$. The effective resistance $R(s,t)$, the force $F(s,t)$, and the commute time* $\mathsf{comm}(s,t)$ *are related by the equations*

$$R(s,t) = \frac{1}{F(s,t)} = \frac{\mathsf{comm}(s,t)}{2m}.$$

PROOF. Consider the function $h \in \mathbb{R}^V$ satisfying $h(s) = 0$, $h(t) = 1$ and harmonic for $v \neq s,t$ (we know that this exists and it is uniquely determined). By (4.8), we have $F(s,t) = \sum_{j\in N(s)} h(j)$.

By the discussion in Section 4.1.3, $h(v)$ is equal to the potential of v if we fix the potential of s and t as above. By Ohm's Law, the current through an edge sj is $h(j)-h(s) = h(j)$. Hence the current through the network is $\sum_{j\in N(s)} h(j) = F(s,t)$, and the effective resistance is $R(s,t) = 1/F(s,t)$.

The assertion about commute time is a bit more difficult to prove. We know that $h(u)$ is the probability that a random walk starting at u visits b before a. Hence $p = \big(1/\deg(s)\big)\sum_{u\in N(s)} h(u) = F(s,t)/\deg(s)$ is the probability that a random walk starting at s hits t before returning to s.

Let $\mathbf{T}$ be the first time when a random walk starting at s returns to s and let $\mathbf{S}$ be the first time when it returns to s after having visited t. We know from the theory of random walks that $\mathsf{E}(\mathbf{T}) = 2m/\deg(s)$ and by definition, $\mathsf{E}(\mathbf{S}) = \mathsf{comm}(s,t)$. Clearly $\mathbf{T} \le \mathbf{S}$ and $\mathbf{T} = \mathbf{S}$ means that the random walk starting at s hits t before returning to s. Thus $\mathsf{P}(\mathbf{T} = \mathbf{S}) = p$. If $\mathbf{T} < \mathbf{S}$, then after the first $\mathbf{T}$ steps, we have to walk from s until we reach t and then return to s. Hence $\mathsf{E}(\mathbf{S}-\mathbf{T} \mid \mathbf{T} < \mathbf{S}) = \mathsf{E}(\mathbf{S})$, and so

$$\mathsf{E}(\mathbf{S}-\mathbf{T}) = p\,\mathsf{E}(\mathbf{S}-\mathbf{T} \mid \mathbf{T} = \mathbf{S}) + (1-p)\,\mathsf{E}(\mathbf{S}-\mathbf{T} \mid \mathbf{T} < \mathbf{S}) = (1-p)\,\mathsf{E}(\mathbf{S}).$$

This gives

$$\frac{2m}{\deg(s)} = \mathsf{E}(\mathbf{T}) = \mathsf{E}(\mathbf{S}) - \mathsf{E}(\mathbf{S}-\mathbf{T}) = p\,\mathsf{E}(\mathbf{S}) = \frac{F(s,t)}{\deg(s)}\mathsf{comm}(s,t).$$

Simplifying, we get the identity in the theorem. □

As a further application of this equivalence between our three models, we describe what can be said if an edge is deleted.

THEOREM 4.7. *Let G be a connected graph, and let us connect two nodes i and j by a new edge. Then for any two nodes $s,t \in V$,*

(a) *The ratio* $\mathsf{comm}(s,t)/m$ *does not increase.*

(b) (Raleigh Monotonicity) *The effective resistance $R(s,t)$ does not increase.*

(c) *If nodes s and t are nailed down at 0 and 1, the force $F(s,t)$ does not decrease.*

PROOF. The three statements are equivalent by Theorem 4.6; we prove (c). By (4.9), it suffices to prove that the equilibrium energy does not decrease. Consider the equilibrium position of G', and delete the new edge. The contribution of this edge to the energy was nonnegative, so the total energy decreases. The current position of the nodes may not be in equilibrium; but the equilibrium position minimizes the energy, so when they move to the equilibrium of G', the energy further decreases. □

4.2. Estimating harmonic functions

4.2.1. Harmonic flows. Suppose (for simplicity) that $S = \{s,t\}$, and let $h:\ V \to \mathbb{R}$ be a function with $h(s) = 0$, $h(t) = 1$, and harmonic outside S. Then

(4.5) tells us that assigning $f(e) = h(v) - h(u)$ to every edge $e = uv$, we get a flow from s to t, which we call a *harmonic flow*. The value of this flow is

$$F(s,t) = \sum_{v \in N(s)} f(sv) = \sum_{v \in N(t)} f(vt). \tag{4.10}$$

It will be convenient to normalize differently and consider the flow $\widehat{f} = f/F(s,t)$; this has value 1.

An s-t flow that is, at the same time, a potential, will be called a *harmonic flow*. The potential constructed from a harmonic function with poles s and t is a harmonic flow, and vice versa. The dimension of the linear space of s-t flows is $m-n+2$, and the dimension of the linear space of potentials is $n-1$. Since $(m-n+2)+(n-1) > m$, these two spaces have a nonzero intersection. This gives yet another proof that harmonic functions with poles s and t exist.

But considering functions on the edges, not only on the nodes, gives more.

THEOREM 4.8. *Let G be a connected graph and $s, t, \in V$. Then*

$$R(s,t) = \min \sum_{e \in E} g(e)^2, \tag{4.11}$$

where $g: E \to \mathbb{R}$ ranges over all functions satisfying $\sum_{e \in C} g(e) \geq 1$ for every s-t cut C. Furthermore,

$$F(s,t) = \min \sum_{e \in E} g(e)^2, \tag{4.12}$$

where $g: E \to \mathbb{R}$ ranges over all functions satisfying $\sum_{e \in E(P)} g(e) \geq 1$ for every s-t path P. In both optimization problems, the optimizing function g is a harmonic s-t flow.

We can restate these formulas as follows. For $g: E \to \mathbb{R}_+$, let $c_g(s,t)$ be the minimum capacity of an s-t cut (viewing g as a capacity function on the edges), and let $d_g(s,t)$ be the minimum length of an s-t path (viewing g as a length function on the edges). Then

$$R(s,t) = \min_{g: \mathsf{E} \to \mathbb{R}_+} \frac{|g|^2}{c_g(s,t)^2} \quad \text{and} \quad F(s,t) = \min_{g: \mathsf{E} \to \mathbb{R}_+} \frac{|g|^2}{d_g(s,t)^2}. \tag{4.13}$$

We can also write this as

$$R(s,t) = \max_{g: \mathsf{E} \to \mathbb{R}_+} \frac{d_g(s,t)^2}{|g|^2} \quad \text{and} \quad F(s,t) = \max_{g: \mathsf{E} \to \mathbb{R}_+} \frac{c_g(s,t)^2}{|g|^2}. \tag{4.14}$$

PROOF. We prove the first identity; the second one can be proved similarly, or by referring to a general duality result (Proposition C.2 in the Appendix). Let g be an optimizer in (4.11). We may assume that $g \geq 0$ (else, we can replace it by $|g|$). We replace each undirected edge e by two, oppositely directed edges e' and e'', and consider $g(e) = g(e') = g(e'')$ as the capacity of each of them. The condition on g says that every s-t cut has total capacity at least 1. By the Max-Flow-Min-Cut Theorem, the graph carries an s-t flow f of value 1 with $0 \leq f \leq g$. We may assume that $\min\{f(e'), f(e'')\} = 0$. Replacing g with $g'(e) = \max\{f(e'), f(e'')\}$ we get a function satisfying the conditions of the theorem, and decrease the objective function, unless $g' = g$. If we delete one of e' and e'' for which $f = 0$, we get an orientation of G on which g defines a flow of value 1.

So g minimizes $\sum_e g(e)^2$ over all s-t flows of value 1. Such flows form an affine subspace $A \subseteq \mathbb{R}^E$. The minimizer is the closest point of A to the origin; hence it is orthogonal to A. Shifting A to the origin we get the space of all circulations; the orthogonal complement of this space is the space A_0 of all potentials. So g is both an s-t flow of value 1 and a potential, i.e., a harmonic flow.

We can write $g(uv) = h(v)-h(u)$ for every edge uv, and h is a harmonic function with poles at s and t. By the electrical networks model, we have $R(s,t) = h(t)-h(s)$. Switching to the rubber band model, $\sum_e g(e)^2 = \sum_e \big(h(v)-h(u)\big)^2$ is twice the energy stored in the rubber bands, when s and t are stretched to a distance of $R(s,t)$. By our discussion of the rubber band model, this is $F(s,t)R(s,t)^2 = R(s,t)$. □

4.2.2. Combinatorial bounds. Choosing appropriate functions g in (4.13) or (4.14), we can prove useful combinatorial bounds. Let us start with an important example.

EXAMPLE 4.9. Consider the grid formed by the lattice points (i,j) with $|i|,|j| \le k$. Identify the boundary to a single node t, and let $s=(0,0)$. We show how to use the inequalities in the previous section to prove

$$R(s,t) \sim \frac{1}{4}\ln k. \tag{4.15}$$

To this end, let us define

$$g(e) = \frac{1}{r+1} \qquad (e \in E),$$

where r is the distance of e from s in the grid graph. (We had to add 1 so that the edges incident with s get finite weight.) This value of $g(e)$ is not the harmonic flow from s to t, which would be more complicated to express. To motivate, notice that (in the whole grid) there are $4r+4$ edges at distance r from s, forming a cut C_r of total weight is 4, independently of r. So from the point of view of these cuts, g "looks like" an s-t-flow of value 4.

We need to do some computations. First, E contains k of the cuts $C_0,\dots,C_{k-1}$, and is covered by the cuts $C_0,\dots,C_{2k-1}$, and hence

$$|g|^2 \ge \sum_{r=0}^{k-1}(4r+4)\frac{1}{(r+1)^2} = 4\sum_{j=1}^{k}\frac{1}{j} = 4H_k,$$

and similarly,

$$|g|^2 \le 4\sum_{j=1}^{2k}\frac{1}{j} \le 4H_k+4.$$

It is clear that every s-t path contains an edge at distance r from s for every r, and the shortest path contains exactly one of each, so

$$d_g(s,t) = H_k.$$

Finally, we have

$$c_g(s,t) = 4. \tag{4.16}$$

The proof of this is elementary, but takes a bit more work, and is left to the reader as Exercise 4.9.

Now by (4.13),

$$R(s,t) \le \frac{|g|^2}{c_g(s,t)^2} \le \frac{4H_k+4}{16} = \frac{H_k+1}{4},$$

and by (4.14),

$$R(s,t) \ge \frac{d_g(s,t)^2}{|g|^2} \ge \frac{H_k^2}{4H_k+4} > \frac{H_k-1}{4}.$$

These estimates prove (4.15). ♦

Let us return to general graphs, and formulate a corollary of Theorem 4.8.

COROLLARY 4.10. *Let $C_1, \dots, C_k$ be edge-disjoint cuts separating s and t. Then*

$$R(s,t) \ge \sum_{j=1}^{k} \frac{1}{|C_j|}. \tag{4.17}$$

PROOF. Let

$$g(e) = \begin{cases} \dfrac{1}{|C_j|}, & \text{if } e \in C_j \text{ for some } j, \\ 0, & \text{otherwise.} \end{cases}$$

Then

$$|g|^2 = \sum_{e\in E} g(e)^2 = \sum_{j=1}^{k} \sum_{e\in C_j} \frac{1}{|C_j|^2} = \sum_{j=1}^{k} \frac{1}{|C_j|}.$$

Since every s-t path intersects every cut C_j, we have

$$d_g(s,t) \ge \sum_{j=1}^{k} \frac{1}{|C_j|},$$

and so

$$R(s,t) \ge \frac{\left(\sum_{j=1}^{k} 1/|C_j|\right)^2}{\sum_{j=1}^{k} 1/|C_j|} = \sum_{j=1}^{k} \frac{1}{|C_j|}. \qquad \square$$

As a further application, we prove a bound on the "smoothness" of a harmonic function.

COROLLARY 4.11. *Let G be a graph, and let $h : V \to \mathbb{R}_+$ be harmonic outside a set $S \subseteq V$. Let $u, v \in V \setminus S$ such that $uv \in E$ and $h(u) \le h(v)$, and let $C_1, \dots, C_t$ be edge-disjoint cuts in $G \setminus uv$ separating u from $S' = \{x \in S : h(x) \le h(u)\}$. Then*

$$h(v) - h(u) \le \Big(\sum_{j=1}^{t} \frac{1}{|C_j|}\Big)^{-1} h(u).$$

PROOF. Recalling the rubber band model, we can think of each node $x \in S$ as nailed at $h(x)$. Then $|h(u) - h(v)|$ is the length to which uv is stretched. We may assume that $h(u) < h(v)$. Let $G' = (V', E')$ be the union of those paths from u to a nailed node along which h is monotone decreasing. We may assume that all nodes of $V \setminus V'$ are nailed; in particular, v is nailed.

We claim that deleting the edges in $E \setminus E'$, the length to which uv is stretched is increasing. This follows if we show that all nodes of V' move down. Suppose that this is not the case, and let $w \in V'$ be node that moves up the most, say by a distance of s. Then all its neighbors in V' move up by at most s, so they will pull down at least as much as before. In addition, any deleted edge pulled w upwards. It follows that equality must hold, so all neighbors of w must move up by the same distance s. Since w is on a path from u to a nailed node x, this implies that x must move up by s, which is impossible as x is nailed.

By a similar argument, we may assume that $h(x) = 0$ for all nodes in S'. It follows by (4.17) that the force with which G' pulls u down is

$$F_{G'}(s,u) = \frac{1}{R_{G'}(s,u)} \le \Big(\sum_{i=1}^{k} \frac{1}{|C_i|}\Big)^{-1} h(u)$$

(note: G' is stretched to a length of $h(u)$). This is counteracted by the force on the edge uv, which is $h(v) - h(u)$. □

We conclude with a special case showing that the bound in the previous example is nontrivial. We return to the planar grid, but more generally than before we consider a simple closed lattice polygon P, and the subgraph G of the grid formed by all edges and nodes on or inside P. Let $h_0 : V(P) \to [0,1]$ be given, and let h be the harmonic extension of h_0 to the interior of P. The following Corollary shows that far from the boundary, h is "smooth". (Such a result does not remain true in dimension 3, see Exercise 4.11.)

COROLLARY 4.12. *Let $uv \in E$, and let t be the graph distance of uv from P. Then*

$$|h(v) - h(u)| \le \frac{6}{\ln t}.$$

PROOF. Let C_j be the cut formed by all edges with one endpoint at Manhattan distance j and the other at Manhattan distance $j-1$ from u. (The Manhattan distance is the graph distance in the whole grid graph, which may be smaller than the graph distance in G.) Then $|C_j| \le 6j$, and the cuts $C_1, \dots, C_t$ separate u from P. So Corollary 4.11 implies that

$$h(v) - h(u) \le \Big(\sum_{j=1}^{t} \frac{1}{|C_j|}\Big)^{-1} \le \Big(\sum_{j=1}^{t} \frac{1}{6j}\Big)^{-1} < \frac{6}{\ln t}. \qquad \square$$

4.3. Inverting the Laplacian

In the proof of the existence of harmonic functions with given poles, we had to invert the Laplacian—in a sense. It will be useful to discuss the solution of an equation

$$Lx = a. \tag{4.18}$$

First, for this to be solvable, we must have $\mathbb{1}^{\mathsf{T}} a = 0$; indeed, $\mathbb{1}^{\mathsf{T}} a = \mathbb{1}^{\mathsf{T}} Lx = 0$. Second, we may assume that $\mathbb{1}^{\mathsf{T}} x = 0$; indeed, if x is a solution, then $x + \alpha\mathbb{1}$ is a solution for every real number α, and from this family of solutions, it suffices to find the one with $\mathbb{1}^{\mathsf{T}} x = 0$. Now if $\mathbb{1}^{\mathsf{T}} x = \mathbb{1}^{\mathsf{T}} a = 0$, or equivalently $Jx = Ja = 0$, then

$$(L+J)^{-1}a = (L+J)^{-1}Lx = (L+J)^{-1}(L+J)x = x, \tag{4.19}$$

thus giving an explicit solution of (4.18).

As an important special case of this simple computation, let G be an electrical network with unit resistances on the edges, and let us send a unit current from node s to node t. Then the potentials x of the nodes will satisfy the equation $Lx = e_s - e_t$, and hence $x = (L+J)^{-1}(e_s - e_t)$. The effective resistance is

$$R(s,t) = x_s - x_t = (e_s - e_t)^{\mathsf{T}}(L+J)^{-1}(e_s - e_t). \tag{4.20}$$

We describe two (related) algebraic methods to produce certain "inverses" of the Laplacian.

4.3.1. Poisson's formula and Green's function. We can rewrite formula (4.7) as follows. Let $\alpha_{v,S}(u)$ denote the probability that the first point in S of a random walk starting at v is u (where $S \subset V$ is a nonempty set). We call $\alpha_{v,S} : S \to \mathbb{R}_+$ the *hitting distribution* (from v on S). Using this distribution, the harmonic extension of a function h_0 on S can be expressed as

$$h(v) = \sum_{u \in S} \alpha_{v,S}(u) h_0(u). \tag{4.21}$$

This equation can be regarded as a discrete version of Poisson's Formula.

Set $T = V \setminus S$. To keep our notation a bit simpler, we write (in this section) f_T for the restriction $f|_T$ of a function f to a set T. We are looking for a function $h : V \to \mathbb{R}$ such that

$$(Lh)_T = 0 \qquad \text{and} \qquad h_S = h_0. \tag{4.22}$$

Let us partition the Laplacian of G as

$$L = \begin{pmatrix} L_T & L_{TS} \\ L_{ST} & L_S \end{pmatrix}, \tag{4.23}$$

(where the first block of rows and columns corresponds to nodes in T; note that $L_{ST} = -A_{ST}$). The matrix L_T will play an important role, and we need two simple properties of it: L_T is invertible and its inverse is nonnegative. Both properties mentioned can be proved in (slightly) larger generality, and we will need them elsewhere in this more general form.

LEMMA 4.13. *Let $M \in \mathbb{R}^{V \times V}$ be a positive semidefinite matrix whose nullspace is one-dimensional and it is spanned by a vector with nonzero coordinates. Then every proper principal minor of M is positive definite.*

PROOF. Let M_T denote the principal minor of M formed by the rows and columns indexed by elements of $T \subset V$. The matrix M_T is trivially positive semidefinite. If $M_T x = 0$ for some $0 \neq x \in \mathbb{R}^T$, then for the vector $y = \binom{x}{0} \in \mathbb{R}^V$, we have $y^{\mathsf{T}} M y = x^{\mathsf{T}} M_T x = 0$. Since M is positive semidefinite, this shows that y is in the nullspace of M. But y cannot be parallel to the vector spanning the nullspace of M, a contradiction. □

LEMMA 4.14. *Let M be a positive definite matrix whose off-diagonal entries are nonpositive. Then $M^{-1} \geq 0$.*

PROOF. We may assume that the largest eigenvalue of M is 1. Then its diagonal entries are at most 1, and hence $B = I - M$ is positive semidefinite and all entries of it are nonnegative. Since M is positive definite, all eigenvalues of B are strictly less than 1, and hence the series expansion

$$M^{-1} = (I-B)^{-1} = I + B + B^2 + \dots$$

is convergent. Since all entries of all terms in the last sum are nonnegative, the same holds for M^{-1}. □

Coming back to solving equation (4.22) for the function h, we see that the restriction h_S is given, and to get the restriction h_T, we only have to solve the equation

$$L_T h_T + L_{TS} h_S = 0.$$

Since L_T is invertible by Lemma 4.13, this system can be solved for h_T, and we get that

$$h_T = -L_T^{-1} L_{TS} h_S \tag{4.24}$$

is the unique harmonic extension of h_S.

The matrix L_T^{-1}, considered as a function in two variables, is a discrete analogue of Green's function in analysis, and we'll call it the *(discrete) Green function* of the graph G, on the set T. (To relate this to the Green function in analysis, think of T as an open domain and of S, as its boundary. It is easy to see that nodes of S that are not adjacent to T play no role in the arguments below.) We denote L_T^{-1} by $\mathcal{G}_T$ or by $\mathcal{G}$ if T is understood.

Formula (4.24) expresses h_T as a linear combination of the values of h_S:

$$h(v) = \sum_{u\in S} \sum_{w\in T\cap N(u)} \mathcal{G}_{vw} h(u) \qquad (v \in T). \tag{4.25}$$

We have seen such an expression using the hitting probabilities on S. Comparing the two, and using the uniqueness of harmonic extensions, we get that the probability that a random walk starting at a node $v \in T$ first hits the set S at node u is given by

$$\alpha_{v,S}(u) = -(\mathcal{G} L_{TS})_{vu} = \sum_{w\in N(u)\cap T} \mathcal{G}_{vw}. \tag{4.26}$$

Generalizing equation (4.25), the Green function provides an important formula to express the values of an arbitrary function in terms of its harmonicity defect on T and its values on S:

LEMMA 4.15. *For every function $f : V \to \mathbb{R}$ and $v \in T$,*

$$f(v) = \sum_{w\in T} \mathcal{G}_{vw}\Big((Lf)(w) + \sum_{u\in S\cap N(w)} f(u)\Big).$$

We can write the formula more compactly:

$$f_T = \mathcal{G}(Lf)_T - \mathcal{G} L_{TS} f_S, \tag{4.27}$$

or less compactly:

$$f(v) = \sum_{w\in T} \mathcal{G}_{vw}(Lf)(w) + \sum_{u\in S} \sum_{w\in T\cap N(u)} \mathcal{G}_{vw} f(u). \tag{4.28}$$

If f is the harmonic extension of f_S, then the first term vanishes, and the second gives (4.25).

PROOF. From the definition of the partition of L, we have

$$(Lf)_T = L_T f_T + L_{TS} f_S.$$

Applying $\mathcal{G}$ from the left,

$$\mathcal{G}(Lf)_T = \mathcal{G} L_T f_T + \mathcal{G} L_{TS} f_S = f_T + \mathcal{G} L_{TS} f_S.$$

Rearranging, we get (4.27). □

Let us conclude with a nice application of the Green function, raised by Grassberger (private correspondence). Let G be a simple graph, and fix a node $v \in V(G)$. For two nodes $i, j \in N(v)$, let σ_{ij} denote the probability that a random walk starting at v and leaving through the edge vi, first returns to v through the edge vj.

COROLLARY 4.16. *For any two distinct nodes $i, j \in N(v)$, we have $\sigma_{ij} < \sigma_{ii}$.*

PROOF. Let $T = V(G) \setminus \{v\}$ and $\mathcal{G} = L_T^{-1}$ the corresponding Green function. We claim that

$$\sigma_{ij} = \mathcal{G}_{ij}. \tag{4.29}$$

Indeed, let us split v into $\deg(v)$ nodes i' ($i \in N(v)$), to get a graph G'. So $ii' \in E(G')$. Let $S = \{i' : i \in N(v)\}$. This operation does not change the Green function on T. The probability σ_{ij} can be interpreted as the probability that a random walk on G' starting at i hits S at j'. By (4.26), we have

$$\sigma_{ij} = \sum_{w \in N(j') \cap T} \mathcal{G}_{iw} = \mathcal{G}_{ij},$$

proving (4.29).

Since $(\sigma_{ij} : j \in N(v))$ is a probability distribution, we have

$$\sum_{j \in N(v)} \mathcal{G}_{ij} = 1 \qquad (i \in V), \tag{4.30}$$

It also follows that the matrix $(\sigma_{ij})_{i,j \in N(v)}$ is symmetric and positive definite.

Let $i \in N(v)$, and let us pivot in $\mathcal{G}$ on the entry (i, i), and then delete the i-th row and column. The resulting matrix is the inverse of the matrix $L_{V \setminus \{v,i\}}$, which is a positive definite matrix with nonpositive offdiagonal entries, and hence its inverse is nonnegative by Lemma 4.14. This implies that

$$\mathcal{G}_{ii}\mathcal{G}_{jk} \geq \mathcal{G}_{ji}\mathcal{G}_{ik} \tag{4.31}$$

for all $i, k, j \in N(v)$. Note that strict inequality holds for $k = j$, since $\mathcal{G}$ is positive definite. Summing over all $k \in N(v)$, and using (4.30), we get the Corollary. □

4.3.2. Generalized inverses, random walks and resistance. The Laplacian L is a singular matrix, and hence it does not have an inverse. But, as every matrix, it has a *pseudoinverse.* While there are different notions of pseudoinverses, in this short section we'll mean that introduced by E.H. Moore, A. Bjerhammar and R. Penrose. The pseudoinverse A^{-1} for an arbitrary matrix A can be defined by the conditions

$$AA^{-1}A = A, \qquad A^{-1}AA^{-1} = A^{-1}, \tag{4.32}$$

and

$$AA^{-1},\ A^{-1}A \text{ are symmetric.} \tag{4.33}$$

It is not very hard to prove that a matrix A^{-1} with these properties always exists, and is uniquely determined (Exercise 4.12). We shall need this construction for the Laplacian of a connected graph G only, in which case it is almost trivial: writing

$$L = \sum_{k=2}^{n} \mu_k v_k v_k^{\mathsf{T}},$$

where $\mu_1 = 0 < \mu_2 \le \dots \le \mu_n$ are the eigenvalues of the Laplacian, and $v_1, \dots, v_n$ are the corresponding eigenvectors, the matrix

$$L^{-1} = \sum_{k=2}^{n} \frac{1}{\mu_k} v_k v_k^{\mathsf{T}}$$

satisfies the conditions of the pseudoinverse, which is straightforward to check. It is also easy to compute that

$$L^{-1}L = LL^{-1} = I - v_1 v_1^{\mathsf{T}} = I - \frac{1}{n}J. \tag{4.34}$$

To compute the pseudoinverse of L, we don't have to find the spectral decomposition of L. It suffices to use again that $L+J$ is an invertible matrix with spectral decomposition

$$L + J = nv_1v_1^{\mathsf{T}} + \sum_{k=2}^{n} \mu_k v_k v_k^{\mathsf{T}},$$

and its inverse is

$$(L+J)^{-1} = \frac{1}{n} v_1 v_1^{\mathsf{T}} + \sum_{k=2}^{n} \frac{1}{\mu_k} v_k v_k^{\mathsf{T}} = \frac{1}{n^2} J + L^{-1}.$$

Hence

$$L^{-1} = (L+J)^{-1} - \frac{1}{n^2} J. \tag{4.35}$$

(We could use in these arguments a matrix $L+cJ$ with any nonzero c instead of $L+J$; it would lead to similar formulas.)

We can go a little further: since L is positive semidefinite, we can raise its nonzero eigenvalues to any real power, and define

$$L^t = \sum_{k=2}^{n} \mu_k^t v_k v_k^{\mathsf{T}}$$

for $t \in \mathbb{R}$. This way we get a family of symmetric matrices, so that $L^1 = L$, L^t is a continuous function of t, and $L^{t+s} = L^t L^s$ for $s, t \in \mathbb{R}$. These matrices can be used to construct an interesting geometric representation, which encodes the effective resistances in the graph (cf. also equation (4.20) above):

PROPOSITION 4.17. *Let G be a connected graph, with a unit resistance on each edge. Let $\mathbf{u}_i \in \mathbb{R}^V$ $(i \in V)$ be the columns of $L^{-1/2}$. Then*

$$R(i,j) = (L^{-1})_{ii} + (L^{-1})_{jj} - 2(L^{-1})_{ij} = |\mathbf{u}_j - \mathbf{u}_i|^2.$$

for any two nodes i and j.

PROOF. Let us push a unit current from i to j, by keeping i and j at appropriate voltages. Let x_k be the voltage of node k. Only the differences of these voltages matter, so we may assume that $\mathbb{1}x = 0$. Then the Kirchhoff node equations say that $Lx = e_j - e_i$, and using (4.34), we get that

$$x = L^{-1}(e_j - e_i).$$

So

$$R(i,j) = x_j - x_i = (e_j - e_i)^{\mathsf{T}} L^{-1} (e_j - e_i) L_{ii}^{-1} + L_{jj}^{-1} - 2L_{ij}^{-1},$$

which we can also write as

$$R(i,j) = |L^{-1/2}(e_j - e_i)|^2 = |\mathbf{u}_j - \mathbf{u}_i|^2. \qquad \square$$

These constructions work for edge-weighted Laplacians as well.

REMARK 4.18. For invertible matrices, their adjoint is a scalar multiple of their inverse. The adjoint can be defined for singular matrices; is the generalized inverse of the Laplacian related to its adjoint? Not really. Since L has rank $n-1$, the adjoint is nonzero. However, from the equation $LL^{\mathrm{adj}} = \det(L)I = 0$ we see that every column and (by symmetry) every row of L^{adj} is a multiple of $\mathbb{1}$, and so $L^{\mathrm{adj}} = cJ$. The only interesting information contained in the adjoint is the constant c, which is the number of spanning trees in the graph (Exercise 4.6).

EXERCISE 4.1. Let G be a connected graph and T, a spanning tree of G. For each edge $e \in E \setminus E(T)$, let C_e denote the cycle of G consisting of e and the path in T connecting the endpoints of e. Prove that if Kirchhoff's Voltage Law holds for these cycles C_e, then it holds for all cycles.

EXERCISE 4.2. Prove that for every connected graph G, the effective resistances $R(i,j)$ as well as their square roots satisfy the triangle inequality: for any three nodes a, b and c, $R(a,c) \le R(a,b) + R(b,c)$, and $\sqrt{R(a,c)} \le \sqrt{R(a,b)} + \sqrt{R(b,c)}$.

EXERCISE 4.3. Let G be a connected graph and let $\mathbf{u}_i \in \mathbb{R}^V$ $(i \in V)$ be the i-th column of $L^{-1/2}$. Prove that for any three nodes $i, j, k \in V$, the triangle $\mathbf{u}_i\mathbf{u}_j\mathbf{u}_k$ is not obtuse. When does it have a right angle?

EXERCISE 4.4. A function $f : V \to \mathbb{R}$ is *subharmonic* at a node $v \in V$ if

$$\frac{1}{\deg(v)} \sum_{u \in N(v)} h(u) \ge h(v). \tag{4.36}$$

Let $S \subseteq V$ be a nonempty set, and let $f : V \to \mathbb{R}$ be subharmonic at every node of $V \setminus S$. Let $v \in V \setminus S$, and let $a_v \in S$ be a random node from the hitting distribution on S (starting at v). Prove that

$$f(v) \le \mathsf{E}_v(f(a_v)).$$

EXERCISE 4.5. Show by examples that the commute time $\mathsf{comm}(a,b)$ can both increase and decrease when we add an edge to the graph.

EXERCISE 4.6. Let G be a connected graph, and let $\mathcal{T}(G)$ denote the number of spanning trees of G. Let L' be the matrix obtained by deleting from L a row and the corresponding column. (a) Prove that $\det(L') = \mathcal{T}(G)$. (b) Prove that $\det(L+J) = n^2\mathcal{T}(G)$. (c) Use these facts to prove Cauchy's Formula: $\mathcal{T}(K_n) = n^{n-2}$.

EXERCISE 4.7. Let G' denote the graph obtained from the graph G by identifying nodes a and b. Prove that $R(a,b) = \mathcal{T}(G')/\mathcal{T}(G)$.

EXERCISE 4.8. Let G be a graph and $e, f \in E$, $e \ne f$. Prove that $\mathcal{T}(G)\mathcal{T}(G \setminus \{e,f\}) \le \mathcal{T}(G \setminus e)\mathcal{T}(G \setminus f)$.

EXERCISE 4.9. To every edge e of the standard grid in the plane, let us assign the weight $g(e) = 1/(r+1)$, where r is the distance of e from the origin. Let S be a set of edges that blocks every infinite path from the origin. Prove that $g(S) \ge 4$.

EXERCISE 4.10. Let A be an irreducible positive definite matrix whose off-diagonal entries are nonpositive. Prove that all entries of A^{-1} are positive.

EXERCISE 4.11. Show by an example that the smoothness property stated in Corollary 4.12 does not remain valid for the 3-dimensional grid.

EXERCISE 4.12. Prove that if the rows of a matrix A are linearly independent, then AA^{T} is nonsingular, and $A^{-1} = A^{\mathsf{T}}(AA^{\mathsf{T}})^{-1}$ satisfies the conditions for the pseudoinverse of A.

EXERCISE 4.13. Prove that every matrix has a unique pseudoinverse.

EXERCISE 4.14. Prove that for a connected d-regular graph G, the pseudoinverse of L can be used to express hitting times by the following formula:

$$H_{ij} = dn\big((L^{-1})_{ij} - (L^{-1})_{ii}\big).$$

Extend this formula to non-regular graphs, using the pseudoinverse of the scaled Laplacian matrix $\widehat{L} = D^{-1/2}LD^{-1/2}$ instead of L^{-1}.

EXERCISE 4.15. We have seen that for every graph G one can define a family of symmetric matrices L^t $(t \in \mathbb{R})$ so that $L^1 = L$, L^t is a continuous function of t, and $L^{t+s} = L^t L^s$ for $s, t \in \mathbb{R}$. Prove that such a family of symmetric matrices is unique.

EXERCISE 4.16. Let M be a positive definite $n \times n$ matrix whose off-diagonal entries are nonpositive. Let N be a submatrix of M^{-1} formed by rows $1, \dots, k, i$ and columns $1, \dots, k, j$, where $k < i, j \le n$. Prove that $\det(N) \ge 0$.

CHAPTER 5

Coin Representation

One of the deepest representation results for planar graphs is the "coin representation" (representation by touching circles) [**Koebe 1936**]. This result has numerous versions, extensions, and applications. It gives a Steinitz representation by a very special polytope (the Cage Theorem), and it has close relations with the Riemann Mapping Theorem in complex analysis.

5.1. Koebe's theorem

We say that a family of (circular) disks is *nonoverlapping*, if their interiors are disjoint. Such a family gives rise to a *tangency graph*: we represent each member of the family by a node, and connect two nodes if the corresponding disks are touching. It is natural to turn this into a geometric representation: we represent each disk by its center, and each edge by the segment connecting the corresponding centers. Trivially, this edge passes trough the point of tangency of the two circles, and different edges do not intersect except if they have a common endpoint. So we get a planar map (Figure 5.1). Koebe's Theorem will provide a converse to this easy observation.

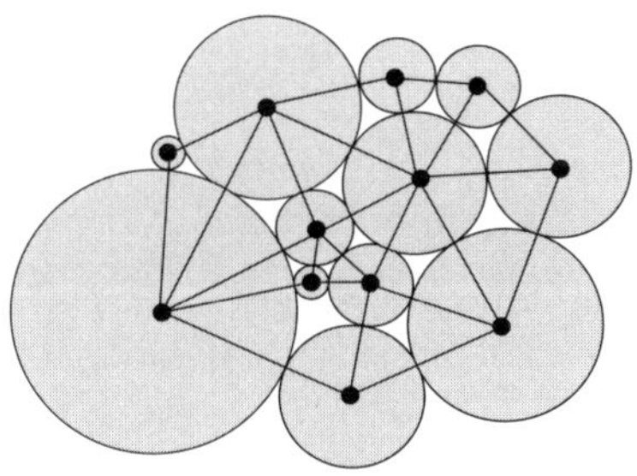

FIGURE 5.1. Tangency graph of a set of disks

The same graph can, of course, be represented by many systems of disks. For example, a cycle can be represented by any necklace of pearls—these may be equal, or different, and laid out in many different ways (Figure 5.2).

If the graph is a triangulation, then the representation is "essentially unique", as we will see. But even in this case, we can modify the set of disks: for example, we can apply inversion to all of them with respect to a further circle C (see Appendix A.4.2 for notions and results concerning circles, relevant throughout this chapter). If the center of C is not covered by any of the disks, then this inversion results in another representation by a family of disks. If the center of C is in the interior of one of the disks (call this D), then we get a similar picture, except that the other

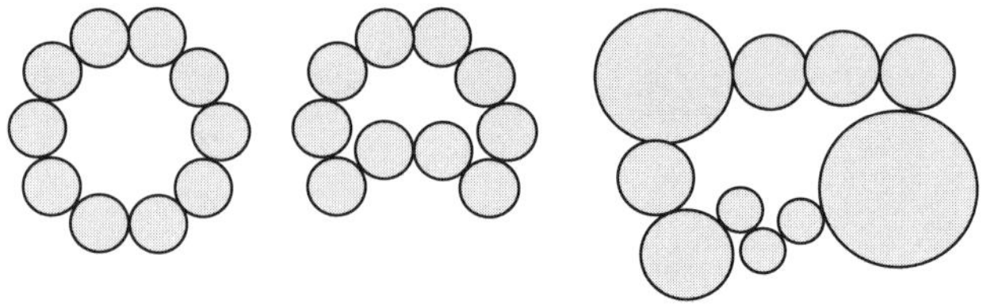

FIGURE 5.2. Different representations of a 10-cycle by touching disks

disks will be contained in the interior of the image D' of D (Figure 5.3). We say that we have a *coin representation with D turned inside out*. We are going to make good use of this version of the coin representation in Section 5.3.

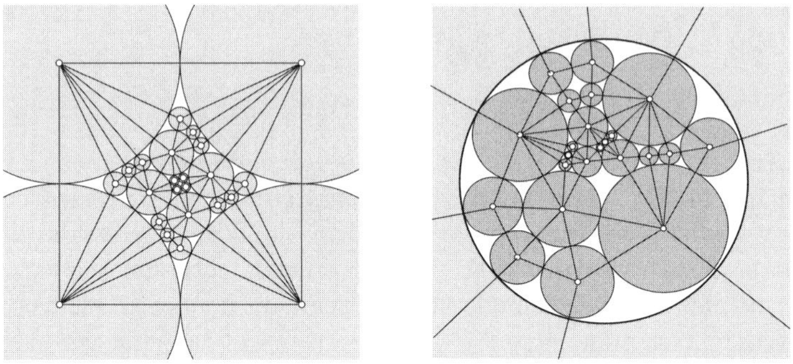

FIGURE 5.3. Representation of a planar graph by touching disks, and a representation of the same graph, where the exterior of one of the disks plays the role of its "body".

After these remarks, we are ready to state the fundamental theorem of Koebe.

THEOREM 5.1 (Koebe's Theorem). *Every planar graph can be represented as the tangency graph of a family of nonoverlapping circular disks.*

Such a family of circular disks is called a *Koebe representation*, or *coin representation* of the graph. We'll prove this theorem after formulating a stronger version.

Koebe's Theorem can be strengthened and generalized in various ways. To motivate the extension we are going to discuss next, consider a bounded triangular face $p = abc$ of the tangency map G of a family $\mathcal{F}$ of non-overlapping circles, where a is the center of the circle C_a etc. Let D_p denote the inscribed circle of p, touching the edges of the triangle at points u, v and w (Figure 5.4). Then $|a-u| = |a-v|$ (as these segments are tangents to D_p from the same point), and similarly $|b-u| = |b-w|$ and $|c-v| = |c-w|$. So the circle C'_a about a with radius $|a-u| = |a-v|$, and the circles C'_b and C'_c defined analogously are mutually tangent. It is easy to see that this implies that $C'_a = C_a$, $C'_b = C_b$ and $C'_c = C_c$. The circle D_p intersects each of these circles orthogonally.

Now consider another bounded triangular country $q = abd$ of G, sharing the ab with abc. The previous argument implies that the inscribed circle D_q of q must also touch the edge ab at u, and it is also tangent to D_p at this point. If all countries are triangular, then the family $\mathcal{F}^*$ of inscribed circles of bounded

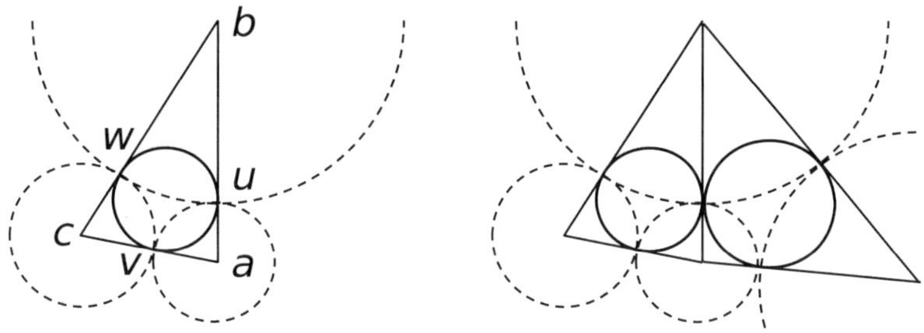

FIGURE 5.4. Left: A triangle in the tangency graph and its inscribed circle. Right: For two triangles sharing an edge, the inscribe circles touch the edge at the same point.

countries is well defined, and its tangency graph is the dual of G (with capital of the unbounded country omitted). We can also include the circle inscribed in the unbounded country, but its "center" should be thought of as the "point at infinity".

What happens if not all countries are triangles? A non-triangular country may not even have an inscribed circle (touching all edges). The main result below will imply that every 3-connected planar graph G has a special representation as the tangency graph of a family of circles in which all countries have inscribed circles.

To be precise, we define a *double circle representation* of a planar map G as two families of circles in the plane, $(C_i : i \in V)$ and $(D_p : p \in V^*)$, so that for every edge ij, bordering countries p and q, the following holds: the circles C_i and C_j are tangent at a point $\mathbf{x}_{ij}$; the circles D_p and D_q are tangent at the same point $\mathbf{x}_{ij}$; and the circles D_p and D_q intersect the circles C_i and C_j at this point orthogonally. Furthermore, the interiors of the circular discs $\widehat{C}_i$ bounded by the circles C_i are disjoint and so are the disks $\widehat{D}_j$, except that the circle D_{p_∞} representing the unbounded country contains all the other circles D_p in its interior.

The conditions in the definition of a double circle representation imply other useful properties. The proof of these facts is left to the reader as an exercise (cf. Figure 5.5).

PROPOSITION 5.2. *Let $(C_i : i \in V;\ D_p : p \in V^*)$ be a double circle representation of a 3-connected planar graph G. Then* (a) *Every country p is bounded by a convex polygon, and the circle D_p touches every edge of this polygon.* (b) *For every bounded country p, the circles D_p and C_i $(i \in V(p))$ cover p.* (c) *If $i \in V$ is not incident with $p \in V^*$, then $\widehat{C}_i$ and $\widehat{D}_p$ are disjoint.*

The main theorem in this chapter asserts that such representations do exist.

THEOREM 5.3. *Every 3-connected planar graph has a double circle representation in the plane.*

The double circle representation is not unique (we can apply circle-preserving transformations to it), but to make it unique, we only need to fix the circles representing the nodes of a triangular country:

COROLLARY 5.4. *Let G be a 3-connected planar graph with a triangular country abc. Fixing three mutually tangent openly disjoint disks in the plane to represent a, b and c, there is a unique extension to a double circle representation of G with abc as the unbounded country.*

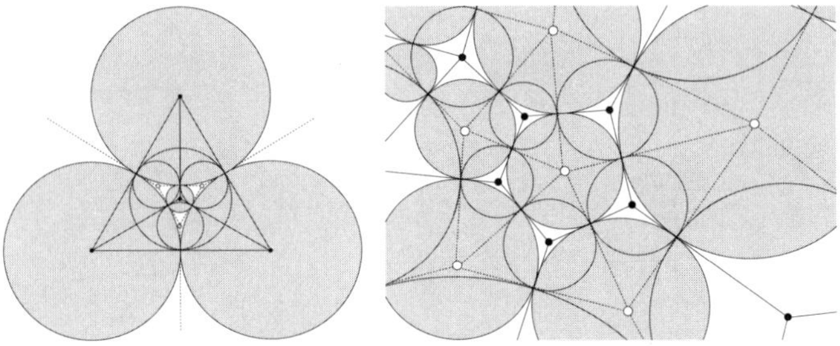

FIGURE 5.5. Left: a double circle representation K_4 and its dual (another complete graph on 4 nodes). For the circle representing the capital of the unbounded country, the exterior domain should be considered as the disk it "bounds". Right: a piece of a double circle representation of a planar graph and its dual.

In the next sections, we will prove these basic results.

5.1.1. Angles and radii. Let G be a 3-connected planar map. We transform G so that a triangular country p_∞ in G becomes the unbounded country; if G has no triangular country, then we interchange the roles of G and G^* (see Proposition 2.1). Let a, b, c be the nodes of p_∞. For every node $i \in V$, let $F(i)$ denote the set of bounded countries containing i, and for every country p, let $V(p)$ denote the set of nodes on the boundary of p. Let $U = V \cup V^* \setminus p_\infty$, and let J denote the set of pairs ip with $p \in V^* \setminus p_\infty$ and $i \in V(p)$.

Let us start with assigning a positive real number r_u to every node $u \in U$. Think of this as a guess for the radius of the circle representing u (we do not guess the radius for the circle representing p_∞; this will be easy to add at the end). For every $ip \in J$, we define

$$\alpha_{ip} = \arctan \frac{r_p}{r_i} \qquad \text{and} \qquad \alpha_{pi} = \arctan \frac{r_i}{r_p}. \tag{5.1}$$

We can think of these numbers as guesses for the angles of the right triangle formed by the centers of circles representing a node, an adjacent dual node, and the intersection point of these circles. It is clear that

$$\alpha_{ip} + \alpha_{pi} = \frac{\pi}{2}. \tag{5.2}$$

Now suppose that the circles with radii r_u form a proper double circle representation, in which the triangle abc is regular. Consider a node $i \in V$. For every country p incident with i, $2\alpha_{ip}$ is the angle of the corner of p at i (Figure 5.6). Since these corners fill out the available angle around i, we have

$$\sum_{p \in F(i)} \alpha_{ip} = \begin{cases} \pi, & \text{if } i \in V \setminus \{a, b, c\}, \\ \pi/6, & \text{if } i \in \{a, b, c\}. \end{cases} \tag{5.3}$$

We can derive similar conditions for the countries:

$$\sum_{i \in V(p)} \alpha_{pi} = \pi \qquad (p \in V^* \setminus p_\infty). \tag{5.4}$$

The key to the construction of a double circle representation is that these conditions are sufficient.

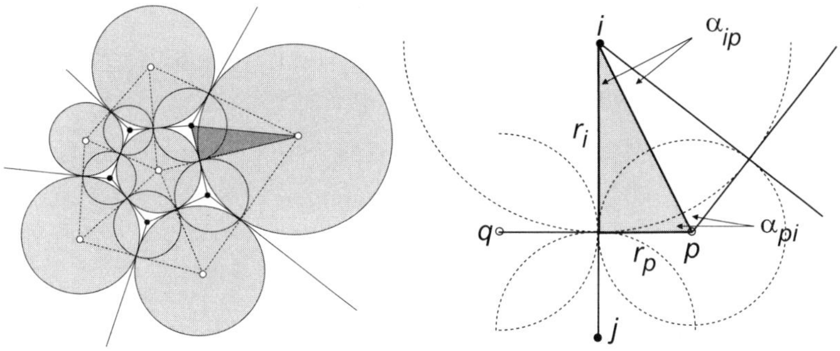

FIGURE 5.6. A right triangle in the double circle representation and its parameters

LEMMA 5.5. *Suppose that the radii* $r_u > 0$ $(u \in U)$ *are chosen so that the angles defined by* (5.1) *satisfy* (5.3)*and* (5.4). *Then there is a double circle representation with these radii.*

PROOF. Construct two right triangles with legs r_i and r_p for every $ip \in J$, one with each orientation, and glue these two triangles together along their hypotenuse to get a *kite* K_{ip}. Starting from node a of p_∞, put down all kites K_{ap} $(p \in F(a))$ in the order of the corresponding countries in a planar embedding of G. By (5.3), these will fill an angle of $\pi/3$ at a. Now proceed to a bounded country p_1 incident with a, and put down all the remaining kites $K_{p_1 i}$ in the order in which the nodes of p_1 follow each other on the boundary of p_1. By (5.4), these triangles will cover a neighborhood of p_1. We proceed similarly to the other bounded countries containing i_1, then to the other nodes of p_1, etc. The conditions (5.3) and (5.4) guarantee that we tile a regular triangle.

Let C_i be the circle with radius r_i about the position of node i constructed above, and define D_p analogously. We still need to define D_{p_∞}. It is clear that p_∞ is drawn as a regular triangle, and hence we necessarily have $r_a = r_b = r_c$. We define D_{p_∞} as the inscribed circle of the regular triangle abc.

It is clear from the construction that we get a double circle representation of G. □

Of course, we cannot expect conditions (5.3) and (5.4) to hold for an arbitrary choice of the radii r_u (equation (5.2) is valid for any choice of positive real numbers r_u). In the next sections we will develop a method to construct radii satisfying the conditions in Lemma 5.5; but first we need two simple lemmas, going in this direction, but not far enough.

For a given assignment of radii r_u $(u \in U)$, consider the violation of the conditions in Lemma 5.5. To be precise, for $u \in U$, we define its *defect* as follows:

$$\delta_u = \begin{cases} \sum_{p \in F(u)} \alpha_{up} - \pi, & \text{if } u \in V \setminus \{a,b,c\}, \\ \sum_{p \in F(u)} \alpha_{up} - \frac{\pi}{6}, & \text{if } u \in \{a,b,c\}, \\ \sum_{i \in V(u)} \alpha_{pu} - \pi, & \text{if } u \in V^* \setminus p_\infty. \end{cases} \tag{5.5}$$

(These defects may be positive or negative.) While of course an arbitrary choice of the radii r_u will not guarantee that all these defects are 0, the following lemma shows that this is true at least "on the average":

LEMMA 5.6. *For every assignment of radii, we have*

$$\sum_{u\in U}\delta_u = 0.$$

PROOF. From the definition,

$$\sum_{u\in U}\delta_u = \sum_{i\in V\setminus\{a,b,c\}}\left(\sum_{p\in F(i)}\alpha_{ip}-\pi\right)+\sum_{i\in\{a,b,c\}}\left(\sum_{p\in F(i)}\alpha_{ip}-\frac{\pi}{6}\right)$$
$$+\sum_{p\in V^*\setminus p_\infty}\left(\sum_{i\in V(p)}\alpha_{pi}-\pi\right).$$

Every pair $ip\in J$ contributes $\alpha_{ip}+\alpha_{pi}=\pi/2$. Since $|J|=2m-3$, we get

$$(2m-3)\frac{\pi}{2}-(n-3)\pi-3\frac{\pi}{6}-(f-1)\pi=(m-n-f+2)\pi$$

By Euler's formula, this proves the lemma. □

Next we prove that conditions (5.3) and (5.4), considered as linear equations for the angles α_{ip}, can be satisfied. (We do not claim here that the solutions are obtained in the form (5.1).)

LEMMA 5.7. *Let G be a planar map with a triangular unbounded country $p_\infty = abc$. Then there are real numbers $0<\beta_{ip}<\pi/2$ $(ij\in J)$ satisfying*

$$\sum_{p\in F(i)}\beta_{ip}=\begin{cases}\pi, & \text{if } i\in V\setminus\{a,b,c\},\\ \pi/6, & \text{if } i\in\{a,b,c\}.\end{cases}$$

and

$$\sum_{i\in V(p)}\left(\frac{\pi}{2}-\beta_{ip}\right)=\pi \qquad (p\in V^*\setminus p_\infty).$$

PROOF. Consider the Tutte rubber band embedding of the graph, with a,b,c nailed to the vertices of a regular triangle. For $i\in V(p)$, let β_{pi} denote the angle of the polygon p at the vertex i. Then the conclusions are easily checked. □

5.1.2. An almost-proof. To prove Theorem 5.3, we want to prove the existence of an assignment of radii so that all defects are zero. Let us start with an arbitrary assignment; we measure its "badness" by its *total defect*

$$\mathcal{D}=\sum_{u\in U}|\delta_u|.$$

How to modify the radii so that we reduce the total defect? The key observation is the following. Let i be a node with positive defect $\delta_i>0$. Suppose that we increase the radius r_i, while keep the other radii fixed. Then α_{ip} decreases for every country $p\in F(i)$, and correspondingly α_{pi} increases, but nothing else changes. Hence δ_i decreases, δ_p increases for every $p\in F(i)$, and all the other defects remain unchanged.

Since the sum of (signed) defects remains the same by Lemma 5.6, we can say that some of the defect of node i is distributed to its neighbors (in the graph (U, J)). (It is more difficult to describe in what proportion this defect is distributed, and we will try to avoid to have to explicitly describe this).

It is clear that the total defect does not increase, and if at least one of the countries incident with i had negative defect, then in fact the total defect strictly decreases.

The same argument applies to nodes in V^*. Thus if the graph (U, J) contains a node with positive defect, which is adjacent to a node with negative defect, then we can decrease the total defect.

Unless we already have a double circle packing, there certainly are nodes in U with positive defect and other nodes with negative defect. What if these positive-defect nodes are isolated from the negative-defect nodes by a sea of zero-defect nodes? This is only a minor problem: we can repeatedly distribute some of the defect of positive-defect nodes, so that more and more zero-defect nodes acquire positive defect, until eventually we reach a negative-defect node.

Thus if we start with an assignment of radii that minimizes the total defect, then all defects must be 0, and (as we have seen) we can construct a double circle representation from this.

Are we done? What remains to be proved is that there is a choice of positive radii that minimizes the total defect. This kind of argument about the existence of minimizers is usually straightforward based on some kind of compactness argument, but in this case it is more awkward to prove this existence directly. Below we give a different proof, whose idea is less transparent, but where the existence of a minimizer will be easy to prove.

5.1.3. A more sophisticated objective function. The proof of Theorem 5.3 to be described here is from [**Colin de Verdière 1991**]. It might help to follow the argument below if we point out that we work with the variables $x_i = \ln r_i$.

Proof of Theorem 5.3. For $x \in \mathbb{R}$, define

$$\phi(x) = \int_{-\infty}^{x} \arctan(e^t)\, dt.$$

It is easy to verify that the function ϕ is monotone increasing, strictly convex, and

$$\frac{\pi}{2}|x|_+ < \phi(x) < 1 + \frac{\pi}{2}|x|_+ \tag{5.6}$$

for all $x \in \mathbb{R}$.

Let $x \in \mathbb{R}^U$. Using the numbers β_{ip} from Lemma 5.7, consider the function

$$\Phi(x) = \sum_{ip \in J} \big(\phi(x_p - x_i) - \beta_{ip}(x_p - x_i)\big).$$

It follows from the convexity of ϕ that Φ is convex; it is not strictly convex, because it is constant if all differences $x_p - x_i$ remain constant. But it is easy to see that this is the only direction in which it is not strictly convex: if we fix $x_a = 0$, then it will be a strictly convex function of the remaining variables.

Furthermore, $\Phi(x)\to\infty$ if $x_a=0$ but $|x|\to\infty$. This follows easily from (5.6):

$$\Phi(x)=\sum_{ip\in J}\big(\phi(x_p-x_i)-\beta_{ip}(x_p-x_i)\big)\ge\sum_{ip\in J}\Big(\frac{\pi}{2}|x_p-x_i|_+-\beta_{ip}(x_p-x_i)\Big)$$
$$=\sum_{ip\in J}\Big(\beta_{pi}|x_p-x_i|_+ +\beta_{ip}|x_p-x_i|_-\Big)\ge\sum_{ip\in J}\min\Big\{\beta_{ip},\beta_{pi}\Big\}\,|x_p-x_i|$$

(where $\beta_{pi}=\pi/2-\beta_{ip}$). Since $\beta_{ip},\beta_{pi}>0$, each term is nonnegative, and any given term tends to infinity if the corresponding difference $|x_i-x_p|$ tends to infinity. If x_a remains 0 but $|x|\to\infty$, then (due to the connectivity of $G^\diamond$), at least one difference $|x_i-x_p|$ must tend to infinity, and so $\Phi(x)\to\infty$.

It follows from these properties of Φ that it attains its minimum at some point $y\in\mathbb{R}^U$, and here

$$\frac{\partial\Phi}{\partial x_u}(y)=0\qquad(u\in U).\tag{5.7}$$

We define the radii $r_i=e^{y_i}$ $(i\in U)$, and the values α_{ip} according to (5.1). To complete the proof, it suffices to check that these radii and angles satisfy the conditions of Lemma 5.5.

Let $i\in V$, then

$$\frac{\partial}{\partial x_i}\Phi(y)=-\sum_{p\in F(i)}\phi'(y_p-y_i)+\sum_{p\in F(i)}\beta_{ip}=-\sum_{p\in F(i)}\arctan(e^{y_p-y_i})+\sum_{p\in F(i)}\beta_{ip},$$

and so by (5.7),

$$\sum_{p\in F(i)}\alpha_{ip}=\sum_{p\in F(i)}\arctan\frac{r_p}{r_i}=\sum_{p\in F(i)}\arctan(e^{y_p-y_i})=\sum_{p\in F(i)}\beta_{ip}.\tag{5.8}$$

Using the definition of the numbers β, this proves (5.3). Condition (5.4) follows by a similar computation. □

Proof of Corollary 5.4. The existence of the representation follows by Theorem 5.3. In the proof above, the fact that Φ is strictly convex (if $x_a=0$ is fixed) implies that the minimizer of Φ is unique, and grad $\Phi(x)=0$ happens at this unique point only. The computation in the proof above shows that every double circle representation (with $r_a=r_b=r_c$ fixed) gives a point x where grad $\Phi(x)=0$, which shows that the double circle representation is essentially unique. It is not hard to extend this argument to the case when we fix three arbitrary mutually tangent disks to represent the nodes of the unbounded country. □

REMARK 5.8. The proof described above provides an algorithm to compute a double circle representation (up to an arbitrary precision). It is in fact quite easy to program: We can use an "off the shelf" optimization algorithm for smooth convex functions to compute the minimizer of F and through it, the representation.

5.2. Formulation in space

5.2.1. Double cap representation. We can lift the Koebe or double circle representation of a graph from the plane to the sphere. The advantage will be that we do not have to deal with the exceptional unbounded country, and there will be a perfect duality between a graph and its dual. The price we pay is that we have to leave the more familiar realm of plane geometry for spherical geometry.

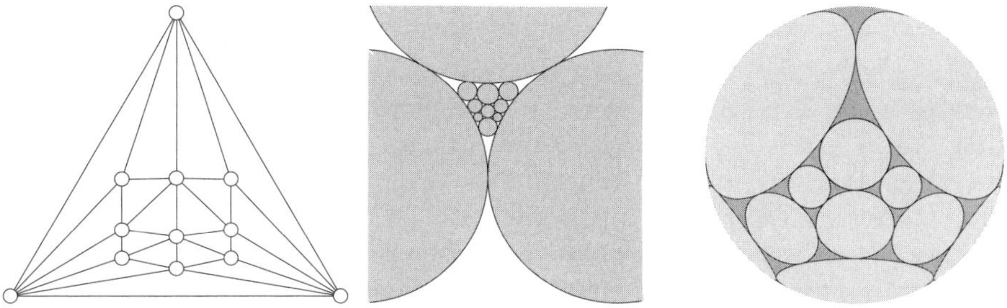

FIGURE 5.7. A planar map, its representation by touching circles, and a stereographic projection of the latter to the sphere.

On the sphere, we represent the nodes of the graph and those of its dual by "caps". It will be convenient to specialize some notions from Appendix C.4 to the ordinary 2-dimensional sphere. A circle on a sphere cuts the sphere into two regions; together with one of the regions, the circle forms a *cap*. The cap may cover a hemisphere or even more; if it covers less than a hemisphere, we call it *proper*. A great circle (like the equator) bounds two hemispheres, which are not proper caps by this definition. Every other circle bounds one proper and one improper cap.

Note that in connection with a cap, we can talk about more than one notions of "center". To be more precise, let S be the unit sphere, and let C be a cap on S with boundary circle C'. The circle C' is the intersection of the sphere with a plane, and in this plane it has a center $\mathbf{a}$. The line through the origin orthogonal to this plane intersects the cap at a point $\mathbf{c}$, which we call the *center of the cap* C. If C is proper, then there is another useful point $\mathbf{p} \in \ell$ to consider, which is the point from which the portion of the sphere that is visible is exactly the cap C. The circle C' is the "horizon" from $\mathbf{p}$. Clearly $\mathbf{a}$, $\mathbf{c}$ and $\mathbf{p}$ are on one line through the origin, and by elementary geometry, $|\mathbf{a}|\cdot|\mathbf{p}| = |\mathbf{c}|^2 = 1$. We call $\mathbf{p}$ the *pole* of the cap C.

We define a *double cap representation* of a planar map G as two families of caps on a sphere S, $(C_i : i \in V)$ and $(D_p : p \in V^*)$, so that for every edge ij, bordering countries p and q, the following holds: the interiors of the caps C_i are disjoint, and so are the interiors of the caps D_p; the caps C_i and C_j are tangent at a point $\mathbf{u}_{ij}$; the caps D_p and D_q are tangent at the same point $\mathbf{u}_{ij}$; and the boundary circles of D_p and D_q intersect the boundary circles of C_i and C_j orthogonally.

Such a double cap representation provides an embedding of the graph in the sphere, by representing each node i by the center of the cap C_i, and connecting adjacent nodes i and j by an arc of a large circle through the tangency point $\mathbf{u}_{ij}$. Similarly we get an embedding of the dual graph. Note that in this spherical setting, the primal and dual graphs are completely interchangeable.

Similarly as in the plane, these conditions imply other useful properties of the sphere and the embedding: Every country p is bounded by a convex spherical polygon, and the cap D_p touches every edge of p. Every country p is covered the caps D_p and C_i $(i \in V(p))$. If $i \in V$ is not incident with $p \in V^*$, then C_i and D_p are disjoint.

We have not excluded caps covering more than a hemisphere in the definition of a double cap representation; if all caps in a double cap representation are proper, we call the representation *proper*.

THEOREM 5.9. *Every 3-connected planar map G has a proper double cap representation on the sphere.*

PROOF. Let G be a 3-connected planar map, and let p_∞ be its unbounded country. Let $(C_i : i \in V)$ and $D_p : p \in V^*)$ be a double circle representation of G in the plane. Take any sphere S touching the plane, and project the plane onto the sphere by inverse stereographic projection. Let $(C_i' : i \in V)$ and $D_p' : p \in V^*)$ be the images of the circles in the double circle representation; these are circles by the fundamental properties of stereographic projection. We define caps $(\widehat{C}_i : i \in V)$ and $\widehat{D}_p : p \in V^*)$ with boundaries C_i' and D_p', respectively: we assign the cap not containing the North pole to every circle except to D_{p_∞}, to which we assign the cap containing the north pole. This way we get two families of caps on the sphere.

Clearly the cap $\widehat{C}_i$ is the image of the disk bounded by C_i, and similar assertion holds for $\widehat{D}_p$, except that $\widehat{D}_{p_\infty}$ is the image of the exterior of D_{p_∞} (plus the north pole). Hence the caps $\widehat{C}_i$ are nonoverlapping, and so are the caps $\widehat{D}_j$. All the properties of the double cap representation follow easily from the corresponding properties of the double circle representation.

If we want a proper representation, all we have to do is to choose the sphere S appropriately: we let it touch the plane at the center $\mathbf{c}$ of D_{p_∞}, in which case $\widehat{D}_{p_\infty}$ will be a cap centered at the north pole. The size of this cap depends on the radius of S. If the radius of S is half of the radius of D_{p_∞}, then $\widehat{D}_{p_\infty}$ is just the northern hemisphere. We choose the radius of S a bit smaller than this, so that $\widehat{D}_{p_\infty}$ be contained in the interior of the northern hemisphere, but just barely. All other caps $\widehat{D}_p$, as well as the caps $\widehat{C}_i$ for $i \notin V(p_\infty)$ are properly contained in the complement of the interior of $\widehat{D}_{p_\infty}$, and so if $\widehat{D}_{p_\infty}$ is sufficiently close to the whole northern hemisphere, then these caps will be proper. A cap $\widehat{C}_i$ for $i \in V(p_\infty)$ contains neither the North pole nor the South pole (since the South pole is the center of the circle D_{p_∞}, which is orthogonal to C_i and hence its center is an exterior point of C_i). This implies that $\widehat{C}_i$ is smaller than a hemisphere, and so the representation is proper. □

5.2.2. The Cage Theorem. One of the nicest consequences of the coin representation (more exactly, the double cap representation in Theorem 5.9) is the following theorem [**Andre'ev 1970a**], which is a far reaching strengthening of the Steinitz Representation Theorem.

THEOREM 5.10 (The Cage Theorem). *Every simple 3-connected planar graph can be represented as the skeleton of a convex* 3*-polytope such that every edge of it touches a given sphere.*

PROOF. The nice picture of a double cap representation translates into polyhedral geometry as follows. Let $(C_i : i \in V)$ and $(D_p : p \in V^*)$ be a proper double cap representation of G. Let $\mathbf{u}_i$ be the pole of C_i, and let $\mathbf{v}_p$ be the pole of D_p. Let $ij \in E$, and let pq be the corresponding edge of G^*. The points $\mathbf{u}_i$ and $\mathbf{u}_j$ are contained in the tangent line of the sphere that is orthogonal to the circles C_i' and C_j' at their common point $\mathbf{u}_{ij}$; this line is clearly the common tangent of the circles D_p' and D_q' at $\mathbf{u}_{ij}$. The plane $\mathbf{v}_p^\mathsf{T}\mathbf{u} = 1$ intersects the sphere in the circle D_p', and hence it contains it tangents, in particular the points $\mathbf{u}_i$ and $\mathbf{u}_j$, along with all points $\mathbf{u}_k$ where k is a node of the facet p. Since the cap $\widehat{D}_p$ is disjoint from $\widehat{C}_k$ if k is not a node of the facet p, we have $\mathbf{v}_p^\mathsf{T}\mathbf{u}_k < 1$ for every such node.

This implies that the polytope $P = \text{conv}\{\mathbf{u}_i : \ i \in V\}$ is contained in the polyhedron $P' = \{\mathbf{u} \in \mathbb{R}^3 : \ \mathbf{v}_p^\mathsf{T}\mathbf{u} \leq 1 \ \forall p \in V^*\}$. Furthermore, every inequality $\mathbf{v}_p^\mathsf{T}\mathbf{u} \leq 1$ defines a facet F_p of P with vertices $\mathbf{u}_i$, where i is a node of p. Every ray from the origin intersects one of the countries p in the drawing of the graph on the sphere, and therefore it intersects F_p. This implies that P has no other facets, and thus $P = P'$. It also follows that every edge of P connects two vertices $\mathbf{u}_i$ and $\mathbf{u}_j$, where $ij \in E$, and hence the skeleton of P is isomorphic to G and every edge is tangent to the sphere. □

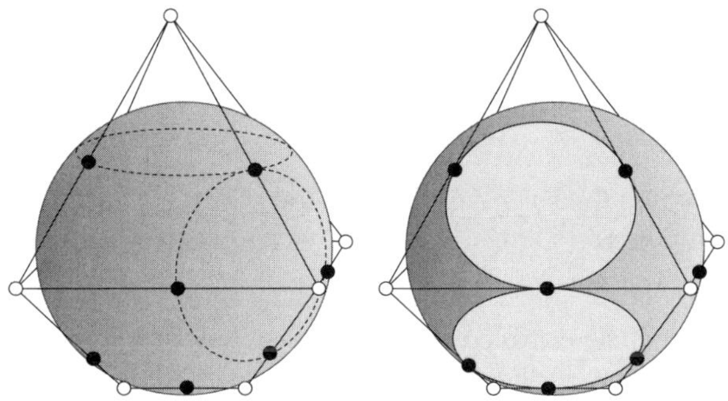

FIGURE 5.8. A convex polytope in which every edge is tangent to a sphere creates two families of circles on the sphere.

Let us note that, conversely, every polytope P whose edges are tangent to the unit sphere creates a double cap representation of its skeleton. It is convenient to assume that the center of the sphere is in the interior of P; this can be achieved by an appropriate projective transformation of the space (we come back to this in Section 5.2.3).

Each node in V is represented by the cap of points you see on the sphere when looking from the corresponding vertex, and each facet is represented by the cap its plane cuts off the sphere (Figure 5.8). Elementary geometry tells us that the caps representing adjacent nodes touch each other at the point where the edge touches the sphere, and the two caps representing the adjacent countries also touch each other at this point, and they are orthogonal to the first two circles. Furthermore, the interiors (smaller sides) of the horizon-circles are disjoint, and the same holds for the facet-circles.

5.2.3. Möbius transformations.

We have seen that the double circle representation of a planar graph is uniquely determined, once a triangular country of G or G^* is chosen as the unbounded country, and the circles representing the nodes of this triangular country are fixed. Similar "essential uniqueness" holds true for double cap representations on the unit sphere.

To formalize what we mean by that, observe that we can apply an arbitrary circle-preserving transformation of the sphere to any double cap representation, to get another one. But these modifications are the only ones to worry about; this is asserted by the following spherical version of Corollary 5.4.

PROPOSITION 5.11. *The double cap representation of a 3-connected planar graph is unique up to circle-preserving transformations of the sphere.*

Proof. Let $(C_i, D_p : i \in V, p \in V^*)$ and $(C'_i, D'_p : i \in V, p \in V^*)$ be two double cap representations of the same 3-connected planar graph G. One of G and G^* has a triangular country; let (say) $p_\infty = abc$ be triangular country of G. We can apply a circle preserving transformation to the representation (C'_i, D'_p) to obtain that $C'_a = C_a$, $C'_b = C_b$ and $C'_c = C_c$ (see Exercise 5.6). This implies that $D_{p_\infty} = D'_{p_\infty}$. Choosing the center of D_{p_∞} as the north pole, and applying stereographic projection, we get two double circle representations of G in the plane with p_∞ as the unbounded country and with the same circles representing a, b and c in both. Uniqueness of the extension to a double circle representation (Corollary 5.4) implies that these are the same, and hence $(C_i, D_p : i \in V, p \in V^*)$ and $(C'_i, D'_p : i \in V, p \in V^*)$ are the same. □

A nice application of this last fact is a proof of the following theorem [**Mani 1971**]:

Corollary 5.12. *Every 3-connected planar graph G has a representation as the skeleton of a polytope P such that every automorphism of the graph extends to an isometry of the polytope.*

Proof. Consider the double cap representation of G. If α is an automorphism of G, then relabeling the nodes according to α gives another representation of G, using the same caps, just representing different nodes. This representation can be obtained by applying a cycle preserving transformation $\overline{\alpha}$ to the sphere, and by Proposition 5.11, this transformation is uniquely determined. It follows that transformations of the form $\overline{\alpha}$, where α is an automorphism of G, form a finite subgroup of the group of cycle preserving transformations. It is well known that every such finite transformation group is conjugate to a group of isometries, which means that after an appropriate circuit-preserving transformation, automorphisms will extend to isometries. □

While the double cap representation of a planar graph is "essentially unique", we can make good use of the richness of Möbius transformations, by "normalizing" the representation in an appropriate way. A natural normalization goal is that the center of gravity of the centers of the caps should be the origin. The reason why such a normalization is nontrivial is that when applying a Möbius transformation, the caps are mapped onto caps, but the centers of the caps are not necessarily mapped onto the centers of their images. In general, it is difficult to compute the appropriate transformation directly; to prove its existence, we have to use topology. The topological argument, on the other hand, has the advantage that it applies to a number of similar normalization demands without any essential change.

As a preparation, let us discuss the topological tool we need. Brouwer's Fixed Point Theorem implies that if we have a continuous map $f : B \to B$ whose restriction to the boundary sphere S is the identity, then the center of B is in the range of f. We need an extension of this fact to the case when f is only defined in the interior B' of B. Suppose that the range of $f : B' \to B$ does not cover the origin. Trivially, the extension of f to the boundary of B as the identity map cannot be continuous; the next lemma say that it is quite a bit discontinuous.

Lemma 5.13. *Let $f : B' \to B$ be a continuous map, and suppose that the origin is not in the range of f. Then there are points $\mathbf{x}_k \in B'$ $(k = 1, 2, \dots)$ and $\mathbf{y} \in S$ such that $\mathbf{x}_k \to \mathbf{y}$ and $f(\mathbf{x}_k)^0 \to -\mathbf{y}$.*

PROOF. Since the range of f does not contain the origin, the mapping $f_t(\mathbf{x}) = -tf(\mathbf{x})^0$ is well defined for $0 < t < 1$ and $\mathbf{x} \in B'$, and it is a continuous map $tB \to tB$. So by Brouwer's Fixed Point Theorem, it has a fixed point $\mathbf{p}_t$, satisfying

$$\mathbf{p}_t = -tf(\mathbf{p}_t)^0.$$

We may select a sequence of numbers $t_k \in (0,1)$ such that $t_k \to 1$ and $\mathbf{p}_{t_k} \to \mathbf{y} \in B$. Since $|\mathbf{p}_{t_k}| = |t_k f(\mathbf{p}_{t_k})^0| = t_k \to 1$, we have $\mathbf{y} \in S$. Furthermore, $f(\mathbf{p}_{t_k})^0 = -(1/t_k)\mathbf{p}_{t_k} \to -\mathbf{y}$. □

LEMMA 5.14. *Let $\{C_1, \dots, C_n\}$ be a family of nonoverlapping caps on the sphere S. Then there is a Möbius transformation τ of the sphere such that $\mathbf{v}_1 + \dots + \mathbf{v}_n = 0$, where $\mathbf{v}_i$ is the center of $\tau(C_i)$.*

PROOF. Excluding trivial cases, we assume that $n \geq 3$. For every interior point $\mathbf{x}$ of the unit ball, we define a conformal (circle-preserving) transformation $\tau_{\mathbf{x}}$ of the unit sphere. For $\mathbf{x} = 0$, we define $\tau_{\mathbf{x}} = \mathrm{id}_S$. If $\mathbf{x} \neq 0$, then we take a tangent plane T at $\mathbf{x}^0$, and project the sphere stereographically onto T, shrink the plane from center $\mathbf{x}^0$ by a factor of $1 - |\mathbf{x}|$, and project it back stereographically to the sphere. It is easy to check that this construction has the following property: if $\mathbf{x}_k \in B$ is a sequence of points such that $\mathbf{x}_k \to \mathbf{y} \in S$, then $\tau_{\mathbf{x}_k}(\mathbf{z}) \to \mathbf{y}$ for every $\mathbf{z} \in S$, $\mathbf{z} \neq \mathbf{y}$.

Let $\mathbf{v}_i(\mathbf{x})$ denote the center of the cap $\tau_{\mathbf{x}}(C_i)$ (warning: this is not the image of the center of C_i in general!), and let $f(\mathbf{x}) = (1/n)\big(\mathbf{v}_1(\mathbf{x}) + \dots + \mathbf{v}_n(\mathbf{x})\big)$. We want to show that the range of f contains the origin. Suppose not, then by Lemma 5.13, there are points $\mathbf{x}_k \in B'$ and $\mathbf{y} \in S$ such that $\mathbf{x}_k \to \mathbf{y}$ and $f(\mathbf{x}_k)^0 \to -\mathbf{y}$ as $k \to \infty$.

Let us think of $\mathbf{y}$ as the north pole and $-\mathbf{y}$ as the south pole. From the properties of $\tau_{\mathbf{x}}$ it follows that if C_i does not contain the south pole, then $\mathbf{v}_i(\mathbf{x}_k) \to \mathbf{y}$ as $k \to \infty$; if the south pole is on the boundary of C_i, then $\mathbf{v}_i(\mathbf{x}_k)$ tends to the equator. So with at most two exceptional values of i, we have $\mathbf{y}^{\mathsf{T}}\mathbf{v}_i(\mathbf{x}_k) \to 1$, and it there are two exceptional values i and j, then $\mathbf{y}^{\mathsf{T}}\mathbf{v}_i(\mathbf{x}_k), \mathbf{y}^{\mathsf{T}}\mathbf{v}_j(\mathbf{x}_k) \to 0$. So $\mathbf{y}^{\mathsf{T}} f(\mathbf{x}_k) > 0$ if k is large enough, which contradicts the fact that $f(\mathbf{x}_k)^0 \to -\mathbf{y}$. □

REMARK 5.15. The topological proof above is quite robust: instead of the center of gravity, it applies to other "centers" satisfying some natural conditions. To be more precise, consider a continuous function $f : S \times \dots \times S \to B$ (n factors of the unit sphere S) with the property that $\mathbf{y}^{\mathsf{T}} f(\mathbf{y}_1, \dots, \mathbf{y}_n) > 0$ if $\mathbf{y}_i = \mathbf{y}$ for at least $n-2$ indices i. Let $\{C_1, \dots, C_n\}$ be a family of nonoverlapping caps on the sphere S ($n \geq 4$). Then there is a Möbius transformation τ of the sphere such that $f(\mathbf{v}_1, \dots, \mathbf{v}_n) = 0$, where $\mathbf{v}_i$ is the center of $\tau(C_i)$. (See Exercise 5.13 for a slightly stronger statement.)

5.3. Circle packing and the Riemann Mapping Theorem

Koebe's Circle Packing Theorem and the Riemann Mapping Theorem in complex analysis are closely related. More exactly, we consider the following generalization of the Riemann Mapping Theorem, also due to Koebe:

THEOREM 5.16 (Circle Domain Theorem). *Every connected open domain on the sphere whose boundary has a finite number of connected components is conformally equivalent to a domain obtained from the sphere by removing a finite number of disjoint disks and points.*

The Circle Packing Theorem and the Circle Domain Theorem are mutually limiting cases of each other [**Koebe 1936**], [**Rodin–Sullivan 1987**]. The exact proof of this fact has substantial technical difficulties, but it is not hard to describe the idea.

1. To see that the Circle Domain Theorem implies the Circle Packing Theorem, let G be a connected map on the sphere. We may assume that G and its dual G^* are 3-connected, and their arcs are smooth curves (these assumptions are not essential, just convenient). Let $\varepsilon > 0$, and let U denote the open ε-neighborhood of G^*. By Theorem 5.16, there is a conformal map of U onto a domain D' which is obtained by removing a finite number of disjoint caps and points from the sphere (the removed points can be considered as degenerate caps). If ε is small enough, then these caps are in one-to-one correspondence with the nodes of G. We normalize using Lemma 5.14 and assume that the center of gravity of the cap centers is 0 (Figure 5.9 shows this correspondence drawn in the plane).

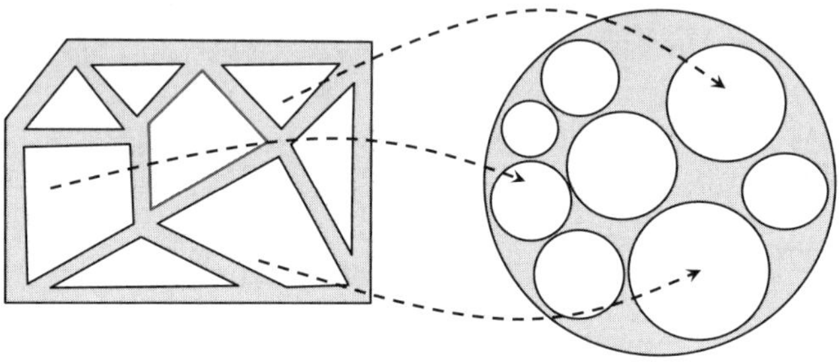

FIGURE 5.9. The Koebe Circle Domain Theorem, applied to a small neighborhood of the edges of a planar map, yields a touching disk representation of the dual map.

Letting $\varepsilon \to 0$, and selecting a suitable subsequence, we may assume that the cap representing any given node $v \in V$ converges to a cap C_v. One can argue that these caps are nondegenerate, caps representing different nodes tend to nonoverlapping caps, and caps representing adjacent nodes tend to caps that are touching.

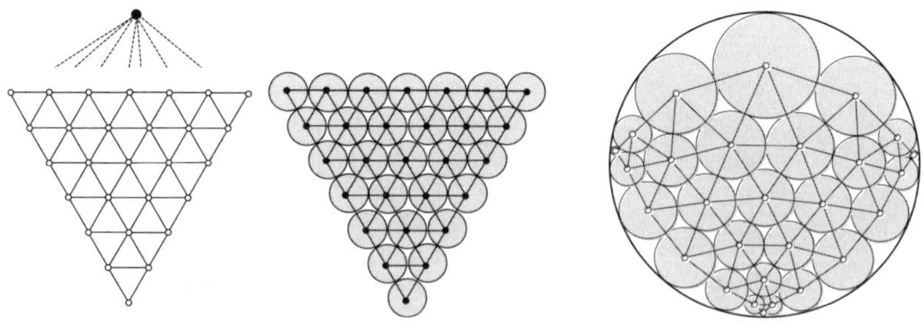

FIGURE 5.10. Coin representation of a triangular grid, and of the graph obtained by adding a new node (the black node on the top), the latter converging to a conformal mapping of a triangle onto a disk.

2. In the other direction, let us warm up by describing how to get a conformal map from a regular triangle to a circular disk. (Here we argue in the planar setting.)

We subdivide the interior of the triangle into "many" small triangles to get piece of a triangular grid (Figure 5.10). This graph G has an obvious representation by touching circular disks. The union of these disks approximates the original triangle, so nothing is gained here.

However, if we add a new node v and connect it to all the boundary nodes of G, we get a triangulation. Representing this triangulation as a tangency graph, with the circle representing v turned inside out, we get the picture on the right: we can think of it as the middle picture distorted so that the outer disks touch the circle C representing the new node. We can map the full triangle into the interior of C by mapping each little triangle affinely to the corresponding triangle on the right. Refining the grid, we get a sequence of maps that (with some care) converge to a conformal map of the interior of the triangle to the interior of the circle.

Turning to the general case, and returning to the spherical formulation, let $U = S^2 \setminus K_1 \setminus \cdots \setminus K_n$, where $K_1, \dots, K_n$ are disjoint closed connected sets which do not separate the sphere. We may assume that no K_i is a single point (these are trivial to handle). Let $\varepsilon > 0$. It is not hard to construct a family $\mathcal{C}(\varepsilon)$ of nonoverlapping caps such that the radius of each cap is less than ε and their tangency graph G is a triangulation of the sphere.

Let H_i denote the subgraph of G consisting of all edges intersecting K_i. If ε is small enough, then the subgraphs H_i are node-disjoint, and each H_i is nonempty. It is also easy to see that the subgraphs H_i are connected.

Let us contract in G each subgraph H_i to a single node w_i. The spherical map G' obtained this way can be represented as the tangency graph of a family of caps $\{D_u(\varepsilon) : u \in V(G')\}$. We can normalize so that the center of gravity of the centers of $D_{w_1}(\varepsilon), \dots, D_{w_n}(\varepsilon)$ is the origin.

Now let $\varepsilon \to 0$. By selecting and appropriate subsequence, we may assume that each $D_{w_i} = D_{w_i}(\varepsilon)$ tends to a cap $D_{w_i}(0)$. Furthermore, we have a map f_ε that assigns to each node u of G_ε the center of the corresponding cap D_u. One can prove (but this is nontrivial) that these maps f_ε, in the limit as $\varepsilon \to 0$, give a conformal map of U onto $S^2 \setminus D_{w_1}(0) \setminus \cdots \setminus D_{w_n}(0)$.

5.4. Applications of coin representations

5.4.1. Planar separators. Koebe's Theorem has several important applications. We start with a simple proof of a version of the Planar Separator Theorem 2.7 [**Lipton–Tarjan 1979**]:

THEOREM 5.17. *Every planar graph G on n nodes contains a set $S \subseteq V$ such that $|S| \le \sqrt{2n}$, and every connected component of $G \setminus S$ has at most $3n/4$ nodes.* □

This version and the proof to be presented is due to [**Miller et al. 1997**]; see [**Spielman–Teng 1996b**] for an improved analysis of the method.

We need the notion of the "statistical center", which is important in many other studies in geometry. Let $S \subseteq \mathbb{R}^d$ be a set of n points in $\mathbb{R}^d$. A point $c \in \mathbb{R}^d$ is a *statistical center* of S if every closed halfspace containing c contains at least $n/(d+1)$ elements of S. (We could say open halfspace instead of closed without changing this notion.) For $d = 1$, this point is just the median of a finite set of real numbers. This is unique if the set has an odd number of elements, otherwise the allowable values form an interval.

LEMMA 5.18. *Every finite set of points in $\mathbb{R}^d$ has a statistical center.*

PROOF. Let B be a ball containing all points in S, and let $\mathcal{H}$ be the family of intersection of B with those closed halfspaces that contain more than $dn/(d+1)$ points of S. Then all members of $\mathcal{H}$ are closed, bounded and convex. Furthermore, the intersection of any $d+1$ of these still contains an element of S, so in particular it is nonempty. Thus by Helly's Theorem, the intersection of all sets in $\mathcal{H}$ is nonempty. We claim that any point $c \in \cap\mathcal{H}$ satisfies the conclusion of the lemma.

If H be any open halfspace containing c, then $\mathbb{R}^d \setminus H \notin \mathcal{H}$, which means that H contains at least $n/(d+1)$ points of S. If H is a closed halfspace containing c, then it can be enlarged slightly to an open halfspace H' that intersects S in exactly the same set. We know already that H' contains at least $n/(d+1)$ elements of S, and hence so does H. □

Proof of Theorem 5.17. Let $(C_i : i \in V)$ be a Koebe representation of G on the unit sphere, and let $\mathbf{u}_i$ be the center of C_i on the sphere, and ρ_i, the spherical radius of C_i.

We want to make the origin a statistical center of the points $\mathbf{u}_i$. This takes a little care, since a set may have more than one statistical center. It is easy to see that statistical centers of the set form a convex set, and we can consider the center of gravity of all statistical centers; let us call this (just for this argument) the *central statistical center*. It is also easy to see that the central statistical center depends continuously on the position of the points of S, and satisfies the conditions formulated in Remark 5.15. So we can transform the Koebe representation into one whose central statistical center is the origin. This also implies that all caps C_i are proper.

Take any plane H through 0. Let S denote the set of nodes i for which C_i intersects H, and let H separate $V \setminus S$ into two sets S_1 and S_2. Clearly there is no edge between S_1 and S_2, and so the subgraphs G_1 and G_2 are disjoint and their union is $G \setminus S$. Since 0 is a statistical center of the $\mathbf{u}_i$, it follows that $|S_1|, |S_2| \le 3n/4$.

It remains to make sure that S is small. To this end, we choose H at random, and estimate the expected size of S.

What is the probability that H intersects C_i? Since the caps C_i are proper, we have $\rho_i < \pi/2$ for every i. By symmetry, instead of fixing C_i and choosing H at random, we can fix H and choose the center of C_i at random. Think of H as the plane of the equator, then C_i will intersect H if and only if its center is at a latitude at most ρ_i (North or South). The area of this belt around the equator is, by elementary geometry, $4\pi \sin\rho_i$, and since the total area of the sphere is 4π, the probability that the center of C_i falls into here is $\sin\rho_i$. It follows that the expected number of caps C_i intersected by H is $\sum_{i\in V} \sin\rho_i$.

To get an upper bound on this quantity, we use the fact that the surface area of the cap C_i is $4\pi\sin(\rho_i/2)^2$, and since these caps are disjoint,

$$\sum_{i\in V} \left(\sin\frac{\rho_i}{2}\right)^2 < 1. \tag{5.9}$$

Using that $\sin\rho_i \le 2\sin\frac{\rho_i}{2}$, we get by Cauchy-Schwarz

$$\sum_{i\in V} \sin\rho_i \le \sqrt{n}\left(\sum_{i\in V}(\sin\rho_i)^2\right)^{1/2} \le 2\sqrt{n}\left(\sum_{i\in V}\left(\sin\frac{\rho_i}{2}\right)^2\right)^{1/2} < 2\sqrt{n}.$$

So the expected size of S is less than $2\sqrt{n}$, and so there is at least one choice of H for which $|S| < 2\sqrt{n}$. □

5.4.2. Laplacians of planar graphs. The Planar Separator theorem, discussed in the last section, was first proved by direct graph-theoretic arguments, and other elegant graph-theoretic proofs are available [**Alon–Seymour–Thomas 1984**]. However, for the following theorem on the eigenvalue gap of the Laplacian (Fiedler value) of planar graphs [**Spielman–Teng 1996a**], there is no proof known avoiding Koebe's theorem. Informally, the theorem says that planar graphs are poor expanders; see Section B.3 for more on the significance of the eigenvalue gap.

THEOREM 5.19. *For every connected planar graph G on n nodes and maximum degree D, the second smallest eigenvalue of L_G is at most $8D/n$.*

PROOF. By Raleigh's formula, the second smallest eigenvalue of L_G is given by

$$\lambda_2 = \min_{\substack{x\neq 0\\ \sum_i x_i=0}} \frac{\sum_{ij\in E}(x_i-x_j)^2}{\sum_{i\in V} x_i^2},$$

Let $C_i : i \in V$ be a Koebe representation of G on the unit sphere, and let $\mathbf{u}_i$ be the center of C_i, and ρ_i, the spherical radius of C_i. By Lemma 5.14 may assume that $\sum_i \mathbf{u}_i = 0$.

Let $\mathbf{u}_i = (u_{i1}, u_{i2}, u_{i3})$, then $x_i = u_{ik}$ must be taken into account in this formula for $k = 1, 2, 3$, which implies that

$$\sum_{ij\in E} (u_{ik}-u_{jk})^2 \ge \lambda_2 \sum_{i\in V} u_{ik}^2$$

holds for every coordinate k, and summing over k, we get

$$\sum_{ij\in E} |\mathbf{u}_i-\mathbf{u}_j|^2 \ge \lambda_2 \sum_{i\in V} |\mathbf{u}_i|^2 = \lambda_2 n. \tag{5.10}$$

On the other hand, we have

$$\begin{aligned}|\mathbf{u}_i-\mathbf{u}_j|^2 &= 4\left(\sin\frac{\rho_i+\rho_j}{2}\right)^2 = 4\left(\sin\frac{\rho_i}{2}\cos\frac{\rho_j}{2}+\sin\frac{\rho_j}{2}\cos\frac{\rho_i}{2}\right)^2\\ &\le 4\left(\sin\frac{\rho_i}{2}+\sin\frac{\rho_j}{2}\right)^2 \le 8\left(\sin\frac{\rho_i}{2}\right)^2+8\left(\sin\frac{\rho_j}{2}\right)^2,\end{aligned}$$

and so by (5.9)

$$\sum_{ij\in E} |\mathbf{u}_i-\mathbf{u}_j|^2 \le 8D\sum_{i\in V}\left(\sin\frac{\rho_i}{2}\right)^2 \le 8D.$$

Comparison with (5.10) proves the theorem. □

Among other applications of Koebe's Theorem of similar nature, we mention a bound on the cover time of the random walk on a planar graph [**Jonasson–Schramm 2000**].

5.5. And more...

5.5.1. General convex domains and bodies. Let $\mathcal{H}$ be a family of closed convex domains in the plane, such that their interiors are disjoint. If we represent each of these domains by a node, and connect two of these nodes if the corresponding domains touch, we get a $G_{\mathcal{H}}$, which we call the *tangency graph* of the family. It is easy to see that if no four of the domains touch at the same point, then this graph is planar.

In what follows, we restrict ourselves to the case when the members of $\mathcal{H}$ are all homothetical copies of a centrally symmetric convex domain. It is natural in this case to represent each domain by its center.

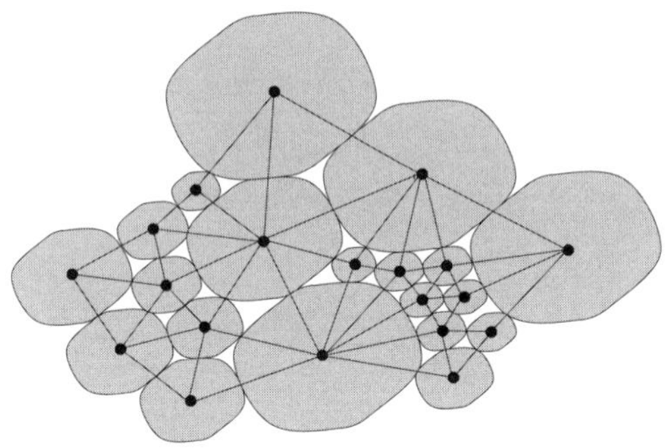

FIGURE 5.11. Straight line embedding of a planar graph from touching centrally symmetric convex figures.

In [**Schramm 1991**] the following deep converse of this construction was proved.

THEOREM 5.20. *For every smooth strictly convex domain D, every maximal planar graph can be represented as the tangency graph of a family of homothetical copies of D.* □

In fact, Schramm proves the more general theorem that if we assign a smooth convex domain D_i to each node i of a triangulation G of the plane, then there is a family $\mathcal{D} = (D'_i : i \in V)$ of nonoverlapping convex domains such that D'_i is positively homothetical to D_i and the tangency graph of $\mathcal{D}$ is G.

The smoothness of the domain is a condition that cannot be dropped. For example, K_4 cannot be represented by touching squares. A strong result about representation by touching squares, also due to Schramm, will be stated and proved in Chapter 6.

The Cage Theorem 5.10 has a far-reaching generalization to general convex bodies [**Schramm 1992**].

THEOREM 5.21 (Caging the Egg). *For every smooth strictly convex body K in $\mathbb{R}^3$, every 3-connected planar graph can be represented as the skeleton of a polytope in $\mathbb{R}^3$ such that every one of its edges touch K.* □

5.5.2. Polyhedra in hyperbolic space. If we have a double cap representation of a planar graph, we can read off another interesting polyhedral representation, not of the graph itself, but of its medial graph, and not in Euclidean space, but in hyperbolic space.

Let $(C_i : i \in V)$ and $(D_p : p \in V^*)$ form a proper double cap representation of G on the sphere S. Let B be the ball bounded by S. We can represent each cap C_i as the intersection of a ball A_i with S, where the boundary sphere A_i' of A_i is orthogonal to S. Similarly, we can write $D_p = B_p \cap S$, where B_p is a ball whose boundary sphere B_p' is orthogonal to S.

Let P denote the set of points in B that are not contained in any A_i and B_p. This set is open and nonempty (since it contains the origin). The balls A_i and B_p cover the sphere S, and even their interiors do so with the exception of the points $\mathbf{x}_{ij}$ where two caps C_i and C_j touch. It follows that P is a domain whose closure $\overline{P}$ contains a finite number of points of the sphere S. It also follows that no three of the balls A_i and B_p have a point in common except on the sphere S. Hence those points in the interior of S that belong to two of the spheres A_i' form circular arcs that go from boundary to boundary, and it is easy to see that these arcs are orthogonal to S. For every incident pair (i, p) $(i \in V, p \in V^*)$ there is such an "edge" of P. The "vertices" of P are the points $\mathbf{x}_{ij}$, which are all on the sphere S, and together with the "edges" as described above they form a graph isomorphic to the medial graph $G^{\bowtie}$ of G.

All this translates very nicely, if we view the interior of S as a Poincaré model of the 3-dimensional hyperbolic space. In this model, spheres orthogonal to S are "planes", and hence P is a polyhedron. All the "vertices" of P are at infinity, and P is a Steinitz representation of G^+ in hyperbolic space. Every dihedral angle of P is $\pi/2$.

Conversely, a representation of G_+ in hyperbolic space as a polyhedron with dihedral angles $\pi/2$ and with all vertices at infinity gives rise to a double cap representation of G.

What about representing not the medial graph, but an arbitrary planar graph by such a polyhedron in hyperbolic space? This is not always possible any more. [**Andre'ev 1970a**, **Andre'ev 1970b**] gave a general necessary and sufficient condition for the existence of a representation of a planar graph by a polyhedron in hyperbolic 3-space with "vertices" at infinity and with prescribed dihedral angles. From this condition, he was able to derive Theorems 5.10 and 5.3.

5.5.3. Other angles.

The double circle representation of G gives a representation of the lozenge graph $G^{\diamond}$ by circles such that adjacent nodes correspond to orthogonal circles. We may want to extend Koebe's Theorem to construct representations of more general graphs by orthogonal circles: circles representing adjacent nodes must intersect at 90°, other pairs should be disjoint or intersect at an angle larger than 90° (i.e., their centers must be farther apart) (Figure 5.12).

More generally, we may want to prescribe the angle at which two circles representing adjacent nodes meet. Circles touching from the outside correspond to 180°, orthogonal circles, to 90°. Most results in this generality concern planar triangulations. We quote the main result, due to [**Andre'ev 1970a**, **Andre'ev 1970b**] and [**Thurston 1997**], without proof. See also [**Bobenko–Springborn 2004**] for a proof along the lines of the proof of Koebe's Theorem given above.

THEOREM 5.22. *Let G be a simple planar triangulation different from K_4. Let an angle $0 \le \psi_{ij} \le \pi/2$ be assigned to every edge ij. Suppose that the following conditions hold:*

(i) *If ijk is a triangle in G such that $\psi_{ij} + \psi_{jk} + \psi_{ki} \ge \pi$, then ijk is a country;*

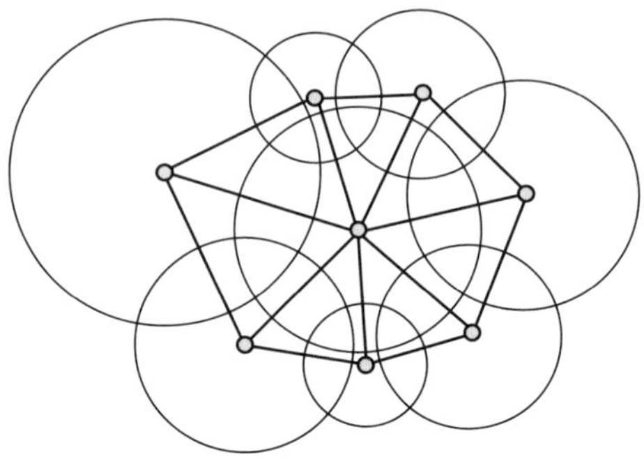

FIGURE 5.12. Representing a planar graph by orthogonal circles.

(ii) *If $ijkl$ is a quadrilateral in G such that $\psi_{ij}+\psi_{jk}+\psi_{il}+\psi_{li}=2\pi$, then $ijkl$ is the boundary of the union of two countries.*

Then G has a representation by circles $(C_i:\ i\in V)$ on the 2-sphere such that for every edge ij, the circles C_i and C_j intersect at an angle of ψ_{ij}. □

In the special case when all prescribed angles are $\pi/2$, a necessary and sufficient condition was proved in [**Kotlov–Lovász–Vempala 1997**].

THEOREM 5.23. *Let G be a simple planar triangulation different from K_4. Then G has a representation by orthogonal proper caps on the sphere if and only if no 3- or 4-cycle separates it into two parts with at least two nodes.* □

EXERCISE 5.1. Let G be a connected planar map with a triangular unbounded country abc. With the notation of Section 5.1.1, define $J[S]=\{ip\in J:\ i,p\in S\}$ $(S\subseteq U)$, and $f(S)=\frac{1}{2}|J[S]|-|S|+\frac{5}{6}|S\cap\{a,b,c\}|$. Prove that (a) $f(\emptyset)=f(U)=0$; (b) $f(S)<0$ for every set $\emptyset\subset S\subset U$.

EXERCISE 5.2 (Continued). Prove that if $(r_u:\ u\in U)$ and $(r'_u:\ u\in U)$ are two assignments of positive radii such that they give rise to the same defect at every node, then there is a constant $C>0$ such that $r'_i=Cr_i$ for all i.

EXERCISE 5.3 (Continued). (a) Prove that $\sum_{u\in S}\delta_u>f(S)\pi$ for every set $\emptyset\subset S\subset U$. (b) Prove that if $(d_u:\ u\in U)$ is a vector of real numbers such that $\sum_u d_u=0$ and $\sum_{u\in S}d_u>f(S)\pi$ for every set $\emptyset\subset S\subset U$, then there is an assignment of positive radii whose defects are $\delta_u=d_u$ for all $u\in U$. (c) Give a new proof of Theorem 5.3 based on these assertions.

EXERCISE 5.4. Prove that every double circle representation of a planar graph is a rubber band representation with appropriate rubber band strengths.

EXERCISE 5.5. Prove that in every double cap representation of a 3-connected planar graph on the sphere, at most one cap is improper.

EXERCISE 5.6. Prove that for three mutually tangent caps C_1, C_2 and C_3 on the sphere, and three other mutually tangent caps C'_1, C'_2 and C'_3, there is a Möbius transformation that maps C_i onto C'_i $(i=1,2,3)$.

EXERCISE 5.7. Show by an example that the bound in Lemma 5.18 is sharp.

EXERCISE 5.8. Let $\mathcal{H}$ be a family of convex domains in the plane with smooth boundaries and disjoint interiors. Then the tangency graph of $\mathcal{H}$ is planar.

EXERCISE 5.9. Construct a planar graph G whose nodes cannot be represented by squares in the plane so that the interiors of the squares are disjoint and two squares share a boundary point if and only if the corresponding nodes of G are adjacent.

EXERCISE 5.10. Prove that every simple planar graph can be represented as the tangency graph of a family of triangles [**de Fraysseix et al. 1994**].

EXERCISE 5.11. Let $\mathcal{H}$ be a family of homothetical centrally symmetric convex domains in the plane with smooth boundaries and disjoint interiors. Prove that the centers of the domains give a straight line embedding in the plane of its tangency graph (Figure 5.11).

EXERCISE 5.12. Prove that in a representation of a 3-connected planar graph by orthogonal circles on the sphere the "interiors" can be chosen so that the corresponding caps cover the whole sphere.

EXERCISE 5.13. Let $f : S\times\cdots\times S \to B$ (n factors of the unit sphere S) be a continuous function with the property that $\mathbf{y}^{\mathsf{T}} f(\mathbf{y}_1,\dots,\mathbf{y}_n) > 0$ if either $\mathbf{y}_i = \mathbf{y}$ for all i, or $\mathbf{y}_i = \mathbf{y}$ with the exception of one index $i = i_1$, for which $\mathbf{y}_{i_1} = -\mathbf{y}$, or $\mathbf{y}_i = \mathbf{y}$ with the exception of two indices $i = i_1, i_2$, for which $\mathbf{y}^{\mathsf{T}}\mathbf{y}_{i_1} = \mathbf{y}^{\mathsf{T}}\mathbf{y}_{i_2} = 0$. Let $\{C_1,\dots,C_n\}$ be a family of nonoverlapping caps on the sphere S ($n \ge 4$). Prove that there is a Möbius transformation τ of the sphere such that $f(\mathbf{v}_1,\dots,\mathbf{v}_n) = 0$, where $\mathbf{v}_i$ is the center of $\tau(C_i)$.

EXERCISE 5.14. Let G be a simple planar map, and consider the graph $H = G\cup G^{\diamond}$, obtained from $G^{\diamond}$ by adding the edges of the original graph. Let us prescribe an angle of 90° to the edges of $G^{\diamond}$, and 0° to the edges of the original graph. Show that a representation of H by circles with these prescribed angles yields a double circle representation of G.

EXERCISE 5.15. Let G be a connected planar graph on n nodes with maximum degree D. Prove that the second largest eigenvalue of the transition matrix of the random walk on G is at least $1-8D/n$.

EXERCISE 5.16. Show by an example that Theorem 5.21 does not remain valid if we drop the conditions of smoothness and strict convexity.

CHAPTER 6

Square Tilings

We can also represent planar graphs by squares, rather then circles, in the plane. There are in fact two quite different ways of doing this: the squares can correspond to the edges, a classic result [**Brooks et al. 1940**], or the squares can correspond to the nodes, a more recent result [**Schramm 1993**].

6.1. Electric current through a rectangle

The classical paper [**Brooks et al. 1940**] used a physical model of electrical currents to show how to relate square tilings to planar graphs. The ultimate goal was to construct tilings of a square with squares whose edge-lengths are all different. This will not be our concern (but see Exercise 6.2); we will allow squares of the same size and also the domain to be tiled can be any rectangle, not necessarily a square.

Consider a tiling $\mathcal{T}$ of a rectangle R with a finite number of squares, whose sides are parallel to the coordinate axes. We can associate a planar map with this tiling as follows. Let us call a maximal horizontal segment composed of edges of the squares a *long edge*. Represent every long edge by a single node. Each square "connects" two horizontal segments, and we can represent it by an edge connecting the two corresponding nodes, directed top-down. We get a directed graph $G_{\mathcal{T}}$ (Figure 6.1), with a single source s (representing the upper edge of the rectangle) and a single sink t (representing the lower edge).

It is easy to see that graph $G_{\mathcal{T}}$ is planar: it can be obtained by first considering the midpoints of the horizontal edges of the squares, connecting two of them horizontally if they are neighbors along a long edge, and vertically if they belong to opposite edges of the same square. This graph is clearly planar, and contracting the horizontal edges, we get $G_{\mathcal{T}}$.

A little attention must be paid to points where four squares meet; we call these points *4-fold corners*. Suppose that squares A, B, C, D share a corner p, where A is the upper left, and B, C, D follow counterclockwise. In this case, we may consider the lower edges of A and B to belong to a single long edge, or to belong to different long edges. In the latter case, we may or may not imagine that there is an infinitesimally small square sitting at p, which may "connect" A with C or B with D (Figure 6.2). What this means is that we have to declare if the four edges of $G_{\mathcal{T}}$ corresponding to A, B, C and D are adjacent to the same node, two nonadjacent nodes, or two adjacent nodes. We can orient this horizontal edge arbitrarily. Deciding between these four possibilities will be called "resolving" the 4-fold corner p.

If we assign the edge length of each square to the corresponding edge, we get a flow f from s to t: If a node v represents a segment I, then the total flow into v is the sum of edge lengths of squares attached to I from the top, while the total flow

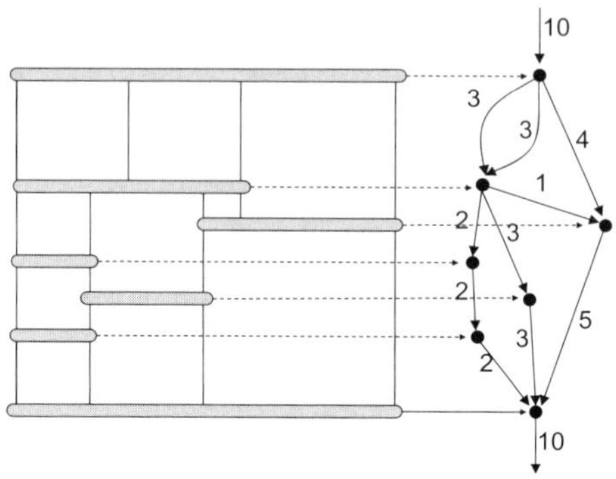

FIGURE 6.1. Constructing a graph with a harmonic function from a tiling of a rectangle by squares.

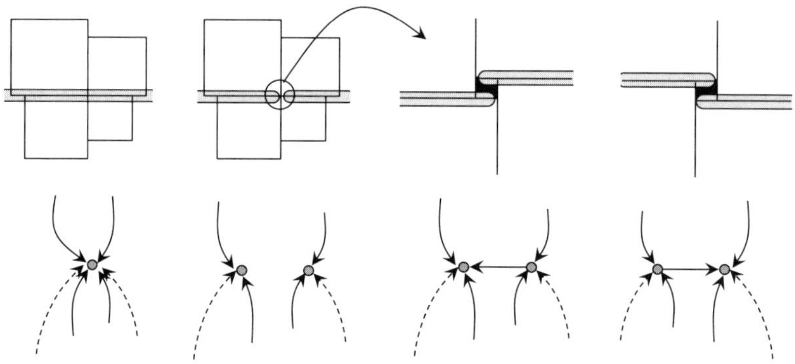

FIGURE 6.2. Possible resolutions of four squares meeting at a point.

out of v is the sum of edge length of squares attached to I from the bottom. Both of these sums are equal to the length of I.

Let $h(v)$ denote the distance of node v from the upper edge of R. Since the edge-length of a square is also the difference between the y-coordinates of its upper and lower edges, the function h is harmonic:

$$h(i) = \frac{1}{\deg(i)} \sum_{j \in N(i)} h(j)$$

for every node different from s and t (Figure 6.1).

It is not hard to see that this construction can be reversed.

THEOREM 6.1. *For every connected planar map G with two specified nodes s and t on the unbounded country, there is a unique tiling $\mathcal{T}$ of a rectangle such that (resolving the 4-fold corners appropriately) $G \cong G_{\mathcal{T}}$.*

PROOF. Consider the harmonic function $f : V \to \mathbb{R}$ with $f(s) = 0$ and $f(t) = 1$ (obtained, say, as the 1-dimensional rubber band embedding with s and t nailed; in Figure 6.3, this is the vertical coordinate of each node). We assign a square to each edge uv with $f(v) > f(u)$, of side length $f(v) - f(u)$. This square will be placed so that its lower edge is at height $f(u)$, and its upper edge, at height $f(v)$. To find the horizontal position of these squares, we start from node s: we

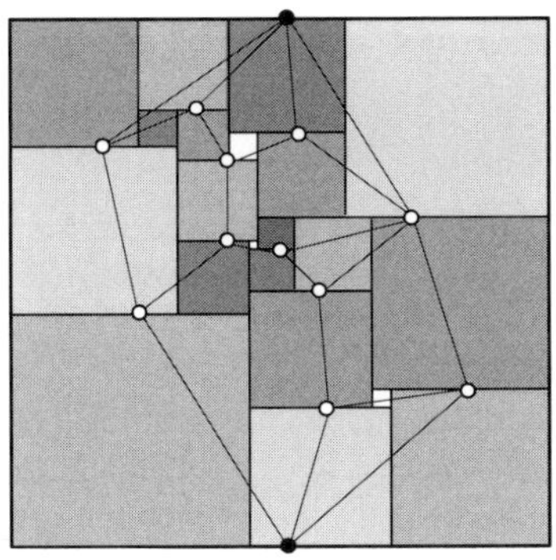

FIGURE 6.3. A planar graph and the tiling generated from it.

line up the squares corresponding to the edges incident with s along the bottom line, in the order as these edges emanate from s. We go through the nodes in the increasing order of the values of f. Getting to a node v, those edges entering v from below have squares assigned to them whose top edges are at height $f(v)$, and these edges form a segment I_v of length

$$|I_v| = \sum_{\substack{u\in N(v)\\ f(u)<f(v)}} \big(f(v)-f(u)\big).$$

Since by the harmonic property of f we also have

$$|I_v| = \sum_{\substack{u\in N(v)\\ f(u)\geq f(v)}} \big(f(u)-f(v)\big),$$

so we can line up the squares corresponding to edges exiting v upwards, along I_v, in the order given by the embedding in the plane. When we get to t, we have filled up the rectangle. □

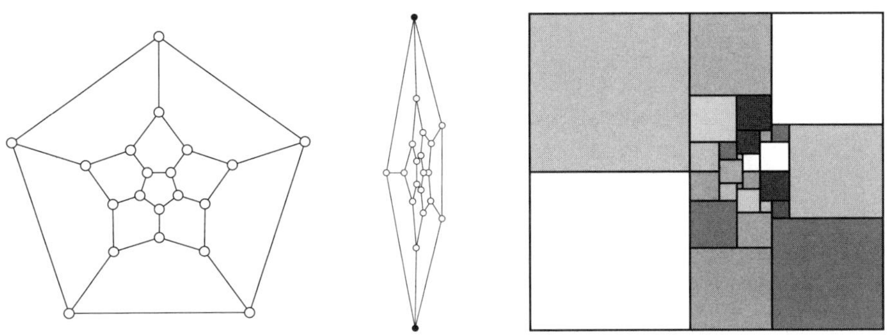

FIGURE 6.4. The dodecahedron graph, its rubber band embedding on the line (horizontally distorted to show the structure), and a square tiling generated from it.

6.2. Tangency graphs of square tilings

Let R be a rectangle in the plane, and consider a tiling of R by squares. Let us add four further squares attached to each edge of R from the outside, sharing the edge with R. We want to look at the tangency graph of this family of squares.

Since the squares do not have a smooth boundary, the conclusion of Exercise 5.8 does not apply, and the tangency graph of the squares may not be planar. Let us try to draw the tangency graph in the plane by representing each square by its center and connecting the centers of touching squares by a straight line segment. Similarly as in the preceding section, we get into trouble when four squares share a vertex. In this case we can specify arbitrarily one diametrically opposite pair as "infinitesimally overlapping", and connect the centers of these two square but not the other two centers. We call this a *resolved tangency graph* of the family of squares.

Every resolved tangency graph is planar, and it is easy to see that it has exactly one country that is a quadrilateral (namely, the unbounded country), and its other countries are triangles; briefly, it is a triangulation of a quadrilateral (Figure 6.5).

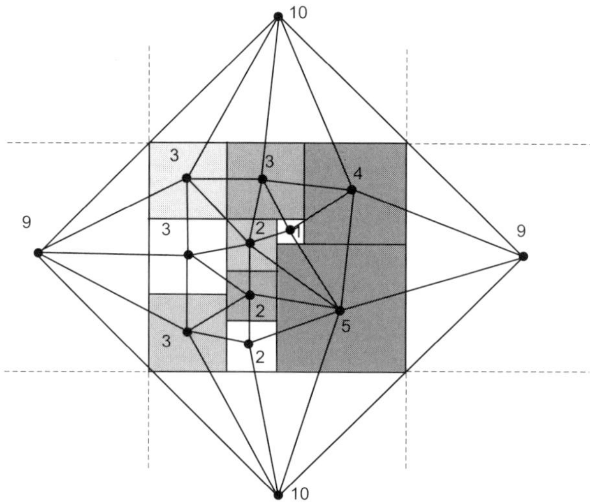

FIGURE 6.5. The resolved tangency graph of a tiling of a rectangle by squares. The numbers indicate the edge lengths of the squares.

Under some connectivity conditions, this fact has the following converse [**Schramm 1993**].

THEOREM 6.2. *Every planar map in which the unbounded country is a quadrilateral, all other countries are triangles, and is not separated by a 3-cycle or 4-cycle, can be represented as a resolved tangency graph of a square tiling of a rectangle.*

Schramm proved a more general theorem, in which separating cycles were allowed; the prize to pay was that degenerate squares with edge-length 0 had to be allowed. It is easy to see that a separating triangle forces everything inside it to degenerate in this sense, and so we do not loose anything interesting by excluding these. Separating 4-cycles may or may not force degeneracy (Figure 6.6), and it does not seem easy to tell when they do.

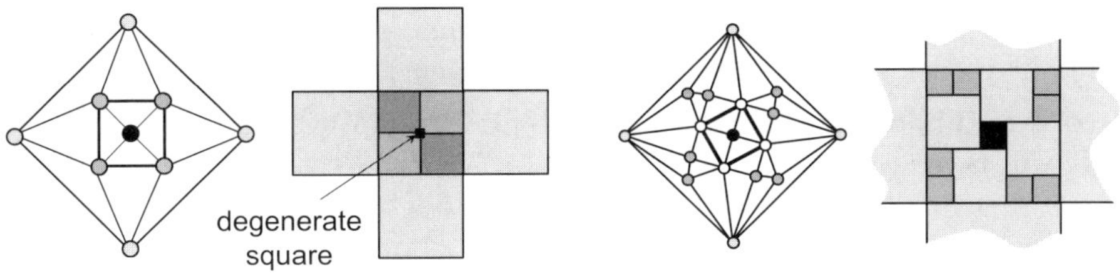

FIGURE 6.6. A separating 4-cycle that causes degeneracy and one that does not.

Before proving Theorem 6.2, let us make a detour to a well-known game. The following folklore fact is easy to prove:

PROPOSITION 6.3. *Let G be a planar map in which the unbounded country is a quadrilateral $abcd$ and all other countries are triangles. Let us 2-color the nodes with black and white, so that a and c are black and b and d are white. Then either there is an all-black a-c path, or an all-white b-d path, but not both.*

PROOF. From the planarity of G it is immediate that if an all-black a-c path exists, then it separates b and d, and so it excludes an all-white b-d path. To prove that one of these paths exists, let us start a walk through the map, avoiding the nodes. We enter through the edge ab, so that we have a black node to the left and a white node to the right, and leave through another edge that has differently colored endpoints (clearly again black on the left and white on the right). Going on similarly, we maintain that we cross only edges having a black endnode to our left and a white endnode to our right. It is easy to see that we never return to a triangle which we left earlier, and so we must return to the unbounded country through one of the edges bc or ad: Exiting through cd is impossible, since then we would have the wrong colors on our left and right (Figure 6.7, left).

Suppose (say) that we exit through bc; then the black-black edges of those triangles that we cross form a walk from a to c, which contains a black a-c path. □

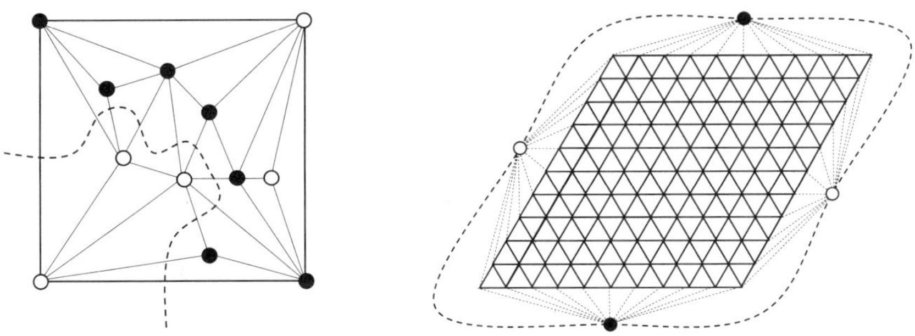

FIGURE 6.7. Left: Walking through a 2-colored triangulation of a square. Right: The game of Hex as a special case of the situation in Proposition 6.3.

The reader familiar with basic algebraic topology will notice that the proof above is very similar to one of the (algorithmic) proofs of Sperner's Lemma. In fact it would be easy to derive Proposition 6.3 from Sperner's Lemma.

A special case of this proposition is the fact that in every game of Hex, one or the other player wins, but not both (Figure 6.7, right).

Proof of Theorem 6.2. Let G be the planar map with unbounded country $abcd$. It will be convenient to set $V' = V \setminus \{a,b,c,d\}$. Recall that $V'(P)$ denotes the set of inner nodes of the path P. For a path P and two nodes u and v on the path, let us denote by $P(u,v)$ the subpath of P consisting of all nodes strictly between u and v. Later on we will also need the notation $P[u,v)$ for the subpath formed by the nodes between u and v, with u included but not v, and $P[u,v]$ for the subpath with both included.

We consider the node-path and node-cut polyhedra, where a and c are the special nodes and b and d are deleted, or vice versa (see Appendix C.2.1). Proposition 6.3 implies that the a-c node-cut polyhedron of $G \setminus \{b,d\}$ is the same as the b-d node-path polyhedron of $G \setminus \{a,c\}$, and vice versa. Another way of saying this is that the b-d node-path polyhedron of $G \setminus \{a,c\}$ is the blocker of the a-c node-path polyhedron of $G \setminus \{b,d\}$.

Let $\mathcal{P}$ be the set of a–c paths in $G \setminus \{b,d\}$ and $\mathcal{Q}$, the set of b–d paths in $G \setminus \{a,c\}$. The a-c node-cut polyhedron is described by the linear constraints

$$x_u \geq 0 \qquad (u \in V') \tag{6.1}$$

$$x\big(V'(P)\big) \geq 1 \qquad (P \in \mathcal{P}). \tag{6.2}$$

Consider the solution $\overline{x}$ of (6.1)–(6.2) minimizing the objective function $\sum_u x_u^2$, and let R^2 be the minimum value. By Proposition C.2, $\overline{y} = (1/R^2)\overline{x}$ minimizes the same objective function over the blocker (which is the b-d node-path polyhedron of $G \setminus \{a,c\}$), and the optimum value is $1/R^2$. Let us rescale the optimizers to get $z = \frac{1}{R}\overline{x} = R\overline{y}$. Then we have

$$z\big(V'(P)\big) \geq \frac{1}{R} \qquad (P \in \mathcal{P}) \tag{6.3}$$

$$z\big(V'(Q)\big) \geq R \qquad (Q \in \mathcal{Q}) \tag{6.4}$$

$$\sum_{u \in V'} z_u^2 = 1. \tag{6.5}$$

It will be convenient to define $z_a = z_b = z_c = z_d = 0$.

We assign the *length* $\widehat{z}_{ij} = \frac{1}{2}(z_i + z_j)$ to every edge ij. Using this, we can define (as usual) the length $\widehat{z}(P)$ of path as the sum of lengths of its edges, and the *distance* $d_z(u,v)$ of two nodes as the minimum length of any path connecting them. Inequalities (6.3) and (6.4) say that $d_z(a,c) \geq 1/R$ and $d_z(b,d) \geq R$, and the minimality of $\overline{x}$ and $\overline{y}$ implies that we have equality here. It is easy to see that $d_z(i,j) = \widehat{z}_{ij}$ for every edge ij, so the edge is a shortest path between its endpoints.

We know that $\overline{y}$ is in the a-c node-path polyhedron, and it is a minimal vector in there, so it is in the a-c node-path polytope, and it can be written as a convex combination of indicator vectors $\mathbb{1}_P$ of sets $V'(P)$, where P is an a–c path. It follows that z can be written as

$$z = \sum_{P \in \mathcal{P}} \lambda_P \mathbb{1}_P, \qquad \sum_P \lambda_P = R, \qquad \lambda_P \geq 0. \tag{6.6}$$

Similarly, we have a decomposition

$$z = \sum_{Q \in \mathcal{Q}} \mu_Q \mathbb{1}_Q, \qquad \sum_Q \mu_Q = \frac{1}{R}, \qquad \mu_Q \geq 0. \tag{6.7}$$

Let $\mathcal{P}' = \{P \in \mathcal{P} : \lambda_P > 0\}$, and define $\mathcal{Q}'$ analogously. Obviously, a node $u \in V'$ has $z_u > 0$ if and only if it is contained in one of the paths in $\mathcal{P}'$, which must then be equivalent to be contained in one of the paths in $\mathcal{Q}'$ (we'll prove later that all nodes have $z_u > 0$).

From conditions (6.3)-(6.5) we can derive some simple but powerful properties of the paths in $\mathcal{P}'$ and $\mathcal{Q}'$. It is trivial from the topology of G that $|V(P) \cap V(Q)| \geq 1$ for $P \in \mathcal{P}'$ and $Q \in \mathcal{Q}'$. On the other hand, (6.5) implies that

$$1 = \sum_{u \in V'} z_u^2 = \Big(\sum_P \lambda_P \mathbb{1}_P\Big)^{\mathsf{T}} \Big(\sum_Q \mu_Q \mathbb{1}_Q\Big) = \sum_{P,Q} \lambda_P \mu_Q |V(P) \cap V(Q)|$$
$$\geq \sum_{P,Q} \lambda_P \mu_Q = \Big(\sum_P \lambda_P\Big)\Big(\sum_Q \mu_Q\Big) = 1.$$

We must have equality here, which implies that

$$|V(P) \cap V(Q)| = 1 \qquad (P \in \mathcal{P}',\ Q \in \mathcal{Q}'). \tag{6.8}$$

For any path $Q \in \mathcal{Q}'$, we have

$$\widehat{z}(Q) = z^{\mathsf{T}} \mathbb{1}_Q = \sum_{P \in \mathcal{P}'} \lambda_P \mathbb{1}_P^{\mathsf{T}} \mathbb{1}_Q = \sum_{P \in \mathcal{P}'} \lambda_P |V'(P) \cap V'(Q)| = \sum_{P \in \mathcal{P}'} \lambda_P = R. \tag{6.9}$$

Similarly, for every $P \in \mathcal{P}'$, we have

$$\widehat{z}(P) = \frac{1}{R}. \tag{6.10}$$

So the paths in $\mathcal{P}'$ and $\mathcal{Q}'$ are shortest paths with respect to the metric d_z. It follows that they are chordless, and also for every path $P \in \mathcal{P}'$ and $u, v \in V(P)$, the subpath $P[u, v]$ is a shortest path between its endpoints. We get by the same kind of argument that the common nodes of any two paths $P, P' \in \mathcal{P}'$ are encountered by the two paths in the same order.

A consequence of (6.9) is that a b-d path Q satisfies $\widehat{z}(Q) = R$ (so it is a shortest b-d path) if and only if $|V(P) \cap V(Q)| = 1$ for every $P \in \mathcal{P}'$.

We say that a $P' \in \mathcal{P}'$ is *to the right* of $P \in \mathcal{P}'$, if every node of P' is either a common point of P and P', or is separated from b by P. Clearly this defines a partial ordering of $\mathcal{P}'$. In this case we say that P and P' *do not cross.* We can similarly define noncrossing paths in $\mathcal{Q}'$, and a partial ordering.

We continue with two somewhat more elaborate properties of the paths in $\mathcal{P}'$ and $\mathcal{Q}'$. The first of these arguments could be omitted at the cost of making the arguments later less transparent. However, under the name of "uncrossing", this method is a standard and powerful step in many proofs in graph theory, and it is worth describing.

Claim 1. *We can choose the decompositions in* (6.6) *and* (6.7) *so that the families $\mathcal{P}'$ and $\mathcal{Q}'$ are pairwise non-crossing.*

In other words, "to the right" defines a total order on $\mathcal{P}'$. Similarly, we get a total order on $\mathcal{Q}'$.

To prove the Claim, let $P, P' \in \mathcal{P}'$ be a pair of paths that are crossing. Using that their common nodes are in the same order along both, we can construct two other a-c paths P_0 and P_0', consisting of the common nodes and those nodes of each path that are to the left (resp. to the right) of the other path (Figure 6.8, left). Clearly $\widehat{z}(P) + \widehat{z}(P') = \widehat{z}(P_0) + \widehat{z}(P_0')$, and since P and P' satisfy (6.3) with equality, so do P_0 and P_0'. Let $\varepsilon = \min\{\lambda_P, \lambda_{P'}\}$, and let us decrease λ_P and $\lambda_{P'}$ by ε and increase λ_{P_0} and $\lambda_{P_0'}$ by ε. This gives another decomposition in (6.6). So one of P and P' (say, P') drops out of $\mathcal{P}'$, and P_0 and P_0' enter it (they may have been there already, in which case only their weight increases).

This describes the main step, but we want to repeat this procedure to get rid of all crossings; to this end, we have to justify that we are making progress. This is not quite obvious, since we have replaced two paths by (possibly) three, creating new crossings with the other paths. Looking more carefully, we see that

(a) the remaining path P and the new paths P_0 and P_0' are mutually noncrossing, and

(b) if a further path in $\mathcal{P}'$ crosses either one of P_0 and P_0', then it crosses at least one of P and P'; and if it crosses both P_0 and P_0', then it crosses both P and P'.

From these observations it follows that the sum

$$\sum_{P_1,P_2\in\mathcal{P}' \text{ crossing}} \lambda_{P_1}\lambda_{P_2}$$

decreases at the above operation, and so if we start with a decomposition minimizing this sum, then the family $\mathcal{P}'$ will consist of non-crossing paths. We argue for $\mathcal{Q}'$ similarly. This proves the Claim.

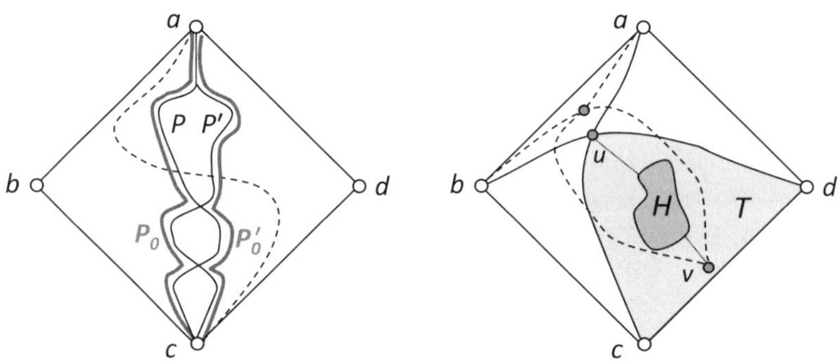

FIGURE 6.8. Left: uncrossing two paths. Right: every node has positive weight.

Let $u \in V'$. Every path in $\mathcal{P}'$ either goes through u or it separates u from exactly one of b or d in the topology of the plane. Let $\mathcal{P}_u^-$, $\mathcal{P}_u$ and $\mathcal{P}_u^+$ denote the sets of paths in $\mathcal{P}'$ separating u from b, passing through u, and separating u from d, respectively. Clearly the sets $\mathcal{P}_u^-$, $\mathcal{P}_u$ and $\mathcal{P}_u^+$ form intervals in the ordering of $\mathcal{P}'$. We define the partition $\mathcal{Q}' = \mathcal{Q}_u^- \cup \mathcal{Q}_u \cup \mathcal{Q}_u^+$ analogously: the paths in $\mathcal{Q}^-$ separate u from a etc. If $Q \in \mathcal{Q}'$ is any path through u, then it is easy to tell which paths in $\mathcal{P}'$ separate u from b: exactly those whose unique common node with Q lies on the subpath $Q[b, u)$.

Clearly $\lambda(\mathcal{P}_u) = \mu(\mathcal{Q}_u) = z_u$. We can express the distance of a node $u \in V'$ with $z_u > 0$ from b as

$$d_z(b, u) = \frac{z_u}{2} + \lambda(\mathcal{P}_u^-). \tag{6.11}$$

Indeed, choosing any $Q \in \mathcal{Q}_u$, a path $P \in \mathcal{P}'$ belongs to $\mathcal{P}_u^-$ if and only if it intersects $Q(b,u)$, and such paths contribute λ_P to z_v for a single node v on $Q(b,u)$.

Claim 2. *Every node $i \in V'$ satisfies $z_i > 0$.*

Suppose that there are nodes with $z_i = 0$, and let H be a connected component of the subgraph induced by these nodes. There are no separating 3- and 4-cycles by the hypothesis of the theorem, hence H has at least five neighbors in $V \setminus V(H)$. Each neighbor v of H has weight $z_v > 0$, and so it is contained in a path $P_v \in \mathcal{P}'$ as well as in a path $Q_v \in \mathcal{Q}'$. The path P_v is disjoint from H and separates H either from b or from d, and similarly, the path Q_v separates H from either a or c. Thus there are two neighbors u and v of H so that both P_u and P_v separate H from (say) b, and both Q_u and Q_v separate h from (say) a. We choose the same path for P_u and P_v if possible, and similarly for Q_u and Q_v, but (6.8) implies that we cannot succeed in both cases. So we may assume that $Q_u \neq Q_v$.

The rest of the argument consists of analyzing the topology around H. If $P_u = P_v$, then we may assume that u comes before v along the path P_u. The path Q_v separates u from a, so P_u meets Q_v before u, and then meets it again at v, contradicting (6.8).

Suppose that $P_u \neq P_v$. Then, as before, it follows that $P_u[a,u)$ meets Q_v. Since P_u and Q_v have only one intersection point, it follows that Q_v is disjoint from $P_u[u,c]$. Similarly, P_v is disjoint from $Q_u[u,d]$.

The paths P_u and Q_u cross each other at u and split the interior of $abcd$ into four open parts (Figure 6.8, right). The subgraph H is contained in the part T incident with the edge cd, and since v is a neighbor of H and not on the paths P_u and Q_u, it must also be in T. Now the path $P_v[a,v]$ must enter T through a node of $P_u(u,c]$, since it is disjoint from $Q_u[u,d]$. Similarly, the path $Q_v[b,v]$ must enter T though $Q_u(u,d]$. But then $P_v[a,v]$ and $Q_v[b,v]$ must have an intersection point outside T, different from v, contradicting (6.8).

After all this preparation, we can describe the squares representing G. We start with the rectangle $\mathfrak{R}$ with one vertex at $(0,0)$ and the opposite vertex at $(R,1/R)$. Every node $u \in V'$ will be represented by a square S_u centered at $\mathbf{p}_u = (d_z(a,u), d_z(b,u))^\mathsf{T}$ and having edge length z_u. We represent node a by the square S_a of side length R attached to the top of $\mathfrak{R}$. We represent b, c and d similarly by squares attached to the other edges of $\mathfrak{R}$.

Claim 3. *Let $i,j \in V$. If i and j are adjacent, then S_i and S_j are tangent. If i and j are nonadjacent, then S_i and S_j are either disjoint or tangent along a single vertex.*

The Claim is easily checked when i and/or j belong to $\{a,b,c,d\}$, so suppose that $i,j \in V'$.

First, let $ij \in E(P)$, where (say) $P \in \mathcal{P}'$. Assume that P encounters i before j. Then $d_z(a,j) = d_z(a,i) + \widehat{z}_{ij}$ (since P is a shortest path). On the other hand, let (say) $d_z(b,j) \geq d_z(b,i)$, and let $Q \in \mathcal{Q}'$ go through i and $Q' \in \mathcal{Q}'$ go through j. Then $Q'' = Q[b,i] \cup \{ij\} \cup Q'[j,d]$ is a b-d path, and since Q'' intersects P in at least two nodes, it is not a shortest b-d path. Thus $z(Q'') > R$, which implies that

$$d_z(b,i) = \widehat{z}(Q[b,i]) > R - \widehat{z}(Q'[j,d]) - \widehat{z}_{ij} = \widehat{z}(Q'[b,j]) - \widehat{z}_{ij} = d_z(b,j) - \widehat{z}_{ij}.$$

So $|d_z(b,j) - d_z(b,i)| < \widehat{z}_{ij}$. This proves that the squares S_i and S_j are tangent along horizontal edges.

Second, let $ij \in E(G)$, and assume that ij is an not an edge of any path in $\mathcal{P}' \cup \mathcal{Q}'$. It is clear that $\mathcal{P}_i \cap \mathcal{P}_j = \emptyset$, since for any path in $\mathcal{P}'$ passing through both i and j, the edge ij would be a chord. Since $\mathcal{P}_i$ and $\mathcal{P}_j$ are intervals in the ordering of $\mathcal{P}'$, it follows that (say) $\mathcal{P}_i$ is completely to the left of $\mathcal{P}_j$. Since i and j are adjacent, no path in $\mathcal{P}'$ can separate i and j, and hence $\cap \mathcal{P}_j^- = \mathcal{P}_i^- \cup \mathcal{P}_i$. Applying (6.11) for both i and j, we see that

$$d_z(b,j) - d_z(b,i) = \frac{z_j}{2} - \frac{z_i}{2} + \lambda(\mathcal{P}_j^-) - \lambda(\mathcal{P}_i^-) = \frac{z_j}{2} - \frac{z_i}{2} + \lambda(\mathcal{P}_i) = \frac{1}{2}(z_i + z_j). \tag{6.12}$$

Finally, assume that i and j are nonadjacent nodes. If there is a path $Q \in \mathcal{Q}'$ going through both i and j, encountering (say) i before j, then

$$d_z(b,j) = d_z(b,i) + d_z(i,j) > d_z(b,i) + \frac{z_i + z_j}{2}$$

and hence S_i and S_j are disjoint. We conclude similarly is there is a path in $\mathcal{P}'$ through both i and j. We are left with the case when $\mathcal{P}_i \cap \mathcal{P}_j = \mathcal{Q}_i \cap \mathcal{Q}_j = \emptyset$. As in the previous proof, we may assume that $\mathcal{P}_i$ is completely to the left of $\mathcal{P}_j$, but we can only conclude that $\cap \mathcal{P}_j^- \subseteq \mathcal{P}_i^- \cup \mathcal{P}_i$, and so the computation in (6.12) gives an inequality only: $d_z(b,j) - d_z(b,i) \geq \frac{1}{2}(z_i + z_j)$. Similarly, $|d_z(a,j) - d_z(a,i)| \geq \frac{1}{2}(z_i + z_j)$. If strict inequality holds in both directions, then S_i and S_j are disjoint; if equality holds in both cases, then S_i and S_j have a vertex in common. This proves Claim 3.

Consider G as drawn in the plane so that node i is at the center $\mathbf{p}_i$ of the square S_i, and every edge ij is a straight segment connecting $\mathbf{p}_i$ and $\mathbf{p}_j$. This gives a planar embedding of G. Indeed, we know from Claim 3 that every edge ij is covered by the squares S_i and S_j. This implies that edges do not cross, except possibly in the degenerate case when four squares S_i, S_j, S_k, S_l share a vertex (in this clockwise order around the vertex), and $ik, jl \in E$. Since the squares S_i and S_j are touching along their vertical edges, we have $ij \in E$. Similarly, $jk, kl, li \in E$, and hence i, j, k, l form a complete 4-graph. But this is impossible in a triangulation of a quadrilateral that has no separating triangles.

Next, we argue that the squares S_i ($i \in V \setminus \{a,b,c,d\}$) tile the rectangle $\mathfrak{R} = [0,R] \times [0,1/R]$. It is easy to see that all these squares are contained in the rectangle $\mathfrak{R}$, and they are non-overlapping by Claim 3. The total area covered by the squares is $\sum_{i \in V'} z_i^2 = 1$, which is just the area of $\mathfrak{R}$, so they must cover $\mathfrak{R}$.

Finally, we show that an appropriately resolved tangency graph of the squares S_i is equal to G. By the above, it contains G (where for edges of type (iii), the 4-corner is resolved so as to get the edge of G). Since G is a triangulation of the outside quadrilateral, the containment cannot be proper, so G is the whole tangency graph. □

EXERCISE 6.1. Figure out how to resolve the common points of four squares in the tiling in Figure 6.2 in order to get the dodecahedron graph.

EXERCISE 6.2. Verify that the graph in Figure 6.9 gives rise to a tiling of a square by different size squares. (This construction provides the minimum number of different squares tiling a square.)

EXERCISE 6.3. Prove that if G is a resolved tangency graph of a square tiling of a rectangle, then every triangle in G is a country.

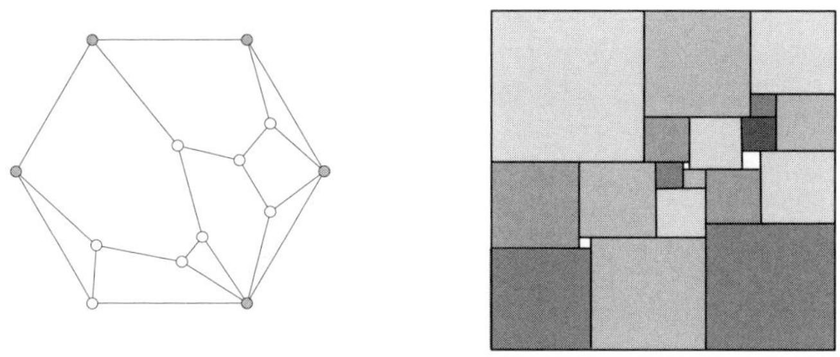

FIGURE 6.9. Tiling a square with the smallest number of different squares.

EXERCISE 6.4. Construct a resolved tangency graph of a square tiling of a rectangle that contains a quadrilateral with a further node in its interior.

EXERCISE 6.5. Let G be a maximal planar graph with at least 4 nodes, with unbounded country abc. Let us 2-color the nodes different from a, b and c with black and white. Prove that either there is a black node connected by black paths to each of a, b and c, or there is a white node connected by white paths to each of a, b and c, but not both.

CHAPTER 7

Discrete Analytic Functions

We have seen in Chapter 4 that the notion of harmonic functions in analysis has an analogue in graph theory, and this analogue has proved very useful in the study of electrical networks, rubber band embeddings, random walks, and elsewhere. In this chapter we discuss an extension of analytic functions to graphs; for this to be meaningful, we have to consider graphs embedded in orientable surfaces (maps).

In analysis, a complex analytic function is just a pair of real harmonic functions connected by the Cauchy-Riemann equations. We know what harmonic functions are on graphs, but what is the "right" analogue of the Cauchy-Riemann equations?

The motivation comes from numerical analysis. In order to do numerical computations with an analytic function, we have to restrict its domain to a finite set, most naturally to a large piece of a fine enough grid. Then we can mimic differentiation by taking the difference of values on adjacent gridpoints, integration by summation along grid paths etc. In this setting, important properties, like the Cauchy–Riemann equations and the Cauchy Integral Theorem, will only remain approximately true. Ferrand and later Duffin suggested to turn this around: use only approximate values of the function on the grid, but require that appropriate forms of (say) the Cauchy Integral Theorem should hold *exactly*. The resulting theory of discrete analytic functions on the grid leads to interesting results, even though some properties of analytic functions seem to be irrevocably lost (for example, the product of analytic functions will be no longer analytic).

In fact, we study two different, but closely related notions: discrete analytic functions, and discrete holomorphic forms (in graph theory terms, these will be called rotation-free circulations). We have seen that no nonconstant function can be harmonic everywhere on a finite graph, and accordingly discrete analytic functions are best defined on countable graphs embedded in the plane; on the other hand, discrete holomorphic forms can be best studied on finite graphs embedded in compact orientable surfaces.

Discrete analytic functions on the square grid were introduced in the 1940's [**Ferrand 1944**] and studied quite extensively over the next decades [**Duffin 1956**], [**Duffin 1968**], [**Duffin–Peterson 1968**], [**Zeilberger 1977b**], [**Zeilberger–Dym 1977**]. For the case of a general map, the notion of discrete analytic functions is implicit in the classic paper [**Brooks et al. 1940**] (cf. also Section 6.1 in this book) and more recent work [**Benjamini–Schramm 1996a**]. These functions were formally introduced by [**Mercat 2001**], and applications were given in [**Benjamini–Lovász 2002**], [**Benjamini–Lovász 2003**] and, perhaps most importantly, in [**Smirnov 2010b**][**Chelkak–Smirnov 2012**], where they were used in the proof of the conformal invariance of the Ising model on the square grid (to be discussed in more detail in Chapter 8). Similar, but different theories of discrete analytic functions (not discussed here) were developed in [**Isaacs 1952**], [**Dynnikov–Novikov 2003**], [**Kiselman 2005**].

7.1. Warmup: discrete analytic functions on the grid

7.1.1. Grid graphs and lattice domains. We consider the square grid as a (countable) map in the complex plane, whose node set is the set $\mathbb{G}$ of Gaussian integers (complex numbers $a+bi$, where $a, b \in \mathbb{Z}$), and two such nodes are connected by an edge if and only if their difference is ± 1 or $\pm i$. We are going to denote, with a little abuse of notation, this graph by $\mathbb{G}$. Its dual graph $\mathbb{G}^*$ can be obtained from $\mathbb{G}$ by shifting it by $(1+i)/2$. Let $\mathbb{G}_r$ $(r = 0, 1)$ denote the set of Gaussian integers $a+bi$ with $a+b \equiv r \pmod 2$. Note that $\mathbb{G}_r$ is a grid itself, where we consider two lattice points adjacent if they are nearest, which in this case means that their difference is $\pm(1+i)$ or $\pm(1-i)$.

The graph $\mathbb{G}$ is bipartite, with bipartition $\mathbb{G}_0 \cup \mathbb{G}_1$; sometimes we are going to call the nodes in $\mathbb{G}_0$ "black", and those in $\mathbb{G}_1$, "white". So the origin is black. We can also 2-color the bounded countries of $\mathbb{G}$ (the nodes of $\mathbb{G}^*$) with black and white (like on a chessboard); to be specific, let each square get the same color as its Southwest corner.

We will also need to define an orientation of these graphs. There are two natural orientations we can consider: we can orient each edge so that the country on its left is black (say), or we can orient it from the black endnode to the white endnode. We call these two orientations the *cyclic* and *acyclic* orientation, respectively. If we start with the acyclic orientation of $\mathbb{G}$, and define the orientation of $\mathbb{G}^*$ as described in Chapter 2, then we get the cyclic orientation of $\mathbb{G}^*$ (Figure 7.1, left).

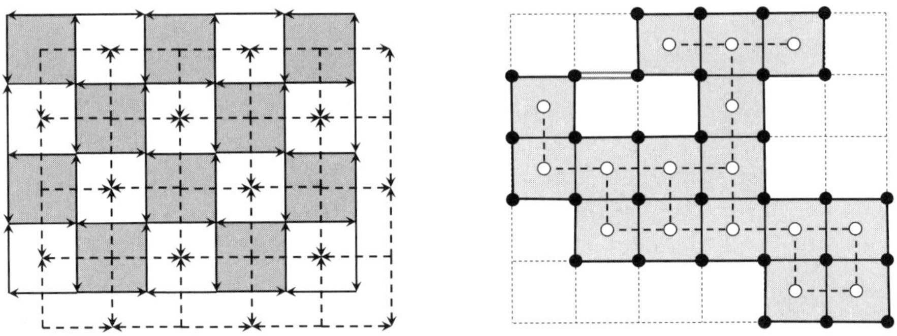

FIGURE 7.1. Left: A cyclic orientation of the planar grid and the corresponding acyclic orientation of its dual. Right: a lattice domain D and the graphs $\mathbb{G}_D$ and $\mathbb{G}^*_D$. Note that the doubled edge is not an edge of $\mathbb{G}_D$, so $\mathbb{G}_D$ is not an induced subgraph of $\mathbb{G}$.

The subset of the plane bounded by a simple closed polygon in $\mathbb{G}$ will be called a *lattice domain* (Figure 7.1, right). The vertices and edges of the lattice graph $\mathbb{G}$ in a lattice domain D form a subgraph $\mathbb{G}_D$. Note that $\mathbb{G}_D$ is not necessarily an induced subgraph of $\mathbb{G}$; but $\mathbb{G}^*_D$, consisting of all nodes representing squares whose interior if contained in the interior of D, is an induced subgraph of $\mathbb{G}^*$. This is the tangency graph of its squares (which we do not resolve into a triangulation, unlike in Chapter 6): we connect the centers of two squares contained in D if and only if they share an edge. This graph $\mathbb{G}_D^*$ can be obtained from the dual graph of G by removing the node representing the unbounded country of G.

7.1.2. Discrete analytic functions. A *discrete analytic function* is a function $f: \mathbb{G} \to \mathbb{C}$, satisfying the equation

$$f(z+i)-f(z+1) = i\big(f(z+1+i)-f(z)\big) \tag{7.1}$$

for every $z \in \mathbb{G}$. This equation has a nice geometric meaning: the diagonals of the image of any lattice square under a discrete analytic function are orthogonal and of equal length. We can write this equation as

$$f(z)+if(z+1)+i^2f(z+i+1)+i^3f(z+i) = 0. \tag{7.2}$$

Dividing (7.1) by $i-1$, we get

$$\frac{f(z+i+1)-f(z)}{i+1} = \frac{f(z+i)-f(z+1)}{i-1}. \tag{7.3}$$

This latter equation can be thought of as a discrete version of the fact that the directional derivative of a complex analytic function is the same in all directions; in other words, it gives a discrete version of the Cauchy–Riemann equation.

We can allow more general discrete analytic functions, namely functions defined on the lattice points in a lattice domain. In this case, we stipulate equation (7.1) for all lattice squares contained in this domain.

Discrete analytic functions form a linear space over $\mathbb{C}$, i.e., linear combinations of them is discrete analytic again. It is trivial to check that every linear function, when restricted to Gaussian integers, is a discrete analytic function. Furthermore, the restriction of any quadratic polynomial to the Gaussian integers is a discrete analytic function. It suffices to verify this for the function z^2, which is straightforward:

$$(z+1)^2-(z+i)^2 = 2(i-1)z+2 = i\big((z+i+1)^2-z^2\big),$$

verifying (7.1) (see Figure 7.2).

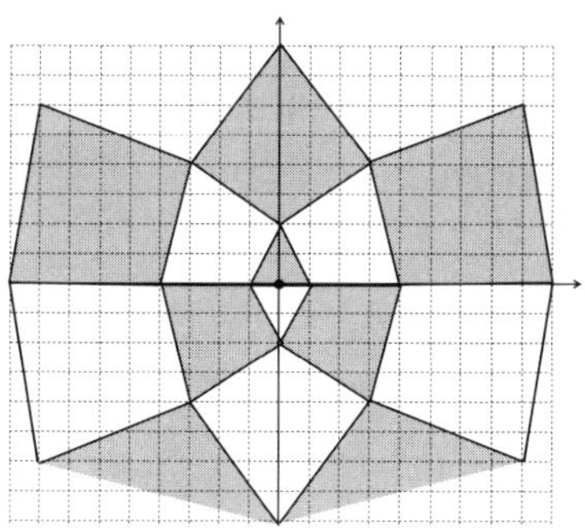

FIGURE 7.2. The image of the standard grid under the discrete analytic function $z \mapsto z^2$. Every vertex other than the origin is the image of two lattice points, negatives of each other.

In contrast to the previous example, the restriction of the function $f(z) = z^3$ to the Gaussian integers is not discrete analytic. In fact, substituting $z = 0$ in (7.1), we get $i^3 - 1^3 = -i-1$ on the left, while $i((1+i)^3-0^3) = -2(1+i)$ on the right. This shows that the product of two discrete analytic functions is not discrete analytic in general. This fact, which is in strong contrast to the continuous case, is a major difficulty in this theory.

Some other operations on "true" analytic functions do work out quite well. Let f be a function on the Gaussian integers. Let $P = (z_0, z_1, \dots, z_k)$ be a walk on the grid, i.e., $z_{k+1} - z_k \in \{\pm 1, \pm i\}$. We define the "line integral"

$$\int_P f(z)\,dz = \sum_{j=0}^{k-1} \frac{f(z_{j+1}) + f(z_j)}{2}(z_{j+1} - z_j).$$

The nice fact about this integral is that for analytic functions, it is independent of the path P, depends on its endpoints only. More precisely, let P and P' be two walks on the grid, with the same beginning node and endnode. Then

$$\int_P f(z)\,dz = \int_{P'} f(z)\,dz. \tag{7.4}$$

This is equivalent to saying that

$$\int_P f(z)\,dz = 0 \tag{7.5}$$

if P is a closed walk. It suffices to verify this for the case when P is a simple polygon, which we can recognize as a discrete version of the Cauchy Integral Theorem. The proof is quite easy: the case when P is the boundary of a single square is just a rearrangement of equation (7.3). For a general lattice polygon P, (7.5) follows by summing over all lattice squares in the interior of P.

It follows that we can write $\int_u^v f(z)\,dz$, without specifying the path from u to v along which we integrate. Furthermore, it is straightforward to verify that for a fixed u, $\int_u^v f\,dz$ is a discrete analytic function of v.

For a "true" analytic function, its real and imaginary parts are harmonic functions. For discrete analytic functions, this connection is a bit more complicated. Let f be a discrete analytic function on the grid $\mathbb{G}$, and let f_r denote its restriction to $\mathbb{G}_r$ $(r = 0, 1)$. It is easy to see that the functions f_0 and f_1 determine each other up to an additive constant.

The function f_r is a harmonic on $\mathbb{G}_r$. Indeed, for $z \in \mathbb{G}_0$ (say) we have

$$\begin{aligned} f_0(z+i+1) - f_0(z) &= i\big(f_0(z+1) - f_0(z+i)\big), \\ f_0(z+i-1) - f_0(z) &= i\big(f_0(z+i) - f_0(z-1)\big), \\ f_0(z-i-1) - f_0(z) &= i\big(f_0(z-1) - f_0(z-i)\big), \\ f_0(z-i+1) - f_0(z) &= i\big(f_0(z-i) - f_0(z+1)\big). \end{aligned}$$

Summing these equations, we get

$$f_0(z+i+1) + f_0(z+i-1) + f_0(z-i-1) + f_0(z-i+1) - 4f_0(z) = 0,$$

which means that f_0 is harmonic at z.

Conversely, let f_0 be a harmonic function on $\mathbb{G}_0$. By the potential argument, we can find a function f_1 on $\mathbb{G}_1$ such that

$$\begin{aligned} f_1(z+1) - f_1(z+i) &= -i\big(f_0(z+i+1) - f_0(z)\big) \\ f_1(z+i) - f_1(z-1) &= -i\big(f_0(z+i-1) - f_0(z)\big) \end{aligned}$$

for every $z \in \mathbb{G}_0$. It is easy to see that f_1 is harmonic as well, and $f_0 \cup f_1$ is a discrete analytic function on the grid.

Let us also notice that a complex valued harmonic function is simply a pair of real valued harmonic functions (its real and imaginary parts). Furthermore, the real part of f_1 is related to the imaginary part of f_0, and vice versa. Hence it suffices to study analytic functions that are real on $\mathbb{G}_0$ and imaginary on $\mathbb{G}_1$. We can forget about the factor i on $\mathbb{G}_1$, and consider a pair of functions $f: \mathbb{G}_0 \to \mathbb{R}$ and $g: \mathbb{G}_1 \to \mathbb{R}$ such that the following version of the Cauchy–Riemann equations holds:

$$
\begin{aligned}
g(z+1) - g(z+i) &= f(z+i+1) - f(z) \\
g(z+i) - g(z-1) &= f(z+i-1) - f(z)
\end{aligned}
\tag{7.6}
$$

We call the pair (f,g) an *analytic pair* on the grid.

We can define discrete analytic functions on the nodes of $\mathcal{G}_D$ for some lattice domain D, instead of all lattice points. Condition (7.1) is then required only for lattice squares contained in D. Properties of discrete analytic functions discussed above extend to this more general setting easily.

Further constructions like analytic polynomials on the grid, and some properties of discrete analytic functions, similar and not similar to properties of "true" analytic functions, are described in Exercises 7.1–7.6.

7.1.3. Strongly analytic functions. Smirnov [**Smirnov 2001**] introduced a class of functions satisfying a stronger condition than analycity. Consider a lattice domain D and the graph $\mathbb{G}_D$ with the cyclic orientation. We say that a function g defined on D is *strongly analytic*, if

$$g(v) - g(u) \parallel \sqrt{\overline{v} - \overline{u}}$$

for every edge $uv \in \mathbb{G}_D$. (Note that the sign of the square root plays no role, but the orientation of the edge is important: $\sqrt{\overline{v}-\overline{u}}$ is orthogonal to $\sqrt{\overline{u}-\overline{v}}$.) Strongly analytic means that $g(z+1)-g(z)$ is real if $z \in \mathbb{G}_0$, and imaginary otherwise; $g(z+i)-g(z)$ is parallel to $1+i$ if $z \in \mathbb{G}_0$, and to $1-i$ otherwise (Figure 7.3, left).

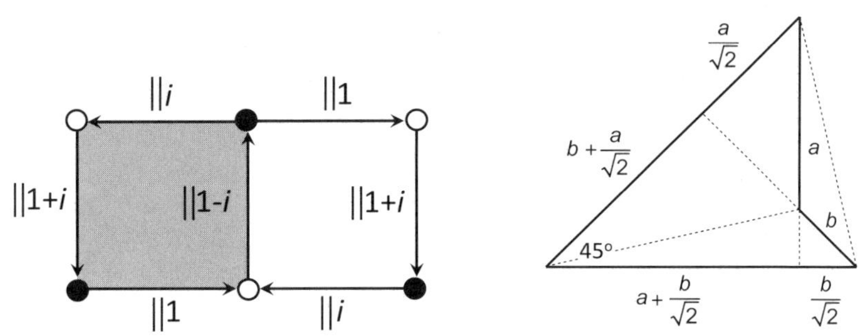

FIGURE 7.3. Left: directions of differences along edges of a strongly analytic function. Right: the image of a lattice square under a strongly analytic function. (The quadrilateral may be rotated by any multiple of 45°.)

It is easy to figure out the shape of the image of a (say, black) square (Figure 7.3, right). The image of a lattice square (assuming it is nondegenerate) is always concave, but never self-intersecting; if its two shorter edges (at the concave angle) have lengths a and b, then the two longer ones have $a+\sqrt{2}b$ (next to b) and $b+\sqrt{2}a$ (next to a). In particular, the sums of squared length of opposite edges are equal: $(a+\sqrt{2}b)^2+a^2 = (b+\sqrt{2}a)^2+b^2 = 2a^2+2\sqrt{2}ab+2b^2$. Three vertices of it form a triangle with one angle of 45°, and the fourth point is the orthocenter of this triangle. It also follows that the diagonals of the quadrilateral are perpendicular and of equal length. Checking the directions of the diagonals too, we see that

$$g(z+i)-g(z+1) = i\big(g(z+i+1)-g(z)\big),$$

which implies that *every strongly analytic function is analytic* (justifying the name).

EXAMPLE 7.1. Consider the function $z \mapsto z$ on $\mathbb{G}$; this is analytic, but not strongly analytic. We get further analytic functions by adding a complex number a at every lattice point in $\mathbb{G}_1$. If $|a|=1$, then it is easy to check that the concave quadrilateral $(0,1+a,1+i,i+a)$ (the image of the lattice square $(0,1,1+i,i)$ has the property that its opposite edges are perpendicular, and so are its diagonals. By elementary geometry, it follows that the angles of this quadrilateral are, in order, 45°, 225°, 45°, 45°. If we divide all function values by $1+a$, then the edges of this quadrilateral will be parallel, in order, to $1, 1-i, i, 1+i$. Thus this function is strongly analytic. In formula:

$$g(z) = \begin{cases} \dfrac{z}{1+a} & \text{if } z \in \mathbb{G}_0, \\ \dfrac{z+a}{1+a} & \text{if } z \in \mathbb{G}_1 \end{cases} \qquad (|a|=1).$$

See Figure 7.4. ♦

FIGURE 7.4. The image of the grid under a strongly analytic function.

Exercise 7.1 tells us that prescribing the values of a discrete analytic function on the coordinate axes determines the function uniquely. Analogously, if we prescribe the value of a function at the diagonal points $x+xi$, then it has a unique extension g to a strongly analytic function. Indeed, if we know the values at the opposite vertices of a square (say at z and $z+i+1$), then $g(z+1)$ and $g(z+i)$ are determined by the directions of the corresponding edges. Hence we can define g recursively farther and farther away from the diagonal. Example 7.1 above is the extension of the function $g(x+xi) = \frac{1+i}{1+a}x$.

Strongly analytic functions form a linear space over $\mathbb{R}$, but not over $\mathbb{C}$. In fact, they generate the linear space of all discrete analytic functions over $\mathbb{C}$ (Exercise 7.9). Many interesting properties of such functions are proved in [**Smirnov 2010b**],

[**Chelkak–Smirnov 2011**] and [**Chelkak–Smirnov 2012**]. See Exercises 7.1 and 7.10 for further important properties of strongly analytic functions.

7.2. Discrete analytic functions on planar graphs

We want to define analytic functions not only on the grid, but on more general planar maps. Definition (7.1) does not generalize (unless all countries are quadrilaterals, a case to which we will come back), but (7.6) does suggest a reasonable generalization.

Let G be a planar map and let G^* be its dual, with a reference orientation. We call a pair of functions $f : V \to \mathbb{R}$ and $g : V^* \to \mathbb{R}$ an *analytic pair*, if the discrete Cauchy–Riemann equation is satisfied for every edge:

$$g\big(\mathsf{ls}(e)\big) - g\big(\mathsf{rs}(e)\big) = f\big(\mathsf{hd}(e)\big) - f\big(\mathsf{tl}(e)\big). \tag{7.7}$$

Relation (7.7) implies a number of further properties; for example, it follows that f is harmonic. Indeed, let $uv_1, \dots, uv_k$ be the edges incident with $u \in V$, in counterclockwise order, and let p_j be the country in the corner of uv_j and uv_{j+1} (we assume, for simplicity of presentation, that the edges are all oriented away from u). Then

$$\sum_{j=1}^{k} \big(f(v_j) - f(u)\big) = \sum_{j=1}^{k} \big(g(p_j) - g(p_{j-1})\big) = 0. \tag{7.8}$$

But this is bad news, since we have seen that no nonconstant function can be harmonic at all nodes of a finite graph. We can get around this obstacle in two ways: either we allow "singularities" at some locations, or we consider infinite graphs. The first approach is quite easy if we disregard the edges of the unbounded country (see Exercise 7.11). We are going to focus on the second.

To this end, we have to redefine (for this section only) what we mean by planar maps. A *planar map* means an (infinite) graph G that is embedded in the plane so that every country is an open topological disk whose closure is compact, and every bounded piece of the plane meets a finite number of edges and nodes only. In particular, this implies that V and E are countable, every node has finite degree, and every country has a finite number of edges. Such a map has a dual $G^* = (V^*, E^*)$, which can be represented as another planar map. The lozenge graph $G^\diamond$ and other simple manipulations on planar graphs can be defined in a straightforward way.

In this more general setting, the definition (7.7) can be required for every edge without forcing the function to be trivial; recall the examples from Section 7.1.

Note that for an analytic pair (f, g), the function f determines g up to an additive constant; furthermore, for such a g to exist it is necessary and sufficient that f is harmonic. Indeed, (7.7) defines a function on the edges of G^*. The computation in (7.8), followed backwards, tells us that this function adds up to zero around each country of G^*, and so the "potential argument" gives us the existence of g.

In this sense, the theory of discrete analytic functions on planar maps is just another formulation of the theory of harmonic functions on such maps. However, the existence of the function on the dual map, which is harmonic as well, leads to a richer theory.

It is often convenient to think of the two functions f and g above as a single function f defined on $V \cup V^* = V^\diamond$, and allow complex values. We say that $f : V^\diamond \to \mathbb{C}$ is *analytic*, if

$$f(\mathsf{ls}(e)) - f(\mathsf{rs}(e)) = i\big(f(\mathsf{hd}(e)) - f(\mathsf{tl}(e))\big) \tag{7.9}$$

for every edge $e \in E$. As we noted for the grid before, the complex version is not substantially more general than the real. Indeed, (7.9) only relates the real part of $f|_V$ to the imaginary part of $f|_{V^*}$ and vice versa. In other words, a complex analytic function f can be decomposed as the sum of two very special analytic functions f_1 and f_2: here f_1 is real on V and imaginary on V^*, and f_2 is the other way around.

EXAMPLE 7.2 (Rubber bands). Let G be a toroidal map. We consider the torus as $\mathbb{R}^2/\mathbb{Z}^2$, endowed with the metric coming from the Euclidean metric on $\mathbb{R}^2$. Let us replace each edge by a rubber band, and let the system find its equilibrium. Topology prevents the map from collapsing to a single point: every non-contractible cycle remains non-contractible. In mathematical terms, we are minimizing the energy $\sum_{e\in E} \ell(e)^2$, where the length $\ell(e)$ of the edge e is measured in the given metric, and we are minimizing over all continuous mappings of the graph into the torus homotopic to the original embedding.

It is not hard to see that the minimum is attained. Clearly, the edges are mapped onto geodesic curves (which become straight line segments if we "roll out" the torus). A nontrivial fact is that if G is a simple 3-connected toroidal map, then the rubber band mapping is an embedding. This can be proved along the lines of the proof of Theorem 3.2, and it also follows from Theorem 7.4 below.

Rolling out the torus, we get a doubly periodic embedding of an infinite graph $\widehat{G}$ in the plane. From the minimality of the energy it follows that every node is at the center of gravity of its neighbors. So considering the points of the plane as complex numbers, the embedding defines a function on $V(\widehat{G})$ that is harmonic at every node, and hence it is one component of an analytic pair on $\widehat{G}$ (Figure 7.5). ♦

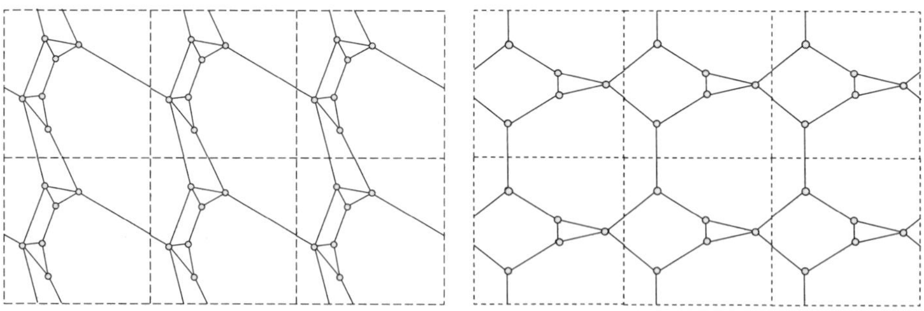

FIGURE 7.5. Periodic rubber band maps obtained from toroidal graphs

It is rather straightforward to extend these notions to planar maps with weighted edges. We weight the edges of G by $\omega \in \mathbb{R}_+^E$; this defines a weighting of the edges of G^* by $\omega_e^* = 1/\omega_{e^*}$. We call a function $f : V^\diamond \to \mathbb{C}$ *analytic for the edge-weighted graph* G, if the weighted version of the discrete Cauchy–Riemann equation

$$f(\mathsf{hd}(e)) - f(\mathsf{tl}(e)) = i\omega_e\big(f(\mathsf{rs}(e)) - f(\mathsf{ls}(e))\big) \tag{7.10}$$

is satisfied for every edge $e \in E$. Again, we can consider instead two real-valued functions $\Re(f) : V \to \mathbb{R}$ and $\Im(f) : V^* \to \mathbb{R}$, and (completely independently) other two, namely $\Im(f) : V \to \mathbb{R}$ and $\Re(f) : V^* \to \mathbb{R}$. Both pairs satisfy the real version of the Cauchy–Riemann equations.

7.2.1. Integration and critical functions. Let f and g be two analytic functions on the nodes of the lozenge graph $G^\diamond$ of a weighted graph in the plane. We define integration along a path $P = (v_0, v_1, \dots, v_k)$ in $G^\diamond$:

$$\int_P f\,dg = \sum_{j=0}^{k-1} \frac{f(v_j)+f(v_{j+1})}{2}\big(g(v_{j+1})-g(v_j)\big).$$

This can also be written as

$$2\int_P f\,dg = f(v_0)g(v_0) + \sum_{j=0}^{k-1} \begin{vmatrix} f(v_j) & f(v_{j+1}) \\ g(v_j) & g(v_{j+1}) \end{vmatrix} - f(v_k)g(v_k).$$

The nice fact about this integral is that it depends on the endpoints of the path only. By the usual argument, it suffices to verify that this integral is zero when P the boundary of a country of $G^\diamond$, which is a straightforward computation. It follows that we can write $\int_u^v f\,dg$. It is also easy to check the rule of integration by parts: for $u, v \in V^\diamond$,

$$\int_u^v f\,dg = f(v)g(v) - f(u)g(u) - \int_u^v g\,df. \tag{7.11}$$

These have been the good news. Now the bad news: for two analytic functions f and g, and a fixed starting node u, the function $F(v) = \int_u^v f\,dg$ is *not* analytic in general. A simple computation gives the "analyticity defect" of an edge uv and dual edge $e = pq$:

(7.12)
$$F(v) - F(u) - i\omega_{uv}(F(p) - F(q)) = i(f(v) - f(u))\big[g(v) + g(u) - g(q) - g(p)\big].$$

Assuming that the function g is such that the factor in brackets in (7.12) is zero for every edge:

$$g\big(\mathsf{ls}(e)\big) + g\big(\mathsf{rs}(e)\big) = g\big(\mathsf{hd}(e)\big) + g\big(\mathsf{tl}(e)\big), \tag{7.13}$$

the integral function F will be analytic. We call such a function g *critical.* What we have found above is that $\int_u^v f\,dg$ is an analytic function of v for every analytic function f if and only if g is a critical analytic function.

This notion was introduced, in a somewhat different setting, in [**Duffin 1968**] under the name of *rhombic lattice. Critical maps* were defined in [**Mercat 2001**]: these are maps which admit a critical analytic function.

Recall condition (7.10) on g being an analytic function:

$$g\big(\mathsf{hd}(e)\big) - g\big(\mathsf{tl}(e)\big) = i\omega_e\big(g(\mathsf{rs}(e)) - g(\mathsf{ls}(e))\big).$$

These equations, together with (7.13), have a nice geometric meaning: the diagonals of the quadrilateral $g\big(\mathsf{ls}(e)\big)g\big(\mathsf{rs}(e)\big)g\big(\mathsf{hd}(e)\big)g\big(\mathsf{tl}(e)\big)$ have the same midpoint, and they are orthogonal. In other words, this quadrilateral is a rhombus. (In the unweighted case it is a square, so in this case, we are back to the study of analytic functions on the square grid!)

Consider the critical analytic function g as a mapping of $G \cup G^*$ into the complex plane $\mathbb{C}$. This defines embeddings of G, G^* and $G^\diamondsuit$ in the plane with following (equivalent) properties:

(a) The countries of $G^\diamondsuit$ are rhombi.

(b) Every edge of $G^\diamondsuit$ has the same length.

(c) Every country of G is inscribed in a circle with the same radius.

(d) Every country of G^* is inscribed in a circle with the same radius.

If you want to construct such maps, you can start with a planar map straight-line embedded in the plane so that all countries are rhombi; this is bipartite, and you can define a critical graph on its white nodes by adding he appropriate diagonal to each rhombus and deleting the original edges (see Figure 7.6).

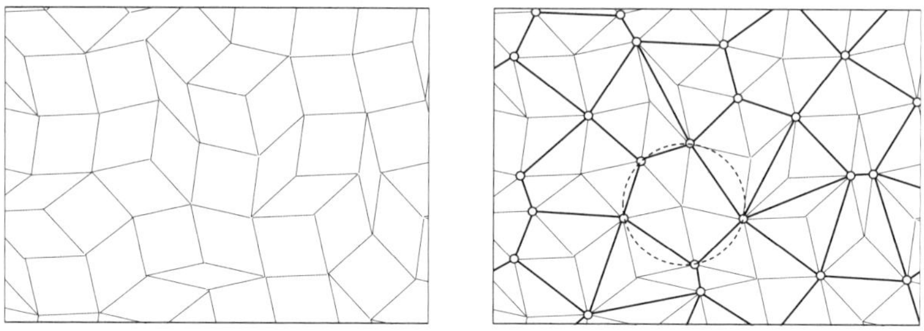

FIGURE 7.6. A map with rhombic countries, and the critical graph constructed from it.

Critical maps have a nice interpretation in terms of the Delaunay diagram: it means that all Delaunay circles of the node set V have the same radius. This is clearly invariant under duality. It follows that the Voronoi graph of V is the Delaunay graph of the dual (Figure 7.7, left).

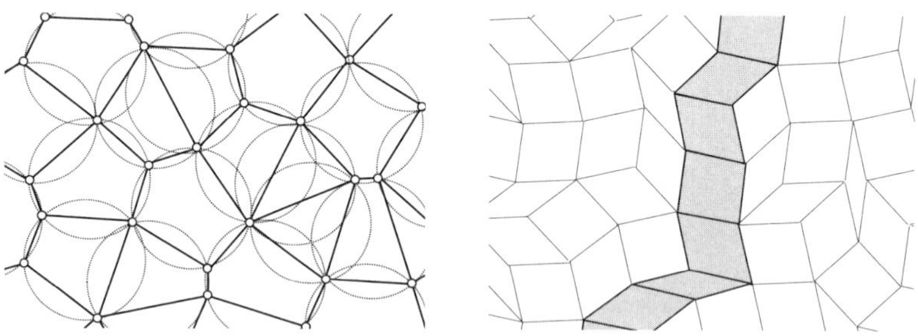

FIGURE 7.7. Left: the Delaunay circles of the node set of a critical map. Right: A track in a rhombic graph.

Whether or not a weighted map in the plane admits a critical analytic function depends on the weighting of the edges. Which maps can be weighted this way? [**Kenyon–Schlenker 2004**] answered this question. Consider any country F_0 of the lozenge graph $G^\diamondsuit$, and a country F_1 incident with it. This is a quadrilateral, so there is a well-defined country F_2 so that F_0 and F_2 are attached to F_1 along opposite edges. Repeating this, we get a sequence of countries $(F_0, F_1, F_2 \dots)$. Using the country attached to F_0 on the opposite side to F_1, we can extend this to a two-way infinite sequence $(\dots, F_{-1}, F_0, F_1, \dots)$. We call such a sequence a *track* (see Figure 7.7, right).

THEOREM 7.3. *A planar map has a rhombic embedding in the plane if and only if every track consists of different countries and any two tracks have at most one country in common.* □

7.2.2. Embedding by analytic functions. Restricting a complex analytic function f on $G^\diamond$ (where G is an infinite planar map) to V, we get a mapping of its nodes into the complex plane. We can extend this map to a drawing of the whole graph by mapping each edge uv onto the segment connecting $f(u)$ and $f(v)$. It turns out that under rather general conditions, this mapping is an embedding in the plane.

THEOREM 7.4. *Let G be a 3-connected map in the plane and let f be a complex analytic function on G. Suppose that there exist a constant c such that*

$$\frac{1}{c} \le \frac{|f(u)-f(v)|}{d_G(u,v)} \le c. \tag{7.14}$$

for every pair of distinct nodes $u, v \in V^\diamond$. Then f defines an embedding of G in the plane such that every country is a convex polygon, and every node is in the center of gravity of its neighbors.

PROOF. Analogously to the proof of Theorem 3.2, the main step in the proof is the following.

Claim 1. *For every (open or closed) halfplane H, the set $S = \{i \in V : \ f(i) \in H\}$ induces a connected subgraph.*

We may assume that H is the halfplane $\{y \ge 0\}$. Let u and v be two nodes in S, we want to show that they can be connected by a path whose image stays in the halfplane H. We use induction on $d_G(u, v)$. Consider a shortest path P in G connecting u and v. We may assume that P is not just an edge, and that all the inner nodes of this path are outside H. Let w be a node on P which is lowest.

We claim that there exists an infinite path $P_u = (u_0 = u, u_1, \dots)$ starting at u such that $f(P_u)$ lies in the upper halfplane and the distance of $f(u_k)$ from the x-axis tends to infinity. Let G' be the connected component of $G[S]$ containing u. It suffices to show that the distance of points of $f(G')$ from the x-axis is unbounded, since then the existence of P_u follows by simple compactness.

Suppose that $f(G')$ lies in a strip $0 \le y \le y_0$, and let U denote the connected component of $\mathbb{C} \setminus f(G')$ containing the halfplane $y > y_0$. The boundary of U is a polygon, whose vertices are points $f(i)$, $i \in V$ and intersection points of images of edges. Neither type of vertices can give a concave angle, and so U is convex; since U contains a halfplane, it follows that U is a halfplane. But its boundary cannot contain a line (in fact, it cannot contain any segment longer than c), which is clearly impossible. This proves that P_u exists as claimed.

Similarly, we can find an infinite path $P_v = (v_0 = v, v_1, \dots)$ such that $f(P_v)$ lies in the upper halfplane and the distance of $f(v_k)$ from the x-axis tends to infinity, and an infinite path $P_w = (w_0 = w, w_1, \dots)$ such that $f(P_w)$ lies in the lower halfplane and the distance of $f(w_k)$ from the x-axis tends to infinity. Clearly P_w is node-disjoint from P_u and P_v, and we may assume that its only intersection node with P is w. If P_u and P_v intersect, then the conclusion is trivial, so assume that they are node-disjoint.

Consider $P \cup P_u \cup P_v \cup P_w$ in the original planar embedding of G. This subgraph splits the plane into three infinite regions; let Ω be the region bounded by $P_u \cup P \cup P_v$. It is easy to see that there are infinitely many disjoint paths $Q_1, Q_2, \dots$ connecting P_u and P_v inside Ω. We claim that if k is large enough, the image of Q_k must stay in H, proving that u and v can be connected by a path in S.

Let D denote the diameter of $f(P)$. Let u_1 be the last node on the path P_u such that the distance of $f(u_1)$ from the x-axis is at most $D+2c^3$, let P'_u be the piece of P_u between u and u_1, and let $P''_u = P_u \setminus P'_u$. We define P'_v, P'_w etc. analogously.

Suppose that $f(P''_u)$ intersects $f(P''_v)$, say edge ij of $f(P''_u)$ intersects edge ab of $f(P''_v)$. By (7.14), the length of the image of any edge is at most c, so $|f(i)-f(a)| \le 2c$, and so $d_G(i,a) \le 2c^2$. Thus there exists a path R of length at most $2c^2$ in G connecting i to a. Again by (7.14), the diameter of $f(R)$ is at most $2c^3$. By the definition of P''_u, the distance of $f(i)$ from the x-axis is more than $2c^3$, so $f(R)$ cannot cross the x-axis. It follows that u and v can be connected by a path whose image stays in the upper halfplane, using paths P_u, R, and P_v.

So we may assume that $f(P''_u)$ and $f(P''_v)$ are disjoint. Let T be the set of all nodes in G at a graph-distance at most c_2 from $P \cup P'_u \cup P'_v \cup P'_w$. Since T is a finite set, there is a k for which Q_k does not intersect T. By (7.14) we get that for every node s of Q_k and every node t of P,

$$|f(s)-f(t)| \ge \frac{d_G(s,t)}{c} \ge \frac{c_1}{c} = c_2.$$

In particular, $f(s)$ cannot be in the convex hull of $f(P)$.

If $f(Q_k)$ does not intersect the lower halfplane, then we are done. Suppose it does, then either it intersects $f(P'_w)$ or else it contains a subpath Q'_k such that $f(Q'_k)$ lies in the upper halfplane and intersects both $f(P''_u)$ and $f(P''_v)$.

Suppose that $f(Q_k)$ intersects $f(P_w)$. Similarly as above, we find a path R of length at most $2c^2$ in G connecting a node a on P_w to a node i on Q_k. This path must intersect the path $P \cup P_u \cup P_v$ at some node z; this means that $d_G(z,a) \le 2c^2$, and so $|f(z)-f(a)| \le 2c^3$. But $f(z)$ is either in the upper halfplane or at a distance at most D from it, and so $f(a)$ is at a distance at most $2c^3+D$ from the upper halfplane. So $a \in V(P'_u)$ and hence $i \in T$, a contradiction, since Q_k avoids T.

Finally, if there is a path Q such that $f(Q)$ lies in the upper halfplane and intersects both $f(P''_u)$ and $f(P''_v)$, then similarly as above, we find two paths R_u and R_v of length at most $2c^2$ connecting Q to P''_u and P''_v, respectively. Similarly as above, these paths must stay in the upper halfplane, and so again we find that u and v can be connected by a path staying in the upper halfplane through P_u, R_u, Q, R_v and P_v. This completes the proof of the Claim.

The assertion that every node is in the center of gravity of its neighbors is just a restatement of the fact that every analytic function is harmonic. This also shows that (assuming that f gives an embedding) no country can have a concave angle, and so the countries are convex polygons. So the main step is to show that f defines an embedding.

We start with observing that the image of every edge of G is a segment of length at most c, and for any two nodes u and v of G (adjacent or not) the distance of $f(u)$ and $f(v)$ is at least $1/c$.

Now we turn to the proof that f defines an embedding. Let us triangulate each country of G arbitrarily, to get a new graph G_1. Let us draw the images of these new edges as straight segments. We claim that even with these new edges, f defines an embedding of G_1.

It is enough to show that (a) two triangular countries of G_1 sharing an edge xy are mapped onto triangles on different sides of the line $f(x)f(y)$, and (b) the images of triangular countries incident with the same node x cover a neighborhood of $f(x)$ exactly once. We describe the proof of (a); the proof of (b) is similar.

So suppose that xyz and xyw are two triangular countries of G_1, and that $f(z)$ and $f(w)$ are on the same side of the line L through $f(x)$ and $f(y)$, say on the right side. By the Lemma, there is a path P in G connecting z and w whose image under f stays on the right side of L. Since x is mapped to the center of gravity of its neighbors, there is a node x' adjacent to x in G such that $f(x')$ lies on the left side of L, and similarly, y has a neighbor y' such that $f(y')$ lies on the left side of L. Again by the Lemma, there is a path Q in G connecting x' to y' such that the image of Q stays on the left side of L. Extend Q with the edges xx' and yy' to get a path Q'.

Now obviously P and Q' are node-disjoint paths. But if we consider them in the planar embedding of G, it is clear that they must cross each other. This contradiction proves the theorem. □

7.2.3. Geometric representations and discrete analytic functions. Harmonic functions are closely related to geometric representations, as we have seen, among others, in rubber band representations. In the setting of analytic functions, this connection is even richer.

Square tiling. We have described the connection between square tilings and harmonic functions on planar graphs, as discovered by [**Brooks et al. 1940**], in Section 6.1. Here we consider the closely related structure of tiling the whole plane with squares, whose sides are parallel to the coordinate axes. Assume that the tiling is discrete, i.e., every bounded region contains only a finite number of squares.

Points where four squares meet must be "resolved" just as in the finite case. Every such point will be either "vertical-through" or "horizontal-through". Let us also call a vertex of a square that is an interior point of a vertical edge of another square a "vertical-through". Also assume that there is no horizontal semiline composed of edges of the squares that contains no "vertical-through" point. We call a horizontal segment *basic*, if it is composed of edges of squares and its endpoints are "vertical-through" points.

We associate a map in the plane with this tiling just as in Section 6.1: We represent each basic segment by a single node (say, positioned at the midpoint of the segment). Each square "connects" two basic segments, and we can represent it by an edge connecting the two corresponding nodes, directed top-down. We get a directed graph G, which is infinite but locally finite.

It is not hard to see that G is planar. If we assign to each node (basic segment) its y coordinate, we get a harmonic function; from this, we get an analytic pair as described above. One can give an explicit description of the other function in this pair by repeating the same construction using the vertical edges of the squares. It is not hard to see that this gives the dual graph, and the x coordinates supply the second function of an analytic pair.

This construction can be reversed. Starting with a planar map G and an analytic pair on it, we can construct a tiling of the plane by squares, where the squares correspond to the edges of G (or, equivalently, to the edges of G^*, or, equivalently, to the countries of $G^\diamond$).

The following special case is worth mentioning. Let G be a 3-connected simple toroidal map, and let us construct its rubber band embedding as in 7.2. We can lift this to a periodic embedding of the universal cover map $\widehat{G}$ in $\mathbb{C}$, where the node positions define a harmonic function. We can construct a square tiling representing this graph as above. This too will be periodic, so the universal covering mapping maps it onto a square tiling of the torus.

Touching polygons. The ideas in [**Kenyon 2002**] give another geometric interpretation of analytic functions.

We need to define one further graph derived from a planar map G: define the *total map* of a planar map as the medial map of its lozenge map. To clarify the name: the lozenge map has a node for each node and for each country of G, and a country for each edge of G. So its medial graph has a country for each node, edge and country of G.

Suppose that a function $f : V \cup V^* \to \mathbb{C}$ gives a straight line embedding of G in the plane with convex countries, and a straight line embedding of G^* with convex countries (we consider the infinite case; in the finite case, we should ignore the unbounded country). The only condition we need connecting the two embeddings is that if uv is an edge of G (with this orientation), and pq is the corresponding edge of G^* (so that p represents the country $\mathsf{ls}(uv)$), then the line $f(p)f(q)$ crosses the line $f(u)f(v)$ from left to right. Let P_u denote the convex polygon representing the country of G (or G^*) corresponding to $u \in V^*$ (or $u \in V$)). Shrink each P_u from the point $f(u)$ by a factor of 2, to get a polygon P'_u. (The point $f(u)$ is not necessarily in the interior of P_u.) This way we get a system of convex polygons, where for every edge $uv \in G^\diamond$, the two polygons P_u and P_v share a vertex at the point $\big(f(u)+f(v)\big)/2$ (Figure 7.8).

There are three kinds of polygons in this picture: the dark and light grey ones correspond to the nodes in V and V^*, respectively, while the white parallelograms Q_e between the polygons correspond to the edges $e \in E$. It is easy to prove that these are indeed rectangles, and the two vectors spanning the parallelogram corresponding to edge e are $\frac{1}{2}\big(f(\mathsf{hd}(e)) - f(\mathsf{tl}(e))\big)$ and $\frac{1}{2}\big(f(\mathsf{ls}(e)) - f(\mathsf{rs}(e))\big)$. It can be shown (using our hypothesis about the relationship between the two straight line embeddings) that the interiors of these polygons will be disjoint. Thus this method provides an embedding of the total graph of G in the plane.

It is clear that *the function f is analytic if and only if every parallelogram Q_e is a square.* If the parallelograms Q_e are rectangles, then we can weight every edge $e \in E$ with the ratio of the sides of Q_e. The function f will be analytic on this weighted graph.

This correspondence with analytic functions is nice enough, but there is more. Let us take another analytic function g on this edge-weighted graph, and construct the polygons $g(u)P'_u$ (multiplication by the complex number $g(u)$ corresponds to blowing up and rotating). The resulting polygons will not meet at the appropriate vertices any more, but let's try to translate them so that they do (Figure 7.9). The computation is quite simple. Let $h(u)$ be the vector with which we try to translate $g(u)P'_u$, where (say) $u \in V$. Let p be a country of G incident with u. Then the

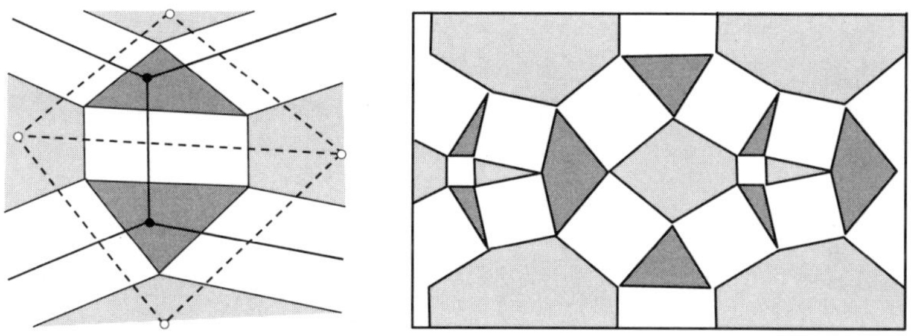

FIGURE 7.8. Left: the total graph of a planar map G, obtained from straight-line embeddings of the map and its dual (only the neighborhood of a single edge is shown). Right: similar representation of the graph in Figure 7.5. Note that since the node positions of the original graph define an analytic function, the white regions representing the edges are squares.

image of u in the image of P_p is

$$\frac{f(u)+f(p)}{2}g(p)+h(p),$$

while the image of p in the image of P_u is

$$\frac{f(u)+f(p)}{2}g(u)+h(u).$$

These two points must be the same, which leads to the equation

$$\frac{f(u)+f(p)}{2}\big(g(p)-g(u)\big) = h(u)-h(p).$$

Recall that f and g are given, and we want to find a function h satisfying this equation for every edge up of $G^\diamondsuit$. Functions h satisfying this condition are precisely of the form

$$h(v) = \int_u^v f\,dg + \text{const}$$

(with an arbitrary but fixed starting node u). We have seen that this integral is well defined for any two analytic functions. If g is, in addition, critical, then this function h is a third analytic function defined by this representation of the total graph.

It is not hard to see that conversely, every "deformation" of the total graph picture such that the polygons P_u remain similar to themselves defines an analytic function on G.

Circle packings. We can also relate discrete analytic functions to circle representations. Let G be a 3-connected toroidal map. It is again best to go to the universal cover map $\widehat{G}$. Then we can construct two (infinite, but discrete) families $\mathcal{F}$ and $\mathcal{F}^*$ of circles in the plane so that they are double periodic modulo a lattice $L = \mathbb{Z}a+\mathbb{Z}b$, $\mathcal{F} \pmod L$ corresponds to the nodes of G, $\mathcal{F}^* \pmod L$ corresponds to the countries of G, and for every edge e, there are two circles C, C' representing $\mathsf{hd}(e)$ and $\mathsf{tl}(e)$, and two circles D and D' representing $\mathsf{rs}(e)$ and $\mathsf{ls}(e)$ so that C, C' are tangent at a point p, D, D' are tangent at the same point p, and C, D are orthogonal. (This can be proved along the lines of the proof of Theorem 5.3.)

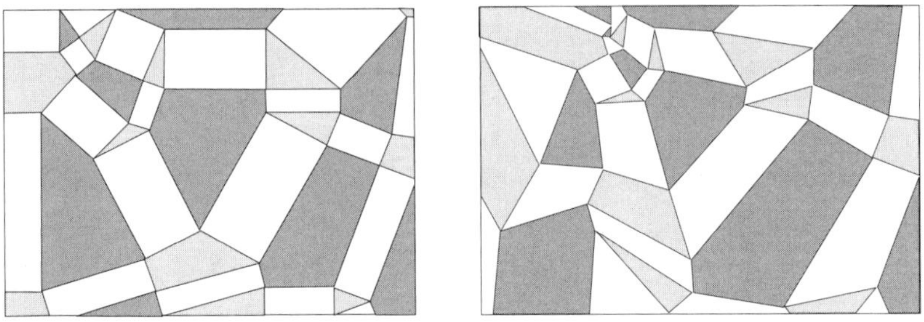

FIGURE 7.9. The deformation of the touching polygon representation given by an additional analytic function.

If we consider the centers of the circles in $\mathcal{F}$ as nodes, and connect two centers by a straight line segment if the circles touch each other, then we get a straight line embedding of the universal cover map in the plane (periodic modulo L). Let $\mathbf{x}_v$ denote the point representing node v of the universal cover map or of its dual.

To get a discrete analytic function out of this representation, consider the plane as the complex plane, and define $f(u) = \mathbf{x}_u$ for every node $u \in V \cup V^*$. By the orthogonality property of the circle representation,

$$f\big(\mathsf{hd}(e)\big) - f\big(\mathsf{tl}(e)\big) = i\omega(e)\big(f(\mathsf{rs}(e)) - f(\mathsf{ls}(e))\big)$$

with an appropriate $\omega(e) > 0$. It follows that if we consider the map G with weights $\omega(e)$, then f is a discrete analytic function on this weighted map.

For more on the subject, see the book [**Stephenson 2005**] and the papers [**Mercat 2002**], [**Mercat 2004**].

7.3. Discrete holomorphic forms, a.k.a. rotation-free circulations

Let us return to finite maps, but this time on surfaces other than the plane. In this section, it will be convenient to allow the graphs to have parallel edges and loops. While no nonconstant analytic function can be defined on any finite graph, we can do a lot better if we consider functions on the edges, analogous to derivatives of "true" analytic functions, namely holomorphic forms. We have to start with some topological considerations concerning maps on surfaces.

7.3.1. Maps, boundaries and coboundaries. Most of the basic definitions concerning planar maps carry over without any difficulty to graphs embedded in other surfaces. A point to pay attention to is the definition of a *proper map* on a compact surface S. By this we mean a finite graph G embedded in S so that each country is homeomorphic to an open disk. Once we have this property, the notions of the cyclic sequence of edges forming the boundary, the dual map, the lozenge map, etc. can be defined. If the surface is orientable, then we can define for each edge a right shore $\mathsf{rs}(e)$ and a left shore $\mathsf{ls}(e)$ just like in the plane. For a proper map on a compact surface, Euler's Formula generalizes as

$$n - m + f = \chi, \tag{7.15}$$

where χ is the *Euler characteristic* of the surface. For a compact orientable surface of genus g (a sphere with g handles), one has $\chi = 2 - 2g$.

Let G be a finite directed map on an orientable surface. For every face $F \in V^*$, we denote by $\partial_F \in \mathbb{R}^E$ the boundary of F:

$$(\partial_F)_e = \begin{cases} 1, & \text{if } \mathsf{ls}(e) = F, \\ -1, & \text{if } \mathsf{rs}(e) = F, \\ 0, & \text{otherwise.} \end{cases}$$

Then $|\partial F|^2$ is the length of the cycle bounding F. Note that ∂F is just the special circulation f_C defined in (B.2), where C is the boundary cycle of F with the counterclockwise orientation.

This construction is dual to vector δ_v and the 0-dimensional coboundary operator, introduced in Appendix B.3 (those can be defined for every graph, not only for maps on surfaces). Similarly as there, we can consider the boundary operator as an $E \times V^*$ matrix $D = D_G$, where $D\mathbb{1}_F = \partial_F$ for every country $F \in V^*$. Its transpose $D^\mathsf{T} \in \mathbb{R}^{V^* \times E}$ is the coboundary operator from dimension 1 to dimension 2, with the rather simple action of $D^\mathsf{T}\mathbb{1}_e = \mathbb{1}_{\mathsf{ls}(e)} - \mathbb{1}_{\mathsf{rs}(e)}$. It is easy to check the fundamental relations

$$B^\mathsf{T} D = 0 \quad \text{and} \quad D^\mathsf{T} B = 0. \tag{7.16}$$

(In topology, one would denote both coboundary operators B and D^T by δ and both boundary operators B^T and D by ∂, so (7.16) could be written as $\delta^2 = 0$ and $\partial^2 = 0$. In the simple cases we need, it will be more convenient to distinguish the boundary and coboundary maps in different dimensions.)

Some basic facts about various subspaces defined by the coboundary operator B are summarized in Appendix B.3. If the graph is embedded in a surface, then we get a little more interesting spaces, relating the graph with its dual. (This still belongs to the almost-trivial part of algebraic topology.)

As a warmup, consider a (finite) 2-connected planar map G. Then the space of circulations $\mathcal{H}$ and its orthogonal complement $\mathcal{A}$, the space of potentials, are related by map duality. Mapping each edge to the corresponding dual edge defines an isomorphism $\phi : \mathbb{R}^E \to \mathbb{R}^{E^*}$. It is not hard to verify that $\phi(\mathcal{A}_G) = \mathcal{H}_{G^*}$ and $\phi(\mathcal{H}_G) = \mathcal{A}_{G^*}$. Boundary cycles of bounded countries of G form a basis of $\mathcal{H}_G$. So $\dim(\mathcal{A}_{G^*}) = m-n+1$; recalling that $\dim(\mathcal{A}) = n-1$, we see that this is just a restatement of Euler's formula.

We get a richer structure if we assume that G is embedded in an orientable surface of higher genus. Let Σ be a closed compact orientable surface of genus g, and suppose that the directed graph G is embedded in Σ as a map. Let $f = |V^*|$ be the number of countries. We say that a vector $x \in \mathbb{R}^E$ is *rotation-free*, if for every country p the equation

$$\sum_{e:\,\mathsf{ls}(e)=p} x_e = \sum_{e:\,\mathsf{rs}(e)=p} x_e \tag{7.17}$$

holds. We can express this as $x \perp \partial_p$. This is also equivalent to saying that the vector x defines a circulation on the dual graph.

Each vector ∂_p ($p \in V^*$) is a circulation; circulations that are linear combinations of these special circulations are called *null-homologous*. Two circulations z and z' are called *homologous* if $z - z'$ is null-homologous. Null-homologous circulations form a linear space $\mathcal{B} = \mathcal{B}_G \subseteq \mathbb{R}^E$ generated by the vectors ∂_p.

We have defined two orthogonal linear subspaces of $\mathcal{R}^E$: the space $\mathcal{A}$ of potentials and the space $\mathcal{B}$ of null-homologous circulations. Interchanging the roles of the map and its dual, the roles of $\mathcal{A}$ and $\mathcal{B}$ are interchanged. The orthogonal complement $\mathcal{A}^\perp = \mathcal{H}$ is the space of all circulations, and $\mathcal{B}^\perp$ is (under the identification of E and E^*) the space of circulations on the dual graph. The intersection $\mathcal{C} = \mathcal{A}^\perp \cap \mathcal{B}^\perp = (\mathcal{A}+\mathcal{B})^\perp$ is the space of rotation-free circulations. This subspace will be our main object of study in this section.

The orthogonal decomposition

$$\mathbb{R}^E = \mathcal{A} \oplus \mathcal{B} \oplus \mathcal{C} \tag{7.18}$$

can be thought of as a discrete version of the Hodge decomposition from differential geometry. We know that $\dim(\mathcal{A}) = n-1$ and $\dim(\mathcal{B}) = \dim(\mathcal{A}_{G^*}) = f-1$. Counting dimensions, we can express the dimension of $\mathcal{C}$ by the genus of the surface:

$$\dim(\mathcal{C}) = m - \dim(\mathcal{A}) - \dim(\mathcal{B}) = m-(n-1)-(f-1) = 2g. \tag{7.19}$$

In particular, we see that for maps on the sphere, the space $\mathcal{C}$ is the null space: no nonzero circulation can be rotation-free.

EXAMPLE 7.5 (Toroidal grid). Figure 7.10 shows a rotation-free circulation on the toroidal grid. The numbers 5 and 17 can be replaced by two arbitrary numbers, so this picture gives a 2-dimensional space of rotation-free circulations. By (7.19), the dimension of $\mathcal{C}$ is 2, so there are no other rotation-free circulations. ♦

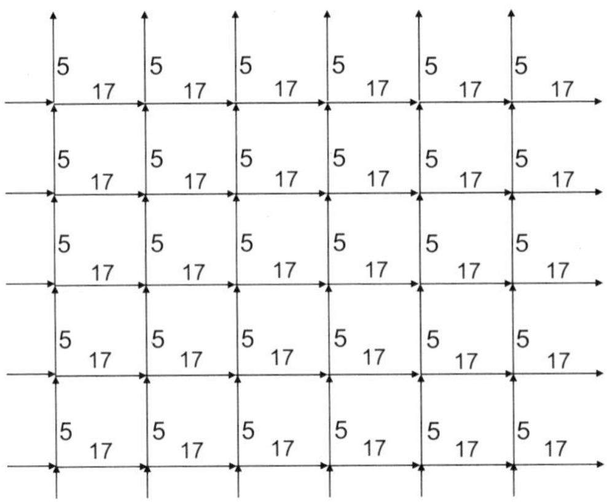

FIGURE 7.10. Rotation-free circulation on the toroidal grid. (Edges leaving on the top are identified with edges entering on the bottom, and similarly for right-left, so that we have a graph embedded in the torus.)

7.3.2. Rotation-free circulations and discrete analytic functions. Let $z \in \mathbb{R}^E$ be a rotation-free circulation on a graph G embedded in a surface. Consider a family $\mathcal{F}'$ of countries whose union is homeomorphic to the disk, and let $G' = (V', E')$ be the graph formed by the boundary edges of these countries. Every cycle in G' is null-homologous, and so rotation-freeness and the potential argument (Corollary 4.5) give us a function $f : V' \to \mathbb{R}$ such that $z_e = f(\mathsf{hd}(e)) - f(\mathsf{tl}(e))$ for every edge $e \in E'$. Similarly, the flow condition implies that there is a function $g : \mathcal{F}' \to \mathbb{R}$ such that $z_e = g(\mathsf{rs}(e)) - g(\mathsf{ls}(e))$ for every edge e between two countries in $\mathcal{F}'$. (To see the analogy with complex function theory better, we shifted our notation to functions on node sets rather than vectors indexed by nodes.)

The functions f and g defined above satisfy

$$f(\mathsf{tl}(e)) - f(\mathsf{hd}(e)) = g(\mathsf{rs}(e)) - g(\mathsf{ls}(e)) \tag{7.20}$$

for every edge between two countries in $\mathcal{F}$. We can think of f and g as the real and imaginary parts of a (discrete) analytic function, and (7.20) can be thought of as a discrete analogue of the Cauchy–Riemann equations.

It is important to emphasize that these functions f and g exist only locally, in the planar piece of Σ; but this is not surprising, since "true" analytic functions may have well-defined values only when restricted to a simply connected region. However, we can use the trick of Riemann surfaces from complex function theory. For a map G on a closed orientable surface Σ different from the sphere, we consider the universal cover mapping $\phi: R^2 \to \Sigma$. Rolling up the plane twice to get a torus is an example.

The graph $\hat{G} = \phi^{-1}(G) = (\hat{V}, \hat{E}, \hat{\mathcal{F}})$ is called the *universal cover* of G. This is an infinite graph embedded in the plane, and it is a planar map in the sense of Section 7.2. Furthermore, it is invariant under the action of an appropriate discrete group of isometries of the Euclidean plane (in the case of the torus) or of the hyperbolic plane (in the case of higher genus). The universal cover of the dual map G^* is the dual map $\hat{G}^*$.

Any rotation-free circulation on G corresponds to a rotation-free circulation on $\hat{G}$. This circulation can be represented as $\hat{a}_e = f(\mathsf{hd}(e)) - f(\mathsf{tl}(e))$, where $f: \hat{V} \to \mathbb{R}$. Note that the function f can *not* be obtained by lifting a function on V. Similarly, we obtain a function $g: V(\hat{G}^*) \to \mathbb{R}$ such that $\hat{a}_e = g(\mathsf{rs}(e)) - g(\mathsf{ls}(e))$. The pair (f, g) is an analytic pair.

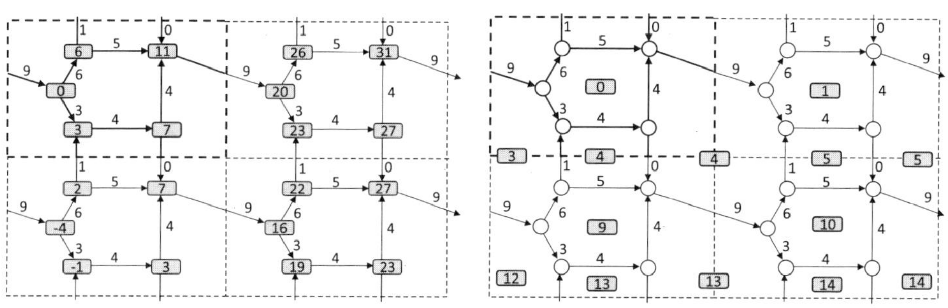

FIGURE 7.11. Left: A rotation-free circulation on the torus (upper left corner) and a corresponding harmonic function on the universal cover map. Right: A corresponding harmonic function on the dual.

Figure 7.11 shows a rotation-free circulation on a graph embedded in the torus. The first figure shows an associated harmonic function on the nodes of the universal cover map, the second, a harmonic function on the dual map.

We can introduce these basic notions in a little larger generality: we associate a positive weight $\omega(e)$ with every edge. In most of this chapter, we restrict our discussion to the unweighted case, because extending this to the weighted case would not amount to much more than inserting '$\omega(e)$' at appropriate places in the equations.

7.3.3. Nondegeneracy properties. We state and prove two key nondegeneracy properties of rotation-free circulations. We start with a simple lemma about maps on surfaces. For every country p, let $\mathsf{rev}(p)$ denote the number of times the orientation of the edges changes if we move along the boundary of p. For every node v, let $\mathsf{rev}(v)$ denote the number of times the orientation of the edges changes in their cyclic order as they emanate from v.

LEMMA 7.6. *Let $G = (V, E, V^*)$ be a proper directed map on a surface of Euler characteristic χ. Then*

$$\sum_{u \in V \cup V^*} (\mathsf{rev}(u) - 2) = -2\chi.$$

PROOF. Every corner means an orientation change either for the node or for the country defining it. Hence

$$\sum_{u \in V \cup V^*} \mathsf{rev}(u) = \sum_{v \in V} \deg(v) = 2m = 2n + 2f - 2\chi.$$

Rearranging, we get the equality in the lemma. □

The first nondegeneracy property [**Benjamini–Lovász 2003**] concerns certain special rotation-free circulations, namely those obtained as projections of basis vectors. Let Ψ denote the orthogonal projection of $\mathbb{R}^E$ to the space $\mathcal{C}$ of rotation-free circulations.

THEOREM 7.7. *Let G be a 3-connected proper map on an orientable surface with genus $g > 0$. Then $\Psi\mathbb{1}_e \neq 0$ for $e \in E$.*

Note that $\Psi\mathbb{1}_e \neq 0$ implies that

$$(\Psi\mathbb{1}_e)_e = \mathbb{1}_e^{\mathsf{T}}\Psi\mathbb{1}_e = \mathbb{1}_e^{\mathsf{T}}\Psi^2\mathbb{1}_e = |\Psi\mathbb{1}_e|^2 > 0, \tag{7.21}$$

so a (seemingly stronger) conclusion of the theorem could be that the vector $\Psi\mathbb{1}_e$ is positive on the edge e. In a similar fashion, for any $c \in \mathcal{C}$, we have

$$c_e = \mathbb{1}_e^{\mathsf{T}}c = \mathbb{1}_e^{\mathsf{T}}(\Psi c) = c^{\mathsf{T}}(\Psi\mathbb{1}_e),$$

so it would suffice to prove that not *every* rotation-free circulation vanishes on e.

For $g = 0$ there is no nonzero rotation-free circulation, and hence $\Psi\mathbb{1}_e = 0$ for every edge e. So the hypothesis that $g > 0$ cannot be omitted. The assumption that G is 3-connected cannot be dropped either, as shown by the example in Figure 7.12. Here every rotation-free circulation is 0 on e, and hence $\Psi\mathbb{1}_e = 0$. The orientability of the surface is also important (Exercise 7.17).

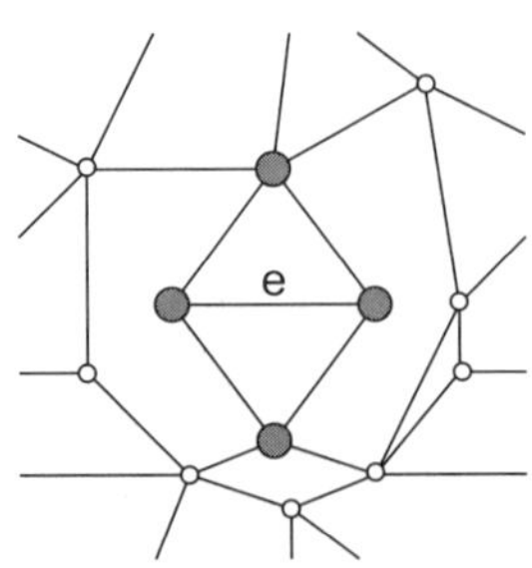

FIGURE 7.12. Every rotation-free circulation is 0 on the edge e.

PROOF. Suppose that $\Psi \mathbb{1}_e = 0$ for some edge e. This means that $\mathbb{1}_e \in \mathcal{A}+\mathcal{B}$, and so we have a decomposition $\mathbb{1}_e = a+b$, where $a \in \mathcal{A}$ and $b \in \mathcal{B}$. So we have $a_f + b_f = 0$ for every edge $f \neq e$, but $a_f + b_f = 1$ for $f = e$. By the definition of $\mathcal{A}$ and $\mathcal{B}$, there are vectors $x \in \mathbb{R}^V$ and $y \in \mathbb{R}^{V^*}$ such that $a = Bx$ and $b = D^\mathsf{T} y$. We may choose the reference orientation so that $a \geq 0$.

Let $S = \{y_p : p \in V^*\}$. For $c \in S$, let U_c denote the union of countries p with $y_p \geq c$. The boundary of U_c is an eulerian subgraph D_c of G. Let D be a cycle contained in D_c. For every edge $f \in E(D) \setminus \{e\}$ we have $a_f > 0$, and all these edges are oriented in the same way around the cycle. So x_i strictly increases as we traverse the cycle D. This can only happen if $e \in E(D)$, and since this holds for every cycle in $D \subseteq D_c$, the whole subgraph D_c is a single cycle through e. It also follows that all the values x_i on D_c are different, and they grow strictly from one endpoint of e to the other.

Let G_0 denote the subgraph formed by those edges f for which $a_f = 0$, and G_1 the subgraph formed by the other edges. Since x_i is constant on every connected component of G_0, any cycle D_c meets any component of G_0 at most once.

Let $a < b$ be two consecutive numbers in S, then the boundary of $U_a \setminus U_b$ is contained in $D_a \cup D_b$. A connected component of G_0 can have at most two nodes in common with $D_a \cup D_b$; hence, by 3-connectivity, this connected component must consist of a single edge that is contained in $U_a \setminus U_b$, connecting D_a and D_b.

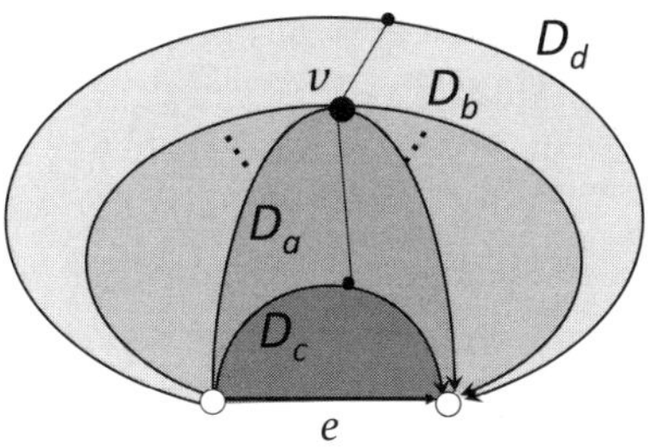

FIGURE 7.13. Level sets of the vector y, and the neighborhood of a node v

This argument implies, in particular, that every node is contained in at least one cycle D_c. Let $v \neq \mathsf{hd}(e), \mathsf{tl}(e)$ be any node, and let $a \in S$ and $b \in S$ be smallest and largest values for which $v \in V(D_a)$ and $v \in D_b$ (Figure 7.13). Let $c < a$ and $d > b$ be a nearest numbers in S (we assume that these numbers exist; the opposite case can be settled similarly). The edges of the cycles D_r $(a \leq r \leq b)$ entering v are consecutive in the cyclic orientation about v, and so are their edges leaving v. Furthermore, G_0 can have at most two edges adjacent to v, one connecting v to D_c, and one to D_d. No matter how these two edges are oriented, the edges incident with v form two consecutive sequences, one entering and one leaving v. In other words,

$$\mathsf{rev}(v) = \begin{cases} 0, & \text{if } v = \mathsf{tl}(e) \text{ or } v = \mathsf{hd}(e), \\ 2, & \text{otherwise.} \end{cases} \tag{7.22}$$

A similar argument shows that $\mathsf{rev}(p) = 2$ for every country p. Substituting in Lemma 7.6 yields $-4 = -2\chi$, so $\chi = 2$, and thus the surface is the sphere, a contradiction. $\square$

Considering a "generic" linear combination of the vectors $\Psi\mathbb{1}_e$ ($e \in E$), we obtain the following corollary.

COROLLARY 7.8. *Every 3-connected proper map on an orientable surface with genus $g > 0$ carries a rotation-free circulation that does not vanish on any edge.* □

The arguments above also yield a more explicit formula for $\Psi\mathbb{1}_e$.

PROPOSITION 7.9. *Let G be an oriented map on an orientable surface Σ, and let $e \in E$. Let $x \in \mathbb{R}^V$ be the vector of voltages of the nodes, if a unit current is sent from $\mathsf{tl}(e)$ to $\mathsf{hd}(e)$, and let $y \in \mathbb{R}^{V^*}$ be defined analogously for G^*. Then*

$$\Psi\mathbb{1}_e = \mathbb{1}_e - Bx - Dy.$$

PROOF. We have the decomposition $\mathbb{1}_e = a+b+c$ with $a \in \mathcal{A}$, $b \in \mathcal{B}$ and $c \in \mathcal{C}$. Using the definitions of $\mathcal{A}$, $\mathcal{B}$ and $\mathcal{C}$, this can be written as

$$\mathbb{1}_e = Bx + Dy + \Psi\mathbb{1}_e$$

for appropriate vectors $x \in \mathbb{R}^V$ and $y \in \mathbb{R}^{V^*}$. Applying the boundary operator B^T, we get

$$\mathbb{1}_{\mathsf{hd}(e)} - \mathbb{1}_{\mathsf{tl}(e)} = B^\mathsf{T}Bx = Lx,$$

since Dy and $\Psi\mathbb{1}_e$ are circulations and hence $B^\mathsf{T}Dy = B^\mathsf{T}\Psi\mathbb{1}_e = 0$. As noted at the end of Section 4.1.3, we can interpret x (and similarly y) as in the Proposition. □

The effective resistance between the endpoints of $e = ij$ can be expressed as $R(i,j) = \mathbb{1}_e^\mathsf{T}Bx$, and similarly the effective resistance of the dual map between the endpoints of the dual edge $e^* = pq$ is $R(p,q) = \mathbb{1}_e^\mathsf{T}Dy$. Taking scalar product of the formula in Proposition 7.9 with $\mathbb{1}_e$, we get

$$(\Psi\mathbb{1}_e)_e = 1 - R(i,j) - R(p,q). \tag{7.23}$$

Using (7.21), we get a corollary for effective resistances:

COROLLARY 7.10. *If G is map on an orientable surface, then for every edge e, $R(i,j) + R(p,q) \le 1$. If the surface is the sphere, then equality holds. If the surface is different from the sphere, and G is 3-connected, then strict inequality holds.* □

The second main nondegeneracy property [**Benjamini–Lovász 2002**] tries to capture the property of "true" analytic functions that if they vanish on an open disk, then they vanish identically. What should correspond to the notion of "open disk" in a discrete setting? It is not hard to guess that it should mean a "large enough connected subgraph", but just specifying the number of nodes will not work.

EXAMPLE 7.11. Let G be a map on an orientable surface with genus $g > 0$, and let X be a connected induced subgraph of G such that at most $2g$ "external" nodes of X are connected to the rest of G (otherwise, X can be arbitrarily large). Suppose that X is embedded in a subset of Σ of the surface that is topologically a disk. Delete the edges of X as well as all the non-external nodes, create a new node x, and connect it to the external nodes (Figure 7.14).

We get a graph G' that is still a map on Σ. Thus this graph has a $2g$-dimensional space of circulations, and hence there is a nonzero rotation-free circulation ψ vanishing on each of the $2g-1$ of the edges incident with x. Since this is a circulation, it must vanish on all the edges incident with x. Delete x, put X back, and extend ψ with 0-s to the edges of X. We get a nonzero rotation-free circulation vanishing on X. ♦

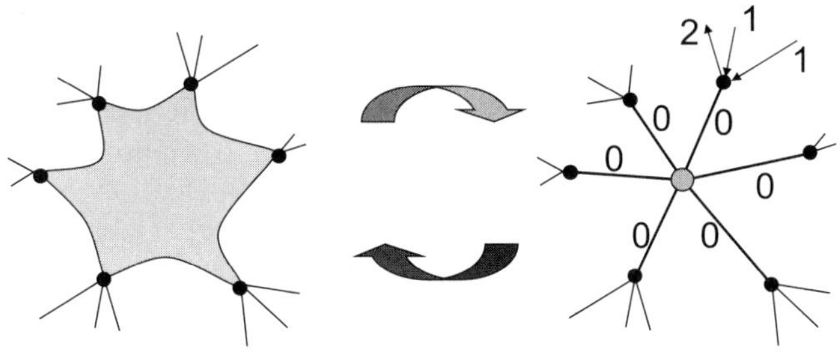

FIGURE 7.14. Contracting a loosely connected piece

This example suggests that the right thing to consider is not the size of the zero-region but its node-boundary. The following theorem asserts that this is indeed so, and the number $2g$ in the example is best possible up to a factor of about 2.

THEOREM 7.12. *Let G be a graph embedded in an orientable surface Σ as a map. Let φ be a nonzero rotation-free circulation on G and let G' be the subgraph of G on which φ does not vanish. Suppose that φ vanishes on all edges induced by, and incident with, a connected subgraph U of G. Then U can be separated from G' by at most $4g-3$ nodes.*

PROOF. In our topological arguments, we have to be careful not to use intuition based on planar maps. In particular, we may have edges both sides of which belong to the same country (even though the graph is 3-connected). We call such an edge *non-bordering*.

Let W be the connected component of $G\setminus V(G')$ containing U, and let Y denote the set of nodes in $V\setminus V(W)$ adjacent to W. So $Y \subseteq V(G')$, every node in Y is adjacent to W, and all edges e between Y and W, as well as in $G[W]$, have $\varphi(e) = 0$.

Consider an edge e with $\varphi(e) = 0$. If e is not a loop, then it can be contracted to get a map on the same surface with a rotation-free circulation. In particular, we can contract W to a single node u (Figure 7.15).

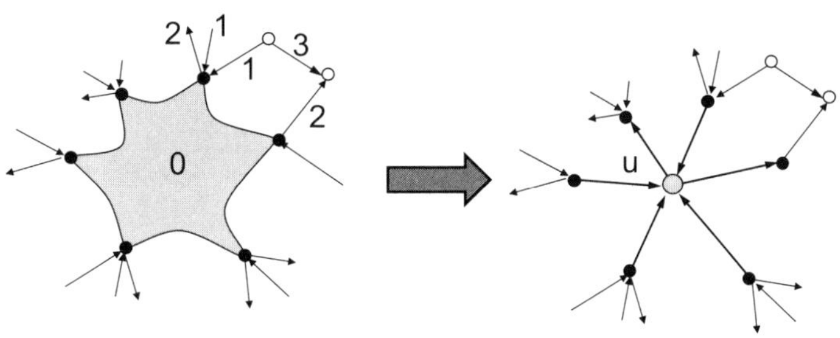

FIGURE 7.15. Contracting a connected component of zero edges

If e is a loop with $\mathsf{ls}(e) \neq \mathsf{rs}(e)$, then $G\setminus e$ is still a map (the interior of every country is topologically an open disk), and φ is a rotation-free circulation on it.

So we can eliminate an edge with $\varphi(e) = 0$ unless it is a non-bordering loop. In this latter case, we can change $\varphi(e)$ to any nonzero value and still have a rotation-free circulation.

We apply this elimination procedure to all edges with $\varphi = 0$ except for the edges between u and Y, and between two nodes of Y. We don't try to contract edges between nodes in Y (we do not want to reduce the size of Y), but we can delete them unless they are non-bordering. If there are parallel edges between u and Y, and at least one of them is a border between distinct countries, then we delete this too. So at the end we may assume that every edge between u and Y is either single, or non-bordering. Let D be the number of edges between u and Y.

Re-orient each edge with $\varphi \neq 0$ in the direction of the flow φ, and orient the edges between u and Y alternatingly in an out from u (there may be one pair of consecutive edges in the same direction, if D is odd). Orient the edges with $\varphi = 0$ between two nodes of Y arbitrarily, to get a digraph G_1.

It is easy to check that G_1 has no sources or sinks, so $\mathsf{rev}(v) \geq 2$ for every node v, and of course $\mathsf{rev}(u) \geq D-1 \geq |Y|-1$. Furthermore, every country either has an edge with $\varphi > 0$ or a non-bordering edge on its boundary. In the first case, it cannot be bounded by a directed cycle, since φ would add up to a positive number on its boundary. In the second, the country boundary goes through the non-bordering edge twice in different directions, so again it is not a directed cycle. Thus we have $\mathsf{rev}(p) \geq 2$ for every country p.

Substituting in Lemma 7.6 (with G_1 in place of G), we get that $|Y|-1 \leq 4g-4$. Since Y separates U from G', this proves the theorem. □

7.4. Global information from local observation

As an application of the nondegeneracy results above, we describe an algorithmic problem; I don't think this problem has any important applications, but it is perhaps interesting for its unusual setting. Suppose that we live on a (finite) map on a compact orientable surface with genus g (the embedding is reasonably dense). Can we determine global properties of the graph, say the genus of the surface, if we only have access to a small neighborhood of our location (Figure 7.16)?

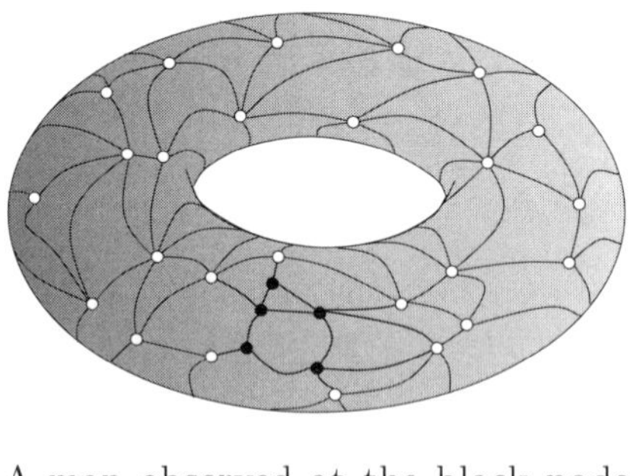

FIGURE 7.16. A map observed at the black nodes. Can you determine the genus of the surface?

Well, clearly not. But suppose that the graph is "active", in the sense that the edges carry weights, which change according to random local transitions, and we can observe the process in that small neighborhood for some time. Let us describe a particular process that allows us to determine the genus of the graph, based on [**Benjamini–Lovász 2002**].

Discrete holomorphic forms motivate a reasonably natural and simple procedure, which we call *random balancing*. Informally, this can be described as follows.

We consider a finite map G, with a reference orientation, on an orientable surface S. Each edge carries a real weight x_{ij}. With some frequency, a node wakes up, and balances the weights on the edges incident with it, so that locally the flow condition is restored. With the same frequency, a country wakes up, and balances the weights on the edges incident with it, so that the rotation around the country is canceled.

To make this precise, we start with any vector $x \in \mathbb{R}^E$. At each step, one of the following two operations is carried out on the current edge-weighting $x \in \mathbb{R}^E$:

(a) NODE BALANCING. Let $v \in V$ and let $s_v = (\delta_v)^\mathsf{T} x$ be the "imbalance" at node v (the value by which the flow condition at v is violated). We modify x by subtracting $\big(s_v/\deg(v)\big)\delta_v$ from it.

(b) COUNTRY BALANCING. Let $p \in V^*$, and let $r_p = (\partial_p)^\mathsf{T} x$ be the "rotation" around p. We modify x by subtracting $\big(r_p/\deg(p)\big)\partial_p$ from it.

Rotation-free circulations are invariant under node and country balancing. Starting with any vector x on the edges and doing country and node balancing of *randomly chosen* nodes and countries, from the uniform distribution on $V \cup V^*$, the weights will converge to Ψx quite fast (Exercise 7.18).

How do we observe the procedure? Let U be a connected subgraph of G, which cannot be separated from any non-zero-homotopic cycle by fewer than $4g-1$ nodes. Let E_0 be the set of edges incident with U (including the edges of U). We think of E_0 as the "window" through which the graph G can be observed.

Let Γx denote the restriction of $x \in R^\mathsf{E}$ to E_0. By Theorem 7.12, $\dim(\Gamma\mathcal{C}) = \dim(\mathcal{C}) = 2g$. For any $x \in \mathbb{R}^E$, the vector $z = \Psi x$ is in $\mathcal{C}$, and Γz can be observed in the window. If we can repeat this with sufficiently many starting edgeweights $x^1, \dots, x^m$ so that their projections $z^i = \Psi x^i$ span $\mathcal{C}$, then $\Gamma z^1, \dots, \Gamma z^m$ span $\Gamma\mathcal{C}$ by Theorem 7.12, and so determining the dimension of the space spanned by $\Gamma z^1, \dots, \Gamma z^m$ tells us the genus of the surface.

How to generate such starting vectors x^i? Since by our model we have access to the edges in $E_0 = \{e_1, \dots, e_m\}$, it is natural to restart from $x^i = \mathbb{1}_{e_i} + z^{i-1}$. (For $i = 1$, we start from $\mathbb{1}_{e_1}$.) We call this a *local excitation*.

Do local excitations yield sufficiently many linearly independent vectors Γz^i? Luckily, they do, as shown by the next lemma.

LEMMA 7.13. *The vectors* $\Psi\mathbb{1}_e$ $(e \in E_0)$ *generate* $\mathcal{C}$.

PROOF. Suppose not, then there is a nonzero vector $\mathbf{v} \in \mathcal{C}$ orthogonal to every vector $\Psi\mathbb{1}_e$. Then

$$\mathbf{v}^\mathsf{T}\mathbb{1}_e = (\Psi\mathbf{v})^\mathsf{T}\mathbb{1}_e = (\Psi\mathbb{1}_e)^\mathsf{T}\mathbf{v} = 0$$

for every edge $e \in E_0$, which means that $\Gamma\mathbf{v} = 0$. By Theorem 7.12, it follows that $\mathbf{v} = 0$, a contradiction. □

The more difficult issue is: how to compute $z = \Psi x$ from x? Starting with $x_0 = x$, let $x_t \in \mathbb{R}^E$ be the vector of edgeweights after t random balancing steps. If $z = \Psi x_0$, then $x_t \to z$ as $t \to \infty$, and so $\Gamma x_t \to \Gamma z$.

Of course, after a local excitation we cannot wait for infinite time to let the system settle to a rotation-free circulation. We have to make an observation of the window E_0 allowing sufficient but finite time T after each local excitation. These observations provide vectors $y_1, \dots, y_m \in \mathbb{R}^{E_0}$ such that $y_i - y_{i-1} \approx \Gamma\Psi\mathbb{1}_{e_i}$ (here $y_0 = 0$). If we had exact equality here, then the rank of $(y_1, \dots, y_m)$ would be exactly $2g$, by Lemma 7.13 and Theorem 7.12.

But we have errors! We can do the following: we construct a linear subspace Σ (approximating $\Gamma\mathcal{C}$) as follows. Start with $\Sigma_0 = 0$. For $i = 1, \dots, m$, check if the distance of $y_i - y_{i-1}$ from Σ_{i-1} is at least a pre-defines $\varepsilon > 0$. If so, let $\Sigma_i = \text{lin}(\Sigma_{i-1} \cup \{y_i\})$, else let $\Sigma_i = \Sigma_{i-1}$. Return the dimension of Σ_m. Let us call this procedure *subspace fitting*.

THEOREM 7.14. *Let* $\varepsilon = (n+f)^{-(n+f)^2}$, $T = (n+f)^7$, *and let* k *be the result of the subspace fitting algorithm. Then with high probability,* $g = k/2$. □

The proof of this theorem involves some rather tedious computation, analogous to that in [**Benjamini–Lovász 2002**], and will not be reproduced here. For the interested reader, we state the main steps as Exercises 7.18 and 7.19.

REMARK 7.15. We defined balancing steps as a node/country chosen uniformly from $V \cup V^*$. While this is perhaps easier to describe, it has an unpleasant feature, in particular if we think of the map G as very large: random choice for the site of the balancing step is a global operation. The following (essentially equivalent) model is therefore more appealing: Let us endow each node and each country with a "Poisson clock", emitting ticks at times according to a Poisson process with parameter 1. When its clock ticks, the site balances. Thus the model becomes truly local. Unfortunately, the waiting time in the algorithm still depends (polynomially) on the size of the graph; this is natural, since information from the distant parts of the map must get to the observation window.

REMARK 7.16. In [**Benjamini–Lovász 2002**] a different method for local excitation was analyzed: In that version, we do not manipulate the edgeweights in the window, but (besides node and country balancing) we allow a third random change with some small probability p:

(c) EXCITATION. Add a random number $X \in \{-1, 1\}$ to the weight of a random edge $e \in E$.

The main result is that there is a constant $c > 0$ such that if $p < 1/(n+f)^c$, then observing the modified balancing process for at least $(n+f)^c/p$ steps, we can determine the genus g with high probability.

EXERCISE 7.1. (a) Prove that if we assign a complex number to every integer point on the real and imaginary axes, then there is a unique analytic function on the Gaussian integers extending these values. (b) Prove that if we assign a complex number to every Gaussian integer on the diagonal $x = y$ in the plane, then there is a unique strongly analytic function on the Gaussian integers extending these values.

EXERCISE 7.2. Let f be an analytic function on $\mathbb{G}$. Prove that $\widehat{f}(x+iy) = (-1)^{x+y}\overline{f}(x+iy)$ is analytic on $\mathbb{G}$.

EXERCISE 7.3. Prove that the restriction of any discrete analytic function to the (tilted) grid $\mathbb{G}_0$ is a (complex valued) harmonic function on this grid graph. Also prove that every harmonic function on the grid graph $\mathbb{G}_0$ arises as the restriction of a discrete analytic function.

EXERCISE 7.4. (a) Prove that if f is an analytic function on the grid, then $\int_0^u f$ is an analytic function of u. (b) If f is a polynomial in $Re(z)$ and $Im(z)$, then $\int_0^u f$ is a polynomial in $Re(u)$ and $Im(u)$.

EXERCISE 7.5. For $u \in \mathbb{G}$, let $u^{:0:} = 1$, and for $n \geq 1$, define

$$u^{:n:} = n \int_0^u z^{:n-1:} dz.$$

Define a *discrete polynomial* of degree n as any linear combination of the discrete analytic functions $1, z, z^{:2:}, \ldots, z^{:n:}$. Prove that the space of polynomials of a given degree is invariant under shifting the origin [**Mercat 2002**].

EXERCISE 7.6. Let f be an analytic function on the grid, and let a be a nonzero Gaussian integer. Prove that the following functions are discrete analytic:

(a) $\partial_a f(z) = \big(f(z+a) - f(z)\big)/a$.

(b) $f^\dagger(z) = \begin{cases} \overline{f}(z) & \text{if } z \in \mathbb{G}_0, \\ -\overline{f(z)} & \text{if } z \in \mathbb{G}_1. \end{cases}$

EXERCISE 7.7. Define the *discrete exponential function* as a discrete analytic function $\mathrm{Exp}(x)$ on $\mathbb{G}$ satisfying

$$\int_u^v f(z) d\,\mathrm{Exp}(z) = \int_u^v f(z)\,\mathrm{Exp}(z) dz.$$

for every discrete analytic functions f and every $u, v \in \mathbb{G}$.

(a) Prove that there is a unique exponential function.

(b) Prove that

$$\mathrm{Exp}(z) = \sum_{n=0}^{\infty} \frac{z^{:n:}}{n!},$$

for values of z for which the series on the right hand side is absolute convergent.

(c) Prove that exponentials form a basis for all discrete analytic functions [**Mercat 2004**].

EXERCISE 7.8. Let $\varepsilon_n \searrow 0$, let Ω be a simply connected bounded open set in the plane, and let $g_n : \Omega \cap (\varepsilon_n \mathbb{G}) \to \mathbb{Z}$ be a function that satisfies the discrete analycity condition on every square contained in Ω. We extend the definition of g_n to every point in Ω, as the value at the nearest point of $\Omega \cap (\varepsilon_n \mathbb{G})$. Let $g : \Omega \to \mathbb{Z}$, and assume that $g_n \to g$ uniformly on every compact subset of Ω. Prove that g is analytic (in the classical sense).

EXERCISE 7.9. Prove that every analytic function on a lattice domain can be written as $f_1 + i f_2$, where f_1 and f_2 are strongly analytic functions.

EXERCISE 7.10. Let g be a strongly analytic function on a lattice domain graph G (with the cyclic orientation). For an internal edge uv, define $\widehat{g}(uv) = g(u) + (\overline{u} - \overline{v})\overline{g(u)}$.

(a) Prove that $\widehat{g}(uv) = g(v) + (\overline{u} - \overline{v})\overline{g(v)}$.

(b) Prove that $\widehat{g}(uv)$ is the orthogonal projection of $g(u)$ onto the line parallel to $\sqrt{u-v}$.

(c) Prove that the function $|\widehat{g}|^2$ satisfies the flow condition at every internal node of G^*.

(d) Show that one can define a function $H : V(G^*) \setminus \infty \to \mathbb{R}$ such that if $uv \in E$, $p = \mathsf{ls}(uv)$ and $q = \mathsf{rs}(uv)$, then $H(q) - H(p) = |\widehat{g}(uv)|^2$.

(e) Prove that H is superharmonic on the lattice graph of $\mathbb{G}_0$.

EXERCISE 7.11. Let G be a finite connected planar map. Let us assign values $f_0(u) \in \mathbb{C}$ to the nodes of the unbounded country p_∞. (a) Prove that there is a pair of functions $f : V \to \mathbb{C}$ and $g : V^* \setminus p_\infty \to \mathbb{C}$ such that f extends f_0 and (7.7) is satisfied at every edge except for the edges of p_∞. (b) Prove that the pair (f, g) is uniquely determined up to adding a constant to g.

EXERCISE 7.12. Let G be a connected planar map with a reference orientation and let $v \in V$ such that v is not a cutpoint. Let us assign values $g_0(uv) \in \mathbb{C}$ to edges exiting v so that $\sum_{u\in N(v)} g_0(uv) = 0$. Prove that there is a unique extension $g:\ E \to \mathcal{C}$ of g_0 that is a circulation on G and rotation-free on $G \setminus v$.

EXERCISE 7.13. Prove that every circulation on a map on any orientable surface is homologous to a unique rotation-free circulation.

EXERCISE 7.14. Let G be a map on an orientable surface Σ and $e \in E$. Prove that if there is a any rotation-free circulation that does not vanish on e, then $\Psi \mathbb{1}_e$ is such a circulation.

EXERCISE 7.15. Show by an example that there is a 3-connected map G on the torus and a nonzero vector $f \in \mathbb{R}^E$ supported on two edges such that $\Psi f = 0$.

EXERCISE 7.16. Let G be a map on an orientable surface S of genus $g > 0$, such that every noncontractible Jordan curve on the surface intersects the map in at least $4g$ points. Prove that if a rotation-free circulation vanishes on all edges of a non-0-homologous cycle and on all edges incident with it, then it is identically 0.

EXERCISE 7.17. Rotation-free circulations can be defined on non-orientable closed surfaces as well. (a) Prove that there is no nonzero rotation-free circulation on any map on the projective plane. (b) Prove that on a map on a non-orientable surface with Euler characteristic χ, the dimension of the space of rotation-free circulations is $1-\chi$. (c) Show by an example that Theorem 7.7 does not remain valid on nonorientable surfaces.

EXERCISE 7.18. Let G be a map on an orientable surface.

(a) Prove that starting with any vector on the edges and doing country and node balancing repeatedly so that each node and each country gets balanced infinitely often, the edgeweights converge to a rotation-free circulation.

(b) For $u \in V \cup V^*$ and $x \in \mathbb{R}^E$, let Φ_u denote the result of node/country balancing at u. Let $u_1, u_2, \dots$ be the sequence of independent uniform random elements of $V \cup V^*$. Starting with any vector $x \in \mathbb{R}^E$, define a sequence of vectors by recurrence as follows: $x_0 = x$ and $x_t = \Phi_{u_t} x_{t-1}$ $(t \geq 1)$. Prove that

$$\mathsf{E}\big(|x_t - \Psi x|^2\big) \leq \Big(1 - \frac{1}{(n+f)^4}\Big)^t |x|^2.$$

EXERCISE 7.19. Prove that if G is a map on an orientable surface with genus $g > 0$, then $(\Psi \mathbb{1}_e)_e \geq n^{-n} f^{-f}$ for every edge e.

CHAPTER 8

Discrete Analytic Functions: Statistical Physics

There are several models in geometric probability and statistical physics which are defined on the part of a lattice in a domain Ω in the plane, and which show a fully unexpected invariance under conformal transformations of Ω, at least in the *scaling limit*, when the edge-length of the lattice tends to zero. This fact (in the cases when proved at all) is not only unexpected, but highly nontrivial. It is perhaps not surprising that discrete analytic functions play a substantial role in several of these proofs.

We describe four such models, and show the discrete analytic functions (or similar functions) that arise from them; then we sketch the proof in one of the cases.

8.1. Conformal invariance

8.1.1. Domino tilings. A *domino* is the union of two adjacent squares. A lattice domain D may or may not be the union of disjoint dominos; the question is equivalent to the existence of a perfect matching in the tangency graph G_D^*. A necessary and sufficient condition for this can be obtained by specializing the Marriage Theorem, but more geometric conditions can be formulated [**Fournier 1995**][**Kenyon 2009**].

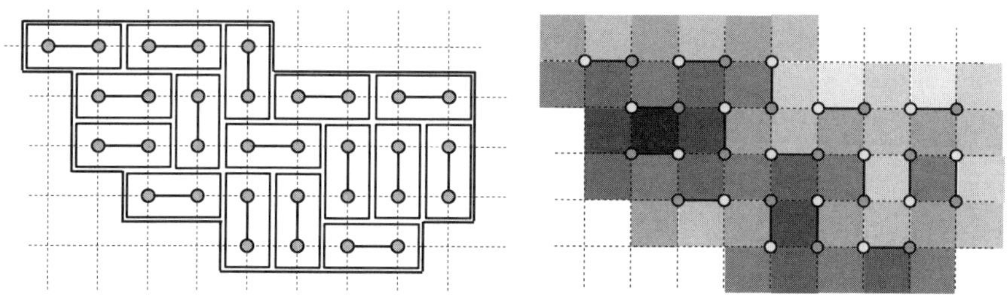

FIGURE 8.1. Left: a domino tiling of a lattice domain and the corresponding perfect matching in the dual graph. Right: the height function of this domino tiling, in grayscale. Note that the edges of the matching correspond to rifts (changes by more than 1) in the height function.

A construction capturing important features of domino tilings was introduced in [**Thurston 1989**]. Given a covering by disjoint dominos, we assign a value of 1 to each edge on the boundary of a domino and -3 to the edge separating two squares on the same domino. It is trivial to see that this assignment is rotation-free with respect to the cyclic orientation, and so (by the potential argument) it can be represented as $\delta\mathbf{h}$, for some vector $\mathbf{h} \in \mathbb{R}^V$. The function $\mathbf{h}$ is called the *height* *function* of the domino tiling. It is uniquely determined up to adding a constant function; to be specific, we set the function value 0 on the leftmost tile among the lowermost tiles. In terms of G_D^*, the function is defined on those squares containing a node of V, and so we can consider it to be defined on the union of these squares (Figure 8.1).

On the boundary of the lattice domain, the height function is uniquely determined, independently from the domino tiling. On the other hand, the height function determines the domino tiling: a pair of squares forms a domino if and only if they share an edge along which the height function changes by 3.

Perhaps the first mathematical result about conformal invariance of the scaling limit of combinatorial quantities defined on lattices was obtained by [**Kenyon 2000**]. He proved that *the variance of the height function is asymptotically conformally invariant.* This means the following. Fix a point p in the interior of the domain Ω. Select a tiling $\mathcal{T}$ of the corresponding lattice domain, then the dual square containing p has a certain height $h(p)$. If we select $\mathcal{T}$ at random, uniformly among all tilings, then $h(p)$ will be a random variable, that has a certain expectation $\overline{h}(p)$. When the edge length ε of the grid tends to zero, then the random variable $h(p) - \overline{h}(p)$ tends to a limit random variable $g(p)$. This random variable is conformally invariant: If $\phi : \Omega \to \Omega_1$ is a conformal mapping of Ω to another domain Ω_1, and g_1 denotes the function corresponding to g on Ω_1 then $g_(\phi(p))$ has the same distribution as $g(p)$. See [**Kenyon 2009**] for a nice introduction to this rich area.

8.1.2. Honeycombs, triangles and hex. Let G be a 2-connected simple planar graph in which all countries are triangles except for the unbounded country p_∞. Let a, b, c and d be four edges of the boundary of p_∞ (labeled in their counterclockwise order), splitting it into four node-disjoint arcs ab, bc, cd and da . Let us 2-color the nodes with black and white, where we require that nodes in $ab \cup cd$ are black and nodes in $bc \cup da$ are white. Let us say briefly that the *boundary is properly colored.*

By Proposition 6.3 and the construction in Figure 6.7(b), there is either an all-black path connecting ab and cd, or and all-white path connecting bc and da, but not both. What is the probability that for a random coloring (subject only to the requirement that the boundary is properly colored), the first possibility, namely ab-cd connectivity occurs?

This number is quite difficult to compute in general, but there is a beautiful asymptotic formula in the case when G is a piece of the triangular grid. More exactly, consider the regular triangular grid in the plane, and a finite set of lattice triangles whose union is a simply connected region Ω. Let G be the graph consisting of the edges and vertices of these triangles. Splitting the boundary into four parts as above, we can ask for the probability that in a random coloring that is proper on the boundary, ab and cd are connected by a black path. We call this the *crossing probability* (from ab to cd).

Next, consider a sequence $G_1, G_2, \dots$ of such graphs, with the edge-lengths of the grids tending to 0. Suppose, furthermore, that the corresponding domains Ω_n tend to a simply connected domain Ω in the Carathéodory sense. Let the boundary polygon of Ω_n be split into four arcs as before by edges a_n, b_n, c_n and d_n, and assume $a_n \to a \in \partial\Omega$, $b_n \to b \in \partial\Omega$ etc.

It can be proved that the crossing probability between the arcs $a_n b_n$ and $c_n d_n$ tends to a value $p = p(\Omega, a, b, c, d)$ which depends on the domain Ω and on the four points $a, b, c, d \in \partial\Omega$ only. The surprising fact, conjectured in [**Cardy 1992**], based on computational experiments and ingenious intuition, is that this value is invariant under conformal transformation. This fact was eventually proved in [**Smirnov 2001**]:

THEOREM 8.1. *Let Ω be a simply connected domain in the plane, let $a, b, c, d, \in \partial\Omega$, and let $f : \overline{\Omega} \to \mathbb{C}$ be a bijective conformal map. Then $p(\Omega, a, b, c, d) = p\big(f(\Omega), f(a), f(b), f(c), f(d)\big)$.*

This allows us to compute the crossing probabilities for any domain (assuming we can compute that map f sufficiently explicitly), when perhaps ad hoc methods can be used.

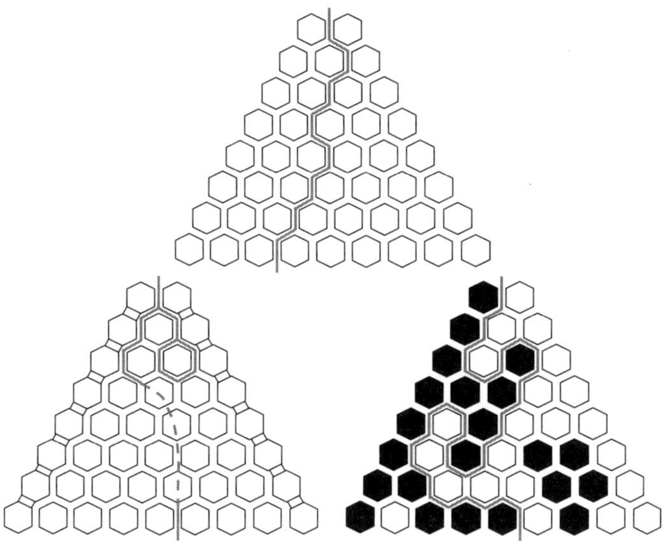

FIGURE 8.2. Three ways of randomly walking down from the top of a triangle to the bottom on the streets amidst hexagonal buildings. Top: a Galton board; left: random walk (not a proper drawing, which would pass too many streets); right: along the line separating a random 2-coloring of the hexagons.

A particularly nice special case was conjectured by [**Cardy 1992**] and proved by [**Smirnov 2001**]. Consider a triangular domain T in the regular triangular lattice, with vertices a, b and c. Take those hexagons in the dual lattice whose center is a lattice point in T; but omit the hexagon containing a. Let all the hexagons along the edge ab be colored black, all hexagons along the edge ac be colored white. Let the other hexagons be colored randomly. Start a polygonal walk between the two hexagons on the top, and keep going so that you always have a black hexagon on your right and a white hexagon on your left. Eventually, you will exit at a point y on the edge bc. Then Cardy's Formula tells us that the point y is asymptotically uniformly distributed along the edge bc, if the edge length of the lattice tends to 0 (Figure 8.2, right). Another way of saying this is that if x is a point on the edge bc of a regular triangle abc, then

$$p(T; a, b, c, x) = \frac{|c-x|}{|c-b|}.$$

The other two figures illustrate other ways of walking down from a to bc: the Galton board (when we move always down, choosing a left or right turn randomly), and a random walk. The Galton board produces a binomial distribution, which is asymptotically Gaussian.

8.1.3. Paving your porch, or the Fortuin–Kasteleyn model. The following construction [**Smirnov 2010a, Chelkak–Smirnov 2012**], providing an interesting class of strongly analytic functions, will be important for its applications in physics. Here we start with an elementary mathematical formulation.

Consider the tiling of a lattice domain (a "porch") with square tiles that have two circular arcs on their face, with radius half of their side length, centered at opposite corners (Figure 8.3). There are two "doors", special tiles marked by arrows.

Such a tile has two possible orientations. We assume that the tiles on the boundary of the domain are oriented alternatingly, except for the doors. More precisely, we may assume that in the 2-coloring of the vertices, entering the door on the left edge of the picture, we see a white vertex on our left. Then the black vertices on the upper boundary have partial circles about them (which may consist of one, two or three quarter-circles), and so do the white vertices on the lower boundary. The arcs that constitute these incomplete circles will be called *ornamental*, and the edges of the tiles incident with ornamental arcs will be called *external*. No matter how the remaining tiles are put down (including two on the boundary, at the doors), the non-ornamental arcs on the tiles form a single open curve γ connecting the two doors, and a number of closed curves (possibly quite long and winding).

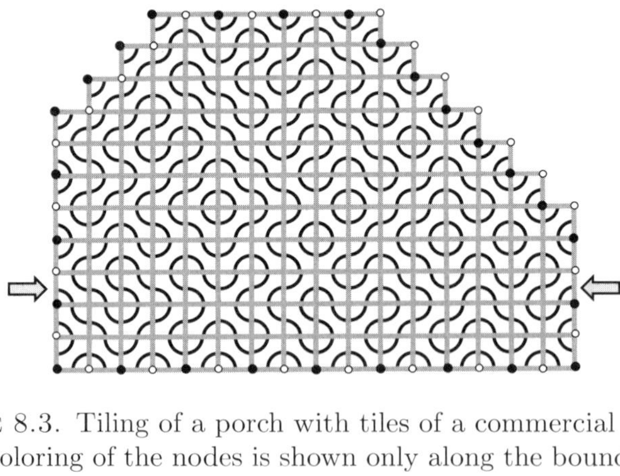

FIGURE 8.3. Tiling of a porch with tiles of a commercial pattern. The 2-coloring of the nodes is shown only along the boundary, the 2-coloring of the squares, only near the doors. Tiles on the boundary are placed so that they form a neat sequence of incomplete circles, interrupted only at the two doors. No matter how we put down the tiles in the middle, we get a single curve connecting the two doors, together with a collection of closed curves.

Suppose that we have a sequence of lattice domains Ω_n, with the size of tiles tending to zero, so that the domains "tend to" a simply connected domain Ω.

Let the two doors (shorter and shorter edges) converge to points a and b on the boundary of Ω.

Let us tile the domains Ω_n randomly, under the boundary conditions as discussed above. There is a complication here: not every tiling has the same probability; we weight each tiling $\mathcal{T}$ with $w(\mathcal{T}) = (\sqrt{2})^{\#\text{ of closed curves}}$, and choose a tiling with probability proportional to its weight. Some explanation for this strange choice of weights will be forthcoming when we discuss the connection of tilings with the Ising model. We call this probability distribution on tilings the *physics distribution* (as opposed to the uniform distribution).

Each tiling will define a random door-to-door curve γ_n. It turns out that the distributions of these random curves will tend to a distribution $\Gamma(\Omega, a, b)$ on Jordan curves inside Ω connecting a and b. This is not trivial, but not too surprising so far. The amazing fact is that this limit distribution will be invariant not only under translations (trivial), similarities (a matter of normalization), rotations (nontrivial), but under all conformal transformations! If we have a conformal map $\phi: \Omega \to \Omega'$, and γ is a random curve from the distribution $\Gamma(\Omega, a, b)$, then the distribution of $\phi(\gamma)$ is $\Gamma\big(\Omega', \phi(a), \phi(b)\big)$. This result was proved in [**Smirnov 2010a**]; see also [**Smirnov 2010b**] for a more informal presentation, and [**Chelkak–Smirnov 2012**] for a generalization to planar graphs with a rhombic embedding (a.k.a. isoradial graphs or critical graphs), but also for more details and applications.

To illustrate why this result is surprising, let us consider a very special case: Ω is the unit disk $\mathbb{D}$, and a and b are diametrically opposite points. The situation looks quite different when the segment ab is horizontal (so parallel to the sides of the tiles), and when it is rotated by some angle. Nevertheless, in the limit the orientation of the lattice plays no role in the limit.

It is not easy to make the above discussion precise. The convergence $\Omega_n \to \Omega$ can be defined in the Carathéodory sense: there are conformal maps $\phi_n: \mathbb{D} \to \Omega_n$ and $\phi: \mathbb{D} \to \Omega$ so that $\phi_n \to \phi$ inside $\mathbb{D}$. The limit distribution of a-b curves is much more difficult to define and describe; this needs the theory of stochastic Löwner evolution (SLE) developed by [**Schramm 2000**], [**Schramm 2007**], whose presentation is outside the possibilities of this book.

Let us formulate one consequence of the method that can be formulated, and its proof can be sketched, without invoking the full machinery of SLE. Let us consider a porch with four doors a, b, c, d (in this counterclockwise order). These mean four special edges on the boundary. Assume that the number of edges of the boundary between consecutive doors is even, and that the tiles incident with these edges are fixed so that the vertices at an odd distance from the nearest door are centers of partial circles (Figure 8.4).

If we lay down the remaining tiles arbitrarily, a door-to-door curve γ_1 will start at a. It is easy to argue (using parity) that γ_1 must exit the porch either at b or d. Another curve γ_2, disjoint from γ_1, will connect the remaining two doors. This tells us that there are two possibilities, and we may wonder what is the probability that the curve starting at a exits at b.

THEOREM 8.2. *Consider a sequence Ω_n of lattice domains, tending to a simply connected domain Ω. Let a_n, b_n, c_n, d_n be four doors on Ω_n, and suppose that they tend to four different points a, b, c, d on the boundary of Ω. Then the probability that for a random tiling a door-to-door curve connects a_n and b_n tends to a value*

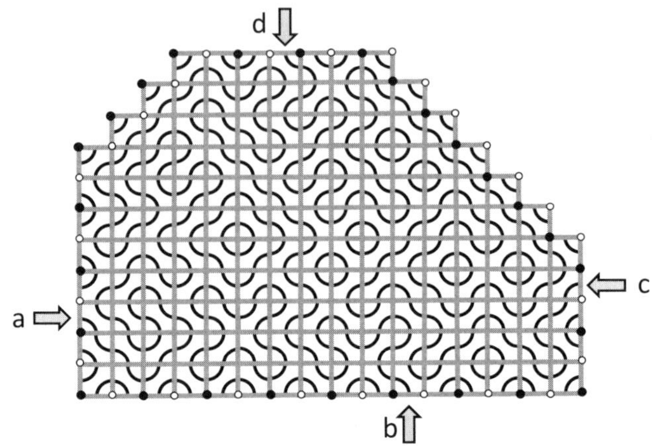

FIGURE 8.4. A tiling with four doors.

$p(\Omega; a, b, c, d)$, and this number is conformally invariant: if $\phi : \Omega \to \Omega'$ is a conformal map, then $p(\Omega; a, b, c, d) = p\big(\Omega'; \phi(a), \phi(b), \phi(c), \phi(d)\big)$.

The proof of this theorem (even in a somewhat sketchy form) will take a lot of preparations, to be done in the Section 8.2.1. But first (to justify the interest in this seemingly quite special question), we discuss its relation to an important problem in statistical physics.

8.1.4. Ising model in the plane. In this most classical model in statistical physics, we consider a lattice domain and its lattice graph G, and assume that there is an atom sitting at each node. Assigning to every node of G (every "site") a "state", which can be UP or DOWN, we get a "configuration". The configuration can be described by the set S of atoms in the UP state (Figure (8.5).

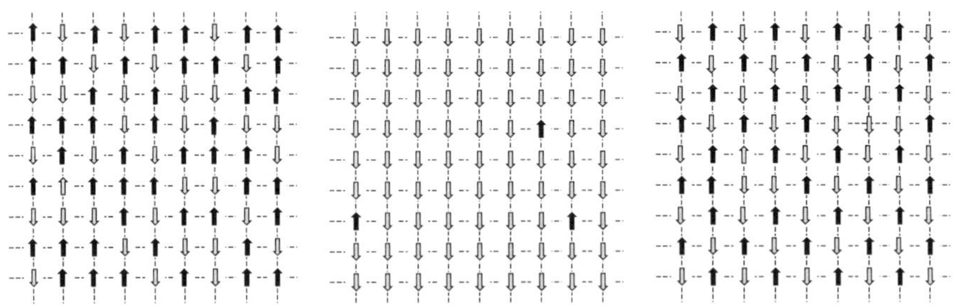

FIGURE 8.5. Three kinds of states in an Ising model: a disordered state, a state close to the homogeneous ground state in the ferromagnetic case, and a state close to the chessboard-like ground state in the antiferromagnetic case.

Two atoms that are adjacent in the grid have an "interaction energy", which depends on their states. In the simplest version of the basic Ising model, the interaction energy depends only on whether or not the atoms are in the same state. Adding the same constant a to the interaction energies shifts the total energy by $a|E|$ independently of the configuration, and such a shift is irrelevant;

so we may assume that the interaction energy between atoms in the same state is 0. Multiplying the other interaction energy by a positive number means only changing the unit of temperature, so we may assume that the interaction energy between atoms in different states is either $J = 1$ or $J = -1$.

The total energy of a given configuration S is

$$H(S) = J|E(S, V \setminus S)|.$$

Basic physics (going back to Boltzmann) tells us that the system is more likely to be in states with low energy. In formula, the probability of a given configuration is proportional to $e^{-H(S)/T}$, where T is the temperature (from our point of view, just a parameter). Since probabilities must add up to 1, these values must be normalized:

$$\mathsf{P}(S) = \frac{e^{-H(S)/T}}{Z},$$

where the normalizing factor Z is called the *partition function* of the system. The partition function is simple to describe:

$$Z = \sum_{S \subseteq V} e^{-H(S)/T},$$

but the number of terms is enormous, and so partition functions can be very hard to compute or analyze.

The behavior of the system depends very much on the sign of J. If $J = 1$, then adjacent pairs that are in the same state contribute less to the total energy than those that are in different states, and so the configuration with the lowest energy is attained when all atoms are in the same state. The typical configuration of the system will be close to this, at least as long as the temperature T is small. This is called the *ferromagnetic* Ising model, because it explains how materials like iron get magnetized. This will be the case we consider in this section. If $J = -1$ (the antiferromagnetic case), then the behavior is different: the chessboard-like pattern minimizes the energy, and no magnetization occurs at any temperature.

One may notice that the temperature T emphasizes the difference between the energy of different configurations when $T \to 0$ (and de-emphasizes it when $T \to \infty$). In the limit when $T \to 0$, all the probability will be concentrated on the states with minimum energy, which are called *ground states*. In the simplest ferromagnetic Ising model, there are two ground states: either all atoms are in state UP, or all of them are in state DOWN. As the temperature increases, disordered states become more likely. The transition from the ordered state to the disordered may be gradual (in dimension 1), or it may happen suddenly at a given temperature (in dimensions 2 and higher, for large graphs G). This temperature is called the *critical temperature*, and the phenomenon itself, a *phase transition*.

8.1.5. Porch tiling and the Ising model. Our next goal is to show the connection between the Ising and the porch tiling models. We start with rephrasing the porch tiling model, at least at the critical temperature. Let us draw, for each tile, the diagonal that does not intersect the quarter-circles. The new edges form a subgraph of the union of the two lattice graphs G_0 and G_1 on $\mathbb{G}_0 \cap V$ and $\mathbb{G}_1 \cap V$. If we specify the subgraph H_0 on $\mathbb{G}_0 \cap V$, then the other subgraph H_1 is determined as the graph of those edges of G_1 that do not cross the edges of H_0. So a porch tiling can be specified by a subgraph H_0 of G_0 (we keep all nodes).

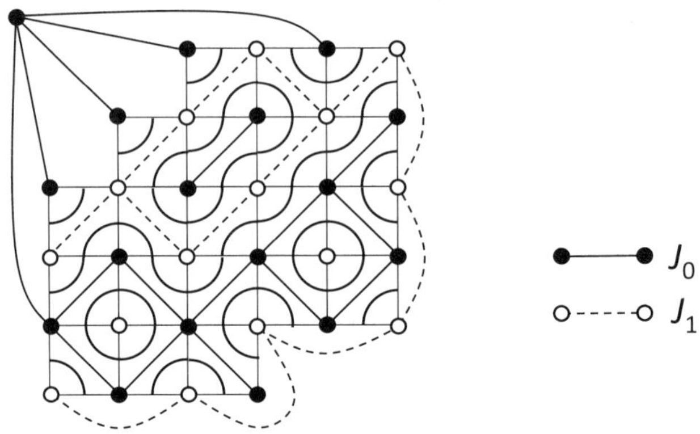

FIGURE 8.6. A porch tiling, and subgraphs of G_0 and G_1 surrounded by closed curves.

Note that each connected component of H_0, as well as each connected component of H_1, is surrounded by a closed curve of the tiling, except for the connected components containing the two boundary arcs. So the number of closed curves is $c(H_0)+c(H_1)-2$ (recall that $c(H)$ denotes the number of connected components of the graph H).

The component numbers $c(H_0)$ and $c(H_1)$ are related. Consider the graph G_0 formed by the black nodes and the diagonals between them. Add a new black node b outside that lattice domain D, and connect it to every black node on the boundary. This way we get a planar graph $\widehat{G}_0$. The dual of this graph is the graph $\widehat{G}_1$ obtained from G_1 by adding edges crossing the edges incident with b. From the graph H_0 (as defined above) we construct a subgraph J_0 of G_0 by adding to it the new node b, all edges connecting b to black nodes on the upper arc between the doors, and (say) to the black endnode of the left door. The corresponding subgraph of G_1 will be obtained from H_1 by adding all edges of the path connecting consecutive white nodes on the lower arc, and ending at the white endnode of the right door (Figure 8.6). It follows that

$$c(J_1) = c(J_0)+|E(J_0)|-|V(J_0)|+1 = c(J_0)+|E(J_0)|-|V(G_0)| \tag{8.1}$$

(recall Exercise 2.3). It is also clear that $c(J_0) = c(H_0)$ and $c(J_1) = c(H_1)$, since the additional part simply connects the ornamental centers to the main part.

Let ℓ_0 be the number of black nodes on the lower arc, then $|E(J_0)| = |E(H_0)|+\ell_0$. Using (8.1), we get that $c(H_1) = c(H_0)+|E(H_0)|+\ell_0-|V(G_0)|$. Thus the weight we assign to a tiling (to define the physical distribution) is

$$w(\mathcal{T}) = (\sqrt{2})^{\#\text{ of closed curves}} = (\sqrt{2})^{c(H_0)+c(H_1)-2} \sim 2^{|E(H_0)|/2+c(H_0)}.$$

The fact that Ising model is closely related to the porch tiling model is not obvious, since the Ising model concerns subsets of the nodes, while the tiling model (after the reformulation above) concerns subsets of the edges. To connect the two models, let $G = (V, E)$ be an arbitrary graph (we do not even need planarity for this argument), and let us consider pairs (S, Y), where $S \subseteq V$ and $Y \subseteq E[S] \cup E[V \setminus S]$. Fix a parameter $0 < p < 1$, and consider a probability distribution on such pairs, where

$$\mathsf{P}(S,Y) \sim p^{|Y|}(1-p)^{|E\setminus Y|} \sim \Big(\frac{p}{1-p}\Big)^{|Y|}$$

(we have to normalize the given values to get a probability distribution). This value is the same for all sets S that are unions of connected components of the graph (V,Y); the number of such sets is $2^{c(V,Y)}$. If we generate a pair (S,Y) from this distribution and then forget S, then we get a probability distribution on subsets of E for which

$$\mathsf{P}(Y) \sim \Big(\frac{p}{1-p}\Big)^{|Y|} 2^{c(V,Y)}.$$

The value $p = 2-\sqrt{2}$ plays a special role: this corresponds to the *critical temperature* in the Ising model. In this case, $p/(1-p) = \sqrt{2}$, and so

$$\mathsf{P}(Y) \sim 2^{|Y|/2+c(V,Y)}.$$

This is the same probability distribution as coming from the porch tiling model.

On the other hand, let us forget about Y. Fixing S and setting $E' = E[S] \cup E[V\setminus S]$, we get

$$\begin{aligned}\mathsf{P}(S) &\sim \sum_{Y\subseteq E'} p^{|Y|}(1-p)^{|E\setminus Y|}\\ &= (1-p)^{|E(S,V\setminus S)|}\sum_{Y\subseteq E'} p^{|Y|}(1-p)^{|E'\setminus Y|} = (1-p)^{|E(S,V\setminus S)|}.\end{aligned}$$

Defining the temperature T so that $1-p = e^{-1/T}$, we get

$$\mathsf{P}(S) \sim e^{-|E(S,V\setminus S)|/T}.$$

This is just the probability distribution in the Ising model.

8.2. Analytic functions and porch tilings

The remainder of this chapter is devoted to sketching the proof of the conformal invariance of the porch tiling model (a.k.a. the Fortuin–Kasteleyn random cluster model).

8.2.1. A strongly analytic function. We consider a lattice domain D, with the acyclic orientation: every edge of G_D is oriented from its black endpoint to its white endpoint. We orient the graph G_D^* (the tangency graph of the squares) in the cyclic sense (the boundary of every black country is a counterclockwise oriented cycle).

Let us start with a simple observation. Tracing the curve γ from the left door to the right door, we cross several edges. A different tiling will create a different curve, crossing different edges. However, if an edge is crossed by both curves, it is crossed in the same direction. Indeed, following the curve, we see that every edge is crossed so that its black endpoint is on the right (so every edge is crossed in the same direction as the oriented edges of G_D^* crossing it).

It is also easy to observe that if we consider a random tiling (say, from the physics distribution, but here we could talk about any distribution), and for an edge e we denote by $p(e)$ the probability that the door-to-door curve crosses e, then for the four edges e_1, e_2, e_3, e_4 of a tile (in this cyclic order), we have

$$p(e_1)+p(e_3) = p(e_2)+p(e_4)$$

(since every curve loads $\{e_1, e_3\}$ by the same amount as it loads $\{e_2, e_4\}$). Our main tool below will be a similar relation for the physics distribution even if we assign a "phase" (a unit complex vector) to each crossing.

Let us remove those boundary tiles that contain two ornamental arcs; they play no role. Those tiles that contain one ornamental arc have a fixed orientation, but their non-ornamental arc can be part of the curve γ. Let us call them *boundary tiles.*

We are going to construct a strongly analytic function on the graph G^*_D of this grid from these tilings. Recall that we define a weight of a tiling as

$$w(\mathcal{T}) = (\sqrt{2})^{\#\text{ of closed cycles}}.$$

The total weight, in physical terms the *partition function*, is $Z = \sum_{\mathcal{T}} w(\mathcal{T})$. The ratio $w(\mathcal{T})/Z$ gives the probability of the tiling in the physics distribution.

Next, we construct a function on the edges of G. Set $\theta = e^{\pi i/4} = (1+i)/\sqrt{2}$ for convenience. Given a tiling $\mathcal{T}$ (we keep the tiles on the boundary fixed, except for the two tiles in front of the doors), we trace out the curve γ connecting the doors, starting (say) at the left door. Along the curve, we count turns with sign: when crossing a tile, we make either a left turn, which counts 1, or a right turn, which counts -1. When an edge e is crossed, this tiling $\mathcal{T}$ contributes

$$f_{\mathcal{T}}(e) = \frac{w(\mathcal{T})}{Z}\,\theta^{-\#\text{ of turns up to } e} \tag{8.2}$$

to the function value of e; edges not crossed by γ receive no contribution from this tiling.

As pointed out above, the curve γ crosses e in the same direction for every tiling $\mathcal{T}$. It follows that $f_{\mathcal{T}}(e)$ is parallel to the same line for all $\mathcal{T}$; $f_{\mathcal{T}}(e)$ is real if e is a vertical edge oriented down (in the acyclic orientation), $f_{\mathcal{T}}(e)$ is a real multiple of θ if e is a horizontal edge oriented right to left etc. We can formalize this as follows: $f_{\mathcal{T}}(uv)$ is parallel to $\sqrt{i(\overline{u}-\overline{v})}$ (when we talk about an edge uv, we assume that it is oriented from u to v; so in this case u is black and v is white).

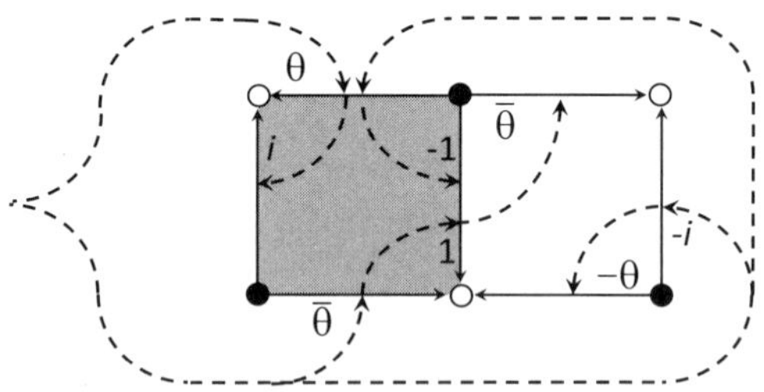

FIGURE 8.7. Directions in the complex plane of the contributions of door-to-door curves to f. They are determined by the edge up to the sign.

We define, for every edge $e = uv$,

$$f(e) = \sum_{\mathcal{T}} f_{\mathcal{T}}(e) = \mathsf{E}\big(\mathbb{1}(e \in \gamma)\theta^{-\#\text{ of turns up to } e}\big), \tag{8.3}$$

where the expectation is taken for a random tiling from the physical distribution. By the observations above,

$$f(uv) \parallel \sqrt{i(\overline{u}-\overline{v})} \quad \text{and} \quad |f(e)| \le \mathsf{P}(e \in \gamma) \le 1. \tag{8.4}$$

Let us consider a tile S, and let e_1, e_2, e_3, e_4 be its edges in counterclockwise order. If S is an internal tile, then we define

$$g(S) = \frac{1}{2}\big(f(e_1)+f(e_2)+f(e_3)+f(e_4)\big). \tag{8.5}$$

For a boundary tile, we have to modify this definition. We may assume that e_1 and e_2 are its internal edges (the other two edges carry no information). We define

$$g(S) = (2-\sqrt{2})\big(f(e_1)+f(e_2)\big). \tag{8.6}$$

Note that if a door-to-door curve γ intersects this tile at all, then it intersects both e_1 and e_2. If the tile is on the upper boundary, then γ must enter through e_1 and make a right turn inside. So $f(e_2) = \theta f(e_1)$. For tiles on the lower boundary, the curve crosses e_1 and e_2 in the reverse order, but the relation $f(e_2) = \theta f(e_1)$ remains valid.

LEMMA 8.3. *The function g is strongly analytic on the graph whose nodes are the tile centers, and in which two centers are connected by an edge if the tiles share an internal edge.*

PROOF. We compare the contributions from a tiling $\mathcal{T}$ and from the tiling $\mathcal{T}'$ obtained by rotating an internal tile S by $90°$. Let γ and γ' be the door-to-door curves in $\mathcal{T}$ and $\mathcal{T}'$, respectively. If γ does not cross S, then neither does γ', and all these contributions are 0. If γ crosses S, then so does γ'; one of them crosses once, and the other crosses twice (Figure 8.8). Let (say) γ cross twice and γ' cross once. We may assume that γ enters S through e_1. There are still two possibilities to consider, namely whether γ turns left or right after entering S. We discuss the case of a left turn; the other case is analogous.

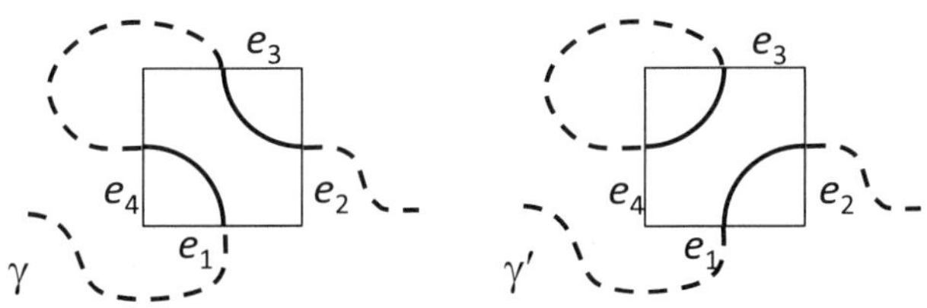

FIGURE 8.8. Rotating a non-boundary tile that is crossed by γ.

We can express all contributions in terms of $a = f_{\mathcal{T}}(e_1)$:

$$f_{\mathcal{T}}(e_2) = \theta a, \quad f_{\mathcal{T}}(e_3) = ia, \quad f_{\mathcal{T}}(e_4) = \overline{\theta} a,$$

and (since there is one more cycle in $\mathcal{T}'$ than in $\mathcal{T}$)

$$f_{\mathcal{T}'}(e_1) = \sqrt{2}a, \quad f_{\mathcal{T}'}(e_2) = \theta\sqrt{2}a, \quad f_{\mathcal{T}'}(e_3) = f_{\mathcal{T}'}(e_4) = 0.$$

Hence

$$f_{\mathcal{T}}(e_1)+f_{\mathcal{T}'}(e_1) = (\sqrt{2}+1)a, \quad f_{\mathcal{T}}(e_2)+f_{\mathcal{T}'}(e_2) = (\sqrt{2}+1)\theta a,$$
$$f_{\mathcal{T}}(e_3)+f_{\mathcal{T}'}(e_3) = ia, \quad f_{\mathcal{T}}(e_4)+f_{\mathcal{T}'}(e_4) = \overline{\theta} a.$$

Notice that

$$f_{\mathcal{T}}(e_1)+f_{\mathcal{T}'}(e_1)+f_{\mathcal{T}}(e_3)+f_{\mathcal{T}'}(e_3) = f_{\mathcal{T}}(e_2)+f_{\mathcal{T}'}(e_2)+f_{\mathcal{T}}(e_4)+f_{\mathcal{T}'}(e_4).$$

Summing over all tilings, we get that

$$f(e_1)+f(e_3) = f(e_2)+f(e_4) = g(S). \tag{8.7}$$

Consider an edge $e = uv$ of G between two non-boundary tiles S and S', so that S' is to its left. Let p_S and $p_{S'}$ denote their centers, and let st and xy be the edges of S and S', respectively, opposite to uv. Then

$$g(S)-g(S') = \big(f(st)+f(uv)\big)-\big(f(xy)+f(uv)\big) = f(st)-f(xy).$$

Since both $f(st)$ and $f(xy)$ are parallel to $\sqrt{i(\overline{t}-\overline{s})} = \sqrt{i(\overline{y}-\overline{x})} = \sqrt{i(\overline{u}-\overline{v})} = \sqrt{\overline{v}_{S'}-\overline{v}_S}$, so is their difference: $g(S)-g(S') \parallel \sqrt{\overline{v}_S-\overline{v}_{S'}}$. Similar computation shows that this relation remains valid for edges between a boundary tile and a non-boundary tile, as well as for non-ornamental edges between boundary tiles. This proves that g is strongly analytic. $\square$

For the special edge at the left door, we have $f_{\mathcal{T}}(e) = \frac{w(\mathcal{T})}{Z}$, and so $f(e) = 1$. The tile S next to the door satisfies $\Re\big(g(S)\big) = 1$.

From the fact that $f(e_1) \perp f(e_3)$ and $f(e_2) \perp f(e_4)$, it follows that (8.7) describes a decomposition of $g(S)$ into the sum of two orthogonal vectors in two different ways. In other words, $f(e_i)$ is the orthogonal projection of $g(S)$ onto the corresponding direction (Figure 8.9, left shows how the values of f on the edges can be expressed if $g(S)$ is given in the standard form). It is easy to check that this relation is also valid for the non-ornamental edges of boundary tiles (the definition of g was adjusted so that this should hold). In particular,

$$|f(e_1)|^2+|f(e_3)|^2 = |f(e_2)|^2+|f(e_4)|^2 = |g(S)|^2 \tag{8.8}$$

for the four edges of an internal tile S.

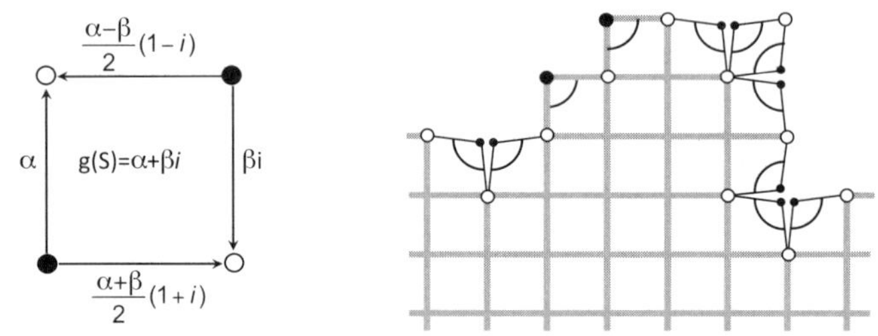

FIGURE 8.9. Left: the values of the function f on the edges of a lattice square, in terms of the value of g. Right: incisions at the ornamental arcs. Only part of the boundary is shown.

For a boundary tile S, where f_1 and f_2 are its internal edges, we can define $f(e_3) = (\sqrt{2}-1)if(e_1)$ and $f(e_4) = -(\sqrt{2}-1)if(e_2)$ for the other two edges. Then $g(S)$ satisfies (8.5). We have to be careful here on two counts: first, this definition of $f(e_3)$ and $f(e_4)$ does not have the same combinatorial meaning as it does for internal edges (which would make these values equal to zero); second, one or even both of the edges e_3 and e_4 may be shared by other boundary tiles, and then the values computed from those tiles might be different. It is best to think of making an incision between the two boundary tiles along such edges (or scratch out the grout, if you wish). A boundary node with two (or three) ornamental arcs around it is split into two (or three) separate nodes of degree two (Figure 8.9, right). A tile containing two ornamental arcs will be completely cut off, and it can be discarded, playing no role in the arguments. This domain with incisions will be denoted by D^{inc}, and the corresponding graph by $G^{\mathrm{inc}} = (V^{\mathrm{inc}}, E^{\mathrm{inc}})$.

The f-values of any two edges of a lattice square S determine the value $g(S)$, as well as the f-values of the other two edges of S. For two opposite edges, this follows by (8.7). It is not hard to calculate that

$$g(S) = (1-i)f(e_1)+(1+i)f(e_2), \tag{8.9}$$

and

$$f(e_3) = -if(e_1)+(1+i)f(e_2), \quad f(e_4) = (1-i)f(e_1)+if(e_2). \tag{8.10}$$

For boundary tiles in the graph G^{inc}, (8.6) expresses g as a linear combination of $f(e_1)$ and $f(e_2)$ with different coefficients; nevertheless, these formulas remain valid for these tiles as well (since $f(e_1)$ and $f(e_2)$ are not independent, but related by $f(e_2) = \theta f(e_1)$.

To formulate a further key property of the function g, we describe another way of reducing the domain: the "irrelevant fringes" can be removed. To be precise, we cut each boundary tile into two triangles by the diagonal separating the arcs on the tile, and delete the triangle containing the ornamental arc. In particular, we delete all tiles with two ornamental arcs (Figure 8.10). This reduced domain will be denoted by D^{red}, and the corresponding graph by $G^{\mathrm{red}} = (V^{\mathrm{red}}, E^{\mathrm{red}})$. Note that the graph G^{red} is no longer a subgraph of the grid, since the diagonals of the tiles where we have cut them are new (longer) edges on the boundary now; we call these edges *diagonal edges*. In the 2-coloring of the nodes, let (say) all the nodes on the lower arc be black, and all the nodes on the upper arc be white. We also need to define the orientation of the edges of the boundary arcs: we orient them away from the left door (along the boundary).

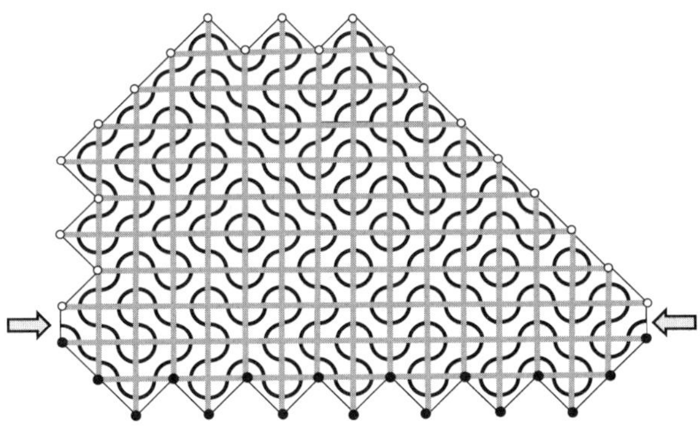

FIGURE 8.10. Cutting off the fringes of a porch tiling.

LEMMA 8.4. *Let uw be a diagonal boundary edge, obtained from the boundary tile S. Then $g(S) \parallel \sqrt{\overline{w}-\overline{u}}$.*

PROOF. Let uw be an edge of (say) the upper boundary arc of the porch between the doors, where this boundary arc passes the edge from u to w. Let v be the third vertex of the triangle containing uw. Then, as we have seen, $f(vw) =$

$\theta f(vu)$, and so

$$g(S) \parallel f(vu)+f(vw) = (1+\theta)f(vu) \parallel (1+\theta)\sqrt{i(\overline{v}-\overline{u})}$$
$$= \sqrt{(1+\theta)^2 i(\overline{v}-\overline{u})} \parallel \sqrt{\overline{\theta}(\overline{v}-\overline{u})} \parallel \sqrt{\overline{w}-\overline{u}}.$$ □

8.2.2. Guessing the limit. Lemma 8.4 suggests what the limit of the discrete analytic functions g should be when we are looking at tiling the same porch D with smaller and smaller tiles. Let us think of the difference $w-u$ as the "discrete tangent vector" of the domain D^{red} (following the boundary from the left door to the other one). Then we have a discrete analytic function on D^{red}, whose value is parallel to the square root of the conjugate of the tangent vector on the boundary. What would the continuous analogue of such a property be?

Let Ω be a simply connected domain in the plane with a smooth (or piecewise smooth) boundary, and let a and b be two points on its boundary. Let $\Phi : \Omega \to \mathbb{C} \cup \{\infty\}$ be a conformal map that maps Ω onto the strip $0 \le \mathfrak{I}(z) \le 1$, so that $\Phi(a) = -\infty$ and $\Phi(b) = +\infty$. The mapping Φ is determined only up to an additive real constant, but this will not matter, since we need its derivative Φ' only. Furthermore, since the map is one-to-one, its derivative is not zero in the interior of Ω, and hence $\sqrt{\Phi'}$ can be defined choosing the branch of the square root with $\mathfrak{R}(\sqrt{\Phi'(c)}) > 0$ for a point c on the counterclockwise lower arc.

Let $\psi = (\Phi|_L)^{-1}$, then $\Phi(\psi(s)) = s$, and differentiating, we get $\Phi'(\psi(s))\psi'(s) = 1$. This means that

$$\Phi'(\psi(s)) = \frac{1}{\psi'(s)} = \frac{1}{|\psi'(s)|^2}\overline{\psi'(s)},$$

where $\psi'(s)$ is a tangent vector to the domain Ω at the boundary point $\psi(s)$. Thus

$$\sqrt{\Phi'(\psi(s))} = \frac{1}{|\psi'(s)|}\sqrt{\overline{\psi'(s)}};$$

the square root on the right is well defined once we fix its sign at (say) $s = 0$ appropriately. This suggests that if Ω is "close" to D, then $(\Phi')^{1/2}$ could be "close" to g.

To make this more precise, let G_n ($n = 1, 2, \dots$) be the graph of a simply connected lattice domain D_n of the lattice $\mathbb{G}_n = \varepsilon_n \mathbb{G}$, where $\varepsilon_n \to 0$. Assume that $D_n \to \Omega$, where Ω is a simply connected bounded domain in the plane, and the convergence is in the Carathéodory sense.

Let f_n and g_n denote the functions defined by (8.3) and (8.5) for the lattice graph G_n of D_n. We can consider $g_n(z)$ to be defined on all points $z \in D_n$ as the value of g_n on the square containing z (ties broken arbitrarily). The main theorem of [**Smirnov 2010b**] and [**Chelkak–Smirnov 2011**] is the following.

THEOREM 8.5. *With the notation above,*

$$\frac{1}{\sqrt{\varepsilon_n}} g_n \to \sqrt{\Phi'}$$

uniformly on every compact subset in the interior of Ω.

The considerations above imply that if the functions $g_n/\sqrt{\varepsilon_n}$ tend to any function, then this must be analytic and in fact equal to $\sqrt{\Phi'}$. But to show that the sequence of functions $g_n/\sqrt{\varepsilon_n}$ is convergent is highly nontrivial, and it takes careful estimates based on analysis and on the theory of random walks.

8.2.3. An almost harmonic function. The proof of Theorem 8.5 depends on the introduction of an auxiliary function on the nodes, that is "almost" harmonic, and which is interesting on its own right. Equation (8.8) means that *the function* $|f(e)|^2$ *is rotation-free* (on the edges of the domain D^{inc}, with respect to the acyclic orientation). Hence it has a potential, which means that there is a function $H : V^{\text{inc}} \to \mathbb{R}$ such that

$$H(v) - H(u) = |f(uv)|^2 \tag{8.11}$$

for every (oriented) edge $uv \in E^{\text{inc}}$.

The difference of H along a diagonal of a square can be expressed in terms of the function g. Let v and z be diametrically opposite vertices of a lattice square S, then

$$H(v) - H(z) = \frac{1}{2}\mathfrak{Re}\big((v-z)g(S)^2\big). \tag{8.12}$$

This follows by direct computation using (8.9) and (8.10). It is easy to derive from this an expression for the analytic defect of H:

$$H(v_1) + iH(v_2) + i^2H(v_3) + i^3H(v_4) = \frac{i-1}{2}g(S)^2 \tag{8.13}$$

for every lattice square $(v_1v_2v_3v_4)$ in D.

LEMMA 8.6. *The function H is constant along both boundary arcs of D^{red}.*

PROOF. Consider a boundary triangle uvw, with vertex v at the right angle. Let, say, v be white. Then by the argument in the proof of Lemma 8.4, we have $f(wv) = \theta f(uv)$ or $f(wv) = \overline{\theta} f(uv)$. In both cases,

$$H(w) - H(u) = H(w) - H(v) + H(v) - H(u) = |f(vw)|^2 - |f(vu)|^2 = 0. \qquad \square$$

Let us add that the values of H on the upper boundary arc of D^{red} are one larger than its values on the lower arc of this domain, since $f(uv) = 1$ for the edge uv at the left door. Since H is determined only up to an additive constant, we may assume that

$$H = \begin{cases} 1 & \text{on the upper arc of } D^{\text{red}}, \\ 0 & \text{on the lower arc of } D^{\text{red}}. \end{cases} \tag{8.14}$$

It follows from the definition of H that if uv is an edge of G (oriented from u to v), then $H(u) \le H(v)$.

This function H is not harmonic in general, but it has the following useful weaker property:

LEMMA 8.7. *Let v be a white internal lattice point, and let $v_1, \dots, v_4$ be the lattice points diagonally opposite to v in the squares containing it. Then*

$$H(v) \le \frac{1}{4}\big(H(v_1) + H(v_2) + H(v_3) + H(v_4)\big),$$

i.e., H is subharmonic on the white grid graph in the interior of D. Analogously, H is superharmonic on the black grid graph in the interior of D.

PROOF. Let $u_1, \dots, u_4$ be (black) neighbors of v, and set $f_k = f(u_kv)$ (Figure 8.11). It suffices to prove the identity

$$H(v_1) + H(v_2) + H(v_3) + H(v_4) - 4H(v) = 2|f_1 - if_2 - f_3 + if_4|^2. \tag{8.15}$$

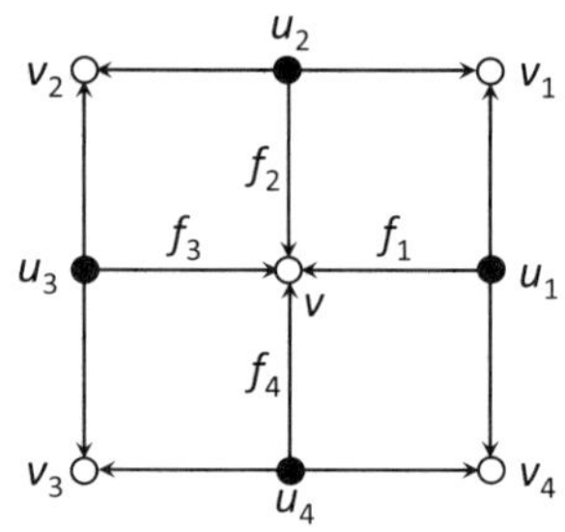

FIGURE 8.11. H is subharmonic.

The proof involves some computation. By (8.10), $f(u_k v_k) = -if_{k+1} + (1+i)f_k$, and hence

$$\begin{aligned} H(v_k) - H(v) &= H(v_k) - H(u_k) + H(u_k) - H(v) = |f(u_k v)|^2 - |f(u_k v_k)|^2 \\ &= |f_{k+1}|^2 + |f_k|^2 - (1+i)\overline{f}_k f_{k+1} - (1-i) f_k \overline{f}_{k+1}. \end{aligned}$$

Note that $f_{k+1} \parallel (1-i)f_k$, and this implies that $(1-i)\overline{f}_k f_{k+1} + (1+i) f_k \overline{f}_{k+1} = 0$, and so we can write

$$H(v_k) - H(v) = |f_{k+1}|^2 + |f_k|^2 - 2i(\overline{f}_k f_{k+1} - f_k \overline{f}_{k+1}). \tag{8.16}$$

On the other hand, expanding $2|f_1 - if_2 - f_3 + if_4|^2$ we get terms $2|f_k|^2$, terms $2i(\overline{f}_k f_{k+1} - f_k \overline{f_{k+1}})$, and terms containing $\overline{f}_k f_{k+2} + f_k \overline{f}_{k+2}$, which is zero since $f_{k+2} \parallel if_k$. So we get the same expression as obtained by adding up the right hand sides of (8.16). This proves the identity (8.15). □

A useful consequence of this lemma is that for every node u,

$$0 \le H(u) \le 1. \tag{8.17}$$

Indeed, to see e.g. the case when u is black, we notice that it holds true if u is a node on the lower arc of D^{red} (when $H(u) = 0$), and also if u is a neighbor of a white node v of the upper arc of D^{red} (because then $H(u) \le H(v) = 1$ and $H(v) - H(u) = |f(uv)|^2 \le 1$). So $H(u) \ge 0$ on all black nodes by its superharmonic property. Similarly, $H(v) \le 1$ for all white nodes. Finally, if u is a black node, v is a white neighbor of u, then $0 \le H(u) \le H(v) \le 1$.

Consider the restriction of H to the black nodes on the boundary of D^{inc}, and extend it to a harmonic function h_B on the black grid inside D^{inc}. We define h_W analogously. It follows from Lemma 8.7 that $h_B(u) \le H(u)$ for every black node u, and $h_W(v) \ge H(v)$ for every white node v.

Let us add that for a black node u, and any (white) neighbor v of u, $H(u) \le H(v)$ by definition, so (informally) we may have in mind the inequalities

$$h_B \le H|_B \text{“} \le \text{”} H|_W \le h_W \tag{8.18}$$

(we have to put quotes around the middle inequality sign, since $H|_B$ and $H|_W$ are defined on different sets). The functions h_B and h_W are harmonic extensions of almost the same values on almost the same sets. So if we can show that $h_B \approx h_W$ on the boundary, then we will know that $H \approx h_B$, so H is approximately harmonic.

8.2.4. Sketch of the proof of Theorem 8.5. Let H_n denote the function defined by (8.11) and (8.14) for the graph G_n, and define $h_{B,n}$ and $h_{W,n}$ analogously.

First, it suffices to prove that there is a subsequence of the sequence $(G_n : n = 1, 2, \dots)$ for which the convergence holds as claimed in the theorem. This follows by a standard "subsequential limit" argument.

Second, there is a subsequence for which $g_n/\sqrt{\varepsilon_n}$ converges, uniformly in the interior of Ω, to some function $g : \Omega \to \mathbb{Z}$, and also H_n, $h_{B,n}$ and $h_{W,n}$ converge to appropriate functions $\widehat{H}, \widehat{h}_B, \widehat{h}_W : \Omega \to \mathbb{R}$. (This step takes some careful estimates of harmonic and sub/superhamonic functions on the grid, for which we refer to the papers of Chelkak and Smirnov.) From the fact that the functions g_n are discrete analytic, it follows that g is analytic (in the classical sense). Using (8.14) and (8.18) (in an exact quantitative form), one can prove that $\widehat{H} = \widehat{h}_B = \widehat{h}_W$, and it is a harmonic function (in the analysis sense). Furthermore, the boundary values of $\widehat{H}$ are 0 on the lower a-b arc of $\partial\Omega$ and 1 on the upper a-b arc.

Third, notice that the function $\Im(\Phi)$ (where Φ is defined in Section 8.2.2) has the same properties: it is harmonic, 0 on the lower arc, and 1 on the upper arc. It follows that $\widehat{H} = \Im(\Phi)$ and so

$$H_n \to \Im(\Phi)$$

in the interior of Ω.

Fourth, (8.12) implies that for any node u and any path $P = (u_0, \dots, u_k = u)$ in G_B connecting u to a (black) node u_0 on the lower arc, we have

(8.19)

$$H_n(u) = \frac{1}{2}\sum_{j=1}^{k}\Re\Big(\frac{u_j - u_{j-1}}{\sqrt{2}\varepsilon_n} g_n(S_j)^2\Big) = \frac{1}{2\sqrt{2}}\Re\Big(\sum_{j=1}^{k}(u_j - u_{j-1})\Big(\frac{g_n(S_j)}{\sqrt{\varepsilon_n}}\Big)^2\Big),$$

where S_j is the lattice square with diagonal $u_{j-1}u_j$. The sum on the right is an approximation of the integral of g_n^2/ε_n along P (taking into account that the lattice $\varepsilon_n\mathcal{G}_0$ has edge length $\sqrt{2}\varepsilon_n$). it follows that

$$H_n(u) \to \frac{1}{2}\Re\Big(\int_v^u g(z)^2\,dz\Big),$$

where v is any point on the lower arc. So it follows that

$$\Re\Big(\frac{1}{2}\int_v^u g(z)^2\,dz - i\Phi(u)\Big) = 0.$$

Since the function in the large parenthesis is analytic, we can conclude that not only its real part is zero, but so is the function itself:

$$\frac{1}{2}\int_v^u g(z)^2\,dz = i\Phi(u).$$

Differentiating and taking square roots, we get the theorem.

8.2.5. Sketch of the proof of Theorem 8.2. Let us add a curve ξ on the outside connecting doors a and b (Figure 8.12). For every tiling, there will be a unique curve from door c to door d, which may or may not go through ξ. In fact, it goes through ξ if and only if the door-to-door curves of the tiling (without ξ) match a with c and b with d. The arguments above can be extended to the porch with the additional curve to show that the probability that the door-to-door curve goes through ξ is conform invariant.

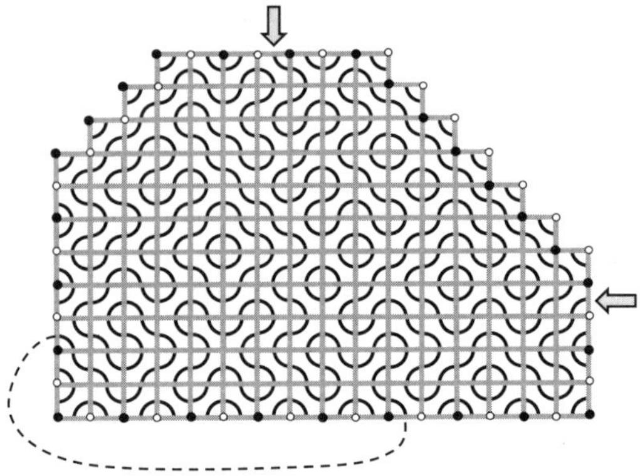

FIGURE 8.12

EXERCISE 8.1. A lattice domain with the same number of black and white squares can be domino tiled unless there are two boundary vertices x, y whose distance in the reduced dual graph is less than their height difference.

EXERCISE 8.2. Let v and z be diametrically opposite vertices of a lattice square S. Prove that $H(v) - H(z) = \frac{1}{2}\Re\big((v-z)g(S)^2\big)$.

EXERCISE 8.3. Let $e_1, \dots, e_4$ be the edges incident with a lattice point in the interior of D, and let $S_1, \dots, S_4$ be the squares incident with it. Prove that

$$|g(S_1)|^2 + \dots + |g(S_4)|^2 \le 2(|f(e_1)|^2 + \dots + |f(e_4)|^2).$$

CHAPTER 9

Adjacency Matrix and its Square

In the previous chapters, the vector-labelings we considered were mostly in low dimensions, and planar graphs played the major role. Now we turn to representations in higher dimensions, where planarity plays no role, or at most a little role.

Perhaps the easiest nontrivial way of assigning a vector to each node is to use the corresponding column of the adjacency matrix. Even this easy construction has some applications, and we will discuss one of them.

However, somewhat surprisingly, it turns out that a more interesting vector-labeling can be obtained by using the columns of the *square* of the adjacency matrix. This construction is related to the Regularity Lemma of Szemerédi, one of the most important tools of graph theory, and it leads to reasonably efficient algorithms to construct (weak) regularity partitions.

9.1. Neighborhoods, rank and size

If we know the rank of the adjacency matrix of a graph, how large can the graph be? Obviously we have to make some assumption, since we can replace any node by an arbitrary number of twins without increasing the rank of the adjacency matrix. However, if we exclude twins, then the question becomes meaningful.

An almost trivial answer is given by the following argument. Let $r = \mathrm{rk}(A_G)$ be the rank of the adjacency matrix of the simple graph G (we will call this number simply the *the rank of* G), and let (say) the first r columns of the adjacency matrix form a basis of the column space. Two rows that agree in their first r positions will agree everywhere, which is impossible if there are no twins. So the initial r-tuples of rows will be all different, and hence there are at most 2^r rows.

This argument works over any field, and no better bound can be given over $GF(2)$, for example (see Exercise 9.2). However, over the real field we can use geometry, not just linear algebra, and prove a substantially better (almost optimal) bound [**Kotlov–Lovász 1996**].

THEOREM 9.1. *Let G be a twin-free graph on n nodes of rank r. Then $n = O(2^{r/2})$.*

Before turning to the proof, we recall a result from discrete geometry. The *kissing number* $s(d)$ is the largest integer N such that there are N nonoverlapping unit balls touching a given unit ball in $\mathbb{R}^d$. For $d = 2$, it is easy to see that this number is 6, but the 3-dimensional case is already difficult: whether the 3-dimensional kissing number is 12 or 13 was the subject of a famous debate between Isaac Newton and David Gregory in the 17-th century (the fact that $s(3) = 12$ was not proved in a watertight way until [**Schütte–v.d.Waerden 1953**]).

The exact value of $s(d)$ is only known for a few other small dimensions, but the following bound, which follows from tighter estimates [**Kabatjanskiĭ–Levenštein 1978**], will be enough for us:

$$s(d) = O(2^{d/2}). \tag{9.1}$$

In other words, every set of more than $s(d)$ vectors in $\mathbb{R}^d$ contains two vectors such that angle between them is less than $\pi/3$. This theorem has an almost immediate consequence for graphs with low rank adjacency matrices:

LEMMA 9.2. *If a graph G of rank r has $n > s(r+1)$ nodes, than it has two nodes i and j such that $|N(i)\triangle N(j)| < n/4$.*

PROOF. Let $\mathbf{a}_i$ denote the column of A_G corresponding to $i \in V$, and let $\mathbf{u}_i = \mathbb{1} - 2\mathbf{a}_i$. Clearly the vectors $\mathbf{u}_i$ are ± 1-vectors, which belong to an (at most) $(r+1)$-dimensional subspace. Applying the kissing number bound to the vectors $\mathbf{u}_i$, we get that there are two vectors $\mathbf{u}_i$ and $\mathbf{u}_j$ forming an angle less than $\pi/3$. For two ± 1 vectors, this means that $\mathbf{u}_i$ and $\mathbf{u}_j$ differ in fewer than $n/4$ positions. The vectors $\mathbf{a}_i$ and $\mathbf{a}_j$ also differ in these positions only. □

We also need a couple of simple facts about twins. Recall that they cannot be adjacent. If we delete a node from a graph G that has a twin, then the rank of the graph is not changed. Furthermore, no new twins are created: if a pair i, j of remaining nodes is distinguished by some node u of G having a twin v, and if u is deleted, then v still distinguishes i and j.

LEMMA 9.3. *Let G be a graph rank r.* (a) *If i and j (nonadjacent) nodes of G that are twins in an induced subgraph H, but not in G, then* $\mathrm{rk}(H) \le r-2$. (b) *If i and j are adjacent nodes in G and no node of a $X \subseteq V\setminus\{i,j\}$ distinguishes i and j, then* $\mathrm{rk}(G[X]) \le r-1$.

PROOF. (a) The matrix A_G must contain a column that distinguishes rows i and j, and so it is not in the linear span of the columns in X. Adding this column to A_H increases its rank. Adding the corresponding row increases the rank further. (b) The columns corresponding to X have the same entry in rows i and j, but column i has different entries in these positions, and so these columns cannot generate column i. So the rank of these columns is at most $r-1$, which implies that $\mathrm{rk}(G[X]) \le r-1$. □

Proof of Theorem 9.1. To facilitate the proof by induction on r, we set $f(r) = C2^{r/2} - 16$, where the constant $C > 16$ is chosen so that $s(r+1) \le f(r)$ for all r. The statement we prove is that $n \le f(r)$ for every twin-free graph on n nodes and of rank r. For $r = 1, 2, 3$ this is easily checked. So we may suppose that $r > 3$.

If $|N(i)\triangle N(j)| \ge n/4$ for all pairs of nodes, then Lemma 9.2 implies that $n \le s(r+1) < f(r)$. So we may assume that there are two nodes i and j such that $|N(i)\triangle N(j)| < n/4$. The set $X = V\setminus\big(N(i)\triangle N(j)\big)$ does not distinguish i and j, and hence by Lemma 9.3, $\mathrm{rk}\,\big(G[X]\big) \le r-1$.

The rest of the proof is a somewhat lengthy argument combining graphs and matrices; geometric representation has done its job. We want to find an induced subgraph that is twin-free and lower rank, so that we can apply induction.

Let $G[Z]$ be a largest induced subgraph of G with $\mathrm{rk}(G[Z]) \le r-1$. Since X induces such a subgraph, we have

$$|Z| \ge |X| > \frac{3}{4}n. \tag{9.2}$$

If $G[Z]$ is twin-free, then by the induction hypothesis applied to $G[Z]$,

$$n < \frac{4}{3}|Z| \le \frac{4}{3}f(r-1) < f(r).$$

So we may assume that $G[Z]$ has twins. By Lemma 9.3, we have the stronger bound $\mathrm{rk}(G[Z]) \le r-2$.

Let $u \in V \setminus Z$. By the maximality of Z, we must have $\mathrm{rk}(G[Z \cup \{u\}]) = r$, and hence $G[Z \cup \{u\}]$ must be twin-free. This implies that u must distinguish every twin pair in Z; in other words, u is connected to exactly one member of every such pair. This in particular implies that $G[Z]$ does not have three mutually twin nodes. Let T be the set of nodes that are twins in $G[Z]$ and are adjacent to u, and let T' be the set of twins of nodes of T. Let $U = Z \setminus (T \cup T')$.

Next consider any $v \in V \setminus Z$ different form u. We claim that it is either connected to all of nodes in T and none in T', or the other way around. Row v of A_G is a linear combination of rows corresponding to Z and u; let, say, the coefficient of row u in this linear combination be positive. Then for every twin pair $\{i,j\}$ (where $i \in T$ and $j \in T'$) we have $a_{ui} > a_{uj}$, but (by the definition of twins) $a_{wi} = a_{wj}$ for all $w \in Z$, and hence $a_{vi} > a_{vj}$. This means that v is adjacent to i, but not to j, and so we get that v is adjacent to all nodes of T but not to any node of T'.

Thus we have a decomposition $V \setminus Z = Y \cup Y'$, where every node of Y is connected to all nodes of T but to no node of T', and for nodes in Y', the other way around. So G has the structure in Figure 9.1.

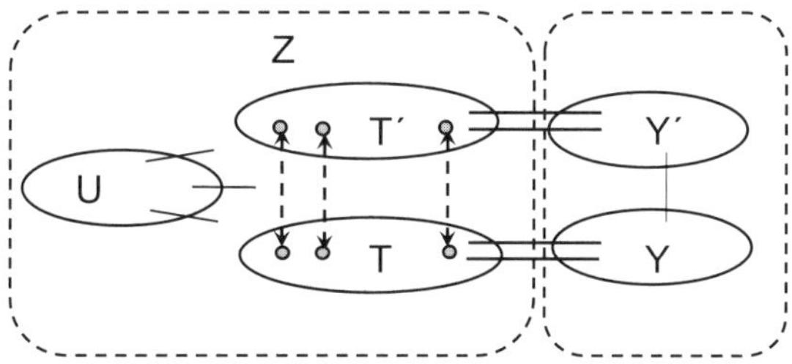

FIGURE 9.1. Node sets in the proof of Theorem 9.1. Dashed arrows indicate twins in $G[Z]$.

The graph $G[T \cup U]$ is obtained from $G[Z]$ by deleting one member of each twin-pair, hence it is twin-free. Since $\mathrm{rk}(G[T \cup U]) = \mathrm{rk}(G[Z]) \le r-2$, we can apply the induction hypothesis to get

$$|T| + |U| \le f(r-2). \tag{9.3}$$

If $|Y \cup Y'| \le 16$, then

$$n = 2|T| + |U| + |Y \cup Y'| \le 2f(r-2) + 16 = f(r),$$

and we are done. So we may assume that (say) $|Y'| > 8$. Since the rows in Y' are all different, this implies that they form a submatrix of rank at least 4. Using that in their columns corresponding to T these rows have zeros, we see that the matrix formed by the rows in $Y'\cup T$ has rank at least $\mathrm{rk}(G[T])+4$. This implies that $\mathrm{rk}(G[T])\le r-4$.

If $G[T]$ is twin-free, then we can apply induction:

$$|T|\le f(r-4). \tag{9.4}$$

Using (9.2), (9.3) and (9.4), we get that

$$n<\frac{4}{3}|Z|=\frac{4}{3}(2|T|+|U|)\le\frac{4}{3}\big(f(r-2)+f(r-4)\big)<f(r).$$

So we may assume that $G[T]$ is not twin-free. Any twins in $G[T]$ remain twins in $G\setminus U$: no node in T' can distinguish them (because its twin does not), and no node in $Y\cup Y'$ can distinguish them, as we have seen. So $G\setminus U$ is not twin-free, and hence $\mathrm{rk}(G\setminus U)\le r-2$ by Lemma 9.3. By the maximality of Z, we have $|V\setminus U|\le|Z|$. Using (9.3), this gives

$$n\le|Z|+|U|=2|T|+2|U|\le 2f(r-2)<f(r).$$

□

The bound in Theorem 9.1 is sharp up to the constant. (This shows that even though the bound on the kissing number could be improved, this would not give an improvement of the whole argument.) Let us define a sequence of graphs G_r recursively: Let $G_1=K_2$, and given G_r ($r\ge 2$), let G_{r+1} be obtained from G_r by doubling each node (creating $|V(G_r)|$ pairs of twins), adding a new node, and connecting it to one member of each twin pair. It is easy to see by induction that G_r is twin-free, $|V(G_r)|=3\cdot 2^{r-1}-1$, and $\mathrm{rk}(A_{G_r})=2r$.

As a corollary, we get that if the rank of the adjacency matrix of a simple graph (which may have twins) is r, then its chromatic number is at most $f(r)$. This argument can be refined [**Kotlov 2000**] to prove a better upper bound of $\chi(G)=O\big(r(4/3)^r\big)$. It was conjectured [**van Nuffelen 1976**] that $\chi(G)\le r$ for every graph G, but this was disproved [**Alon–Seymour 1989**]; the worst example known so far [**Nisan–Wigderson 1984**], gives a sequence of graphs with $\chi(G)=\Omega(2^{(\log r)^{1.09\dots}})$.

9.2. Regularity partitions

9.2.1. Regularity Lemmas. In order to formulate this application of geometric representations, we have to include a little introduction to regularity lemmas. The original Regularity Lemma [**Szemerédi 1975**],[**Szemerédi 1978**] has been the key to many proofs in extremal graph theory, graph algorithms, number theory, the theory of graph limits, and more; the range of its applications is so broad that we do not attempt to describe any of them in this book. It has formulations not only in graph theory but in analysis and information theory as well, and its applications often depend on different versions of the Lemma, which are not equivalent. Stronger versions have been proved [**Alon at al. 2000**], [**Tao 2006**], [**Lovász–Szegedy 2007**]. Another version is weaker than the original but gives much better error bounds [**Frieze–Kannan 1993**], and this is the version we will be dealing with in this section.

Let G be a graph on n nodes. For two sets $S,T\subseteq V$, let $e_G(S,T)$ denote the number of edges $ij\in E$ with $i\in S$ and $j\in T$. We do not assume here that S and T are disjoint; edges induced by $S\cap T$ should be counted twice. We also use this notation in the case when G is edge-weighted, when the weights of edges connecting

S and T should be added up. (An unweighted edge can be considered as an edge with weight 1.)

We come to an important definition of this section, introduced by [**Frieze–Kannan 1993**]. We define, for two edge-weighted graphs G and H on the same set V of n nodes, their *cut-distance* by

$$d_\square(G,H) = \frac{1}{n^2} \max_{S,T\subseteq V} |e_G(S,T) - e_H(S,T)|.$$

This is, of course, not the only way to define a meaningful distance between two graphs; for example, the *edit distance* $|E(G)\triangle E(H)|$ is often used. For us the cut-distance, which measures a certain global similarity, will be more important. Note that the normalization guarantees that $0 \le d_\square(G,H) < 1$ for any two simple graphs G and H. It is straightforward to check that $d_\square$ satisfies the triangle inequality.

REMARK 9.4. One may wonder why we divide by n^2, and why not by $|S|\,|T|$, which would also guarantee that the distance of two simple graphs is at most one. It is not hard to see that in this case the maximum would be attained when $|S| = |T| = 1$, and the distance would be 0 if the two graphs are identical, and 1 otherwise.

REMARK 9.5. The notion of the cut-distance can be modified to apply to graphs that are not defined on the same node set; they may even have different number of nodes. We don't need these extensions in this book; see [**Lovász 2012**] for a detailed discussion and applications.

Let $\mathcal{P} = \{V_1, \dots, V_k\}$ be a partition of V into nonempty sets. We define the edge-weighted graph $G_\mathcal{P}$ on V by taking the complete graph and weighting its edge uv by $e_G(V_i,V_j)/(|V_i|\,|V_j|)$ if $u \in V_i$ and $v \in V_j$. The case $i = j$ takes some care: we then count edges twice, so $e_G(V_i,V_i) = 2|E(G[V_i])|$, and we include the case $u = v$, so $G_\mathcal{P}$ will have loops. Informally, $G_\mathcal{P}$ is obtained by averaging the adjacency matrix over sets $V_i \times V_j$.

The Regularity Lemma says, roughly speaking, that the node set of every graph has a partition $\mathcal{P}$ into a "small" number of classes such that $G_\mathcal{P}$ is "close" to G. Figure 9.2 shows a schematic explanation of the Regularity Lemma.

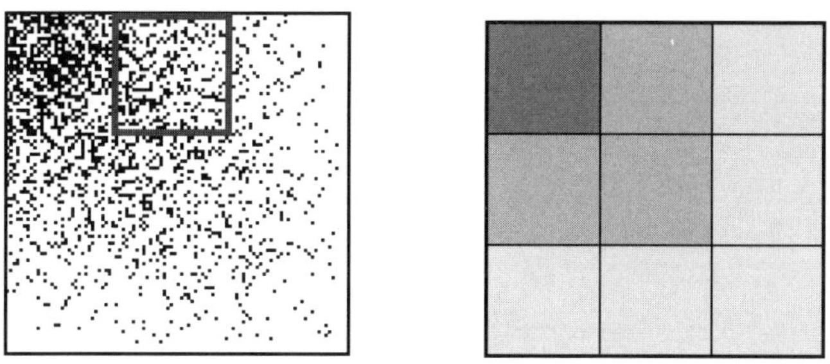

FIGURE 9.2. Left: adjacency matrix of a graph G, where 1's are replaced by black squares, and 0's, by white squares. Right: similar representation of $G_\mathcal{P}$ for a 3-partition $\mathcal{P}$. The partition is a regularity partition if the submatrix in (almost) every rectangle $V_i \times V_j$ is random-like.

LEMMA 9.6 (Weak Regularity Lemma). *For every simple graph G and every $k \ge 1$, the node set V has a partition $\mathcal{P}$ into k classes such that*

$$d_\square(G, G_\mathcal{P}) \le \frac{4}{\sqrt{\log k}}.$$

The main point here (and in subsequent related results) is that the error bound on the right tends to 0 as $k \to \infty$, independently of the graph.

A proof of this lemma will follow from its matrix form (Lemma 9.7). We do not require here that $\mathcal{P}$ be an equitable partition. It is not hard to see, however, that this version implies that there is also a k-partition in which all partition classes are almost equal (in the sense that each class has $\lfloor n/k \rfloor$ or $\lceil n/k \rceil$ elements), and which satisfies the same inequality as in the lemma, just we have to double the error bound.

9.2.2. Cut norm. It will be useful to reformulate the Regularity Lemma in terms of matrices, and for this, we need to define norms on the space of $n \times n$ matrices. For $S, T \subseteq [n]$, we define the special matrix $\mathbb{1}_{S\times T}$ by

$$(\mathbb{1}_{S\times T})_{i,j} = \begin{cases} 1, & \text{if } i \in S \text{ and } j \in T, \\ 0, & \text{otherwise.} \end{cases}$$

This is a 0-1 matrix of rank 1. For a matrix $M \in \mathbb{R}^{n\times n}$ and $S, T \subseteq [n]$, we define

$$M(S,T) = \sum_{i\in S,\, j\in T} M_{i,j} = M \cdot \mathbb{1}_{S\times T}.$$

Now we come to defining various norms on matrices. We need the standard Frobenius norm:

$$\|M\|_2 = \Big(\sum_{i,j=1}^n M_{i,j}^2\Big)^{1/2}.$$

The *cut-norm* is less standard, and is used mostly in combinatorics. It is defined as follows:

$$\|M\|_\square = \max_{S,T\subseteq[n]} \big|M \cdot \mathbb{1}_{S\times T}\big| = \max_{S,T\subseteq[n]} |M(S,T)|. \tag{9.5}$$

It is easy to verify that this is indeed a norm (most importantly, it satisfies the triangle inequality). Applying this notation to the adjacency matrices of two graphs G and H on the same set of n nodes, we get

$$d_\square(G,H) = \frac{1}{n^2}\|A_G - A_H\|_\square.$$

There are many alternative ways to define the cut-norm, or norms that are closely related. (The proofs are left to the reader as exercises.) We could define it by

$$\|M\|_\square = \max\big\{|x^\mathsf{T} M y| : \ x, y \in [0,1]^n\big\}. \tag{9.6}$$

If we maximize instead over all vectors in $[-1,1]^n$, we obtain a norm that may be different, but only by a factor of 4:

$$\|M\|_\square \le \max\{|x^\mathsf{T} M y| : \ x, y \in [-1,1]^n\} \le 4\|M\|_\square. \tag{9.7}$$

It will be useful to state the upper bound here in a form valid for all vectors $x, y \in \mathbb{R}^n$:

$$|x^\mathsf{T} M y| \le 4\|M\|_\square\, |x|_\infty\, |y|_\infty. \tag{9.8}$$

We can also play with the sets S and T in the definition, without changing the value too much. For example,

$$\frac{1}{4}\|M\|_\square \le \max_{S\subseteq[n]} M(S,S) \le \|M\|_\square, \tag{9.9}$$

$$\frac{1}{4}\|M\|_\square \le \max_{|S|,|T|\ge n/2} M(S,T) \le \|M\|_\square, \tag{9.10}$$

$$\frac{1}{4}\|M\|_\square \le \max_{T=[n]\setminus S} M(S,T) \le \|M\|_\square. \tag{9.11}$$

(The last formula is the origin of the name "cut norm", since the edges counted form a cut in the graph.)

9.2.3. Averaging. For a partition $\mathcal{P} = \{V_1, \dots, V_k\}$ of $[n]$, vector $x \in \mathbb{R}^n$, and matrix $M \in \mathbb{R}^{n\times n}$, we define the vector $x_\mathcal{P}$ and matrix $M_\mathcal{P} \in \mathbb{R}^{n\times n}$ by

$$(x_\mathcal{P})_u = \frac{x(V_i)}{|V_i|} \qquad \text{and} \qquad (M_\mathcal{P})_{u,v} = \frac{M(V_i,V_j)}{|V_i|\,|V_j|} \qquad (u\in V_i,\ v\in V_j).$$

In words, the matrix $M_\mathcal{P}$ is obtained by averaging the entries of M over every block $V_i \times V_j$ (and analogously for $x_\mathcal{P}$). If A is the adjacency matrix of a simple graph G, then $A_\mathcal{P}$ is the weighted adjacency matrix of $G_\mathcal{P}$.

Let us say that $B \in \mathbb{R}^{n\times n}$ is a $\mathcal{P}$-matrix, if B is constant on every block $V_i \times V_j$, i.e., $B_\mathcal{P} = B$. Then the operator $M \mapsto M_\mathcal{P}$ is the orthogonal projection of $\mathbb{R}^{n\times n}$ to the linear space of $\mathcal{P}$-matrices. This operator is symmetric and idempotent:

$$M_\mathcal{P}\cdot N = M\cdot N_\mathcal{P} = M_\mathcal{P}\cdot N_\mathcal{P}, \qquad (M_\mathcal{P})_\mathcal{P} = M_\mathcal{P}.$$

With respect to many matrix norms, it is also contractive:

$$\|M_\mathcal{P}\|_2 \le \|M\|_2, \quad \text{and} \quad \|M_\mathcal{P}\|_\square \le \|M\|_\square. \tag{9.12}$$

(The latter inequality takes some work to prove, see Exercise 9.5.)

9.2.4. The Weak Regularity Lemma. Using the cut norm and this averaging operator, we can re-state the Weak Regularity Lemma. In fact, we state it in three versions, to make the proof easier to follow.

LEMMA 9.7. *Let $M \in \mathbb{R}^{n\times n}$.*

(a) *For every $r \ge 1$ there are $2r$ sets $S_1, \dots, S_r, T_1, \dots, T_r \subseteq [n]$ and r real numbers a_i such that*

$$\Big\|M - \sum_{i=1}^r a_i \mathbb{1}_{S_i\times T_i}^\mathsf{T}\Big\|_\square \le \frac{1}{\sqrt{r}}\|M\|_2.$$

(b) *For every $k \ge 2$ there is a partition $\mathcal{P}$ of $[n]$ into k classes for which*

$$\|M - M_\mathcal{P}\|_\square \le \frac{4}{\sqrt{\log k}}\|M\|_2.$$

(c) *For every $k \ge 2$ there is a partition $\mathcal{P}$ of $[n]$ into k classes, and a $\mathcal{P}$-matrix B, for which*

$$\|M - B\|_\square \le \frac{2}{\sqrt{\log k}}\|M\|_2,$$

Lemma 9.7(a) gives a low-rank approximation of M. The important fact about version (a) of the lemma is that the number of terms r and the error bound $1/\sqrt{r}$ are polynomially related. The order of magnitude of the errors cannot be improved (see Exercise 9.10).

PROOF. We start with proving (a). Let $S,T \subseteq [n]$ be two nonempty sets such that $\|M\|_\square = |M(S,T)|$. For every real number a, we have

$$\begin{aligned}\|M - a\mathbb{1}_{S\times T}\|_2^2 &= \|M\|^2 + a^2\|\mathbb{1}_{S\times T}\|_2^2 - 2aM\cdot\mathbb{1}_{S\times T}\\ &= \|M\|^2 + a^2|S|\,|T| - 2aM(S,T),\end{aligned}$$

which is minimized when $a = M(S,T)/(|S|\,|T|)$, and for this choice

$$\|M - a\mathbb{1}_{S\times T}\|_2^2 = \|M\|^2 - \frac{1}{|S|\,|T|}M(S,T)^2 \le \|M\|_2^2 - \|M\|_\square^2. \tag{9.13}$$

Repeated application of this inequality gives a sequence of matrices

$$M_r = M - \sum_{i=1}^{r} a_i\mathbb{1}_{S_i\times T_i},$$

where

$$\|M_r\|_2^2 \le \|M\|_2^2 - \sum_{j=0}^{r-1}\|M_j\|_\square^2.$$

Since the left side is nonnegative, there must be an index $j \le r-1$ for which

$$\|M_j\|_\square^2 = \Big|M - \sum_{i=1}^{j} a_i\mathbb{1}_{S_i\times T_i}\Big|_\square^2 \le \frac{1}{r}\|M\|_2^2.$$

Replacing the remaining coefficients $a_{j+1},\dots,a_{r-1}$ by 0, we get the decomposition in (a).

Next we prove (c). Consider the decomposition as in (a) with $r = \lfloor(\log k)/2\rfloor$, and let $\mathcal{P}$ be the partition into the atoms of the Boolean algebra generated by all the sets S_i and T_i. Then $|\mathcal{P}| \le 2^{2r} \le k$, and the $\mathcal{P}$-matrix $B = \sum_{i=1}^{r} a_i\mathbb{1}_{S_i\times T_i}$ satisfies

$$\|M - B\|_\square \le \frac{1}{\sqrt{r}}\|M\|_2 \le \frac{2}{\sqrt{\log k}}\|M\|_2.$$

(This last step holds for $k \ge 4$; for $k = 2,3$ we can just take $B = 0$.)

To prove (b), we use (9.12) and the fact that $B_{\mathcal{P}} = B$:

$$\begin{aligned}\|M - M_{\mathcal{P}}\|_\square &\le \|M - B\|_\square + \|B - M_{\mathcal{P}}\|_\square = \|M - B\|_\square + \|B_{\mathcal{P}} - M_{\mathcal{P}}\|_\square\\ &= \|M - B\|_\square + \|(B - M)_{\mathcal{P}}\|_\square \le 2\|M - B\|_\square.\end{aligned}$$ □

REMARK 9.8. In (a), one could require in addition that $\sum_i a_i^2 \le 4$, at the cost of increasing the error bound to $(4/\sqrt{r})\|M\|_2$. The trick is to use the maximizing sets in (9.10).

REMARK 9.9. The original Regularity Lemma measures the error of the approximation $A \approx A_{\mathcal{P}}$ differently. Again there are several versions, closest to ours is the following: For a matrix $M \in \mathbb{R}^{n\times n}$ and partition $\mathcal{P} = \{V_1,\dots,V_k\}$ of $[n]$, and for every $1 \le i, j \le k$, we choose the sets $S_{ij} \subseteq V_i$ and $T_{ij} \subseteq V_j$ maximizing the "local error" $|(M - M_{\mathcal{P}})(S_{ij}, T_{ij})|$. The "total error" is the sum of these, normalized as before:

$$e(\mathcal{P}) = \frac{1}{n^2} \sum_{i,j=1}^{k} |(M - M_{\mathcal{P}})(S_{ij}, T_{ij})|.$$

Appropriate modification of the proof above gives the following: *There is a partition $\mathcal{P}$ of $[n]$ into k classes for which*

$$e(\mathcal{P}) = O\Big(\frac{1}{\sqrt{\log^* k}}\Big).$$

(Here $\log^* k$ is a very slowly decreasing function, defined as the number of times we have to apply the log function to k to get a negative number.) This bound is also known to be best possible up to a constant factor [**Conlon–Fox 2012**], [**Fox–L.M.Lovász 2014**].

9.3. A-squared representation

9.3.1. Voronoi cells and regularity partitions. Now we are ready to tie regularity partitions to geometric representations. We define the *A-squared representation* of a graph G as the map $i \mapsto \mathbf{u}_i$, where $\mathbf{u}_i = A^2\mathbf{e}_i$ is the column of A^2 corresponding to node i (where $A = A_G$ is the adjacency matrix of G). Squaring the matrix seems unnatural, but it is crucial. We define the *2-neighborhood distance* of two nodes by

$$\mathsf{d}_2(s,t) = \frac{1}{n^{3/2}} |\mathbf{u}_s - \mathbf{u}_t|.$$

This normalization guarantees that the distance of any two nodes is at most 1. We need further notation: For a nonempty set $S \subseteq V$, we consider the average distance from S:

$$\overline{\mathsf{d}_2}(S) = \frac{1}{n} \sum_{i \in V} \mathsf{d}_2(i, S) = \frac{1}{n} \sum_{i \in V} \min_{j \in S} \mathsf{d}_2(i,j).$$

EXAMPLE 9.10. To illustrate the substantial difference between the vector representations by the columns of the adjacency matrix and by the columns of its square, let us consider a random graph with a very simple structure: Let $V = V_1 \cup V_2$, where $|V_1| = |V_2| = n/2$, and let any node in V_1 be connected to any node in V_2 with probability $1/2$. With high probability, the distance between any two columns of the adjacency matrix is of the order $\sqrt{n}$ (approximately $\sqrt{n}/\sqrt{2}$ for two nodes in different classes, and $\sqrt{n}/2$ for two nodes in the same class). But if we square the matrix, the distance of two columns in different classes will be approximately $n^{3/2}/8$, while for two columns in the same class it will be $O(n^{5/4})$. With the normalization above, the two classes will be collapsed to single points (asymptotically), but the distance of these two points will remain constant. So the 2-neighborhood distance reflects the structure of the graph very nicely! ♦

EXAMPLE 9.11. Let G be obtained by selecting n random points on the d-dimensional unit sphere S^d, and connecting two of these points x and y if $\measuredangle(x,y) \le 90°$. Then with high probability $\mathsf{d}_2(x,y) \sim \sqrt{d}\,\measuredangle(x,y)$. So, even if we don't use any information except the graph structure, the 2-neighborhood distance of nodes is, up to a little perturbation tending to 0 as $n \to \infty$, is proportional to the geometric distance of the original points on the sphere. In other words, the A-squared representation is the very high dimension n is almost isometric with the d-dimensional representation we started with.

One may object that this is due to the very special construction of the graph. However, we can modify it an connect any two points depending on their distance, just making sure that we will have an edge density bounded away from both 0 and 1.

Another objection can be that the same conclusion would hold (with an even simpler proof) for the 1-neighborhood distance. We can further generalize the construction, by connecting two nodes randomly, where the probability of connecting them is an arbitrary continuous function of their distance. (Decisions about connecting pairs of nodes must be made independently, and as before, we have to choose this function so that we get an edge density bounded away from 0 and 1.) In this case the 1-neighborhood distance between two nodes will be of constant order even if the original points are close on the sphere.

This graph can be used to show that the bound in the Weak Regularity Lemma 9.6 is sharp, up to a constant factor. This will follow from Theorem 9.12(b) below, noticing that for any subset S of $k = 2^d$ points on S^d, at least half of the surface will be at a distance larger that $1/\sqrt{d}$ from S. ♦

The following connection between Voronoi diagrams and regularity partitions was proved in [**Lovász–Szegedy 2007**] (in a somewhat weaker form).

THEOREM 9.12. *Let G be a simple graph, and let $\mathsf{d}_2(.,.)$ be its 2-neighborhood distance.*

(a) *The Voronoi cells of a nonempty set $S \subseteq V$ define a partition $\mathcal{P}$ of V such that $d_\square(G, G_\mathcal{P}) \le 2\overline{\mathsf{d}_2}(S)^{1/2}$.*

(b) *For every partition $\mathcal{P} = \{V_1, \dots, V_k\}$ we can select elements $s_i \in V_i$ so that $S = \{s_1, \dots, s_k\}$ satisfies $\overline{\mathsf{d}_2}(S) \le 4 d_\square(G, G_\mathcal{P})^{1/2}$.*

PROOF. In both parts of the proof we work with linear algebra, using the adjacency matrix $A = A_G$. In both parts we consider a particular partition $\mathcal{P} = \{V_1, \dots, V_k\}$. We will be interested in the "error" matrix $R = A - A_\mathcal{P}$, for which $\|R\|_\square = n^2 d_\square(G, G_\mathcal{P})$.

(a) Let $S = \{s_1, \dots, s_k\}$, let $\mathcal{P} = \{V_1, \dots, V_k\}$ be the partition of V defined by the Voronoi cells of S (where $s_i \in V_i$), and let $\phi(v) = s_t$ for $v \in V_t$. To bound $\|R\|_\square$, we use (9.9):

$$\|R\|_\square = \max_{x,y \in \{0,1\}^V} |x^\mathsf{T} R y|.$$

Let x, y be the maximizers on the right, and let $\widehat{x} = x - x_\mathcal{P}$ and $\widehat{y} = y - y_\mathcal{P}$. The crucial equation is

$$x^\mathsf{T} R y = x^\mathsf{T} A y - x^\mathsf{T} A_\mathcal{P} y = x^\mathsf{T} A y - x_\mathcal{P}^\mathsf{T} A y_\mathcal{P} = x^\mathsf{T} A \widehat{y} + \widehat{x}^\mathsf{T} A y_\mathcal{P},$$

which implies that

$$|x^\mathsf{T} R y| \le |x|\,|A\widehat{y}| + |y_\mathcal{P}|\,|A\widehat{x}| \le n^{1/2}\big(|A\widehat{x}| + |A\widehat{y}|\big). \tag{9.14}$$

To connect with the A-squared representation $\mathbf{u}$, we use that

$$A^2 \widehat{y} = \sum_v \widehat{y}_v \mathbf{u}_v = \sum_v \widehat{y}_v (\mathbf{u}_v - \mathbf{u}_{\phi(v)})$$

(since $\sum_{v\in V_t}\widehat{y}_v = 0$ for every partition class V_i). The fact that we have a Voronoi partition means that $|\mathbf{u}_v - \mathbf{u}_{\phi(v)}| = n^{3/2}\mathsf{d}_2(v,S)$ for every node v. Hence

$$|A^2\widehat{y}| \le \sum_v |\widehat{y}_v||\mathbf{u}_v - \mathbf{u}_{\phi(v)}| \le \sum_v |\mathbf{u}_v - \mathbf{u}_{\phi(v)}| = n^{3/2}\sum_v \mathsf{d}_2(v,S) = n^{5/2}\overline{\mathsf{d}_2}(S),$$

and

$$|A\widehat{y}|^2 = \widehat{y}^{\mathsf{T}}A^2\widehat{y} \le |\widehat{y}|\,|A^2\widehat{y}| \le n^{1/2}n^{5/2}\overline{\mathsf{d}_2}(S) = n^3\overline{\mathsf{d}_2}(S).$$

We get the same upper bound for $|A\widehat{x}|$. Combining with (9.14), we get

$$d_\square(G,G_{\mathcal{P}}) = \frac{1}{n^2}|x^{\mathsf{T}}Ry| \le \frac{1}{n^{3/2}}(|A\widehat{x}|+|A\widehat{y}|) \le 2\overline{\mathsf{d}_2}(S)^{1/2}.$$

(b) Let i,j be two nodes in the same partition class of $\mathcal{P}$, then $A_{\mathcal{P}}e_i = A_{\mathcal{P}}e_j$, and hence $A(e_i-e_j) = R(e_i-e_j)$. Thus

$$|\mathbf{u}_i - \mathbf{u}_j| = |A^2e_i - A^2e_j)| = |AR(e_i-e_j)| \le |ARe_i| + |ARe_j|.$$

For every set $V_t \in \mathcal{P}$, choose a point $s_t \in V_t$ for which $|ARe_j|$ is minimized over $j\in V_t$ by $j = s_t$, let $S=\{s_1,\dots,s_k\}$, and let $\phi(i) = s_t$ for $i\in V_t$. Then

$$|\mathbf{u}_i - \mathbf{u}_{\phi(i)}| \le |ARe_i| + |ARe_{\phi(i)}| \le 2|ARe_i|,$$

and using the Cauchy–Schwarz Inequality,

$$\begin{aligned}\text{(9.15)}\qquad \overline{\mathsf{d}_2}(S) &= \frac{1}{n}\sum_{i\in V}\mathsf{d}_2(i,S) \le \frac{1}{n}\sum_{i\in V}\mathsf{d}_2(i,\phi(i)) = \frac{1}{n^{5/2}}\sum_{i\in V}|\mathbf{u}_i - \mathbf{u}_{\phi(i)}|\\ &\le \frac{2}{n^{5/2}}\sum_{i\in V}|ARe_i| \le \frac{2}{n^2}\Big(\sum_{i\in V}|ARe_i|^2\Big)^{1/2}.\end{aligned}$$

Here by (9.8),

$$\begin{aligned}\sum_{i\in V}|ARe_i|^2 &= \operatorname{tr}(RAAR) = \operatorname{tr}(ARRA) = \sum_{i\in V}(e_i^{\mathsf{T}}A)R(RAe_i)\\ &\le 4\|R\|_\square\sum_{i\in V}|Ae_i|_\infty|RAe_i|_\infty \le 4n^2\|R\|_\square,\end{aligned}$$

and so by (9.15),

$$\overline{\mathsf{d}_2}(S) \le \frac{4}{n}\|R\|_\square^{1/2} = 4d_\square(G,G_{\mathcal{P}})^{1/2}.$$ □

Combining Theorem 9.12 with the Weak Regularity Lemma, it follows that every graph has an "average representative set" in the following sense.

COROLLARY 9.13. *For every simple graph G and every $k\ge 1$, there is a set $S\subseteq V$ of k nodes such that $\overline{\mathsf{d}_2}(S) \le 8/(\log k)^{1/4}$.*

REMARK 9.14. In [**Lovász 2012**], a version of Theorem 9.12 was formulated using the ℓ_1-distance between the representing points rather than the ℓ_2 (Euclidean) distance. This gives a sharper result in the sense that we don't have to take the square root in (b); as a consequence, we get $4/(\log k)^{1/2}$ instead of $8/(\log k)^{1/4}$ in Corollary 9.13. On the other hand, the Voronoi diagram is much nicer, and more efficiently computable, for the Euclidean distance (recall Figure 2.10).

REMARK 9.15. Theorem 9.12 suggests to define the *dimension* of a family $\mathcal{G}$ of graphs as the infimum of real numbers $d > 0$ for which the following holds: for every $\varepsilon > 0$ and $G \in \mathcal{G}$ the node set of G can be partitioned into a set of at most $\varepsilon|V(G)|$ nodes and into at most ε^{-d} sets of d_2-diameter at most ε. (This number d can be infinite.) In the cases when the graphs have a natural dimensionality, this dimension tends to give the right value (recall Example 9.11).

REMARK 9.16. It would be tempting to reverse the above treatment: prove Corollary 9.13 by some more direct geometric argument, and then use Theorem 9.12 to prove the (weak) Regularity Lemma. However, a direct proof of Corollary 9.13 has been elusive so far.

REMARK 9.17. Does Corollary 9.13 remain valid if we want to find a set of k nodes such that the maximum distance from the set, rather than the average distance, can be bounded uniformly by a function of k tending to zero as $k \to \infty$? Alon (unpublished) proved that the answer is in the affirmative, if we consider the ℓ_1 distance instead of the Euclidean distance in the A-square representation, and allow a slightly weaker bound of $O\big((\log\log k)/\sqrt{\log k}\big)$ for the distance from the k-set (see [**Lovász 2012**] for a proof). It is not known at this time whether an analogous statement remains true for the Euclidean distance. Again, a geometric proof would be very desirable.

9.3.2. Algorithms for nets and regularity partitions. The results above can be applied to the computation of (weak) regularity partitions in the property testing model for dense graphs. This topic is treated in detail in [**Lovász 2012**], so we only sketch the model and the application.

We need to survey some notions and elementary arguments about packing and covering. Let $\varepsilon > 0$. A set of nodes $S \subseteq V$ is an *ε-cover*, if $\mathsf{d}_2(v, S) \le \varepsilon$ for every point $v \in V$. (Here, as usual, $\mathsf{d}_2(v, S) = \min_{s\in S} \mathsf{d}_2(s, v)$.) We say that S is an *ε-packing*, if $\mathsf{d}_2(u, v) \ge \varepsilon$ for every $u, v \in S$. An *ε-net* is a set that is both an ε-packing and an ε-cover. It is clear that a maximal ε-packing must be an ε-cover (and so, and ε-net). It is clear that if S is an ε-cover, then the Voronoi cells of S have diameter at most 2ε.

An *average ε-cover* is a set $S \subseteq V$ such that $\sum_{v\in V} \mathsf{d}_2(v, S) \le \varepsilon n$. An *average ε-net* is an average (2ε)-cover that is also an ε-packing. (It is useful to allow this relaxation by a factor of 2 here.) For every average ε-cover, a maximal subset that is an ε-packing is an average ε-net. Corollary 9.13 implies that every graph has an average ε-net with $K(\varepsilon) = 2^{O(1/\varepsilon^2)}$ elements.

We will need the following simple lemma.

LEMMA 9.18. *Let $T, S \subseteq V$. Then there is a subset $S' \subseteq S$ such that $|S'| \le |T|$ and $\overline{d}(S') \le \overline{d}(S) + 2\overline{d}(T)$.*

PROOF. For every point in T choose a nearest point of S, and let S' be the set of points chosen this way. Clearly $|S'| \le |T|$. For every $x \in V$, let $y \in S$ and $z \in T$ be the points nearest to x. Then $\mathsf{d}_2(z, S) \le \mathsf{d}_2(z, y) \le \mathsf{d}_2(x, z) + \mathsf{d}_2(x, y) = \mathsf{d}_2(x, T) + \mathsf{d}_2(x, S)$, and hence, by its definition, S' contains a point y' with $\mathsf{d}_2(z, y') \le \mathsf{d}_2(x, T) + \mathsf{d}_2(x, S)$. Hence

$$\mathsf{d}_2(x, S') \le \mathsf{d}_2(x, y') \le \mathsf{d}_2(x, z) + \mathsf{d}_2(z, y') \le 2\mathsf{d}_2(x, T) + \mathsf{d}_2(x, S).$$

Averaging over x, the lemma follows. □

Now we turn to algorithms. We assume, as before, that we can generate independent, uniformly distributed random points from V. The 2-neighborhood metric

$$\mathsf{d}_2(i,j)^2 = \frac{1}{n^2}|A^2(e_i - e_j)| = \frac{1}{n^3}\sum_k |(A^2)_{ik} - (A^2)_{jk}|^2$$
$$= \frac{1}{n^3}\sum_k \Big|\sum_r (A_{ir} - A_{jr})A_{rk}\Big|^2$$

(where A is the adjacency matrix of the graph G) can be written as

$$\mathsf{d}_2(i,j) = \mathsf{E}_k \big|\mathsf{E}_r\big((A_{ir} - A_{jr})A_{rk}\big)\big|^2,$$

where k and r are chosen randomly and uniformly from V. This shows that we can approximately compute $\mathsf{d}_2(i,j)$ by random sampling from V. (We suppress the analysis using Laws of Large Numbers and concentration inequalities.)

How to construct ε-nets? If V is not too large, then we can go through the nodes in any order, and build up the ε-net S, by adding a new node v to S if $\mathsf{d}_2(v,S) \geq \varepsilon$. In other words, S is a greedily selected maximal ε-packing, and therefore an ε-net.

Average ε-nets can be constructed by a simple randomized algorithm even if V is large. We build up S randomly: at each step we generate a new random node, and add it to S if its distance from S is at least ε (otherwise, we throw it away). For some pre-specified $A \geq 1$, we stop if for A/ε consecutive steps no new point has been added to S. The set S formed this way is trivially an ε-packing, but not necessarily maximal. However, it is likely to be an average (2ε)-cover. Indeed, let t be the number of points x with $\mathsf{d}_2(x,S) > \varepsilon$. These points are at distance at most 1 from S, and hence

$$\sum_{x\in V} \mathsf{d}_2(x,S) \leq (n-t)\varepsilon + t < n\varepsilon + t.$$

So if S is not an average (2ε)-cover, then $t > \varepsilon n$. The probability that we have not hit this set for A/ε steps is

$$\Big(1 - \frac{t}{n}\Big)^{A/\varepsilon} < e^{-\frac{tA}{n\varepsilon}} < e^{-A}.$$

The time cost of this algorithm is at most $K(\varepsilon)A/\varepsilon$.

We can modify this procedure to find an almost optimal average $O(\varepsilon)$-cover. Consider the smallest average ε-cover T (we do not know which points belong to T, of course). Suppose that we can construct a (possibly much larger) average 2ε-cover S by the procedure described above. By Lemma 9.18, S contains an average (3ε)-cover S' of size $|S'| \leq |T|$.

How to find this subset of S? We can try all $|T|$-subsets of S. This clearly inefficient last step is nevertheless useful: In $2^{O(K(\varepsilon))}$ (randomized) time, we can construct a set that is an average (3ε)-cover with probability $1-\varepsilon$. (The point is that the time bound depends only on ε, not on the size of V.)

Once this set S is computed, we can consider the task of computing a weak regularity partition solved: by Theorem 9.12, we can take the Voronoi cells of the set S, with respect to the metric d, as the classes of this partition. This partition $\mathcal{P}$ satisfies $d_\square(G, G_\mathcal{P}) \leq 8\varepsilon^{1/2}$.

Note, however, that "computing the partition $\mathcal{P}$" does not mean that we compute a list of all nodes, specifying the partition class it belongs to: we assume that the graph is very large, so such a list would be too long. What we want is a way to determine about any given node which partition class it belongs to, in other words, which element of S is closest to it. Since the metric d is computable by sampling, this can be done in the property testing model. (Of course, using sampling we can only compute an approximation of the metric d, with high probability. We do not discuss the issue of estimating the errors here.)

REMARK 9.19. The considerations in Section 9.3.2 work in general for every metric, assuming that we have an upper bound on the size of an ε-packing.

EXERCISE 9.1. Find all twin-free simple graphs with rank at most 4.

EXERCISE 9.2. Show that for every even $r \geq 2$, there is a twin-free simple graph on 2^{r-1} nodes whose adjacency matrix has rank r over $GF(2)$.

EXERCISE 9.3. Prove that every graph on n nodes contains two nodes i and j with $|N(i)\triangle N(j)| < n/2$. Show that this bound is best possible.

EXERCISE 9.4. Let G_1 and G_2 be two independently generated random graphs on the same set V of nodes, obtained by connecting any two elements of V with probability 1/2, independently for different pairs. Prove that with high probability, $d_\square(G_1, G_2) = o(1)$.

EXERCISE 9.5. Prove (9.12). Prove that similar inequality holds for any matrix norm that is invariant under reordering of the rows and columns.

EXERCISE 9.6. Let M be a real $n\times n$ matrix, and define $\|M\|_1 = \sum_{i,j=1}^n |M_{ij}|$. (a) Prove that $\|M\|_1 \leq 2n\|M\|_\square$. (b) Prove that $\|M\|_1 \leq 2\sqrt{n}\|M\|_\square$.

EXERCISE 9.7. If a metric space has an ε-cover with N elements, then no (2ε)-packing can have more than N elements.

EXERCISE 9.8. Every ε-cover of a metric space has a subset that is both an ε-packing and a (2ε)-cover.

EXERCISE 9.9. Let (V, d) be a finite metric space and let T and S be ε-covers. Then there is a subset $S' \subseteq S$ such that $|S'| \leq |T|$ and S' is a (4ε)-cover.

EXERCISE 9.10. Let S^d denote the d-dimensional unit sphere. Let V be a random subset of S^d with $|V| = n$, where we let $n \to \infty$ in the statements below. Connect two points of V if their spherical distance is larger than $\pi/2$, to get a graph G. (a) Prove that with high probability, $(A_G^2)_{uv} \sim \measuredangle(u,v)/2\pi$ for all $u, v \in V$. (b) Prove that with high probability, $\mathsf{d}_2(u,v) = \Omega(\measuredangle(u,v)/\sqrt{d})$ for all $u, v \in V$. (c) Prove that with high probability, every set $S \subseteq V$ with $\overline{d}(S) \leq 1/(8\sqrt{d})$ has at least 2^d elements. (d) Prove that with high probability, every partition $\mathcal{P}$ for which $d_\square(G, G_\mathcal{P}) \leq 1/(64\sqrt{d})$ has at least 2^d classes.

EXERCISE 9.11. Describe how to solve the maximum cut problem in the property testing model in the following sense: There is a randomized algorithm $\mathcal{A}$ which, given a graph G, an error bound $\varepsilon > 0$, and a node $v \in V$, returns LEFT or RIGHT. For a fixed G and ε, the partition into nodes found on the LEFT and nodes found on the RIGHT determine a cut containing at least $(1-\varepsilon)$ times as many edges as the maximum cut. All this happens with probability $1-\varepsilon$, and the computation time is bounded by a function of ε.

CHAPTER 10

Orthogonal Representations: Dimension

We have used orthogonal representations, (labelings of the nodes by vectors in which nonadjacent nodes are labeled by orthogonal vectors of graphs) in Section 1.4 of the Introduction. In this and the next chapter we study them from two different aspects. One can be concerned with the smallest dimension in which such a representation exists, or with representations that are "economical" in some other sense. It turns out that orthogonal representations are related to a number of fundamental properties of graphs, like connectivity, tree-width, stability number and chromatic number.

Perhaps the most natural way to be "economic" in constructing an orthogonal representation is to minimize the dimension, which will be the main topic of this chapter. We can say only a little about the minimum dimension of orthogonal representations overall, but we get interesting results if we impose "nondegeneracy" conditions. We will study three nondegeneracy conditions: general position, faithfulness, and transversality.

10.1. Basics

An *orthogonal representation* of a simple graph G in $\mathbb{R}^d$ assigns to each $i \in V$ a vector $\mathbf{u}_i \in \mathbb{R}^d$ such that $\mathbf{u}_i^\mathsf{T}\mathbf{u}_j = 0$ whenever $ij \in \overline{E}$. An *orthonormal representation* is an orthogonal representation in which all the representing vectors have unit length. Clearly we can always scale the nonzero vectors in an orthogonal representation to unit length, and zero vectors (which are orthogonal to everything) can be either excluded or ignored in most cases.

There is something arbitrary in requiring that nonadjacent nodes be represented by orthogonal vectors; why not adjacent nodes? We use here the more standard convention, but sometimes we need to talk about an orthogonal representation of the complement, which we also call a *dual orthogonal representation.*

Note that we did not insist that adjacent nodes are mapped onto nonorthogonal vectors. If this condition also holds, then we call the orthogonal representation *faithful.*

EXAMPLE 10.1. For $d = 1$, the vector labels are just real numbers u_i, and the constraints $u_i u_j = 0$ ($ij \in \overline{E}$) mean that no two nodes labeled by nonzero numbers are adjacent; in other words, the support of u is a stable set of nodes. Scaling the nonzero numbers to 1, we obtain incidence vectors of stable sets. (The connection between stable sets and orthonormal representations in higher dimension will be the central topic of the next chapter.)

Since very simple problems about stable sets are NP-hard (for example, their maximum size), this example should warn us that orthogonal representations can be very complex. ♦

EXAMPLE 10.2. Every graph has a trivial orthonormal representation in $\mathbb{R}^V$, in which node i is represented by the standard basis vector $\mathbf{e}_i$. This representation is not faithful unless the graph has no edges. However, it is easy to perturb this representation to make it faithful. Of course, we are interested in "nontrivial" orthogonal representations, which are more "economical" than this trivial one. ◆

EXAMPLE 10.3. Every graph G has a faithful orthogonal representation in $\mathbb{R}^E$, in which we label a node by the indicator vector of the set of edges incident with it. It is perhaps natural to expect that this simple representation will be rather "uneconomical" for most purposes. ◆

EXAMPLE 10.4. Figure 10.1 below shows a simple orthogonal representation in 2 dimensions of the graph obtained by adding a diagonal to the pentagon. ◆

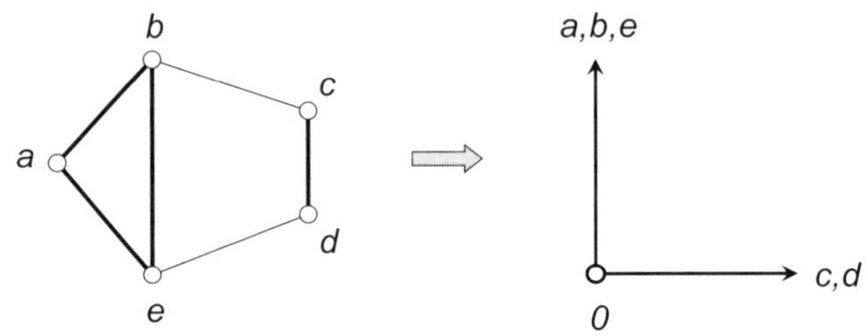

FIGURE 10.1. An (almost) trivial orthogonal representation

EXAMPLE 10.5. The previous example can be generalized. In an orthogonal representation, we can label a set of nodes with the same nonzero vector if and only if these nodes form a clique. Let $k = \chi(\overline{G})$ (the chromatic number of the complement of G), then there is a family $\{B_1, \dots, B_k\}$ of disjoint complete subgraphs covering all the nodes. Mapping every node of B_i to $\mathbf{e}_i$ ($i = 1, \dots, k$) is an orthonormal representation. ◆

EXAMPLE 10.6. In the introduction we have seen an orthogonal representation with a more interesting geometric content. The previous example gives an orthogonal representation of C_5 in 3-space (Figure 10.2, left). The "umbrella" representation defined in the introduction gives another orthogonal representation of the graph C_5 in 3-space. ◆

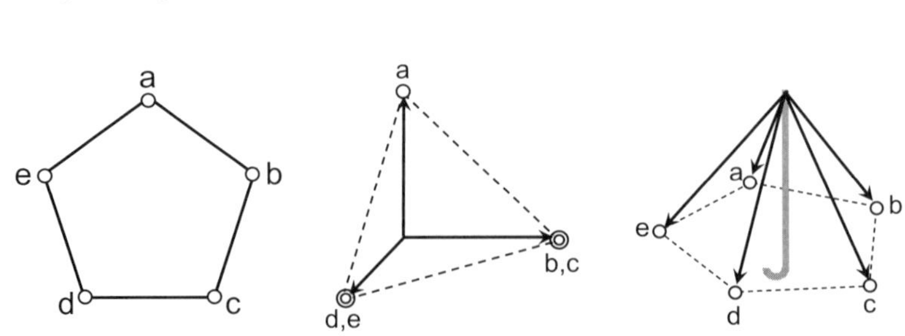

FIGURE 10.2. The graph C_5 and two orthonormal representations of it.

EXAMPLE 10.7. Let L be the Laplacian of a graph G, and let $\mathbf{u}_i$ $(i \in V)$ be the columns of the matrix $L^{1/2}$. Then $\mathbf{u}_i^{\mathsf{T}}\mathbf{u}_j = L_{ij}$, and hence $\mathbf{u}$ is a faithful orthogonal representation of G. ♦

10.2. Minimum dimension of orthonormal representations

Our first topic is to construct orthogonal representations in a dimension as low as possible. Of course, we can represent every node by the zero vector, to get an orthogonal representation in dimension 0. So we come to a very weak nondegeneracy assumption: we exclude zero vectors. Nonzero vectors can be normalized to unit vectors, so in other words, we want to study the minimum dimension in which a given graph G has an *orthonormal* representation. It is trivial that this dimension is at least $\alpha(G)$, since we have to use at least $\alpha(G)$ mutually orthogonal vector labels. The construction in Example 10.5 shows that this minimum dimension is at most $\chi(\overline{G})$.

The minimum dimension of an orthonormal representation is (loosely) tied to the chromatic number of the complement. Let us consider the (infinite) graph H_d, whose nodes are all unit vectors in $\mathbb{R}^d$, two of them being adjacent if and only if the vectors are orthogonal. The chromatic number of this graph is not known precisely, but (for large d) rather good bounds follow from the results of [**Larman–Rogers 1972**] (upper bound) and [**Frankl–Wilson 1981**], improved by [**Raigorodsky 2012**] . Not stating the most precise bounds, the following will suffice for us:

$$1.2^d \le \chi(H_d) \le 4^d. \tag{10.1}$$

Using this, it will be easy to prove the following.

PROPOSITION 10.8. *The minimum dimension d in which a graph G has an orthonormal representation satisfies*

$$\frac{1}{2}\log\chi(\overline{G}) \le d \le \chi(\overline{G}).$$

PROOF. We have already shown the upper bound. On the other hand, any orthonormal representation of G in $\mathbb{R}^d$ gives a homomorphism $\overline{G} \to H_d$, and hence

$$\chi(\overline{G}) \le \chi(H_d) \le 4^d,$$

proving the lower bound in the proposition. □

The bounds above are essentially tight. More exactly, the theorem of [**Erdős–de Bruijn 1951**] implies that H_d has a finite subgraph H such that $\chi(H) = \chi(H_d) \ge 1.2^d$, showing that the lower bound is tight up to a constant factor. The graph with d nodes and no edges shows the tightness of the upper bound.

Stated in a different language, [**Rosenfeld 1991**] proved that no triangle-free graph with n nodes can have an orthonormal representation in dimension less than $n/2$ (this is clearly tight, for example, for a complete bipartite graph; see Exercise 10.3). The case of general clique number was considered in [**Füredi–Stanley 1992**] and [**Alon–Szegedy 1999**], but the upper and lower bounds are rather far apart.

10.3. General position and connectivity

The first nontrivial nondegeneracy condition we study is *general position*: we assume that any d of the representing vectors in $\mathbb{R}^d$ are linearly independent. This implies that neither one of them is the zero vector. Another way of expressing this property is that every subset of the vectors is either linearly independent or spans the whole space.

The main result of this section [**Lovász–Saks–Schrijver 1989**] finds an exact condition for this type of geometric representability.

THEOREM 10.9. *A graph G has an orthogonal representation in $\mathbb{R}^d$ in general position if and only if G is $(n-d)$-connected.*

It is useful to remark that a graph is $(n-d)$-connected if and only if its complement does not contain a complete bipartite graph with more than d nodes (where for a complete bipartite graph we always assume that its bipartition classes are nonempty). So we could formulate this theorem as follows: *A graph has a dual orthonormal representation in general position in $\mathbb{R}^d$ if and only if it does not contain any complete bipartite subgraph with $d+1$ nodes.*

The condition that the given set of representing vectors is in general position is not easy to check (it is NP-hard). A weaker, but very useful condition will be that the vectors representing the nodes nonadjacent to any node are linearly independent. We say that such a representation is in *locally general position*. This condition implies that every node is represented by a nonzero vector unless it is connected to all the other nodes. In this case there is no condition on the representing vector, so we may assume that all vectors are nonzero (equivalently, unit vectors).

Theorem 10.9 will be proved in the following slightly more general form:

THEOREM 10.10. *For every graph G, the following conditions are equivalent:*

(i) *G has an orthogonal representation in $\mathbb{R}^d$ in general position;*

(ii) *G has an orthogonal representation in $\mathbb{R}^d$ in locally general position;*

(iii) *G is $(n-d)$-connected.*

Two steps of the proof are easy. Consider an orthogonal representation in general position, then the vectors representing the non-neighbors of a node v cannot span the whole space (since they are all orthogonal to the vector label of v), so they are linearly independent. Thus (i)⇒(ii). To illustrate the connection between connectivity and orthogonal representations, we prove that (ii)⇒(iii). Let V_0 be a cutset of nodes of G, then $V = V_0 \cup V_1 \cup V_2$, where $V_1, V_2 \neq \emptyset$, and no edge connects V_1 and V_2. This implies that the vectors representing V_1 are linearly independent, and similarly, the vectors representing V_2 are linearly independent. Since the vectors representing V_1 and V_2 are mutually orthogonal, all vectors representing $V_1 \cup V_2$ are linearly independent. Hence $d \geq |V_1 \cup V_2| = n - |V_0|$, and so $|V_0| \geq n-d$ (Figure 10.3).

The difficult part of the proof will be (iii)⇒(i): this requires the construction of a general position orthogonal representation for $(n-d)$-connected graphs. We describe and analyze the algorithm constructing the representation. As a matter of fact, to describe the construction is quite easy, the difficulty lies in the proof of its validity.

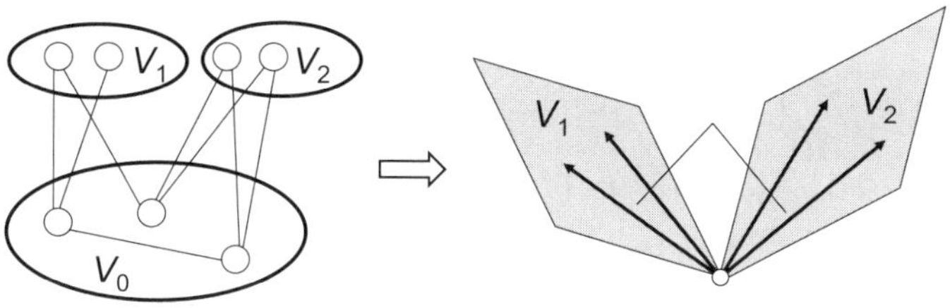

FIGURE 10.3. Locally general position orthogonal representation in low dimension implies high connectivity.

10.3.1. *G*-orthogonalization. The following procedure can be viewed as an extension of the Gram-Schmidt orthogonalization algorithm. Let G be any simple graph, where $V = [n]$, and let $\mathbf{u} : V \to \mathbb{R}^d$ be any vector labeling. Let us choose vectors $\mathbf{f}_1, \mathbf{f}_2, \dots$ consecutively as follows. Let $\mathbf{f}_1 = \mathbf{u}_1$. Supposing that the vectors $\mathbf{f}_i$ $(i < j)$ are already chosen, we define $\mathbf{f}_j$ as the orthogonal projection of $\mathbf{u}_j$ onto the subspace $L_j = \mathrm{lin}\{\mathbf{f}_i : i < j, ij \in \overline{E}\}^\perp$. The sequence $(\mathbf{f}_1, \dots, \mathbf{f}_n)$ is a vector labeling of G, which we call the *G-orthogonalization* of the vector labeling $\mathbf{u}$. It is trivial that $\mathbf{f}$ is an orthogonal representation of G. If $E = \emptyset$, then $(\mathbf{f}_1, \dots, \mathbf{f}_n)$ is just the Gram–Schmidt orthogonalization $(\mathbf{u}_1, \dots, \mathbf{u}_n)$. It follows by straightforward induction on j that

$$\mathrm{lin}\{\mathbf{u}_1, \dots, \mathbf{u}_j\} = \mathrm{lin}\{\mathbf{f}_1, \dots, \mathbf{f}_j\}$$

for every $j \in [n]$.

It is important to note that the G-orthogonalization of a vector labeling depends on the ordering of V. In this section we will be concerned with G-orthogonalizations obtained through different orderings of a fixed graph G. If $\sigma : V \to [n]$ defines an ordering of the nodes and $\mathbf{u} : V \to \mathbb{R}^d$ is a vector labeling, then we denote its G-orthogonalization by $\mathbf{f}^\sigma$ (we omit σ if the ordering of the nodes is understood).

From now on, we consider a *random* vector labeling $\mathbf{u}$: by this we mean that the vectors $\mathbf{u}_i$ are random vectors whose entries are chosen independently from the standard Gaussian distribution. Note that the length of the vector $\mathbf{u}_i$ influences the vector $\mathbf{f}_i$ only, and only its length; so we could replace the (multidimensional) standard Gaussian distribution of $\mathbf{u}_i$ by any rotationally symmetric distribution in the lemmas below (for example, by the uniform distribution on S^{d-1}).

LEMMA 10.11. *Let G be a graph in which every node has degree at least $n-d$, and let $\mathbf{u} : V \to \mathbb{R}^d$ be a random vector labeling. Then the vectors in the G-orthogonalization of $\mathbf{u}$ (with respect to any ordering of the nodes) are almost surely nonzero.*

PROOF. For any j, there are at most $d-1$ vectors $\mathbf{f}_i$ $(i < j, ij \notin E)$, hence they do not span $\mathbb{R}^d$, and hence almost surely $\mathbf{u}_j$ is linearly independent of them, and then its projection onto L_j is nonzero. □

The main fact we want to prove is the following (which implies Theorem 10.10 immediately).

LEMMA 10.12. *Let G be an $(n-d)$-connected graph, and let $\mathbf{u} : V \to \mathbb{R}^d$ be a random vector labeling. Then the vectors in the G-orthogonalization of $\mathbf{u}$ (with respect to any ordering of the nodes) are almost surely in general position.*

To motivate our approach to the proof, let us consider a simple example.

EXAMPLE 10.13. Let $V = \{a,b,c,d\}$ and $E = \{ac, bd\}$. Consider a random vector labeling $\mathbf{u}$ in $\mathbb{R}^3$ and compute its G-orthogonalization $\mathbf{f}$, associated with the given ordering. Since every node has degree 1, the vectors $\mathbf{f}_i$ are almost surely nonzero. We have $\mathbf{f}_a = \mathbf{u}_a$, $\mathbf{f}_b \in \mathbf{f}_a^\perp$, and $\mathbf{f}_c \in \mathbf{f}_b^\perp$; almost surely, $\mathbf{f}_c$ will not be parallel to $\mathbf{f}_a$, so together they span the plane $\mathbf{f}_b^\perp$. This means that $\mathbf{f}_d$, which is orthogonal to both $\mathbf{f}_a$ and $\mathbf{f}_c$, must be parallel to $\mathbf{f}_b$, no matter what $\mathbf{u}_d$ is.

Now suppose that we reverse the order of the nodes. Let $\mathbf{f}'_d, \mathbf{f}'_c, \mathbf{f}'_b, \mathbf{f}'_a$ be the vectors obtained by the G-orthogonalization in this order. Almost surely $\mathbf{f}'_d$ will not be parallel to $\mathbf{f}'_b$, but $\mathbf{f}'_c$ will be parallel with $\mathbf{f}'_a$. So not only are the two distributions different, but a particular event, namely $\mathbf{f}_b \| \mathbf{f}_d$, occurs with probability 0 in one and with probability 1 in the other.

Let us modify this example by connecting b and c by an edge. Processing the vectors in the order (a,b,c,d) again, the vectors $\mathbf{f}_a = \mathbf{u}_a$ and $\mathbf{f}_b$ will be orthogonal again. No condition on $\mathbf{f}_c$, so almost surely $\mathbf{f}_c = \mathbf{u}_c$ will be linearly independent of $\mathbf{f}_a$ and $\mathbf{f}_b$, but not orthogonal to either one of them. The direction of $\mathbf{f}_d$ is still determined, but now it will not be parallel to $\mathbf{f}_b$; in fact, depending on $\mathbf{f}_c$, it can have any direction orthogonal to $\mathbf{f}_a$. Doing this in the reverse order, we get vectors $\mathbf{f}'_d, \mathbf{f}'_c, \mathbf{f}'_b, \mathbf{f}'_a$ that have similar properties. The distributions of $(\mathbf{f}_d, \mathbf{f}_c, \mathbf{f}_b, \mathbf{f}_a)$ and $(\mathbf{f}'_d, \mathbf{f}'_c, \mathbf{f}'_b, \mathbf{f}'_a)$ are still different, but any event that occurs with probability 0 in one will also occur with probability 0 in the other (this is not quite obvious). ♦

This example motivates the following considerations. As noted, the distribution of the G-orthogonalization may depend on the ordering of the nodes; the key to the proof will be that this dependence is not too strong. To define what this means, consider two random variables X and Y with values in $\mathbb{R}^M$. We say that these are *mutually absolutely continuous*, if for every Borel set B,

$$\mathsf{P}(X \in B) = 0 \quad \Leftrightarrow \quad \mathsf{P}(Y \in B) = 0.$$

Informally, if a "reasonable" property of X almost never holds (for example, its coordinates almost surely do not satisfy an algebraic equation), then this is also true for Y, and vice versa.

We will repeatedly use the following easy property of being mutually absolutely continuous.

LEMMA 10.14. *Consider two random variables, each of which consist of two components: $X = (U, V)$ and $X' = (U', V')$, where $U, U' \in \mathbb{R}^M$ and $V, V' \in \mathbb{R}^N$. Assume that U and U' are mutually absolutely continuous, and for every Borel set $A \in \mathbb{R}^M$ with $\mathsf{P}(U \in A) > 0$, the random variables $V|A$ and $V'|A$ are mutually absolutely continuous. Then X and Y are mutually absolutely continuous.* □

The main lemma (which will be used later as well) is the following.

LEMMA 10.15 (Main Lemma). *Let G be an $(n-d)$-connected graph, and let $\mathbf{u}: V \to \mathbb{R}^d$ be a random vector labeling. Then for any two orderings σ and τ of V, the random variables $\mathbf{f}^\sigma$ and $\mathbf{f}^\tau$ are mutually absolutely continuous.*

This implies Lemma 10.12, since the first d vectors in the G-orthogonalization of a random vector labeling are almost surely in general position, and we can start the ordering with any d-tuple.

PROOF. It suffices to prove this lemma in the case when τ is the ordering obtained from σ by swapping the nodes in positions j and $j+1$ ($1 \le j \le n-1$). Let us label the nodes so that $\sigma = (1, 2, \dots, n)$.

Clearly $\mathbf{f}_i^\sigma = \mathbf{f}_i^\tau$ for $i < j$. The main part of the proof is to show that the distributions of $(\mathbf{f}_1^\sigma, \dots, \mathbf{f}_{j+1}^\sigma)$ and $(\mathbf{f}_1^\tau, \dots, \mathbf{f}_{j+1}^\tau)$ are mutually absolutely continuous. We prove this by induction on j, distinguishing several cases.

Case 1. j and $j+1$ are adjacent in G. In this case the vector $\mathbf{f}_{j+1}$ does not depend on $\mathbf{u}_j$ and vice versa, so $\mathbf{f}_j^\sigma = \mathbf{f}_j^\tau$ and $\mathbf{f}_{j+1}^\sigma = \mathbf{f}_{j+1}^\tau$.

Case 2. j and $j+1$ are not adjacent, but they are connected by a path that lies entirely in $\{1, \dots, j, j+1\}$. Let P be a shortest such path and let t be its length (number of edges), so $2 \le t \le j$. We argue by induction on t (and on j). Let i be any internal node of P. We swap j and $j+1$ by the following steps (Figure 10.4):

(1) Interchange i and j, by several swaps of consecutive elements among the first j.

(2) Swap i and $j+1$.

(3) Interchange $j+1$ and j, by several swaps of consecutive elements among the first j.

(4) Swap j and i.

(5) Interchange $j+1$ and i, by several swaps of consecutive elements among the first j.

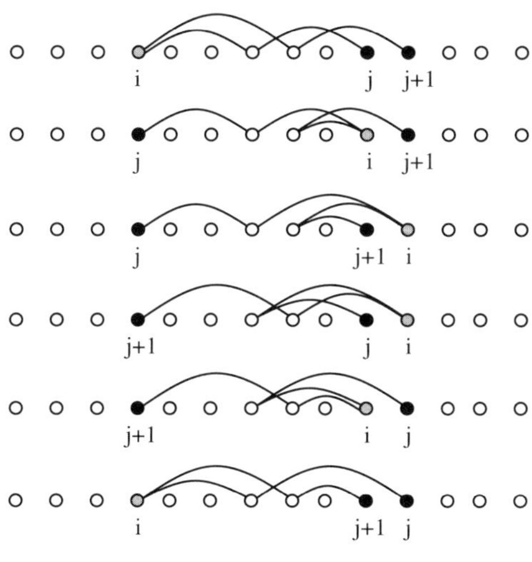

FIGURE 10.4. Interchanging j and $j+1$.

In each step, the new and the previous distributions of the G-orthogonalized vectors are mutually absolutely continuous: In steps (1), (3) and (5) this is so because the swaps take place among the first j nodes, and we can invoke induction on j; in steps (2) and (4), because the nodes swapped are at a smaller distance than t in the graph distance, and we can invoke induction on t.

Case 3. There is no path connecting j to $j+1$ within $\{1, \dots, j+1\}$. This case is more tedious but not particularly deep. It suffices to show that the distributions of the pairs of vectors $(\mathbf{f}_j^\sigma, \mathbf{f}_{j+1}^\sigma)$ and $(\mathbf{f}_j^\tau, \mathbf{f}_{j+1}^\tau)$, conditioned on the previous vectors $\mathbf{u}_1, \dots, \mathbf{u}_{j-1}$, are mutually absolutely continuous. The remaining vectors $\mathbf{f}_k^\sigma$ and $\mathbf{f}_k^\tau$ $(k > j+1)$ are generated in the same way from these two sequences, and hence the distributions of $\mathbf{f}^\sigma$ and $\mathbf{f}^\tau$ are mutually absolutely continuous.

Clearly $V \setminus \{1, \dots, j+1\}$ is a cutset, whence by our hypothesis on the connectivity of G it follows that $n-j-1 \geq n-d$ and so $j \leq d-1$. Let $B = \text{lin}\{\mathbf{u}_1, \dots, \mathbf{u}_{j-1}\}$ and $C = B^\perp$, then $t = \dim(C) = d - \dim(B) \geq d-(j-1) \geq 2$.

The random vector $\mathbf{u}_j$ can be decomposed as $\mathbf{u}_j = \mathbf{b}_j + \mathbf{c}_j$, where $\mathbf{b}_j$ and $\mathbf{c}_j$ are random Gaussian vectors in B and C, respectively, and $\mathbf{b}_j$ and $\mathbf{c}_j$ are independent as random variables. The subspace which is orthogonal to all previous nonneighbors of j (in the ordering σ) is $L_j = \big(\overline{N}^\sigma(j)\big)^\perp$. Thus $C \subseteq L_j$. Let $\mathbf{b}_j'$ be the orthogonal projection of $\mathbf{b}_j$ onto L_j. Since $\mathbf{b}_j$ is orthogonal to C, so is $\mathbf{b}_j'$, thus $\mathbf{b}_j' \in B$.

We define L_{j+1} similarly for the node $j+1$ and ordering τ, along with the decomposition $\mathbf{u}_{j+1} = \mathbf{b}_{j+1} + \mathbf{c}_{j+1}$, and the orthogonal projection $\mathbf{b}_{j+1}'$ of $\mathbf{b}_{j+1}$ onto L.

The following observation will be important:

$$\mathbf{b}_j' \perp \mathbf{b}_{j+1}'. \tag{10.2}$$

Here we have to use that there is no path in J connecting j and $j+1$. This implies that the set $\{1, \dots, j-1\}$ has a partition $W' \cup W''$ so that $N^\sigma(j) \subseteq W'$, $N^\tau(j+1) \subseteq W''$, and there is no edge between W' and W''. Let $B' = \text{lin}(W')$ and $B'' = \text{lin}(W'')$, then $B' \perp B''$, and $L_j \perp B''$, which implies that $\mathbf{b}_j' \in B'$. Similarly, $\mathbf{b}_{j+1}' \in B''$, which implies (10.2).

To get $\mathbf{f}_j^\sigma$, we have to project $\mathbf{u}_j$ onto L_j; using that $C \subseteq L_j$, the component of $\mathbf{u}_j$ in C remains unchanged, and this projection can be expressed as

$$\mathbf{f}_j^\sigma = \mathbf{b}_j' + \mathbf{c}_j \tag{10.3}$$

Similarly, we have

$$\mathbf{f}_{j+1}^\tau = \mathbf{b}_{j+1}' + \mathbf{c}_{j+1}. \tag{10.4}$$

To describe the other two orthogonalized vectors, notice that the only difference between $\mathbf{f}_{j+1}^\tau$ and $\mathbf{f}_{j+1}^\sigma$ is that we have to make $\mathbf{f}_{j+1}^\sigma$ orthogonal to $\mathbf{f}_j^\sigma$ as well. Since the vectors $\mathbf{b}_j'$ and $\mathbf{b}_{j+1}'$ are already orthogonal to each other, as well as to all vectors in C, this simply means that we have to modify (10.4) by replacing $\mathbf{c}_{j+1}$ by its projection $\mathbf{c}_{j+1}'$ onto the orthogonal complement of $\mathbf{c}_j$. So we can write

$$\mathbf{f}_{j+1}^\sigma = \mathbf{b}_{j+1}' + \mathbf{c}_{j+1}'. \tag{10.5}$$

Similarly

$$\mathbf{f}_j^\tau = \mathbf{b}_j' + \mathbf{c}_j', \tag{10.6}$$

where $\mathbf{c}_j'$ is the projection of $\mathbf{c}_j$ onto the orthogonal complement of $\mathbf{c}_{j+1}$.

We can generate these vectors as follows. We condition on the vectors $\mathbf{u}_1, \dots, \mathbf{u}_{j-1}$, determining the subspaces B and C. Then we generate independent Gaussian vectors $\mathbf{b}_j, \mathbf{b}_{j+1} \in B$, determining the vectors $\mathbf{b}_j'$ and $\mathbf{b}_{j+1}'$. Finally, we generate two independent Gaussian vectors $\mathbf{c}_j, \mathbf{c}_{j+1} \in C$. Both pairs of vectors $(\mathbf{c}_j, \mathbf{c}_{j+1}')$ and $(\mathbf{c}_j', \mathbf{c}_{j+1})$ belong to the manifold $\{(\mathbf{x}, \mathbf{y}) : \mathbf{x}, \mathbf{y} \in \mathbb{R}^d, \ \mathbf{x} \perp \mathbf{y}\}$, and they have positive density functions, which implies that they are mutually absolutely continuous. Hence $(\mathbf{f}_j^\sigma, \mathbf{f}_{j+1}^\sigma)$ and $(\mathbf{f}_j^\tau, \mathbf{f}_{j+1}^\tau)$ are mutually absolutely continuous.

This completes the proof of Lemma 10.15 and of Theorem 10.10. □

10.4. Faithfulness

It is probably difficult to decide whether a given graph has a faithful orthogonal representation in a given dimension. In other words, we do not know how to determine the minimum dimension of a faithful orthogonal representation. What we can do is to give almost-trivial upper and lower bounds, and—as an application of the results in the previous section—a nontrivial upper bound.

Let us start with some examples.

EXAMPLE 10.16. It is easy to construct a faithful orthogonal representation in dimension $n-1$ of the path P_n with n nodes: we label the nodes in order by $\mathbf{e}_1, \mathbf{e}_1+\mathbf{e}_2, \dots, \mathbf{e}_{n-2}+\mathbf{e}_{n-1}, \mathbf{e}_{n-1}$. The path P_n has no faithful orthogonal representation in lower dimension. Indeed, in any faithful orthogonal representation $(\mathbf{v}_i : i \in V(P_n))$, for every $1 \le k \le n-1$, the vector $\mathbf{v}_{k+1}$ is orthogonal to $\mathbf{v}_1, \dots, \mathbf{v}_{k-1}$, but not orthogonal to $\mathbf{v}_k$, which implies that $\mathbf{v}_k$ is linearly independent of $\mathbf{v}_1, \dots, \mathbf{v}_{k-1}$. Hence the vectors $\mathbf{v}_1, \dots, \mathbf{v}_{n-1}$ are linearly independent, and so the ambient space must have dimension at least $n-1$. ♦

This example motivates the following upper and lower bounds on the dimension of a faithful orthogonal representation. We have constructed orthogonal representations from cliques covering the nodes of a graph (Example 10.5). These representations are far from being faithful in general. However, using a family of cliques covering all *edges*, we can construct a faithful representation.

PROPOSITION 10.17. *If the edges of a graph G can be covered by k cliques, then it has a faithful orthogonal representation in $\mathbb{R}^k$.*

PROOF. Let $\{B_1, \dots, B_k\}$ be any family of complete subgraphs covering all edges of a graph G. We construct the vector labeling $\mathbf{u}$ in $\mathbb{R}^k$, defined by $(\mathbf{u}_i)_j = \mathbb{1}(i \in B_j)$. Then $\mathbf{u}$ is a faithful orthogonal representation of G: if i and j are adjacent nodes, then they are contained in one of the cliques B_i, and hence $\mathbf{u}_i^\mathsf{T}\mathbf{u}_j > 0$; if they are nonadjacent, then the supports of $\mathbf{u}_i$ and $\mathbf{u}_j$ are disjoint, and so $\mathbf{u}_i^\mathsf{T}\mathbf{u}_j = 0$. □

It is trivial that $\alpha(G)$ is a lower bound on the dimension of any orthogonal representation by nonzero vectors (faithful or not). To strengthen this bound in the case of faithful representations, we say that a subset $S \subseteq V$ is *almost stable*, if for every connected subgraph H of $G[S]$ with at least one edge there is a node $j \in V \setminus V(H)$ that is adjacent to exactly one node in $V(H)$. (The node j may or may not belong to S.) It is trivial that every stable set is almost stable.

PROPOSITION 10.18. *The dimension of any faithful orthogonal representation of a graph G is at least as large as its largest almost stable set.*

PROOF. Let $\mathbf{v}$ be a faithful orthogonal representation of G in dimension d, and let S be an almost stable set. We prove by induction on $|S|$ that the vectors in $\mathbf{v}(S)$ are linearly independent. If S is a stable subset of V, then this is trivial. Else, let H be a connected component of $G[S]$ with at least one edge, and let $j \notin V(H)$ be a node adjacent to a unique node $i \in V(H)$. The set $S' = S \setminus i$ is almost stable, and hence the vectors in $\mathbf{v}(S')$ are linearly independent by the induction hypothesis.

Next, observe that the vectors $\mathbf{v}(V(H)\setminus i)$ cannot span $\mathbf{v}_i$; this follows from the fact that $\mathbf{v}_j$ is orthogonal to each of these vectors, but not to $\mathbf{v}_i$. Since the vectors $\mathbf{v}_k$ with $k \in S\setminus V(H)$ are all orthogonal to every vector $\mathbf{v}_k$ with $k \in V(H)$, it follows that S' cannot span $\mathbf{v}_i$. This proves that $\mathbf{v}(S) = \mathbf{v}(S') \cup \{\mathbf{v}_i\}$ is a linearly independent set of vectors. □

From the results in Section 10.3.1, it is easy to derive an upper bound on the minimum dimension of faithful orthogonal representations.

PROPOSITION 10.19. *Every $(n-d)$-connected graph on n nodes has a faithful orthogonal representation in $\mathbb{R}^d$.*

PROOF. It suffices to show that in a G-orthogonalized random representation, the probability that two adjacent nodes are represented by orthogonal vectors is zero. By the Lemma 10.15, it suffices to prove this for the G-orthogonalization based on an ordering starting with these two nodes. But then the assertion is obvious. □

Using that a graph with minimum degree D cannot contain a complete bipartite graph with more than $2D$ nodes, Proposition 10.19 implies the following bound.

COROLLARY 10.20. *If the maximum degree of a graph G is D, then G has a faithful dual orthogonal representation in $2D$ dimensions.*

This was proved in [**Maehara–Rödl 1990**]. They conjecture that the bound on the dimension can be improved to $D+1$. Proposition 10.19 shows that the conjecture is true if we strengthen its assumption by requiring that $\overline{G}$ is $(n-D-1)$-connected.

We conclude this section with two examples illustrating the use (and insufficiency) of the upper and lower bounds above.

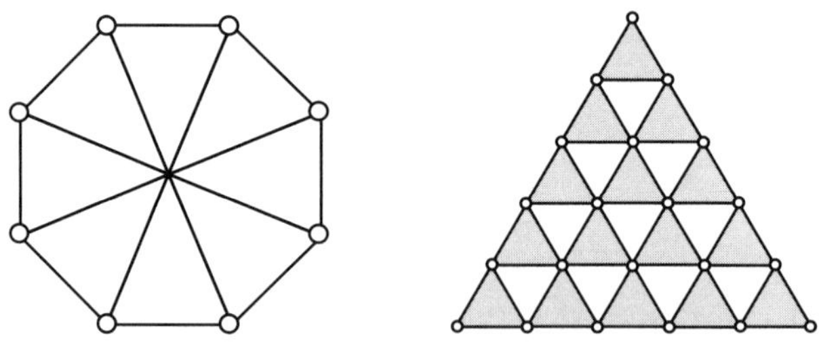

FIGURE 10.5. The graph V_8 and the triangular grid Δ_6.

EXAMPLE 10.21. Let V_8 denote the graph obtained from the cycle C_8 by adding its longest diagonals (Figure 10.5, left). This graph is 3-connected, and hence by Proposition 10.19, it has a faithful orthogonal representation in $\mathbb{R}^5$. It does not have a faithful orthogonal representation in $\mathbb{R}^4$ [**van der Holst 1996**]. But to show this is not easy, because it does not have an almost stable set of size 5. (The last two facts need some case analysis, not reproduced here.) So the lower bound in Proposition 10.18 does not always give the right value. ♦

EXAMPLE 10.22 (Triangular grid I). Consider the triangular grid Δ_k with $k > 2$ nodes along the bottom line Figure 10.5, right). This graph has $n = \binom{k+1}{2}$ nodes.

The dark triangles form a family of cliques covering all edges, so the construction in Example 10.16 yields a faithful orthogonal representation in dimension $n-k=\binom{k}{2}$, the number of dark triangles. This is much better than the upper bound from Proposition 10.19, which is $n-2$.

This dimension is in fact the smallest possible; this follows by Proposition 10.18, since it is easy to see that the set of nodes above the bottom line is almost stable, and hence the minimum dimension of a faithful representation is $n-k$.

We note that Δ_k is planar, so all of its minors are at most 5-connected. So we could not hope to salvage the upper bound provided by Proposition 10.17 by looking at highly connected subgraphs or minors. The significance of this observation will be made more clear in the next section. ♦

10.5. Transversality

We have seen various nondegeneracy conditions (general position, faithfulness), and how they turn a complicated and probably untractable question like the existence of an orthogonal representation in a given dimension into a question where some sort of answer can be obtained by combinatorial means. In this section we introduce a more complicated, deeper nondegeneracy condition, which was first used in this spirit by Colin de Verdière. We spend a bit more time on it than what we would need in this chapter, as a preparation for its use later on.

Consider a system of polynomial equations with real coefficients

$$p_1(x)=0,\dots,p_m(x)=0 \quad (x\in\mathbb{R}^n). \tag{10.7}$$

Each of these equations defines a hypersurface in $\mathbb{R}^n$. We say that a solution u is *transversal*, if these surfaces intersect transversally at this point. This means that their normal vectors at this point, the vectors grad $p_1(u),\dots,$ grad $p_m(u)$, are linearly independent.

We could also rephrase this property as follows: for a linear combination $p=\sum_i\alpha_ip_i$ of the left sides of (10.7), the point $\mathbf{u}$ is a stationary point of p (i.e., the gradient of p at $\mathbf{u}$ is nonzero) only if the linear combination is trivial.

The most important consequence of this property is the following: if a solution y of (10.7) is transversal, then for every $\varepsilon>0$ there is a $\delta>0$ such that changing each coefficient of each polynomial by at most δ, the modified system

$$\widehat{p}_1(x)=0,\dots,\widehat{p}_m(x)=0 \tag{10.8}$$

has a solution z such that $|z-y|<\varepsilon$. Furthermore, the vector z, as a solution of the system (10.8), is transversal as well. This follows by the Implicit Function Theorem.

Informally, this means that the solvability of the system of equations (10.7) is not just a numerical coincidence, but it is a "robust" consequence of the structure of the equations. We will see that this fact leads to important combinatorial applications of this nondegeneracy notion.

10.5.1. Transversal orthogonal representations.

We can view an orthogonal representation $i\mapsto\mathbf{u}_i\in\mathbb{R}^d$ of a graph G as a system of dn variables, entries of a matrix $U\in\mathbb{R}^{d\times V}$, whose columns satisfy the quadratic equations

$$\mathbf{u}_i^{\mathsf{T}}\mathbf{u}_j=0 \qquad (ij\in\overline{E}). \tag{10.9}$$

We say that the representation of G is *transversal*, if this solution of (10.9) is transversal. Computing the gradient, this says that there are no real numbers X_{ij} $(ij \in \overline{E})$ such that

$$\sum_{j\in\overline{N}(i)} X_{ij}\mathbf{u}_j = 0 \qquad \text{for every } i \in V. \tag{10.10}$$

Using terminology to be introduced in Chapter 14, *the orthogonal representation* $\mathbf{u}$ *of G is transversal if and only if* $(\overline{G}, \mathbf{u})$ *carries no nonzero homogeneous stress.*

It will be convenient to think of X as a nonzero symmetric $V \times V$ matrix such that $X_{ij} = 0$ if $i = j$ or $ij \in E$. In other words, X is a nonzero $\overline{G}$-matrix with zero diagonal. Note that X has a zero-nonzero pattern complementary to $A = \mathsf{Gram}(\mathbf{u})$, and condition (10.10) is equivalent to the matrix equation $XA = 0$.

For a row of X that is not all 0, equation (10.10) means a linear dependence between the nonneighbors of i, and if $X \neq 0$ then at least one of these linear dependencies is nontrivial. It follows that every orthogonal representation of a graph G in locally general position is transversal. This shows that transversality can be thought of as some sort of weakened, symmetrized version of locally general position. But the two conditions are not equivalent, as the following example shows.

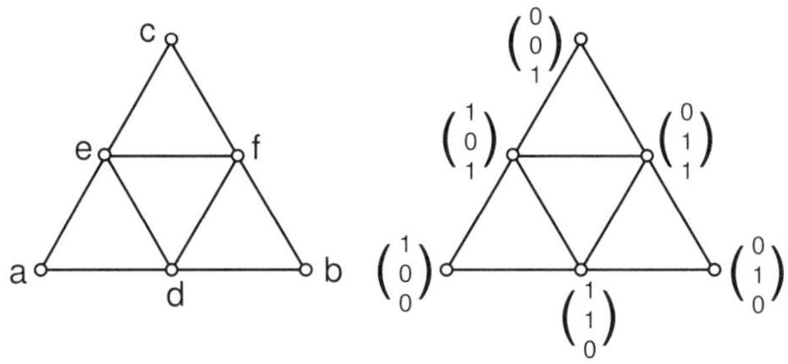

FIGURE 10.6. The graph Δ_3 with an orthogonal representation that is transversal but not in locally general position.

EXAMPLE 10.23 (Triangular grid II). Consider the graph Δ_3 obtained by attaching a triangle to each edge of a triangle (Figure 10.6). We have constructed an orthogonal representation of this graph in $\mathbb{R}^3$ (Example 10.22), which is shown in the figure. This representation is not in locally general position, since the nodes nonadjacent to (say) node a are represented by linearly dependent vectors. But it is transversal. Indeed, suppose that a nonzero $\overline{G}$-matrix X with zero diagonal satisfies (10.10). Since node d is adjacent to all the other nodes except node c, we have $X_{d,j} = 0$ for $j \neq c$, and therefore (10.10) implies that $X_{d,c} = 0$. By symmetry, $X_{c,d} = 0$, and hence (10.10) implies that $X_{c,j} = 0$ for all j. Going on similarly, we get that all entries of X must be zero. Note that the symmetry of X was important in this argument. ♦

10.5.2. Strong Arnold Property. We can forget about the meaning of A (in particular, about its positive semidefiniteness) and define the *Strong Arnold Property* of any G-matrix A: this means that $XA \neq 0$ for every nonzero $\overline{G}$-matrix X with zero diagonal.

For a general G-matrix A, this definition of the Strong Arnold Property does not correspond to a transversality property of a system like (10.9). Nevertheless, we can describe it as transversality with respect to another, more complicated system of algebraic manifolds. First, we have to extend the definition of transversality to more general manifolds. We say that smooth manifolds (not necessarily hypersurfaces) in $\mathbb{R}^n$ *intersect transversally* at a point x, if their normal subspaces at this point are linearly independent. For $m = 2$, this means that their normal subspaces at x intersect at the zero subspace.

Consider the manifold $\mathcal{R}_k$ of all symmetric $V \times V$ matrices with rank k. It is not hard to see that the dimension of $\mathcal{R}_k$ is $kn - \binom{k}{2}$. We need the following lemma describing the tangent space of $\mathcal{R}_k$ and its orthogonal complement.

LEMMA 10.24. (a) *The tangent space $\mathcal{T}_A$ of $\mathcal{R}_k$ at $A \in \mathcal{R}^k$ consists of all matrices of the form $AU + U^\mathsf{T}A$, where U ranges through all $V \times V$ matrices.*

(b) *The normal space $\mathcal{N}_A$ of $\mathcal{R}_k$ at A consists of all symmetric $V \times V$ matrices X such that $AX = 0$.*

Note that U in (a) is not necessarily symmetric, but $AU + U^\mathsf{T}A$ is.

PROOF. (a) Consider any matrix of the form $AU + U^\mathsf{T}A$, and the family

$$A_t = (I + tU)^\mathsf{T}A(I + tU).$$

Clearly $\mathrm{rk}(A_t) \le \mathrm{rk}(A) = k$ and equality holds if $|t|$ is small enough. Furthermore, $(\frac{d}{dt}A_t)_{t=0} = AU + U^\mathsf{T}A$, showing that $AU + U^\mathsf{T}A \in \mathcal{T}_A$.

Conversely, let $Y \in \mathcal{T}_A$. Then there is a one-parameter differentiable family A_t of symmetric matrices, defined in a neighborhood of $t = 0$, so that $A_t \in \mathcal{R}_k$, $A_0 = A$ and $\dot{A}_0 = (\frac{d}{dt}A_t)_{t=0} = Y$. In a sufficiently small neighborhood of $t = 0$ all matrices A_t have the same signature of eigenvalues, which implies that they can be written in the form $A_t = U_t^\mathsf{T}DU_t$, where D is a diagonal matrix in which the first k_1 diagonal entries are 1, the next $k - k_1$ diagonal entries are -1, and the rest are zero, and U_t is a nonsingular matrix which depends on t in a smooth way. Differentiating,

$$Y = \dot{A}_0 = \dot{U}_0^\mathsf{T}DU_0 + U_0^\mathsf{T}D\dot{U}_0 = (U_0^{-1}\dot{U}_0)^\mathsf{T}A + A(U_0^{-1}\dot{U}_0).$$

This shows that Y has the form as stated.

(b) We have $X \in \mathcal{N}(A)$ if and only if $X \cdot Y = 0$ for every $Y \in \mathcal{T}_A$. By (a), this means that $X \cdot (AU + U^\mathsf{T}A) = 0$ for every $U \in \mathbb{R}^{n \times n}$. This can be written as

$$\begin{aligned} 0 = \mathrm{Tr}\big(X(AU + U^\mathsf{T}A)\big) &= \mathrm{Tr}(XAU) + \mathrm{Tr}(XU^\mathsf{T}A) = \mathrm{Tr}(XAU) + \mathrm{Tr}(AUX) \\ &= 2\mathrm{Tr}(XAU). \end{aligned}$$

This holds for every U if and only if $XA = 0$. □

Using this description, we can easily prove the following lemma giving a geometric meaning to the Strong Arnold Property in the case of general G-matrices. Recall that $\mathcal{M}_G$ denotes the linear space of G-matrices.

LEMMA 10.25. *For a graph G, a G-matrix A has the Strong Arnold Property if and only if the manifolds $\mathcal{R}_k$ ($k = \mathrm{rk}(A)$) and $\mathcal{M}_G$ intersect transversally at A.*

PROOF. Let X be a nonzero matrix in the normal spaces of $\mathcal{M}_G$ and $\mathcal{R}_k$ at A. The first condition says that $X_{ij} = 0$ for every $ij \in E \cup \Delta$; the second says that $XA = 0$ (by Lemma 10.24). These are just the conditions on X needed to disprove the Strong Arnold property. □

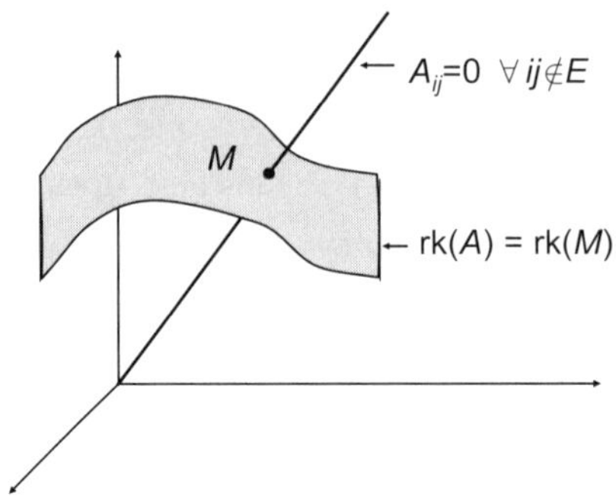

FIGURE 10.7. The Strong Arnold Property

We note that every matrix A with corank at most 1 has the Strong Arnold Property automatically. Indeed, if $AX = 0$, then X has rank at most 1, but since X is symmetric with 0's in the diagonal, this implies that $X = 0$.

It is not hard to prove that whether or not a G-matrix A has the Strong Arnold Property depends only on the graph G and on the nullspace of A (see Exercise 10.10). A useful combinatorial property of the nullspace is the following.

LEMMA 10.26. *Let G be a graph, let A be a G-matrix with the Strong Arnold Property. Let $x, y \in \mathrm{Ker}(A)$. Then either* $\mathrm{supp}(x)$ *and* $\mathrm{supp}(y)$ *intersect, or there is an edge of G connecting them.*

As a special case, we see that G has at most one connected component such that the corresponding diagonal block of A is singular.

PROOF. Suppose not, then $X = xy^{\mathsf{T}} + yx^{\mathsf{T}}$ is a nonzero $\overline{G}$-matrix with zero diagonal satisfying $MX = 0$, contradicting the Strong Arnold Property. □

Coming back to orthogonal representations, we have seen that transversality is weaker than locally general position; but a condition in the spirit of locally general position does follow from the previous lemma.

COROLLARY 10.27. *Let G be a simple graph and let $S, T \subseteq V$ be disjoint subsets such that no edge of G connects them. Let $\mathbf{u}$ be a transversal orthogonal representation of G. Then either $\mathbf{u}(S)$ or $\mathbf{u}(T)$ is linearly independent.*

As a special case, two nonadjacent nodes cannot be both represented by the zero vector.

10.5.3. Algebraic width. Based on transversality, an interesting graph invariant related to connectivity was introduced in [**Colin de Verdière 1998a**] (this is different from the better known Colin de Verdière number to be discussed in Chapter 16). He formulated this notion in terms of matrices, but from our perspective, working with orthogonal representations is more natural.

It will be useful to introduce an extension of the notion of faithful representations, based on [**van der Holst 1996**]. We would like to "interpolate" between faithful and nonfaithful orthogonal representations, by allowing "weak" edges for which the inner product may be zero or nonzero. Van der Holst suggests to encode this requirement by considering these edges as edges with multiplicity larger than one. This seems to be somewhat arbitrary, but it will turn out to be quite useful.

Given a multigraph G, let $A_G = (a_{ij})$ be its adjacency matrix, where a_{ij} denotes the number of edges connecting nodes i and j. We define a faithful orthogonal representation of it as a vector labeling $\mathbf{u}: V \to \mathbb{R}^d$, such that

$$\mathbf{u}_i^\mathsf{T}\mathbf{u}_j \begin{cases} = 0, & \text{if } i \neq j \text{ and } a_{ij} = 0, \\ \neq 0, & \text{if } a_{ij} = 1. \end{cases} \tag{10.11}$$

There is no condition when $i = j$ or when i and j are connected by more than one edge.

If the graph is simple, this gives the previously studied notion of faithfulness. If all edges have multiplicity larger than one, then we get all orthogonal representations. (The value of the multiplicity does not matter, once it is larger than one.) For every graph G, we define $G^=$ as the graph obtained from G by doubling all edges. Then faithful orthogonal representations of $G^=$ are just all orthogonal representations of G.

Let d be the smallest dimension in which a multigraph G has a faithful transversal orthogonal representation, and define $\mathsf{w}_{\text{alg}}(G) = n - d$. We call $\mathsf{w}_{\text{alg}}(G)$ the *algebraic width* of the graph (the name refers to its connection with tree width, see below). This definition is meaningful, since it is easy to construct a faithful orthogonal representation in $\mathbb{R}^n$, in which the representing vectors are almost orthogonal and hence linearly independent, which implies transversality. The algebraic width measures how much we can save in the dimension of a faithful transversal orthogonal representation, relative to the trivial dimension n.

This definition can be rephrased in terms of the Gram matrix of the representation. For any faithful transversal orthogonal representation $(\mathbf{u}_i : i \in V)$, the matrix $N = \mathsf{Gram}(\mathbf{u})$ has the following properties:

(N1) $N_{ij} \begin{cases} = 0 & \text{if } i \neq j \text{ and } a_{ij} = 0, \\ \neq 0 & \text{if } a_{ij} = 1; \end{cases}$

(N2) N is positive semidefinite;

(N3) If X is a nonzero $\overline{G}$-matrix with zero diagonal, then $NX \neq 0$.

Conversely, every matrix N with these properties can be written as the Gram matrix of a vector labeling (by (N2)), and it is easy to check that this vector labeling is a faithful transversal orthogonal representation. The rank of the matrix is the dimension of the representation. Hence we get the following reformulation:

LEMMA 10.28. *The algebraic width of a graph G is equal to the maximum corank of a matrix with properties* (N1)–(N3).

EXAMPLE 10.29 (Complete graphs). The Strong Arnold Property is void for complete graphs, and every representation is orthogonal, so we can use the same vector to represent every node. This shows that $\mathsf{w}_{\text{alg}}(K_n) = n-1$. For every noncomplete graph G, $\mathsf{w}_{\text{alg}}(G) \leq n-2$, since a faithful orthogonal representation requires at least two dimensions.

Doubling edges makes a difference: to get a faithful orthogonal representation of $K_n^=$, we can represent all nodes by the null vector. The Strong Arnold Property is void again, which shows that $\mathsf{w}_{\text{alg}}(K_n^=) = n$. ♦

EXAMPLE 10.30 (Edgeless graphs). To have a faithful orthogonal representation, all representing vectors must be mutually orthogonal, hence $\mathsf{w}_{\text{alg}}(\overline{K}_n) = 0$. It is not hard to see that every other multigraph G has $\mathsf{w}_{\text{alg}}(G) \geq 1$. ♦

EXAMPLE 10.31 (Paths II). Let P_n denote the path with n nodes, and assume that $n \geq 2$. Every matrix N satisfying (N1) has an $(n-1)\times(n-1)$ nonsingular (triangular) submatrix, and hence by Lemma 10.28, $\mathsf{w}_{\mathrm{alg}}(P_n) \leq 1$. Since P_n has an edge, we know that equality holds here. ♦

EXAMPLE 10.32 (Triangular grid II). To see a more interesting example, let us have a new look at the graphs Δ_k in Figure 10.5 (Example 10.22). For $k=3$, we have seen that Δ_3 has a faithful transversal orthogonal representation in $\mathbb{R}^3$, and this is clearly best possible, so $\mathsf{w}_{\mathrm{alg}}(\Delta_3)=3$.

For general k, the construction in Example 10.22 yields a faithful orthogonal representation $\mathbf{u}$. Recall that each coordinate of $\mathbf{u}_i$ corresponds to one of the dark triangles B, and it is 1 if $i \in B$, and 0 otherwise. We show that this representation is transversal. Let X be a nonzero $\overline{G}$-matrix with zero diagonal disproving the transversality of $\mathbf{u}$. This means that $\sum_j X_{ij}\mathbf{u}_j = 0$ for every node i, which implies that

$$\sum_{j\in B} X_{ij} = 0 \tag{10.12}$$

for every dark triangle B and node i.

We prove that $X_{ij}=0$ for all $i,j \in V = V(\Delta_k)$ by induction on the distance of i and j (in the sense of 2-dimensional Euclidean distance in the figure). If i and j are neighbors, then this is clear, since X is supported on $\overline{E} = E(\overline{\Delta}_k)$. Suppose that $ij \in \overline{E}$, then the interior segment connecting i and j intersects a dark triangle incident with either i or j; say it intersects triangle $B = jkl$. Then both k and l are closer to i than j is, and hence we already know that $X_{ik} = X_{il} = 0$. By (10.12), this implies that $X_{ij}=0$.

So $\mathbf{u}$ is a faithful transversal orthogonal representation of Δ_k. The dimension of this representation is the number of dark triangles, which is clearly $\binom{k-1}{2}$. We have argued that this is the smallest possible dimension (this follows just from faithfulness, not using transversality). Thus

$$\mathsf{w}_{\mathrm{alg}}(\Delta_k) = n - \binom{k-1}{2} = \binom{k}{2} - \binom{k-1}{2} = k.$$

♦

We continue with some easy bounds on the algebraic width. The condition that the representation must be faithful implies that the vectors representing a largest stable set of nodes must be mutually orthogonal nonzero vectors, and hence the dimension of the representation is at least $\alpha(G)$. This implies that

$$\mathsf{w}_{\mathrm{alg}}(G) \leq n - \alpha(G) = \tau(G). \tag{10.13}$$

By Theorem 10.10, every k-connected graph G has a faithful general position orthogonal representation in $\mathbb{R}^{n-k}$, and hence

$$\mathsf{w}_{\mathrm{alg}}(G) \geq \kappa(G). \tag{10.14}$$

We may also use the Strong Arnold Property to bound the algebraic width. There are $\binom{n}{2} - m$ orthogonality conditions, and in an optimal representation they involve $(n - \mathsf{w}_{\mathrm{alg}}(G))n$ variables. If their normal vectors are linearly independent, then $\binom{n}{2} - m \leq (n - \mathsf{w}_{\mathrm{alg}}(G))n$, and hence

$$\mathsf{w}_{\mathrm{alg}}(G) \leq \frac{n+1}{2} + \frac{m}{n}. \tag{10.15}$$

The most important consequence of the Strong Arnold Property is the following.

THEOREM 10.33. *The algebraic width* $\mathsf{w}_{\rm alg}(G)$ *is minor-monotone: If* H *is a minor of* G*, then* $\mathsf{w}_{\rm alg}(H) \le \mathsf{w}_{\rm alg}(G)$.

PROOF. It suffices to consider three cases: when H arises from G by deleting an isolated node, by deleting an edge, or by contracting an edge. We consider a faithful transversal orthogonal representation $\mathbf{u}$ of H in dimension $d = |V(H)| - \mathsf{w}_{\rm alg}(H)$, and modify it to get a faithful transversal orthogonal representation of G.

The case when H arises by deleting an isolated node i is trivial: Starting with an optimal representation of H, and representing i by the $(d+1)$-st unit vector, we get a representation of G in dimension $d+1$. It is straightforward to see that this faithful representation of G is transversal. Hence $\mathsf{w}_{\rm alg}(G) \ge n-(d+1) = |V(H)|-d = \mathsf{w}_{\rm alg}(H)$.

Suppose that $H = G \setminus ab$, where $ab \in E$. We start with an optimal representation $i \mapsto \mathbf{u}_i$ of H, this satisfies the conditions

$$\mathbf{u}_i^{\mathsf{T}}\mathbf{u}_j = 0 \qquad \big(ij \in \overline{E}(H)\big). \tag{10.16}$$

By transversality, for a sufficiently small $\varepsilon > 0$, the system of equations

$$\mathbf{x}_i^{\mathsf{T}}\mathbf{x}_j = 0 \qquad (ij \in \overline{E}(H) \setminus ab) \tag{10.17}$$

$$\mathbf{x}_a^{\mathsf{T}}\mathbf{x}_b = \varepsilon \tag{10.18}$$

has a solution in $\mathbf{x}_i \in \mathbb{R}^d$, $i \in V$ such that $|\mathbf{u}_i - \mathbf{x}_i| \le \varepsilon$ for all $i \in V$, and which is transversal with respect to the system (10.17)-(10.18). If ε is small enough, then this representation is a faithful orthogonal representation of G, and it is transversal with respect to (10.17), showing that $\mathsf{w}_{\rm alg}(G) \ge n-d = \mathsf{w}_{\rm alg}(H)$.

The most complicated case is obtained when H arises from G by contracting an edge ab to a node. It will be convenient to assume that $N(a) \cap N(b) = \emptyset$ (we can delete the edges connecting b to $N(a)$; this does not change the contracted graph, and does not increase $\mathsf{w}_{\rm alg}(G)$, as we have just proved). Let $A = N(a) \setminus b$, $B = N(b) \setminus a$ and $R = V(H) \setminus A \setminus B$. We identify b with the node of H obtained by the contraction, and add a new isolated node a to H, to obtain a graph G'. We have $V(G') = V$, and G' is obtained from G by shifting all ends of edges from a to b (Figure 10.8).

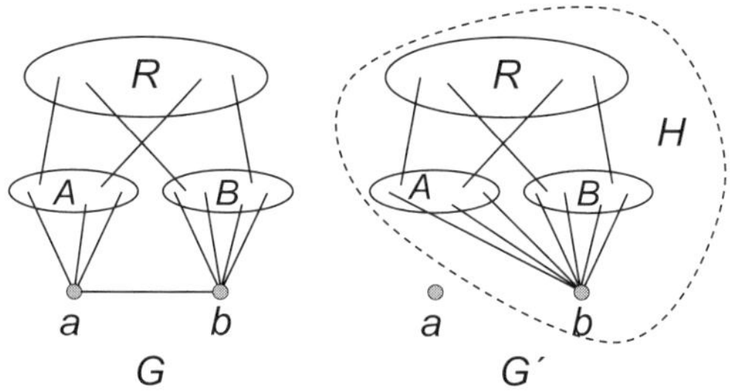

FIGURE 10.8. Minor monotonicity of the algebraic width

We start again with a faithful transversal orthogonal representation $\big(\mathbf{u}_i : i \in V(H)\big)$ in dimension $d = |V(H)| - \mathrm{w}_{\mathrm{alg}}(H)$ of H, and want to construct a faithful transversal orthogonal representation of G in dimension $d+1$. We start with constructing a vector labeling of G' in $\mathbb{R}^{d+1}$:

$$\mathbf{v}_i = \begin{cases} \begin{pmatrix} \mathbf{u}_i \\ 0 \end{pmatrix}, & \text{if } i \neq a, \\ \mathbf{e}_{d+1} & \text{if } i = a. \end{cases}$$

We have seen above that this is a faithful transversal orthogonal representation of G'. By the Implicit Function Theorem, for a sufficiently small $\varepsilon > 0$ there is a vector labeling $(\mathbf{z}_i : i \in V)$ such that $\mathbf{z}_i$ is arbitrarily close to $\mathbf{v}_i$, and $\mathbf{z}$ is a transversal solution of the system of equations

$$\mathbf{z}_i^{\mathsf{T}}\mathbf{z}_j = 0 \qquad (ij \in \overline{E}(H) \text{ or } i = a,\ j \in B \cup R), \tag{10.19}$$

$$\mathbf{z}_a^{\mathsf{T}}\mathbf{z}_j - \varepsilon\mathbf{z}_b^{\mathsf{T}}\mathbf{z}_j = 0 \qquad (j \in A). \tag{10.20}$$

Define another vector labeling of V by

$$\mathbf{w}_i = \begin{cases} \mathbf{z}_a - \varepsilon\mathbf{z}_b, & \text{if } i = b, \\ \mathbf{z}_i, & \text{otherwise.} \end{cases}$$

We claim that $(\mathbf{w}_i : i \in V)$ is an orthogonal representation of G. Let $ij \in \overline{E}(G)$. If $i, j \neq b$, then $\mathbf{w}_i^{\mathsf{T}}\mathbf{w}_j = 0$ follows immediately from (10.19). If (say) $i = b$ and $j \in R$, then

$$\mathbf{w}_b^{\mathsf{T}}\mathbf{w}_j = (\mathbf{z}_a - \varepsilon\mathbf{z}_b)^{\mathsf{T}}\mathbf{z}_j = \mathbf{z}_a^{\mathsf{T}}\mathbf{z}_j - \varepsilon\mathbf{z}_b^{\mathsf{T}}\mathbf{z}_j = 0$$

by (10.19). Finally, if $i = b$ and $j \in A$, then

$$\mathbf{w}_b^{\mathsf{T}}\mathbf{w}_j = (\mathbf{z}_a - \varepsilon\mathbf{z}_b)^{\mathsf{T}}\mathbf{z}_j = 0$$

by (10.20).

Second, we show that $\mathbf{w}$ is faithful, if ε is small enough, i.e., $\mathbf{w}_i^{\mathsf{T}}\mathbf{w}_j \neq 0$ for all $ij \in E$. If $i, j \neq a, b$, then this is clear, since $\mathbf{w}_i^{\mathsf{T}}\mathbf{w}_j \approx \mathbf{v}_i^{\mathsf{T}}\mathbf{v}_j = \mathbf{u}_i^{\mathsf{T}}\mathbf{u}_j \neq 0$ (the error at the $\approx$ sign can be made arbitrarily small). If $ij = ab$, then $\mathbf{w}_i^{\mathsf{T}}\mathbf{w}_j = \mathbf{w}_a^{\mathsf{T}}\mathbf{w}_b = \mathbf{z}_a^{\mathsf{T}}\mathbf{z}_a - \varepsilon\mathbf{z}_a^{\mathsf{T}}\mathbf{z}_b \neq 0$ if ε is small enough. If $i = a$ and $j \in A$, then $\mathbf{w}_a^{\mathsf{T}}\mathbf{w}_j = \varepsilon\mathbf{z}_i^{\mathsf{T}}\mathbf{z}_b \neq 0$ (since $ic \in E(H)$). If $i = b$ and $j \in B$, then $\mathbf{w}_b^{\mathsf{T}}\mathbf{w}_j = (\mathbf{z}_a - \varepsilon\mathbf{z}_c)^{\mathsf{T}}\mathbf{z}_j = \varepsilon\mathbf{z}_c^{\mathsf{T}}\mathbf{z}_j \approx \varepsilon\mathbf{v}_c^{\mathsf{T}}\mathbf{v}_j = \varepsilon\mathbf{u}_c^{\mathsf{T}}\mathbf{u}_j \neq 0$.

Finally, we show that $(\mathbf{w}_i : i \in V)$ is transversal (as an orthogonal representation of G), if ε is small enough. Suppose that there exists a nontrivial linear combination

$$f(\mathbf{x}) = \sum_{ij \in \overline{E}} X_{ij}\mathbf{x}_i^{\mathsf{T}}\mathbf{x}_j$$

of the orthogonality conditions such that $\mathbf{w}$ is a stationary point of the function f. The substitution

$$\mathbf{x}_i = \begin{cases} \mathbf{y}_a - \varepsilon\mathbf{y}_b, & \text{if } i = b, \\ \mathbf{y}_i, & \text{otherwise} \end{cases}$$

maps $\mathbf{w}$ onto $\mathbf{z}$, and it turns f into the function

$$g(\mathbf{y}) = \sum_{\substack{ij\in\overline{E}\\ i,j\neq b}} X_{ij}\mathbf{y}_i^\mathsf{T}\mathbf{y}_j + \sum_{j\in A\cup R} X_{bj}(\mathbf{y}_a - \varepsilon\mathbf{y}_b)^\mathsf{T}\mathbf{y}_j$$

$$= \sum_{\substack{ij\in\overline{E}\\ i,j\neq b}} X_{ij}\mathbf{y}_i^\mathsf{T}\mathbf{y}_j + \sum_{j\in R} X_{bj}\mathbf{y}_a^\mathsf{T}\mathbf{y}_j - \varepsilon\sum_{j\in R} X_{bj}\mathbf{y}_b^\mathsf{T}\mathbf{y}_j + \sum_{j\in A} X_{bj}(\mathbf{y}_a - \varepsilon\mathbf{y}_b)^\mathsf{T}\mathbf{y}_j.$$

The function g is a linear combination of the left hand sides of the equations (10.19)–(10.20), and $\mathbf{z}$ is a stationary point of this linear combination. Since $\mathbf{z}$ is transversal, this linear combination must be trivial. The last sum is the only one where the functions $(\mathbf{y}_a - \varepsilon\mathbf{y}_b)^\mathsf{T}\mathbf{y}_j$ ($j \in A$) are used, and hence $X_{bj} = 0$ for $j \in A$. Similarly, the third sum gives that $X_{bj} = 0$ for $j \in R$. This means that all coefficients in the second sum are 0. This leaves us with the first sum, so all coefficients there must be 0 as well. But this is a contradiction, since $f(\mathbf{x})$ was assumed to be a nontrivial linear combination. □

10.5.4. Algebraic width, treewidth and connectivity. We start with surveying several minor-monotone graph parameters related to connectivity.

The connectivity of a minor of a graph can be larger than the connectivity of the graph itself. (A trivial example is a subdivision of a highly connected graph, which is only 2-connected, but by edge contraction we get the original graph as a minor.) Thus it makes sense to define the *monotone connectivity* $\kappa_{\text{mon}}(G)$ of a graph G as the maximum connectivity of any minor of G.

Tree-width was introduced in [**Robertson–Seymour 1984**] as an important element in their graph minor theory. We try to decompose our graph G as the union of subgraphs G_i, which are indexed by the nodes of a tree T, with the following property: if i, j, k are three nodes of T and j lies on the path between i and k, then $V(G_i)\cap V(G_k) \subseteq V(G_j)$. Such a family is called a *tree-decomposition* of the graph G. The *tree-width* of a graph G is the smallest integer k for which G has a tree-decomposition into parts with at most $k+1$ nodes. It is not hard to see that $\kappa(G) \le \mathsf{w}_{\text{tr}}(G)$, which implies

$$\kappa_{\text{mon}}(G) \le \mathsf{w}_{\text{tr}}(G). \tag{10.21}$$

A closely related parameter, which we call the *product-width* $\mathsf{w}_{\text{pr}}(G)$ of a graph G, was also defined in [**Colin de Verdière 1998a**]. This is the smallest positive integer r for which G is a minor of a Cartesian product $K_r\square T$, where T is a tree. The reason why we consider the product width a "connectivity-related" parameter is that it is almost equal to treewidth:

$$\mathsf{w}_{\text{tr}}(G) \le \mathsf{w}_{\text{pr}}(G) \le \mathsf{w}_{\text{tr}}(G)+1. \tag{10.22}$$

The lower bound was proved in [**van der Holst 1996**],, the upper was established in [**Colin de Verdière 1998a**].. The proof of these inequalities is left to the reader as a (not quite easy) exercise.

Monotone connectivity, product width and treewidth are minor-monotone graph parameters; this is trivial for the first two, and well known for treewidth. It is easy to verify that a graph is a forest if and only if either one of these parameters is at most 1.

So far, we have discussed connectivity-type graph parameters; now we relate them to the algebraic width, which will be sandwiched between two of them:

THEOREM 10.34. *For every graph G,*

$$\kappa_{\rm mon}(G) \le \mathsf{w}_{\rm alg}(G) \le \mathsf{w}_{\rm pr}(G).$$

The upper bound was proved in [**Colin de Verdière 1998a**], while the lower bound is an easy consequence of the results in Section 10.3.

PROOF. Let H be a minor of G with largest connectivity. Using (10.14) and the fact that $\mathsf{w}_{\rm alg}(G)$ is minor-monotone, we get the first inequality:

$$\kappa_{\rm mon}(G) = \kappa(H) \le \mathsf{w}_{\rm alg}(H) \le \mathsf{w}_{\rm alg}(G).$$

To prove the second, we use again the minor-monotonicity of $\mathsf{w}_{\rm alg}(G)$: it suffices to prove that

$$\mathsf{w}_{\rm alg}(K_r\Box T) \le r \tag{10.23}$$

for any $r \ge 1$ and tree T. Note that if T has at least two nodes, then K_{r+1} can be obtained as a minor of $K_r\Box T$, and hence $\mathsf{w}_{\rm alg}(K_r\Box T) \ge \mathsf{w}_{\rm alg}(K_{r+1}) = r$ by the minor monotonicity of algebraic width. So for every tree T with at least one edge, inequality (10.23) holds with equality.

Inequality (10.23) says that the minimum dimension in which $K_r\Box T$ has a faithful transversal orthogonal representation is at least $(n-1)r$, where $n = |V(T)|$. In the argument below, we only need the faithfulness of the representation, transversality is not used in this part. Using Proposition 10.18, it suffices to show that $K_r\Box T$ has an almost stable set of size $(n-1)r$. We claim that for any node $u \in V(T)$, the subset $S = (V(T)\setminus u)\times V(K_r)$ is almost stable.

Let H be a connected subgraph of $(K_r\Box T)[S]$. Considering u as the root of T, let v be a node of T closest to u such that H contains a node from $\{v\}\times V(K_r)$. Note that all nodes of H are contained in sets $\{w\}\times V(K_r)$ where $w = v$ or w is a descendant of v. Let (v,i) be a node of H, and let v' be the parent of v in T. Then (v',i) is a node of $K_r\Box T$ adjacent to (v,i), but to no other node of H. This proves that S is almost stable, and thus completes the proof of the theorem. □

For small values, equality holds in Theorem 10.34; this was proved in [**van der Holst 1996**] and in [**Kotlov 2000**].

PROPOSITION 10.35. *Let G be a graph with at least one edge. If $\mathsf{w}_{\rm alg}(G) \le 2$, then $\kappa_{\rm mon}(G) = \mathsf{w}_{\rm alg}(G) = \mathsf{w}_{\rm pr}(G)$.*

PROOF. Using Theorem 10.34 (an easy special case) and the remark above about forests, we get

$$\mathsf{w}_{\rm alg}(G)=1 \Longrightarrow \kappa_{\rm mon}(G)=1 \Longrightarrow G \text{ is a forest} \Longrightarrow \mathsf{w}_{\rm pr}(G)=1 \Longrightarrow \mathsf{w}_{\rm alg}(G)=1,$$

and this proves the assertion for $\mathsf{w}_{\rm alg}(G) = 1$. Suppose that $\mathsf{w}_{\rm alg}(G) = 2$, then $\kappa_{\rm mon}(G) \le 2$ by Theorem 10.34. Furthermore, neither K_4 nor the graph Δ_3 in Example 10.23 is a minor of G, since $\mathsf{w}_{\rm alg}(K_4) = \mathsf{w}_{\rm alg}(\Delta_3) = 3$.

By Exercise 10.13, G satisfies $\mathsf{w}_{\rm pr}(G) \le 2$. We must have equality, since otherwise $\mathsf{w}_{\rm alg}(G) = 1$ would follow. □

A planar graph has $\kappa_{\rm mon}(G) \le 5$ (since every simple minor of it is planar and hence it has a node with degree at most 5). Since planar graphs can have arbitrarily large algebraic width (see Exercise 10.15), the first inequality in Theorem 10.34 can be arbitrarily far from equality: no function of $\kappa_{\rm mon}$ bounds $\mathsf{w}_{\rm alg}$ from above.

The following example from [**van der Holst 1996**] shows that in general, equality does not hold in the upper bound in Theorem 10.34 either.

EXAMPLE 10.36 (Wagner graph). Consider the *Wagner graph* W_8 obtained from the cycle C_8 by adding its longest diagonals. Then $\kappa_{\text{mon}}(W_8) = \mathsf{w}_{\text{alg}}(W_8) = 3$ and $\mathsf{w}_{\text{tr}}(W_8) = \mathsf{w}_{\text{pr}}(W_8) = 4$. These facts take some work to verify, especially the value of the algebraic width. ♦

As a weak converse of the inequality $\mathsf{w}_{\text{alg}} \le \mathsf{w}_{\text{pr}}$ in Theorem 10.34, the product width (or tree-width) is bounded by *some* function of the algebraic width: For every graph G, we have

$$\mathsf{w}_{\text{pr}}(G) = O(\mathsf{w}_{\text{alg}}(G)^{20}). \tag{10.24}$$

indeed, Example 10.32 implies that G cannot contain the triangular grid $\Delta_{\mathsf{w}_{\text{alg}}(G)+1}$ as a minor. A basic result of graph minor theory [**Robertson–Seymour 1986**] says that the treewidth of such a graph is bounded by an appropriate function of the grid size. The bound used in (10.24) was proved in [**Chuzoy 2016**].

10.6. The variety of orthogonal representations

In this section we take a global view of all orthogonal representations of a graph in $\mathbb{R}^d$. For small values of d, several of our considerations might be easier to follow in terms of the complementary graph $\overline{G}$. In particular, G is $(n-d)$-connected if and only if $\overline{G}$ does not contain a complete bipartite graph on $d+1$ nodes (i.e., a complete bipartite graph $K_{a,b}$ with $a, b \ge 1$, $a+b = d+1$).

Let G be a simple graph on n nodes, which are labeled $1, \dots, n$. Every vector labeling $i \mapsto \mathbf{v}_i \in \mathbb{R}^d$ can be considered as a point in $\mathbb{R}^{dn}$, and orthogonal representations of G in $\mathbb{R}^d$ form an algebraic variety $\mathcal{OR}_d(G) \subseteq \mathbb{R}^{dn}$, defined by the equations

$$\mathbf{u}_i^{\mathsf{T}}\mathbf{u}_j = 0 \quad (ij \in \overline{E}). \tag{10.25}$$

We could consider orthonormal representations, whose variety is defined by the equations

$$\mathbf{u}_i^{\mathsf{T}}\mathbf{u}_i = 1 \quad (i \in V), \qquad \mathbf{u}_i^{\mathsf{T}}\mathbf{u}_j = 0 \quad (ij \in \overline{E}). \tag{10.26}$$

The set of orthogonal representations of G that are *not* in general position also forms an algebraic variety: we only have to add to (10.25) the equation

$$\prod_{1\le i_1<\dots<i_d\le n} \det(\mathbf{u}_{i_1}, \dots, \mathbf{u}_{i_d}) = 0. \tag{10.27}$$

Similarly, the set of orthogonal representations that are not in locally general position form a real algebraic variety. The set $\mathcal{GOR}_d(G)$ of general position orthogonal representations of G in $\mathbb{R}^d$, as well as the set $\mathcal{LOR}_d(G)$ of locally general position orthogonal representations of G, are semialgebraic.

Let us a look at some examples showing the complications (or interesting features, depending on your point of view) of the variety of orthogonal representations of a graph.

EXAMPLE 10.37. Consider the graph $\overline{K_2}$ consisting of two isolated nodes. Every orthogonal representation of this graph consists of two orthogonal vectors. It follows that in dimension $d = 1$, one of the labels must be 0; the other one is arbitrary. So the variety of orthogonal representations decomposes into two subvarieties, which intersect at the point $(0,0)$. This graph has no orthonormal representation in dimension 1.

For $d=2$, we look at pairs $(\mathbf{u}_1,\mathbf{u}_2)$ of orthogonal vectors. In the *orthonormal* case, this implies that $\mathbf{u}_2$ arises from $\mathbf{u}_1$ by rotation of $90°$ either in the positive or in the negative direction. In other words, either $(\mathbf{u}_1)_1=(\mathbf{u}_2)_2$ and $(\mathbf{u}_1)_2=-(\mathbf{u}_2)_1$, or $(\mathbf{u}_1)_1=-(\mathbf{u}_2)_2$ and $(\mathbf{u}_1)_2=(\mathbf{u}_2)_1$. So the variety of orthonormal representations decomposes into two algebraic varieties. It is not hard to see that these two are irreducible.

Considering, more generally, *orthogonal* representations in the plane, their variety decomposes into two semialgebraic sets in a natural way, depending on the sign of $\det(\mathbf{u}_1,\mathbf{u}_2)$. These two semialgebraic sets intersect in the set of degenerate labelings $\{(\mathbf{u}_1,\mathbf{u}_2):\ \mathbf{u}_1=0 \text{ or } \mathbf{u}_2=0\}$. These two semialgebraic sets are not algebraic however.

For $d>2$, both varieties (orthogonal and orthonormal representations) are indecomposable. Orthonormal representations of $\overline{K}_n$ in $\mathbb{R}^d$ form the Stiefel manifold of n-frames in $\mathbb{R}^d$. ♦

EXAMPLE 10.38. Let us look at the structure of orthogonal representations of an arbitrary graph G in one dimension. We have a single variable x_i assigned to each node i, and the orthogonality conditions say that $x_ix_j=0$ for $ij\in\overline{E}$. This is trivially equivalent to saying that the support of the vector $x=(x_i:\ i\in V)$ is a clique. So this variety decomposes into as many subvarieties as the number of maximal cliques.

This shows that the variety of orthogonal representations can have a very complex structure even in the smallest dimension. For example, the dimension of this variety is $\omega(G)$, an NP-hard quantity. ♦

EXAMPLE 10.39. Our last example is nontrivial, and it will play an important role later in this section. Let Q^3 denote the graph of the (ordinary) 3-dimensional cube. Note that Q^3 is bipartite; let $A=\{a_1,a_2,a_3,a_4\}$ and $B=\{b_1,b_2,b_3,b_4\}$ be its color classes. The indices are chosen so that a_i is opposite to b_i (Figure 10.9 (a)).

Since Q^3 contains neither $K_{1,4}$ nor $K_{2,3}$, the graph $\overline{Q^3}$ is 4-connected, and hence it has a general position orthogonal representation in $\mathbb{R}^4$. This is in fact easy to construct. Choose any four linearly independent vectors $\mathbf{x}_1,\dots,\mathbf{x}_4$ to represent $a_1,\dots,a_4$, then the vector $\mathbf{y}_i$ representing b_i is uniquely determined (up to scaling) by the condition that it has to be orthogonal to the three vectors $\mathbf{x}_j$, $j\neq i$. It follows from Theorem 10.10 that for almost all choices of $\mathbf{x}_1,\dots,\mathbf{x}_4$, the orthogonal representation obtained this way is in general position.

For bipartite graphs G, we can think of orthogonal representations of $\overline{G}$ in $\mathbb{R}^d$ in another way. Vectors representing one bipartition class can be thought of as homogeneous coordinates of points in projective $(d-1)$-space, while vectors in the other class can be thought of as homogeneous coordinates of planes. Orthogonality is then translated to incidence. So to construct a general position orthogonal representation of $\overline{Q^3}$ in $\mathbb{R}^4$, we need four points in general position in 3-space, and four planes, each going through exactly three of these points. In other words, we get the vertices and faces of a tetrahedron (Figure 10.9 (b)).

We show that every orthogonal representation of this graph in general position satisfies a certain algebraic equation. Let $\mathbf{x}_i'$ denote the orthogonal projection of $\mathbf{x}_i$ onto the first two coordinates, and let $\mathbf{y}_i'$ denote the orthogonal projection of $\mathbf{y}_i$

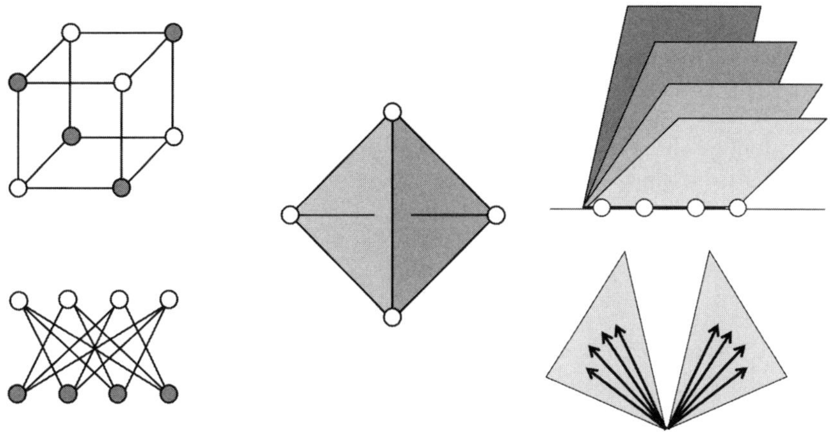

(a) (b) (c)

FIGURE 10.9. (a) The cube as bipartite graph. (b) An orthogonal representation in general position, depicted as a tetrahedron in projective 3-space. The homogeneous coordinates of the vertices represent one color class, those of the facets represent the other. (c) A degenerate orthogonal representation.

onto the third and fourth coordinates. We claim that

$$\frac{\det(\mathbf{x}'_1, \mathbf{x}'_3)\det(\mathbf{x}'_2, \mathbf{x}'_4)}{\det(\mathbf{x}'_1, \mathbf{x}'_4)\det(\mathbf{x}'_2, \mathbf{x}'_3)} = \frac{\det(\mathbf{y}'_1, \mathbf{y}'_3)\det(\mathbf{y}'_2, \mathbf{y}'_4)}{\det(\mathbf{y}'_1, \mathbf{y}'_4)\det(\mathbf{y}'_2, \mathbf{y}'_3)}. \tag{10.28}$$

Before proving (10.28), we remark that in a geometric language, this equation says the following: *Let ℓ be a line in projective 3-space that intersects the planes of the faces of a tetrahedron T at four distinct points $\mathbf{x}_1, \mathbf{x}_2, \mathbf{x}_3, \mathbf{x}_4$. Let $\Sigma_1, \Sigma_2, \Sigma_3, \Sigma_4$ be the four planes spanned by ℓ and the vertices of T, indexed so that the face defining $\mathbf{x}_i$ is opposite to the vertex defining Σ_i. Then the cross ratios $(\mathbf{x}_1 : \mathbf{x}_2 : \mathbf{x}_3 : \mathbf{x}_4)$ and $(\Sigma_1 : \Sigma_2 : \Sigma_3 : \Sigma_4)$ are equal* (see Appendix A.4).

To prove (10.28), we may assume that $\mathbf{x}_i^\mathsf{T}\mathbf{y}_i = 1$ for $i = 1, \dots, 4$. This can be done, since $\mathbf{x}_i^\mathsf{T}\mathbf{y}_i \neq 0$ (else, $\mathbf{x}_1, \dots, \mathbf{x}_4$ would all be contained in the 3-space orthogonal to $\mathbf{y}_i$, contradicting the assumption that the representation is in general position), and the equation is invariant under scaling. This means that the 4×4 matrices $X = (\mathbf{x}_1, \mathbf{x}_2, \mathbf{x}_3, \mathbf{x}_4)$ and $Y = (\mathbf{y}_1, \mathbf{y}_2, \mathbf{y}_3, \mathbf{y}_4)$ satisfy $X^\mathsf{T}Y = I$. By the well-known relationship between the subdeterminants of inverse matrices, we have

$$\det(\mathbf{x}'_1, \mathbf{x}'_3) = \det(X)\det(\mathbf{y}'_2, \mathbf{y}'_4),$$

and similarly for the other determinants in (10.28). Substituting these values, we get (10.28) after simplification.

There is another type of orthogonal representation of this graph, in which the elements of U are represented by vectors $\mathbf{x}_1, \dots, \mathbf{x}_4$ in a linear subspace L of $\mathbb{R}^4$ and the elements of W are represented by vectors $\mathbf{y}_1, \dots, \mathbf{y}_4$ in the orthogonal complement $L^\perp$ of L. Clearly, this is an orthogonal representation for any choice of the representing vectors in these subspaces. This representation is never faithful.

As a special case, we can choose 4 vectors $\mathbf{x}_i$ in the plane spanned by the first two coordinates, and 4 vectors $\mathbf{y}_i$ in the plane spanned by the third and fourth coordinates. Since these choices are independent, equation 10.28 will not hold in general. (In the projective representation, we get four collinear points and four planes incident with their line; Figure 10.9 (c)).

It follows from these arguments that the equation

$$\det(\mathbf{x}_1,\mathbf{x}_2,\mathbf{x}_3,\mathbf{x}_4)\Big(\det(\mathbf{x}'_1,\mathbf{x}'_3)\det(\mathbf{x}'_2,\mathbf{x}'_4)\det(\mathbf{y}'_1,\mathbf{y}'_4)\det(\mathbf{y}'_2,\mathbf{y}'_3) \\ -\det(\mathbf{x}'_1,\mathbf{x}'_4)\det(\mathbf{x}'_2,\mathbf{x}'_3)\det(\mathbf{y}'_2,\mathbf{y}'_4)\det(\mathbf{y}'_1,\mathbf{y}'_3)\Big)=0$$

is satisfied by every orthogonal representation, but the first factor is nonzero for the orthogonal representations in general position, while the second is nonzero for almost all of those of the second type. So the variety $\mathcal{OR}_d(\overline{Q^3})$ is reducible. ♦

10.6.1. G-orthogonalization revisited. For any graph G with ordered nodes, and any vector labeling $\mathbf{x}\in\mathbb{R}^{d\times V}$, the G-orthogonalization procedure described in Section 10.3.1 assigns an orthogonal representation $\Phi\mathbf{x}\in\mathbb{R}^{d\times V}$. The mapping $\Phi:\ \mathbb{R}^{d\times V}\to\mathbb{R}^{d\times V}$ is well-defined, and it is a retraction: if $\mathbf{x}$ itself is an orthogonal representation, then $\Phi\mathbf{x}=\mathbf{x}$. In particular, the range of Φ is just $\mathcal{OR}_d(G)$. But the map Φ is not necessarily continuous (check Example 10.37).

We can modify this procedure so that we get vectors whose coordinates are polynomials of the original coordinates. Given $\mathbf{x}\in\mathbb{R}^{d\times V}$, we define the vectors $\mathbf{z}_j\in\mathbb{R}^d$ recursively by

$$\mathbf{z}_j=\begin{vmatrix}\mathbf{x}_j & \mathbf{z}_{i_1} & \dots & \mathbf{z}_{i_r}\\ \mathbf{z}_{i_1}^{\mathsf T}\mathbf{x}_j & \mathbf{z}_{i_1}^{\mathsf T}\mathbf{z}_{i_1} & \dots & \mathbf{z}_{i_1}^{\mathsf T}\mathbf{z}_{i_r}\\ \vdots & \vdots & & \vdots\\ \mathbf{z}_{i_r}^{\mathsf T}\mathbf{x}_j & \mathbf{z}_{i_r}^{\mathsf T}\mathbf{z}_{i_1} & \dots & \mathbf{z}_{i_r}^{\mathsf T}\mathbf{z}_{i_r}\end{vmatrix}, \tag{10.29}$$

where $\overline{N}(j)\cap[j]=\{i_1,\dots,i_r\}$. (In this compact notation, the first row consists of vectors, the later rows consist of numbers. Expanding the determinant gives a meaningful expression.) This defines a map $\Psi:\ \mathbf{x}\mapsto\mathbf{z}$.

Obviously, $\mathbf{z}_j$ is a linear combination

$$\mathbf{z}_j=\alpha_j\mathbf{x}_j+\sum_{t=1}^{r}\beta_{j,t}\mathbf{z}_{i_t}, \tag{10.30}$$

and it is orthogonal to $\mathbf{z}_{i_1},\dots,\mathbf{z}_{i_r}$, since taking inner product with any of these vectors, we get a determinant with two equal rows. It follows that $\mathbf{z}$ is an orthogonal representation of G for every $\mathbf{x}\in\mathbb{R}^{d\times V}$. It also follows by induction on j that each entry of each $\mathbf{z}_j$ is a polynomial in the coordinates of $\mathbf{x}$.

We want to compare the vectors $\mathbf{z}$ with the vectors $\mathbf{y}=\Phi\mathbf{x}$. The following lemma sums up some simple properties of this modified G-orthogonalization procedure.

Lemma 10.40. *With the notation above, $\mathbf{z}_j=0$ if and only if $\mathbf{x}_j,\mathbf{z}_{i_1},\dots,\mathbf{z}_{i_r}$ are linearly dependent. The vector $\mathbf{z}_j$ is a nonnegative scalar multiple of $\mathbf{y}_j$.*

Proof. If $\mathbf{x}_j,\mathbf{z}_{i_1},\dots,\mathbf{z}_{i_r}$ are linearly dependent (in particular, if either one of them is zero), then $\mathbf{z}_j=0$, since the columns of the determinant (10.29) are linearly dependent. If $\mathbf{x}_j, \mathbf{z}_{i_1}, \dots, \mathbf{z}_{i_r}$ are linearly independent, then the linear combination as in (10.30) is unique, and

$$\alpha_j = \begin{vmatrix} \mathbf{z}_{i_1}^\mathsf{T}\mathbf{z}_{i_1} & \dots & \mathbf{z}_{i_1}^\mathsf{T}\mathbf{z}_{i_r} \\ \vdots & & \vdots \\ \mathbf{z}_{i_r}^\mathsf{T}\mathbf{z}_{i_1} & \dots & \mathbf{z}_{i_r}^\mathsf{T}\mathbf{z}_{i_r} \end{vmatrix} > 0,$$

so $\mathbf{z}_j \neq 0$.

For the second assertion, we prove by induction on j that if $\mathbf{z}_j \neq 0$, then $\mathbf{z}_j = \alpha_j \mathbf{y}_j$. We know by the first assertion of the lemma that $\mathbf{z}_{i_t} \neq 0$ for $1 \le t \le r$, and hence $\mathbf{z}_{i_t} = \alpha_{i_t}\mathbf{y}_{i_t}$ by induction. It follows that $\mathbf{z}_j$ is orthogonal to $\mathbf{y}_{i_1}, \dots, \mathbf{y}_{i_r}$, and

$$\frac{1}{\alpha_j}\mathbf{z}_j = \mathbf{x}_j + \sum_{t=1}^{r} \frac{\beta_{j,t}\alpha_{i_t}}{\alpha_j}\mathbf{y}_{i_t}.$$

Since these properties determine $\mathbf{y}_j$, we get that $(1/\alpha_j)\mathbf{z}_j = \mathbf{y}_j$. □

Suppose that we scale one of the vectors $\mathbf{x}_j$ by a scalar $\lambda_j \neq 0$. From the orthogonalization procedure we see that $\mathbf{y}_i$ does not change for $i \neq j$, and $\mathbf{y}_j$ is scaled by λ_j. Trivially, the vectors $\mathbf{z}_i$ do not change for $i < j$, and $\mathbf{z}_j$ will be scaled by λ_j. The vectors $\mathbf{z}_i$ with $i > j$ change by appropriate factors that are powers of λ_j. It follows that if we want to scale each $\mathbf{z}_j$ by some given scalar $\lambda_j \neq 0$, then we can scale $\mathbf{x}_1, \dots, \mathbf{x}_n$ one by one to achieve that. As a consequence, we see that the range $\mathrm{Rng}(\Psi)$ is closed under rescaling each vector independently by a nonzero scalar.

Every orthogonal representation is a fixed point of Φ, and so $\mathrm{Rng}(\Phi) = \mathcal{OR}(G)$. The range of Ψ depends on the ordering of the nodes: an orthogonal representation is in the range of Ψ if and only if for every node with $\mathbf{z}_j \neq 0$ the vectors representing the nodes in $\overline{N}(j) \cap \{1, \dots, j-1\}$ are linearly independent. We have

$$\mathrm{Rng}(\Psi) \subseteq \mathrm{Rng}(\Phi) = \mathcal{OR}(G), \tag{10.31}$$

and Example 10.39 shows that equality may not hold even for $(n-d)$-connected graphs.

Every orthogonal representation in locally general position is in the range of Ψ, so we have $\mathcal{GOR}(G) \subseteq \mathcal{LOR}(G) \subseteq \mathrm{Rng}(\Psi)$. This is uninteresting if G is not $(n-d)$-connected (in this case $\mathcal{GOR}(G) = \mathcal{LOR}(G) = \emptyset$). Let us assume that G is $(n-d)$-connected, then $\Phi\mathbf{x}$ is in general position for almost all $\mathbf{x}$ by Lemma 10.12, and hence so is $\Psi\mathbf{x}$. Since Ψ is continuous, it follows that $\mathcal{GOR}(G)$ is dense in the range of Ψ for every ordering of the nodes, and so for every $(n-d)$-connected graph G,

$$\overline{\mathcal{GOR}(G)} = \overline{\mathcal{LOR}(G)} = \mathrm{Rng}(\Psi). \tag{10.32}$$

10.6.2. Irreducibility. One may wonder whether various algebraic properties of the varieties $\mathcal{OR}_d(G)$, $\mathcal{GOR}_d(G)$ etc. have any combinatorial significance. The dimension of this variety may be interesting, but the fact that this is an NP-hard quantity even in the simplest case when $d = 1$ (Example 10.1) should warn as that this dimension my be intractable. In this section we address only one question, namely the irreducibility of $\mathcal{OR}_d(G)$, which does have some interesting combinatorial aspects.

A set $A \subseteq \mathbb{R}^N$ is *reducible* if there exist two polynomials $p, q \in \mathbb{R}[x_1, \dots, x_N]$ such that their product pq vanishes on A, but neither p nor q vanishes on A;

equivalently, the polynomial ideal $\{p: \; p \text{ vanishes on } A\}$ is not a prime ideal. If an algebraic variety A is reducible, then it is the union of two algebraic varieties that are both proper subvarieties, and therefore they can be considered "simpler".

PROPOSITION 10.41. *If G is not $(n-d)$-connected, then the algebraic variety $\mathcal{OR}_d(G)$ of its orthogonal representations is reducible.*

PROOF. In this case G has no general position orthogonal representation in $\mathbb{R}^d$ by Theorem 10.9. Hence equation (10.27) holds on $\mathcal{OR}_d(G)$. On the other hand, none of the equations

$$\det(\mathbf{v}_{i_1}, \dots, \mathbf{v}_{i_d}) = 0 \quad (1 \le i_1 < \dots < i_d \le n)$$

holds everywhere on $\mathcal{OR}_d(G)$, because

$$\mathbf{v}_j = \begin{cases} \mathbf{e}_k, & \text{if } j = i_k,\ k = 1, \dots, d, \\ 0, & \text{otherwise}, \end{cases}$$

is an orthogonal representation with $\det(\mathbf{v}_{i_1}, \dots, \mathbf{v}_{i_d}) \neq 0$. □

In the opposite direction, we show the following.

PROPOSITION 10.42. *For every graph G, the semialgebraic variety $\mathcal{GOR}_d(G)$ of its orthogonal representations in general position is irreducible.*

PROOF. Recall that $\mathcal{GOR}_d(G)$ is nonempty if and only if G is $(n-d)$-connected, so the assertion is trivial unless G is $(n-d)$-connected. Let f and g be two polynomials such that fg vanishes on $\mathcal{GOR}_d(G)$. Then fg vanishes on the closure of $\mathcal{GOR}_d(G)$; by (10.32) this means that fg vanishes on the range of Ψ. In other words, $f(\Psi\mathbf{x})g(\Psi\mathbf{x})$ vanishes for all $\mathbf{x} \in \mathbb{R}^{d\times V}$. Since $f(\Psi\mathbf{x})$ and $g(\Psi\mathbf{x})$ are polynomials in $\mathbf{x}$, it follows that either $f(\Psi\mathbf{x})$ or $g(\Psi\mathbf{x})$ vanishes identically; say the first occurs, then f vanishes on $\text{Rng}(\Psi)$, and hence also on $\mathcal{GOR}_d(G)$ by (10.32). □

From now on, we assume that G is $(n-d)$-connected. To exclude some trivial complications, we assume that $d < n$, so that G is connected. Since this connectivity condition does imply a lot about orthogonal representations of G in $\mathbb{R}^d$ by Theorem 10.9, we might expect that $\mathcal{OR}_d(G)$ will be irreducible in this case. Example 10.39 shows that this is false in general, so Proposition 10.41 cannot be reversed. Our goal is to prove that, at least for $d \le 4$, this is essentially the only counterexample. The case of higher dimensions remains open.

THEOREM 10.43. *Let G be an $(n-d)$-connected graph, where $d \le 4$. Then $\mathcal{GOR}_d(G)$ is dense in $\mathcal{OR}_d(G)$, and hence $\mathcal{OR}_d(G)$ is irreducible, unless $d = 4$ and $\overline{G}$ has a connected component that is the skeleton of the 3-cube.*

To give the proof, we need some definitions (just for this section). We say that an orthogonal representation of a graph G in $\mathbb{R}^d$ is *exceptional* if it does not belong to the topological closure of $\mathcal{GOR}_d(G)$. A graph G is *exceptional*, if it is has an exceptional orthogonal representation. (If G is not $(n-d)$-connected, then every representation in $\mathcal{OR}_d(G)$ is exceptional, but we are interested in the $(n-d)$-connected case.) By this definition, the set of exceptional orthogonal representations of G in $\mathbb{R}^d$ is relatively open in $\mathcal{OR}_d(G)$.

If an induced subgraph of a graph G is exceptional, then so is G itself: we can extend the exceptional representation of the induced subgraph to an orthogonal representation of the whole graph, and clearly this remains exceptional. (Note that every induced subgraph G' is $(|V(G')|-d)$-connected.) A graph G is *minimally exceptional*, if it is exceptional, but no proper induced subgraph of it is exceptional.

Let $\mathbf{x}$ and $\mathbf{y}$ be orthogonal representations of G in $\mathbb{R}^d$. If we keep $\mathbf{x}$ fixed and make $\sum_i |\mathbf{y}_i - \mathbf{x}_i|$ small enough, then for every $S \subseteq V$ for which $\mathbf{x}(S)$ is linearly independent, so is $\mathbf{y}(S)$, while some of the linear dependencies among the $\mathbf{x}_i$ may be broken. We say that the orthogonal representation $\mathbf{x}$ is *locally freest*, if for every orthogonal representation $\mathbf{y}$ for which $\sum_i |\mathbf{y}_i - \mathbf{x}_i|$ is small enough, and every subset $S \subseteq V$, $\mathbf{x}(S)$ is linearly dependent if and only if $\mathbf{y}(S)$ is. Clearly every graph having an exceptional representation in $\mathbb{R}^d$ has a locally freest exceptional representation arbitrarily close to it. Equation (10.32) implies that if a locally freest orthogonal representation is in locally general position, then it is in general position.

The following lemma summarizes some useful properties of minimally exceptional graphs and their locally freest exceptional representations. When a graph G and an orthogonal representation $\mathbf{x}$ are understood, we define $A_u = \mathrm{lin}\big(\mathbf{x}(\overline{N}(u))\big)$ for $u \in V$.

LEMMA 10.44. *Let G be a minimally exceptional graph, and let $\mathbf{x}$ be a locally freest exceptional orthogonal representation of it.*

(a) *The complementary graph $\overline{G}$ is connected.*

(b) *For every $u \in V$, the set $\mathbf{x}(\overline{N}(u))$ is linearly dependent.*

(c) *Let $S \subseteq V$ and $u \in V \setminus S$. If $\mathbf{x}_u \in \mathrm{lin}\big(\mathbf{x}(S)\big)$, then $A_u^\perp \subseteq \mathrm{lin}\big(\mathbf{x}(S)\big)$.*

(d) *Any two vectors $\mathbf{x}_u$ are linearly independent, and $2 \le \dim(A_u) \le d-2$ for every node u. In particular, no node is represented by the zero vector.*

PROOF. (a) Suppose that $V(G) = V_1 \cup V_2$ is a partition into nonempty sets so that every node in V_1 is adjacent to every node in V_2 in G. Consider any orthogonal representation $\mathbf{x}$ of G. Then the orthogonal representations of $\mathbf{x}|_{V_i}$ of $G[V_i]$ are not exceptional, and so they can be approximated by general position orthogonal representations $\mathbf{y}_i$. Then $\mathbf{y}_1 \cup \mathbf{y}_2$ is an orthogonal representation of G which is in locally general position. By (10.32), $\mathbf{y}_1 \cup \mathbf{y}_2$ can be approximated by orthogonal representations in general position, so $\mathbf{x}$ is not exceptional.

It follows that no node of $\overline{G}$ is isolated, and since every node of G has degree at least $n-d$, we get that $n \ge d+2$.

(b) Call a node $v \in V$ *fat* if $\mathbf{x}\big(\overline{N}(v)\big)$ is linearly independent and *slim* otherwise. We want to prove that all nodes are slim. There must be at least one slim node (else, the representation is in locally general position and hence in general position, and not exceptional).

Suppose that G has a slim node i and a fat node j. Since G is connected, we may choose i and j so that they are adjacent.

The representation $\mathbf{x}'$ obtained by restricting $\mathbf{x}$ to $V(G \setminus j)$, is an orthogonal representation of $G' = G \setminus j$, and hence there exists a $\mathbf{y}' \in \mathcal{GOR}_d(G')$ arbitrarily close to $\mathbf{x}'$. We can extend $\mathbf{y}'$ to an orthogonal representation $\mathbf{y}$ of G; since $\mathbf{x}'\big(\overline{N}(j)\big)$ and $\mathbf{y}'\big(\overline{N}(j)\big)$ are linearly independent, we can choose $\mathbf{y}_j$ so that it is arbitrarily close to $\mathbf{x}_j$ if $\mathbf{y}'$ is sufficiently close to $\mathbf{x}'$.

This extended $\mathbf{y}$ is strictly locally freer than $\mathbf{x}$: indeed, a fat node remains fat (if $\mathbf{x}'$ and $\mathbf{y}'$ are close enough), and (since any d nodes different from j are represented by linearly independent vectors) j is fat in $\mathbf{y}$. This contradicts the assumption that $\mathbf{x}$ is locally freest.

(c) Clearly $\mathbf{x}_u \in A_u^\perp$. If $A_u^\perp \not\subseteq \operatorname{lin}(\mathbf{x}(S))$, then $A_u^\perp$ contains a vector $\mathbf{y}$ arbitrarily close to $\mathbf{x}_u$ but not in $\operatorname{lin}(\mathbf{x}(S))$. Replacing $\mathbf{x}_u$ by $\mathbf{y}$ we obtain another orthogonal representation of G which is strictly freer than $\mathbf{x}$, contradicting the definition of locally freest representations.

(d) For every node u, $|\overline{N}(u)| \le d-1$ and the vectors in $\mathbf{x}(\overline{N}(u))$ are linearly dependent by (a), hence $\dim(A_u) \le d-2$. Any two representing vectors must be linearly independent: if $\mathbf{x}_u = \lambda \mathbf{x}_v$, then by (b), $A_u^\perp$ is contained in the linear span of $\mathbf{x}_v$, which is impossible, since $\dim(A_u^\perp) = d - \dim(A_u) \ge 2$. Since $|N(u)| \ge n-d \ge 2$, this shows that $\dim(A_u) \ge 2$. $\square$

Using these facts about minimally exceptional graphs, we turn to the proof of the main theorem.

Proof of Theorem 10.43. We may assume that G is a minimally exceptional graph with an exceptional orthogonal representation $\mathbf{x}$. We may assume that $\mathbf{x}$ is locally freest. Lemma 10.44(d) implies that $2 \le d-2$, which implies that in the case $d \le 3$, no exceptional representations exist. Suppose that $d = 4$, then Lemma 10.44(d) implies that $\overline{G}$ is 3-regular, and the subspaces A_v defined above have $\dim(A_v) = 2$. Any two vectors $\mathbf{x}_u$ $(u \in \overline{N}(v))$ generate A_v, and thus the third element of $\mathbf{x}(\overline{N}(v))$ is linearly dependent on them.

Let u and v be nonadjacent nodes. Then $\mathbf{x}_u \in \operatorname{lin}(\mathbf{x}(\overline{N}(v)\setminus u)) = A_v$, and so Lemma 10.44(c) implies that $A_u^\perp \subseteq A_v$. Since these are 2-dimensional subspaces, it follows that $A_u = A_v^\perp$. Since $\overline{G}$ is connected, fixing any node v, the rest of the nodes fall into two classes: those with $A_u = A_v$ and those with $A_u = A_v^\perp$. This defines a bipartition $V = U \cup W$ of $\overline{G}$. By this argument, $\mathbf{x}(U)$ and $\mathbf{x}(W)$ span orthogonal 2-dimensional subspaces L and $L^\perp$.

Let us try to replace $\mathbf{x}_i$ by $\mathbf{x}_i + \varepsilon \mathbf{y}_i$, where

$$\mathbf{y}_i \in \begin{cases} L^\perp, & \text{if } i \in U, \\ L, & \text{if } i \in W \end{cases} \tag{10.33}$$

(so the modifying vector $\mathbf{y}_i$ is orthogonal to $\mathbf{x}_i$), and ε is sufficiently small. Let us assume that the vectors $\mathbf{y}_i$ satisfy the equations

$$\mathbf{x}_i^\mathsf{T} \mathbf{y}_j + \mathbf{x}_j^\mathsf{T} \mathbf{y}_i = 0 \quad \text{for all } ij \in \overline{E}. \tag{10.34}$$

Then for every $ij \in \overline{E}$ and real ε, we have

$$(\mathbf{x}_i + \varepsilon \mathbf{y}_i)^\mathsf{T} (\mathbf{x}_j + \varepsilon \mathbf{y}_j) = \mathbf{x}_i^\mathsf{T} \mathbf{x}_j + \varepsilon (\mathbf{x}_i^\mathsf{T} \mathbf{y}_j + \mathbf{x}_j^\mathsf{T} \mathbf{y}_i) + \varepsilon^2 \mathbf{y}_i^\mathsf{T} \mathbf{y}_j = 0,$$

so the modified vectors form an orthogonal representation of G. Choosing ε small enough, the orthogonal representation $\mathbf{x} + \varepsilon \mathbf{y}$ will be arbitrarily close to $\mathbf{x}$, and hence it is a locally freest exceptional representation. This implies that sets $\{\mathbf{x}_i + \varepsilon \mathbf{y}_i : i \in U\}$ and $\{\mathbf{x}_i + \varepsilon \mathbf{y}_i : i \in W\}$ span orthogonal 2-dimensional subspaces. Thus $(\mathbf{x}_i + \varepsilon \mathbf{y}_i)^\mathsf{T} (\mathbf{x}_j + \varepsilon \mathbf{y}_j) = 0$ for all $i \in U$ and $j \in W$ (not just for nonadjacent pairs). Expanding this equation, we get that (10.34) holds for all $i \in U$ and $j \in W$.

One way to construct vectors $\mathbf{y}_i$ satisfying (10.33) and (10.34) is the following: we take any linear map $B : L \to L^\perp$, and define

$$\mathbf{y}_i = \begin{cases} B\mathbf{x}_i & \text{if } i \in U, \\ -B^\mathsf{T} \mathbf{x}_i & \text{if } i \in W. \end{cases} \tag{10.35}$$

It is easy to check that these vectors satisfy (10.33) and (10.34). We claim that this is all that can happen.

Claim. *If the vectors* $\mathbf{y}_i \in \mathbb{R}^4$ *satisfy* (10.33) *and* (10.34), *then they can be obtained as in* (10.35).

To prove this, consider any linear relation $\sum_{i\in U} \alpha_i \mathbf{x}_i = 0$ between the vectors in $\mathbf{x}(U)$. Since (10.34) holds for all $i \in U$ and $j \in W$, it follows that

$$\sum_{i\in U} \alpha_i \mathbf{x}_j^{\mathsf{T}} \mathbf{y}_i = -\sum_{i\in U} \alpha_i \mathbf{x}_i^{\mathsf{T}} \mathbf{y}_j = 0$$

for every $j \in W$. Since the vectors in $\mathbf{x}(W)$ span $L^\perp$, this implies that $\sum_{i\in U} \alpha_i \mathbf{y}_i = 0$. So every linear equation satisfied by the vectors $\mathbf{x}_i$ $(i \in U)$ is also satisfied by the vectors $\mathbf{y}_i$. This implies by elementary linear algebra that there is a 2×2 matrix B for which $\mathbf{y}_i = B\mathbf{x}_i$. Substituting in (10.34), we get that $\mathbf{x}_i^{\mathsf{T}}(\mathbf{y}_j + B^{\mathsf{T}}\mathbf{x}_j) = 0$. So the vector $\mathbf{y}_j + B^{\mathsf{T}}\mathbf{x}_j \in L$ is orthogonal to $\mathbf{x}_i$ for every $i \in \overline{N}(j)$. This implies that $\mathbf{y}_j + B^{\mathsf{T}}\mathbf{x}_j = 0$, proving the Claim.

Now it is easy to complete the proof. Equations (10.33) and (10.34), viewed as a system of linear equations for the $\mathbf{y}_i$, have $4n$ unknowns, and with $2n+m = 7n/2$ equations, so it has a solution set that is at least $n/2$-dimensional. By the Claim, all these solutions can be represented by (10.35). This gives a 4-dimensional space of solutions, so $n \le 8$. Thus $\overline{G}$ is a 3-regular bipartite graph on at most 8 nodes; it is easy to see that this means that $G = \overline{K_{3,3}}$ or $G = \overline{Q^3}$. The former is ruled out since G is connected. Thus the only minimally exceptional graph for $d = 4$ is $\overline{Q^3}$.

It follows that every exceptional graph G contains $\overline{Q^3}$ as an induced subgraph. Since all degrees of G are at least $n-4$, it follows that Q^3 is a component of $\overline{G}$. □

10.6.3. More on algebraic properties. The algebraic structure of orthogonal representations has, quite probably, further interesting connections with the combinatorial structure of the underlying graph.

For a subset $S \subseteq \mathbb{R}^n$, we denote by $\mathcal{I}(S)$ the set of polynomials in n variables that vanish on all points of S. This is an ideal, whose algebraic properties are closely related to geometric properties of S. We are interested in the ideals $\mathcal{I}_d = \mathcal{I}(\mathcal{OR}_d(G))$ and $\mathcal{I}_d' = \mathcal{I}(\mathcal{GOR}_d(G))$, both of which consist of polynomials in variables x_{ir}, where $i \in V$ and $r \in [d]$.

Theorem 10.42, Proposition 10.41, and Theorem 10.43 can be recast in this algebraic language. It follows that the ideal $\mathcal{I}_d'$ is a prime ideal if and only if G is $(n-d)$-connected. If G is not $(n-d)$-connected, then $\mathcal{I}_d$ is not prime. If G is $(n-d)$-connected, then for $d \le 3$ the ideal $\mathcal{I}_d$ is prime, and this also holds for $d = 4$ unless $\overline{G}$ has a connected component isomorphic to the 3-cube.

Another interesting ideal is $\mathcal{L}_d$, generated by the polynomials $\mathbf{x}_i^{\mathsf{T}}\mathbf{x}_j = \sum_{r=1}^d x_{ir}x_{jr}$, where $ij \in \overline{E}$. It is clear that $\mathcal{L} \subseteq \mathcal{I}_d \subseteq \mathcal{I}_d'$ for every dimension $d \ge 1$.

The ideal $\mathcal{L}_1$ is called an *edge ideal*, and many properties of it have been studied; see e.g. [**Morey–Villarreal 2012**]. Their results easily imply that $\mathcal{L}_1 = \mathcal{I}_1' = \mathcal{I}_1$ for any graph. The structure of the ideal $\mathcal{L}_2$ was studied in [**Herzog et al. 2015**]. Among others, they prove that this ideal is radical.

EXERCISE 10.1. Which graphs have an orthonormal representation in the plane?

EXERCISE 10.2. Prove the edges of every graph on n nodes can be covered by $\lfloor n^2/4 \rfloor$ cliques, and show that this is best possible.

EXERCISE 10.3. Let G be a triangle-free graph and let $\mathbf{u}$ be an orthonormal representation of G in $\mathbb{R}^d$. Consider the matrix $B \in \mathbb{R}^{V\times V}$ defined by

$$B_{ij} = \begin{cases} \mathbf{u}_i^\mathsf{T}\mathbf{u}_j, & \text{if } i \neq j, \\ 0, & \text{if } i = j. \end{cases}$$

Let s_k be the sum of k-th powers of the eigenvalues of B, and let a_k be the k-th elementary symmetric polynomial of these eigenvalues. Prove that (a) $s_1 = a_1 = 0$, (b) $a_3 = 0$, (c) $s_3 = 0$, (d) $n \le 2d$.

EXERCISE 10.4. Prove the inequalities (10.22) and (10.21).

EXERCISE 10.5. Let Σ and Δ be two planes in $\mathbb{R}^3$ that are neither parallel nor weakly orthogonal. Select a unit vector $\mathbf{a}_1$ uniformly from Σ, and a unit vector $\mathbf{b}_1 \in \Delta\cap\mathbf{a}_1^\perp$. Let the unit vectors $\mathbf{a}_2$ and $\mathbf{b}_2$ be defined similarly, but selecting $\mathbf{b}_2 \in \Delta$ first (uniformly over all unit vectors), and then $\mathbf{a}_2$ from $\Sigma\cap\mathbf{b}_2^\perp$. Prove that the distributions of $(\mathbf{a}_1, \mathbf{b}_1)$ and $(\mathbf{a}_2, \mathbf{b}_2)$ are different, but mutually absolutely continuous.

EXERCISE 10.6. Let $\mathbf{u}: V \to \mathbb{R}^d$ be an orthogonal representation of the graph G, and let S be a minimal set such that $\mathbf{u}(S)$ is linearly dependent. Prove that $G[S]$ is connected.

EXERCISE 10.7. Prove that for a graph G, the following properties are equivalent: (i) G is $(n-d)$-connected; (ii) there is a G-matrix such that all $d\times d$ submatrices are nonsingular; (iii) there is a positive semidefinite G-matrix of rank d such that all $d\times d$ submatrices are nonsingular.

EXERCISE 10.8. Let G be a graph and let G' be obtained from G by creating a new node and connecting it to all nodes of G. Prove that $\mathsf{w}_{\rm tr}(G') = \mathsf{w}_{\rm tr}(G)+1$, $\mathsf{w}_{\rm pr}(G') = \mathsf{w}_{\rm pr}(G)+1$ and $\mathsf{w}_{\rm alg}(G') = \mathsf{w}_{\rm alg}(G)+1$.

EXERCISE 10.9. If $\mathbf{x}$ is transversal as a solution of a system of algebraic equations, then it also has this property as a solution of any system obtained by dropping some of the equations.

EXERCISE 10.10. A G-matrix A has the Strong Arnold Property if and only if for every symmetric matrix B there exists a G-matrix B' such that $x^\mathsf{T}(B'-B)x = 0$ for all vectors x in the nullspace of A.

EXERCISE 10.11. Let $\widehat{\mathcal{R}}_k$ be the manifold of all (not necessarily symmetric) $V\times V$ matrices with rank k. Determine (a) dimension of $\widehat{\mathcal{R}}_k$, (b) the tangent space of $\widehat{\mathcal{R}}_k$, and (c) the normal space $\widehat{\mathcal{R}}_k$ at a matrix $A \in \widehat{\mathcal{R}}^k$.

EXERCISE 10.12. Let G be a graph, and let $G = G_1\cup G_2$ where $|V(G_1)\cap(G_2)| = 1$. Prove that $\mathsf{w}_{\rm alg}(G) = \max\{\mathsf{w}_{\rm alg}(G_1), \mathsf{w}_{\rm alg}(G_2)\}$.

EXERCISE 10.13. Suppose that G has no K_4 minor. Prove that $\mathsf{w}_{\rm pr}(G) \le 3$, and if $\mathsf{w}_{\rm pr}(G) = 3$, then G contains Δ_3 as a minor.

EXERCISE 10.14. (a) For every bipartite graph G, the graph $G\boxtimes K_2$ is a minor of $G\square C_4$. (b) $K_{2^{m+1}}$ is a minor of Q^{2m+1}.

EXERCISE 10.15. (a) Prove that if a graph G has node-expansion c, then its treewidth is at least $cn/(c+3)-1$. (b) Prove that the treewidth of an $n\times n$ grid graph is n. (c) Prove that the product-width of the n-cube is at least 2^{n-1}.

EXERCISE 10.16. Determine all convex polytopes with an edge-transitive isometry group for which $\mathbf{u}^\mathsf{T}\mathbf{v} \le 0$ for every edge $\mathbf{uv}$.

EXERCISE 10.17. Prove that the variety $\{(\mathbf{u}_1, \mathbf{u}_2) : \mathbf{u}_i \in \mathbb{R}^2,\ \mathbf{u}_1^\mathsf{T}\mathbf{u}_2 = 0\}$ is indecomposable.

EXERCISE 10.18. Given a graph G, construct a real algebraic variety $\mathcal{V}$ that consists of general position orthogonal representations, and every general position orthogonal representation, after being scaled by a positive scalar, belongs to $\mathcal{V}$.

CHAPTER 11

Orthogonal Representations: The Smallest Cone

In this chapter we generalize the discussion in the introductory Section 1.4, and study orthogonal representations that are "economical" in a sense different from dimension. It turns out that the smallest half-angle of a rotational cone (in arbitrary dimension) that contains all vectors in an orthogonal representation contains interesting information about the graph [**Lovász 1979b**]. To be precise, we will work with a transformed version of this quantity, the "theta-function". This quantity will be related to the stability number, the clique number and the chromatic number of graphs.

We describe several applications of this graph parameter, of which we mention two problems in which this parameter yields the only known solution. One of these is the problem of the Shannon capacity (or zero-error capacity) of a graph from Section 1.4, which is a tough parameter to compute, and whose evaluation even for a small graph like the pentagon needs the theta-function. The other application is the computation of the clique number and chromatic number for perfect graphs in polynomial time.

11.1. The theta-function

Let us start with the definition:

$$\vartheta(G) = \min_{\mathbf{u},\mathbf{c}} \max_{i\in V} \frac{1}{(\mathbf{c}^{\mathsf{T}}\mathbf{u}_i)^2}, \tag{11.1}$$

where the minimum is taken over all orthonormal representations $(\mathbf{u}_i : i \in V)$ of G and all unit vectors $\mathbf{c}$. We call $\mathbf{c}$ the "handle" of the representation (for the origin of the name, see Example 1.2). If ϕ is the half-angle of the smallest rotational cone with axis in the direction of c, containing all vectors in an orthonormal representation, then clearly

$$\vartheta(G) = \frac{1}{(\cos\phi)^2}. \tag{11.2}$$

Of course, we could fix $\mathbf{c}$ to be (say) the standard basis vector $\mathbf{e}_1$, but this is not always convenient. We may always assume that the dimension d of the ambient space is at most n. We could require that $\mathbf{c}^{\mathsf{T}}\mathbf{u}_i \geq 0$ for all nodes, since we can replace any $\mathbf{u}_i$ by its negative. With just a little more complicated argument, we may even require that

$$\mathbf{c}^{\mathsf{T}}\mathbf{u}_i = \frac{1}{\sqrt{\vartheta(G)}} \tag{11.3}$$

for every node i without changing the optimum value in (11.1). Indeed, let $\alpha_i = 1/(\sqrt{\vartheta(G)}\mathbf{c}^\mathsf{T}\mathbf{u}_i)$, then $0 \le \alpha_i \le 1$, and so we can replace $\mathbf{u}_i$ by the $(d+n)$-dimensional vector

$$\mathbf{u}'_i = \begin{pmatrix} \alpha_i \mathbf{u}_i \\ \sqrt{1-\alpha_i^2}\mathbf{e}_i \end{pmatrix}$$

Extending $\mathbf{c}$ with n zeroes to get a unit vector $\mathbf{c}'$, we get an orthonormal representation in which $(\mathbf{c}')^\mathsf{T}\mathbf{u}'_i = 1/\sqrt{\vartheta(G)}$ for all i.

The following rather easy bounds will be very important.

THEOREM 11.1. *For every graph G,*

$$\alpha(G) \le \vartheta(G) \le \chi(\overline{G}).$$

PROOF. First, let $S \subseteq V$ be a maximum stable set of nodes in G. Then in every orthonormal representation $(\mathbf{u}_i)$, the vectors $\{\mathbf{u}_i : i \in S\}$ are mutually orthogonal unit vectors. Hence

$$1 = |\mathbf{c}|^2 \ge \sum_{i\in S}(\mathbf{c}^\mathsf{T}\mathbf{u}_i)^2 \ge |S| \min_i (\mathbf{c}^\mathsf{T}\mathbf{u}_i)^2, \tag{11.4}$$

and so

$$\max_{i\in V} \frac{1}{(\mathbf{c}^\mathsf{T}\mathbf{u}_i)^2} \ge |S| = \alpha(G).$$

This implies the first inequality. The second follows from the orthogonal representation constructed in Example 10.5, using $\mathbf{c} = (1/\sqrt{m})\mathbb{1}$ as the handle. □

Our examples in Section 10.1 give further upper bounds on ϑ. Example 10.3 leads to the following construction: Assuming that there are no isolated nodes, we assign to each node $i \in V$ the vector $\mathbf{u}_i \in \mathbb{R}^E$ defined by

$$(\mathbf{u}_i)_e = \begin{cases} \dfrac{1}{\sqrt{\deg(i)}} & \text{if } e \text{ is incident with } i, \\ 0, & \text{otherwise,} \end{cases}$$

and define the handle $\mathbf{c} = (1/\sqrt{m})\mathbb{1}$. Then $\mathbf{c}^\mathsf{T}\mathbf{u}_i = \sqrt{\deg(i)/m}$, and so we get the bound

$$\vartheta(G) \le \max_i \frac{m}{\deg(i)} = \frac{m}{d_{\min}}.$$

(The upper bound $m/d_{\min}$ for the stability number $\alpha(G)$ is easy to prove by counting edges.) From Example 10.6 we get by elementary trigonometry that

$$\vartheta(C_5) \le \sqrt{5}. \tag{11.5}$$

Soon we will see that equality holds here.

It is clear that if G' is an induced subgraph of G, then $\vartheta(G') \le \vartheta(G)$ (an optimal orthonormal representation of G, restricted to $V(G')$, is an orthonormal representation of G'). It is also clear that if G' is a spanning subgraph of G (i.e., $V(G') = V(G)$ and $E(G') \subseteq E$), then $\vartheta(G') \ge \vartheta(G)$ (every orthonormal representation of G' is an orthonormal representation of G).

11.2. Duality for theta

11.2.1. Alternative definitions. The graph parameter ϑ has many equivalent definitions. We are going to state some, which lead to an important dual formulation of this quantity.

Vector chromatic number. The following geometric definition was proposed by Karger, Motwani and Sudan. In terms of the complementary graph, this value is called the "vector chromatic number". As a motivation for this name, consider a t-colorable graph ($t \geq 2$), and let us color its nodes by t unit vectors $\mathbf{f}_1, \dots, \mathbf{f}_t \in \mathbb{R}^{t-1}$, pointing to the vertices of a regular simplex centered at the origin. We get a vector labeling $\mathbf{w}$ (not an orthogonal representation of $\overline{G}$, but closely related). It is not hard to compute that $\mathbf{f}_k^\mathsf{T}\mathbf{f}_l = -1/(t-1)$ for $k \neq l$, and so $\mathbf{w}_i^\mathsf{T}\mathbf{w}_j = -1/(t-1)$ for $ij \in E$.

Now let us forget about the condition that $\mathbf{w}_i$ must be one of the vectors $\mathbf{f}_k$: define a (strict) *vector t-coloring* ($t > 1$) of the graph G as a vector labeling $i \mapsto \mathbf{w}_i \in \mathbb{R}^n$ such that $|\mathbf{w}_i| = 1$ for all $i \in V$, and $\mathbf{w}_i^\mathsf{T}\mathbf{w}_j = -1/(t-1)$ for all $ij \in E$. (The dimension n in the definition above is just chosen to be large enough; allowing a higher dimension would not make any difference.)

If G has at least one edge, then the last condition implies that $t \geq 2$. The smallest t for which the graph G has a vector t-coloring is called its (strict) *vector chromatic number*, and is denoted by χ_{vec}. If $E = \emptyset$, so G is edgeless and $\chi(G) = 1$, then we define $\chi_{\text{vec}} = 1$. For $t = 2$, the endpoints of any edge must be labeled by antipodal unit vectors. It follows that every connected component of G is labeled by two antipodal vectors only, and so G is bipartite. Conversely, every bipartite graph has a strict vector 2-coloring.

It is clear from the construction above that $\chi_{\text{vec}}(G) \leq \chi(G)$ for every graph G. Equality does not hold in general. Vector-labeling the nodes of a pentagon by the vertices of a regular pentagon inscribed in the unit circle, so that the edges are mapped onto the diagonals, we see that $\chi_{\text{vec}}(C_5) \leq \sqrt{5} < \chi(C_5) = 3$. It is also easy to see that $\chi_{\text{vec}}(G) \geq \omega(G)$.

Semidefinite matrices. Next, we give a couple of formulas for ϑ in terms of semidefinite matrices. Let

$$\vartheta_{\text{diag}} = \min\{1 + \max_{i\in V} Y_{ii} : Y \in \mathbb{R}^{V\times V},\ Y \succeq 0,\ Y_{ij} = -1\ (ij \in \overline{E})\} \tag{11.6}$$

and

$$\vartheta_{\text{sum}} = \max\Big\{\sum_{i,j\in V} Z_{ij} : Z \in \mathbb{R}^{V\times V},\ Z \succeq 0,\ Z_{ij} = 0\ (ij \in E),\ \text{tr}(Z) = 1\Big\}. \tag{11.7}$$

It will turn out that these two values are equal. This equality is in fact a special case of the Duality Theorem of semidefinite optimization. It is not hard to check that (11.6) and (11.7) are dual semidefinite programs, and the first one has a strictly feasible solution. So the Duality Theorem 13.5 of semidefinite programming applies, and asserts that the two programs have the same objective value. However, we are going to include a proof, to make our treatment self-contained.

Dual orthogonal representation. We use orthonormal representations of the complementary graph to define

$$\vartheta_{\text{dual}} = \max \sum_{i\in V} (\mathbf{d}^{\mathsf{T}}\mathbf{v}_i)^2, \tag{11.8}$$

where the maximum extends over all orthonormal representations $(\mathbf{v}_i : i \in V)$ of the complementary graph $\overline{G}$ and all unit vectors (handles) $\mathbf{d}$.

11.2.2. The main duality result. The main theorem of this section asserts that all these definitions lead to the same value.

THEOREM 11.2. *For every graph G,*

$$\vartheta(G) = \chi_{\text{vec}}(\overline{G}) = \vartheta_{\text{diag}}(G) = \vartheta_{\text{sum}}(G) = \vartheta_{\text{dual}}(G).$$

PROOF. We prove the circle of inequalities

$$\vartheta(G) \le \chi_{\text{vec}}(\overline{G}) \le \vartheta_{\text{diag}}(G) \le \vartheta_{\text{sum}}(G) \le \vartheta_{\text{dual}}(G) \le \vartheta(G). \tag{11.9}$$

To prove the first inequality, let $t = \chi_{\text{vec}}(\overline{G})$, and let $(\mathbf{w}_i : i \in V)$ be an optimal vector t-coloring. Let $\mathbf{c}$ be a vector orthogonal to all the $\mathbf{w}_i$ (we increase the dimension of the space if necessary). Let

$$\mathbf{u}_i = \frac{1}{\sqrt{t}}\mathbf{c} + \sqrt{\frac{t-1}{t}}\mathbf{w}_i.$$

Then $|\mathbf{u}_i| = 1$ and $\mathbf{u}_i^{\mathsf{T}}\mathbf{u}_j = 0$ for $ij \in \overline{E}$, so $(\mathbf{u}_i)$ is an orthonormal representation of G. Furthermore, with handle $\mathbf{c}$ we have $\mathbf{c}^{\mathsf{T}}\mathbf{u}_i = 1/\sqrt{t}$, which implies that $\vartheta(G) \le \chi_{\text{vec}}(\overline{G})$.

Second, let Y be an optimal solution of (11.6). We may assume that all diagonal entries Y_{ii} are the same number t, since we can replace all of them by the largest without violating the other constraints. The matrix $1/(t-1)Y$ is positive semidefinite, and so it can be written as $\mathsf{Gram}(\mathbf{w}_i : i \in V)$ with appropriate vectors $(\mathbf{w}_i \in \mathbb{R}^n)$. These vectors form a strict vector t-coloring. Since $\chi_{\text{vec}}(\overline{G})$ is the smallest t for which this exists, this proves that $\chi_{\text{vec}}(\overline{G}) \le t = \vartheta_{\text{diag}}$.

The main step in the proof is to show that $\vartheta_{\text{diag}} \le \vartheta_{\text{sum}}$. Fix any $t > \vartheta_{\text{sum}}$; it is easy to see that $\vartheta_{\text{sum}} \ge 1$ and hence $t > 1$. Let $\mathcal{L}$ denote the linear space of symmetric $V\times V$ matrices satisfying $Z_{ij} = 0$ $(ij \in E)$ and $(tI - J)\cdot Z = 0$, and let $\mathcal{P}$ denote the cone of positive semidefinite $V\times V$ matrices.

We claim that $\mathcal{P}\cap\mathcal{L} = \{0\}$. Suppose, to the contrary, that there is a symmetric matrix $Z \ne 0$ such that $Z \in \mathcal{P}\cap\mathcal{L}$. Clearly $\mathrm{tr}(Z) > 0$, and so, by scaling, we may assume that $\mathrm{tr}(Z) = 1$. Then Z satisfies the conditions in the definition of ϑ_{sum}, and so $\vartheta_{\text{sum}} \ge J\cdot Z = tI\cdot Z = t$, contradicting the choice of t.

So the linear space $\mathcal{L}$ touches the convex cone $\mathcal{P}$ at its apex only, and hence there is a hyperplane $\mathcal{H}$ such that $\mathcal{L} \subseteq \mathcal{H}$ and $\mathcal{H}\cap\mathcal{P} = \{0\}$. Let $Y\cdot X = 0$ be the equation of $\mathcal{H}$ (where $Y \ne 0$ is a symmetric $V\times V$ matrix); we may assume that $Y\cdot X \ge 0$ for all $X \in \mathcal{P}$. This means that Y is in the dual cone of $\mathcal{P}$. This dual cone is $\mathcal{P}$ itself, which means that $Y \succeq 0$. Furthermore, $\mathcal{L} \subseteq \mathcal{H}$ means that the equation of $\mathcal{H}$ is a linear combination of the equations defining $\mathcal{L}$, i.e., there are real numbers a_{ij} $(ij \in E)$ and b such that

$$Y = \sum_{ij\in E} a_{ij}E_{ij} + b(tI - J).$$

Considering a positive diagonal entry of Y, we see that $b > 0$, and since we are free to scale Y by positive scalars, we may assume that $b = 1$. But this means that Y satisfies the conditions in the definition of ϑ_{diag}, and so $\vartheta_{\text{diag}} \le 1+\max_i Y_{ii} = t$. Since this holds for every $t > \vartheta_{\text{sum}}$, this implies that $\vartheta_{\text{diag}} \le \vartheta_{\text{sum}}$.

To prove the fourth inequality in (11.9), let Z be an optimum solution of the program (11.7) with objective function value ϑ_{sum}. We can write Z as $\mathsf{Gram}(\mathbf{z}_i : i \in V)$ where $\mathbf{z}_i \in \mathbb{R}^n$. Let us rescale the vectors $\mathbf{z}_i$ to get the unit vectors $\mathbf{v}_i = \mathbf{z}_i^0$ (if $\mathbf{z}_i = 0$, then we take a unit vector orthogonal to everything else as $\mathbf{v}_i$). Define $\mathbf{d} = (\sum_i \mathbf{z}_i)^0$.

By the properties of Z, the vectors $\mathbf{v}_i$ form an orthonormal representation of $\overline{G}$, and hence

$$\vartheta_{\text{dual}} \ge \sum_i (\mathbf{d}^{\mathsf{T}}\mathbf{v}_i)^2.$$

To estimate the right side, we use the equations

$$\sum_i |\mathbf{z}_i|^2 = \sum_i \mathbf{z}_i^{\mathsf{T}}\mathbf{z}_i = \operatorname{tr}(Z) = 1, \qquad \Big|\sum_i \mathbf{z}_i\Big|^2 = \sum_{i,j} \mathbf{z}_i^{\mathsf{T}}\mathbf{z}_j = \sum_{i,j} Z_{ij} = \vartheta_{\text{sum}},$$

and the Cauchy–Schwarz Inequality:

$$\begin{aligned}\sum_i (\mathbf{d}^{\mathsf{T}}\mathbf{v}_i)^2 &= \Big(\sum_i |\mathbf{z}_i|^2\Big)\Big(\sum_i (\mathbf{d}^{\mathsf{T}}\mathbf{v}_i)^2\Big) \ge \Big(\sum_i |\mathbf{z}_i| \mathbf{d}^{\mathsf{T}}\mathbf{v}_i\Big)^2 \\ &= \Big(\sum_i \mathbf{d}^{\mathsf{T}}\mathbf{z}_i\Big)^2 = \Big(\mathbf{d}^{\mathsf{T}}\sum_i \mathbf{z}_i\Big)^2 = \Big|\sum_i \mathbf{z}_i\Big|^2 = \vartheta_{\text{sum}}.\end{aligned}$$

This proves that $\vartheta_{\text{dual}} \ge \vartheta_{\text{sum}}$.

Finally, to prove the last inequality in (11.9), it suffices to prove that if $(\mathbf{u}_i : i \in V)$ is an orthonormal representation of G in $\mathbb{R}^n$ with handle $\mathbf{c}$, and $(\mathbf{v}_i : i \in V)$ is an orthonormal representation of $\overline{G}$ in $\mathbb{R}^m$ with handle $\mathbf{d}$, then

$$\sum_{i\in V} (\mathbf{d}^{\mathsf{T}}\mathbf{v}_i)^2 \le \max_{i\in V} \frac{1}{(\mathbf{c}^{\mathsf{T}}\mathbf{u}_i)^2}. \tag{11.10}$$

The tensor product vectors $\mathbf{u}_i \circ \mathbf{v}_i$ $(i \in V)$ are mutually orthogonal unit vectors. Indeed, $(\mathbf{u}_i \circ \mathbf{v}_i)^{\mathsf{T}}(\mathbf{u}_j \circ \mathbf{v}_j) = (\mathbf{u}_i^{\mathsf{T}}\mathbf{u}_j)(\mathbf{v}_i^{\mathsf{T}}\mathbf{v}_j) = 0$, since either $\mathbf{u}_i$ is orthogonal to $\mathbf{u}_j$ or $\mathbf{v}_i$ is orthogonal to $\mathbf{v}_j$. Hence

$$\sum_i (\mathbf{c}^{\mathsf{T}}\mathbf{u}_i)^2(\mathbf{d}^{\mathsf{T}}\mathbf{v}_i)^2 = \sum_i \big((\mathbf{c}\circ\mathbf{d})^{\mathsf{T}}(\mathbf{u}_i\circ\mathbf{v}_i)\big)^2 \le 1. \tag{11.11}$$

On the other hand,

$$\sum_i (\mathbf{c}^{\mathsf{T}}\mathbf{u}_i)^2(\mathbf{d}^{\mathsf{T}}\mathbf{v}_i)^2 \ge \min_i (\mathbf{c}^{\mathsf{T}}\mathbf{u}_i)^2 \sum_i (\mathbf{d}^{\mathsf{T}}\mathbf{v}_i)^2,$$

which implies that

$$\sum_i (\mathbf{d}^{\mathsf{T}}\mathbf{v}_i)^2 \le \frac{1}{\min_i (\mathbf{c}^{\mathsf{T}}\mathbf{u}_i)^2} = \max_i \frac{1}{(\mathbf{c}^{\mathsf{T}}\mathbf{u}_i)^2}.$$

This proves (11.10) and completes the proof of Theorem 11.2. □

We can state the theorem more explicitly as the following sequence of formulas.

$$\vartheta(G) = \min\Big\{\max_{i\in V} \frac{1}{(\mathbf{c}^{\mathsf T}\mathbf{u}_i)^2} : \mathbf{u} \text{ ONR of } G,\ |\mathbf{c}| = 1\Big\} \tag{11.12}$$

$$= \min\Big\{t > 1 : |\mathbf{w}_i| = 1,\ \mathbf{w}_i^{\mathsf T}\mathbf{w}_j = -\frac{1}{t-1}\ (ij \in \overline{E})\Big\} \tag{11.13}$$

$$= \min\big\{1 + \max_{i\in V} Y_{ii} : Y \succeq 0,\ Y_{ij} = -1\ (ij \in \overline{E})\big\} \tag{11.14}$$

$$= \max\Big\{\sum_{i,j\in V} Z_{ij} : Z \succeq 0,\ Z_{ij} = 0\ (ij \in E),\ \operatorname{tr}(Z) = 1\Big\} \tag{11.15}$$

$$= \max\Big\{\sum_{i\in V} (\mathbf{d}^{\mathsf T}\mathbf{v}_i)^2 : \mathbf{v} \text{ ONR of } \overline{G},\ |\mathbf{d}| = 1\Big\}. \tag{11.16}$$

From this form it is clear (and we have seen this in the proof as well) that the powerful step in this sequence of formulas is the equality (11.14)=(11.15), where an expression as a minimum switches to an expression as a maximum. Note that before this equality we have conditions on the edges of $\overline{G}$, which then get replaced by conditions on the edges of G for the last two rows.

REMARK 11.3. If we want to get a better upper bound on $\alpha(G)$ than $\vartheta(G)$, we can relax the conditions in (11.12), (11.13) or (11.14). This is equivalent to tightening the conditions in the dual optimization problems (11.15) or (11.16).

[**McEliece–Rodemich–Rumsey 1978**] and [**Schrijver 1979**] proposed to tighten the conditions in (11.15) by requiring that $Z_{ij} \geq 0$ for all i and j. This is equivalent to relaxing the conditions in (11.14) by requiring only $Y_{ij} \geq -1$ for $ij \in \overline{E}$. The resulting quantity $\vartheta^-(G)$ is still a well-characterized (and polynomial time computable) upper bound on $\alpha(G)$, tighter than $\vartheta(G)$. The other characterizations of $\vartheta(G)$ have natural modifications to yield $\vartheta^-(G)$.

If we want to get a better bound on $\chi(\overline{G})$, we need to relax the conditions in (11.14), or tighten the conditions in (11.15). [**Szegedy 1994**] studied the quantity obtained by adding the conditions $A_{ij} \geq -1$ $(ij \in E)$ in (11.14). This is equivalent to relaxing the conditions in (11.15) by requiring only $Z_{ij} \leq 0$ for $ij \in E$. This results in a quantity $\vartheta^+(G)$, which is again well-characterized and polynomial time computable.

These quantities satisfy the inequalities

$$\alpha(G) \leq \vartheta^-(G) \leq \vartheta(G) \leq \vartheta^+(G) \leq \chi(\overline{G}).$$

The two modified graph parameters lack some of the more algebraic properties of ϑ (for example, Theorem 11.8 below), but they give better bounds on $\alpha(G)$ and $\chi(\overline{G})$, respectively. See Exercises 11.5–11.8 for more on these parameters.

11.2.3. Consequences of duality. From the fact that equality holds in (11.9), it follows that equality holds in all of the arguments above. Let us formulate some consequences. From the fact that equality must hold in (11.11), and from the derivation of this inequality, we see that for an optimal orthonormal representation $(\mathbf{u}, \mathbf{c})$ and an optimal dual orthonormal representation $(\mathbf{v}, \mathbf{d})$, the unit vector $\mathbf{c} \circ \mathbf{d}$ must be a linear combination of the mutually orthogonal unit vectors $\mathbf{u}_i \circ \mathbf{v}_i$. The coefficients are easy to figure out, and we get

$$\mathbf{c} \circ \mathbf{d} = \sum_i (\mathbf{c}^{\mathsf T}\mathbf{u}_i)(\mathbf{d}^{\mathsf T}\mathbf{v}_i)(\mathbf{u}_i \circ \mathbf{v}_i), \tag{11.17}$$

or in a matrix form:

$$\mathbf{c}\mathbf{d}^{\mathsf{T}} = \sum_i (\mathbf{c}^{\mathsf{T}}\mathbf{u}_i)(\mathbf{d}^{\mathsf{T}}\mathbf{v}_i)\mathbf{u}_i\mathbf{v}_i^{\mathsf{T}}. \tag{11.18}$$

This equation gives (upon multiplying by $\mathbf{d}$ from the right and by $\mathbf{c}^{\mathsf{T}}$ from left, respectively) two equations expressing the handles as linear combinations of the (primal and dual) orthonormal representations:

$$\mathbf{c} = \sum_i (\mathbf{c}^{\mathsf{T}}\mathbf{u}_i)(\mathbf{d}^{\mathsf{T}}\mathbf{v}_i)^2\mathbf{u}_i, \quad \mathbf{d} = \sum_i (\mathbf{c}^{\mathsf{T}}\mathbf{u}_i)^2(\mathbf{d}^{\mathsf{T}}\mathbf{v}_i)\mathbf{v}_i. \tag{11.19}$$

Using an optimal orthonormal representation with $\mathbf{c}^{\mathsf{T}}\mathbf{u}_i = 1/\sqrt{\vartheta}$, we get the simpler formulas

$$\mathbf{c} = \frac{1}{\sqrt{\vartheta(G)}} \sum_{i\in V} (\mathbf{d}^{\mathsf{T}}\mathbf{v}_i)^2\mathbf{u}_i, \quad \mathbf{d} = \frac{1}{\vartheta(G)} \sum_{i\in V} (\mathbf{d}^{\mathsf{T}}\mathbf{v}_i)\mathbf{v}_i. \tag{11.20}$$

There is a simple relationship between the theta value of a graph and of its complement.

PROPOSITION 11.4. *For every graph G, we have $\vartheta(G)\vartheta(\overline{G}) \ge n$.*

PROOF. Let $(\mathbf{u}, \mathbf{c})$ be an optimal orthogonal representation of G. Then applying (11.16) to the complementary graph, we get

$$\vartheta(\overline{G}) \ge \sum_i (\mathbf{c}^{\mathsf{T}}\mathbf{u}_i)^2 \ge n \min_i (\mathbf{c}^{\mathsf{T}}\mathbf{u}_i)^2 = \frac{n}{\vartheta(G)}. \qquad \square$$

Equality does not hold in 11.4 in general, but it does when G has a node-transitive automorphism group. To prove this, we need an important fact about the symmetries of orthogonal representations. We say that an orthonormal representation $(\mathbf{u}_i, \mathbf{c})$ in $\mathbb{R}^d$ of a graph G is *automorphism invariant*, if every automorphism $\gamma \in \mathrm{Aut}(G)$ can be lifted to an orthogonal transformation O_γ of $\mathbb{R}^d$ such that $O_\gamma\mathbf{c} = \mathbf{c}$ and $\mathbf{u}_{\gamma(i)} = O_\gamma\mathbf{u}_i$ for every node i. An optimal orthonormal representation (say, in the sense of (11.12)) is not necessarily invariant under automorphisms, but there is always one that is (see Figure 11.1). The representation on the left is also optimal with respect to minimizing the dimension, and it is not hard to see that C_4 has no automorphism invariant orthonormal representation in $\mathbb{R}^2$. So minimizing the dimension and minimizing the cone behave differently from this point of view.

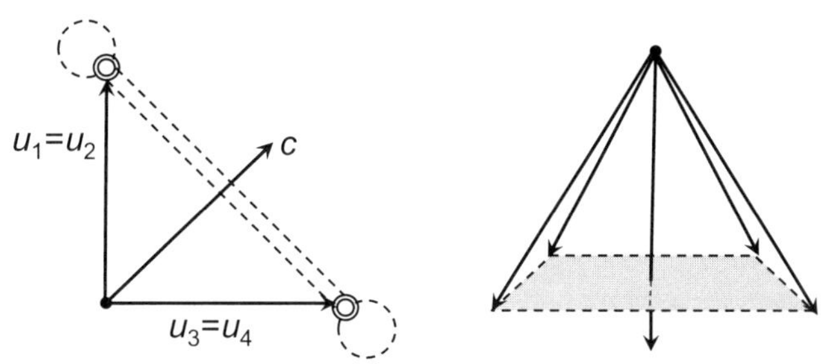

FIGURE 11.1. An optimal orthonormal representation of C_4 that is not invariant under its automorphisms, and one that is.

THEOREM 11.5. *Every graph G has an optimal orthonormal representation and an optimal dual orthonormal representation that are both automorphism invariant.*

PROOF. We give the proof for the dual orthonormal representation. The optimum solutions of the semidefinite program in (11.6) form a bounded convex set, which is invariant under the transformations $Z \mapsto P_\alpha^\mathsf{T} Z P_\alpha$, where P_α is the permutation matrix defined by an automorphism α of G. If Z is an optimizer in (11.15), then so is $P_\alpha^\mathsf{T} Z P_\alpha$ for every automorphism $\alpha \in \operatorname{Aut}(G)$, and hence also

$$\widehat{Z} = \frac{1}{|\operatorname{Aut}(G)|} \sum_{\alpha \in \operatorname{Aut}(G)} P_\alpha^\mathsf{T} Z P_\alpha.$$

This matrix satisfies $P_\alpha^\mathsf{T} Z P_\alpha = Z$ for all automorphisms α.

The construction of an orthonormal representation of $\overline{G}$ in the proof of $\vartheta_{\text{diag}} \le \vartheta_{\text{sum}}$ in Theorem 11.2 can be done in a canonical way: we choose the columns of $Z^{1/2}$ as the vectors $\mathbf{z}_i$, and use them to construct the dual orthonormal representation with $\mathbf{v}_i = \mathbf{z}_i^0$ and $\mathbf{d} = (\sum_i \mathbf{z}_i)^0$. The optimal dual orthonormal representation constructed this way will be invariant under the automorphism group of G. □

COROLLARY 11.6. *If G has a node-transitive automorphism group, then*

$$\vartheta(G)\vartheta(\overline{G}) = n.$$

PROOF. It follows from Theorem 11.5 that $\overline{G}$ has an orthonormal representation $(\mathbf{v}_i, \mathbf{d})$ in $\mathbb{R}^n$ such that $\sum_i (\mathbf{d}^\mathsf{T}\mathbf{v}_i)^2 = \vartheta(G)$, and $\mathbf{d}^\mathsf{T}\mathbf{v}_i$ is the same for every i. So $(\mathbf{d}^\mathsf{T}\mathbf{v}_i)^2 = \vartheta(G)/n$ for all nodes i, and hence

$$\vartheta(\overline{G}) \le \max_i \frac{1}{(\mathbf{d}^\mathsf{T}\mathbf{v}_i)^2} = \frac{n}{\vartheta(G)}.$$

Since we already know the reverse inequality (Proposition 11.4), this proves the Corollary. □

COROLLARY 11.7. *If G is a self-complementary graph with a node-transitive automorphism group, then $\vartheta(G) = \sqrt{n}$. In particular, $\vartheta(C_5) = \sqrt{5}$.* □

A further important feature of the theta-function is its nice behavior with respect to graph product; we will see that this is what underlies its applications in information theory.

There are many different ways of multiplying two simple graphs G and H, of which we need one in this chapter. The *strong product* $G \boxtimes H$ is defined on the underlying set $V(G) \times V(H)$. Two nodes (u_1, v_1) and (u_2, v_2) are adjacent if and only if either $ij \in E$ and $uv \in E(H)$, or $ij \in E$ and $u = v$, or $i = j$ and $uv \in E(H)$. It is easy to see that this multiplication is associative and commutative (up to isomorphism). The product of two complete graphs is a complete graph.

THEOREM 11.8. *For any two graphs G and H, we have $\vartheta(G \boxtimes H) = \vartheta(G)\vartheta(H)$.*

PROOF. Let $(\mathbf{u}_i : i \in V)$ be an optimal orthogonal representation of G with handle $\mathbf{c}$ ($\mathbf{u}_i, \mathbf{c} \in \mathbb{R}^n$), and let $(\mathbf{v}_j : j \in V(H))$ be an optimal orthogonal representation of H with handle $\mathbf{d}$ ($\mathbf{v}_j, \mathbf{d} \in \mathbb{R}^m$). Recalling the identity $(\mathbf{u} \circ \mathbf{v})^\mathsf{T}(\mathbf{x} \circ \mathbf{y}) = (\mathbf{u}^\mathsf{T}\mathbf{x})(\mathbf{v}^\mathsf{T}\mathbf{y})$ (used in the Introduction), we see that the vectors

$\mathbf{u}_i \circ \mathbf{v}_j$ $((i,j) \in V(G) \times V(H))$ form an orthogonal representation of $G \boxtimes H$. Furthermore, taking $\mathbf{c} \circ \mathbf{d}$ as its handle, we have

$$\big((\mathbf{c}\circ\mathbf{d})^\mathsf{T}(\mathbf{u}_i\circ\mathbf{v}_j)\big)^2 = (\mathbf{c}^\mathsf{T}\mathbf{u}_i)^2(\mathbf{d}\circ\mathbf{v}_j)^2 \geq \frac{1}{\vartheta(G)}\cdot\frac{1}{\vartheta(H)},$$

and hence

$$\vartheta(G\boxtimes H) \leq \max_{i,j} \frac{1}{\big((\mathbf{c}\circ\mathbf{d})^\mathsf{T}(\mathbf{u}_i\circ\mathbf{v}_j)\big)^2} \leq \vartheta(G)\vartheta(H).$$

To prove that equality holds, we use the duality established in Section 11.2. Let $(\mathbf{v}_i, \mathbf{d})$ be an orthonormal representation of $\overline{G}$ which is optimal in the sense that $\sum_i (\mathbf{d}^\mathsf{T}\mathbf{v}_i)^2 = \vartheta(G)$, and let $(\mathbf{w}_j, \mathbf{e})$ be an orthonormal representation of $\overline{H}$ such that $\sum_i (\mathbf{e}^\mathsf{T}\mathbf{w}_i)^2 = \vartheta(H)$. It is easy to check that the vectors $\mathbf{v}_i \circ \mathbf{w}_j$ form an orthonormal representation of $\overline{G\boxtimes H}$, and so using handle $\mathbf{d}\circ\mathbf{e}$ we get

$$\vartheta(G\boxtimes H) \geq \sum_{i,j}\big((\mathbf{d}\circ\mathbf{e})^\mathsf{T}(\mathbf{v}_i\circ\mathbf{w}_j)\big)^2 = \sum_{i,j}(\mathbf{d}^\mathsf{T}\mathbf{v}_i)^2(\mathbf{e}^\mathsf{T}\mathbf{w}_j)^2 = \vartheta(G)\vartheta(H).$$

We already know the reverse inequality, which completes the proof. □

11.2.4. Eigenvalues and theta. To motivate the identities and inequalities to be proved in this section, let us survey some of the classical results which use spectral properties of graphs, or more generally linear algebra techniques, to bound quantities like the stability number $\alpha = \alpha(G)$, the clique number $\omega = \omega(G)$, or the chromatic number $\chi = \chi(G)$ in terms of the eigenvalues of the adjacency matrix $A = A_G$. It turns out that several of these results could be used to define $\vartheta(G)$, if generalized appropriately.

Let us start with an almost trivial inequality:

$$\omega \leq 1 + \lambda_{\max}(A). \tag{11.21}$$

In terms of the complementary graph,

$$\alpha \leq 1 + \lambda_{\max}(A_{\overline{G}}). \tag{11.22}$$

Indeed, the matrix $A+I$ contains an $\omega\times\omega$ submatrix J_ω of 1's, so $1+\lambda_{\max}(A) = \lambda_{\max}(A+I) \geq \lambda_{\max}(J_\omega) = \omega$. Note that in this argument, only the matrix entries in adjacent and diagonal positions are used. We could consider any symmetric matrix A' obtained from A by substituting arbitrary real numbers for the remaining entries (which are originally zeroes), to minimize the bound $\lambda_{\max}(A')+1$. (We keep A' symmetric, to have real eigenvalues.) What is the best bound on ω we can obtain this way?

The following lower bound on the chromatic number of a graph is more difficult to prove [**Hoffman 1970**]:

$$\chi \geq 1 - \frac{\lambda_{\max}(A)}{\lambda_{\min}(A)} \tag{11.23}$$

(note that $\lambda_{\min}(A) < 0$ if G has at least one edge, which we may assume). We will not go through the proof; but if you do, you realize that it uses only the 0's in the adjacency matrix, so we can play with the 1's to get the sharpest possible lower bound. What is the best bound on χ we can obtain this way?

Hoffman (unpublished) proved the following upper bound on α, somewhat analogous to the bound (11.23): If G is a d-regular graph, then

$$\alpha(G) \le \frac{-n\lambda_{\min}(A)}{d-\lambda_{\min}(A)} = \frac{-n\lambda_{\min}(A)}{\lambda_{\max}(A)-\lambda_{\min}(A)}. \tag{11.24}$$

Looking at the proof, one realizes that we use here where the 0's of A are, and also the fact that all row sums are the same; but not the actual values of the entries corresponding to edges. What is the best bound we can obtain by playing with the entries in adjacent positions?

Perhaps it is not surprising that the answer to the first two questions posed above is $\vartheta(\overline{G})$, and the answer to the third is closely related. This will follow from the next identities. The first and third formulas below were proved in [**Lovász 1979b**], the second, in [**Knuth 1994**].

Proposition 11.9. *For every graph G,*

$$\vartheta(G) = \min_U \lambda_{\max}(U), \tag{11.25}$$

where U ranges over all $V\times V$-matrices with $U_{ij}=1$ for $ij\in\overline{E}$ and also for $i=j$. Furthermore,

$$\vartheta(G) = \max_W \lambda_{\max}(W), \tag{11.26}$$

where W ranges over all positive semidefinite $V\times V$-matrices with $W_{ij}=0$ for $ij\in E$ and $W_{ii}=1$ for $i\in V$. Furthermore,

$$\vartheta(G) = 1+\max_T \frac{\lambda_{\max}(T)}{|\lambda_{\min}(T)|}, \tag{11.27}$$

where T ranges over all symmetric nonzero $V\times V$-matrices with $T_{ij}=0$ for $ij\in E$ and also for $i=j$.

Note that (11.26) can be written as

$$\vartheta(G) = \max_{\mathbf{v}} \lambda_{\max}\big(\mathsf{Gram}(\mathbf{v})\big), \tag{11.28}$$

where $\mathbf{v}$ ranges over dual orthonormal representations of G (see Exercise 11.12 for the connection of this formula with the definition 11.16 of ϑ).

Proof. Let Y be a minimizer matrix in (11.14), let D be the diagonal matrix obtained from Y by changing all off-diagonal entries to 0, and define $U=I+D-Y$. Then U satisfies the conditions in the Proposition, and (using that $Y\succeq 0$)

$$\lambda_{\max}(U) \le \lambda_{\max}(I+D) = 1+\max_{i\in V} Y_{ii} = \vartheta(G).$$

The reverse inequality follows similarly, by starting with a minimizer in (11.25), and considering $Y=\lambda_{\max}(U)I-U$.

To prove (11.26), let Z be a minimizer in (11.15), and define a $V\times V$ matrix B by

$$W_{ij} = \frac{Z_{ij}}{\sqrt{Z_{ii}Z_{jj}}}.$$

Then W satisfies the conditions in (11.26). Define a vector $x\in\mathbb{R}^V$ by $x_i=\sqrt{Z_{ii}}$, then x is a unit vector and

$$\lambda_{\max}(B) \ge x^{\mathsf{T}}Wx = \sum_{i,j} Z_{ij} = \vartheta(G).$$

The reverse inequality follows similarly, by starting with an optimizer W in (11.26), and scaling its rows and columns by the entries of an eigenvector belonging to $\lambda_{\max}(W)$.

Finally, to prove (11.27), consider an optimizer T in it, then $W = I - \frac{1}{\lambda_{\min}(T)}T$ is positive semidefinite, has $0's$ in adjacent positions and 1's on the diagonal. Hence by (11.26), we have

$$\vartheta(G) \geq \lambda_{\max}(W) = 1 - \frac{\lambda_{\max}(T)}{\lambda_{\min}(T)}.$$

The reverse inequality follows by a similar argument. □

Proposition 11.9 can be combined with different known estimates for the largest eigenvalue of a matrix. As an example, using that the largest eigenvalue of a matrix is bounded above by the largest ℓ_1-norm of its rows, we get that for every optimal dual orthogonal representation $\mathbf{v}$,

$$\vartheta(G) \leq \max_i \sum_j |\mathbf{v}_i^{\mathsf{T}}\mathbf{v}_j|. \tag{11.29}$$

Finally, we show how a strengthening of Hoffman's bound (11.24) can be derived from Proposition 11.9.

PROPOSITION 11.10. *Let G be a d-regular graph. Then for every symmetric nonzero $V \times V$-matrix M such that $M_{ij} = 0$ for $ij \in \overline{E}$ and also for $i = j$, and M has equal row-sums, we have*

$$\vartheta(G) \leq \frac{-n\lambda_{\min}(M)}{\lambda_{\max}(M) - \lambda_{\min}(M)}.$$

If the automorphism group of G is transitive on the nodes, then there is such a matrix M attaining equality. If the automorphism group is transitive on the edges, then equality holds for $M = A_G$.

PROOF. Let M have eigenvalues $\lambda_1 = d \geq \lambda_2 \geq \cdots \geq \lambda_n$. The matrix $J - tM$ satisfies the conditions in (11.25) for every value of t. Using the condition that all row-sums of M are the same, we see that $\mathbb{1}$ is a common eigenvector of J and M, and it follows that all eigenvectors of M are eigenvectors of J as well. Hence the eigenvalues of $J - tM$ are $n - td, -t\lambda_2, \dots, -t\lambda_n$. From the fact that the trace of M is zero, and hence $\lambda_n < 0$, it follows that the largest eigenvalue of $J - tM$ is either $n - td$ or $-t\lambda_n$, and we get the best bound if we choose t so that these two are equal: $t = n/(d - \lambda_n)$, giving the bound in the Proposition.

We can see just as in the proof of Theorem 11.5 that there is an optimizing matrix U in (11.25) that is invariant under the automorphisms. So if G has a node-transitive automorphism group, then the row-sums of this matrix are equal, and the same holds for $M = J - U$. This matrix M satisfies the conditions in the Proposition, and attains equality.

If G has an edge-transitive automorphism group, then all nonzero entries of M are the same, and hence $M = tA_G$ for some $t \neq 0$. The value of t cancels from the formula, so $M = A_G$ also provides equality. □

11.3. Computing the theta-function

Perhaps the most important consequence of the formulas proved in Section 11.2 is that the value of $\vartheta(G)$ is polynomial time computable, as proved in [**Grötschel–Lovász–Schrijver 1981**]. More precisely,

THEOREM 11.11. *There is an algorithm that computes, for every graph G and every $\varepsilon > 0$, a real number t such that*

$$|\vartheta(G) - t| < \varepsilon.$$

The running time of the algorithm is polynomial in n and $\log(1/\varepsilon)$.

Algorithms proving this theorem can be based on almost any of our formulas for ϑ. The simplest is to refer to Theorem 11.2 giving a formulation of $\vartheta(G)$ as the optimum of a semidefinite program (11.6), and the polynomial time solvability of semidefinite programs (see Chapter 13).

The significance of this fact is underlined if we combine it with Theorem 11.1: The two important graph parameters $\alpha(G)$ and $\chi(\overline{G})$ are both NP-hard, but they have a polynomial time computable quantity sandwiched between them.

Computability of the theta-function in an approximate sense, with an arbitrary small error, is an important fact, but not always satisfactory. Often we want explicit algebraic expressions, or at least explicit bounds. For example, it is important to know that $\vartheta(C_5) = \sqrt{5}$, not just that it is between 2.23606 and 2.23607. In the rest of this section we compute the theta-function of several classes of graphs, to illustrate the use of the formulas in the previous sections.

EXAMPLE 11.12 (Cycles). Even cycles are trivial: If n is even, then $\alpha(C_n) = \vartheta(C_n) = \chi(\overline{C}_n) = n/2$ and $\alpha(\overline{C}_n) = \vartheta(\overline{C}_n) = \chi(C_n) = 2$. To derive the theta-function on odd cycles, we can use Proposition 11.10: The eigenvalues of C_n are $2\cos(2k\pi/n)$ $(k = 0, 1, \ldots, n-1)$, of which $k = 0$ gives the largest (which is 2) and $k = (n-1)/2$ gives the smallest one (which is $2\cos\big((n-1)\pi/n\big) = -2\cos(\pi/n)$. Hence

$$\vartheta(C_n) = \frac{n\cos(\pi/n)}{1+\cos(\pi/n)}.$$

Since C_n has a node-transitive automorphism group, this implies that

$$\vartheta(\overline{C}_n) = 1 + \frac{1}{\cos(\pi/n)}.$$

In particular, $\vartheta(\overline{C}_n) \to 2$ as $n \to \infty$. ♦

EXAMPLE 11.13 (Kneser graphs). The Kneser graph K_k^n is defined on node set $\binom{[n]}{k}$, by connecting two k-sets if and only if they are disjoint $(1 \le k \le n)$. Let us assume that $n \ge 2k$ to exclude the trivial case of a graph with no edges. The set of k-sets containing any fixed element of $[n]$ is stable, hence $\alpha(K_k^n) \ge \binom{n-1}{k-1}$. The Erdős–Ko–Rado Theorem asserts that this is the exact value; this fact will follow from our considerations below.

To compute the theta-function of this graph, we use the eigenvalues of its adjacency matrix. These are well known from coding theory:

$$\lambda_t = (-1)^t \binom{n-k-t}{k-t}, \quad (t = 0, 1, \ldots, k). \tag{11.30}$$

The multiplicity of eigenvalue λ_t is $\binom{n}{t} - \binom{n}{t-1}$ (but this will not be important here). The largest eigenvalue in $\binom{n-k}{k}$ (the degree of each node), while the smallest is the next one, $-\binom{n-k-1}{k-1}$.

We apply the formula in Proposition 11.10, and get

$$\vartheta(K_k^n) \le \frac{n\binom{n-k-1}{k-1}}{\binom{n-k}{k} - \binom{n-k-1}{k-1}} = \binom{n-1}{k-1}. \tag{11.31}$$

Comparing this upper bound with the lower bound on $\alpha(K_n^k)$, we see that they are equal, and so

$$\vartheta(K_k^n) = \alpha(K_k^n) = \binom{n-1}{k-1}. \tag{11.32}$$

In particular, the Petersen graph K_2^5 has $\vartheta(K_2^5) = 4$.

Since K_k^n has a node-transitive automorphism group, it follows that

$$\vartheta(\overline{K}_k^n) = \binom{n}{k} \Big/ \binom{n-1}{k-1} = \frac{n}{k}.$$

This is quite close to $\alpha(\overline{K}_k^n) = \lfloor n/k \rfloor$, but can be arbitrarily far from the chromatic number of K_k^n, which is known to be $n-2k+2$. ♦

EXAMPLE 11.14 (Paley graphs). The *Paley graph* Pal_p is defined for a prime $p \equiv 1 \pmod 4$. We take the $\{0, 1, \dots, p-1\}$ as nodes, and connect two of them if their difference is a quadratic residue. It is clear that these graphs have a node-transitive automorphism group, and it is easy to see that they are self-complementary. So Corollary 11.7 applies, and gives that $\vartheta(\mathsf{Pal}_p) = \sqrt{p}$. To determine the stability number of Paley graphs is an unsolved number-theoretic problem; it is conjectured that $\alpha(\mathsf{Pal}_p) = O\big((\log p)^2\big)$. ♦

EXAMPLE 11.15 (Cycles with diagonals). For graphs with a node-transitive automorphism group, the existence of automorphism-invariant optima can be very useful. We illustrate this on the graph W_n obtained of an even cycle C_n with its longest diagonals added. The graph W_6 is just the Kuratowski graph $K_{3,3}$, W_8 is the Wagner graph from Example 10.36. In fact, we can restrict our attention to the case when $n = 4k$ is a multiple of 4, since otherwise W_n is bipartite and $\vartheta(W_n) = \alpha(W_n) = n/2$. We can observe the easy bounds $2k-1 \le \vartheta(W_n) \le 2k$, since W_n has $2k-1$ nonadjacent nodes and can be covered by $2k$ edges.

Let $V(W_n) = \{0, 1, \dots, n-1\}$, where the nodes are labeled in the order of the original cycle. There exists an optimizing matrix U in (11.25) that is invariant under rotation, which means that it has only three different entries:

$$U_{ij} = \begin{cases} 1+a, & \text{if } j-i \equiv \pm 1 \pmod{4k}, \\ 1+b, & \text{if } j-i \equiv 2k \pmod{4k}, \\ 1, & \text{otherwise.} \end{cases}$$

It is easy to see that for every $(4k)$-th root of unity ε, the vector $(1, \varepsilon, \varepsilon^2, \dots \varepsilon^{n-1})$ is an eigenvector of U. (This is a complex vector, so if we want to stay in the real field, we have to consider its real and imaginary parts; but it is more convenient here to do computations with complex vectors.) The eigenvalue λ_r corresponding to $\varepsilon = e^{2\pi i r/n}$ is easy to compute:

$$\lambda_0 = 4k + 2a + b,$$

and

$$\lambda_r = 1+a\varepsilon+\varepsilon^2+\dots+b\varepsilon^{2k}+\varepsilon^{2k+1}+\dots+a\varepsilon^{n-1} = a(\varepsilon+\overline{\varepsilon})+b\varepsilon^{2k}$$
$$= 2a\cos\frac{r\pi}{2k}+b(-1)^r \qquad (r>0).$$

We want to minimize $\max_r \lambda_r$. Each $\lambda_r = \lambda_r(a,b)$ is a linear function of a and b, so we can find ϑ as the optimum of the linear program in 3 variables a, b and t:

$$\vartheta(W_n) = \min_{a,b}\max_{r}\lambda_r(a,b) = \begin{cases}\text{minimize} & t\\ \text{subject to} & \lambda_r(a,b)\le t,\ r=1,\dots,n-1.\end{cases} \tag{11.33}$$

There are many ways to do a back-of-the-envelope computation here; one gets that $a=-k$, $b=-k+k\cos(\pi/k)$ is an optimal solution, giving

$$\vartheta(W_n) = k+k\cos\frac{\pi}{k}. \tag{11.34}$$

The main point in this example is to illustrate that for graphs with a node-transitive automorphism group, the value of the theta-function can be computed by a linear program, analogous to (11.33), where the number of unknowns is the number of orbits of the automorphism group on the edges. This may or may not lead to simple formula like in this case, but the computation is easy to perform, often even by hand. ♦

EXAMPLE 11.16 (Self-polar polytopes). A polytope $P\subseteq\mathbb{R}^d$ is called *self-polar*, if $P^* = -P$. Note that this condition implies that for each vertex $\mathbf{v}$, the inequality $(-\mathbf{v})^{\mathsf T}\mathbf{x}\le 1$ defines a facet $F_{\mathbf{v}}$ of P, and we obtain all facets this way. We call two vertices $\mathbf{v}$ and $\mathbf{v}'$ of P *opposite*, if $\mathbf{v}'$ lies on $F_{\mathbf{v}}$. In other words, $\mathbf{v}^{\mathsf T}\mathbf{v}'=-1$, which shows that this is a symmetric relation. We call the polytope *strongly self-polar*, if it is inscribed in a ball centered at the origin, in other words, there is an $r>0$ such that $|\mathbf{v}|=r$ for all vertices $\mathbf{v}$. For two opposite vertices we have $1=\mathbf{v}^{\mathsf T}\mathbf{u}<|\mathbf{v}|\,|\mathbf{u}|=r^2$, and hence $r>1$. It also follows that the distance of any facet from the origin is $1/r$, so the sphere with radius $1/r$ about the origin is contained in P and touches every facet.

In dimension 2, regular polygons with an odd number of vertices, with appropriate edge length, are strongly self-polar. It was proved in [**Lovász 1983b**] that for every dimension d and $\varepsilon>0$ there exist strongly self-polar polytopes inscribed in a sphere with radius $r<1+\varepsilon$.

Let P be a strongly self-polar polytope in $\mathbb{R}^d$, and let G be the graph on $V=V(P)$, in which two vertices are connected if and only of they are opposite. It was proved in [**Lovász 1983b**] that $\chi(G)\ge d+1$.

We can estimate $\vartheta(G)$ and $\vartheta(\overline{G})$ as follows. Let us label each vertex $\mathbf{v}$ of P with the vector

$$\mathbf{u}_{\mathbf{v}} = \frac{1}{\sqrt{r^2+1}}\begin{pmatrix}\mathbf{v}\\ 1\end{pmatrix}.$$

This is trivially a unit vector in $\mathbb{R}^{d+1}$, and $\mathbf{u}_{\mathbf{v}}\perp\mathbf{u}_{\mathbf{v}'}$ for opposite vertices $\mathbf{v}$ and $\mathbf{v}'$. So $\mathbf{u}$ is an orthonormal representation of $\overline{G}$. Using the vector $\mathbf{e}_{d+1}$ as handle, we see that

$$\vartheta(\overline{G}) \le \max_{\mathbf{v}}\frac{1}{(\mathbf{e}_{d+1}^{\mathsf T}\mathbf{u}_{\mathbf{v}})^2} = r^2+1,$$

and

$$\vartheta(G) \geq \sum_{\mathbf{v}} (\mathbf{e}_{d+1}^{\mathsf{T}} \mathbf{u}_{\mathbf{v}})^2 = \frac{n}{r^2+1}.$$

In particular, we see that $\chi(G)$ can be arbitrarily large while $\vartheta(\overline{G})$ can be arbitrarily close to 2. ♦

Example 11.17 (Random graphs). Consider the most basic random graph $G(n, 1/2)$ (as usual, $G(n,p)$ denotes a random graph on n nodes with edge probability p). It is a nontrivial problem to determine the theta-function of a random graph, and not completely solved. To start with a heuristic, recall Corollary 11.7: a self-complementary graph G with a node-transitive automorphism group has $\vartheta(G) = \sqrt{n}$. As a special case, the Paley graph Pal_p is quasirandom (informally, it behaves like a random graph in many respects) with edge-density 1/2 and $\vartheta(\mathsf{Pal}_p) = \sqrt{p}$.

The graph $G(n, 1/2)$ is, of course, not self-complementary, and its automorphism group is trivial, with high probability. However, its distribution is invariant under complementation and also under all permutations of the nodes. Informally, it is difficult to distinguish it from its complement (as it is difficult to distinguish any two random graphs with the same edge-density), and apart from a little variance in the degrees, it is difficult to distinguish any two nodes. So perhaps it does behave like a self-complementary graph with a node-transitive automorphism group would!

This heuristic predicts the right order of magnitude of $\vartheta\big(G(n,1/2)\big)$, namely it is of the order $\sqrt{n}$. However, no proof is known that would build on the heuristic above. It was proved in [**Juhász 1982**] that with high probability,

$$\frac{1}{3}\sqrt{n} < \vartheta\Big(G\big(n, \tfrac{1}{2}\big)\Big) < 3\sqrt{n}. \tag{11.35}$$

The method extends to estimating the theta-function for random graphs with other edge-densities. If p is a constant and $n \to \infty$, then with high probability,

$$\frac{1}{3}\sqrt{\frac{pn}{1-p}} \leq \vartheta\big(G(n,p)\big) \leq 3\sqrt{\frac{(1-p)n}{p}}.$$

The analysis can be extended to the case when $(\ln n)^{1/6}/n \leq p \leq 1-(\ln n)^{1/6}/n$ [**Coja-Oghlan–Taraz 2004**] (see [**Coja-Oghlan 2003**] for more results about the concentration of this value).

We sketch the proof in the case $p = 1/2$. First, consider the upper bound. The proof of Juhász refers to a result of [**Füredi–Komlós 1981**] bounding the eigenvalues of random matrices. Let A be the matrix defined by

$$A_{ij} = \begin{cases} -1, & \text{if } ij \in E, \\ 1, & \text{otherwise.} \end{cases}$$

Note that $\mathsf{E}(A_{ij}) = 0$ and $A_{ij}^2 = 1$ for $i \neq j$. The matrix A satisfies the conditions in Proposition 11.9, and hence the results of [**Füredi–Komlós 1981**] provide the bound

$$\vartheta(G) \leq \lambda_{\max}(A) \leq 3\sqrt{n}.$$

To prove the lower bound, it suffices to invoke Lemma 11.4 and apply the upper bound to the complementary graph:

$$\vartheta\Big(G(n,\tfrac{1}{2})\Big) = \vartheta\Big(\overline{G}(n,\tfrac{1}{2})\Big) \ge \frac{n}{\vartheta\Big(G(n,\tfrac{1}{2})\Big)} \ge \frac{n}{3\sqrt{n}} = \frac{1}{3}\sqrt{n}$$

with high probability. ♦

11.4. Stable sets

11.4.1. Stability number and theta. We have introduced $\vartheta(G)$ as an upper bound on the stability number $\alpha(G)$. How good an approximation of the stability number is obtained this way? We have seen in Theorem 11.1 that

$$\alpha(G) \le \vartheta(G) \le \chi(\overline{G}).$$

But $\alpha(G)$ and $\chi(\overline{G})$ can be very far apart, and unfortunately, the approximation of α by ϑ can be quite as poor. We have seen that for a random graph G with edge density 1/2, we have $\alpha(G) = O(\ln n)$, but $\vartheta(G) = \Theta(\sqrt{n})$ (with high probability as $n \to \infty$). Even worse examples can be constructed [**Feige 1995**]: sequences of graphs G for which $\alpha(G) = n^{o(1)}$ and $\vartheta(G) = n^{1-o(1)}$; in other words, ϑ/α can be larger than $n^{1-\varepsilon}$ for every $\varepsilon > 0$. (The existence of such graphs also follows from the $P \ne NP$ hypothesis and the results of [**Håstad 1999**], asserting that it is NP-hard to determine $\alpha(G)$ with a relative error less than $n^{1-\varepsilon}$.) By [**Szegedy 1994**], this also implies that $\vartheta(\overline{G})$ does not approximate the chromatic number within a factor of $n^{1-\varepsilon}$.

These constructions do leave a little room for something interesting, namely in the cases when either α is very small, or if ϑ is very large. There are indeed (rather weak, but useful) results in both cases.

First, consider the case when α is very small. [**Konyagin 1981**] proved that if $\alpha(G) = 2$ then $\vartheta(G) = O(n^{1/3})$, and constructed a graph with $\alpha(G) = 2$ and $\vartheta(G) = \Omega(n^c)$ for $c > 0$. The matching lower bound of $\Omega(n^{1/3})$ was proved by Alon [**Alon 1994**], improving a slightly weaker lower bound of [**Kashin–Konyagin 1981**]). (The construction, which is quite involved, is not reproduced here.) For general $\alpha(G)$, [**Alon–Kahale 1998**] proved the following theorem:

THEOREM 11.18. *Let G be a graph with n nodes and $\alpha(G) = k$. Then*

$$\vartheta(G) \le 16 n^{\frac{k-1}{k+1}}.$$

To prove this theorem, we need a lemma that facilitates recurrence when bounding the theta-function.

LEMMA 11.19. *Let G be a graph, and for $i \in V$, let $G_i = G[\overline{N}(i)]$. Then*

$$\vartheta(G) \le 1 + \max_{i\in V} \sqrt{|\overline{N}(i)|\,\vartheta(G_i)}.$$

PROOF. Let $(\mathbf{v}_i, \mathbf{d})$ be an optimal dual orthonormal representation of G. By inequality (11.29),

$$\vartheta(G) \le \max_i \sum_j |\mathbf{v}_i^\mathsf{T}\mathbf{v}_j| = 1 + \max_i \sum_{j\in\overline{N}(i)} |\mathbf{v}_i^\mathsf{T}\mathbf{v}_j|.$$

For any given i, we use the Cauchy–Schwarz Inequality:

$$\Big(\sum_{j\in\overline{N}(i)} |\mathbf{v}_i^{\mathsf{T}}\mathbf{v}_j|\Big)^2 \le |\overline{N}(i)| \sum_{j\in\overline{N}(i)} (\mathbf{v}_i^{\mathsf{T}}\mathbf{v}_j)^2 \le |\overline{N}(i)|\vartheta(G_i),$$

since we can consider $(\mathbf{v}_j:\ j\in\overline{N}(i))$ as a dual orthonormal representation of G_i with handle $\mathbf{v}_i$. Combining these two inequalities, we get the Lemma. □

Proof of Theorem 11.18. The case $k=1$ is trivial, so assume that $k>1$. We use induction on the number of nodes. Let $\Delta=\max_i|\overline{N}(i)|$ and $G_i=G[\overline{N}(i)]$, then $\alpha(G_i)\le k-1$ for every node i, and hence $\vartheta(G_i)\le 16\Delta^{(k-2)/k}$. So Lemma 11.19 implies that

$$\vartheta(G)\le 1+\max_i\sqrt{|\overline{N}(i)|\,\vartheta(G_i)}\le 1+\sqrt{\Delta\cdot(16\Delta^{\frac{k-2}{k}})}=1+4\Delta^{\frac{k-1}{k}}.$$

If $\Delta\le 3n^{k/(k+1)}$, then

$$1+4\Delta^{\frac{k-1}{k}}\le 1+4\cdot 3^{\frac{k-1}{k}}n^{\frac{k}{k+1}\frac{k-1}{k}}<1+12n^{\frac{k-1}{k+1}}<16n^{\frac{k-1}{k+1}},$$

and we are done. If $\Delta>3n^{k/(k+1)}$, then let i be a node with $|\overline{N}(i)|=\Delta$, and consider the partition of the nodes into the sets $S_1=\overline{N}(i)$ and $S_2=\{i\}\cup N(i)$. Note that $\alpha(G[S_2])\le\alpha(G)-1=k-1$. So by induction on the number of nodes,

$$\begin{aligned}\vartheta(G)&\le\vartheta(G[S_1])+\vartheta(G[S_2])\le 16\Delta^{\frac{k-2}{k}}+16(n-\Delta)^{\frac{k-1}{k+1}}\\&=16\Delta^{\frac{k-2}{k}}+16n^{\frac{k-1}{k+1}}\Big(1-\frac{\Delta}{n}\Big)^{\frac{k-1}{k+1}}<16\Delta^{\frac{k-2}{k}}+16n^{\frac{k-1}{k+1}}\Big(1-\frac{k-1}{k+1}\cdot\frac{\Delta}{n}\Big)\\&=16n^{\frac{k-1}{k+1}}+16\Delta\Big(\Delta^{-\frac{2}{k}}-\frac{k-1}{k+1}n^{-\frac{2}{k+1}}\Big)\\&\le 16n^{\frac{k-1}{k+1}}+16\Delta n^{-\frac{2}{k+1}}\Big(3^{-\frac{2}{k}}-\frac{k-1}{k+1}\Big)\le 16n^{\frac{k-1}{k+1}}.\end{aligned}$$

□

To motivate the next theorem, let us think of the case when the chromatic number is small, say $\chi(G)\le t$, where we think of t as a constant. Trivially, $\alpha(G)\ge n/t$, and nothing better can be said. From the weaker assumption that $\omega(G)\le t$ we get a much weaker bound: using the result of [**Ajtai–Komlós–Szemerédi 1980**] in the theory of off-diagonal Ramsey numbers, one gets (essentially) a power of n as a lower bound:

$$\alpha(G)=\Omega\left(\Big(\frac{n}{\log n}\Big)^{\frac{1}{t-1}}\right). \tag{11.36}$$

The inequality $\vartheta(\overline{G})\le t$ is a condition that is between the previous two, and the following theorem of [**Karger–Motwani–Sudan 1994**] does give a lower bound on $\alpha(G)$ that is better than (11.36) (for $t>2$). In addition, the proof provides a polynomial time randomized algorithm to construct a stable set of the appropriate size, whose use we will explain later.

THEOREM 11.20. *Let G be a graph and $t=\vartheta(\overline{G})$. Then*

$$\alpha(G)\ge\frac{n^{\frac{3}{t+1}}}{10\sqrt{\ln n}}.$$

PROOF. The case $t = 2$ is easy (cf. Exercise 11.1), so we assume that $t > 2$. By Theorem 11.2, t is the strict vector chromatic number of G, and hence there are unit vectors $\mathbf{u}_i \in \mathbb{R}^k$ (in some dimension k) such that $\mathbf{u}_i^\mathsf{T}\mathbf{u}_j = -1/(t-1)$ whenever $ij \in E$.

Let $\mathbf{w}$ be a random point in $\mathbb{R}^d$ whose coordinates are independent standard Gaussian random variables. Fix an $s > 0$, and consider the set $S = \{i : \mathbf{w}^\mathsf{T}\mathbf{u}_i \geq s\}$. The inner product $\mathbf{w}^\mathsf{T}\mathbf{u}_i$ has standard Gaussian distribution, and hence the probability that a given node belongs to S is

$$Q(s) = \frac{1}{\sqrt{2\pi}} \int_s^\infty e^{-x^2/2}\, dx,$$

and the expected size of S is $\mathsf{E}|S| = Q(s)n$.

Next we show that S does not induce too many edges. Let $ij \in E$, then $|\mathbf{u}_i + \mathbf{u}_j|^2 = (2t-4)/(t-1)$, and so the probability that both nodes $\mathbf{u}_i$ and $\mathbf{u}_j$ belong to S can be estimated as follows:

$$\mathcal{P}(\mathbf{w}^\mathsf{T}\mathbf{u}_i \geq s,\ \mathbf{w}^\mathsf{T}\mathbf{u}_j \geq s) \leq \mathcal{P}(\mathbf{w}^\mathsf{T}(\mathbf{u}_i + \mathbf{u}_j) \geq 2s) = Q\Big(\sqrt{\frac{2t-2}{t-2}}s\Big).$$

Hence the expected number of edges spanned by S satisfies

$$\mathsf{E}|E[S]| \leq Q\Big(\sqrt{\frac{2t-2}{t-2}}s\Big)m,$$

where $m = |E|$. We can delete at most $|E[S]|$ nodes from S (one from each edge it induces) to get a stable set T with expected size

$$\mathsf{E}|T| \geq \mathsf{E}(|S| - |E[S]|) = \mathsf{E}|S| - \mathsf{E}|E[S]| \geq Q(s)n - Q\Big(\sqrt{\frac{2t-2}{t-2}}s\Big)m.$$

We want to choose s so that it maximizes the right hand side. By elementary computation we get that

$$s = \sqrt{\frac{2t-4}{t}\ln\frac{m}{n}}$$

is an approximately optimal choice. Using the well known estimates

$$\frac{1}{\sqrt{2\pi}}\frac{1}{s}e^{-s^2/2} < Q(s) < \frac{1}{\sqrt{2\pi}}\frac{s}{s^2+1}e^{-s^2/2}, \tag{11.37}$$

we get

$$\alpha(G) \geq \frac{1}{10\sqrt{\ln(m/n)}} m^{-\frac{t-2}{t}} n^{\frac{2t-2}{t}}.$$

If $m < n^{(2t+1)/(t+1)}$, then this proves the theorem.

If $m \geq n^{(2t+1)/(t+1)}$, then there is a node i with degree $\Delta \geq 2m/n \geq 2n^{t/(t+1)}$. Clearly $\vartheta(G[N(i)]) \leq t-1$ (see Exercise 11.3), and hence by induction on n, $G[N(i)]$ has a stable set of size at least $\Delta^{3/t}/(10\sqrt{\ln\Delta}) > n^{3/(t+1)}/(10\sqrt{\ln n})$. This proves the theorem in this case as well. □

11.4.2. The stable set polytope. Stable sets and cliques give rise to important polyhedra associated with graphs. After summarizing some basic properties of these polyhedra, we show that orthogonal representations provide an interesting related convex body, with nice duality properties.

The *stable set polytope* STAB(G) of a graph G is the convex hull of incidence vectors of all stable sets. This gives us a polytope in $\mathbb{R}^V$. The stability number $\alpha(G)$ can be obtained by maximizing the linear function $\sum_{i\in V} x_i$ over this polytope, which suggests that methods from linear programming can be used here.

With this goal in mind, we have to find a system of linear inequalities whose solution set is exactly the polytope STAB(G). It would be best to find a minimal such system, which is unique. If we can find this system, then the task of computing the stability number $\alpha(G)$ of G reduces to maximizing $\sum_{i\in V} x_i$ subject to these constraints, which means solving a linear program. Unfortunately, this system of linear inequalities is in general exponentially large and very complicated. But if we find at least some linear inequalities valid for the stable set polytope, then solving the linear program we get an upper bound on $\alpha(G)$, and for special graphs, we get the exact value.

So we want to find linear inequalities (constraints) valid for the incidence vector of every stable set. We start with the trivial *nonnegativity constraints*:

$$x_i \ge 0 \quad (i \in V). \tag{11.38}$$

The fact that the set is stable is reflected by the *edge constraints*:

$$x_i + x_j \le 1 \quad (ij \in E). \tag{11.39}$$

Inequalities (11.38) and (11.39) define the *fractional stable set polytope* FSTAB(G). Integral points in FSTAB(G) are exactly the incidence vectors of stable sets, but FSTAB(G) may have other (nonintegral) vertices, and is in general larger than STAB(G) (cf. Exercise 11.18). The case of equality has a nice characterization.

PROPOSITION 11.21. STAB(G) = FSTAB(G) *if and only if* G *is bipartite.* □

Let $\alpha^f(G)$ denote the maximum of $\sum_i x_i$ over $\mathbf{x} \in$ FSTAB(G). Trivially $\alpha(G) \le \alpha^f(G)$, and $\alpha^f(G)$ is computable in polynomial time (since (11.38) and (11.39) describe a linear program defining α^f). The difference $\alpha^f - \alpha$ will play a role in Chapter 18.

We can strengthen the edge constraints if the graph has larger cliques. Every clique B gives rise to a *clique constraint*:

$$\sum_{i\in B} x_i \le 1. \tag{11.40}$$

Inequalities (11.38) and (11.40) define a polytope QSTAB(G), the *clique-constrained fractional stable set polytope* of G. Since cliques in G correspond to stable sets in $\overline{G}$ and vice versa, it is easy to see that QSTAB(G) is just the antiblocker of STAB($\overline{G}$) (cf. Section C.3).

Again, we can introduce a corresponding relaxation of the stability number, namely the quantity $\alpha^*(G)$ defined as the maximum of $\sum_i x_i$ over $\mathbf{x} \in$ QSTAB(G). This quantity is a sharper upper bound on $\alpha(G)$ then α^f, but it is NP-hard to compute [**Grötschel–Lovász–Schrijver 1984**]. See also Exercise 11.20.

The polytope QSTAB(G) is contained in FSTAB(G), but is still larger than STAB(G) in general. The case of equality leads us to an interesting and rich class of graphs, of which we give a very brief survey.

11.4.3. Perfect graphs. Recall again the basic inequality

$$\alpha(G) \le \vartheta(G) \le \chi(\overline{G}).$$

For graphs with $\alpha(G) = \chi(\overline{G})$, we have equality here, so ϑ is an integer. To know this is useful; just to mention one consequence, calling the approximation algorithm in Theorem 11.11 with an error bound of 1/3 gives the exact value of ϑ.

But which graphs have this nice property? It turns out that the condition $\alpha(G) = \chi(\overline{G})$ does not say much about the structure of G, but a strengthened version of it leads to a very interesting class of graphs. A graph G is called *perfect*, if for every induced subgraph G' of G, we have $\omega(G') = \chi(G')$. Every bipartite graph is perfect, since they satisfy $\omega(G) = \chi(G) = 2$ (if they have an edge) or $\omega(G) = \chi(G) = 1$ (if they have no edge), and their induced subgraphs are also bipartite. Figure 11.2 shows some perfect and nonperfect graphs.

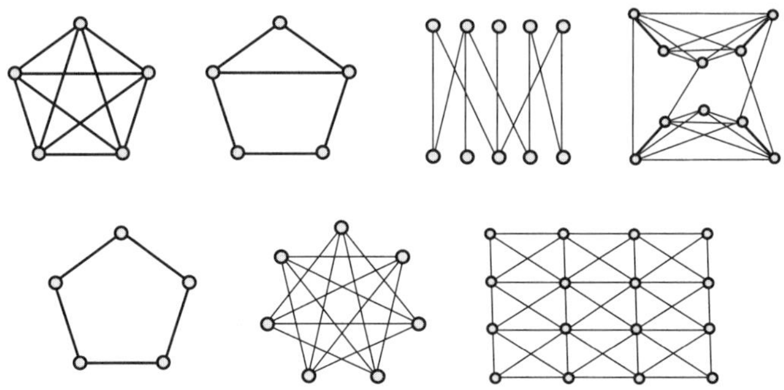

FIGURE 11.2. Some perfect graphs (first row) and some nonperfect graphs (second row).

To be perfect is a rather strong structural property; nevertheless, many interesting classes of graphs are perfect (bipartite graphs, their complements and their linegraphs, interval graphs, comparability and incomparability graphs of posets, chordal graphs, split graphs, etc.).

The following deep characterization of perfect graphs was conjectured by Berge in 1961 and proved by Chudnovski, Robertson, Seymour and Thomas [**Chudnovsky et al. 2006**].

THEOREM 11.22 (Strong Perfect Graph Theorem). *A graph is perfect if and only if neither the graph nor its complement contains a chordless odd cycle longer than* 3. □

As a corollary we can state the "Weak Perfect Graph Theorem" proved earlier [**Lovász 1972**]:

THEOREM 11.23. *The complement of a perfect graph is perfect.* □

From this theorem it follows that in the definition of perfect graphs we could replace the equation $\omega(G') = \chi(G')$ by $\alpha(G') = \chi(\overline{G'})$. Perfectness can also be characterized in terms of the stable set polytope [**Chvátal 1975**]:

THEOREM 11.24. $\mathrm{STAB}(G) = \mathrm{QSTAB}(G)$ *if and only if G is perfect.* □

Based on our remark above, the condition $\mathrm{STAB}(G) = \mathrm{QSTAB}(G)$ is equivalent to saying that $\mathrm{STAB}(\overline{G})$ and $\mathrm{STAB}(G)$ are antiblockers, which is a condition symmetric in G and $\overline{G}$. So Theorem 11.24 implies Theorem 11.23.

Turning to algorithms, Theorem 11.11 implies:

COROLLARY 11.25. *The stability number and the chromatic number of a perfect graph are polynomial time computable.*

Using the algorithms of Corollary 11.25 one can compute more than just these values: one can compute a maximum stable set and an optimal coloring in a perfect graph in polynomial time (see Exercise 11.19).

It is interesting (and somewhat frustrating) that the only other way of computing the stability number of a perfect graph in polynomial time is to use the very advanced structure theory of perfect graphs, developed more recently by Chudnovski, Robertson, Seymour and Thomas [**Chudnovsky et al. 2006**]. This deep theory is combinatorial, but quite involved.

Theorem 11.11 extends to the weighted version of the theta-function. Maximizing a linear function over $\mathrm{STAB}(G)$ or $\mathrm{QSTAB}(G)$ is NP-hard; but, surprisingly, TSTAB behaves much better: *Every linear objective function can be maximized over* $\mathrm{TSTAB}(G)$ *(with an arbitrarily small error) in polynomial time.* This applies in particular to $\vartheta(G)$, which is the maximum of $\sum_i x_i$ over $\mathrm{TSTAB}(G)$. See [**Grötschel–Lovász–Schrijver 1988, Knuth 1994**] for more detail.

11.4.4. Orthogonality constraints.

For every orthonormal representation $(\mathbf{u}_i, \mathbf{c})$ of G, we consider the linear constraint

(11.41)
$$\sum_{i\in V} (\mathbf{c}^\mathsf{T}\mathbf{u}_i)^2 x_i \le 1,$$

which we call an *orthogonality constraint.* The solution set of nonnegativity and orthogonality constraints is denoted by $\mathrm{TSTAB}(G)$. It is clear that TSTAB is a closed, full-dimensional, convex set. The orthogonality constraints are valid if x is the indicator vector of a stable set of nodes (cf. (11.4)), and therefore they are valid for $\mathrm{STAB}(G)$. Furthermore, every clique constraint is an orthogonality constraint. Indeed, for every clique B, the constraint $\sum_{i\in B} x_i \le 1$ is obtained from the orthogonal representation

$$i \mapsto \begin{cases} \mathbf{e}_1, & i \in B, \\ \mathbf{e}_i, & \text{otherwise}, \end{cases} \qquad \mathbf{c} = \mathbf{e}_1.$$

Hence

(11.42)
$$\mathrm{STAB}(G) \subseteq \mathrm{TSTAB}(G) \subseteq \mathrm{QSTAB}(G)$$

for every graph G.

There are several other characterizations of TSTAB. These are based on an extension of the theta-function to the case when we are also given a weighting $w:\ V \to \mathbb{R}_+$. Generalizing the formulas (11.12)-(11.16), the quantity $\vartheta(G, w)$ can be defined by any of the following formulas [**Grötschel–Lovász–Schrijver 1986**]:

$$\vartheta(G,w) = \min\Big\{\max_{i\in V} \frac{w_i}{(\mathbf{c}^\mathsf{T}\mathbf{u}_i)^2} : \mathbf{u} \text{ ONR of } G,\ |\mathbf{c}| = 1\Big\} \tag{11.43}$$

$$= \min\Big\{t \ge 2 : |\mathbf{y}_i|^2 = t - w_i,\ \mathbf{y}_i^\mathsf{T}\mathbf{y}_j = -\sqrt{w_i w_j}\ (ij \in \overline{E})\Big\} \tag{11.44}$$

$$= \min\{\max_{i\in V}(Y_{ii} + w_i) : Y \succeq 0,\ Y_{ij} = -\sqrt{w_i w_j}\ (ij \in \overline{E})\} \tag{11.45}$$

$$= \max\Big\{\sum_{i,j\in V} w_i w_j Z_{ij} : Z \succeq 0,\ Z_{ij} = 0\ (ij \in E),\ \sum_i Z_{ii} = 1\Big\} \tag{11.46}$$

$$= \max\Big\{\sum_{i\in V} w_i (\mathbf{d}^\mathsf{T}\mathbf{v}_i)^2 : \mathbf{v} \text{ ONR of } \overline{G},\ |\mathbf{d}| = 1\Big\}. \tag{11.47}$$

The equivalence of (11.43)–(11.47) can be obtained extending the proof Theorem 11.2 to the node-weighted version (at the cost of a little more computation), which is described in the book [**Grötschel–Lovász–Schrijver 1988**], and we do not reproduce the proof here.

Similarly as in the unweighted case, the optima in (11.43) and (11.47) satisfy the relations

$$\mathbf{c}\mathbf{d}^\mathsf{T} = \sum_i (\mathbf{c}^\mathsf{T}\mathbf{u}_i)(\mathbf{d}^\mathsf{T}\mathbf{v}_i)\mathbf{u}_i\mathbf{v}_i^\mathsf{T} \tag{11.48}$$

and its consequences formulated there. (The node-weights don't enter this relation.)

For every orthonormal representation $\mathbf{u} = (\mathbf{u}_i : i \in V)$ with handle $\mathbf{c}$, we call the vector $\big((\mathbf{c}^\mathsf{T}\mathbf{u}_i)^2 : i \in V\big)$ the *profile* of the representation. We can state two further characterizations of TSTAB(G):

PROPOSITION 11.26. (a) $x \in \text{TSTAB}(G)$ *if and only if* $\vartheta(\overline{G}, x) \le 1$.

(b) *The body* TSTAB(G) *is exactly the set of profiles of dual orthonormal representations of* G.

PROOF. (a) follows from (11.47).

(b) The profile of every dual orthonormal representation belongs to TSTAB(G); this is equivalent to (11.11). Conversely, let $x \in \text{TSTAB}(G)$. Then $\vartheta(\overline{G}, x) \le 1$ by (a), so (11.43) implies that there is a dual orthonormal representation $\mathbf{v}$ of G with handle $\mathbf{d}$ for which $x_i \le \mathbf{d}^\mathsf{T}\mathbf{v}_i$ for all nodes $i \in V$. Thus the vectors $\mathbf{v}_i' = (x_i/\mathbf{d}^\mathsf{T}\mathbf{v}_i)\mathbf{v}_i$ satisfy $\mathbf{d}^\mathsf{T}\mathbf{v}_i' = x_i$. The vectors $\mathbf{v}_i'$ are not of unit length, but the vectors

$$\mathbf{v}_i'' = \begin{pmatrix} \mathbf{v}_i' \\ \sqrt{1 - |\mathbf{v}_i'|^2}\mathbf{e}_i \end{pmatrix} \quad \text{and} \quad \mathbf{d}'' = \begin{pmatrix} \mathbf{d} \\ 0 \end{pmatrix}$$

form a dual orthonormal representation of G with profile x. □

The last characterization of TSTAB(G) is equivalent to the following duality result.

COROLLARY 11.27. TSTAB($\overline{G}$) *is the antiblocker of* TSTAB(G). □

Next we determine the vertices of $TSTAB(G)$. Recall that a *vertex* of a convex body K is a boundary point $\mathbf{v}$ that is the unique point of intersection of all hyperplanes supporting K at $\mathbf{v}$. This means that there is a pointed convex cone containing K with $\mathbf{v}$ as its vertex. This is to be distinguished from an *extreme point*, which is the unique point of intersection of a hyperplane supporting K with K.

THEOREM 11.28. *The vertices of* TSTAB(G) *are exactly the incidence vectors of stable sets in* G.

Does this imply that TSTAB(G) = STAB(G)? Of course not, since TSTAB(G) (as every convex body) is the convex hull of its extreme points, but not necessarily of its vertices.

PROOF. The vector $\mathbb{1}_A$, where A is a stable set of nodes, is the unique common point of n supporting hyperplanes $x_i = 1$ $(i \in A)$ and $x_i = 0$ $(i \in V \setminus A)$, and so it is a vertex of TSTAB(G).

Conversely, let $z = (z_i : i \in V)$ be a vertex of TSTAB(G). If $z_i = 0$ for some node i, then we can delete i: We get a graph G' for which $\mathrm{TSTAB}(G') = \mathrm{TSTAB}(G) \cap \{z_i = 0\}$, and so $z|_{V\setminus i}$ is a vertex of TSTAB(G'), and we can proceed by induction.

So we may assume that $z_i > 0$ for all $i \in V$. Since $z \in \mathrm{TSTAB}(G)$, we can write $z_i = (\mathbf{d}^\mathsf{T}\mathbf{v}_i)^2$ for some dual orthonormal representation $(\mathbf{v}_i : i \in V)$ of G and unit vector $\mathbf{d}$.

Let $a^\mathsf{T}x \le 1$ be a hyperplane that supports TSTAB(G) at z. Then $a \in \mathrm{TSTAB}(\overline{G})$ by Corollary 11.27, and hence there is an orthonormal representation $(\mathbf{u}_i : i \in V)$ of G and unit vector $\mathbf{c}$ such that $a_i = (\mathbf{c}^\mathsf{T}\mathbf{u}_i)^2$. From (11.48), we get

$$\mathbf{d} = \sum_i (\mathbf{c}^\mathsf{T}\mathbf{u}_i)^2(\mathbf{d}^\mathsf{T}\mathbf{v}_i)\mathbf{v}_i.$$

Multiplying by any vector $\mathbf{y}$, we get

$$\mathbf{d}^\mathsf{T}\mathbf{y} = \sum_i (\mathbf{c}^\mathsf{T}\mathbf{u}_i)^2(\mathbf{d}^\mathsf{T}\mathbf{v}_i)(\mathbf{v}_i^\mathsf{T}\mathbf{y}) = \sum_i a_i(\mathbf{d}^\mathsf{T}\mathbf{v}_i)(\mathbf{v}_i^\mathsf{T}\mathbf{y}).$$

Thus if $\mathbf{y}$ is not orthogonal to $\mathbf{d}$, then the point z' defined by $z'_i = (\mathbf{d}^\mathsf{T}\mathbf{v}_i)(\mathbf{v}_i^\mathsf{T}\mathbf{y})/(\mathbf{d}^\mathsf{T}\mathbf{y})$ is contained in the supporting hyperplane $\sum_i a_i x_i = 1$. This holds for every supporting hyperplane at z. Since z is a vertex, the only common point of hyperplanes supporting TSTAB(G) at z is z itself (here we use that z is a vertex, not just an extreme point). Thus $z' = z$, which means that $\mathbf{v}_i^\mathsf{T}\mathbf{y} = (\mathbf{d}^\mathsf{T}\mathbf{v}_i)(\mathbf{d}^\mathsf{T}\mathbf{y})$ for all i (here we use that $z_i \ne 0$ for all i). This relation holds for almost all $\mathbf{y}$, which implies that $\mathbf{v}_i = (\mathbf{d}^\mathsf{T}\mathbf{v}_i)\mathbf{d}$, and since $\mathbf{v}_i$ is a unit vector, we get $\mathbf{v}_i = \mathbf{d}$. So no two vectors $\mathbf{v}_i$ are orthogonal, and thus G has no edges. But then $z = \mathbb{1}_V$ is the incidence vector of a stable set as claimed. □

EXAMPLE 11.29. Consider the graph C_5, with node set $\{1,\dots,5\}$. The polytope STAB(C_5) has 11 vertices (the origin, the basic unit vectors, and the incidence vectors of nonadjacent pairs of nodes). The facets are defined by the nonnegativity constraints, edge constraints, and single further inequality

$$x_1+x_2+x_3+x_4+x_5 \le 2. \tag{11.49}$$

Since C_5 has no triangles, we have QSTAB = FSTAB. This polytope has a single vertex $(\frac{1}{2},\dots,\frac{1}{2})$ in addition to the incidence vectors of stable sets.

Turning to TSTAB, we know by Theorem 11.28 that it has 11 vertices, the same vertices as STAB. The umbrella construction in Example 10.6 gives a point

$$\Big(\frac{1}{\sqrt{5}},\dots,\frac{1}{\sqrt{5}}\Big)^\mathsf{T} \in \mathrm{TSTAB}(C_5) \tag{11.50}$$

which is not in STAB(G) by (11.49). Applying the umbrella construction to the complement, and scaling, we get an orthogonality constraint

$$x_1 + \cdots + x_5 \leq \sqrt{5}, \tag{11.51}$$

showing that the special vertex of FSTAB(C_5) does not belong to TSTAB(C_5). ♦

This example shows that not every orthogonality constraint follows from the clique constraints. In fact, the number of essential orthogonality constraints is infinite unless the graph is perfect.

PROPOSITION 11.30. TSTAB(G) *is polyhedral if and only if the graph is perfect.*

PROOF. If G is perfect, then STAB(G) = QSTAB(G) by Theorem 11.24, and (11.42) implies that TSTAB(G) = STAB(G) = QSTAB(G), so TSTAB(G) is polyhedral. To prove the converse, suppose that TSTAB(G) is polyhedral, then Theorem 11.28 implies that TSTAB(G) = STAB(G). We can apply this argument to $\overline{G}$, since the antiblocker $\text{TSTAB}(G)^{\text{abl}} = \text{TSTAB}(\overline{G})$ is also polyhedral; we get that

$$\text{TSTAB}(G) = \text{TSTAB}(\overline{G})^{\text{abl}} = \text{STAB}(\overline{G})^{\text{abl}} = \text{QSTAB}(G).$$

So STAB(G) = TSTAB(G) = QSTAB(G), which implies that G is perfect by Theorem 11.24. □

11.5. Applications

11.5.1. Shannon capacity. In the introduction, we have described how to use orthogonal representations to determine the Shannon zero-error capacity of the pentagon. What happens with other confusion graphs?

Let V be an alphabet with confusion graph $G = (V, E)$. To describe the confusion graph of longer messages, we use the strong product of two graphs. In these terms, $\alpha(G^{\boxtimes k})$ is the maximum number of non-confusable words of length k: words composed of elements of V, so that for every two words there is at least one i ($1 \leq i \leq k$) such that the i-th letters are different and nonadjacent in G, i.e., non-confusable. It is easy to see that

$$\alpha(G \boxtimes H) \geq \alpha(G)\alpha(H). \tag{11.52}$$

This implies that

$$\alpha(G^{\boxtimes(k+l)}) \geq \alpha(G^{\boxtimes k})\alpha(G^{\boxtimes l}), \tag{11.53}$$

and hence

$$\alpha(G^{\boxtimes k}) \geq \alpha(G)^k. \tag{11.54}$$

The *Shannon capacity* (zero-error capacity, if we want to be pedantic) of a graph G is the value

$$\Theta(G) = \lim_{k\to\infty} \alpha(G^{\boxtimes k})^{1/k}. \tag{11.55}$$

Inequality (11.53) implies, via Fekete's Lemma, that the limit exists, and (11.54) implies that

$$\Theta(G) \geq \alpha(G). \tag{11.56}$$

Rather little is known about this graph parameter for general graphs. For example, it is not known whether $\Theta(G)$ can be computed for all graphs by any algorithm (polynomial or not), although there are several special classes of graphs for which this is not hard. The behavior of $\Theta(G)$ and the convergence in (11.55) are rather erratic; see [**Alon 1998, Alon–Lubetzky 2006**].

Let us describe a few facts we do know. First, let us generalize the argument from the Introduction bounding $\Theta(C_4)$. Let $\overline{\chi}(G)$ denote the minimum number of complete subgraphs covering the nodes of G (this is the same as the chromatic number of the complementary graph.) Trivially

$$\alpha(G) \le \overline{\chi}(G). \tag{11.57}$$

Any covering of G by $\overline{\chi}(G)$ cliques and of H by $\overline{\chi}(H)$ cliques gives a "product covering" of $G \boxtimes H$ by $\overline{\chi}(G)\overline{\chi}(H)$ cliques, and so

$$\overline{\chi}(G \boxtimes H) \le \overline{\chi}(G)\overline{\chi}(H). \tag{11.58}$$

Hence

$$\alpha(G^{\boxtimes k}) \le \overline{\chi}(G^{\boxtimes k}) \le \overline{\chi}(G)^k,$$

and thus

$$\Theta(G) \le \overline{\chi}(G). \tag{11.59}$$

It follows that if $\alpha(G) = \overline{\chi}(G)$, then $\Theta(G) = \alpha(G)$; for such graphs, nothing better can be done than reducing the alphabet to the largest mutually non-confusable subset. In particular, this answers the Shannon capacity problem for perfect graphs.

Instead of $\overline{\chi}$, we can use ϑ to bound the Shannon capacity:

$$\alpha(G^k) \le \vartheta(G^k) \le \vartheta(G)^k,$$

which implies

PROPOSITION 11.31. *For every graph G,*

$$\Theta(G) \le \vartheta(G).$$

Since $\vartheta(C_5) = \sqrt{5}$, we get that equality holds in Example 1.2: $\Theta(C_5) = \sqrt{5}$. This argument can be generalized to an infinite class of graphs:

COROLLARY 11.32. *If G is a self-complementary graph with a node-transitive automorphism group, then $\Theta(G) = \sqrt{n}$.*

PROOF. The diagonal in $G \boxtimes \overline{G}$ is stable, so $\alpha(G \boxtimes G) = \alpha(G \boxtimes \overline{G}) \ge n$, and hence $\Theta(G) \ge \sqrt{n}$. On the other hand, $\Theta(G) \le \vartheta(G) = \sqrt{n}$ by Corollary 11.7. □

EXAMPLE 11.33 (Paley graphs II). Paley graphs form a class of graphs to which this corollary applies, and whose Shannon capacity can be determined exactly: $\Theta(\mathrm{Pal}_p) = \vartheta(\mathrm{Pal}_p) = \sqrt{p}$. Assuming that the stability number of a Paley graph is polylogarithmic in p (as conjectured), for this infinite family of graphs the Shannon capacity is much higher than the stability number. ♦

The tensor product construction in the proof of Theorem 11.8 shows that if G has an orthonormal representation in dimension c, and H has an orthonormal representation in dimension d, the $G \boxtimes H$ has an orthonormal representation in dimension cd. It follows that the minimum dimension of any orthonormal representation is an upper bound on $\Theta(G)$. Exercise 11.2 shows that this bound is never better than $\vartheta(G)$.

However, if we consider orthogonal representations over fields of finite characteristic, then the analogue of ϑ is not defined, but the minimum dimension in which an orthogonal representation exists may provide a better bound on the Shannon capacity than ϑ. More generally, consider a G-matrix A with nonzero diagonal entries. Haemers [**Haemers 1979**] proved that $\mathrm{rk}(A)$ is an upper bound on the Shannon capacity of G, and constructed matrices (over an appropriate finite field) for which this bound is better that ϑ (see Exercise 11.17). There are examples where such a matrix has rank 3 while $\vartheta = \Omega(n^{1/4})$ [**Alon 2018**]. Random graphs behave in the opposite way: by Example 11.17, $\vartheta\big(G(n,1/2)\big) = O(n^{1/2})$ with high probability, while for every fixed field (finite or infinite), every matrix A with the properties above has rank $\Omega(n/\log n)$ [**Alon at al. 2018**].

We'll return to the Shannon capacity in a quantum communication setting in Chapter 12.

11.5.2. Approximate coloring. Suppose that somebody gives us a graph and guarantees that the graph is 3-colorable, without telling us the 3-coloring itself. Can we find this 3-coloring? (This "hidden 3-coloring problem"may sound artificial, but this kind of situation does arise in cryptography and other data security applications; one can think of the hidden 3-coloring as a "watermark" that can be verified if we know where to look.)

It is easy to argue that knowing that the graph is 3-colorable does not help: it is still NP-hard to find the 3-coloration. But suppose that we would be satisfied with finding a 4-coloration, or 5-coloration, or $(\log n)$-coloration; is this easier? It is known that to find a 4-coloration is still NP-hard [**Khanna–Linial–Safra 2000**] (see also [**Garey–Johnson 1976**] for inapproximability results about larger chromatic number), but little is known above this. Improving earlier results, [**Karger—Motwani–Sudan 1994**] gave a polynomial time algorithm that, given a 3-colorable graph, computes a coloring with $O(n^{1/4}(\ln n)^{3/2})$ colors. This was improved in several papers; the current best exponent is $O(n^{0.19996})$ [**Kawarabayashi–Thorup 2014**].

We sketch the algorithm of [**Karger–Motwani–Sudan 1994**]. Theorem 11.20 implies that G contains a stable set of size at least $n^{3/4}/(10\sqrt{\ln n})$, and the proof can be converted to a randomized polynomial time algorithm to find such a set. Color this set with one color, delete its nodes, and repeat the procedure with the remaining graph G_1.

We claim that this procedure results in a coloring of G with $40n^{1/4}\sqrt{\ln n}$ colors. Indeed, the number of nodes in G_1 is

$$n_1 \le n - 10\frac{n^{3/4}}{10\sqrt{\ln n}} = n\Big(1 - \frac{1}{10n^{1/4}\sqrt{\ln n}}\Big),$$

then by induction, the number of colors we use is at most

$$\begin{aligned} 1 + 40n_1^{1/4}\sqrt{\ln n_1} &\le 1 + 40n^{1/4}\Big(1 - \frac{1}{10n^{1/4}\sqrt{\ln n}}\Big)^{1/4}\sqrt{\ln n} \\ &\le 1 + 40n^{1/4}\Big(1 - \frac{1}{40n^{1/4}\sqrt{\ln n}}\Big)\sqrt{\ln n} = 40n^{1/4}\sqrt{\ln n}. \end{aligned}$$

Exercise 11.1. If $\vartheta(\overline{G}) = 2$, then G is bipartite.

Exercise 11.2. Prove that the minimum dimension in which a graph G has an orthonormal representation is at least $\vartheta(G)$.

EXERCISE 11.3. Let G be a graph and $v \in V$. (a) $\vartheta(G \setminus v) \geq \vartheta(G) - 1$. (b) If v is an isolated node, then $\vartheta(G \setminus v) = \vartheta(G) - 1$. (c) If v is adjacent to all other nodes, then $\vartheta(G \setminus v) = \vartheta(G)$.

EXERCISE 11.4. Let G be a graph and let $V = S_1 \cup \cdots \cup S_k$ be a partition of V. (a) $\vartheta(G) \leq \sum_i \vartheta(G[S_i])$. (b) If no edge connects nodes in different sets S_i, then equality holds. (c) Suppose that any two nodes in different sets S_i are adjacent. How can $\vartheta(G)$ be expressed in terms of the $\vartheta(G[S_i])$?

EXERCISE 11.5. Prove that the graph parameter $\vartheta^-(G)$ introduced in Remark 11.3 could be defined in any of the following ways: (a) tightening the conditions in (11.16) by requiring that $\mathbf{v}_i^\mathsf{T}\mathbf{v}_j \geq 0$ for all i and j; (b) relaxing the conditions in (11.13) by requiring only $\mathbf{w}_i^\mathsf{T}\mathbf{w}_j \leq -1/(t-1)$ for $ij \in \overline{E}$; (c) relaxing the orthogonality conditions in (11.12) by requiring only $\mathbf{u}_i^\mathsf{T}\mathbf{u}_j \leq 0$ for $ij \in \overline{E}$.

EXERCISE 11.6. Formulate and prove the analogue of Exercise 11.5 for the graph parameter $\vartheta^+(G)$.

EXERCISE 11.7. Prove that $\vartheta^-(G)\vartheta^+(\overline{G}) \geq n$ for every graph G.

EXERCISE 11.8. Prove the following weak converse to Proposition 11.4 [**Szegedy 1994**]: Every graph G has a nonempty subset $S \subseteq V$ such that

$$\vartheta(\overline{G}) \leq \Big(1 + \frac{1}{2} + \cdots + \frac{1}{n}\Big) \frac{|S|}{\vartheta(G[S])}.$$

EXERCISE 11.9. Prove that allowing orthogonal representations in the complex Hilbert space (define!) does not change the value defined by formula (11.16).

EXERCISE 11.10. Prove the formula (11.34) for even cycles with their longest diagonals added, by showing that the solution given above it is optimal.

EXERCISE 11.11. Construct an optimal dual orthogonal representation for the Kneser graph K_k^n.

EXERCISE 11.12. Let $\mathbf{v}_1, \dots, \mathbf{v}_n \in \mathbb{R}^n$ be any set of vectors, and let $A = \mathsf{Gram}(\mathbf{v})$. Prove that

$$\lambda_{\max}(A) = \max\Big\{\sum_i (\mathbf{d}^\mathsf{T}\mathbf{v}_i)^2 : \ \mathbf{d} \in \mathbb{R}^d,\ |\mathbf{d} = 1|\Big\}.$$

EXERCISE 11.13. Let a be the profile of an orthogonal representation $(\mathbf{u}, \mathbf{c})$ of G in $\mathbb{R}^d$, and let z be the profile of an orthogonal representation $(\mathbf{v}, \mathbf{d})$ of $\overline{G}$ in $\mathbb{R}^e$. Suppose that $a^\mathsf{T} z = 1$.

(a) Prove that such orthogonal representations exist.

(b) Prove that $\sum_i (\mathbf{c}^\mathsf{T}\mathbf{u}_i)(\mathbf{d}^\mathsf{T}\mathbf{v}_i)(\mathbf{u}_i \circ \mathbf{v}_i) = \mathbf{c} \circ \mathbf{d}$.

(c) Defining $\widehat{\mathbf{u}}_i = (\mathbf{d}^\mathsf{T}\mathbf{v}_i)\mathbf{u}_i$ and $\widehat{\mathbf{v}}_i = (\mathbf{c}^\mathsf{T}\mathbf{u}_i)\mathbf{v}_i$, Prove that the matrices $U = (\widehat{\mathbf{u}}_1, \dots, \widehat{\mathbf{u}}_n, \mathbf{c})$ and $W = (\widehat{\mathbf{v}}_1, \dots, \widehat{\mathbf{u}}_n, -\mathbf{d})$ satisfy $UW^\mathsf{T} = 0$. (So the columns of W almost form a Gale dual of the columns of U: only "almost", because ranks of U and W do not necessarily add up to $n+1$.)

EXERCISE 11.14. With the notation of Lemma 11.19, every graph G satisfies

$$\vartheta(G)(\vartheta(G) - 1)^2 \leq \sum_{i \in V} \vartheta(G_i)^2.$$

EXERCISE 11.15. Prove that the length of the sum of n unit vectors such that among any three of them some two are orthogonal is at most $4n^{2/3}$. Prove that this is best possible up to the constant.

EXERCISE 11.16. (a) Show that any stable set S provides a feasible solution of the dual program in (11.6). (b) Show that any k-coloring of $\overline{G}$ provides a feasible solution of the primal program in (11.6). (c) Give a new proof of the Sandwich Theorem 11.1 based on (a) and (b).

EXERCISE 11.17. Let G be a graph and let A be a G-matrix with nonzero diagonal entries. (a) Prove that $\alpha(G) \le \mathrm{rk}(A)$; (b) prove that $\Theta(G) \le \mathrm{rk}(A)$; (c) these estimates are valid for matrices A over any field. (d) Construct a graph G and a matrix A as above (over some field) for which $\mathrm{rk}(A) < \vartheta(G)$.

EXERCISE 11.18. Prove that the vertices of FSTAB(G) are half-integral, and show by an example that they are not always integral.

EXERCISE 11.19. Show that for a perfect graph, a maximum stable set and a coloring with minimum number of colors can be computed in polynomial time.

EXERCISE 11.20. The *fractional chromatic number* $\chi^*(G)$ is defined as the least real number t for which there exists a family $(A_j : j = 1, \dots, p)$ of stable sets in G, and nonnegative weights $(\tau_j : j = 1, \dots, p)$ such that $\sum_j \tau_j = t$ and $\sum_j \tau_j \mathbb{1}_{A_j} \ge \mathbb{1}_V$. The *fractional clique number* $\omega^*(G)$ is the largest real number s for which there exist nonnegative weights $(\sigma_i : i \in V)$ such that $\sum_i \sigma_i = s$ and $\sum_{i \in A} \sigma_i \le 1$ for every stable set A.

(a) Prove that $\omega(G) \le \omega^*(G)$ and $\chi(G) \ge \chi^*(G)$.

(b) Prove that $\chi^*(G) = \omega^*(G)$.

(b) Prove that $\vartheta(\overline{G}) \le \chi^*(G)$.

CHAPTER 12

Orthogonal Representations: Quantum Physics

As the basic setup in quantum physics, the state of a physical system can be described by a vector of unit norm in a (complex) Hilbert space. (In the simple systems we need, this space will be finite dimensional.) A *measurement* on a system in quantum state is performed by applying a self-adjoint linear operator to it. The simplest measurement operator is the orthogonal projection of the state vector onto a one-dimensional subspace; such an experiment can be thought of as checking a particular property of the state $\mathbf{x}$. If $\mathbf{u}$ is the vector generating this subspace (which is unique up to a scalar of absolute value 1), then the probability that the property checks out is the squared inner product of $\mathbf{u}$ and $\mathbf{x}$.

The fact that quantum physics assigns vectors to discrete objects like "properties" suggests analogies with orthogonal representations and other geometric representations treated in this book. As it turns out, this is more than just an analogy.

The most successful area of applying quantum physics in computer science, at least so far, has been quantum information theory. Using phenomena of quantum physics (for example, entanglement of particles), one can create communication channels that are more efficient than classical communication channels. We will discuss a quantum version of the Shannon capacity problem from the Introduction, along with two further interesting applications of orthogonal representations, to the theory of hidden variables, and in the construction of strangely entangled states.

12.1. Preliminaries: quantum states and entanglement

As it happens, the applications we are going to discuss are all about entanglement, so let us introduce this notion first.

We write $\mathbf{u}\cdot\mathbf{v}$ for the inner product of two vectors $\mathbf{u},\mathbf{v}\in\mathbb{C}^d$, to emphasize that it is used in the Hilbert space sense: $\mathbf{u}\cdot\mathbf{v}=\sum_i \overline{u}_i v_i$. However, allowing complex entries will not play any important role.[1]

The simplest example of a quantum physical system is a *qubit*: a qubit is a unit vector in $\mathbb{C}^2$. We fix a basis $\{\mathbf{e}_0,\mathbf{e}_1\}$ in $\mathbb{C}^2$. The vector $\mathbf{e}_0$ corresponds (in a sense) to the logical value "NO" of the bit, while $\mathbf{e}_1$, to the value "TRUE". Qubits are the basic units of quantum computation and quantum information theory, just like bits are for the classical versions.

Consider two quantum physical systems A and B. Separately, their states can be described by unit vectors $\mathbf{x}\in\mathbb{C}^d$ and $\mathbf{y}\in\mathbb{C}^e$. The state of the union of the two systems can be described by a vector in the tensor product space $\mathbb{C}^d\otimes\mathbb{C}^e$. If the two systems in states $\mathbf{x}$ and $\mathbf{y}$ are "independent" (unentangled), their joint state

[1] We digress from the standard physics notation, where a row vector is denoted by $\langle x|$, a column vector, by $|y\rangle$, and their inner product by $\langle x|y\rangle$. This would make reading this chapter easier for physicists, but not for mathematicians. Since most of this book is about mathematics, I do not change notation for this single chapter.

is $\mathbf{x} \circ \mathbf{y}$, which is then called a *product state*. However, this will not be the case in general, as a vector $\mathbf{z} \in \mathbb{C}^d \otimes \mathbb{C}^e$ cannot generally be written as the tensor product of two vectors in $\mathbb{C}^d$. A state that cannot be written as such a tensor product is called *entangled*. Every state in $\mathbb{C}^d \otimes \mathbb{C}^e$ can be written as

$$\mathbf{z} = \sum_{i=1}^{d} \sum_{j=1}^{e} S_{ij}(\mathbf{e}_i \circ \mathbf{f}_j),$$

where $\{\mathbf{e}_1, \dots, \mathbf{e}_d\}$ is the standard basis in $\mathbb{C}^d$, and $\mathbf{f}_1, \dots, \mathbf{f}_e$ is the standard basis in $\mathbb{C}^e$. The coefficients S_{ij} are complex numbers with $\sum_{i,j} |S_{ij}|^2 = 1$. Sometimes it is convenient to describe $\mathbf{z}$ by the $d \times e$ complex matrix $S = (S_{ij})$. In the case of two isomorphic systems (so $e = d$), the state

$$\mathbf{z} = \sum_{i=1}^{d} \frac{1}{\sqrt{d}} \mathbf{e}_i \circ \mathbf{e}_i.$$

is *maximally entangled*: in a sense, farthest from a product state. The rank of the matrix S is 1 in the case of a product state, and d in the case of the above maximally entangled state (when the matrix is the identity).

Measurement in a quantum system is a complicated issue, but we need only a very simple special case. We have mentioned that checking a property of the quantum state $\mathbf{x}$ corresponds to projecting it onto a one-dimensional subspace. Given mutually orthogonal vectors $\mathbf{a}_1, \dots, \mathbf{a}_q \in \mathbb{C}^d$, we can think of them as representing q different values of the same property, which are mutually exclusive (this is not hard to see). We can check these properties simultaneously (since projections to these one-dimensional subspaces commute), and tell which of them holds: the property described by $\mathbf{a}_i$ will be found to hold with probability $|\mathbf{a}_i \cdot \mathbf{x}|^2$. If $q = d$, then exactly one of them will hold.

12.2. Capacity of quantum channels

Entanglement leads to rather paradoxical behavior of particles; this was pointed out by Einstein, Podolsky and Rosen in 1935. Consider a pair of maximally entangled particles with a 2-dimensional state space: say two electrons, whose spin can be either "up" or "down". The maximally entangled state to consider here is $\frac{1}{\sqrt{2}}\mathbf{e}_1 \circ \mathbf{e}_1 + \frac{1}{\sqrt{2}}\mathbf{e}_2 \circ \mathbf{e}_2$. Such a pair of particles is often called an Einstein–Podolsky–Rosen pair or EPR pair. Both elements of such a pair represent a qubit: the same qubit for each.

Suppose that Alice and Bob split an EPR pair between themselves (while it remains in the same entangled state), and they travel to different far away places. If Alice measures the state of her electron, she will find it in one of the states $\mathbf{e}_1$ and $\mathbf{e}_2$ with the same probability. Say it is in state $\mathbf{e}_1$, then the entangled state collapses to $\mathbf{e}_1 \circ \mathbf{e}_1$ immediately. These long range correlated events could be interpreted from the point of view of Alice as an action apparently faster than light, so it would contradict special relativity - but only apparently. If Bob measures the state of his electron, then it will be in state $\mathbf{e}_1$; but if they do the measurements in different order what they observe will come out the same. You can play around with more EPR pairs to convince yourself that no information can be transmitted between Alice and Bob using only such pairs.

While splitting an EPR pair cannot be used to transmit information faster than light, such a strange behavior can be utilized in information theory and computer science. One important feature of sharing such an EPR pair is that Alice and Bob obtain the same random bit, which is very secure, since by the principles of quantum physics, nobody can learn this bit without destroying the entanglement.

Here we utilize EPR pairs in a different way. One can ask for analogues of the Shannon zero-error capacity in quantum information theory. Let us start with an example showing that using entanglement does improve the efficiency of communication in certain cases.

EXAMPLE 12.1. While entanglement does not allow to send information directly (faster than the speed of light) between two people, it does allow to make better use of classical communication channels. The following example is due to [**Cubitt et al. 2010**].

We need a graph G with the following properties:

(a) its edges can be covered by r complete d-subgraphs $H_1, \dots, H_r$;

(b) some q of these, say $H_1, \dots, H_q$, partition the node set;

(c) $\alpha(G) < q$;

(d) G has a dual orthonormal representation $\mathbf{v}$ in $\mathbb{R}^d$.

Several constructions of such graphs are known, it is perhaps easiest to describe such a graph with $d = 4$ and $q = 6$ [**Peres 1991**]. The node set of this graph can be defined as the set of vectors in $\mathbb{R}^4$ with coordinates $0, 1$ or -1, where the number of nonzero coordinates is $1, 2$ or 4, and the first nonzero coordinate is 1. We connect two of these nodes by an edge if they are orthogonal. By definition, this graph has a dual orthonormal representation in $\mathbb{R}^4$. To verify that properties (a)–(c) above are satisfied as well is a tedious but routine exercise.

Next we construct a (noisy) classical communication channel: Its input alphabet is V, its output alphabet is $[r]$, and on input $a \in V$ it outputs one of the indices i for which $a \in V(H_i)$. The output i is chosen randomly and uniformly from all such i. Thus two inputs a and b are confusable if and only if there is a chance that they lead to the same output, i.e., if there is an $i \in [r]$ such that $a, b \in V(H_i)$. This is equivalent to $ab \in E$, so the confusability graph of the channel is G. The number of inputs that can be used without any danger of confusion is $\alpha(G) < q$.

Now we endow the channel with an extra feature: we construct two copies a quantum system with a d-dimensional state space $\mathbb{C}^d$, and prepare them by bringing them to the above maximally entangled state. Alice and Bob get one of the two copies each.

We use the dual orthonormal representation to show that in the presence of such entanglement, one can safely transmit q different one-letter messages. Alice wants to send message $x \in [q]$ to Bob. She measures the state of her side in the orthonormal basis $\{\mathbf{v}_s : s \in V(H_x)\}$. As discussed above, the result of the measurement will be a random element $\mathbf{v}_s$ of this basis. After that, the entangled system will be in the state $\mathbf{v}_s \circ \mathbf{v}_s$.

Alice sends s to Bob through the noisy channel. Bob receives an index i for which $V(H_i) \ni s$. (If $i \le q$, this would be the intended information x, but the noisy channel may have output any such index i.) Since H_i is complete, Bob can measure the state of his side in the orthonormal basis $\{\mathbf{v}_t : t \in V(H_i)\}$, and determine s. Since there is a unique $x \in [q]$ for which $s \in V(H_x)$, this determines x.

It is interesting to notice that while in Section 11.5.1, as well as in the rest of this section, orthogonal representations are used to prove upper bounds on the capacity of a channel, here we have used dual orthogonal representations to design better protocols. ♦

So we see that using an entangled state can improve the zero-error capacity of a noisy channel. How much is the improvement? [**Beigi 2010**] and [**Cubitt et al. 2011**] prove that the theta-function is still an upper bound. One can use an entanglement-assisted channel repeatedly to gain in its capacity just like in the classical case. Since the theta-function is multiplicative, it remains an upper bound on the zero-error capacity.

THEOREM 12.2. *Suppose that Alice and Bob are connected by a classical noisy channel with confusability graph G. In addition, there is an entangled state $\mathbf{u} \in \mathbb{C}^d \circ \mathbb{C}^d$, where Alice has access to the first factor and Bob has access to the second factor. Then the maximum number of one-letter messages that Alice can transmit without the possibility of confusion is bounded by $\vartheta(G)$.*

PROOF. Let the noisy channel have input alphabet X and output alphabet Z. For each input x, let $Z_x \subseteq Z$ be the set of outputs that occur with positive probability. (Since we are interested in zero-error, the actual probabilities do not matter.) Two elements x and y are confusable if and only if $Z_x \cap Z_y \neq \emptyset$.

The other part of the equipment is the shared entangled state $\mathbf{u} \in \mathbb{C}^d \otimes \mathbb{C}^d$, which we consider as a $d \times d$ complex matrix S such that $S \cdot S = \operatorname{tr}(S\overline{S}^{\mathsf{T}}) = \sum_{u,v} |S_{uv}|^2 = 1$.

The use of this entanglement-assisted channel can be described like this. Alice wants to transmit a message $i \in [m]$. She performs a measurement on her half of $\mathbf{u}$; this collapses the joint state to a state $\mathbf{u}'$. She transmits a message $x \in X$ that depends on i and the result of the measurement. Once Bob gets the message (more exactly, he gets some message $z \in Z_x$), he performs a measurement (depending on z) on his half of $\mathbf{u}'$, which has m possible outcomes; the outcome of this measurement is supposed to coincide with Alice's intended message i.

To set up the linear algebra for the analysis of this protocol, let Alice apply an operator $A_i \otimes I$ to $\mathbf{u}$. As the result of the measurement she gets an eigenvector of A_i. Let A_i^x be the orthogonal projection onto the subspace of $\mathbb{C}^d$ generated by those eigenvectors that lead to message x. The fact that always exactly one message $x \in X$ must be sent means, by the laws of quantum physics, that $A_i^x A_i^y = 0$ for $x \neq y$ and $\sum_i A_i^x = I$ for each $x \in X$. This measurement collapses $\mathbf{u}$ to the state $\mathbf{u}'$ described by the matrix $A_i^x S$.

Bob's measurements consist of m orthogonal projections $B_1^z, \dots, B_m^z$ onto appropriate subspaces of $\mathbb{C}^d$. Similarly as before, we have $B_i^z B_j^z = 0$ for $i \neq j$ and $\sum_i B_i^z = I$. Measurement B_j^z collapses the state $\mathbf{u}'$ to $A_i^x S B_j^z$. The fact that Bob is able to recover i means that $A_i^x S B_j^z = 0$ if $z \in Z_x$ and $i \neq j$.

We claim that

$$(A_i^x S) \cdot (A_j^y S) = 0 \tag{12.1}$$

in the following cases: (a) $i = j$, $x \neq y$; (b) $i \neq j$, $x = y$; (c) $i \neq j$, $xy \in E$. Case (a) is easy, since we can write $(A_i^x S) \cdot (A_i^y S) = \operatorname{tr}(\overline{S}^{\mathsf{T}} A_i^x A_i^y S)$, and, as we have seen, $A_i^x A_i^y = 0$. In cases (b) and (c) there is an element $z \in Z_x \cap Z_y$, and using this we get

$$(A_i^x S)\cdot(A_j^y S) = \mathrm{tr}(A_j^y S\overline{S}^{\mathsf{T}} A_i^x) = \mathrm{tr}\big(A_j^y S(\sum_k B_k^z)\overline{S}^{\mathsf{T}} A_i^x\big)$$
$$= \sum_k \mathrm{tr}\big(A_j^y S B_k^z \overline{S}^{\mathsf{T}} A_i^x\big) = 0,$$

since for every k, either $A_j^y S B_k^z = 0$ or $B_k^z \overline{S}^{\mathsf{T}} \overline{A}_i^x = 0$. This proves (12.1).

Define an $X \times X$ matrix M by

$$M_{x,y} = \sum_{i,j=1}^{m} (A_i^x S)\cdot(A_j^y S).$$

If $xy \in E$, then $M_{x,y} = 0$, since every term is zero by conditions (a) and (c) above. It is easy to see that M is positive semidefinite. Furthermore, using (b),

$$\mathrm{tr}(M) = \sum_x \sum_{i,j=1}^{m} (A_i^x S)\cdot(A_j^x S) = \sum_x S\cdot S = m,$$

and

$$\mathrm{tr}(JM) = \sum_{x,y,i,j} \mathrm{tr}(\overline{S}^{\mathsf{T}} A_i^x A_j^y S) = \sum_{i,j} \mathrm{tr}\Big(\overline{S}^{\mathsf{T}}(\sum_x A_i^x)(\sum_y A_j^y)S\Big)$$
$$= m^2 \mathrm{tr}(\overline{S}^{\mathsf{T}} S) = m^2.$$

This shows that the matrix $\frac{1}{m}M$ almost fulfils definition (11.15) of ϑ, except that it is not necessarily real. But $Z = \frac{1}{2m}(M + \overline{M}^{\mathsf{T}})$ is real, and has the same trace and sum of entries, so $\vartheta(G) \geq m$. □

One can consider quantum physical communication channels more general than entanglement-assisted classical channels. The theta-function can be generalized to such channels, so that it remains an upper bound on the zero–error capacity of the channel [**Duan–Severini–Winter 2013**]; see also [**Cubitt et al. 2011**].

12.3. Hidden variables

Let us return to the Einstein–Podolsky–Rosen objection to quantum physics. One way out of the paradox of sustained correlations between the results of independently performed, spatially separated quantum measurements is the theory of *hidden variables*. Perhaps the states of the electrons of Alice and Bob are already determined when the pair is formed? Then there is nothing surprising in the fact that the two particles are found in the same state. In the well known parable of Schrödinger's cat, this argument would mean that the cat in the closed chamber at a given moment in time is either dead or alive—we just do not know which.

This interpretation, which arose from the objection to the nondeterministic–random interpretation of quantum events, suggests that if we knew the exact state of each particle (its "hidden parameters"), then we could predict quantum events with certainty. A consequence would be that even if we are not able to observe certain events simultaneously, they do actually occur.

Based on the work in [**Cabello–Severini–Winter 2014**], we describe a connection between the theta-function and this fundamental issue in quantum physics. Consider a quantum system, and let $e_1, \dots, e_n$ be observable events. Construct a graph G on $V = [n]$ in which $ij \in E$ if and only if e_i and e_j are exclusive (cannot occur simultaneously). We call G the *exclusivity graph* of the system $e_1, \dots, e_n$ of events.

How many of these events occur?

We can observe any particular event e_i, but this observation changes the state of the system, so we cannot observe all of the events. If we repeat the experiment, even if we can create the same starting state and observe the same event, we may get a different answer. But if we choose the event e_i uniformly at random from the set $\{e_1, \dots, e_n\}$, and we repeat the experiment with the same starting state many times, then we can find, experimentally, the probability p that the observed event does occur. The expected number of events that occur is then pn.

What can we say about p? In the classical setting, when $e_1, \dots, e_n$ are observable events in a probability space (no quantum effects), we would have a probability distribution on the stable subsets of nodes of G. The number pn would be the expected size of this set, i.e., the expected number of events that occur simultaneously. It is trivial that this number is at most $\alpha(G)$, so we get the inequality

$$p \le \frac{\alpha(G)}{n}. \tag{12.2}$$

The same inequality can be derived in quantum physics, if we assume that it makes sense to talk about the number of events e_i that actually hold in the given experiment (even though we can only check one of them). In the usual "hidden variable" interpretation of quantum physics this is the case.

John Bell was the first to suggest simple measurement schemes and derive for them inequalities, which are related to (12.2), and could be experimentally verified (or rather falsified) [**Bell 1964**]. Since their derivation depends on the theory of hidden variables, disproving such an inequality disproves the hidden variable theory (at least in its simplest form).

From basic quantum physical principles (not using hidden variables) one can only derive the weaker inequality

$$p \le \frac{\vartheta(G)}{n}. \tag{12.3}$$

Indeed, state $\mathbf{x}$ is a unit vector in a complex Hilbert space H. Observing the event e_i means projecting $\mathbf{x}$ to a one-dimensional subspace, spanned by a unit vector $\mathbf{u}_i \in H$; the probability that the event occurs is just the squared length of the projection, $p_i = |\mathbf{x} \cdot \mathbf{u}_i|^2$. Thus

$$p = \frac{1}{n}(p_1 + \dots + p_n) = \frac{1}{n} \sum_i |\mathbf{x} \cdot \mathbf{u}_i|^2.$$

Two events that exclude each other must correspond to projections onto orthogonal vectors, and hence $\mathbf{u}$ is a dual orthogonal representation of G in a Hilbert space. This is a complex Hilbert space, but it is not hard to see that the value of the maximum in definition (11.8) of the theta-function does not change if we take complex Hilbert space instead of the real (Exercise (11.9)). Hence we get

$$p = \frac{1}{n} \sum_i |\mathbf{x} \cdot \mathbf{u}_i|^2 \le \frac{\vartheta(G)}{n}.$$

From even weaker (simpler) principles, one gets an even weaker inequality. The *exclusivity principle* says that the sum of probabilities of mutually exclusive events is at most 1. This implies the inequality

$$\sum_{i\in B} p_i \le 1$$

for every clique B in G; in the language of Section 11.4.2, the vector $(p_1,\dots,p_n)$ must belong to the clique-constrained fractional stable set polytope QSTAB(G), and so it follows that

$$p = \frac{1}{n}(p_1+\dots+p_n) \le \frac{\alpha^*(G)}{n}. \tag{12.4}$$

Two special cases have been studied extensively. In the Clauser–Horne–Shimony–Holt version of the Bell experiment, one creates two entangled particles, say electrons, whose spins can be "up" or "down"; so the state of each particle is described by a 2-dimensional unit vector. These are sent to two far away observers, called (as usual) Alice and Bob. Alice has two possible settings a and a' of her equipment measuring the spin, which means that she can choose one of these, and perform a measurement which returns "up" or "down". Bob has, similarly, two settings b and b' (not the same settings as Alice). Let a_+ denote the event that Alice measures "up" in setting a. We have 8 analogous events.

To reformulate this in our setting, consider the following 8 events: $a_+b_+, a_-b_-, a'_+b_+, a'_-b_-, a_+b'_+, a_-b'_-, a'_+b'_-, a'_-b'_+$ (note the twist in the last two terms!). The exclusivity graph of these events is the Wagner graph W_8 familiar from Example 11.15 (Figure 12.1(b)). Looking at the picture, it is easy to see that $\alpha(G) = 3$. Let p denote the probability that selecting one of them uniformly at random and performing the corresponding measurements (one by Alice, one by Bob), the event does occur. Repeating the experiment many times, the value of p can be determined experimentally.

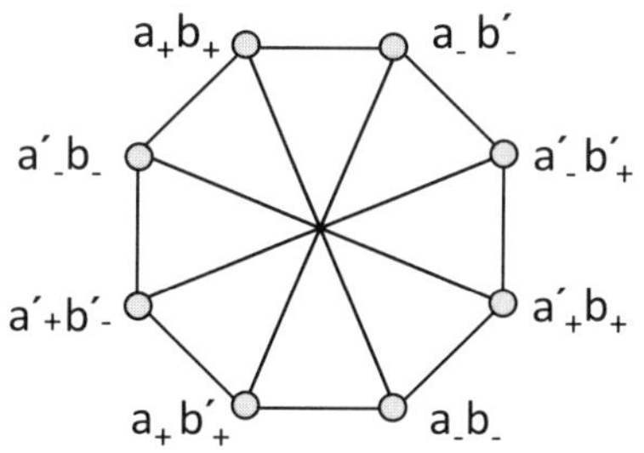

FIGURE 12.1. The exclusivity graph V_8 of the Clauser–Horne–Shimony–Holt experiment. The edge between (say) $a'_-b'_+$ and $a_-b'_-$ indicates that they cannot occur simultaneously: for these two events, test b' returns different results.

Inequality (12.2) says in this case that

$$p \le \frac{\alpha(G)}{n} = \frac{3}{8} = 0.375. \tag{12.5}$$

Recall that this inequality follows from the theory of hidden variables. Since there are trivial linear equations relating the probabilities of the events a_+b_+ etc., this inequality can be written in several equivalent forms, of which (12.5) is the most convenient for us.

We have $\vartheta(G) = 2+\sqrt{2}$ (see Example 11.15), so the inequality

$$p \le \frac{\vartheta(G)}{n} = \frac{2+\sqrt{2}}{8} \approx 0.427\dots \tag{12.6}$$

follows by (12.3) without the hypothesis of hidden variables, just by the laws of quantum physics.

REMARK 12.3. Traditionally, the inequality in the Clauser–Horne–Shimony–Holt experiment is expressed in terms of the *quantum correlations*, defined by

$$E(a_+, b_+) = \mathsf{E}\big((-1)^{\mathbb{1}(a_+)+\mathbb{1}(b_+)}\big),$$

and similarly for the other events. The CHSH inequality is then

$$E(a_+, b_+) + E(a_+, b_-) + E(a_-, b_+) - E(a_-, b_-) \le 2. \tag{12.7}$$

This inequality can be expressed in terms of probabilities as

$$\mathsf{P}(a_-b_-) \le \mathsf{P}(a_+b'_-) + \mathsf{P}(a'_-b_+) + \mathsf{P}(a'_+b'_+), \tag{12.8}$$

which in turn is equivalent to (12.5) (see Exercise 12.1).

Returning to the bounds (12.5) and (12.6), which of these bounds is the "truth"? After a long line of increasingly sophisticated experiments, recent reports [**Hensen et al. 2015**], [**Giustina et al. 2015**], [**Shalm et al. 2015**] claim to have eliminated all the implicit assumptions ("loopholes"), and show that the bound (12.5) does not hold in general. The experiment in [**Hensen et al. 2015**] provides the value $p \approx 0.401$. This value is about half way between the bounds in (12.5) and (12.6), and it can be considered as a disproof of the "hidden variable" interpretation of quantum physics (at least in its basic form). Whether other experimental setup will get even closer to the bound in (12.6) is an open question.

The experiment described in [**Klyachko et al. 2008**] works in a similar spirit, and leads to a more complicated measurement setup but to a simpler graph, namely C_5; see also [**Cabello–Severini–Winter 2014**]. In this latter paper it was also shown that (12.3) is best possible in a sense: for every simple graph G one can construct mathematical models of quantum physical systems in which equality holds. [**Howard et al. 2014**] describes potential applications of these ideas to quantum computing.

12.4. Unextendible product bases

This application of orthogonal representations leads to the construction of highly entangled states.

Our introductory discussion of two entangled states easily extends to a quantum physical system consisting of m parties. If the state of the i-th party is described by a unit vector in a complex Hilbert space $\mathbb{C}^{d_i}$, then a state of the whole system is described by a vector in the tensor product $\mathbb{H} = \mathbb{C}^{d_1} \otimes \dots \otimes \mathbb{C}^{d_m}$. A *product state* is a vector of the form $\mathbf{v}_1 \circ \dots \circ \mathbf{v}_m$, where $\mathbf{v}_i \in \mathbb{C}^{d_i}$, $|\mathbf{v}_i| = 1$. Physically, this corresponds to the states of the individual parties to be "independent", more exactly *disentangled*. We exclude trivial cases only if we assume that $m, d_1, \dots, d_m \ge 2$.

For every product state $\mathbf{v}$, let's write $\mathbf{v} = \mathbf{v}^1 \circ \dots \circ \mathbf{v}^m$, where $\mathbf{v}^i \in \mathbb{C}^{d_i}$ and $|\mathbf{v}_i| = 1$. This decomposition is unique up to scalars of absolute value 1, which we can choose arbitrarily. The inner product of two product states can be computed as follows:

$$\mathbf{u}\cdot\mathbf{v} = (\mathbf{u}^1\cdot\mathbf{v}^1)\cdots(\mathbf{u}^m\cdot\mathbf{v}^m). \tag{12.9}$$

An *incomplete orthogonal product basis*, or shortly *product basis* is a system of mutually orthogonal product states in $\mathbb{H}$. A product basis B is *unextendible*, if it is not the subset of a larger product basis. In other words, the orthogonal complement $B^\perp$ of the linear span of the basis vectors contains no product state.

Unextendible product bases were introduced in [**Bennett et al. 1999**] in order to construct highly entangled states. In fact, they proved that such a state leads to high entanglement in two different senses. We state these consequences informally, without giving the exact definition and explaining their significance, which would lead us too far into physics.

• From the mixed state obtained as the uniform distribution over all states (unit vectors) in $B^\perp$ no product space can be "distilled" by local measurements and classical communication.

• The states in an unextendible product basis are *locally immeasurable*, i.e., none of them can be distinguished from the others by local measurements and classical communication.

There are trivial unextendible product bases: taking any orthonormal basis $\mathbf{b}^i_1,\dots,\mathbf{b}^i_{d_i}$ in each $\mathbb{C}^{d_i}$, the set of all vectors of the form $\mathbf{b}^1_{j_1}\circ\cdots\circ\mathbf{b}^m_{j_m}$ forms a full orthonormal basis in $\mathbb{H}$, of cardinality $d_1\cdots d_n$. Our goal is to construct an unextendible product basis as small as possible.

LEMMA 12.4. *If $\{\mathbf{v}_1,\dots,\mathbf{v}_n\}$ is any unextendible product basis in $\mathbb{H}$, then $n\ge 1+\sum_{i=1}^m(d_i-1)$. If equality holds, then the vectors $\{\mathbf{v}^i_s:\ s=1,\dots,n\}$ are in general position in $\mathbb{C}^{d_i}$ for every $1\le i\le m$.*

PROOF. If $n\le\sum_{i=1}^m(d_i-1)$, then we can partition the basis into m sets $B_1,\dots B_m$, where $|B_i|\le d_i-1$. Then there is a unit vector $\mathbf{u}_i$ that is orthogonal to the $|B_i|$ vectors $\mathbf{v}^i_s$ $(s\in B_i)$, and hence the vector $\mathbf{u}_1\circ\cdots\circ\mathbf{u}_m$ is orthogonal to every $\mathbf{v}\in B$. This contradicts the assumption that B is unextendible.

In the case of equality, the same argument can be applied provided one of the sets B_i has d_i elements but the vectors $\{\mathbf{v}^i_s:\ s\in B_i\}$ do not span $\mathbb{C}^{d_i}$. □

The following result of [**Alon–Lovász 2001**] translates the question whether equality can be attained in Lemma 12.4 into a purely graph-theoretic question.

LEMMA 12.5. *Let $n=1+\sum_{i=1}^m(d_i-1)$. Then there exists an unextendible product basis with n elements in $\mathbb{H}$ if and only if there exists a decomposition $K_n=G_1\cup\cdots\cup G_m$ into edge-disjoint graphs such that G_i does not contain a complete bipartite subgraph with d_i+1 nodes $(i=1,\dots,m)$.*

As we have used before, the condition on G_i is equivalent to saying that its complement $\overline{G}_i$ is $(n-d_i)$-connected.

PROOF. I. Suppose that there is an unextendible product basis $\{\mathbf{v}_1,\dots,\mathbf{v}_n\}$ in $\mathbb{H}$. For $i=1,\dots,m$, we define a graph G_i on $V=[n]$ as follows: we connect two nodes $s,t\in V$ by an edge if $\mathbf{v}^i_s\cdot\mathbf{v}^i_t=0$. The condition that $\mathbf{v}_s\cdot\mathbf{v}_t=0$ implies by (12.9) that every pair (s,t) of nodes will be connected in at least one graph G_i. (If the pair gets connected in more than one graph G_i, we can delete this edge from all but one of the graphs.)

We have to show that G_i does not contain a complete bipartite subgraph on d_i+1 nodes. Suppose it does, and let $U, W \subseteq V$ be the color classes, where $|U|, |W| \geq 1$ and $|U|+|W| \geq d_i+1$. For every $s \in U$ and $t \in W$, we have $\mathbf{v}_s^i \cdot \mathbf{v}_t^i = 0$. So the subspaces $A = \text{lin}\{\mathbf{u}_s^i : s \in U$ and $B = \text{lin}\{\mathbf{u}_s^i : s \in W$ (over $\mathbb{C}$) are orthogonal to each other, which implies that $\dim(A)+\dim(B) \leq d_i$. It follows that one of U and W, say W, consists of linearly dependent vectors, contradicting Lemma 12.4.

II. Suppose that K_n has a decomposition as described. Theorem 10.9 implies that $\overline{G}_i$ has an orthogonal representation $(\mathbf{u}_t^i : t \in V)$ in $\mathbb{R}^{d_i}$ in general position. The vectors $\mathbf{v}_t = \mathbf{u}_t^1 \circ \cdots \circ \mathbf{u}_t^m$ $(t = 1, \ldots, n)$ form a product basis: if $1 \leq s < t \leq n$, then

$$\mathbf{v}_s \cdot \mathbf{v}_t = (\mathbf{u}_s^1 \cdot \mathbf{u}_t^1) \cdots (\mathbf{u}_s^m \cdot \mathbf{u}_t^m) = 0,$$

since $st \in E(G_i)$ for some i, and then $\mathbf{u}_s^i \cdot \mathbf{u}_t^i = 0$.

We show that the product basis $\{\mathbf{v}_1, \ldots, \mathbf{v}_n\}$ is unextendible. Indeed, suppose that it could be extended by a product state $\mathbf{w}$. Then $\mathbf{w}$ is orthogonal to every vector $\mathbf{v}_t$, and hence

$$(\mathbf{w}^1 \cdot \mathbf{u}_t^1) \cdots (\mathbf{w}^m \cdot \mathbf{u}_t^m) = 0 \qquad (t = 1, \ldots, n).$$

Hence for every $1 \leq t \leq n$ there is an index $1 \leq i \leq m$ for which $\mathbf{w}^i \cdot \mathbf{u}_t^i = 0$. But since the vectors $\mathbf{u}_1^i, \ldots, \mathbf{u}_n^i$ are in general position, one and the same index i could serve at most $d_i - 1$ choices of t. Hence $n \leq \sum_i (d_i - 1)$, a contradiction. □

Unfortunately, the condition given in Lemma 12.4 is not always sufficient for the existence of a decomposition of K_n as in the lemma. Suppose that equality holds:

$$n = 1 + \sum_i (d_i - 1), \tag{12.10}$$

and there is a decomposition $K_n = G_1 \cup \cdots \cup G_m$ as in the lemma. Since G_i contains no K_{1,d_i}, the maximum degree in G_i is at most $d_i - 1$. So (12.10) implies that G_i is regular of degree $d_i - 1$. This is impossible if n is odd and d_i is even. Another, more trivial exception is when $m = 2$ and $d_1 = 2 \leq d_2$ (or the other way around): Then G_1 must be a perfect matching, and so G_2 contains a complete bipartite graph $K_{2,n-2} = K_{2,d_2-1}$.

It was shown in [**Alon–Lovász 2001**] that these are the only exceptional cases:

THEOREM 12.6. *Let $m, d_1, \ldots, d_m \geq 2$, and $n = 1 + \sum_{i=1}^m (d_i - 1)$. Assume that if n is odd then every d_i is odd, and if $m = 2$ then $d_1, d_2 \geq 3$. Then there exists an unextendible product basis with n elements in $\mathbb{C}^{d_1} \otimes \cdots \otimes \mathbb{C}^{d_m}$.* □

The proof uses Lemma 12.5, by constructing an appropriate decomposition of K_n. For the details, we refer to [**Alon–Lovász 2001**]. Other constructions, settling several of the exceptional cases, were given by [**Chen–Johnston 2014**].

EXERCISE 12.1. Prove that (12.5) is equivalent to the simpler inequality $P(a_+b_+) \leq P(a_+b'_-) + P(a'_-b_+) + P(a'_+b'_+)$.

CHAPTER 13

Semidefinite Optimization

In our treatment of orthogonal representations, a major tool was the duality theory for the theta-function (Section 11.2). We have remarked there that the main step could be considered as a special case of semidefinite duality. Since semidefinite optimization has other important applications in discrete mathematics, foremost in the area of approximation algorithms, but also in constructing geometric representations, it is now time to survey some of the key results and illustrate the general technique that leads to these applications.

13.1. Linear and semidefinite programs

Linear programming has been perhaps the most fundamental and successful tools in optimization and discrete mathematics. We assume some basic familiarity with this theory; see for example [**Vanderbei 2001**], [**Schrijver 2003**], [**Korte–Vygen 2008**].linebreakKorte, Bernhard

Linear programs are special cases of convex programs; *semidefinite programs* are more general than linear programs but still convex programs, to which many of the useful properties of linear programs extend. For more comprehensive presentations of semidefinite optimization, see [**Vandeberghe–Boyd 1996**], [**Wolkowicz–Saigal–Vandenberghe 2000**].

13.1.1. Different forms of semidefinite programs.

A semidefinite program is an optimization problem of the following form:

$$\text{(13.1)} \qquad \begin{aligned} &\text{minimize} && c^{\mathsf{T}}x \\ &\text{subject to} && x_1A_1+\dots+x_nA_n-B\succeq 0. \end{aligned}$$

Here $A_1,\dots,A_n,B$ are given symmetric $m\times m$ matrices, and $c\in\mathbb{R}^n$ is a given vector. Recall that $M\succeq 0$ means that M is positive semidefinite. We can think of $\sum_i x_iA_i-B$ as a matrix whose entries are linear functions of the variables.

As usual, any choice of the values x_i that satisfies the given constraint is called a *feasible solution*. A solution is *strictly feasible*, if the matrix $\sum_i x_iA_i-B$ is positive definite. We denote by v_{primal} the infimum of the objective function.

The special case when $A_1,\dots,A_n,B$ are diagonal matrices is just an arbitrary linear program, and it is very fruitful to think of semidefinite programs as generalizations of linear programs. But there are important technical differences. Unlike in the case of linear programs, the infimum may be finite but not a minimum, i.e., not attained by any feasible solution.

EXAMPLE 13.1. Consider the following simple semidefinite program in two real variables x and y:

$$\begin{aligned} &\text{minimize} \quad x \\ &\text{subject to} \quad \begin{pmatrix} x & 1 \\ 1 & y \end{pmatrix} \succeq 0. \end{aligned} \tag{13.2}$$

The constraint means that $x \geq 0$, $y \geq 0$ and $xy \geq 1$, which implies that $x > 0$, but x can be arbitrarily close to 0. ♦

As in the theory of linear programs, there are a large number of equivalent formulations of a semidefinite program. Of course, we could consider minimization instead of maximization. We could allow additional linear constraints on the variables x_i (inequalities and/or equations). These could be incorporated into the form above by extending the A_i and B with new diagonal entries.

We could introduce the entries of the matrix $X = \sum_i x_i A_i - B$ as variables, in which case the fact that they are linear functions of the original variables translates into linear equations between them. Straightforward linear algebra can be used to transform (13.1) into an optimization problem of the form

$$\begin{aligned} \text{minimize} \quad & C \cdot X \\ \text{subject to} \quad & X \succeq 0 \\ & D_i \cdot X = d_i \qquad (i = 1, \dots, k) \end{aligned} \tag{13.3}$$

where $C, D_1, \dots, D_k$ are symmetric $m \times m$ matrices and $d_1, \dots, d_k \in \mathbb{R}$. Note that $C \cdot X$ is the general form of a linear combination of entries of X, and so $D_i \cdot X = d_i$ is the general form of a linear equation in the entries of X. It is easy to see that we would not get any substantially more general problem if we allowed linear inequalities in the entries of X in addition to the equations.

With these variants in mind, it should be clear that equations (11.6) and (11.7) express the theta-function as optima of semidefinite programs. The relationship between these programs will be discussed next.

13.1.2. Duality. Duality is one of the most important aspects of linear programs, and we continue our treatment with generalizing duality to semidefinite programs. We start with generalizations of the Farkas Lemma, which in its original form characterizes the solvability of systems of linear inequalities. Let $\mathcal{S}$ denote the linear space of real symmetric $m \times m$ matrices, and $\mathcal{P} \subseteq \mathcal{S}$, the set of positive semidefinite matrices. The set $\mathcal{P}$ is a convex pointed cone. Let $\mathcal{P}'$ be the set of positive definite matrices, which is just the interior of $\mathcal{P}$.

LEMMA 13.2 (Farkas Lemma, homogeneous version). *Let $\mathcal{L} \subseteq \mathcal{S}$ be a linear space of symmetric matrices. Then exactly one of the following alternatives holds:*

(i) *There is a matrix $X \in \mathcal{L}$, $X \succ 0$.*

(ii) *There is a matrix $Y \in \mathcal{L}^\perp$, $Y \neq 0$, $Y \succeq 0$.*

We could state this fact as follows: either both $\mathcal{L}$ and $\mathcal{L}^\perp$ contain nonzero positive semidefinite matrices, or one of them contains a positive definite matrix.

PROOF. Suppose that both alternatives hold. Then the conditions $X \in \mathcal{L}$ and $Y \in \mathcal{L}^\perp$ imply that $X \cdot Y = 0$, while the conditions $X \succ 0$, $Y \succeq 0$ and $Y \neq 0$ imply that $X\dot{Y} > 0$. (To see this, use an orthonormal basis in which Y is diagonal.)

Conversely, suppose that $\mathcal{L}\cap\mathcal{P}' = \emptyset$. Then there is a hyperplane $\mathcal{H} \subseteq \mathcal{S}$ containing $\mathcal{L}$ but disjoint from $\mathcal{P}'$. The equation of this hyperplane is $Y\cdot X = 0$ with some symmetric matrix $Y \neq 0$. Clearly $Y \in \mathcal{L}^\perp$. Assuming that $\mathcal{P}'$ is on the positive side of $\mathcal{H}$, we have $Y\cdot X > 0$ for every positive definite matrix X, which implies that $Y\cdot X \geq 0$ for every positive semidefinite matrix X. This implies that $Y \succeq 0$. □

Depending on how the subspace $\mathcal{L}$ is given, the Lemma can be reformulated in several ways. For example, let $A_1,\dots,A_n$ be symmetric $m\times m$ matrices. Then exactly one of the following alternatives holds:

(i) There are $x_1,\dots,x_n \in \mathbb{R}$ such that $x_1A_1+\dots+x_nA_n \succ 0$.

(ii) There exists a symmetric matrix $Y \neq 0$, $Y \succeq 0$ such that $A_1\cdot Y = \dots = A_n\cdot Y = 0$.

The Semidefinite Farkas Lemma has several inhomogeneous versions; here we state one (see the exercises for others).

We need some notation. Let $\mathcal{A}$ be an affine subspace of a Euclidean space $\mathcal{S}$ that does not contain 0, and let $\mathcal{A}^* = \{Y \in \mathcal{S} :\ Y\cdot X = -1\ \forall X \in \mathcal{A}\}$. It is easy to see that $(\mathcal{A}^*)^* = \mathcal{A}$. Let $\widehat{\mathcal{A}}$ denote the linear subspace obtained by translating $\mathcal{A}$ to the origin, and let $\widehat{\mathcal{A}^*}$ be obtained similarly from $\mathcal{A}^*$. It is easy to see that $\widehat{\mathcal{A}}$ and $\widehat{\mathcal{A}^*}$ are orthogonal linear subspaces with $\dim(\widehat{\mathcal{A}})+\dim(\widehat{\mathcal{A}^*}) = \dim(\mathcal{S})-1$, and $\widehat{\mathcal{A}^*} = \mathrm{lin}(\mathcal{A})^\perp$.

Suppose that $\mathcal{A}$ is defined by the set of equations $A_i\cdot X = c_i$, $i = 1,\dots,n$. Then elementary linear algebra gives the following expressions for the associated spaces:

$$\begin{aligned} \widehat{\mathcal{A}} &= \Big\{X \in \mathcal{S} :\ A_1\cdot X = \dots = A_n\cdot X = 0\Big\},\\ \widehat{\mathcal{A}^*} &= \Big\{\sum_i x_iA_i :\ \sum_i x_ic_i = 0\Big\},\\ \mathcal{A}^* &= \Big\{\sum_i x_iA_i :\ \sum_i x_ic_i = -1\Big\}. \end{aligned} \tag{13.4}$$

Now let $\mathcal{S}$ be the set of symmetric $m\times m$ matrices. It is trivial that $\mathcal{A}$ and $\mathcal{A}^*$ cannot both contain positive semidefinite matrices. It would be nice if one of them always did, but this is not true, as the following example shows.

EXAMPLE 13.3. Let

$$\mathcal{A} = \left\{\begin{pmatrix} 0 & 1\\ 1 & t\end{pmatrix} :\ t \in \mathbb{R}\right\},$$

then

$$\mathcal{A}^* = \left\{\begin{pmatrix} s & -\frac{1}{2}\\ -\frac{1}{2} & 0\end{pmatrix} :\ s \in \mathbb{R}\right\},$$

and these subspaces contain no positive semidefinite matrices (Figure 13.1). Both $\mathcal{A}$ and $\mathcal{A}^*$ are disjoint from $\mathcal{P}$, but they get arbitrarily close. ♦

But we can state the following.

LEMMA 13.4. *Let $\mathcal{A}$ be an affine space of symmetric $m\times m$ matrices that does not contain the 0 matrix. Then either $\mathcal{A}$ or $\mathcal{A}^*$ contains a nonzero positive semidefinite matrix, or both $\widehat{\mathcal{A}}$ and $\widehat{\mathcal{A}^*}$ contain nonzero positive semidefinite matrices.*

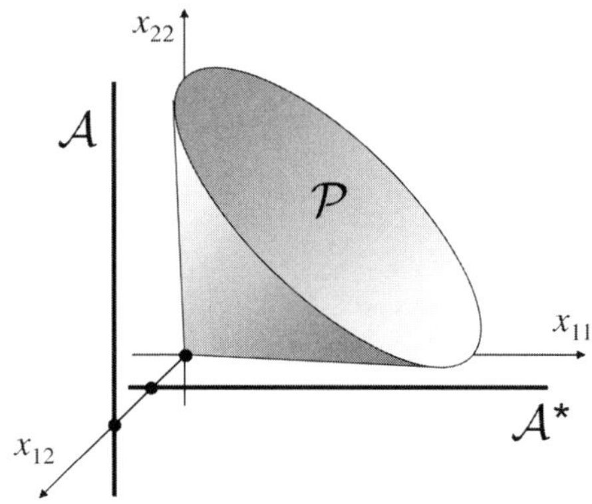

FIGURE 13.1. The space of 2×2 symmetric matrices, the positive semidefinite cone $\mathcal{P}$, and an affine subspace $\mathcal{A}$ getting arbitrarily close to $\mathcal{P}$. The subspace $\mathcal{A}^*$ has similar properties.

If the last alternative occurs, then Lemma 13.2 implies that $\mathrm{lin}(\mathcal{A})$ and $\mathrm{lin}(\mathcal{A}^*)$ contain no positive definite matrices.

PROOF. Clearly $\widehat{\mathcal{A}}$ is a linear hyperplane in $\mathrm{lin}(\mathcal{A})$. If there is a positive semidefinite matrix $Z \in \mathrm{lin}(\mathcal{A})\setminus\widehat{\mathcal{A}}$ that is on the same side of $\widehat{\mathcal{A}}$ as $\mathcal{A}$, then Z has a positive multiple in $\mathcal{A}$ and we are done. So we may assume that $\mathcal{P}\cap\mathrm{lin}(\mathcal{A})$ is contained in the closed halfspace of $\mathrm{lin}(\mathcal{A})$ bounded by $\widehat{\mathcal{A}}$ not containing $\mathcal{A}$.

This implies that $\widehat{\mathcal{A}}$ contains no positive definite matrix, and hence there is a matrix $Y \neq 0$ such that the hyperplane $Y^\perp = \{X \in \mathcal{S} :\ X\cdot Y = 0\}$ contains $\widehat{\mathcal{A}}$ but no positive definite matrix. We may assume that $X\cdot Y > 0$ for all $X \in \mathcal{P}'$, then $X\cdot Y \geq 0$ for all $X \in \mathcal{P}$, and hence $Y \succeq 0$.

If $Y \notin \mathrm{lin}(\mathcal{A})^\perp = \widehat{\mathcal{A}^*}$, then $Y^\perp\cap\mathrm{lin}(\mathcal{A}) = \widehat{\mathcal{A}}$, and so $Y^\perp\cap\mathcal{A} = \emptyset$, which implies that $Y\cdot A = a < 0$ for all $A \in \mathcal{A}$. Then $-(1/a)Y$ is positive semidefinite matrix in $\mathcal{A}^*$, and we are done. If $Y \in \widehat{\mathcal{A}^*}$, then $\widehat{\mathcal{A}^*}$ contains a positive semidefinite matrix. The conclusion follows for $\widehat{\mathcal{A}}$ similarly. □

Now we turn to duality. Recall the meaning of the dual variables in linear programming: they are multipliers for the constraints so that a linear combination of the constraints with these coefficients gives the sharpest possible bound on the objective function. With this view in mind, it is clear that equalities can have arbitrary real multipliers, and inequalities must have nonnegative multipliers (or nonpositive ones, if we want to reverse the direction of the inequality). If we want to construct a dual program in semidefinite programming, the only (but nontrivial!) addition is that the multiplier for a semidefiniteness constraint $X \succeq 0$ is a positive semidefinite matrix $Y \succeq 0$. This is based on the fact that $X \succeq 0$ is equivalent to $X\cdot Y \geq 0$ for every $Y \succeq 0$.

Consider a semidefinite program, which we recall for convenience:

$$\text{(13.5)} \qquad \begin{aligned} \text{minimize} \quad & \sum_{i=1}^{n} c_i x_i \\ \text{subject to} \quad & \sum_{i=1}^{n} x_i A_i \succeq B. \end{aligned}$$

We can follow the "rules" above for defining multipliers for the constraints, and formulate the *dual program*:

$$\text{(13.6)} \qquad \begin{aligned} \text{maximize} \quad & B \cdot Y \\ \text{subject to} \quad & A_i \cdot Y = c_i \quad (i = 1, \dots, n) \\ & Y \succeq 0. \end{aligned}$$

Note that this too is a semidefinite program in the general sense. We denote by v_{dual} the supremum of the objective function.

With this notion of dual programs, the Duality Theorem holds under additional conditions (which cannot be omitted!); see [**Wolkowitz 1981**], [**Vandeberghe–Boyd 1996**], [**Ramana 1997**].

THEOREM 13.5. *Assume that both the primal program* (13.5) *and the dual program* (13.6) *have feasible solutions. Then* $v_{\text{primal}} \geq v_{\text{dual}}$. *If, in addition, the dual program has a strictly feasible solution, then the primal optimum is attained and* $v_{\text{primal}} = v_{\text{dual}}$.

It is easy to see that the roles of primal and dual could be interchanged: if the dual program has a strictly feasible solution, then the primal optimum is attained, and the two optima are equal. In particular, if both programs have strictly feasible solutions, then the supremum, respectively infimum, of the objective functions are attained and are equal.

PROOF. Let x be a feasible solution of the primal program (13.5) and Y, a feasible solution of the dual program (13.6). Then

$$\text{(13.7)} \qquad \sum_i c_i x_i = \sum_i (A_i \cdot Y) x_i = \Big(\sum_i x_i A_i\Big) \cdot Y = \Big(\sum_i x_i A_i - B\Big) \cdot Y + B \cdot Y \geq B \cdot Y.$$

This proves that $v_{\text{primal}} \geq v_{\text{dual}}$.

Suppose that the primal program (13.5) has a strongly feasible solution, but either $v_{\text{primal}} \neq v_{\text{dual}}$ or the dual optimum is not attained. In either case, the dual has no solution with objective value $t = v_{\text{primal}}$. This means that the affine subspace $\mathcal{A}$ of $\mathcal{S}$ defined by

$$\text{(13.8)} \qquad B \cdot Y = t, \qquad A_i \cdot Y = c_i \quad (i = 1, \dots, n)$$

contains no positive semidefinite matrix.

Next, assume that the subspace $\mathcal{A}^*$ contains a positive semidefinite matrix X. By the representation (13.4), this has the form

$$\text{(13.9)} \qquad X = x_0 B + \sum_{i=1}^{n} x_i A_i, \quad \text{where} \quad x_0 t + \sum_{i=1}^{n} x_i c_i = -1.$$

For any dual solution Y, we have

$$X\cdot Y = x_0 B\cdot Y + \sum_{i=1}^{n} x_i A_i\cdot Y = x_0 B\cdot Y + \sum_{i=1}^{n} x_i c_i = x_0(B\cdot Y - t) - 1,$$

and since $B\cdot Y < t$, we must have $x_0 < 0$. Hence $-(1/x_0)X$ is a solution of the primal program (13.6) with objective value $t + 1/x_0 < t$, contradicting the definition of t.

Thus neither $\mathcal{A}$ nor $\mathcal{A}^*$ contains a positive semidefinite matrix. By Lemma 13.4, this implies that neither $\text{lin}(\mathcal{A})$ nor $\text{lin}(\mathcal{A}^*)$ can contain a positive definite matrix. But for a strongly feasible primal solution, the matrix $\sum_i x_i A_i - B$ is a positive definite matrix in $\text{lin}(\mathcal{A}^*)$, a contradiction. □

The following complementary slackness condition also follows from (13.7).

PROPOSITION 13.6. *Let x be a feasible solution of the primal program* (13.5) *and Y, a feasible solution of the dual program* (13.6). *Then both x and Y are optimal solutions if and only if* $(\sum_i x_i A_i - B)\cdot Y = 0$.

Using that for two positive semidefinite matrices X and Y, we have $X\cdot Y = 0$ if and only if $XY = 0$ (matrix product!), we see that the complementary slackness condition is equivalent to $(\sum_i x_i A_i - B)Y = 0$.

13.1.3. Algorithms for semidefinite programs. There are two essentially different algorithms known that solve semidefinite programs in polynomial time: the *ellipsoid method* and *interior point/barrier methods*. Both of these have many variants, and the exact technical descriptions are quite complicated; we refer to [**Porkoláb–Khachiyan 1997**, **Ramana 1997**] for discussions of these.

The first polynomial time algorithm to solve semidefinite optimization problems in polynomial time was the ellipsoid method. This is based on the general fact that if we can test membership in a convex body $K \subseteq \mathbb{R}^d$ (i.e., we have a subroutine that, for a given point $x \in \mathbb{R}^d$, tells us whether or not $x \in K$), then we can use the ellipsoid method to optimize any linear objective function over K [**Grötschel–Lovász–Schrijver 1988**]. The precise statement of this fact needs nontrivial side-conditions.

For any semidefinite program (13.1), the set K of feasible solutions is convex. With rounding and other tricks, we can make it a bounded, full-dimensional set in $\mathbb{R}^n$. To test membership, we have to decide whether a given point x belongs to K; ignoring numerical problems, we can use Gaussian elimination to check whether the matrix $Y = \sum_i x_i A_i - B$ is positive semidefinite. Thus using the ellipsoid method we can compute, in polynomial time, a feasible solution x that is approximately optimal.

Unfortunately, the above argument gives an algorithm which is polynomial, but hopelessly slow, and practically useless. Semidefinite programs can be solved in polynomial time and also *practically reasonably efficiently* by interior point methods [**Overton 1988**, **Alizadeh 1992**, **Alizadeh 1995**]. The algorithm can be described very informally as follows. We consider the form (13.3), denote by K the set of its feasible solutions (these are symmetric matrices), and want to minimize a linear function $C\cdot X$ over $X \in K$. The set K is convex, but the minimum will be attained on the boundary of K, and this boundary is neither smooth nor polyhedral in general. Therefore, neither gradient-type methods nor simplex-type methods of linear programming can be used.

The main idea of barrier methods is that instead of minimizing a linear objective function $C^\mathsf{T}X$, we minimize the convex function $F_\lambda(x) = -\log\det(X) + \lambda C^\mathsf{T}X$ for some parameter $\lambda > 0$. Since F_λ tends to infinity on the boundary of K, the minimum will be attained in the interior. Since F_λ is convex and analytic in the interior, the minimum can be very efficiently computed by a variety of numerical methods (conjugate gradient etc.)

Of course, the point we obtain this way is not what we want, but if λ is large it will be close. If we do not like it, we can increase λ and use the minimizer for the old F_λ as the starting point for a new gradient type algorithm. One can show that (under some technical assumptions about the feasible domain) this algorithm gives a good approximation of the optimum in polynomial time (see [**Alizadeh 1995**],[**Vandeberghe–Boyd 1996**] and the book [**Nesterov–Nemirovsky 1994**]) .

13.2. Geometric representations from semidefinite optimization

13.2.1. Unit distance representation. Given a graph G, how can we represent it by unit distances? To be precise, a *unit distance representation* of a graph G is a geometric representation $\mathbf{u} : V \to \mathbb{R}^d$ for some $d \geq 1$ such that $|\mathbf{u}_i - \mathbf{u}_j| = 1$ for every $ij \in E$ (we allow that $|\mathbf{u}_i - \mathbf{u}_j| = 1$ for some nonadjacent nodes i, j).

Every finite graph has a unit distance representation in a sufficiently high dimension. (For example, we can map its nodes onto the vertices of a regular simplex with edges of length 1.) The first problem that comes to mind, raised in [**Erdős–Harary–Tutte 1965**], is to find the minimum dimension d in which a graph has a unit distance representation. Figure 13.2 shows a 2-dimensional unit distance representation of the Petersen graph. Similarly to orthogonal representations of the complement (and by a very similar argument) the minimum dimension d of a unit distance representation is linked to the chromatic number of the graph:

$$\frac{1}{2}\log\chi(G) \leq d \leq \chi(G) - 1. \tag{13.10}$$

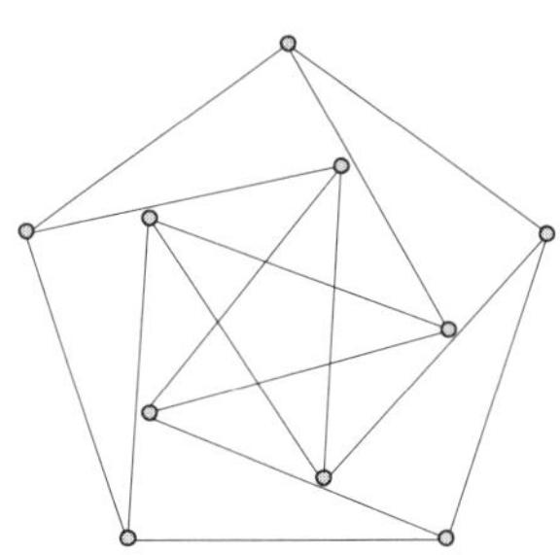

FIGURE 13.2. A unit distance representation of the Petersen graph in the plane.

There are many other senses in which we may want to find the most economical unit distance representation. We only discuss one: what is the smallest radius of a ball containing a unit distance representation of G (in any dimension)?

Considering the Gram matrix $A = \mathsf{Gram}(\mathbf{u})$ of a vector labeling $\mathbf{u}$, it is easy to obtain the following: A graph G has a unit distance representation in a ball of radius R (in some appropriately high dimension) if and only if there exists a positive semidefinite matrix A such that $A_{ii} \le R^2$ for all $(i \in V)$ and $A_{ii} - 2Aij + A_{jj} = 1$ for all $(ij \in E)$. In other words, the smallest radius R is the square root of the optimum value of the following semidefinite program (and so it is polynomial time computable with arbitrary precision):

$$\begin{aligned} \text{(13.11)} \qquad & \text{minimize} \quad w \\ & \text{subject to} \quad A \succeq 0 \\ & \qquad A_{ii} \le w \qquad\qquad (i \in V) \\ & \qquad A_{ii} - 2Aij + A_{jj} = 1 \qquad (ij \in E). \end{aligned}$$

EXAMPLE 13.7. The unit distance embedding of the Petersen graph in Figure 13.2 is optimal if we want to minimize the dimension, but not if we want to minimize the circumradius. Let us illustrate how semidefinite optimization can find the optimal embedding by determining this for the Petersen graph. In the formulation above, we have to find a 10×10 positive semidefinite matrix A satisfying the given linear constraints. For a given w, the set of feasible solutions is convex, and it is invariant under the automorphisms of the Petersen graph. Hence there is an optimum solution which is invariant under these automorphisms (in the sense that if we permute the rows and columns by the same automorphism of the Petersen graph, we get back the same matrix).

Now we know that the Petersen graph has a very rich automorphism group: not only can we transform every node into every other node, but also every edge into every other edge, and every nonadjacent pair of nodes into every other nonadjacent pair of nodes. A matrix invariant under these automorphisms has only 3 different entries: one number in the diagonal, another number in positions corresponding to edges, and a third number in positions corresponding to nonadjacent pairs of nodes. This means that this optimal matrix A can be written as

$$A = xP + yJ + zI,$$

where P is the adjacency matrix of the Petersen graph. So we only have these 3 unknowns x, y and z to determine.

The linear conditions above are easily translated into the variables x, y, z. But what to do with the condition that A is positive semidefinite? Luckily, the eigenvectors of P are also eigenvectors of J and I, and hence the eigenvalues of A are the corresponding linear combinations of the eigenvalues of P, J and I. The eigenvalues of P are well known (and easy to compute): they are 3, 1 (5 times) and -2 (4 times). So the nonnegativity of the eigenvalues of A, together with the linear conditions gives the linear inequalities in (13.11) give us the linear program

$$\text{minimize} \quad y + z$$

$$\text{subject to} \quad \begin{cases} 3x + 10y + z & \ge 0, \\ x + z & \ge 0, \\ -2x + z & \ge 0, \\ 2z - 2x & = 1. \end{cases}$$

It is easy to solve this: $x = -1/4$, $y = 1/20$ and $z = 1/4$. Thus the smallest radius of a ball in which the Petersen graph has a unit distance representation is $\sqrt{y+z} = \sqrt{3/10}$. The corresponding matrix A has rank 4, so this representation is in $\mathbb{R}^4$.

It would be difficult to draw a picture of this representation, but we can offer the following nice matrix, whose columns realize this representation:

$$\frac{1}{2}\begin{pmatrix} 1 & 1 & 1 & 1 & 0 & 0 & 0 & 0 & 0 & 0 \\ 1 & 0 & 0 & 0 & 1 & 1 & 1 & 0 & 0 & 0 \\ 0 & 1 & 0 & 0 & 1 & 0 & 0 & 1 & 1 & 0 \\ 0 & 0 & 1 & 0 & 0 & 1 & 0 & 1 & 0 & 1 \\ 0 & 0 & 0 & 1 & 0 & 0 & 1 & 0 & 1 & 1 \end{pmatrix} \tag{13.12}$$

This matrix reflects the fact that the Petersen graph is the complement of the line-graph of K_5. The center of the smallest ball containing it is not the origin; translating it to the origin, we get a representation in 4-space, but with a less transparent matrix. ♦

Unit distance representations are not only analogous to orthogonal representations, but in fact closely related; see Exercises 13.2, 13.3. They are also special cases of bar-and-joint frameworks, to be discussed in Chapter 15.

13.2.2. Boolean variables and semidefinite optimization. Discrete optimization problems typically ask for an optimal subset of a set with given properties, and with maximum or minimum weight, or some other extremal property. A solution of such a problem can be encoded as a Boolean vector. Often the constraints and the objective function can be expressed as quadratic functions of these variables. If this is the case, we have a chance to use semidefinite optimization to obtain good bounds on the optimum.

In the following general description, we consider the Boolean values of the variables as real variables $x_1, \dots, x_n$, where $x_i = 1$ means TRUE and $x_i = 0$ means FALSE. The quadratic equation $x_i^2 = x_i$ expresses that x is 0-1 valued. For this discussion, assume that the problem is a maximization problem with some (typically NP-hard) optimum value $v_{\rm opt}$. It is often useful to homogenize the problem; for this, we introduce $x_0 = 1$. The condition on the x_i becomes $x_i^2 - x_i x_0 = 0$, and we also have $x_0^2 = 1$.

The symmetric matrix $X = (x_i x_j)_{i,j=0}^n$ is positive semidefinite. There are some further general conditions following just from the definition of X:

$$X_{00} = 1, \tag{13.13}$$

$$X_{i0} = X_{0i} = X_{ii} \qquad (i = 1, \dots, n). \tag{13.14}$$

This is not all; as an example, the following inequality is often useful (it can be verified by checking the possible values of x_i, x_j and x_k; it does not follow from positive semidefiniteness):

$$X_{ij} + X_{jk} \le X_{jj} + X_{ik} \qquad (1 \le i, j, k \le n). \tag{13.15}$$

To obtain a semidefinite relaxation of the problem, we express the conditions as linear constraints on the entries of X. This step is of course problem-specific, and we will discuss some important applications in the sect sections.

Then we solve the semidefinite optimization problem with these constraints, and obtain an optimizer matrix X and an optimum value $v_{\rm sd}$. Since the semidefinite problem is a relaxation, we have $v_{\rm opt} \le v_{\rm sd}$. How good is this approximation? And, equally importantly, how to get an optimum solution of the original problem from X?

We can write $X = \mathsf{Gram}(\mathbf{u})$ for some vector labeling $\mathbf{u}$. Equation (13.13) implies that $\mathbf{u}_0$ is a unit vector; the vectors $\mathbf{u}_i$ $(i > 0)$ satisfy $\mathbf{u}_i^\mathsf{T}(\mathbf{u}_0 - \mathbf{u}_i) = 0$, so the vectors $\mathbf{u}_i$, $\mathbf{u}_0 - \mathbf{u}_i$ and $\mathbf{u}_0$ form a right triangle. We can also say that $\mathbf{u}_i$ lies on the Thales sphere with diameter $[0, \mathbf{u}_0]$. Condition (13.15) translates into $(\mathbf{u}_j - \mathbf{u}_i)^\mathsf{T}(\mathbf{u}_j - \mathbf{u}_k) \ge 0$, which in the geometric language means that no three of the labeling vectors form an obtuse triangle.

More generally, the vector labels satisfy algebraic conditions that are formally similar to original quadratic constraints. For example, a constraint $x_i x_j = 0$ implies $X_{ij} = 0$, which implies $\mathbf{u}_i^\mathsf{T}\mathbf{u}_j = 0$. If we are lucky and the vectors $\mathbf{u}_i$ all belong to the same 1-dimensional space, then we get an optimal solution of the discrete problem right away.

But in general we are not so lucky, and we need to extract and approximate solution from the vectors $\mathbf{u}_i$. The simplest method is to generate a random unit vector $\mathbf{v}$, and define

$$\overline{x}_i = \begin{cases} 1, & \text{if } \mathbf{v}^\mathsf{T}\mathbf{u}_i > \frac{1}{2}\mathbf{v}^\mathsf{T}\mathbf{u}_0, \\ 0, & \text{otherwise.} \end{cases}$$

Geometrically, we take a random hyperplane through the center of the Thales sphere, and define $\overline{x}$ as the incidence vector of the set of nodes represented in the halfspace not containing the origin. The difficult question is, how far is this solution from the optimum? The analysis of a few important applications in the next sections shows that the answer to this question is often not easy, but the method is quite powerful.

This method can augmented with several nontrivial tricks, and other extraction methods have also been developed; and refer to [**Feige–Langberg 2006**] and [**Laurent–Rendl 2005**] for more.

While it is more common to use $x_i = 0$ for FALSE, one could also use $x_i = -1$ to encode this. This means only a linear transformation, and sometimes one, sometimes the other is more useful. In the examples below, both approaches come up.

13.2.3. Stable sets. We want to find a maximum stable set S in the graph G. We can assign 0-1 valued variables x_i to the nodes, where $x_i = 1$ means that $i \in S$. The condition that S is stable means that $x_i x_j = 0$ for all edges ij, and the cardinality of S is just the sum $\sum_i x_i$. It will be convenient to homogenize by introducing a new variable $x_0 = 1$. Then our conditions are $x_i^2 = x_0 x_i$ for every node i, and $x_i x_j = 0$ for every edge ij. The objective function to maximize can be written as $\sum_{i \in V} x_i^2$.

In terms of the $(n+1) \times (n+1)$ matrix $X = (x_i x_j)_{i,j=0}^n$, we have

$$\begin{aligned} \text{maximize} \quad & \sum_{i \in V} X_{ii} \\ \text{subject to} \quad & A \succeq 0, \\ & X_{00} = 1, \\ & X_{i0} = X_{0i} = X_{ii} \quad (i \in V), \\ & X_{ij} = 0 \quad (ij \in E). \end{aligned} \tag{13.16}$$

Writing $X = \mathsf{Gram}(\mathbf{w})$, we get an orthogonal representation together with a unit vector $\mathbf{w}_0$. The vectors $\mathbf{v}_i = \mathbf{w}_i^0$ form a dual orthonormal representation with handle $\mathbf{d} = \mathbf{w}_0$. We know that $\mathbf{w}_i$, $\mathbf{d}-\mathbf{w}_i$ and $\mathbf{d}$ form a right triangle, which means that $\mathbf{w}_i$ is the orthogonal projection of $\mathbf{d}$ onto $\mathbf{v}_i$, and hence $|\mathbf{w}_i| = \mathbf{d}^\mathsf{T}\mathbf{v}_i$. Thus the value of the objective function is

$$\sum_{i\in V} X_{ii} = \sum_{i\in V} |\mathbf{w}_i|^2 = \sum_{i\in V} (\mathbf{d}^\mathsf{T}\mathbf{v}_i)^2.$$

Since $\vartheta(G)$ is the maximum of the right hand side over all orthonormal representations, it follows that $v_{\rm sd} \le \vartheta(G)$. It is easy to check that we have equality here. So ϑ can be defined as the most basic semidefinite relaxation of the stability number.

A next step is to extract a reasonably large stable subset from the optimum solution of (13.16). We have done so in Section 11.4.1 (with a very large error, but as remarked there, one cannot expect to do better).

13.3. Maximum cut

Next we describe the application of semidefinite optimization combined with geometric techniques to another basic algorithmic problem, due to Goemans and Williamson [**Goemans–Williamson 1995**].

13.3.1. The Maximum Cut Problem. A *cut* in a graph G is the set of edges connecting a set $S \subseteq V$ to $V\setminus S$, where $\emptyset \subset S \subset V$. The *Maximum Cut Problem* is to find a cut with maximum cardinality. We denote by $\mathsf{maxcut}(G)$ this maximum. (More generally, we can be given a weighting $w: V \to \mathbb{R}_+$, and we could be looking for a cut with maximum total weight. To keep things simple, however, we restrict our discussion to the unweighted case.)

The Maximum Cut Problem is NP-hard; it was one of the first problems shown to be NP-hard by Karp. A natural next step is to find an "approximately" maximum cut. Formulated in different terms, [**Erdős 1967**] described the following simple heuristic algorithm for the Maximum Cut Problem: for an arbitrary ordering $(v_1,\dots,v_n)$ of the nodes, we color $v_1, v_2, \dots, v_n$ successively red or blue. For each i, v_i is colored blue if and only if the number of edges connecting v_i to blue nodes among $v_1,\dots,v_{i-1}$ is less than the number of edges connecting v_i to red nodes in this set; otherwise, it is colored blue. Then the cut formed by the edges between red and blue nodes contains at least half of all edges. In particular, we get a cut that is at least half as large as the maximum cut.

There is an even easier randomized algorithm to achieve this approximation, at least in expected value. Let us 2-color the nodes of G randomly, so that each node is colored red or blue independently, with probability $1/2$. Then the probability that an edge belongs to the cut between red and blue is $1/2$, and so the expected number of edges in this cut is $m/2$.

These algorithms show that the maximum cut can be approximated from below in polynomial time with a multiplicative error of at most $1/2$. Can we do better? The fundamental paper [**Goemans–Williamson 1995**], building on results of [**Mohar–Poljak 1990**, **Poljak–Rendl 1992**, **Delorme–Poljak 1993a**], gives a polynomial time algorithm that approximates the maximum cut in a graph with a relative error of less than 13%:

THEOREM 13.8. *One can find a cut with at least* $.878\,\mathsf{maxcut}(G)$ *edges in polynomial time.*

What is this strange constant? The proof below will show that its exact value is $2c/\pi$, where c is the largest positive number for which

$$\arccos t \geq c(1-t) \tag{13.17}$$

for $-1 \leq t \leq 1$ (Figure 13.3). It can be shown that $2c/\pi$ is an irrational number, of which the value .878 in the theorem is only an approximation.

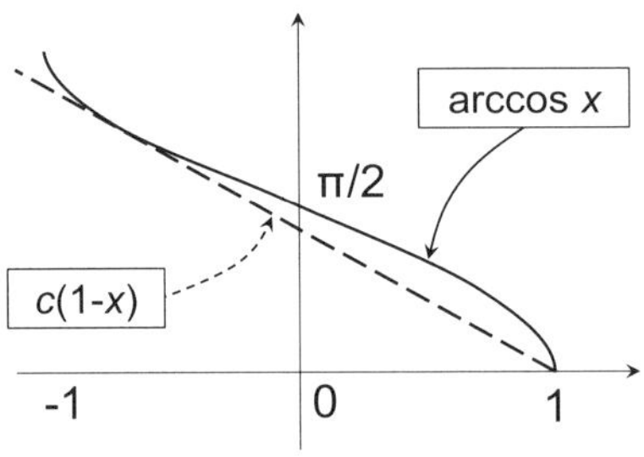

FIGURE 13.3. The constant c in the Goemans–Williamson algorithm

Can we do even better? Not likely. [**Håstad 1997**] showed that it is NP-hard to find a cut with more than $(16/17)\mathsf{maxcut}(G) \approx .941\,\mathsf{maxcut}(G)$ edges, so we cannot get arbitrarily close to the optimum. The constant $2c/\pi \approx .878$ would seem like a "random" byproduct of a particular proof, and one would expect that the "truth" (the best achievable constant) is somewhere between .878... and .941..., perhaps some nice number like 9/10. However, it turns out that if we assume the complexity theoretic hypothesis called the "Unique Games Conjecture" (stronger than $P \neq NP$), then the constant $2c/\pi$ is optimal [**Khot et al. 2007**]: no polynomial time randomized algorithm can produce a cut that for every graph would contain at least $a\,\mathsf{maxcut}(G)$ edges, where $a > 2c/\pi$.

13.3.2. Semidefinite formulation. Let us assign a ± 1 variable x_i to each node, where the sign encodes the two sides of the cut we want to find. These variables satisfy $x_i^2 = 1$. There are no instance-dependent constraints. For an edge ij, the difference $x_i - x_j$ is zero if i and j are on the same side of the cut, and ± 2 otherwise. Hence the number of edges in the cut can be expressed as

$$\frac{1}{4}\sum_{ij\in E}(x_i - x_j)^2 = \frac{1}{4}x^{\mathsf{T}}Lx \tag{13.18}$$

(where L is the Laplacian of the graph). In terms of the symmetric $V \times V$ matrix $X = xx^{\mathsf{T}} = (x_i x_j)_{i,j\in V}$, we have the semidefinite optimization problem

$$\begin{aligned} \text{maximize} \quad & \frac{1}{4}L \cdot X \\ \text{subject to} \quad & X \succeq 0 \\ & X_{ii} = 1 \qquad (i \in V) \end{aligned} \tag{13.19}$$

It is not hard to see that the maximum is attained, since the conditions $X_{ii} = 1$ imply that $|X_{ij}| \leq 1$ for all i and j and so the set of solutions is compact. This semidefinite program is only a relaxation of the maximum cut problem: the matrix X has the additional property that its rank is 1. This property, however, cannot be expressed as a linear condition on the matrix entries. Let $\gamma(G)$ denote optimum value in (13.19).

We can write the optimizer in (13.19) as $X = \mathsf{Gram}(\mathbf{u})$, then the vector labeling $\mathbf{u}$ consists of unit vectors, and the objective function is

$$\frac{m}{2} - \sum_{ij\in E} \mathbf{u}_i^\mathsf{T}\mathbf{u}_j. \tag{13.20}$$

13.3.3. The dual problem. During our treatment of orthogonal representations and the theta-function, we have seen that semidefinite duality has played a crucial role, leading not only to the technical tools of Theorem 11.2, but also to duality relations between the graph and its complement (Proposition 11.4, Corollary 11.6 and Corollary 11.27). Let us conclude with describing the duality of the semidefinite relaxation of the maximum cut problem. It is not hard to see that the dual of (13.19) can be written as

$$\begin{aligned} &\text{minimize} && \frac{1}{4}\sum_{i\in V} z_i \\ &\text{subject to} && \operatorname{diag}(\mathbf{z}) - L \succeq 0 \end{aligned} \tag{13.21}$$

In words, we want to add to the diagonal entries of $-L$ as little as possible to make it positive semidefinite. Clearly both (13.19) and (13.21) have strictly feasible solutions, and both optima are attained and equal to $\gamma(G)$.

In terms of the adjacency matrix, we can rewrite this as

$$\gamma(G) = \frac{m}{2} + \frac{1}{4}\min\{\mathbb{1}^\mathsf{T}\mathbf{x} :\ \operatorname{diag}(\mathbf{x}) + A_G \succeq 0\} \tag{13.22}$$

So we want to increase the diagonal entries of the adjacency matrix A_G to make it positive semidefinite, and want to minimize the total increase. This minimum is just $\gamma(G) - m/2$.

We can write a solution as $\operatorname{diag}(\mathbf{z}) - L = \mathsf{Gram}(\mathbf{v})$, then the vector labeling $\mathbf{v}$ satisfies

$$\mathbf{v}_i^\mathsf{T}\mathbf{v}_j = \begin{cases} 1, & \text{if } ij \in E, \\ 0, & \text{if } ij \in \overline{E}, \end{cases} \tag{13.23}$$

and, letting $\mathbf{v}$ run over all vector labelings satisfying (13.23),

$$\gamma(G) = \min_{\mathbf{v}} \frac{1}{4}\sum_{i\in V} \big(|\mathbf{v}_i|^2 + \deg(i)\big) = \frac{m}{2} + \frac{1}{4}\min_{\mathbf{v}} \sum_{i\in V} |\mathbf{v}_i|^2. \tag{13.24}$$

We see that this is always at least $m/2$, as it should be. Note that $\mathbf{v}$ is a particularly nice orthogonal representation.

An argument similar to the proof of Theorem 11.5 shows that if the graph has a node-transitive automorphism group, then adding the same value in the diagonal is optimal. This implies that for a graph with a node-transitive automorphism group,

$$\gamma(G) = \frac{m}{2} - \frac{n}{4}\lambda_{\min}. \tag{13.25}$$

In particular,

$$\gamma(K_n) = \frac{1}{2}\binom{n}{2} + \frac{n}{4} = \frac{n^2}{4}, \quad \gamma(K_{n/2,n/2}) = \frac{n^2}{4}.$$

EXAMPLE 13.9. Bipartite graphs G are trivial examples for the Maximum Cut Problem, since all edges can be included in a cut. How is the semidefinite programming bound? We claim that adding the degrees $\deg(i)$ in the diagonal of the adjacency matrix, we get a positive semidefinite matrix. Indeed, multiplying the rows and columns corresponding to nodes in one color class (which does not change the signs of the eigenvalues), we get L_G, which is positive semidefinite.

This implies, by (13.22), that

$$\gamma(G) \le \frac{m}{2} + \frac{1}{4}\sum_i \deg(i) = \frac{m}{2} + \frac{m}{2} = m.$$

Since a cut of size m can be attained, it follows that $\gamma(G) = m$, and that $x_i = \deg(i)$ is an optimizer in (13.22). ♦

EXAMPLE 13.10. Consider the line-graph $H = L(G)$ of a simple connected graph G. The maximum cut of H is not hard to compute [**Guruswami 1999**]. A partition of $V(H) = E$ corresponds to a partition $E' \cup E'' = E$. Let $\deg(i)$ and $\deg'(i)$ denote the degree of node i in G and in (V, E'), respectively. Then the number of edges of H between E' and E'' is the number of pairs of edges $e' \in E'$ and $e'' \in E''$ that share an endpoint i, and so it is $\sum_{i\in V} \deg'(i)\big(\deg(i) - \deg'(i)\big)$. Clearly $\deg'(i)\big(\deg(i) - \deg'(i)\big) \le \lfloor \deg(i)^2/4 \rfloor$, and hence

$$\mathsf{maxcut}(H) \le \sum_{i\in V} \left\lfloor \frac{\deg(i)^2}{4} \right\rfloor.$$

This upper bound can be attained except in one case, when it can be improved by 1 (see Exercise 13.8):

$$\mathsf{maxcut}(H) = \sum_{i\in V} \left\lfloor \frac{\deg(i)^2}{4} \right\rfloor - \varepsilon, \tag{13.26}$$

where $\varepsilon = 1$ if G is an eulerian graph with an odd number of edges, and $\varepsilon = 0$ otherwise.

To estimate the upper bound γ, we assign to each edge ij the vector $\mathbb{1}_{\{i,j\}} \in \mathbb{R}^V$. Then we get an orthogonal representation with properties (13.23), and so

$$\gamma(H) \le \frac{|E(H)|}{2} + \frac{m}{2} = \sum_{i\in V} \frac{\deg(i)^2}{4}. \tag{13.27}$$

Equality does not hold in all cases, but it does when G has a node-transitive automorphism group. We see that for a line-graph H, the value $\gamma(H)$ is a reasonable bound on $\mathsf{maxcut}(H)$; it is equal to $\mathsf{maxcut}(H)$, among others, when G is an eulerian graph with an even number of edges, and asymptotically equal if the degrees of G are large. ♦

13.3.4. Extracting a cut. The semidefinite optimization problem (13.19) can be solved in polynomial time (with an arbitrarily small relative error; let's not worry about this error here). The optimum value $\gamma(G)$ is a polynomial time computable upper bound on the size of the maximum cut.

Following the general method, we write the optimal matrix as $X = \mathsf{Gram}(\mathbf{u})$, to get a vector labeling with unit vectors in some space $\mathbb{R}^d$, for which the objective function (13.20) is maximized. To define the cut, we generate a random hyperplane

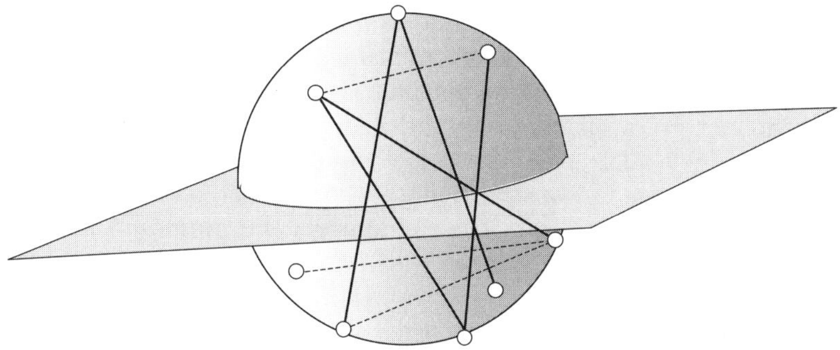

FIGURE 13.4. A cut in the graph given by a random hyperplane

in $\mathbb{R}^d$ (its normal vector should be uniformly distributed over S^d), and define the two sides of the cut as the two sides of this hyperplane (Figure 13.4).

How good is this cut? Let $ij \in E$ and let $\mathbf{u}_i, \mathbf{u}_j \in S^{n-1}$ be the corresponding vectors in the representation constructed above. It is easy to see that the probability that a random hyperplane H through 0 separates $\mathbf{u}_i$ and $\mathbf{u}_j$ is $(\arccos \mathbf{u}_i^\mathsf{T}\mathbf{u}_j)/\pi$. Using the constant c introduced above, the expected number of edges intersected by H is

$$\sum_{ij\in E} \frac{\arccos \mathbf{u}_i^\mathsf{T}\mathbf{u}_j}{\pi} \geq \sum_{ij\in E} c\,\frac{1-\mathbf{u}_i^\mathsf{T}\mathbf{u}_j}{\pi} = -\frac{2c}{\pi}\gamma(G) \geq \frac{2c}{\pi}\,\mathsf{maxcut}(G). \tag{13.28}$$

This completes the analysis of the algorithm, and proves Theorem 13.8.

REMARK 13.11. Often the Max Cut Problem arises in a more general setting, where the edges of a the graph have nonnegative weights (or values), and we want to find a cut with maximum total weight. The algorithm described above is easily extended to this case, by using the weights as coefficients in the sums (13.18), (13.19) and (13.28).

REMARK 13.12. It is not quite obvious how to generate a random hyperplane in a high-dimensional linear space. We want that the random hyperplane does not prefer any direction, which means the normal vector of the hyperplane should be uniformly distributed over the unit sphere in $\mathbb{R}^n$. Such a vector can be generated by normalizing a standard Gaussian (see (C.15) in the Appendix).

Another objection to the above algorithm could be that it uses random real numbers. This can be replaced by random integers from a polynomial-size set, similarly as at the end of Section 3.3.3. In fact, the algorithm can be fully *derandomized* by well established (but nontrivial) techniques. We do not discuss this issue here; see e.g. [**Alon–Spencer 1992**], Chapter 15 for a survey of derandomization methods.

13.3.5. Springs. A different path leading to the same algorithm reaches back to the idea of rubber band representations—with a twist. Suppose that we have somehow placed the nodes of the graph in space, conveniently on the unit sphere in some Euclidean space. Take a random hyperplane H through the origin. This partitions the nodes into two classes (Figure 13.4). We "hope" that this will give a good partition.

How justified is this hope? If the nodes are placed in a random position (say, their location is independently selected uniformly from the unit ball centered at the origin), then it is easy to see that we get half of the edges in expectation.

Can we do better by placing the nodes in a more sophisticated way? Still heuristically, an edge that is longer has a better chance of being cut by the random hyperplane. So we want a construction that pushes adjacent nodes apart; one expects that for such a placement, the random cut will intersect many edges. In other words, we want to replace the edges not by rubber bands (as in Chapter 3), but by springs! Our springs will be, however, very peculiar springs, since the farther the endpoints are, the larger the force with which they push their endpoints apart. (This is certainly counterintuitive, but mathematically definable. In a small neighborhood of a given length, the "spring" image can be helpful, and will be used in Chapter 15 again.)

Clearly we have to constrain the nodes to a bounded region, and the unit sphere S^{d-1} is a natural choice. (The unit ball would work just as well.) Mathematically speaking, we want to find a representation $i \mapsto \mathbf{u}_i$ ($i \in V$) of the nodes of the graph on the unit sphere in $\mathbb{R}^d$ so that the following "energy" is minimized:

$$-\sum_{ij \in E} (\mathbf{u}_i - \mathbf{u}_j)^2. \tag{13.29}$$

What is the minimum of this energy? If we work in $\mathbb{R}^1$, then the problem is equivalent to the Maximum Cut problem: each node is represented by either 1 or -1, which means that a placement is equivalent to a partition of the node set; the energy is the negative of the number of edges connecting the two partition classes. So at least we have been able to relate our placement problem to the Maximum Cut problem.

Unfortunately, the argument above also implies that for $d = 1$, the optimal representation is NP-hard to find. For $d > 1$, the negative of the minimum energy $-\mathcal{E}_{\text{opt}}$ may not be equal to the size of the maximum cut, but it is still an upper bound. While I am not aware of a proof of this, it is probably NP-hard to find the placement with minimum energy for $d = 2$, and more generally, for any fixed d. The surprising fact is that if we do not fix d (equivalently, for $d = n$), such a representation can be found in polynomial time using semidefinite optimization! This has been the content of our discussion in this section.

13.4. The MAX-3SAT problem

The Satisfiability Problem is perhaps the most central algorithmic problem in computer science. We have Boolean variables $x_1, \dots, x_n$. A *literal* is a variable or its negation. A *clause* is a disjunction of literals: $c = y_1 \vee \dots \vee y_k$. A *conjunctive normal form* is a conjunction of clauses: $\Phi(x_1, \dots, x_n) = c_1 \wedge \dots \wedge c_m$. A *$k$-form* is a conjunctive normal form in which every clause contains k literals.

The Satisfiability Problem asks for finding values for the variables making $\Phi(x_1, \dots, x_n)$ true (in other words, making every clause true). This is an NP-hard problem (in a sense, the quintessential NP-hard problem), and therefore many versions and special cases have been studied. It is known that the Satisfiability Problem is polynomially solvable for 2-forms, but NP-hard for 3-forms.

The MAX-3SAT problem is a variant in which, given a 3-form Φ, we want to find an assignment of values to the variables which makes as many clauses true as possible. We denote by $\mathsf{maxsat}(\Phi)$ this maximum number. This is, of course, NP-hard as well, but in this setting we may be interested in good approximation algorithms.

Here we sketch an algorithm that solves the MAX-3SAT Problem within a factor of 7/8 of the optimum, using semidefinite optimization and the geometric representation obtained from it. We follow the general path outlined above: we consider the Boolean variables as real variables of value 0 or 1. Each clause c has a boolean value, indicating the truth of it: $c = 1$ if c is satisfied and $c = 0$ otherwise. The objective function is simply $\sum_c c$.

As before, we use a new variable $x_0 = 1$ to make the system homogeneous, and want to make use of the semidefiniteness of the matrix $X = xx^\mathsf{T} = (x_i x_j)_{0 \le i,j \le n}$. The validity of a clause $c = y_i \vee y_j \vee y_k$ can be expressed in an algebraic form:

$$c = 1 - (1-y_i)(1-y_j)(1-y_k).$$

Here $y_i = x_i$ if x_i is unnegated, and $y_i = 1-x_i$ otherwise. Unfortunately, this inequality is cubic, which does not lead to linear conditions on the entries of the matrix X. But we are only looking for a relaxation, so we might use upper bounds instead of an exact expression of c. A simple bound is

$$c \le y_i + y_j + y_k,$$

but this is not powerful enough. We go sort of half way, using a quadratic upper bound:

$$c \le y_i + y_j + y_k - y_i y_j - y_i y_k, \tag{13.30}$$

which is easy to check. There are three ways to separate one of the variables y_i from a given clause, so every clause gives rise to three inequalities of the form (13.30).

Defining the symmetric $(n+1)\times(n+1)$ matrix $X = (x_i x_j)_{i,j=0}^n$, we want to

$$\begin{aligned} &\text{maximize} && \sum_c (X_{0i}^c + X_{0j}^c + X_{0j}^c - X_{ij}^c - X_{ik}^c) \\ &\text{subject to} && X \succeq 0, \\ &&& X_{00} = 1, \\ &&& X_{0i} = X_{i0} = X_{ii} \qquad (1 \le i \le n), \end{aligned} \tag{13.31}$$

where

$$X_{ij}^c = \begin{cases} X_{ij}, & \text{if } i \text{ and } j \text{ are not negated in } c, \\ X_{ii} - X_{ij}, & \text{if } i \text{ is not negated in } c \text{ but } j \text{ is}, \\ X_{jj} - X_{ij}, & \text{if } j \text{ is not negated in } c \text{ but } i \text{ is}, \\ X_{00} - X_{ii} - X_{jj} + X_{ij}, & \text{if both } i \text{ and } j \text{ are negated in } c. \end{cases}$$

The optimum value of this semidefinite program, along with the optimizing matrix X, is polynomial time computable. This optimization problem is a relaxation of the MAX-3SAT problem: every solution of the MAX-3SAT problem yields a solution of (13.31) (but unfortunately not the other way around). So the optimum value of (13.31) is at least $\mathsf{maxsat}(\Phi)$.

The rest of the algorithm is similar to the maximum cut algorithm. We write $X = \mathsf{Gram}(\mathbf{u})$ with an appropriate vector labeling $\mathbf{u}$ in $\mathbb{R}^d$. Note that $\mathbf{u}_0$ is a unit vector, and all the other vectors $\mathbf{u}_i$ are contained in the sphere with radius 1/2 about $(1/2)\mathbf{u}_0$ (the "Thales sphere" of $[0, \mathbf{u}_0]$). Taking a random hyperplane H

through $(1/2)\mathbf{u}_0$, and assigning $x_i = 1$ if $\mathbf{u}_i$ is on the same side of H as $\mathbf{u}_0$, and $x_i = 0$ otherwise, we get an assignment of boolean values.

It can be proved that the expected number of satisfied clauses is at least $\frac{7}{8}\mathsf{maxsat}(\Phi)$; this was conjectured, and proved in some special cases in [**Karloff–Zwick 1997**], and proved completely in [**Zwick 2002**]. This latter proof is computer-assisted, and we do not reproduce it here. The constant 7/8 is best possible [**Håstad 1997**].

EXERCISE 13.1. How do the bounds (13.10) change if we insist that different nodes be represented by different points of $\mathbb{R}^d$?

EXERCISE 13.2. Let d_1 denote the smallest dimension in which a given graph G has a unit distance representation, and let d_2 denote the smallest dimension in which $\overline{G}$ has an orthonormal representation. Prove that $d_1 \le d_2 \le 4^{d_1}$.

EXERCISE 13.3. Prove that every graph G has a unit distance representation (in some dimension) contained in a ball with radius

$$\sqrt{\frac{1}{2} - \frac{1}{2\vartheta(\overline{G})}}.$$

If the graph has a node-transitive automorphism group, then this is the smallest possible radius.

EXERCISE 13.4. Construct the dual of the semidefinite program (13.2). Does it have an optimal solution? Does it have a strictly feasible solution?

EXERCISE 13.5. Let $A_1, \dots, A_n$ be symmetric $m \times m$ matrices. Then exactly one of the following alternatives holds: (i) There are $x_1, \dots, x_n \in \mathbb{R}$ such that $x_1 A_1 + \dots + x_n A_n \succeq 0$ and $\sum_{i=1}^n x_i A_i \ne 0$. (ii) There exists a matrix $Y \succ 0$ such that $A_i \cdot Y = 0$ $(i = 1, \dots, n)$.

EXERCISE 13.6. Let $\mathcal{A}$ be an affine subspace of $\mathcal{S}$. Then exactly one of the following alternatives holds: (i) There is an $X \in \mathcal{A}$, $X \succ 0$. (ii) There exists a matrix $Y \ne 0$, $Y \succeq 0$ such that either $X \cdot Y \le 0$ for all $X \in \mathcal{L}$.

EXERCISE 13.7. Let $A_1, \dots, A_n$ be symmetric $m \times m$ matrices and $b_i, \dots, b_n \in \mathbb{R}$. Then exactly one of the following alternatives holds: (i) There exist $x_1, \dots, x_n \in \mathbb{R}$ such that $x_1 A_1 + \dots + x_n A_n \succeq 0$, $\sum_i x_i b_i \le 0$, and if $\sum_i x_i b_i = 0$ then $\sum_i |b_i| > 0$ and $\mathbf{y}^\mathsf{T} A_i y = 0$ for all $y \in \mathbb{R}^m$ with $(x_1 A_1 + \dots + x_n A_n)\mathbf{y} = 0$. (ii) There exists a matrix $Y \succeq 0$ such that $A_i \cdot Y = b_i$ for all i.

EXERCISE 13.8. Let G be a connected simple graph. (a) If G is eulerian with an even number of nodes, then it has a spanning subgraph G' such that every node i is incident with $\deg(i)/2$ edges of G'. (b) If G is eulerian with an odd number of nodes, then it has a spanning subgraph G' such that every node except one i is incident with $\deg(i)/2$ edges of G'. (c) If G is not eulerian, then it has a spanning subgraph G' such that every node i is incident with either $\lfloor \deg(i)/2 \rfloor$ or $\lceil \deg(i)/2 \rceil$ edges of G'.

EXERCISE 13.9. Show that the Maximum Cut Problem is a special case of the MAX-2SAT problem.

EXERCISE 13.10. Find infinitely many linegraphs for which equality does not hold in (13.27).

CHAPTER 14

Stresses

We can introduce this chapter as picking up the topic of Chapter 3, studying frameworks in equilibrium in which the edges are rubber bands and also springs and rigid bars; mathematically, all that matters are the forces that act along the edges, which we describe as equilibrium stresses on frameworks. The physical interpretation of the edges does not matter here: in this setting, stresses (as well as braced stresses, to be introduced below) are simply solutions of appropriate linear equations.

Physical experience helps to visualize several of the arguments. Depending on what carries the force along the edges, we can think of various physical models that can be described by the above mathematical definition.

— We can think of an edge ij with a positive stress as a rubber band with this strength. As we have done before, we think of an edge with negative stress as a peculiar "spring" that pushes its endpoints apart.

— We can think of all edges as rigid bars, to get a *bar-and-joint framework*. In a resting position, such a framework will not carry any stress, but if we replace, say, one edge by a rubber band, then it will either move or carry a stress. Stresses are important tools in understanding the rigidity of the framework.

— One can extend the previous model by taking the signs of stresses into account. An edge that can carry a positive stress, but not a negative stress, is called a *cable*; if it is the other way around, it is called a *strut*. A framework with its edges classified as "cables" and "struts" is called a *tensegrity framework*.

From an engineering point of view, this chapter studies static properties of (say) bar-and-joint frameworks; dynamic properties will be the subject matter of the next chapter. Our main goal is to point out the many interesting connections between graph theory and geometry this setup leads to, so we do not attempt to be "practical".

14.1. Stresses and stress matrices

Let us fix a vector-labeling $\mathbf{u}: V \to \mathbb{R}^d$ of the graph G. A function $S: E \to \mathbb{R}$ will called an *equilibrium stress* (or simply a *stress*) on $(G, \mathbf{u})$ if

$$\sum_{j \in N(i)} S_{ij}(\mathbf{u}_j - \mathbf{u}_i) = 0 \qquad \text{(for every } i \in V). \tag{14.1}$$

Since the stress is defined on edges, rather than oriented edges, we have tacitly assumed that $S_{ij} = S_{ji}$. We will extend the notation to nonadjacent pairs i, j by $S_{ij} = 0$ (the values S_{ii} play no role here; we will return to a useful way of defining them soon.)

In the language of physics, we want to keep node i in position $\mathbf{u}_i$ (think of the case $d = 3$), and we can do that through forces acting along the edges of G. Then the force acting along edge ij is parallel to $\mathbf{u}_j - \mathbf{u}_i$, and so wit can be written as $S_{ij}(\mathbf{u}_j - \mathbf{u}_i)$. Newton's Third Law implies that $S_{ij} = S_{ji}$. The stress on an edge can be zero, negative or positive. A positive stress pulls each endpoint in the direction of the other, a negative stress pushes them apart. Equation (14.1) means that every node is in equilibrium. The framework is *stressed*, if it carries a not-identically-zero equilibrium stress, and *stress-free* otherwise.

Let us emphasize the importance of the symmetry $S_{ij} = S_{ji}$. Without this, equations (14.1) would be independent for different nodes i, and a nonzero stress would simply mean that at least one node and its neighbors are affine dependent.

We can arrange a stress on the given framework into a matrix $S \in \mathbb{R}^{V\times V}$, where $S_{ij} = 0$ if $ij \in \overline{E}$, and the diagonal entries are defined by

$$S_{ii} = -\sum_{k\in N(i)} S_{ik} \tag{14.2}$$

(so that each row-sum is zero). This matrix is symmetric. Combining the vectors $\mathbf{u}_i$ into a $d\times V$ matrix U, the definition of a stress and the definition of the diagonal entries can be expressed as

$$US = 0 \qquad \text{and} \qquad \mathbb{1}^{\mathsf{T}} S = 0. \tag{14.3}$$

If the representation is affinely full-dimensional, then the vector $\mathbb{1}$ and the rows of U are $d+1$ linearly independent vectors in the nullspace of S, giving an upper bound on its corank:

$$\mathrm{cork}(S) \geq d+1. \tag{14.4}$$

In a rubber band representation, the strengths of the edges "almost" form a stress: (14.1) holds for all nodes that are not nailed. In this language, Exercise 3.5 says the following: *Every rubber band representation of a simple* 3*-connected planar graph with an infinite triangular country has a stress that is* 1 *on the internal edges and negative on the boundary edges.*

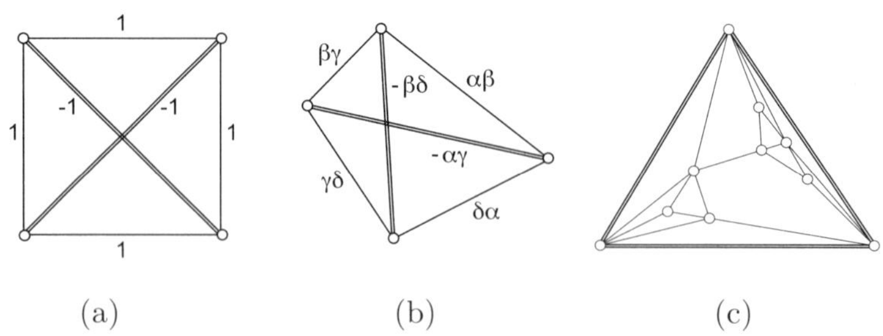

FIGURE 14.1. (a) A framework consisting of a square with diagonals and a stress on it. Simple edges represent positive stress values (pulling), double edges represent negative ones. (b) Stress on a "general" quadrilateral. The intersection point of the diagonals divides them in the ratio of $\beta : \delta$ and $\alpha : \gamma$, respectively. (c) Stress from a rubber band representation: 1 on the interior edges, negative on the boundary edges.

Every stress on a framework $(G, \mathbf{u})$ defines a G-matrix. Clearly stresses on a given graph G with a given vector-labeling $\mathbf{u}$ form a linear space, which we can consider as a space $\mathsf{Str}_{G,\mathbf{u}} \subseteq \mathbb{R}^E$.

For this chapter and the next, we assume that the graphs we consider have no isolated nodes; this excludes some trivial complications here and there.

14.1.1. Stability and the spectrum of the stress matrix. Consider a framework with a stress S on the edges. As before, we consider S as a symmetric matrix, whose diagonal entries are defined by (14.2), so that (14.3) holds. Let us interpret the edges as rubber bands and springs, where rubber bands correspond to positive stresses and springs correspond to negative stresses. The stress S_{ij} describes the strength of the edge ij. The definition of a stress means that the framework is in equilibrium. Is this equilibrium stable? In other words, can we move the nodes so as to decrease the energy of the framework?

For any vector labeling $\mathbf{u}$, and stresses S_{ij}, define the *energy* as

$$\mathcal{E}(u) = \sum_{ij \in E} S_{ij} |\mathbf{u}_j - \mathbf{u}_i|^2.$$

If the equilibrium condition (14.1) holds, then

$$\mathcal{E}(\mathbf{u}) = \sum_{ij \in E} S_{ij} |\mathbf{u}_j - \mathbf{u}_i|^2 = -\sum_i \sum_{j \in N(i)} S_{ij} \mathbf{u}_i^\mathsf{T} (\mathbf{u}_j - \mathbf{u}_i) = 0. \tag{14.5}$$

A similar computation gives more. Suppose that we move each node i to a new position $\mathbf{u}_i + \mathbf{v}_i$; then the energy of the new framework is

$$\mathcal{E}(\mathbf{u}+\mathbf{v}) = \sum_{ij \in E} S_{ij} |\mathbf{u}_j - \mathbf{u}_i|^2 + \sum_{ij \in E} S_{ij} |\mathbf{v}_j - \mathbf{v}_i|^2 + 2 \sum_{ij \in E} S_{ij} (\mathbf{v}_j - \mathbf{v}_i)^\mathsf{T} (\mathbf{u}_j - \mathbf{u}_i).$$

Using the equilibrium condition, we see that the last term vanishes:

$$\begin{aligned} &2 \sum_{ij \in E} S_{ij} (\mathbf{v}_j - \mathbf{v}_i)^\mathsf{T} (\mathbf{u}_j - \mathbf{u}_i) \\ &\quad = \sum_i \sum_{j \in N(i)} S_{ij} \mathbf{v}_i^\mathsf{T} (\mathbf{u}_j - \mathbf{u}_i) + \sum_j \sum_{i \in N(j)} S_{ij} \mathbf{v}_j^\mathsf{T} (\mathbf{u}_i - \mathbf{u}_j) = 0. \end{aligned}$$

Thus we get that if in position $\mathbf{u}$ the framework is in equilibrium, then

$$\mathcal{E}(\mathbf{u}+\mathbf{v}) = \mathcal{E}(\mathbf{v}). \tag{14.6}$$

It follows that the framework is in stable equilibrium if and only if $\mathcal{E}(\mathbf{v}) \geq 0$ for every assignment of vectors $\mathbf{v}_i$ to the nodes. It is easy to check that for every vector $x \in \mathbb{R}^V$,

$$x^\mathsf{T} S x = -\sum_{ij \in E} S_{ij} (x_i - x_j)^2.$$

Thus $\mathcal{E}(\mathbf{v}) \geq 0$ for every vector labeling $\mathbf{v}$ if and only if S is negative semidefinite. To sum up,

PROPOSITION 14.1. *A framework $(G, \mathbf{u})$ in $\mathbb{R}^d$ with stress matrix S is in stable equilibrium if and only if it is in equilibrium and S is negative semidefinite.* □

Notice that if S is a stress on a framework, then so is $-S$; but if $S \neq 0$ is stable, then $-S$ is not stable, since the negation of a negative semidefinite nonzero matrix is never negative semidefinite.

Let S be a negative semidefinite G-matrix with zero row sums (and column sums). Then the corresponding energy function satisfies

$$\mathcal{E}(\mathbf{u}) = \sum_{i,j} S_{ij}|\mathbf{u}_i - \mathbf{u}_j|^2 = \sum_{i,j} S_{ij}|\mathbf{u}_i|^2 + \sum_{i,j} S_{ij}|\mathbf{u}_i|^2 - 2\sum_{i,j} S_{ij}\mathbf{u}_i^\mathsf{T}\mathbf{u}_j.$$

The first two terms are zero, and hence $\mathcal{E}(\mathbf{u}) \geq 0$; furthermore, equality holds if and only if $\sum_j S_{ij}\mathbf{u}_j = 0$ for every node i. This is clearly equivalent to saying that S is a stress on $(G, \mathbf{u})$.

A further consequence of equation (14.6) characterizes equilibrium positions as stationary positions for the energy function (cf. Exercise 14.5).

14.1.2. Stresses on convex polygons. We study planar frameworks that consist of a convex polygon and some of its diagonals.

EXAMPLE 14.2. Let us consider the first nontrivial case, the complete graph K_4 represented in the plane by the vertices of a convex quadrilateral P. This framework must carry a stress; if we delete an edge, and realize the remaining edges by rigid bars, then we get a rigid framework, so if we put back the edge as a rubber band, the geometry does not change. This means that the force with which this last edge acts must be counterbalanced by forces acting along the other edges, which yields a stress S. (This stress is explicitly computed in Figure 14.1.)

The equilibrium condition (14.1) implies that if S is positive on one of the boundary edges, then it must be positive on all other boundary edges, and negative on the diagonals. This implies that the stress is uniquely determined up to scaling (from two essentially different equilibrium stresses we could create one that is zero on one edge but not all by taking linear combination). By (14.4), the rank of S is 1. It is not hard to see that the sum of stresses on the three edges incident with a given node is positive, so the diagonal entries are negative. This implies that the one nonzero eigenvalue of S is negative. Thus S is negative semidefinite. ♦

It is clear that if a framework consisting of a convex polygon and it carries a nowhere-zero stress, then every vertex must be incident with a diagonal (else, the forces acting along the two edges incident with this node could not be in equilibrium). The following lemma generalizes this observation for stresses with a specific signature (see Figure 14.2).

LEMMA 14.3. *Let $(G, \mathbf{u})$ be a framework in the plane consisting of a convex polygon and some of its diagonals, which carries a stress that is positive on the boundary edges and negative on the diagonals. Then G is 3-connected.*

PROOF. Suppose that G can be separated as $G = G_1 \cup G_2$, where $V(G_1) \cap V(G_2) = \{a, b\}$. One boundary edge incident with a belongs to G_1, the other belongs to G_2; indeed, if both belong to (say) G_1, then the whole boundary belongs to G_1 (walking around we cannot switch from G_1 to G_2 and back), and hence the only diagonal belonging to G_2 is ab, but this is not a proper 2-separation.

Let ac be a boundary edge belonging to G_1, and let F be the force with which G_1 acts on the point $\mathbf{u}_a$ (Figure 14.3(a)). Since a is in equilibrium, G_2 acts on $\mathbf{u}_a$ with a force of $-F$. The moment of F relative to the point $\mathbf{u}_b$ must be zero, since the other forces acting on G_2 act at $\mathbf{u}_b$. So F is parallel to $\mathbf{u}_b - \mathbf{u}_a$. We may assume that F points from $\mathbf{u}_a$ in the direction of $\mathbf{u}_b$ (else, we can interchange the roles of G_1 and G_2).

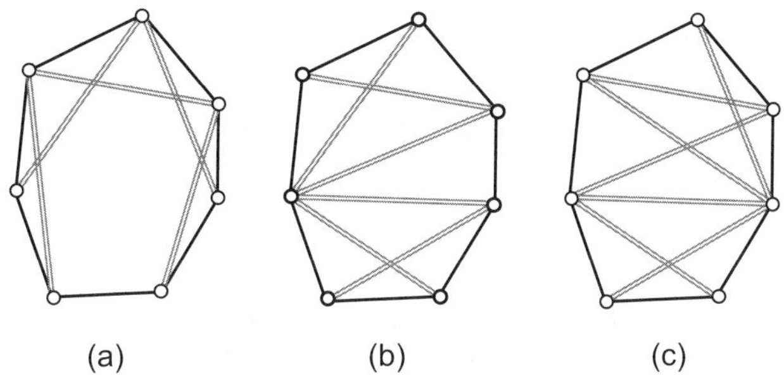

FIGURE 14.2. Structure (a) carries a nowhere-zero stress in the plane, positive on the edges of the polygon, but negative on the diagonals (see the proof of Theorem 14.4). Structures (b) and (c) carry no such stress. In the case of (b), the rightmost edge could pull its endpoints closer without changing the length of any other edge. For both of (b) and (c), Lemma 14.3 implies this, since they are not 3-connected.

Let e be a line through $\mathbf{u}_a$ separating $\mathbf{u}_c$ from the other vertices of P. Then all the forces $S_{ia}(\mathbf{u}_i - \mathbf{u}_a)$, where $i \in V(G_1)$, point to one side of e (due to the assumption about the signs of the stresses S_i). This means that their sum, F, cannot point in the direction of $\mathbf{u}_b$, a contradiction. $\square$

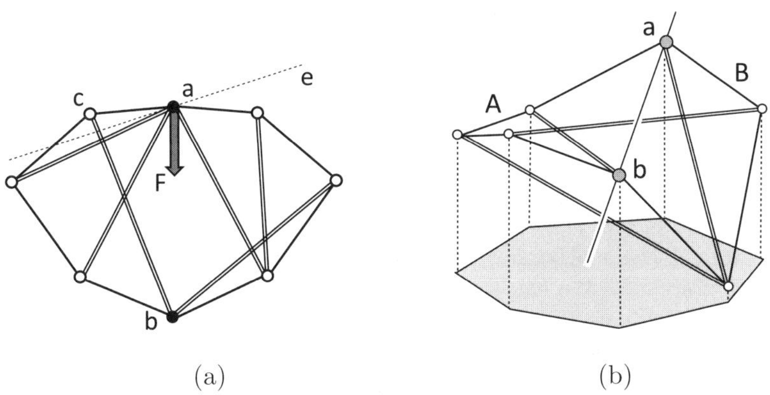

FIGURE 14.3. (a) A convex polygon with positive stress on the edges and negative stress on some diagonals cannot have a 2-separation. (b) Such a stress must have corank 3.

Now we are able to prove the following important theorem on stresses on convex polygons [**Connelly 1982**]:

THEOREM 14.4. *Let $(G, \mathbf{u})$ be a framework in $\mathbb{R}^2$ whose nodes are the vertices of a convex polygon P; its edges are the edges of P and some diagonals of P. Suppose that G has a stress S that is positive on the boundary edges and negative on the diagonals. Then S is negative semidefinite and has corank* 3.

PROOF. The proof consists of three main steps.

Claim 1. *The corank of S is 3.*

We have seen already that it is at least 3. Suppose that there is a vector x in the nullspace of S that is linearly independent from the coordinate functions and the vector $\mathbb{1}$. Consider the 3-dimensional vector labeling

$$i \mapsto \mathbf{v}_i = \begin{pmatrix} \mathbf{u}_i \\ x_i \end{pmatrix} \quad (i \in V).$$

Our choice of x implies that the vectors $\mathbf{v}_i$ are not contained in a plane (Figure 14.3(b)). Thus their convex hull is a 3-polytope Q. The stress S will be a stress on the 3-dimensional framework $(G, \mathbf{v})$.

Projecting Q to the plane of the first two coordinates, we get the polygon P. Every vertex of P is the projection of a unique vertex of Q. The polytope Q must have an edge e connecting vertices $\mathbf{v}_a$ and $\mathbf{v}_b$ that projects onto a proper diagonal of P (not necessarily onto an edge of G). Nodes a and b split the boundary of P into two arcs A and A'.

All outside forces acting on A come from (a) diagonals connecting the interior of A to the interior of A', or (b) from edges of G incident with i or j. Every force of type (a) tries to rotate A about the edge $(\mathbf{u}_a, \mathbf{u}_b)$ in the same direction; forces of type (b) have no moment relative to this axis. Since there is at least one diagonal of type (a) by Lemma 14.3, the framework cannot be in equilibrium. This contradiction proves Claim 1.

Claim 2. *There is at least one graph $G' = (V, E')$ and a stress S' on $(G', \mathbf{u})$ such that S' is positive on the edges of P, negative on the diagonal edges, and negative semidefinite.*

Let K_n be the complete graph on $[n]$, where i is labeled by the vector $\mathbf{u}_i$. For any three consecutive vertices $\mathbf{u}_i, \mathbf{u}_{i+1}, \mathbf{u}_{i+2}$ $(2 \le i \le n-2)$, consider a stress S_i on the complete 4-graph with vertices $\mathbf{u}_1, \mathbf{u}_i, \mathbf{u}_{i+1}, \mathbf{u}_{i+2}$, that is positive on the edges of the quadrilateral and negative on the diagonals (as in Example 14.2). We can consider S_i as a stress on $(K_n, \mathbf{u})$. The sum of these stresses S_i gives a stress that is positive on the edges of P, negative on diagonals connecting second neighbors along the arc from 2 to n, and zero on the other diagonals—except for those diagonals incident with node 1. We can scale these stresses before adding them up, and we are going to scale them so that their sum is zero on all diagonals incident with 1 except $(1, n-1)$. We start with S_2; then we scale S_3 so that adding S_3 kills the stress on $(1, 3)$. Then we add an appropriately scaled version of S_4 to kill the stress on $(1, 4)$ etc. This way we can kill all the stresses on diagonals from 1 except on $(1, n-1)$ (Figure 14.4).

But we have to prove that we scale with positive numbers! We prove this by induction on $i = 3, 4, \dots$. Consider the stress $S_i' = S_2 + S_3 + \dots S_{i-1}$. This is negative on the boundary edges of the convex polygon $(\mathbf{u}_1, \mathbf{u}_2, \dots, \mathbf{u}_{i+1})$, and zero on the diagonals incident with 1 except for $(1, i)$. Since $\mathbf{u}_1$ is in equilibrium, S_i' must be negative on $(1, i)$. So to kill the stress on $(1, i)$, we must add a positive multiple if S_i. This proves Claim 2.

Claim 3. *S is negative semidefinite.*

We know that there is another graph G' on $V(P)$ whose edges are the edges of P and some set of diagonals of P, that has a stress S' that is positive on the boundary edges and negative on the diagonals, and that is negative semidefinite.

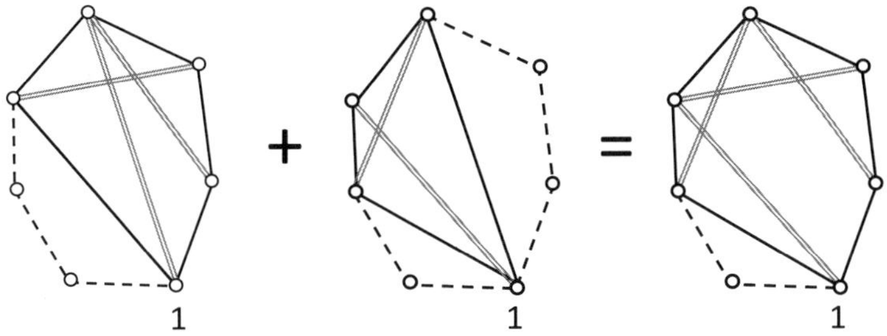

FIGURE 14.4. Making the stress zero on all but one diagonals from node 1.

The matrices $S_t = tS + (1-t)S'$ $(0 \le t \le 1)$ are stresses on the graph $G \cup G'$, of the type in the theorem. By Claim 1, they all have corank 3.

Since the eigenvalues are continuous functions of the matrix, the sets

$$A = \{t : \ S_t \text{ negative semidefinite}\}$$
$$B = \{t : \ S_t \text{ has} \ge 4 \text{ nonnegative eigenvalues}\}$$

are closed and their union is $[0,1]$. They are also disjoint, since a matrix in both of them would have corank 4. So one of them is empty; since $0 \in A$, it follows that $1 \in A$, and so $S = S_1$ is negative semidefinite. This completes the proof of the theorem. $\square$

14.1.3. Stresses on convex polytopes. Let us move up to the 3-dimensional space. An important result about stresses is a classic indeed: it was proved by Cauchy in 1822.

THEOREM 14.5. *The skeleton of a 3-polytope carries no nonzero stress.*

PROOF. Suppose that the edges of a convex polytope P carry a nonzero stress S. Let G' be the subgraph of its skeleton formed by those edges which have a nonzero stress, together with the vertices they are incident with. As in Figure 14.5, we draw an edge ij of G' solid, if $S_{ij} > 0$, and dashed if $S_{ij} < 0$. Edges of P with zero stress are dotted.

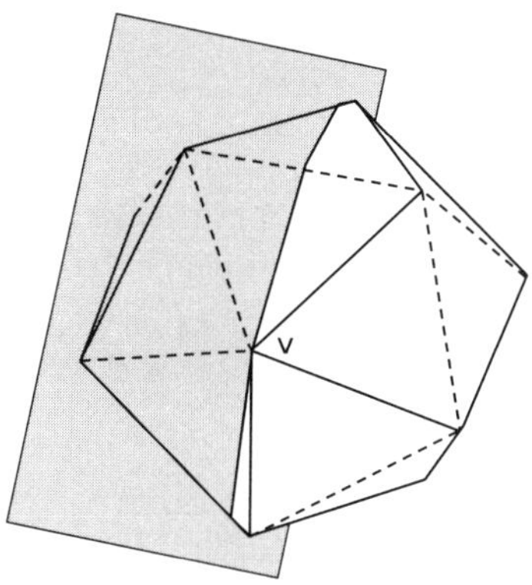

FIGURE 14.5. Why a polytope with an equilibrium stress cannot have a "quiet" node.

The graph G' is planar, and in fact it comes embedded in the surface of P, which is homeomorphic to the sphere. By Lemma 2.3, G' has a "quiet" node i, such that the solid edges (and then also the dashed edges) incident with i are consecutive in the given embedding of G'. This implies that we can find a plane Σ through $\mathbf{u}_i$ which separates (in their embedding on the surface of P) the solid and dashed edges incident with i. Let $\mathbf{e}$ be the normal unit vector of Σ pointing in the halfspace which contains the solid edges. Then for every solid edge ij we have $\mathbf{e}^\mathsf{T}(\mathbf{u}_j - \mathbf{u}_i) > 0$, and for every dashed edge ij we have $\mathbf{e}^\mathsf{T}(\mathbf{u}_j - \mathbf{u}_i) < 0$. This means that we have $S_{ij}\mathbf{e}^\mathsf{T}(\mathbf{u}_j - \mathbf{u}_i) > 0$ for every edge ij of G'. Also by definition, we have $S_{ij}\mathbf{e}^\mathsf{T}(\mathbf{u}_j - \mathbf{u}_i) = 0$ for the dotted edges. Since i is a node of G' and hence it is incident with at least one edge with nonzero stress, we have

$$\sum_{j\in N(i)} S_{ij}\mathbf{e}^\mathsf{T}(\mathbf{u}_j - \mathbf{u}_i) > 0.$$

But

$$\sum_{j\in N(i)} S_{ij}\mathbf{e}^\mathsf{T}(\mathbf{u}_j - \mathbf{u}_i) = \mathbf{e}^\mathsf{T}\Big(\sum_{j\in N(i)} S_{ij}(\mathbf{u}_j - \mathbf{u}_i)\Big) = 0$$

by the definition of a stress, a contradiction. □

The most interesting consequence of Cauchy's Theorem is that if we make a convex polyhedron out of cardboard, then it will be rigid (see Section 15.2).

14.2. Braced stresses

Let G be a simple graph, and let $\mathbf{u} : V \to \mathbb{R}^d$ be a vector-labeling of G. We say that a function $S : E \to \mathbb{R}$ is a *braced stress* on $(G, \mathbf{u})$, if there are appropriate values $S_{ii} \in \mathbb{R}$ such that for every node i,

$$S_{ii}\mathbf{u}_i + \sum_{j\in N(i)} S_{ij}\mathbf{u}_j = 0 \tag{14.7}$$

As before, we define $S_{ij} = 0$ for $ij \in \overline{E}$, to get a symmetric G-matrix S, so that we have the equation

$$US = 0 \tag{14.8}$$

(where U is the matrix whose columns give the vector-labeling $\mathbf{u}$). It follows that if the labeling vectors span the space $\mathbb{R}^d$ linearly, then the corank of the matrix S is at least d.

Clearly "proper" stresses are special braced stresses: if S is a stress on $(G, \mathbf{u})$, then

$$\sum_{j\in N(i)} S_{ij}\mathbf{u}_j = \Big(\sum_{j\in N(i)} S_{ij}\Big)\mathbf{u}_i.$$

The name "braced stress" can be explained by the following physical interpretation of (14.7): We add a new node 0 at position $\mathbf{u}_0 = 0$ to G, and connect it to every old node, to get a graph G' (Figure 14.6). We define a new (proper) stress

S' by

$$\begin{aligned} S'_{ij} &= S_{ij}, \qquad (i,j \in V,\ i \neq j),\\ S'_{i0} &= S'_{0i} = S_{ii}, \qquad (i \in V),\\ S'_{ii} &= -\sum_{k\in\{i\}\cup N(i)} S_{ik}, \qquad (i \in V \cup \{0\}). \end{aligned}$$

Then S' will be a stress on the edges of this representation of G': the equilibrium condition at the old nodes is satisfied by the definition of S, and this easily implies that it is also satisfied at the new node.

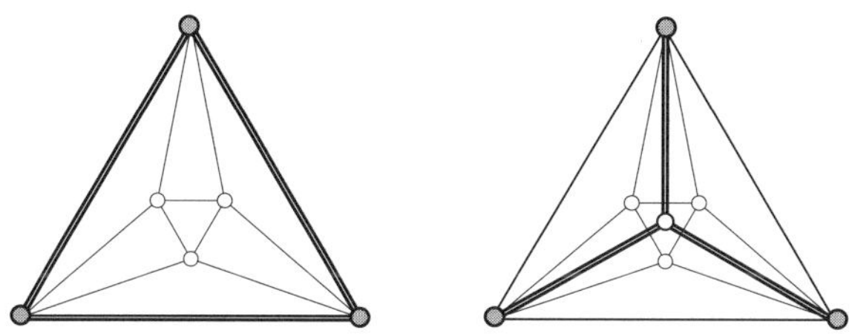

FIGURE 14.6. Rubber band representations of the octahedron graph, framed and braced.

In the case when all stresses (on the original edges) have the same sign, we can imagine this framework as follows: the old edges are strings or rubber bands of appropriate strength, the new edges are rods attached to the origin, and the whole framework is in equilibrium.

REMARK 14.6. We can study equation (14.8) from two different points of view: either we are given the vector-labeling U, and want to understand which G-matrices M satisfy it; this has lead us to "stresses", the main topic of this chapter. Or we can fix M and ask for properties of the vector-labelings that satisfy it; this will lead us to "nullspace representations", which will be important tools in Chapter 16. (It is perhaps needless to say that these two points of view cannot be strictly separated.)

REMARK 14.7. Sometimes we consider a version where instead of (14.1), we assume that

$$\sum_{j\in N(i)} S_{ij}\mathbf{u}_j = 0. \tag{14.9}$$

Such an assignment is called a *homogeneous stress*. Thus we have defined three kinds of stress matrices: "ordinary", braced, and homogeneous. Each of these versions can be described by a G-matrix satisfying the stress equation $US = 0$. Homogeneous stress matrices are those that, in addition, have 0's in their diagonal. "Ordinary" stress matrices have zero row and column sums. "Ordinary" stresses remain stresses when the geometric representation is translated, but homogeneous and braced stresses do not have this property.

14.3. Discrete Nodal Theorems

The model for the main results in this section is the Nodal Theorem of Courant. Informally, this theorem states that the support of an eigenfunction of Laplacian-like operators on a bounded domain, belonging to the k-th smallest eigenvalue, has at most k connected components. We'll see that this result about partial differential equations has very natural and useful analogues for graphs.

14.3.1. Eigenvectors of well-signed G-matrices. Let us fix a connected graph G, let M be a well-signed G-matrix of corank d. We study the spectrum of M, starting with an easy observation. The Perron–Frobenius Theorem easily implies the following fact:

LEMMA 14.8. *The smallest eigenvalue of M has multiplicity* 1, *and all entries of the corresponding eigenvector have the same sign.*

PROOF. For an appropriately large constant $C > 0$, the matrix $CI - M$ is nonnegative and irreducible. Hence it follows from the Perron-Frobenius Theorem that the largest eigenvalue of $CI - M$ has multiplicity 1, and all entries of the corresponding eigenvector have the same sign. This implies that the same holds for the smallest eigenvalue of M. □

Turning to the second smallest eigenvalue, we may assume that it is 0 (this can be achieved by adding an appropriate constant to the diagonal, which changes neither the order and multiplicities of the eigenvalues nor the eigenvectors).

A key lemma concerning well-signed G-matrices, which can be considered as a discrete version of Courant's Nodal Theorem, was proved in [**van der Holst 1995**]. In its simplest form, the Discrete Nodal Theorem says that *if the second smallest eigenvalue of a well-signed G-matrix has multiplicity* 1, *then the positive and negative supports of the corresponding eigenvector induce connected subgraphs.*

Subsequently, various new versions and generalizations of this lemma were obtained; see [**van der Holst et al. 1995b**], [**van der Holst et al. 1999**] and [**Lovász–Schrijver 1999**]; for higher eigenvectors [**Colin de Verdière 1998b**], [**Davies–Gladwell–Leytold 2001**] and [**Duval–Reiner 1999**]; for embedded graphs [**Lin et al. 2010**].

This lemma does not remain true, unfortunately, if the eigenvalue has multiplicity larger than 1 (and this will be an important situation for us):

EXAMPLE 14.9 (Claw). Consider the graph $G = K_{1,3}$ (the 3-star or "claw"), and the following well-signed G-matrix:

$$M = \begin{pmatrix} 2 & -1 & -1 & -1 \\ -1 & 0 & 0 & 0 \\ -1 & 0 & 0 & 0 \\ -1 & 0 & 0 & 0 \end{pmatrix}.$$

This matrix has eigenvalues $(-1, 0, 0, 3)$, and its nullspace consists of all vectors of the form $(0, x, y, z)^{\mathsf{T}}$, where $x+y+z = 0$. So for most vectors in the nullspace, either the positive or negative support consists of two nonadjacent nodes, and so it is disconnected. Note, however, that those vectors in the nullspace with minimal support look like $(0, 0, x, -x)$, for which both the positive and negative supports are connected in a trivial way. ♦

However, these exceptional eigenvectors have a very specific form, which we describe in detail for the purposes of later applications. Some simpler conditions that exclude these exceptions are stated as corollaries.

THEOREM 14.10 (Discrete Nodal Theorem). *Let G be a connected graph, let M be a well-signed G-matrix with one negative eigenvalue, and let $0 \neq x \in \text{Ker}(M)$. Let $N_x = \{i \in V : x_i = 0\}$. Then the sets $\text{supp}_+(x)$ and $\text{supp}_-(x)$ are nonempty, and either*

(i) *both graphs $G[\text{supp}_+(x)]$ and $G[\text{supp}_-(x)]$ are connected, or*

(ii) *$G[\text{supp}(x)]$ has connected components $H_1, \dots, H_k$ $(3 \leq k \leq \text{cork}(M)+1)$, each of which is contained in either $\text{supp}_+(x)$ or $\text{supp}_-(x)$. Furthermore, if z_i denotes the restriction of x onto H_i, extended by 0's so that it is a vector in $\mathbb{R}^V$, then $\text{supp}(Mz_i) \subseteq N_x$ and $Mz_i \parallel Mz_j$ for all $1 \leq i \leq j \leq k$.*

For the graph structure it follows that $N(H_i) \setminus V(H_i)$ is the same subset of N_x, independently of i.

PROOF. By the Perron–Frobenius Theorem, M has a positive eigenvector y belonging to the negative eigenvalue. From the orthogonality relation $y^\mathsf{T} x = 0$, it follows that the sets $\text{supp}_+(x)$ and $\text{supp}_-(x)$ are nonempty. Suppose that at least one of them, say $\text{supp}_+(x)$, induces a disconnected graph. Let $H_1, \dots, H_a$ and $H_{a+1}, \dots, H_k$ be the connected components of $\text{supp}_+(x)$ and $\text{supp}_-(x)$, respectively, then $a \geq 2$ and $k \geq 3$. Clearly $z_i \geq 0$ for $i \leq a$, $z_i \leq 0$ for $i > a$, $z_1, \dots, z_k$ are linearly independent, and $x = \sum_i z_i$. Let $Z = (z_1, \dots, z_k) \in \mathbb{R}^{V \times k}$.

The main step in the proof is to understand the symmetric $k \times k$ matrix $A = Z^\mathsf{T} M Z$. (It will turn out that $A = 0$.) It is easy to figure out the signs of the entries of A. Clearly $A_{ij} = z_i^\mathsf{T} M z_j = 0$ for $1 \leq i < j \leq a$ and $a+1 \leq i < j \leq k$, since there is no edge between $V(H_i)$ and $V(H_j)$ in this case. We have $A_{ij} = z_i^\mathsf{T} M z_j \geq 0$ for $1 \leq i \leq a < j \leq k$, since $M_{uv} \leq 0$ and $x_u x_v < 0$ for $u \in V(H_i), v \in V(H_j)$ in this case. Furthermore, $\sum_j z_i^\mathsf{T} M z_j = z_i^\mathsf{T} M x = 0$, and hence $A_{ii} = z_i^\mathsf{T} M z_i \leq 0$ for each $1 \leq i \leq k$.

To sum up, A has nonpositive diagonal, nonnegative entries off the diagonal, and zero row-sums. Thus the matrix $-A$ is diagonally dominant, and hence positive semidefinite, which implies that A is negative semidefinite. By the Inertia Theorem, the number of nonpositive eigenvalues of A is not larger than the number of nonpositive eigenvalues of M, and hence $k \leq d+1$.

Let $1 \leq i < j \leq a$, and consider the vector $w = (z_j^\mathsf{T} y) z_i - (z_i^\mathsf{T} y) z_j$. Clearly $w \neq 0$, but $w^\mathsf{T} y = 0$. Hence w is a linear combination of eigenvectors belonging to the nonnegative eigenvalues, and so $w^\mathsf{T} M w \geq 0$. On the other hand, using that A is negative semidefinite,

$$w^\mathsf{T} M w = A_{ii} (z_j^\mathsf{T} y)^2 - 2A_{ij} (z_i^\mathsf{T} y)(z_j^\mathsf{T} y) + A_{jj} (z_i^\mathsf{T} y)^2 \leq 0.$$

This implies that $w^\mathsf{T} M w = 0$, and so w is a linear combination of eigenvectors belonging to the zero eigenvalues; this means that $Mw = 0$. Hence $Mz_i \parallel Mz_j$, and so $\text{supp}(Mz_i) = \text{supp}(Mz_j)$.

Let U denote this common support. Since there is no edge connecting H_1 and H_2, the set $U = \text{supp}(Mz_1)$ is disjoint from H_2, and similarly $U \cap V(H_i) = \emptyset$ for $1 \leq i \leq a$. But this also implies that $U \cap V(H_j) = \emptyset$ for $a+1 \leq j \leq k$: else, there would be an edge connecting H_j $(j > a)$ and H_1, showing that $U = \text{supp}(Mz_j)$ contains a node of H_1. □

We note that if, in addition, M has the Strong Arnold Property, then $k = 3$ in (ii) of Theorem 14.10. Indeed, if $k \geq 4$, then $u = (z_2^{\mathsf{T}}y)z_1 - (z_1^{\mathsf{T}}y)z_2$ and $w = (z_3^{\mathsf{T}}y)z_4 - (z_4^{\mathsf{T}}y)z_3$ are vectors in the nullspace of M with disjoint supports, and no edge connects their supports. This contradicts Lemma 10.26.

The following corollary states some conditions under which the awkward possibility (ii) can be excluded, facilitating an easier application of the lemma.

COROLLARY 14.11. *Let G be a connected graph, let M be a well-signed G-matrix with one negative eigenvalue, and let $x \in \mathrm{Ker}(M)$, $x \neq 0$.. Assume that one of the following conditions hold:*

(a) *x has minimal support among all nonzero vectors in* $\mathrm{Ker}(M)$*;*

(b) *x has full support;*

(c) *G is a 3-connected planar graph.*

Then $G[\mathrm{supp}_+(x)]$ and $G[\mathrm{supp}_-(x)]$ are nonempty connected subgraphs.

PROOF. It is straightforward to see that case (ii) in Theorem 14.10 cannot occur under conditions (a) or (b). Suppose that (c) holds and alternative (ii) occurs. Since $V_0 = N\big(V(H_i)\big) \setminus V(H_i)$ is a cutset, we have $|V_0| \geq 3$. Contracting each H_i to a single node, we get a minor containing $K_{3,3}$, which is impossible as G is planar. $\square$

Some further versions of this lemma are stated in Exercises 14.10–14.11.

We can re-state the nodal lemma in a geometric form (see Figure 14.7). Let M be a well-signed G-matrix with one negative eigenvalue, let $\mathbf{u}: V \to \mathbb{R}^d$ be a nullspace representation belonging to M, and U, the corresponding matrix. The nullspace of M consists of all vectors of the form $U^{\mathsf{T}}\mathbf{s}$, where $\mathbf{s} \in \mathbb{R}^d$.

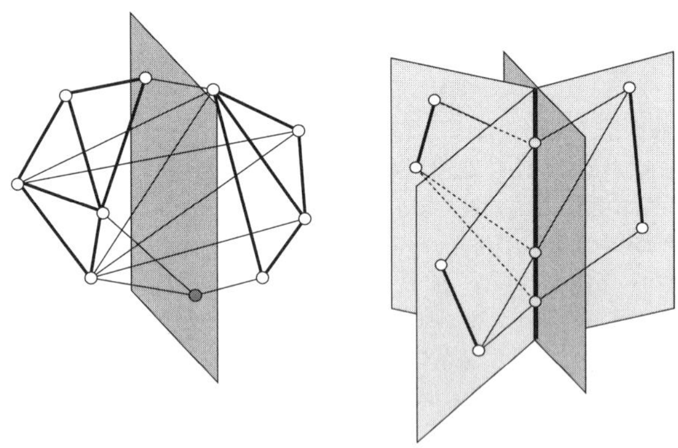

FIGURE 14.7. Two possibilities how the nullspace representation of a connected graph can be partitioned by a hyperplane through the origin.

Let H be a linear hyperplane in $\mathbb{R}^d$. We denote by V_0, V^+ and V^- the set of nodes represented by vectors on H, on one side of H, and on the other side of H, respectively. In terms of the vector $x = U\mathbf{s}$ in the nullspace of M, where $\mathbf{s}$ is a normal vector of H, we have $V^+ = \mathrm{supp}_+(x)$ etc. So it follows that V^+ and V^- are nonempty, and "usually" induce connected subgraphs.

In the exceptional case when one of $G[V^+]$ and $G[V^-]$ is disconnected, no edge connects V^+ and V^-. As further geometric information, there is a subspace $S \subseteq H$ of dimension $d-2$, and there are open half-hyperplanes $T_1, \dots, T_k$ ($k \geq 3$) with boundary S, so that

• all nodes in V_0, and only those, are represented in S, and these vectors generate S;

• the nodes represented in T_i induce a connected subgraph H_i;

• all neighbors of H_i outside H_i belong to V_0;

• every node in V_0 is connected either to every H_i, or to none of them.

Let us rotate the hyperplane H about the subspace S. In each position, both halfspaces bounded by H will contain at least one of the half-hyperplanes T_i, and so they can never contain all of them. In each position where one of these halfspaces contains at least two of the half-hyperplanes T_i we get an exceptional case. This gives the following geometric descriptions of the two cases covered by Corollary 14.11: the subspace H is linearly spanned by some of the vectors $\mathbf{u}_i$; H does not contain any of the vectors $\mathbf{u}_i$.

As a further application of the Discrete Nodal Theorem, we prove a fact that will be important in several chapters of this book.

THEOREM 14.12. *Let G be a 3-connected planar graph and let M be a well-signed G-matrix with exactly one negative eigenvalue. Then* $\mathrm{cork}(M) \leq 3$.

PROOF. Suppose not. Let a_1, a_2 and a_3 be three nodes of the same country (say, the unbounded country), then there is a nonzero vector $x = (x_i : i \in V)$ such that $Mx = 0$ and $x_{a_1} = x_{a_2} = x_{a_3} = 0$. We may assume that x has minimal support.

By 3-connectivity, there are three paths P_i connecting the support S of $\mathbf{u}$ to a_i ($i = 1, 2, 3$), such that they have no node in common except possibly their endpoints b_i in S. Let c_i be a node on P_i next to b_i, then clearly $x_{b_i} \neq 0$ but $x_{c_i} = 0$. From the equation $Mx = 0$ it follows that c_i must have a neighbor d_i such that $x_{d_i} \neq 0$, and x_{b_i} and x_{d_i} have opposite signs. We may assume without loss of generality that $x_{b_i} > 0$ but $x_{d_i} < 0$ for $i = 1, 2, 3$ (Figure 14.8).

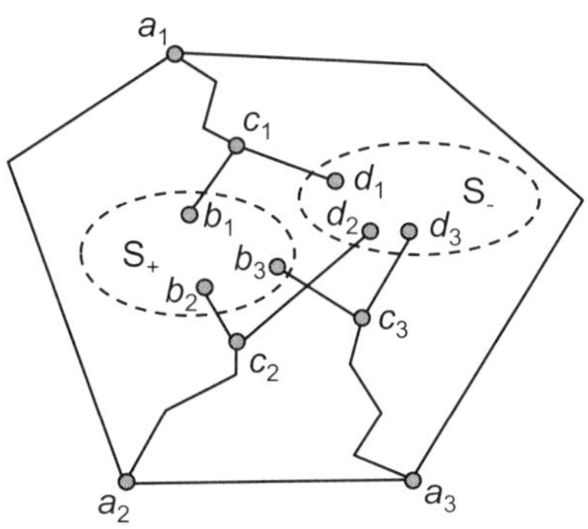

FIGURE 14.8. Paths connecting three nodes of the unbounded country to the negative and positive supports

Create a new node in the interior of the country F and connect it to a_1, a_2 and a_3. Using that the positive support S_+ as well as the negative support S_- of

x induce connected subgraphs by Corollary 14.11, we can contract them to single nodes. We get a subdivision of $K_{3,3}$ embedded in the plane, which is impossible. □

14.4. Braced stresses on polytopes

14.4.1. Existence. In contrast to Cauchy's Theorem 14.5, 3-dimensional polytopes do have *braced* stresses.

THEOREM 14.13. *The skeleton of every convex* 3*-polytope containing the origin in its interior has a braced stress that is positive on every edge of the polytope.*

Of course, we could require the braced stress to be negative on all edges.

PROOF. Let P be a convex polytope in $\mathbb{R}^3$ such that the origin is in the interior of P, and let G be the skeleton of P. Let P^* be its polar, and let $G^* = (V^*, E^*)$ denote the skeleton of P^* (see Figure 14.9).

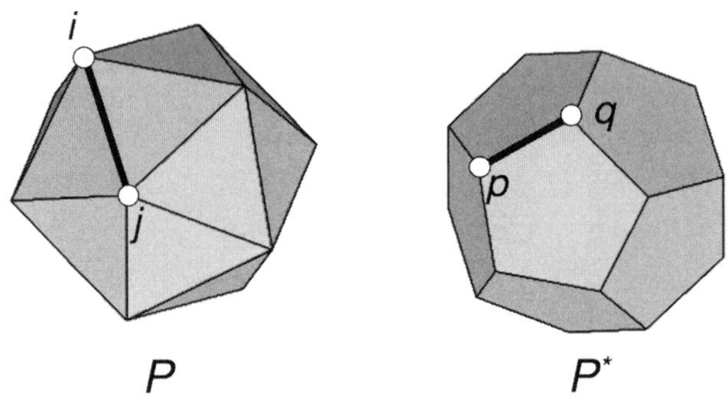

FIGURE 14.9. An edge of a polytope and the corresponding edge of its polar.

Let $\mathbf{u}_i$ and $\mathbf{u}_j$ be two adjacent vertices of P, and let $\mathbf{w}_p$ and $\mathbf{w}_q$ be the endpoints of the corresponding edge $pq \in E^*$, where the p is on the right of the edge ij, oriented from i to j, and viewed from outside. Then by the definition of polarity, we have

$$\mathbf{u}_i^\mathsf{T}\mathbf{w}_p = \mathbf{u}_i^\mathsf{T}\mathbf{w}_q = \mathbf{u}_j^\mathsf{T}\mathbf{w}_p = \mathbf{u}_j^\mathsf{T}\mathbf{w}_q = 1.$$

This implies that the vector $\mathbf{w}_p - \mathbf{w}_q$ is orthogonal to both vectors $\mathbf{u}_i$ and $\mathbf{u}_j$. Here $\mathbf{u}_i$ and $\mathbf{u}_j$ are not parallel, since they are the endpoints of an edge and the origin is not on this line. Hence $\mathbf{w}_q - \mathbf{w}_p$ is parallel to the nonzero vector $\mathbf{u}_i \times \mathbf{u}_j$, and we can write

$$\mathbf{w}_q - \mathbf{w}_p = S_{ij}(\mathbf{u}_i \times \mathbf{u}_j) \tag{14.10}$$

with some real numbers S_{ij}. Clearly $S_{ij} = S_{ji}$ for every edge ij, and our choice for the orientation of the edges implies that $S_{ij} > 0$.

We claim that the values S_{ij} form a braced stress on $(G, \mathbf{u})$. Let $i \in V$, and consider the vector

$$\mathbf{u}_i' = \sum_{j \in N(i)} S_{ij}\mathbf{u}_j.$$

Then

$$\mathbf{u}_i \times \mathbf{u}_i' = \sum_{j \in N(i)} S_{ij}(\mathbf{u}_i \times \mathbf{u}_j) = \sum (\mathbf{w}_q - \mathbf{w}_p),$$

where the last sum extends over all edges pq of the facet of P^* corresponding to i, oriented counterclockwise. Hence this sum vanishes, and we get that $\mathbf{u}_i \times \mathbf{u}'_i = 0$. This means that $\mathbf{u}_i$ and $\mathbf{u}'_i$ are parallel, and there are real numbers S_{ii} such that $\mathbf{u}'_i = -S_{ii}\mathbf{u}_i$. This proves that the values S_{ij} form a braced stress. □

We call the special braced stress constructed above the *canonical braced stress* of the polytope P, and denote the corresponding matrix by $S = S_P$. We express the canonical stress as

$$S_{ij} = \frac{|\mathbf{w}_q - \mathbf{w}_p|}{|\mathbf{u}_i \times \mathbf{u}_j|} = \frac{|\mathbf{w}_q - \mathbf{w}_p|}{|\mathbf{u}_i|\,|\mathbf{u}_j| \sin \measuredangle(\mathbf{u}_i, \mathbf{u}_j)}. \tag{14.11}$$

Hence we have a simple geometric interpretation: S_{ij} is the length of the dual edge, divided by twice the area of the triangle formed by the edge and the origin.

Taking the cross product of (14.10) with $\mathbf{w}_p$ we get the equation

(14.12)

$$\mathbf{w}_q \times \mathbf{w}_p = S_{ij}\big((\mathbf{u}_i \times \mathbf{u}_j) \times \mathbf{w}_p\big) = S_{ij}\big((\mathbf{u}_i^\mathsf{T}\mathbf{w}_p)\mathbf{u}_j - (\mathbf{u}_j^\mathsf{T}\mathbf{w}_p)\mathbf{u}_i\big) = S_{ij}(\mathbf{u}_j - \mathbf{u}_i).$$

This shows that the canonical braced stresses on the edges of the polar polytope are simply the reciprocals of the canonical braced stresses on the original.

The matrix S_P depends on the choice of the origin. We can even choose the origin outside P, as long as it is not on the plane of any facet. The only step in the argument above that does not remain valid is that the stresses on the edges are positive.

The existence, and these elementary properties, of the canonical braced stress are nice facts in themselves, but we will need to spend some pages on the study of its highly nontrivial properties, after presenting some general results about spectra and eigenvectors we need.

14.4.2. Spectra. Returning to the canonical braced stresses, our main goal is to determine the signs of their eigenvalues, and more generally, the signs of eigenvalues of braced stresses on convex polytopes. To this end, we relate the canonical braced stress matrix to volumes and mixed volumes.

It is not hard to prove (see Exercise 14.7) that the canonical stress matrix and volume are related:

$$\mathrm{vol}(P^*) = \frac{1}{6}\sum_{i,j} S_{ij}. \tag{14.13}$$

This simple identity is not the last word on this issue. Let P be a convex 3-polytope containing the origin in its interior, and let $\mathbf{u}_i$ ($i \in V = V(P)$) be its vertices. For $y \in \mathbb{R}^V$, $y > 0$, define

$$Q_y = \{\mathbf{x} \in \mathbb{R}^3 : \ \mathbf{u}_i^\mathsf{T}\mathbf{x} \le y_i \ \forall i \in V\}.$$

Clearly Q_y is a convex 3-polytope, and 0 is in its interior. In particular, $Q_{\mathbb{1}} = P^*$. We compute the gradient and Hessian of the function $f(y) = \mathrm{vol}(Q_y)$. (Note that f is not everywhere an analytic function of y; it is not analytic at the values of y where Q_y has vertices of degree larger than 3. But the proof below will show that it is twice continuously differentiable.)

Let F_y^i denote the facet of Q_y corresponding to $i \in V$. It is easy to see that

$$\frac{\partial}{\partial y_i} f(y) = \frac{\mathrm{area}(F_y^i)}{|\mathbf{u}_i|}. \tag{14.14}$$

Hence the outer normal of F^i_y of length area(F^i_y) is

$$\text{area}(F^i_y)\mathbf{u}^0_i = \frac{\partial}{\partial y_i} f(y)\mathbf{u}_i.$$

The sum of these outer normals is zero; hence we get the equation

$$\sum_i \frac{\partial}{\partial y_i} f(y)\mathbf{u}_i = 0. \tag{14.15}$$

The next important step is the following equation expressing the canonical stress matrix S as the Hessian of the volume of Q_y:

$$\frac{d^2}{dy^2} f(\mathbb{1}) = S. \tag{14.16}$$

To prove identity (14.16), let H denote the matrix on the left. Continuing from (14.14), we want to find

$$H_{ij} = \frac{\partial^2}{\partial y_i\, \partial y_j} f(\mathbb{1}) = \lim_{t\to 0} \frac{\text{area}(F^i_{\mathbb{1}+t\mathbf{e}_j}) - \text{area}(F^i_{\mathbb{1}})}{t|\mathbf{u}_i|} \tag{14.17}$$

for $j \neq i$. We consider the case when $t \searrow 0$ (the case of negative values of t is similar). If j is not adjacent to i, then $F^i_{\mathbb{1}+t\mathbf{e}_j} = F^i_{\mathbb{1}}$, and so $H_{ij} = 0$. Assume that ij is an edge of P, and let pq be the corresponding edge in P^*. The difference in the numerator in (14.17) is the area of a strip along the edge of $F^i_{\mathbb{1}}$ connecting $\mathbf{w}_p$ and $\mathbf{w}_q$. Elementary computation shows that the width of this strip is $t/\big(|\mathbf{u}_j| \sin\measuredangle(\mathbf{u}_i, \mathbf{u}_j)\big)$. It follows that

$$\text{area}(F^i_{\mathbb{1}+t\mathbf{e}_j}) - \text{area}(F^i_{\mathbb{1}}) = \frac{t\,|\mathbf{w}_p - \mathbf{w}_q|}{|\mathbf{u}_j| \sin\measuredangle(\mathbf{u}_i, \mathbf{u}_j)} + O(t^2).$$

(The error term $O(t^2)$ comes from the ends of the strip.) Substituting in (14.17) and using (14.11), we get

$$H_{ij} = \frac{|\mathbf{w}_p - \mathbf{w}_q|}{|\mathbf{u}_i|\,|\mathbf{u}_j| \sin\measuredangle(\mathbf{u}_i, \mathbf{u}_j)} = S_{ij}.$$

So the off-diagonal entries of H and S are the same, and hence $H - S$ is a diagonal matrix. Differentiating with respect to y_j, we get the equation $\sum_j H_{ij}\mathbf{u}_j = 0$, or more compactly $UH = 0$. We know that $US = 0$, and hence $U(H - S) = 0$. Since no column of U is 0, it follows that $H - S = 0$. This completes the proof of (14.16).

Our next goal is to relate the canonical stress matrix to the mixed volumes of these polytopes Q_y (for the definition and properties of mixed volumes, we refer to Appendix C.5 and for a presentation of greater depth, to [**Schneider 1993**], in particular Sections 5.1 and 6.3). We call a vector $y \in \mathbb{R}^V$ *proper*, if for every $\mathbf{w} \in \mathbb{R}^3$ the faces of $Q_{\mathbb{1}}$ and Q_y maximizing the objective function $\mathbf{w}^\mathsf{T}\mathbf{x}$ have the same dimension. (This can be phrased as "$Q_{\mathbb{1}}$ and Q_y are strongly isomorphic.") If $Q_{\mathbb{1}}$ is simple, then every vector y that is close enough to $\mathbb{1} \in \mathbb{R}^V$ is proper. For two proper vectors y and z, the vector $\mathbb{1} + sy + tz$ is proper, and

$$Q_{\mathbb{1}+sy+tz} = Q_{\mathbb{1}} + sQ_y + tQ_z.$$

(This takes some work to check; see Exercise 14.14.) From this, we can determine the mixed volume $V(Q_{\mathbb{1}}, Q_y, Q_z)$. Indeed, by the definition of the mixed volume, the volume of $Q_{\mathbb{1}+sy+tz}$ can be expressed as a polynomial in s and t, in which the coefficient of st is $6V(Q_{\mathbb{1}}, Q_y, Q_z)$. On the other hand, (14.16) implies that this coefficient is

$$\frac{\partial^2}{\partial s\, \partial t} \operatorname{vol}(Q_{\mathbb{1}+sy+tz}) = y^{\mathsf{T}}\Big(\frac{d^2}{dx^2} \operatorname{vol}(Q_x)\Big|_{x=\mathbb{1}}\Big)z = y^{\mathsf{T}}Sz.$$

Hence

$$V(Q_{\mathbb{1}}, Q_y, Q_z) = \frac{1}{6}y^{\mathsf{T}}Sz. \tag{14.18}$$

After this preparation, we are able to prove one of the main results in this section, describing the signature of the matrix S_P of the canonical braced stress on a 3-polytope P.

PROPOSITION 14.14. *If P is a 3-polytope containing the origin in its interior, then its canonical stress matrix $S = S_P$ has exactly three zero eigenvalues and exactly one positive eigenvalue.*

PROOF. From the stress equation $US = 0$ (where U is the matrix whose columns are the vertices of P), it follows that S has at least three zero eigenvalues. By Lemma 14.8, the largest eigenvalue of S is positive with multiplicity 1. Thus it cannot be the zero eigenvalue, which means that S has at least one positive eigenvalue.

The difficult part of the proof is to show that S has only one positive eigenvalue. We may assume that P is simplicial (every facet is a triangle), since every nonsimplicial polytope can be approximated by simplicial ones, and S_P depends continuously on P. (It takes some work to make this argument precise; see Exercise 14.13). Then $P^* = Q_{\mathbb{1}}$ is simple.

We invoke the Alexandrov–Fenchel Inequality (C.7) for mixed volumes:

$$V(Q_{\mathbb{1}}, Q_{\mathbb{1}}, Q_y)^2 \geq V(Q_{\mathbb{1}}, Q_{\mathbb{1}}, Q_{\mathbb{1}})V(Q_{\mathbb{1}}, Q_y, Q_y).$$

By (14.18), this gives that

$$(y^{\mathsf{T}}S\mathbb{1})^2 \geq (y^{\mathsf{T}}Sy)(\mathbb{1}^{\mathsf{T}}S\mathbb{1})$$

for every proper vector y. This holds in particular for every vector sufficiently close to $\mathbb{1}$. Choose $y = \mathbb{1}+\varepsilon z$, where $z \in \mathbb{R}^V$ is orthogonal to $S\mathbb{1}$ and $\varepsilon > 0$ is small enough, then $y^{\mathsf{T}}S\mathbb{1} = \mathbb{1}^{\mathsf{T}}S\mathbb{1}$ and $y^{\mathsf{T}}Sy = \mathbb{1}^{\mathsf{T}}S\mathbb{1}+\varepsilon^2 z^{\mathsf{T}}Sz$. Substituting and simplifying, using that $\mathbb{1}^{\mathsf{T}}S\mathbb{1} < 0$ by (14.13), we get that $z^{\mathsf{T}}Sz \leq 0$. Since this holds for all vectors in a subspace of dimension $n-1$, it follows that S has at most one positive eigenvalue.

Once we know this, Theorem 14.12 implies that S has exactly three zero eigenvalues. □

The previous proposition is a "self-refining" result: by a relatively simple argument (basically, repeating the last step in the proof of Theorem 14.4) it extends to every braced stress.

THEOREM 14.15. *If P is a convex polytope in $\mathbb{R}^3$ containing the origin in its interior, then every braced stress matrix with positive stresses on its skeleton has exactly one positive eigenvalue and exactly three zero eigenvalues.*

PROOF. It follows just as in the proof of Proposition 14.14 that every braced stress matrix Q with positive values on the edges has at least one positive and at least three zero eigenvalues. Furthermore, if it has exactly one positive eigenvalue, then it has exactly three zero eigenvalues by Theorem 14.12.

Suppose that Q has more than one positive eigenvalues. The braced stress matrix S_P has exactly one. All the matrices $Q_t = tS_P + (1-t)Q$ are braced stress matrices for the same vector labeling and with positive values on the edges, and hence they have at least four nonnegative eigenvalues. Since the eigenvalues are continuous functions of the matrix, there is a smallest number $t > 0$ for which Q_t has at least 5 nonnegative eigenvalues. Then M_t must have exactly one positive and at least four zero eigenvalues, contradicting Theorem 14.12. □

REMARK 14.16. Most of our discussion in the last section extends to higher dimensional polytopes. The canonical stress matrix of the skeleton of a polytope can be defined in every dimension. For us, however, only the 3-dimensional case will be relevant.

EXERCISE 14.1. Prove that a stress matrix of a framework with at least two different vector labels is never well-signed.

EXERCISE 14.2. Suppose that a framework has a stress. (a) Prove that applying an affine transformation to the positions of the nodes, the framework obtained has a stress. (b) Prove that applying a projective transformation to the positions of the nodes, so that no node is mapped onto an ideal point, the framework obtained has a stress.

EXERCISE 14.3. Let P be a convex polyhedron in $\mathbb{R}^3$, and let G be a planar graph embedded in the surface of P, with straight edges. Prove that every stress on G is a sum of stresses, each of which is supported on edges of G on a single face of P [**Whiteley 1984**].

EXERCISE 14.4. Let $S = \{\mathbf{u}_1, \dots, \mathbf{u}_n\} \subseteq \mathbb{R}^d$. We say that S is in *isotropic position*, if $\sum_i \mathbf{u}_i = 0$ and $\sum_i \mathbf{u}_i\mathbf{u}_i^\mathsf{T} = I$.

(a) Prove that every finite point set not in an affine hyperplane can be transformed to isotropic position by an appropriate affine transformation.

(b) Let the nodes of K_n be labeled by a set $\{\mathbf{u}_1, \dots, \mathbf{u}_n\} \subseteq \mathbb{R}^d$ in isotropic position. Define $S_{ij} = \mathbf{u}_i^\mathsf{T}\mathbf{u}_j + 1/n$ $(i \neq j)$. Prove that (with appropriate diagonal entries) S defines a stress matrix on $(K_n, \mathbf{u})$, and it is negative semidefinite.

EXERCISE 14.5. Let $(G, \mathbf{u})$ be a rubber-band-and-spring framework, and let S be a G-matrix. Prove that S is a stress on $(G, \mathbf{u})$ if and only if changing each position $\mathbf{u}_i$ by at most ε, the energy changes by $O(\varepsilon^2)$. Use this assertion to give an alternate argument for the conclusion in the proof of Theorem 14.4.

EXERCISE 14.6. Let P be a simple 3-polytope (every vertex has degree 3) containing the origin in its interior. Prove that the braced stress on its edges is unique up to scaling.

EXERCISE 14.7. Let P be a polytope and let S be the canonical stress on P. (a) Prove that the volume of the convex hull of the face i of P^* and the origin is $\frac{1}{6}\sum_j S_{ij}$. (b) Prove identity (14.13).

EXERCISE 14.8. Let P be a convex polygon in the plane, containing the origin in its interior. (a) Prove that its edges carry a braced stress that is negative on the edges. (b) Find a formula for this stress. (c) Prove that the stress matrix has exactly one negative eigenvalue.

EXERCISE 14.9. We have seen that the square with diagonals with the stress shown in Figure 14.1(a) is stable with respect to the positions of the vertices. It is however, very unstable with respect to the stress: if we consider the edges as rubber bands and springs, and change any of their strengths arbitrarily little, it will either collapse or fly off to infinity. If the diagonals are bars (or struts), then changing the strength of any of the rubber bands (representing the edges of the square) by a little amount, changes the equilibrium position only a little.

EXERCISE 14.10. Let G be a connected graph, let M be a well-signed G-matrix with exactly one negative eigenvalue, and let $x \in \mathbb{R}^V$ be a vector satisfying $Mx < 0$. Then $\text{supp}^+(x)$ is nonempty and induces a connected subgraph of G.

EXERCISE 14.11. Let G be a connected graph, let M be a well-signed G-matrix with exactly one negative eigenvalue, and let $x \in \mathbb{R}^V$ be a vector satisfying $Mx \leq 0$. Then $\text{supp}^+(x)$ is nonempty, and either it induces a connected subgraph of G or there is a subset $S \subset V$, such that $G \setminus S$ has at least two connected components $G_1, \dots, G_h$, and there exist nonzero, nonnegative vectors $z_1, \dots, z_r, y \in \mathbb{R}^V$ such that $\text{supp}(x_i) = S_i$, $\text{supp}(y) \subseteq S$, $Mx_1 = Mx_2 = \cdots = Mx_r = -y$, and $x = \sum_i \alpha_i z_i$, where $\sum_i \alpha_i \geq 0$ and at least two α_i are positive.

EXERCISE 14.12. Let P be a convex polygon in the plane, such that the origin is contained in its exterior, and no edge is collinear with the origin. (a) Prove that its edges carry a braced stress. (b) On which edges is it negative? (c) Prove that (switching the sign of all stress values if necessary) the number of negative eigenvalues is equal to the number of edges of P visible from the origin.

EXERCISE 14.13. Let P^t $(t = 0, 1, 2, \dots)$ be convex 3-polytopes with vertices $\mathbf{v}_1^t, \dots, \mathbf{v}_n^t$, containing the origin in their interiors. Suppose that $\mathbf{v}_i^t \to \mathbf{v}_i^0$ as $t \to \infty$ for every $1 \leq i \leq n$. Prove that $S_{P^t} \to S_{P^0}$ entry-by-entry.

EXERCISE 14.14. Let P be a 3-polytope containing the origin in its interior, and let y and z be proper vectors in the sense of Section 14.4.2. Prove that $y+z$ is proper, and $Q_{y+z} = Q_y + Q_z$.

EXERCISE 14.15. If a braced stress on a 3-polytope is nonnegative on every edge (but may be zero), then the corresponding stress matrix has at most one positive eigenvalue.

CHAPTER 15

Rigidity and Motions of Frameworks

After the previous, somewhat special and technical chapter, we turn to a practically important as well as graphically inspiring topic, namely rigidity of frameworks. We replace the edges of a graph by rigid bars (instead of rubber bands and springs as before). This way we obtain a physical model related to rubber bands, but more important from the point of view of applications. This topic has a vast literature, dealing with problems in architecture, engineering, robotics, and (replacing the reference to rigid bars by distance measurements) in the theory of sensor networks. We have to restrict ourselves to a few basic results of this rich theory; see e.g. [**Recski 1989**] or [**Graver–Servatius–Servatius 1993**] for more.

In the introduction to the previous chapter we have seen a whole arsenal of physical gadgets that we can imagine in place of the edges: rubber bands, cables, bars, struts and springs. The difference between these with respect to static properties was little (the sign of the stress). We could study dynamic properties of structures composed of any combinations of these, but we will restrict our attention to bar-and-joint frameworks.

We will look at these frameworks with two (slightly) different goals in mind. From one aspect, we specify the locations of the nodes in space (which determines the lengths of the edges geometrically). We then ask whether the resulting framework is rigid. This is the main use of these ideas in architecture: we specify the geometry of the framework and the rigidifying beams, and then want to make sure that the framework does not collapse; we want to compute the forces acting along the beams etc. From the other aspect, we are given the nodes and edges, along with the lengths of the edges, and then we are interested in all possible ways this framework can move. This aspect was very important in the 19th century, in the design of mechanisms (see Example 15.1); these days, it is extensively studied for further applications, among others, in the design of robot arms.

Example 15.1 (Peaucellier–Lipkin linkage). Perhaps the first time when frameworks and their motions received a lot of interest came after the discovery of the steam engine and other mechanical devices in the 18th and 19th century. One important goal was to guide a point back and forth on a straight line segment. Figure 15.1 shows one of the first solutions of this problem, the Peaucellier–Lipkin linkage. Note the sophistication needed for the solution of this simply sounding problem. ♦

A graph G with a vector-labeling in $\mathbb{R}^d$ will be called a *framework*, where the label is considered as the "position" of the node. A graph with an edge-length assigned to each edge will be called a *linkage*. Every framework $(G, \mathbf{u})$ defines a linkage (G, ℓ); we can say that to get a linkage from a framework we forget the positions of the nodes and only remember the lengths of the edges. The linkage (G, ℓ) is *realizable in* $\mathbb{R}^d$ if it is defined by a d-dimensional linkage, i.e., there is

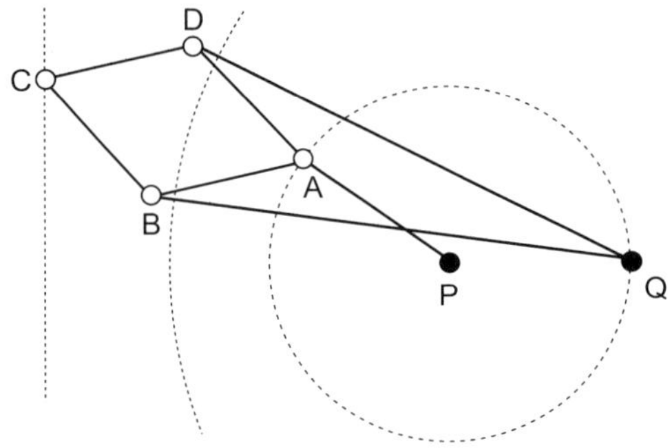

FIGURE 15.1. The Peaucellier–Lipkin linkage. The black joints are nailed, the white joints are free to move. The lengths of the bars satisfy $AB = BC = CD = DA$ and $QB = QD$. Nodes A and C are inverses relative to the fixed circle centered at Q with radius $\sqrt{QB^2 - AB^2}$ (prove!). As A moves along a circle through the center of the inversion, its inverse C moves along a straight line.

a vector-labeling $\mathbf{u} : V \to \mathbb{R}^d$ so that $\ell(ij) = |\mathbf{u}_i - \mathbf{u}_j|$ for every edge ij. The framework $(G, \mathbf{u})$ is called a *realization* of the linkage (G, ℓ).

Most of the time we assume that the framework is not contained in a lower dimensional affine subspace, in which case we call it *full-dimensional.* Two frameworks $(G, \mathbf{u})$ and $(G, \mathbf{v})$ with the same underlying graph G in d-space are called *congruent*, if there is a congruence (isometric transformation) of the whole space that maps $\mathbf{u}_i$ onto $\mathbf{v}_i$ $(i \in V)$. Basically, we do not want to distinguish congruent frameworks; but a warning is in order that congruence will be a nontrivial issue in some considerations.

Sometimes we want to "factor out" congruences. This can be defined as identifying congruent realizations of the same linkage, but a more explicit description of the resulting factor space will be useful. Let us assume that the framework is full-dimensional, and consider an ordered set of nodes, say $S = (0, 1, \dots, d)$, whose positions are affinely independent. Shift $\mathbf{u}_0$ to the origin; rotate the space about 0 so that $\mathbf{u}_1$ is moved to the positive half of the x_1-axis; then rotate the space about the x_1 axis so that $\mathbf{u}_2$ is moved to the upper halfplane of the (x_1, x_2)-plane etc. We end up with a vector labeling that is congruent to the original, and for $0 \le i \le d$ the i-th coordinate of $\mathbf{u}_i$ is positive, and its later coordinates are 0. We call this procedure *fitting S to the coordinate system.* This way we may assume that altogether $d+(d-1)+\dots+1 = \binom{d+1}{2}$ coordinates of the node positions are zero. (This number is the same as the dimension of the manifold of congruences of the Euclidean d-space.)

15.1. Versions of rigidity

There are several versions of the notion of rigidity of a framework $(G, \mathbf{u})$ $(\mathbf{u} : V \to \mathbb{R}^d)$.

• Is there a "proper motion" or "flexing" in the sense that we can specify a smooth orbit $\mathbf{u}_i(t)$ $(t \in [0, 1])$ for each node i so that $\mathbf{u}_i(0) = \mathbf{u}_i$, and $|\mathbf{u}_i(t) - \mathbf{u}_j(t)|$ is constant for every pair of adjacent nodes, but not for all pairs of nodes? If no such motion exists, we say that the framework is *locally rigid* or simply *rigid*. Structures which are not locally rigid will be called *flexible*; sometimes they are also called *mechanisms*.

We have to make a technical remark here: We have defined a flexing as a smooth motion of the nodes, but it would be natural to be more general and allow continuous but not differentiable motions, or to be more restrictive and allow only those motions that are analytic functions of time. As a very weak condition, we could ask for the existence of a noncongruent realization of the same linkage arbitrarily close to the original. It turns out that all these notions of flexibility are equivalent (see e.g. [**Connelly 1993**]).

• Is there a different realization of the same linkage: another representation $\mathbf{v} : V \to \mathbb{R}^d$ such that $|\mathbf{v}_i - \mathbf{v}_j| = |\mathbf{u}_i - \mathbf{u}_j|$ for every pair of adjacent nodes, but not for all pairs of nodes? (The representation $\mathbf{v}$ may or may not be obtained from the representation $\mathbf{u}$ by a continuous motion preserving the lengths of the edges.) If the framework has no different realization, it is called *globally rigid*. It is clear that this implies local rigidity. Further (non)-implications between these notions are illustrated in the examples below.

• Suppose that a framework is rigid in $\mathbb{R}^d$; does it remain rigid, if we embed it (isometrically) in a higher dimensional space? We say that such a framework is *universally rigid*. A single triangle in the plane has this property; two triangles attached along an edge form a rigid framework in the plane but not in 3-space. Universal rigidity is in various ways closely related to global rigidity; we refer to [**Gortler–Thurston 2014**] and [**Connelly–Gortler–Theran 2016a**].

• It may happen that we can assign velocities to the joints so that they do not seem to stretch or shrink any edge, but such velocities cannot be continued to an actual motion (unless they correspond to a rigid motion). We call such an assignment of velocities an "infinitesimal motion", and if no such motion exists, we say that the framework is *infinitesimally rigid*.

We can give an engineering interpretation of infinitesimal motions: Suppose that we move each node i by a (small) distance of at most ε, so that this is not a rigid motion of the framework. Trivially, the length of any edge ij changes by $O(\varepsilon)$. It can happen that for such a nonrigid motion the length of each edge changes by $O(\varepsilon^2)$ only; in this case, we cannot discover "in first order" that the length changes at all. From a practical point of view, this may be a problem, since in practice the bars are never totally rigid, and they may not hold against an infinitesimal motion.

EXAMPLE 15.2. Consider a complete graph K_n. Trivially every representation of it is globally rigid, and also locally rigid. However, it may have infinitesimal motions other than congruences: if all nodes are contained in an affine hyperplane, assign to each node an arbitrary velocity vector orthogonal to this hyperplane. ♦

EXAMPLE 15.3. Have a look at Figure 15.2, which shows 2-dimensional realizations of the same graph, rigid in one sense but not in the other. ♦

EXAMPLE 15.4 (Triangular Prism II). Let us return to the triangular prism in Example 3.4, represented in the plane. From (15.11) we get that $\dim(\mathsf{Inf}) - \dim(\mathsf{Str}) = 2 \cdot 6 - 9 = 3 = \dim(\mathsf{Rig})$ (assuming that not all nodes are collinear in the representation), and hence a representation of the triangular prism in the plane has a noncongruent infinitesimal motion if and only if it has a stress. This happens

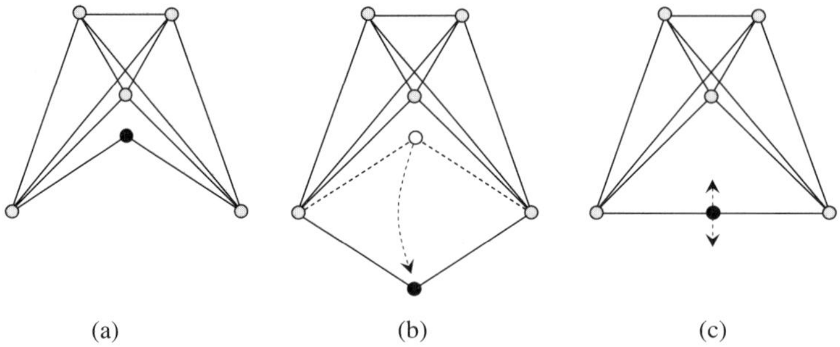

FIGURE 15.2. Structure (a) is infinitesimally rigid and locally rigid in the plane, but has a different realization (b), and so it is not globally rigid. Considering this as a motion in 3-space, we see that this framework is not universally rigid. Structure (c) has no other realization, so it is locally, globally and universally rigid, but it is not infinitesimally rigid, since vertical motion of the middle point is an infinitesimal motion. (Note that motion by ε results in stretching the edges by $O(\varepsilon^2)$.

if and only if the lines of the three edges connecting the two triangles pass through one and the same point (which may be at infinity; Figure 15.3, left and right).

Not all of these representations allow flexing, but some do. In the first, fixing the dark nodes, the white triangle can "circle" the dark triangle. This framework has a stress (can you find it?). The second is infinitesimally rigid and stress-free, but it is not globally rigid (reflect the white triangle in a symmetry axis of the black triangle); the third has an infinitesimal motion (equivalently, with this number of edges, it has a stress, as we will see), but it is globally rigid (and so it has no flexing). ◆

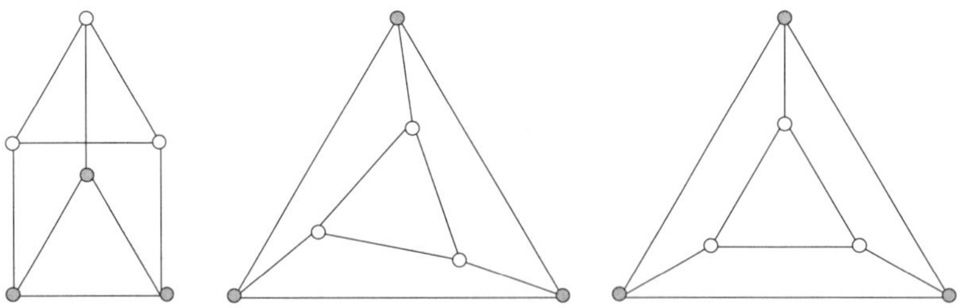

FIGURE 15.3. Rigid and nonrigid representations of the triangular prism in the plane.

15.2. Infinitesimal motions

We will continue with the exact definition and study of infinitesimal rigidity, which was perhaps the most awkward to describe, but easiest to handle.

Suppose that the nodes of the graph move smoothly (in continuously differentiable way) in d-space; let $\mathbf{u}_i(t)$ denote the position of node i at time t, where $\mathbf{u}_i = \mathbf{u}_i(0)$ is the starting position. The fact that the bars are rigid says that for every edge $ij \in E$, the edge-lengths $\ell_{ij} = |\mathbf{u}_i - \mathbf{u}_j|$ are fixed. Squaring these equations leads to polynomial equations

$$(\mathbf{u}_i(t) - \mathbf{u}_j(t))^{\mathsf{T}}(\mathbf{u}_i(t) - \mathbf{u}_j(t)) = \ell_{ij}^2. \tag{15.1}$$

Differentiating, we get

$$(\mathbf{u}_i(t) - \mathbf{u}_j(t))^{\mathsf{T}}(\dot{\mathbf{u}}_i(t) - \dot{\mathbf{u}}_j(t)) = 0. \tag{15.2}$$

In particular,

$$(\mathbf{u}_i - \mathbf{u}_j)^{\mathsf{T}}(\dot{\mathbf{u}}_i(0) - \dot{\mathbf{u}}_j(0)) = 0. \tag{15.3}$$

This motivates the following definition. Given a vector-labeling $\mathbf{u}: V \to \mathbb{R}^d$ of G, a map $\mathbf{v}: V \to \mathbb{R}^d$ is called an *infinitesimal motion* of the framework $(G, \mathbf{u})$, if the equation

$$(\mathbf{u}_i - \mathbf{u}_j)^{\mathsf{T}}(\mathbf{v}_i - \mathbf{v}_j) = 0 \tag{15.4}$$

holds for every edge $ij \in E$. The vector $\mathbf{v}_i$ can be thought of as the "velocity" of node i.

Infinitesimal motions form a linear space $\mathsf{Inf} = \mathsf{Inf}_{G,\mathbf{u}} \subseteq \mathbb{R}^{d\times V}$. There are some trivial infinitesimal motions, called (infinitesimal) *rigid motions* or *congruences*, which are velocities of smooth families of congruences of the whole space. In this case, (15.4) holds for every pair of nodes. Every infinitesimal congruence is the sum of two kinds of special solutions of equations (15.4) (see Exercise 15.3): we can have $\mathbf{v}_i = \mathbf{v}_j$ for all i and j (geometrically, this means translation of the framework), and $\mathbf{v}_i = A\mathbf{u}_i$, where $A \in \mathbb{R}^{d\times d}$ is a skew symmetric matrix (this corresponds to applying the smooth family of orthogonal transformations).

Infinitesimal congruences of a framework form a subspace $\mathsf{Rig} = \mathsf{Rig}_{G,\mathbf{u}} \subseteq \mathsf{Inf}$. By the remarks above, Rig consists of all systems of velocities of the form

$$\mathbf{v}_i = A\mathbf{u}_i + \mathbf{v}, \tag{15.5}$$

where A is a skew symmetric matrix and $\mathbf{v} \in \mathbb{R}^d$. If the node positions are not contained in an affine hyperplane, then this representation is unique. This implies that in this case $\dim(\mathsf{Rig}) = d + \binom{d}{2} = \binom{d+1}{2}$.

In the language of differential geometry, congruences of the d-space form a manifold of dimension $\binom{d+1}{2}$, and infinitesimal congruences form the tangent space of this manifold, which is a linear space of the same dimension.

15.2.1. Infinitesimal and finite motions.

We say that the framework is *infinitesimally rigid*, if $\mathsf{Inf} = \mathsf{Rig}$, i.e., every infinitesimal motion is an infinitesimal congruence. The argument above, motivating the definition of infinitesimal motions, suggests the following:

LEMMA 15.5. *If a framework is flexible, then it is not infinitesimally rigid.*

This looks trivial, since it would seem that if $\mathbf{u}(t)$ $(0 \le t \le 1)$ is a proper flexing, then $\mathbf{v}_i = \dot{\mathbf{u}}_i(0)$ gives an infinitesimal motion by (15.3). However, there is technical difficulty here: (15.3) may be a trivial equation if all the velocities $\dot{\mathbf{u}}_i(t)$ are zero at $t = 0$. It could also happen that $\dot{\mathbf{u}}(0)$ is an infinitesimal congruence, even though the flexing is proper (see Example 15.6).

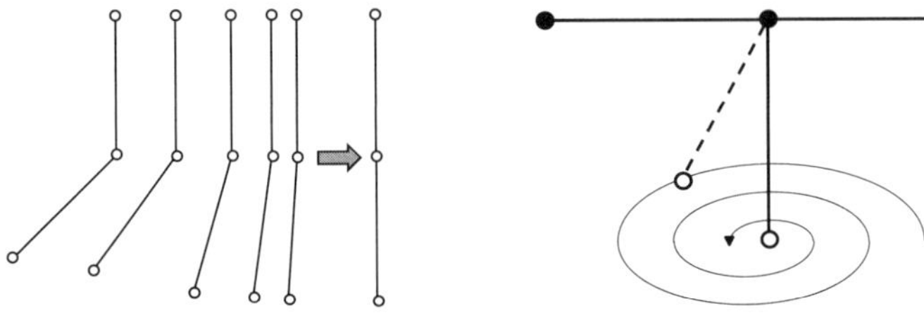

FIGURE 15.4. Left: a proper flexing whose velocities at time 0 form an infinitesimal congruence. Right: A proper flexing where velocities at positions at times converging to 0, when normalized, can converge to an infinitesimal congruence, but also to a proper infinitesimal motion, depending on when the framework is observed.

EXAMPLE 15.6. Consider a smooth proper flexing $(G, \mathbf{u}(t))$ as above, but let us modify it by slowing down and translation: we consider $(G, \mathbf{v}(t))$, where

$$\mathbf{v}_i(t) = \mathbf{u}_i(t^2) + t\mathbf{v}$$

with some fixed vector $\mathbf{v}$. Then

$$\dot{\mathbf{v}}_i(0) = \mathbf{v},$$

which is an infinitesimal congruence (Figure 15.4, left).

Perhaps we should construct an infinitesimal motion not as the velocities at time 0, but as a limit of velocities at times converging to 0, appropriately scaled? The example on the right of Figure 15.4, as a framework in 3-space, should be another warning. We can think of this as a pendulum circling its vertical position and slowing down, starting at time -1. The geometry and timing are arranged so that the k-th arc has length $\sim 4^{-k}$, and it is traversed in time 2^{-k}, so the velocity tends to 0, which is attained at time 0. (If you wish, you can continue to positive times just reversing the process.)

If we look at velocities when the pendulum is closest to us, then we see velocities of a noncongruent infinitesimal motion. However, if we look at the velocities when the pendulum is (say) in a leftmost position, then we see the velocities of a rigid motion. How to choose the "right" moments in time? ♦

Proof of Lemma 15.5. Our first step is to modify the flexing $(G, \mathbf{u}(t))$ in order to eliminate rigid motions as much as possible. Fixing a time t, let $\mathbf{u}_i = \mathbf{u}_i(t)$ and $\overline{\mathbf{u}}_i = \mathbf{u}_i(0)$, and consider $(G, \mathbf{u})$ and $(G, \overline{\mathbf{u}})$ as two different frameworks. We fix $(G, \overline{\mathbf{u}})$, and rigidify $(G, \mathbf{u})$, but let it float freely. Connect node $\mathbf{u}_i$ to node $\overline{\mathbf{u}}_i$ by a rubber band for each node i, and let the resulting structure find its equilibrium. In other words, apply an isometric transformation to $(G, \mathbf{u})$ so that the energy $\sum_i |\mathbf{u}_i - \overline{\mathbf{u}}_i|^2$ is minimized.

The forces keeping $(G, \mathbf{u})$ in equilibrium (as a rigid body) are the vectors $\mathbf{v}_i = \mathbf{v}_i(t) = \mathbf{u}_i - \overline{\mathbf{u}}_i$, acting at the points $\mathbf{u}_i$. By classical physics,

$$\sum_{i \in V} \mathbf{v}_i = 0, \tag{15.6}$$

and

$$\sum_{i\in V} \mathbf{v}_i^{\mathsf{T}} A\mathbf{u}_i = 0 \tag{15.7}$$

for every skew symmetric matrix $A \in \mathbb{R}^{d\times d}$. Also note that since $(G, \mathbf{u})$ and $(G, \overline{\mathbf{u}})$ are the same as linkages, we have $|\overline{\mathbf{u}}_i - \overline{\mathbf{u}}_j|^2 = |\mathbf{u}_i - \mathbf{u}_j|^2$ for every edge ij. This translates into the equation

$$2(\overline{\mathbf{u}}_i - \overline{\mathbf{u}}_j)^{\mathsf{T}}(\mathbf{v}_i - \mathbf{v}_j) = -|\mathbf{v}_i - \mathbf{v}_j|^2. \tag{15.8}$$

Let $\omega_t = \max_{i\in V} |\mathbf{v}_i|$. If $(G, \overline{\mathbf{u}})$ is not isometric to $(G, \mathbf{u})$, then $\omega_t > 0$, and the vectors $(1/\omega_t)\mathbf{v}_i$ have length at most 1. Let us select a sequence of such times converging to 0, then $\omega_t \to 0$. By selecting a subsequence, we may assume that $(1/\omega_t)\mathbf{v}_i(t) \to \mathbf{w}_i$ for some vectors $\mathbf{w}_i$. Clearly $\max_i |\mathbf{w}_i| = 1$ and so $\mathbf{w} \neq 0$.

We claim that $\mathbf{w}$ is an infinitesimal motion of $(G, \overline{\mathbf{u}})$. Indeed, (15.8) implies that

$$(\overline{\mathbf{u}}_i - \overline{\mathbf{u}}_j)^{\mathsf{T}}\Big(\frac{1}{\omega_t}\mathbf{v}_i - \frac{1}{\omega_t}\mathbf{v}_j\Big) = -\frac{1}{2\omega_t}|\mathbf{v}_i - \mathbf{v}_j|^2 = O(\omega_t),$$

and so in the limit we have

$$(\overline{\mathbf{u}}_i - \overline{\mathbf{u}}_j)^{\mathsf{T}}(\mathbf{w}_i - \mathbf{w}_j) = 0.$$

We have to argue that $\mathbf{w}$ is not an infinitesimal congruence of $(G, \overline{\mathbf{u}})$. Suppose it is, then there is a vector $\mathbf{v} \in \mathbb{R}^d$ and a skew symmetric matrix $A \in \mathbb{R}^{d\times d}$ such that $\mathbf{w}_i = \mathbf{v} + A\overline{\mathbf{u}}_i$. We may assume that $\sum_i \overline{\mathbf{u}}_i = 0$, since we can choose the origin at the center of gravity of the points $\overline{\mathbf{u}}_i$. Scaling by ω_t and going to the limit, we see that the vectors $\mathbf{w}_i$ satisfy the equilibrium conditions (15.6) and (15.7). The first condition gives us that

$$0 = \sum_{i\in V} \mathbf{w}_i = n\mathbf{v} + \sum_{i\in V} A\overline{\mathbf{u}}_i = n\mathbf{v},$$

so $\mathbf{v} = 0$ and $\mathbf{w}_i = A\mathbf{u}_i$. Applying the second condition with the same matrix A, we get

$$0 = \sum_{i\in V} \mathbf{w}_i^{\mathsf{T}} A\overline{\mathbf{u}}_i = \sum_{i\in V} \overline{\mathbf{u}}_i^{\mathsf{T}} A^{\mathsf{T}} A\overline{\mathbf{u}}_i = \sum_{i\in V} |A\overline{\mathbf{u}}_i|^2.$$

This implies that $A\overline{\mathbf{u}}_i = 0$ for every i, and so $\mathbf{w}_i = 0$. But we know that $\mathbf{w} \neq 0$, a contradiction. $\square$

The converse of this lemma is not true in general, as Figure 15.2 shows. However, the converse does hold for stress-free frameworks:

LEMMA 15.7. *If a stress-free framework is not infinitesimally rigid, then it is flexible.*

PROOF. Let $(G, \mathbf{u})$ be a stress-free framework in $\mathbb{R}^d$, and let $\mathbf{v}$ be a non-rigid infinitesimal motion. Since $\mathbf{v}$ is not a rigid infinitesimal motion, there is a pair (a, b) of nonadjacent nodes for which $(\mathbf{u}_a - \mathbf{u}_b)^{\mathsf{T}}(\mathbf{v}_a - \mathbf{v}_b) \neq 0$. We may assume that

$$(\mathbf{u}_a - \mathbf{u}_b)^{\mathsf{T}}(\mathbf{v}_a - \mathbf{v}_b) = 1. \tag{15.9}$$

Equations (15.4) give linearly independent equations for $\mathbf{v}$, and (15.9) is linearly independent from them, since the other equations have zero right hand side. In the solution space of (15.4) and (15.9), we can fix some entries of $\mathbf{v}$ to zero to have a unique solution. By Cramer's Rule, the remaining entries of $\mathbf{v}$ can be expressed as rational functions of the entries of $\mathbf{u}$, and these rational functions will satisfy (15.4) and (15.9) in a neighborhood of $\mathbf{u}$. Scaling up with the common denominator, we get polynomials $\mathbf{w}_i(\mathbf{u})$ in the entries of $\mathbf{u}$ such that

$$(\mathbf{x}_i - \mathbf{x}_j)^\mathsf{T}\big(\mathbf{w}_i(\mathbf{x}) - \mathbf{w}_j(\mathbf{x})\big) = 0$$

identically in $\mathbf{x} \in \mathbb{R}^{d\times V}$ for every edge $ij \in E$, and

$$(\mathbf{x}_a - \mathbf{x}_b)^\mathsf{T}\big(\mathbf{w}_a(\mathbf{x}) - \mathbf{w}_b(\mathbf{x})\big) \neq 0$$

if $\mathbf{x}$ is in a neighborhood of $\mathbf{u}$.

Consider the differential equation system

$$\frac{d}{dt}\mathbf{x}_i(t) = \mathbf{w}_i\big(\mathbf{x}(t)\big), \qquad \mathbf{x}(0) = \mathbf{u}.$$

Since $\mathbf{w}(\mathbf{x})$ is a continuous function of $\mathbf{x}$, this system has a solution in an interval $0 \le t \le T$. This defines a nonrigid motion of the graph. Indeed, for every edge ij,

$$\begin{aligned}\frac{d}{dt}|\mathbf{x}_i(t) - \mathbf{x}_j(t)|^2 &= \big(\mathbf{x}_i(t) - \mathbf{x}_j(t)\big)^\mathsf{T}\big(\dot{\mathbf{x}}_i(t) - \dot{\mathbf{x}}_j(t)\big)\\ &= \big(\mathbf{x}_i(t) - \mathbf{x}_j(t)\big)^\mathsf{T}\big(\mathbf{w}_i(\mathbf{x}(t)) - \mathbf{w}_j(\mathbf{x}(t))\big) = 0,\end{aligned}$$

so $|\mathbf{x}_i(t) - \mathbf{x}_j(t)|$ remains constant. On the other hand, this is not a rigid motion, since

$$\frac{d}{dt}|\mathbf{x}_a(t) - \mathbf{x}_b(t)|^2 \,|_{t=0} = (\mathbf{u}_a - \mathbf{u}_b)^\mathsf{T}\big(\mathbf{w}_i(\mathbf{u}) - \mathbf{w}_j(\mathbf{u})\big) \neq 0,$$

so $|\mathbf{x}_a(t) - \mathbf{x}_b(t)|$ cannot remain constant. □

15.2.2. Infinitesimal motions and stresses. Which subsets of the equations in (15.4) are linearly independent? Let λ_{ij} $(ij \in E)$ be multipliers such that combining the equations with them we get 0. Considering the coefficient of $\mathbf{v}_i$, this means that

$$\sum_{j\in N(i)} \lambda_{ij}(\mathbf{u}_j - \mathbf{u}_i) = 0$$

for every node i. In other words, λ is a stress!

Recalling the notion of transversality, and that equations (15.4) were obtained by differentiating equations (15.1), we can assert that *a solution of* (15.1) *is transversal if and only if the framework* $(G, \mathbf{u})$ *is stress-free.* This means that stress-free frameworks can be studied by similar methods as transversal orthogonal representations. For example, the Implicit Function Theorem can be applied to show that *if a linkage* (G, ℓ) *has a stress-free realization, then so does every linkage* (G, ℓ') *for edge-lengths* ℓ' *sufficiently close to* ℓ.

There is a certain "dual" to this observation about linear dependencies between the equations defining infinitesimal motions. What are the linear dependencies between the equations (14.1) defining a stress (where the values S_{ij} are considered as unknowns)? Combining the d equations belonging to the same node i means taking inner product with a vector $\mathbf{w}_i$. Combining all results in one equation, we obtain

$$\sum_{i\in V}\sum_{j\in N(i)} S_{ij}(\mathbf{u}_j - \mathbf{u}_i)^\mathsf{T}\mathbf{w}_i = 0,$$

or

$$\sum_{ij\in E} S_{ij}(\mathbf{u}_j - \mathbf{u}_i)^\mathsf{T}(\mathbf{w}_j - \mathbf{w}_i) = 0.$$

If this is a linear dependency between the stress equations, then the coefficient of every S_{ij} must be zero. Thus $\mathbf{w}$ is an infinitesimal motion!

Let us re-write the basic equations (15.4) in a form that is sometimes more useful. We combine the velocity vectors $\mathbf{v}_i$ into a single vector $\mathbf{v} \in \mathbb{R}^{nd}$, then (15.4) provides a system of homogeneous linear equations for the entries of $\mathbf{v}$. The matrix $R = R_{G,\mathbf{u}} \in \mathbb{R}^{m\times nd}$ of this system is defined for an edge ij, a node k and a coordinate $1 \le \nu \le d$ by

$$R_{ij,k\nu} = \begin{cases} u_{i\nu} - u_{j\nu}, & \text{if } k = i, \\ u_{j\nu} - u_{i\nu}, & \text{if } k = j, \\ 0, & \text{otherwise.} \end{cases} \tag{15.10}$$

We call R the *rigidity matrix* of the framework $(G, \mathbf{u})$ (Figure 15.5).

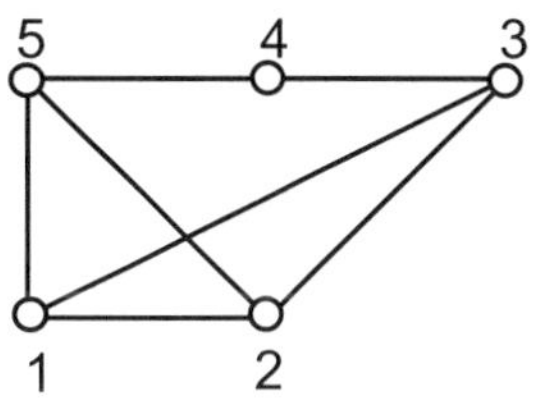

$$\begin{array}{c} 12 \\ 23 \\ 13 \\ 34 \\ 45 \\ 15 \\ 25 \end{array}\left[\begin{array}{cc:cc:cc:cc:cc} 1 & 0 & -1 & 0 & 0 & 0 & 0 & 0 & 0 & 0 \\ 0 & 0 & 1 & 1 & -1 & -1 & 0 & 0 & 0 & 0 \\ 2 & 0 & 0 & 0 & -1 & 0 & 0 & 0 & 0 & 0 \\ 0 & 0 & 0 & 0 & -1 & 0 & 1 & 0 & 0 & 0 \\ 0 & 0 & 0 & 0 & 0 & 0 & -1 & 0 & 0 & 0 \\ 0 & 1 & 0 & 0 & 0 & 0 & 0 & 0 & 0 & -1 \\ 0 & 0 & -1 & 1 & 0 & 0 & 0 & 0 & 1 & -1 \end{array}\right]$$

FIGURE 15.5. A framework in dimension 2 and its rigidity matrix. The framework has 5 nodes and $7 = 2\cdot 5 - 3$ edges. It has both a nonzero stress and a nonrigid infinitesimal motion.

Infinitesimal motions (combined into single vectors $\mathbf{v} \in \mathbb{R}^{nd}$) form the right null space of R, while stresses on $(G, \mathbf{u})$ form the left null space of R. This implies that

$$\mathrm{rk}(R) = dn - \dim(\mathsf{Inf}) = m - \dim(\mathsf{Str}). \tag{15.11}$$

From this we see that if a framework spanning $\mathbb{R}^d$ is stress-free, then $m \le dn - \binom{d+1}{2}$, and if it is infinitesimally rigid, then $m \ge dn - \binom{d+1}{2}$. In the (most interesting) case when $m = dn - \binom{d+1}{2}$, the framework is infinitesimally rigid if and only if it is stress-free. Such frameworks are minimally infinitesimally rigid as well as maximally stress-free (corresponding to the subsets of edges that are bases of the row space of $R(K_n, \mathbf{u})$).

There is another useful meaning of the rigidity matrix. Fix a graph G, and consider the squared-edge-length map $f = f_G : \mathbb{R}^{d\times V} \to \mathbb{R}^E$ defined by

$$f(\mathbf{u}) = \big(|\mathbf{u}_i - \mathbf{u}_j|^2 : ij \in E\big). \tag{15.12}$$

It is easy to compute that *the Jacobian $Df(\mathbf{u})$ of f is just the rigidity matrix $R_{G,\mathbf{u}}$*.

15.2.3. Rigidity of 3-polytopes. To conclude this section, let us illustrate the connection between stresses and infinitesimal rigidity by an important example. Let G be the skeleton of a convex 3-polytope P, with the representation given by the positions of the vertices. Cauchy's Theorem 14.5 tells us that the space of stresses has dimension 0, so we get that the space of infinitesimal motions has dimension $\dim(\mathsf{Inf}) = 3n - m$. We know that $\dim(\mathsf{Inf}) \ge \binom{4}{2} = 6$, and hence $m \le 3n - 6$. Of course, this simple inequality we know already (recall (2.2)). It also follows that if equality holds, i.e., when all facets of P are triangles, then the space of infinitesimal motions is 6-dimensional; in other words,

PROPOSITION 15.8. *The framework consisting of the vertices and edges of a convex 3-polytope with triangular facets is infinitesimally rigid.* □

If not all facets of the polytope P are triangles, then it follows by the same computation that the skeleton is not infinitesimally rigid. However, if the facets are "made out of cardboard", which means that they are forced to preserve their shape, then the polytope will be infinitesimally rigid. To prove this, one can put a pyramid over each facet (flat enough to preserve convexity), and apply Proposition 15.8 to this new polytope P' with triangular faces. Every nontrivial infinitesimal motion of P preserving the shape of the facets would extend to a nontrivial infinitesimal motion of P', but we know already that P' has no such motion.

Proposition 15.8 does not remain true if we only assume that the graph is isomorphic to the skeleton of a convex 3-polytope with triangular facets; see Exercise 15.6 for a classical example [**Bricard 1897**]. It even fails to hold when the framework consists of the vertices and edges of a simple polytope (the surface of the polytope is homeomorphic to the sphere), as shown by a more involved counterexample [**Connelly 1977**].

15.3. Generic frameworks

In Chapter 3 we have met vector-labelings that are in general position; in Chapter 10 another related condition, transversality was introduced. Here we need a further nondegeneracy condition: we say that a vector-labeling is *generic*, if all coordinates of the representing vectors are algebraically independent over $\mathbb{Q}$.

This assumption is more restrictive than general position, and it is usually an overkill, since we could always replace it by *"the coordinates of the representing vectors do not satisfy any algebraic relation that they do not have to, and which is important to avoid in the proof"*. The hypothesis of general position will be used in the next section on global rigidity again. Transversality is equivalent to the basic condition of stress-freeness, and so both of these related but weaker notions have a role to play in rigidity theory.

In a generic framework, all entries of the vector labels are algebraically independent, by definition. However, while isometric transformations of the whole space do not change stress-freeness, and do not change the edge-lengths, they may create algebraic dependencies between the coordinates. For example, if we fit a set of $d+1$ nodes to the coordinate system, then out of dn coordinates, $\binom{d+1}{2}$ become zeros, which means that we are left with $dn - \binom{d+1}{2}$ algebraically independent coordinates.

We say that a graph is *generically stress-free in* $\mathbb{R}^d$, or simply *stress-free in* $\mathbb{R}^d$, if every generic representation of it in $\mathbb{R}^d$ is stress-free; similarly, we say that G is *generically rigid* in $\mathbb{R}^d$, if every generic representation of it in $\mathbb{R}^d$ is infinitesimally rigid. It is enough to check these properties for a single generic representation: if one generic labeling is stress-free or infinitesimally rigid, then so is every other generic labeling. More generally, the following fact was observed by Gluck [**Gluck 1974**]:

PROPOSITION 15.9. *If a graph G has a vector labeling in $\mathbb{R}^d$ that is infinitesimally rigid [stress-free], then G is generically rigid [stress-free] in $\mathbb{R}^d$.*

PROOF. Both infinitesimal rigidity and stress-freeness can be expressed in terms of the nonvanishing of certain determinants composed of the coordinates of the representation. If such a determinant vanishes for one algebraically independent choice of the coordinates, then it vanishes for every choice. □

The assumption that the coordinates are algebraically independent numbers seems very far from any real-life situation; it is quite technical and unverifiable, and for this reason it seems useless. However, another way of handling the genericity assumption brings genericity quite close to applications: this is to choose independent random values for the coordinates uniformly from an interval (or indeed, from any distribution that is absolutely continuous with respect to the Lebesgue measure, like Gaussian). We call the vector labeling obtained this way a *random vector labeling* of G. The random labeling is generic with probability 1, and we can re-state our results as probabilistic statements holding with probability 1. (Mathematically, the definition in terms of algebraic independence yields slightly stronger results.)

By this argument, we can state:

PROPOSITION 15.10. *If G is generically rigid [stress-free] in $\mathbb{R}^d$, then a random vector labeling of G is almost surely infinitesimally rigid [stress-free] in $\mathbb{R}^d$.*

15.3.1. Stress-free and rigid graphs in the plane.

Whether a graph is generically rigid or stress-free is a purely combinatorial property, and one would like to obtain a purely combinatorial characterization of these properties. It is surprising that this basic question is only solved in the planar case, which we present here. We start with stress-freeness.

It is clear that if a framework $(G, \mathbf{u})$ has more than $2n-3$ edges, then it cannot be stress-free (recall (15.11)). We can apply this observation to induced subgraphs: if some subset of $k \geq 2$ nodes spans more than $2k-3$ edges, then the framework cannot be stress-free in the plane. So a natural criterion for a graph G to have a stress-free representation is the following:

(GL) *Every set of k nodes ($k \geq 2$) spans at most $2k-3$ edges.*

Note that this condition implies that the graph has no multiple edges. The necessity of this criterion was recognized already in the 19-th century, and it was stated as a "rule of thumb" for the existence of a stress-free realization in textbooks [**Föppl 1880**]. The sufficiency was proved by [**Pollaczek-Geiringer 1927**], and rediscovered by [**Laman 1970**]. We call (GL) the *Geiringer–Laman condition.*

The Geiringer–Laman condition tells us why a graph is *not* stress-free. To formulate a (general) reason why a graph *is* stress-free, we introduce the *Henneberg construction*, [**Henneberg 1911**] which is an iterative procedure, starting with a single node, and increasing the number of nodes of a graph G in one of two ways (Figure 15.6):

(H1) *Create a new node and connect it to at most two old nodes.*

(H2) *Subdivide an edge and connect the new node to any third node.*

The following important theorem, combining the results of [**Henneberg 1911**], [**Pollaczek-Geiringer 1927**] and [**Laman 1970**] describes stress-free graphs in the plane.

THEOREM 15.11. *For a graph G, the following are equivalent:*

(a) *G is stress-free in the plane;*

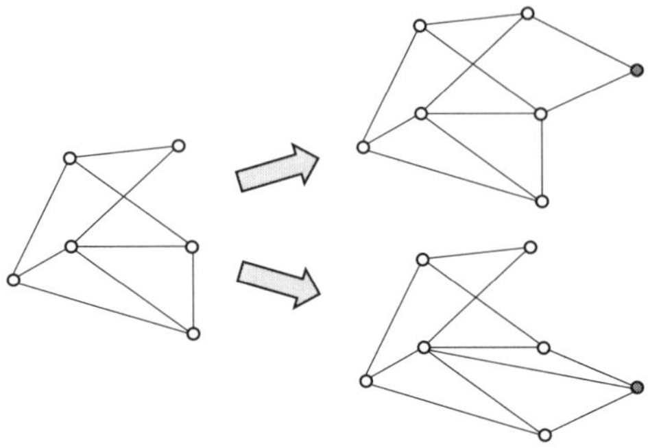

FIGURE 15.6. The Henneberg construction

(b) *G satisfies the Geiringer–Laman condition;*

(c) *G can be obtained by the Henneberg construction.*

Conditions (b) and (c) show that the property of a graph of being (generically) stress-free in the plane is in NP∩co-NP. The fact that it is in P follows by Corollary 15.17 below, using algorithms for the decomposition of a graph into connected spanning subgraphs. (See also Section 19.3.2.)

PROOF. (a)⇒(b): This we have argued above.

(b)⇒(c): We prove by induction on the number of edges that every graph G satisfying the Geiringer–Laman condition can be built up by the Henneberg construction. Since the average degree of G is $2m/n \le (4n-6)/n < 4$, there is a node i of degree at most 3. If i has degree at most 2, then we can delete i and proceed by induction and (H1). So we may assume that i has degree 3. Let a, b, c be its neighbors.

Call a set $S \subseteq V$ spanning exactly $2|S|-3$ edges a *tight set*. Clearly $|S| \ge 2$. The key observation is:

Claim. *If two tight sets have more than one node in common, then their union is also tight.*

Indeed, suppose that S_1 and S_2 are tight, and let $E_1 = E(G[S_i])$. Then $E_1 \cap E_2$ is the set of edges spanned by $S_1 \cap S_2$, and so $|E_1 \cap E_2| \le 2|S_1 \cap S_2| - 3$ (using the assumption that $|S_1 \cap S_2| \ge 2$). The set $E_1 \cup E_2$ is a subset of the set of edges spanned by $S_1 \cup S_2$, and so the number of edges spanned by $S_1 \cup S_2$ is at least

$$|E_1 \cup E_2| = |E_1| + |E_2| - |E_1 \cap E_2| \ge (2|S_1| - 3) + (2|S_2| - 3) - (2|S_1 \cap S_2| - 3)$$
$$= 2|S_1 \cup S_2| - 3.$$

Since G satisfies (GL), we must have equality here, implying that $S_1 \cup S_2$ is tight.

We want to prove that we can delete i and create a new edge between two of its neighbors to get a graph G' satisfying (GL); then G arises from G' by Henneberg construction (H2). Let us try to add the edge ab to $G \setminus i$; if the resulting graph G' violates (GL), then there is a set $S \subseteq V \setminus \{i\}$ with $|S| \ge 2$ spanning more than $2|S| - 3$ edges of G'. Since S spans at most $2|S| - 3$ edges of G, this implies that S is tight, and $a, b \in S$. (If a and b are adjacent in G, then $\{a, b\}$ is such a tight set.) If there are several sets S with this property, then Claim 1 implies that their union is also tight. We denote by S_{ab} this union. Similarly we get the tight sets S_{bc} and S_{ca}.

The set S_{ab} cannot contain c, since then adding i to S_{ab}, condition (GL) would be violated. Similarly, $b \notin S_{ac}$ and $a \notin S_{bc}$. In particular, the sets S_{ab}, S_{bc} and S_{ca} are distinct, and hence the Claim implies that any two of them have only one node in common. Thus their union $S = S_{ab} \cup S_{bc} \cup S_{ca} \cup \{i\}$ spans at least

$$(2|S_{ab}|-3)+(2|S_{bc}|-3)+(2|S_{ca}|-3)+3 = 2|S|-3$$

edges, and so it is tight. This contradicts the maximality of S_{ab}, and completes the proof of (b)⇒(c).

(c)⇒(a): We use induction on the number of edges. Let G arise from G' by one step of the Henneberg construction, and let S be a nonzero stress on a generic framework $(G, \mathbf{u})$. We want to construct a nonzero stress on $(G', \mathbf{u})$ (which will be a contradiction by induction, and so it will complete the proof).

First suppose that we get G by step (H1), and let i be the new node. It suffices to argue that S is 0 on the new edges, and so it gives a stress on $(G', \mathbf{u})$. If i has degree 0 or 1, then this is trivial. Suppose that i has two neighbors a and b. Then by the assumption that the representation is generic, we know that the points $\mathbf{u}_i$, $\mathbf{u}_a$ and $\mathbf{u}_b$ are not collinear, so from the stress condition

$$S_{ia}(\mathbf{u}_i - \mathbf{u}_a) + S_{ib}(\mathbf{u}_i - \mathbf{u}_b) = 0$$

it follows that $S_{ia} = S_{ib} = 0$.

Suppose that G arises by the Henneberg construction (H2), by subdividing the edge $ab \in E(G')$ and connecting the new node i to $c \in V(G') \setminus \{a, b\}$. Let us modify the representation $\mathbf{u}$ as follows:

$$\mathbf{u}'_j = \begin{cases} \frac{1}{2}(\mathbf{u}_a + \mathbf{u}_b) & \text{if } j = i, \\ \mathbf{u}_j & \text{otherwise.} \end{cases}$$

This representation for G is not generic any more, but it is generic if we restrict it to G'.

If the generic representation $\mathbf{u}$ of G admits a nonzero stress, then by Proposition 15.9 so does every other representation, in particular $\mathbf{u}'$ admits a nonzero stress S'. Consider the stress condition for i:

$$S'_{ia}(\mathbf{u}'_i - \mathbf{u}'_a) + S'_{ib}(\mathbf{u}'_i - \mathbf{u}'_b) + S'_{ic}(\mathbf{u}'_i - \mathbf{u}'_c) = 0.$$

Here $\mathbf{u}'_i - \mathbf{u}'_a = \frac{1}{2}(\mathbf{u}_b - \mathbf{u}_a)$ and $\mathbf{u}'_i - \mathbf{u}'_b = \frac{1}{2}(\mathbf{u}_a - \mathbf{u}_b)$ are parallel but $\mathbf{u}'_i - \mathbf{u}'_c$ is not parallel to them. So it follows that $S'_{ic} = 0$ and $S'_{ia} = S'_{ib}$. Defining $S'_{ab} = S'_{ia}$, we get a nonzero stress on $(G', \mathbf{u})$, a contradiction. □

Using these results, we turn to characterizing rigid graphs in the plane. Equation (15.11) implies that minimally rigid graphs (i.e., rigid graphs that cease to be rigid when any edge is deleted) are exactly the maximal graphs satisfying (GL), and they all have $2n-3$ edges. In other words,

COROLLARY 15.12. *A graph with n nodes and $2n-3$ edges is rigid in the plane if and only if it satisfies the Geiringer–Laman condition.*

Minimal stressed graphs (i.e., graphs that are not stress-free but all their proper subgraphs are) are also worth considering.

COROLLARY 15.13. *A graph G is minimal stressed in the plane if and only if it has $2n-2$ edges, but every proper subset $S \subset V$ spans at most $2|S|-3$ edges.*

The Geiringer–Laman Theorem can be used to prove the following characterization of rigid graphs (not just the minimal ones) in the plane. Let us say that G satisfies the *partition condition*, if for every decomposition $G = G_1 \cup \dots \cup G_k$ into the union of graphs with at least one edge, we have

$$\sum_i (2|V(G_i)|-3) \geq 2n-3. \tag{15.13}$$

Note that this condition implies that G has at least $2n-3$ edges (considering the partition into single edges).

THEOREM 15.14. *A graph G is rigid in the plane if and only if it satisfies the partition condition.*

PROOF. To prove that the partition condition is necessary, consider any decomposition $G = G_1 \cup \dots \cup G_k$ into the union of graphs with at least one edge. Let $E' \subset E$ be a smallest rigid subset. We know that E' is stress-free, and hence $E' \cap E(G_i)$ is stress-free, which implies that $|E' \cap E(G_i)| \leq 2V(G_i)-3$ by Theorem 15.11 (here we use the easy implication (a)$\Rightarrow$(b)). It follows that $2n-3 = |E'| \leq \sum_i |E' \cap E(G_i)| \leq \sum_i (2|V(G_i)|-3)$.

To prove the converse, suppose that G is not rigid, and fix any generic representation $\mathbf{u}$ of it. Let H be a maximal spanning subgraph of G satisfying (GL). Since H is not rigid, it has fewer than $2n-3$ edges. Similarly as in the proof of Theorem 15.11, we call a subset $S \subseteq V$ *tight*, if it spans $2|S|-3$ edges of H (by (GL), S cannot span more edges of H than that). Let $S_1, \dots, S_k$ be the maximal tight subsets of H. Since every adjacent pair of nodes is tight, the subgraphs $H[S_i]$ cover H. Just like in the proof of Theorem 15.11, it follows that any two subsets S_i have at most one node in common; in particular, the induced subgraphs $G[S_i]$ are edge-disjoint.

For every edge $e \in E \setminus E(H)$, the graph $H+e$ does not satisfy (GL) any more, and so it contains a subset $S \subseteq V$ spanning more than $2|S|-3$ edges of $H+e$. It follows that S is tight, and e is spanned by S. So $S \subseteq S_i$ for some i, and e is spanned by S_i. This means that $G[S_1] \cup \dots G[S_k] = G$; since

$$\sum_{i=1}^{k} (2|S_i|-3) = \sum_{i=1}^{k} |E(H[S_i])| = |E(H)| < 2n-3,$$

this partition of G violates the partition condition. □

COROLLARY 15.15. *Every minimal stressed graph is rigid in the plane.*

We can derive more graph-theoretic conditions on generic stress-freeness and rigidity in the plane by combining the Geiringer–Laman condition and the partition condition with classical theorems in [**Tutte 1961**] and [**Nash-Williams 1964**]. We state these theorems here without proof; a proof will be easy using more general results in Section 19.3.2.

THEOREM 15.16. *Let G be a multigraph on n nodes. (a) The edges of G can be covered by k subtrees of G if and only if every subset $S \subseteq V$ spans at $k(|S|-1)$ edges. (b) G contains k edge-disjoint spanning trees of G if and only if for every partition $\mathcal{P}$ of V, the number of edges connecting different classes is at least $k(|\mathcal{P}|-1)$.* □

Using this theorem, we obtain the following corollary.

COROLLARY 15.17. *A graph G is generically stress-free in the plane if an only if every graph obtained by doubling an edge of G can be covered by two forests. A graph G is generically rigid in the plane if an only if every graph obtained by doubling an edge of G contains two edge-disjoint spanning trees.*

We give a simpler necessary condition for rigidity in the plane as an application of Theorem 15.14 [**Lovász–Yemini 1982**].

COROLLARY 15.18. *Every 6-connected graph is rigid in the plane.*

PROOF. Suppose that G is 6-connected but not rigid. Then by Theorem 15.14 there exist subgraphs $G_i = (V_i, E_i)$ $(i = 1, \dots, k)$ such that $G = G_1 \cup \dots \cup G_k$ and

$$\sum_{i=1}^{k} (2|V_i| - 3) < 2n - 3. \tag{15.14}$$

We may assume that every G_i is complete, since adding edges inside a set V_i does not change (15.14). The graph itself is trivially not complete. We may also assume that every node belongs to at least two of the V_i: deleting a node contained in one V_i only preserves (15.14) and it is easy to see that it preserves 6-connectivity.

We claim that

$$\sum_{V_i \ni v} \Big(2 - \frac{3}{|V_i|}\Big) \geq 2 \tag{15.15}$$

for each node v. Let (say) $V_1, \dots, V_r$ be those V_i containing v, $|V_1| \leq |V_2| \leq \dots$ The graph is 6-connected, hence v has degree at least 6, and so

$$\sum_{i=1}^{r} (|V_i| - 1) \geq 6. \tag{15.16}$$

Each term on the left side is at least 1/2, so if $r \geq 4$, we are done. If $r = 3$, then $|V_3| \geq 3$, and so the left side of (15.15) is at least $1/2 + 1/2 + 1 = 2$. If $r = 2$ and $|V_1| \geq 3$, then the left side of (15.15) is at least $1 + 1 = 2$. Finally, if $r = 2$ and $|V_1| = 2$, then $|V_2| \geq 6$ by (15.16), and so the left side of (15.15) is at least $1/2 + 3/2 = 2$.

Now summing (15.15) over all nodes v, we get

$$\sum_{v \in V} \sum_{V_i \ni v} \Big(2 - \frac{3}{|V_i|}\Big) = \sum_{i=1}^{k} \Big(2 - \frac{3}{|V_i|}\Big) \sum_{v \in V_i} 1 = \sum_{i=1}^{k} (2|V_i| - 3)$$

on the left side and $2n$ on the right hand side, which contradicts (15.14). □

It is not enough to assume that G is 5-connected. Indeed, consider any 5-connected 5-regular graph H on $k > 6$ nodes, split each node into 5 nodes of degree 1, and attach a copy of K_5 to these nodes, to get a graph G (Figure 15.7). Clearly G is 5-connected. We claim it is not rigid in the plane. Indeed, from each copy of K_5 we can remove 3 edges so that it remains rigid, and so the remaining graph $G' = (V, E')$ is rigid if G is. But $n = 5k$, and

$$|E'| = 7k + |E(H)| = 10k - \frac{1}{2}k < 10k - 3 = 2n - 3,$$

showing that G' is not rigid in the plane.

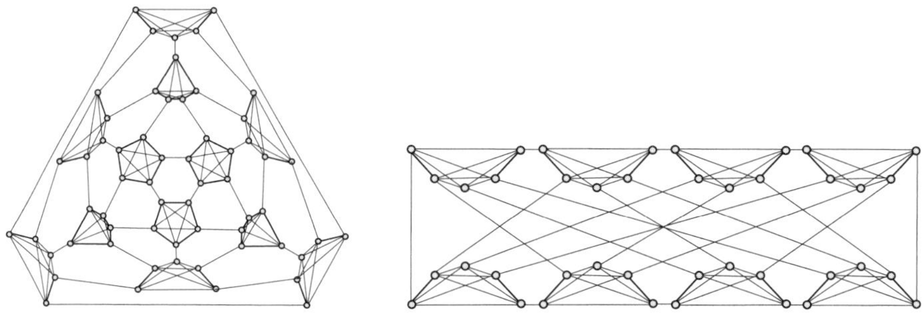

FIGURE 15.7. Two 5-connected graphs that are not rigid in the plane.

15.3.2. Generic rigidity in higher dimension. A 3-dimensional analogue of the Geiringer–Laman Theorem, or of Theorem 15.14, is not known. It is not even known whether there is a positive integer k such that every k-connected graph is rigid in 3-space. One indication why an approach similar to the planar one fails can be illustrated by the following example.

EXAMPLE 15.19. Consider the graph G obtained by gluing together two copies of K_5 along an edge ab, and then removing this edge (Figure 15.8). This graph is clearly not rigid in 3-space: in every generic representation $\mathbf{u}$ we can rotate one copy of K_5 along the line $\mathbf{u}_a\mathbf{u}_b$ while keeping the other fixed. On the other hand, its number of edges is $18 = 3\cdot 8 - \binom{4}{2}$, which is "just right", so this graph is stressed. But all of its subgraphs H have at most $3|V(H)|-6$ edges, and so the existence of a stress on $(G, \mathbf{u})$ does not follow by counting dimensions. This graph is a minimal stressed graph, showing that Corollary 15.15 does not remain valid in 3-space. ♦

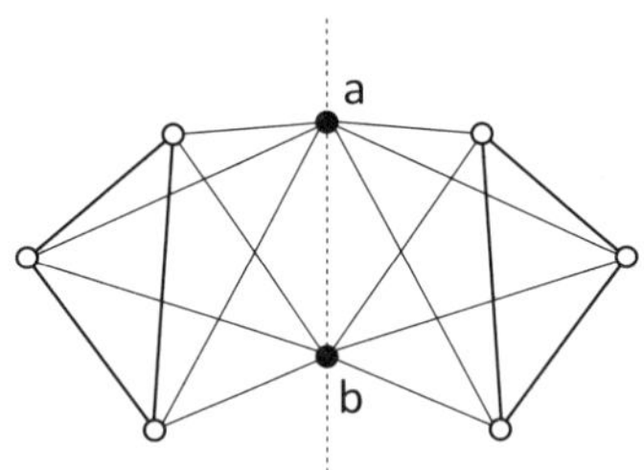

FIGURE 15.8. A nonrigid graph in 3-space.

Condition (GL) generalizes to any dimension, as a *necessary* condition for stress-freeness:

(GL-d) *Every set $S \subseteq V$ with $k = |S| \geq d$ spans at most $dk - \binom{d+1}{2}$ edges.*

Example 15.19 shows that this condition is not sufficient for $d = 3$ (and it is easy to generalize this counterexample to all $d \geq 3$).

While a complete characterization of stress-free or rigid generic frameworks in higher dimension is not available (it may not exist at all), there are a number of interesting and highly nontrivial results, which we will discuss in the rest of this section.

REMARK 15.20. Let us add a discussion of the complexity theoretic aspects of these characterizations. Violation of the Geiringer–Laman condition (GL) gives a simple certificate that a graph *is not* generically stress-free in the plane. Any sequence of Henneberg's construction building up the graph is a certificate that the graph *is* generically stress-free. So generic stress-freeness is in NP$\cap$co-NP. One can use basic matroid algorithms to show that generic stress-freeness is in fact in P. Similar arguments can be given for generic rigidity of a graph in the plane.

We noted that stress-freeness of a framework can be easily tested (in polynomial time) using elementary linear algebra. This is correct if the coordinates of the points are given as rational numbers. For a generic framework, "algebraically independent transcendentals" cannot be specified in a way to facilitate computations with them, and it is not known whether the problem of deciding whether a graph is generically stress-free in dimension 3 (or in any larger dimension) is in P.

Generic stress-freeness of a graph is a property that is in NP for every dimension. One can use the Schwartz–Zippel Lemma 3.12, similarly is in Section 3.3.3 to show that a generically stress-free graph has a stress-free realization with rational coordinates whose numerators and denominators have polynomially bounded numbers of digits. This realization can serve as certificate, whose validity can be easily checked. It is not known, however, how to certify that a generic realization is *not* stress-free; more generally, it is not known whether generic stress-freeness of a graph is in co-NP for dimensions 3 or larger.

15.3.3. Genericity and finite motions. For generic frameworks, there is no difference between infinitesimal nonrigidity and flexibility [**Asimow–Roth 1978**]:

THEOREM 15.21. *A generic framework is infinitesimally rigid if and only if it is locally rigid.*

PROOF. Lemma 15.5 implies the "only if" part. The proof of the "if" part is quite similar to the proof of Lemma 15.7. Suppose that the generic framework $(G, \mathbf{u})$ is not infinitesimally rigid in $\mathbb{R}^d$, we want to show that it has a proper flexing. The key step is to show that there is a vector labeling $\mathbf{v}$ whose entries can be expressed as polynomials of the node positions so that $\mathbf{v}(\mathbf{x})$ is an infinitesimal motion for every vector labeling $\mathbf{x}$ in a neighborhood of $\mathbf{u}$, and $\mathbf{v}(\mathbf{u})$ is not a rigid infinitesimal motion.

For every minimal set C of linearly dependent columns of the rigidity matrix $Rf = R_{G,\mathbf{u}}$, there is a nonzero infinitesimal motion $\mathbf{v}_C$ whose support is C. Note that $\mathbf{v}_C$ is uniquely determined up to a scalar factor, and every infinitesimal motion is a linear combination of the vector labelings $\mathbf{v}_C$ belonging to minimal dependent sets C of columns. Therefore there is a minimal dependent set C for which $\mathbf{v}_C$ is not a rigid motion. From Cramer's Rule it follows that the entries of $\mathbf{v}_C$ can be chosen as polynomials $\mathbf{w}_i(\mathbf{u})$ of the entries of $\mathbf{u}$ with integral coefficients.

The equation $(\mathbf{u}_i - \mathbf{u}_j)^\mathsf{T}(\mathbf{w}_i(\mathbf{u}) - \mathbf{w}_j(\mathbf{u})) = 0$ is an algebraic equation in the entries of $\mathbf{u}$, and so the assumption that these entries are algebraically independent implies that it holds identically in $\mathbf{u}$. In other words, if we define $\mathbf{w}_j(\mathbf{x})$ for every labeling $\mathbf{x} : V \to \mathbb{R}^d$ using the same polynomials as in $\mathbf{v}_C$, then and so $\mathbf{w}(\mathbf{x})$ is an infinitesimal motion of $(G, \mathbf{x})$ for every $\mathbf{x}$. Trivially, $\mathbf{v} = \mathbf{w}(\mathbf{u})$ is not a rigid infinitesimal motion.

From here, we conclude just as in the proof of Lemma 15.7. $\square$

15.3.4. Genericity and edge lengths. The following theorem, adopted from [**Connelly 2005**], connects edge-lengths and stress-freeness.

THEOREM 15.22. (a) *A generic framework is stress-free if and only if its edge-lengths are algebraically independent.* (b) *Every framework with algebraically independent edge-lengths is stress-free.*

PROOF. (a) Assume that the edge-lengths are algebraically dependent; then there is a polynomial $F(x_e:\ e\in E)$ with rational coefficients that is not identically zero, but

$$F(...,|\mathbf{u}_i-\mathbf{u}_j|^2,...)=0 \tag{15.17}$$

(where $|\mathbf{u}_i-\mathbf{u}_j|^2$ is substituted for the variable x_{ij}, $ij\in E$). Consider such a polynomial F with minimum degree. Since the coordinates of the $\mathbf{u}_i$ are algebraically independent, (15.17) holds identically for any choice of representing vectors $\mathbf{u}$. Differentiating with respect to the coordinates of $\mathbf{u}_i$, we get the equation

$$\sum_{j\in N(i)} F_{ij}(...,|\mathbf{u}_s-\mathbf{u}_t|^2,...)(\mathbf{u}_i-\mathbf{u}_j)=0 \tag{15.18}$$

(here F_{ij} is the derivative of F according to the variable x_{ij}). Since F is a nonconstant polynomial, at least one derivative F_{ij} is not identically 0. Then $F_{ij}(...,|\mathbf{u}_s-\mathbf{u}_t|^2,...)\neq 0$, since otherwise we could replace F by F_{ij}. But this means that $S_{ij}=F_{ij}(...,|\mathbf{u}_s-\mathbf{u}_t|^2,...)$ defines a nonzero stress on $(G,\mathbf{u})$.

Conversely, assume that the edge-length are algebraically independent. We may also assume that the representation of the graph does not lie on any affine hyperplane, since otherwise we can proceed by induction on the dimension. We may also assume that we cannot add any further edge so as to preserve this property. We can fit the framework to the coordinate system, and thereby reduce the number of algebraically independent entries in $\mathbf{u}$ to $nd-\binom{d+1}{2}$; this implies that the number of edges is at most $nd-\binom{d+1}{2}$.

We claim that the framework is infinitesimally rigid. Suppose not, then by Theorem 15.21 it has a flexing. During this, the distance of at least one nonadjacent pair of nodes (i,j) must change, and since it changes continuously, at some time i and j will have a distance that is algebraically independent of the edge lengths. Let us add this edge to create a graph G', and let $\mathbf{v}:\ V\to\mathbb{R}^d$ be the node positions at this time. Since we could not add the edge ij to G in its original position, the distance $|\mathbf{u}_i-\mathbf{u}_j|$ must be algebraically dependent on the edge-lengths of G. This determines an algebraic relation between the entries of $\mathbf{u}$, and since it is generic, this algebraic relation must hold for all node positions, in particular when substituting $\mathbf{v}$ for $\mathbf{u}$, a contradiction.

Thus the framework is infinitesimally rigid. This implies that the number of edges must be exactly $dn-\binom{d+1}{2}$, and they cannot carry a stress by (15.11). This completes the proof of (a).

(b) This part of the proof uses deeper tools than those used before. Fix the graph G, and consider the squared-edge-length map $f:\ \mathbb{R}^{d\times V}$ defined by (15.12). The framework $(G,\mathbf{u})$ is stress-free if and only if the rank of $Df(\mathbf{u})$ is m. We know by (a) that this occurs when $\mathbf{u}$ is generic. So the maximum rank of $Df(\mathbf{u})$, over all vector labelings $\mathbf{u}:\ V\to\mathbb{R}^d$, is equal to m. Furthermore, a vector $\mathbf{w}\in\mathbb{R}^E$ of squared edge-lengths is realized by some framework $(G,\mathbf{u})$ carrying a nonzero stress if and only if it is a critical value of the map f. By Sard's Theorem [**Sard 1942**], the set of such vectors $\mathbf{w}$ has measure 0 in $\mathbb{R}^E$.

On the other hand, consider the set of "bad" realizations (critical points of f) $K = \{\mathbf{u} : \operatorname{rk}(Df(\mathbf{u})) < m\}$. It is clear that this is a real algebraic variety over the rationals: the condition $\operatorname{rk}(Df(\mathbf{u})) < m$ is equivalent to saying that $\det(N) = 0$ for all $m \times m$ subdeterminants N of R. Hence the set $\{(\mathbf{u}, f(\mathbf{u})) : \mathbf{u} \in K\}$ is an algebraic variety over the rationals as well. By the Tarski–Seidenberg theorem on semialgebraic sets, the set $f(K)$ is semialgebraic over the rationals, i.e., it can be described as $L_1 \cup \dots \cup L_s$, where each L_i is the solution set of a finite system of polynomial equations and strict polynomial inequalities with rational coefficients.

Now assume that the given linkage (G, ℓ) has a realization $(G, \mathbf{u})$ with a nonzero stress. This means that there is a $\mathbf{w} \in f(K)$, where $\mathbf{w} \in \mathbb{R}^E$ is defined by $w_e = \ell(e)^2$. Since the entries of ℓ are algebraically independent, so are the entries of $\mathbf{w}$, and it follows that they cannot satisfy any nontrivial polynomial equation with rational coefficients. So $\mathbf{w} \in L_i$ for some L_i that is defined by strict polynomial inequalities. But this means that ℓ is an interior point of $f(K)$, which implies that $f(K)$ has positive measure, a contradiction. □

This theorem implies a purely algebraic characterization of generic stress-freeness of a graph G. Consider dn variables $x_{i\nu}$ $(i = 1, \dots, n; \nu = 1, \dots, d)$ and the polynomials

$$p_{ij} = \sum_{\nu=1}^{d} (x_{i\nu} - x_{j\nu})^2 \quad (ij \in E).$$

COROLLARY 15.23. *A graph G is generically stress-free in $\mathbb{R}^d$ if and only if the polynomials p_{ij} $(ij \in E)$ are algebraically independent.* □

Combining parts (a) and (b) of Theorem 15.22, we see that if a generic framework is stress-free, then every other realization of the same linkage is stress-free as well. An important and surprising generalization of this fact was proved in [**Connelly 2005**]: If we consider a generic framework $(G, \mathbf{u})$ and another (not necessarily generic) realization $(G, \mathbf{v})$ of the same linkage, then not only are their stress spaces simultaneously nonempty, and (as it also follows) not only have they the same dimension, but these stress spaces are equal as linear subspaces of $\mathbb{R}^E$. (Figure 15.9 shows that this conclusion does not remain valid without the genericity assumption.)

LEMMA 15.24. *Let $(G, \mathbf{u})$ and $(G, \mathbf{v})$ be two realizations of the same linkage in $\mathbb{R}^d$, and assume that one of them is generic. Then $\mathsf{Str}(G, \mathbf{u}) = \mathsf{Str}(G, \mathbf{v})$.*

PROOF. Let $(G, \mathbf{u})$ be generic, and let $(G', \mathbf{u})$ be a maximal stress-free subgraph of $(G, \mathbf{u})$. Then $\dim(\mathsf{Str}(G' + e, \mathbf{u})) = 1$ for every $e \in E \setminus E(G')$; let S^e be a nonzero stress on $(G' + e, \mathbf{u})$. Trivially, S^e is not zero on the edge e; this implies that the stresses S^e $(e \in E \setminus E(G'))$ are linearly independent. It also follows that they generate $\mathsf{Str}(G, \mathbf{u})$: Let S be any stress on $(G, \mathbf{u})$, then there is a linear combination $\sum_e a_e S^e$ that matches S on the edges in $E \setminus E(G')$, and so the stress $S - \sum_e a_e S^e$ is supported on $E(G')$. Since $(G', \mathbf{u})$ is stress-free, it follows that $S = \sum_e a_e S^e$.

Our next goal is to show that every S^e is a stress on $(G' + e, \mathbf{v})$ as well. Theorem 15.22 implies that the edge lengths of $(G' + e, \mathbf{u})$ are algebraically dependent. Using this, the stress S^e can be constructed on $(G' + e, \mathbf{u})$ just like in the proof of

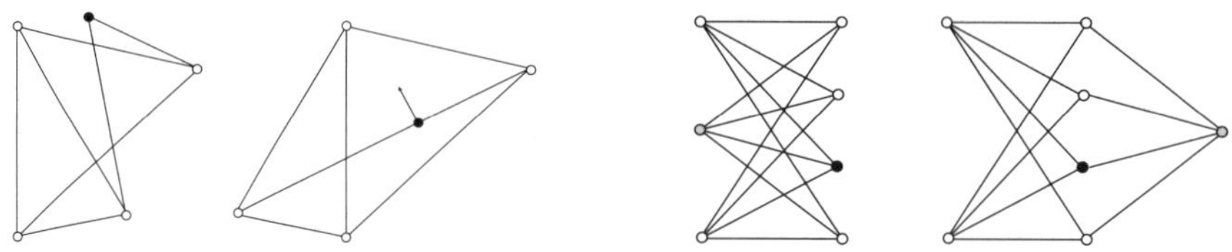

(a) (b) (c) (d)

FIGURE 15.9. (a)–(b) Two realizations of the same linkage in the plane, only one of which has a stress: (a) is stress-free; (b) has an infinitesimal motion as shown, and therefore it carries a stress. (c)–(d) Two realizations of the same linkage in the plane with different stress spaces. The graph $K_{4,3}$ has 7 nodes and $12 > 2\cdot 7-3$ edges, and so every realization in the plane has a stress. The values of any stress on the three edges incident with the black node must have the same sign in (d), but not in (c).

Theorem 15.22(a), using (15.18). Note, however, that (15.18) is an identity, and so it holds true when we replace the coordinates of $\mathbf{u}$ by the coordinates of $\mathbf{v}$:

$$\sum_{j:\ ij\in E(G'+e)} F_{ij}(..., |\mathbf{v}_s-\mathbf{v}_t|^2, ...)(\mathbf{v}_i-\mathbf{v}_j) = 0.$$

But $\mathbf{v}$ realizes the same linkage as $\mathbf{u}$, and so $|\mathbf{v}_i-\mathbf{v}_j| = |\mathbf{u}_i-\mathbf{u}_j|$ for every edge ij of G. So this equation gives the same nonzero stress S^e on $(G'+e,\mathbf{v})$ as on $(G'+e,\mathbf{u})$. As mentioned above, the framework $(G',\mathbf{v})$ is stress-free, so this is a unique stress on $(G'+e,\mathbf{v})$. By the same argument as above, the stresses S^e generate $\mathsf{Str}(G,\mathbf{v})$, which proves the lemma. □

15.4. The realization space

What can we say about the space of all realizations of a linkage, at least in the generic case? Sharpening results of [**Hendrickson 1992**], the following theorem [**Jackson–Keevash 2011**] asserts that the realization space of a generic framework is a smooth manifold. To see that this statement is far from obvious, let us have a look at the simple example of a parallelogram in the plane with a triangle attached (Figure 15.10). To factor out congruences, let us nail down the vertices of the bottom triangle. (The space of all realizations could be obtained as the product of this space with $\mathbb{R}^2\times S^1\mathbb{Z}_2$, taking into account the isometric transformations.) In a typical position, the parallelogram can be deformed in two directions. But in the special position when it is collinear, there are four possible directions we can deform it. So the realization space is not a manifold, but the union of four arcs connecting the two collinear realizations.

THEOREM 15.25. *The space of realizations of a linkage defined by a generic framework* $(G,\mathbf{u})$ *in* $\mathbb{R}^d$ *is a smooth manifold of dimension* $\dim(\mathsf{Inf}_{G,\mathbf{u}})$.

PROOF. Let $t = \mathrm{rk}(R_{G,\mathbf{u}})$. Since $(G,\mathbf{u})$ is generic, we have $\mathrm{rk}(R_{G,\mathbf{v}}) \le \mathrm{rk}(R_{G,\mathbf{u}}) = t$ for every vector labeling $\mathbf{v}$. By Lemma 15.24 and (15.11), we have $\mathrm{rk}(R_{G,\mathbf{v}}) = t$ for every vector labeling $\mathbf{v}$ realizing the linkage $(G,\ell_{\mathbf{u}})$. The matrix

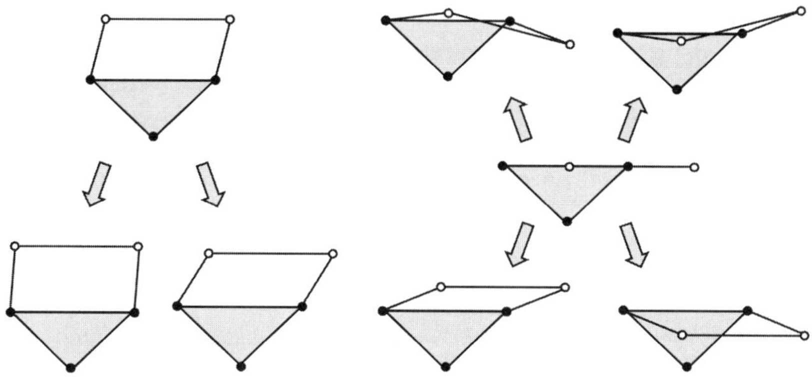

FIGURE 15.10. Motions of a parallelogram with a triangle attached do not form a manifold. The triangle is fixed; this factors out isometries.

$R_{G,\mathbf{v}}$ is the Jacobian of the map $f(\mathbf{v}) = (|\mathbf{v}_i - \mathbf{v}_j|^2 : ij \in E)$, and so it has maximum rank t for every $\mathbf{v} \in f^{-1}(\ell_{\mathbf{u}})$. By basic differential geometry, this implies that $f^{-1}(\ell_{\mathbf{u}})$ is a smooth manifold of dimension $nd - t$. □

Theorem 15.25 is not quite what we would like to see: we don't want to distinguish between congruent frameworks. Define a *reduced realization* as an equivalence class of congruent realizations. Is the space of reduced realizations (endowed with a metric in the natural way) still a manifold? We must be careful, as shown by the following example.

EXAMPLE 15.26. Consider a path of length 2 with nodes a, b, c (in this order), where the two edges have algebraically independent lengths α and β. Any 2-dimensional realization of this linkage can be encoded by specifying the position of b, and the two angles formed by the edges with the positive axis. So the realization space is $\mathbb{R}^2 \times S^1 \times S^1$, a nice 4-dimensional manifold.

If we identify congruent realizations, then we can uniquely describe a reduced realization $\mathbf{u}$ by the distance of $\mathbf{u}_a$ and $\mathbf{u}_c$, which fill the interval $[|\alpha - \beta|, \alpha + \beta]$. So the reduced realization space is homeomorphic to a segment, which is not a manifold. Notice that the trouble (violation of the manifold property) is caused by those realizations that are collinear. ♦

THEOREM 15.27. *The space of reduced full-dimensional realizations of a linkage defined by a generic framework* $(G, \mathbf{u})$ *in* $\mathbb{R}^d$ *is a differentiable manifold of dimension* $\dim(\mathsf{Inf}_{G,\mathbf{u}}) - \binom{d+1}{2}$.

The linkage may or may not have realizations in a lower dimension, but if it does, these are excluded from consideration.

PROOF. The proof combines the proof method of Theorem 15.25 with standard techniques in differential geometry, and we give a sketch only. Select an ordered set $S = (s_0, \dots, s_d)$ of $d+1$ nodes, and consider all vector labelings in which these nodes are affine independent. We can fit S to the coordinate system; this selects a particular labeling from each equivalence class of congruent labelings. The set of all vector labelings with S fitted to the coordinate system is described by $dn - \binom{d+1}{2}$ variables, d of which are restricted to the positive semiline.

By the same argument as above, the realizations $\mathbf{v}$ of $(G, \ell_{\mathbf{u}})$ in which $\mathbf{v}(S)$ is affinely independent is a t-dimensional manifold M_S. For realizations in which both of two $(d+1)$-sets of nodes S and S' are labeled by affinely independent vectors, congruent realizations in the manifolds M_S and $M_{S'}$ can be identified, and the identification maps are smooth. So the manifolds M_S form an atlas of charts, endowing the reduced realization space with a differentiable structure. □

Since the space of all realizations is trivially compact, we get a stronger statement if every realization is full-dimensional.

COROLLARY 15.28. *If the linkage defined by a generic framework* $(G, \mathbf{u})$ *in* $\mathbb{R}^d$ *has no realization in* $\mathbb{R}^{d-1}$, *then the space of all reduced realizations in* $\mathbb{R}^d$ *is a compact differentiable manifold of dimension* $\dim(\mathsf{Inf}_{G,\mathbf{u}}) - \binom{d+1}{2}$.

15.5. Global rigidity

Let us recall that a framework in $\mathbb{R}^d$ is *globally rigid*, if every other realization of the same linkage in $\mathbb{R}^d$ is congruent to it. In other words, the distances between adjacent nodes determine the distances between nonadjacent nodes.

There are a large number of applied problems which can be modeled using the concept of global rigidity rather than local or infinitesimal rigidity. As an example, suppose we have a number of sensors scattered in the plane or in space. These could be ships on the ocean, or animals with embedded chips, or perhaps, in the future, diagnostic nano-devices in the body. We are able to measure the distance between some pairs (say, by the delay of radio signals), and would like reconstruct their geometric positions. This can be done if and only if the framework of the devices as nodes and the measured distances as edges is globally rigid.

One expects that global rigidity is more difficult to handle than local rigidity, since two different realizations of the same linkage may not be continuously movable into each other, and so analytic tools like differentiation are not available here. This is indeed true, but there is a surprising variety of nontrivial results on global rigidity, of which we give a sampler.

Let us warm up with the 1-dimensional case. Trivially, a disconnected graph is not rigid on the line in any sense, and it is not hard to see that every connected graph is locally and infinitesimally rigid with any representation. Global rigidity is more complicated. Our first observation is that if G has a cutnode a, and so we can write $G = G_1 \cup G_2$ where $V(G_1) \cap V(G_2) = \{a\}$, $|V(G_i)| > 1$, then from any representation $(G, \mathbf{u})$ we can create another representation $(G, \mathbf{u}')$ by reflecting the nodes of G_1 in u_a. This shows that no representation can be globally rigid unless either all nodes of G_1 or all nodes of G_2 are represented by the same point.

Assume that G is 2-connected. Even if the nodes are represented by different points, global rigidity does not follow.

EXAMPLE 15.29. Consider a quadrilateral on nodes $V(C_4) = \{a, b, c, d\}$ (in this cyclic order), and its 1-dimensional representation $u_a = 0, u_b = 1, u_c = 3, u_d = 2$ (Figure 15.11). Since C_4 is connected, this representation is locally and infinitesimally rigid. Mapping the nodes to $v_a = 2, v_b = 3, v_c = 1, v_d = 0$, we get another realization of the same linkage, which is not congruent to the first one, since $|u_a - u_c| = 3$ but $|v_a - v_c| = 1$. Thus $(G, \mathbf{u})$ is not globally rigid on the line. It is easy to construct frameworks on the linea that are arbitrarily highly connected but not globally rigid (Exercise 15.13). ♦

FIGURE 15.11. Even a 2-connected framework may not be globally rigid on the line. The pictures are vertically distorted to make the structure visible.

However, if G is 2-connected and the labeling $\mathbf{u} : V \to \mathbb{R}$ is generic, then the framework will be globally rigid on the line. The proof of this fact is left to the reader as Exercise 15.12.

One might expect that if a linkage is not globally rigid, i.e., it has two incongruent realizations, then one can obtain a second realization by slightly deforming the first one; in other words, the framework is flexible. This is, however, not the case, as 1-dimensional representations of any graph with a cutnode show. (It is easy to generalize this example to higher dimensions.)

Which are the globally rigid representations of a graph? In the case of infinitesimal rigidity, generic representations (vector labelings) are the "best", in the sense that if *any* representation is infinitesimally rigid, then *every* generic representation is rigid as well. Nothing like this holds for global rigidity: for example, labeling each node of a connected graph with the same vector gives a globally (even universally) rigid representation, but of course the graph need not be globally rigid (or even locally rigid) in a generic representation. We have seen examples of globally rigid but not infinitesimally rigid graphs in Figures 15.2 and 15.3; Figure 15.12 shows that the 3-cube graph is not globally rigid in generic position in the plane, but it has a nontrivial realization that is globally rigid.

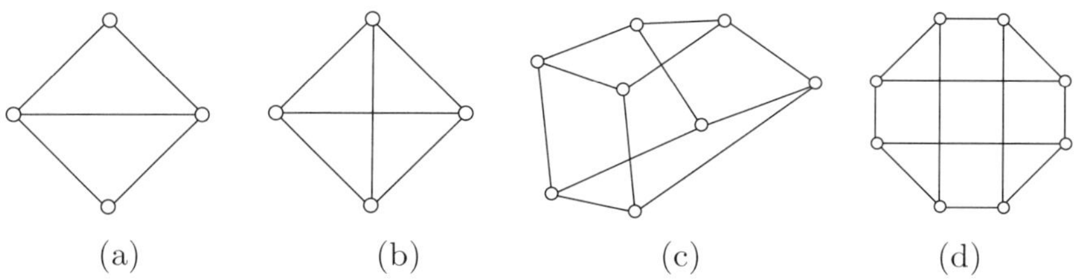

FIGURE 15.12. Structure (a) is not globally rigid in the plane (flip one triangle), but adding an edge (b) we get a globally rigid framework. (c) shows a generic representation of the cube graph, which is not infinitesimally rigid (just count degrees of freedom), and hence it is flexible and so not globally rigid. Representation (d) of the cube is globally rigid; see Exercise 15.16.

So to make the question "Does G have a globally rigid representation?" meaningful, we have to impose some sort of nondegeneracy condition. Genericity is perhaps the most natural, but general position will be another interesting case.

15.5.1. Global rigidity up to an affine transformation. Before starting with the analysis of global rigidity, let us discuss a weaker version of it. We say that a framework $(G, \mathbf{u})$ in $\mathbb{R}^d$ is *globally rigid up to an affine transformation*, if for every vector labeling $\mathbf{v} : V \to \mathbb{R}^d$ realizing the same linkage, there is an affine transformation A of the space such that $\mathbf{v}_i = A\mathbf{u}_i$. Note that the definition of global rigidity only differs in that it requires A to be an isometry.

EXAMPLE 15.30. Figure 15.13 shows a rather trivial example in the plane. Note, however, that a noncongruent affine transformation in the plane preserves the length of segments in at most two directions. So all examples in the plane are similar to this one. ♦

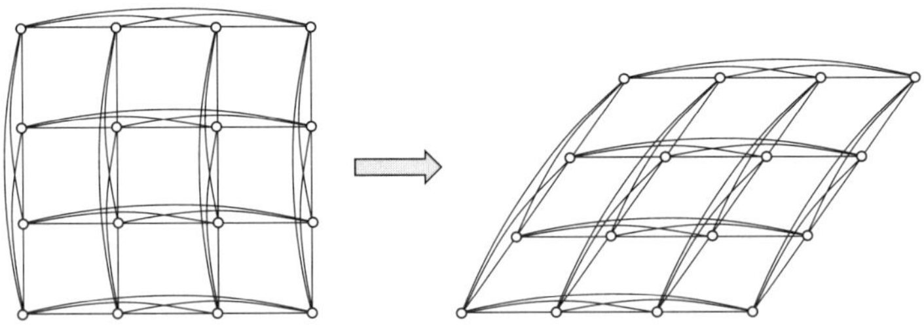

FIGURE 15.13. A planar framework whose edge lengths are preserved by a noncongruent affine transformation. The edges are curved in order to show the structure.

One feels that rigidity up to affine transformations is quite close to global rigidity, but it is not so easy to handle the connection. We start with a rather simple characterization [**Connelly 1982**]. Let us say that in a framework $(G, \mathbf{u})$ in $\mathbb{R}^d$ *the edge directions lie on a conic at infinity*, if there is a symmetric nonzero matrix $Q \in \mathbb{R}^{d\times d}$ such that $(\mathbf{u}_i - \mathbf{u}_j)^{\mathsf{T}} Q(\mathbf{u}_i - \mathbf{u}_j) = 0$ for every edge $ij \in E$. We say that this conic is defined by Q.

LEMMA 15.31. *A framework $(G, \mathbf{u})$ has a noncongruent affine image that preserves the edge lengths if and only if the edge directions lie on a conic at infinity.*

Another way of interpreting this condition is that there is a noncongruent infinitesimal motion where the velocities $Q\mathbf{u}_i$ depend linearly on the positions $\mathbf{u}_i$.

PROOF. Suppose that $(G, \mathbf{v})$ is an affine image of $(G, \mathbf{u})$, realizing the same linkage, but not congruent to $(G, \mathbf{u})$. In other words, there is an affine transformation A of $\mathbb{R}^d$ such that $\mathbf{v}_i = A\mathbf{u}_i$ for all nodes i, but A is not a congruence. We may assume (by translation) that A is a homogeneous linear transformation. Since A preserves the lengths of edges, we have $|A(\mathbf{u}_i - \mathbf{u}_j)| = |\mathbf{u}_i - \mathbf{u}_j|$ for every edge ij, which can be written as $(\mathbf{u}_i - \mathbf{u}_j)^{\mathsf{T}}(A^{\mathsf{T}}A - I)(\mathbf{u}_i - \mathbf{u}_j) = 0$. Since A is not an orthogonal matrix, $Q = A^{\mathsf{T}}A - I \neq 0$ defines the conic at infinity.

The converse follows by a similar argument. □

The matrix Q, defining the conic at infinity, has some further useful properties. Let S be any stress on $(G, \mathbf{u})$. For every node i, we have

$$\sum_j S_{ij}(\mathbf{u}_i - \mathbf{u}_j)^{\mathsf{T}} Q(\mathbf{u}_i - \mathbf{u}_j) = 0,$$

since every term is zero. Expanding, we have

$$\Big(\sum_j S_{ij}\Big)\mathbf{u}_i^{\mathsf{T}} Q\mathbf{u}_i - 2\mathbf{u}_i^{\mathsf{T}} Q\Big(\sum_j S_{ij}\mathbf{u}_j\Big) + \sum_j S_{ij}\mathbf{u}_j^{\mathsf{T}} Q\mathbf{u}_j.$$

The first two terms are zero as S is a stress, so the last term must be zero as well:

$$\sum_{j\in V} S_{ij}\mathbf{u}_j^{\mathsf T} Q\mathbf{u}_j = 0. \tag{15.19}$$

Using a similar expansion, we get that

$$\sum_{j} S_{ij}(\mathbf{u}_j - \mathbf{u}_k)^{\mathsf T} Q(\mathbf{u}_j - \mathbf{u}_l) = 0 \tag{15.20}$$

for any three nodes i, k and l. Also note that if $\mathbf{u}$ is full-dimensional, then $(\mathbf{u}_i - \mathbf{u}_j)^{\mathsf T} Q(\mathbf{u}_i - \mathbf{u}_j) \neq 0$ for at least one nonadjacent pair $i, j \in V$ (this is left to the reader as an exercise).

The following lemma gives some simpler conditions under which affine images that preserve the lengths of edges are automatically congruent images. The first condition was formulated in [**Connelly 2005**], the second in [**Alfakih–Ye 2013**]. As a motivation for condition (b) below, let us recall (14.4): if $(G, \mathbf{u})$ spans $\mathbb{R}^d$, then every stress on it has corank at least $d+1$.

LEMMA 15.32. *Let $(G, \mathbf{u})$ be a framework in $\mathbb{R}^d$, and suppose that one of the following conditions hold:*

(a) *$(G, \mathbf{u})$ is generic and all nodes have degree at least d.*

(b) *$(G, \mathbf{u})$ is in general position and carries a stress of corank $d+1$.*

Then the edge directions do not lie on a conic at infinity.

PROOF. Suppose that the edge directions do lie on a conic at infinity, defined by Q. Since the conditions are invariant under translation, we may assume that $\mathbf{u}_a = 0$ for an appropriately chosen node a.

(a) We have a system of homogeneous linear equations for the entries of Q, and hence the existence of such a matrix Q is equivalent to the vanishing of certain determinants composed of quadratic polynomials of the vectors $\mathbf{u}_i$. If these determinants vanish for the generic vector labeling $\mathbf{u}$, then they vanish for every vector labeling in $\mathbb{R}^d$. So for every other vector labeling $\mathbf{v}$, there is a symmetric nonzero matrix $M \in \mathbb{R}^{d\times d}$ such that $(\mathbf{v}_i - \mathbf{v}_j)^{\mathsf T} M(\mathbf{v}_i - \mathbf{v}_j) = 0$ for every edge $ij \in E$.

Let $\mathbf{v}_i = \binom{1}{\mathbf{u}_i'}$ be obtained by replacing the first entry of $\mathbf{u}_i$ by 1 for $i > 1$, and let $\mathbf{v}_1 = 0$. Consider the matrix M as above, and let M' be obtained from M by deleting its first row and column. For every edge ij with $i, j > 1$, we have

$$(\mathbf{u}_i' - \mathbf{u}_j')^{\mathsf T} M'(\mathbf{u}_i' - \mathbf{u}_j') = (\mathbf{v}_i - \mathbf{v}_j)^{\mathsf T} M(\mathbf{v}_i - \mathbf{v}_j) = 0.$$

Since the graph $G\setminus 1$ has degrees at least $d-1$, and $\mathbf{u}'$ is a generic vector labeling of it in dimension $d-1$, it follows by induction that $M' = 0$. This means that we can write $M = \mathbf{b}\mathbf{e}_1^{\mathsf T} + \mathbf{e}_1^{\mathsf T}\mathbf{b}$ with some nonzero vector $\mathbf{b}$, and so $\mathbf{v}_i^{\mathsf T} M\mathbf{v}_i = 2(\mathbf{b}^{\mathsf T}\mathbf{v}_i)(\mathbf{e}_1^{\mathsf T}\mathbf{v}_i) = 2(\mathbf{b}^{\mathsf T}\mathbf{v}_i)$. For $i \in N(1)$, we have

$$\mathbf{v}_i^{\mathsf T} M\mathbf{v}_i = (\mathbf{v}_i - \mathbf{v}_1)^{\mathsf T} M(\mathbf{v}_i - \mathbf{v}_1) = 0,$$

and hence $\mathbf{b}^{\mathsf T}\mathbf{v}_i = 0$. Since any d of the vectors $\mathbf{v}_i$ $(i \in N(1))$ are linearly independent, this is a contradiction.

(b) Let S be a stress matrix with corank $d+1$. First, note that S cannot have an all-zero row; indeed, deleting such a row (say, row i) and the corresponding column, we would get a stress matrix on $(G\setminus i, \mathbf{u})$ of corank d, while the vectors $\mathbf{u}_j$ $(j \neq i)$ would still span the affine d-space; this would contradict (14.4).

It follows that for every node i, the equation $\sum_j S_{ij}(\mathbf{u}_j - \mathbf{u}_i) = 0$ provides a linear dependence between the vectors $\mathbf{u}_j - \mathbf{u}_i$ ($j \in N(i)$), and hence (by the assumption that the representation is in affinely general position) there must be at least $d+1$ terms with nonzero coefficients. Hence the degree of each node is at least $d+1$. In particular, the vectors $\mathbf{u}_j - \mathbf{u}_i$ ($j \in N(i)$) must span $\mathbb{R}^d$.

Now suppose that the edge directions lie on a conic at infinity defined by the symmetric nonzero matrix Q. Define $\mathbf{q} \in \mathbb{R}^V$ by $q_i = \mathbf{u}_i^\mathsf{T} Q \mathbf{u}_i$, then we have $S\mathbf{q} = 0$ by (15.19). By assumption that S has corank $d+1$, the vector $\mathbb{1}$ and the columns of U^T generate the kernel of S, which means that there is a vector $\mathbf{b} \in \mathbb{R}^d$ and a real number α such that $q_i = \mathbf{u}_i^\mathsf{T} Q \mathbf{u}_i = \mathbf{b}^\mathsf{T}\mathbf{u}_i + \alpha$ for every $i \in V$.

Substituting these expressions, we get for every i and j

$$(\mathbf{u}_i + \mathbf{u}_j)^\mathsf{T} Q (\mathbf{u}_i - \mathbf{u}_j) = \mathbf{b}^\mathsf{T}(\mathbf{u}_i - \mathbf{u}_j).$$

Adding the equation $(\mathbf{u}_i - \mathbf{u}_j)^\mathsf{T} Q (\mathbf{u}_i - \mathbf{u}_j) = 0$ (valid for $ij \in E$), we get

$$2\mathbf{u}_i^\mathsf{T} Q (\mathbf{u}_i - \mathbf{u}_j) = \mathbf{b}^\mathsf{T}(\mathbf{u}_i - \mathbf{u}_j).$$

Since the vectors $\mathbf{u}_j - \mathbf{u}_i$ (i fixed, $j \in N(i)$) span the space, it follows that $2Q\mathbf{u}_i = \mathbf{b}$ for every $i \in V$. But then the representation lies in a proper affine subspace, a contradiction. □

For more on frameworks whose edge directions lie on a conic at infinity, including connections with the Strong Arnold Property and with the completion problem of semidefinite matrices, see [**Connelly–Gortler–Theran 2016b**] and [**Laurent–Varvitsiotis 2014**].

The following lemma provides a way to prove that only affine (but not necessarily congruent) images preserve the lengths of edges.

LEMMA 15.33. *Let* $\mathbf{u} : V \to \mathbb{R}^d$ *and* $\mathbf{v} : V \to \mathbb{R}^d$ *be two vector labelings of the same graph* G, *and suppose that* $(G, \mathbf{u})$ *and* $(G, \mathbf{v})$ *carry the same stress* S *with corank* $d+1$. *Also suppose that* $(G, \mathbf{u})$ *spans* $\mathbb{R}^d$. *Then* $\mathbf{v}$ *is an affine image of* $\mathbf{u}$.

PROOF. Consider a third labeling $\mathbf{w}_i = \binom{\mathbf{u}_i}{\mathbf{v}_i} \in \mathbb{R}^{2d}$. Clearly S remains a stress on this framework too, so by (14.4), the effective dimension of $(G, \mathbf{w})$ is at most d, and so there is a d-dimensional affine subspace $L \subseteq \mathbb{R}^{2d}$ containing all of the vectors $\mathbf{w}_i$. The projection π_1 of L to the first d coordinates must be bijective. Let π_2 be the projection of L to the last d coordinates. It follows that $\pi_1^{-1}\pi_2$ is an affine map from $(G, \mathbf{u})$ onto $(G, \mathbf{v})$. □

THEOREM 15.34. *Assume that a framework* $(G, \mathbf{u})$ *spanning* $\mathbb{R}^d$ *carries a stress* S *whose matrix is negative semidefinite and has corank* $d+1$. *Then* $(G, \mathbf{u})$ *is globally rigid up to an affine transformation.*

Of course, assuming that S is positive semidefinite would lead to the same conclusion: just replace S by $-S$.

PROOF. Let $(G, \mathbf{v})$ be another realization of the framework $(G, \mathbf{u})$ in $\mathbb{R}^d$. Our goal is to show that S is a stress on $(G, \mathbf{v})$ as well; then Lemma 15.33 will complete the proof. Since S is a stress, the energy satisfies $\mathcal{E}(\mathbf{u}) = 0$ by (14.5), and since all edge-lengths are preserved, we have $\mathcal{E}(\mathbf{v}) = \mathcal{E}(\mathbf{u}) = 0$. Since S is negative semidefinite, the labeling $\mathbf{v}$ minimizes the energy, and so S must be a stress on the framework $(G, \mathbf{v})$ as well. By Lemma 15.33, this implies that $\mathbf{v}$ is an affine image of $\mathbf{u}$. □

15.5.2. Global rigidity in general position. The following condition for a graph to have globally rigid vector labelings in general position was proved in [**Alfakih 2017**]:

THEOREM 15.35. *A graph has a globally rigid representation in $\mathbb{R}^d$ in affinely general position if and only if it is either complete or $(d+1)$-connected.*

PROOF. The case of complete graphs is trivial, so suppose that the graph G is not complete. First, suppose that G has a cutset T with $|T| \le d$. Consider any general position vector labeling $\mathbf{u}$ of V, and let H be any hyperplane containing the vectors representing T but no other vector label (by the general position of the labeling, we can choose such a hyperplane H). Reflecting one of the components of $G \setminus T$ in H, we get another representation in which the edge lengths are the same as before, but which is not congruent. So $(G, \mathbf{u})$ is not globally rigid.

To prove the converse, let G be a $(d+1)$-connected graph. By Theorem 10.9, it has an orthogonal representation in general position in $\mathbb{R}^{n-d-1}$. Considering the Gram matrix of these vectors, we get a positive semidefinite G-matrix S of rank $n-d-1$ such that every $(n-d-1)\times(n-d-1)$ submatrix of S is nonsingular.

Next, we construct a vector $\mathbf{v} \in \mathrm{Ker}(S)$ with nonzero entries. It suffices to show that for every $i \in V$ there is a vector $\mathbf{w} \in \mathrm{Ker}(S)$ with $w_i \neq 0$; a "generic" linear combination of such vectors is a vector in $\mathrm{Ker}(S)$ with all coordinates different from 0. If no such vector $\mathbf{w}$ exits, then elementary linear algebra shows that $S\mathbf{a} = \mathbf{e}_i$ for some vector $\mathbf{a} \in \mathbb{R}^V$. S is positive semidefinite, hence it can be written as $S = X^{\mathsf{T}}X$ for some $X \in \mathbb{R}^{(n-d-1)\times V}$, where any $n-d-1$ columns of X are linearly independent. Let Y be any $(n-d-1)\times(n-d-1)$ submatrix not containing column i, then $\mathbf{e}_i = X^{\mathsf{T}}X\mathbf{a}$ implies that $0 = Y^{\mathsf{T}}X\mathbf{a}$. Since Y is nonsingular, this implies that $X\mathbf{a} = 0$, whence $S\mathbf{a} = X^{\mathsf{T}}X\mathbf{a} = 0$, a contradiction.

Scaling the rows and columns of S if necessary, we may assume that $\mathbf{v} = \mathbb{1}$, so $S\mathbb{1} = 0$. We can extend $\mathbb{1}$ by d further column vectors in $\mathbb{R}^n$ to a basis of $\mathrm{Ker}(S)$, and combine these to a single matrix W such that $SW = 0$. The columns of W^{T} have the form $\mathbf{w}_i = \binom{1}{\mathbf{u}_i}$ $(i \in V,\ \mathbf{u}_i \in \mathbb{R}^d)$, and so $SW = 0$ means that S is a stress matrix on the vector labeling $\mathbf{u}$ of G. So $(G, \mathbf{u})$ carries a positive semidefinite stress of rank $n-d-1$, and hence it is globally rigid up to an affine transformation.

If $(G, \mathbf{u})$ is not globally rigid, then its edge directions lie on a conic at infinity, defined by a matrix $Q \in \mathbb{R}^{d\times d}$. By (15.20),

$$\sum_j S_{ij}(\mathbf{u}_j - \mathbf{u}_k)^{\mathsf{T}} Q(\mathbf{u}_j - \mathbf{u}_k) = 0$$

for every i and k. Choose k so that $(\mathbf{u}_j - \mathbf{u}_k)^{\mathsf{T}} Q(\mathbf{u}_j - \mathbf{u}_k) \neq 0$ for at least one j, then the columns of S corresponding $j \in \overline{N}(k)$ are linearly dependent. Since the number of these entries is at most $n-d-1$, this contradicts the choice of S. Thus $(G, \mathbf{u})$ is globally rigid.

What remains is to argue that any $d+1$ of the vectors $\mathbf{u}_i$ are affine independent, or equivalently, any $d+1$ of the vectors $\mathbf{w}_i$ are linearly independent. Suppose not, then there is a nonzero linear combination $\mathbf{b} = W\mathbf{a}$ of the columns of W that has $d+1$ zeroes; but then $S\mathbf{b} = SW\mathbf{a} = 0$, which shows that S has $n-d-1$ linearly dependent columns, a contradiction. $\square$

15.5.3. Generic global rigidity. We say that a graph is *generically globally rigid*, if it is globally rigid for every generic vector labeling. It turns out that it would suffice to require this for a single generic vector labeling; this is a nontrivial fact [**Gortler–Healy–Thurston 2010**], whose proof we do not reproduce here. Figure 15.14 illustrates some of the difficulties.

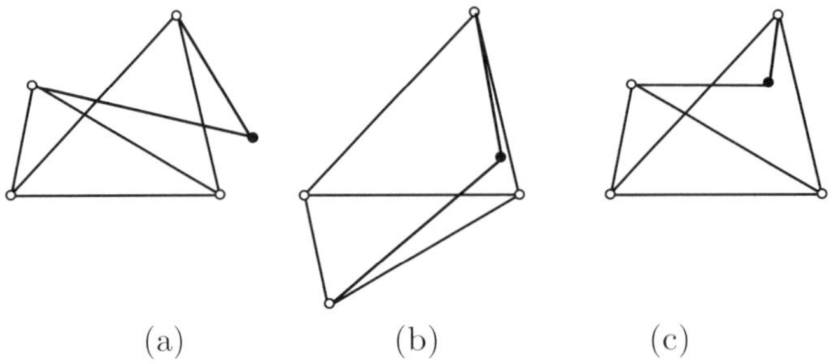

FIGURE 15.14. Three generic frameworks in the plane with the same graph; (a) and (b) realize the same linkage. Three more frameworks are obtained by flipping the black node over the line connecting its neighbors, so neither one of these is globally rigid. (c) cannot be realized so that the two triangles do not overlap. So (a) has four noncongruent realizations, (c) has only two.

In the case of generic frameworks, one can give a full characterization of global rigidity in terms of the existence of appropriate stresses. The "if" part of the following theorem was proved in [**Connelly 2005**], the "only if" part in [**Gortler–Healy–Thurston 2010**]. The proof of the "only if" part, which extends the methods of the proof of Hendrickson's Theorem below, will not be reproduced here.

THEOREM 15.36. *A generic framework $(G, \mathbf{u})$ in $\mathbb{R}^d$ with at least $d+2$ nodes is globally rigid if and only if it carries a stress S whose matrix has corank $d+1$.*

PROOF. (Sufficiency.) Again, we consider another realization $(G, \mathbf{v})$ of the framework $(G, \mathbf{u})$ in $\mathbb{R}^d$. By Lemma 15.24, S is a stress on the framework $(G, \mathbf{v})$ as well, and so by Lemma 15.33, $(G, \mathbf{v})$ is an affine image of $(G, \mathbf{u})$. Lemma 15.32(b) completes the proof. □

While this theorem gives an interesting and nontrivial condition for generic global rigidity, this condition is not easy to verify. It is related to two simpler algorithmic problems that are notorious for their unknown complexity: as we have seen, no polynomial time algorithm is known to decide whether a generic framework carries a stress, and no polynomial time algorithm is known to determine the maximum rank in a linear space of matrices (in this case, in the space of stresses). However, the methods developed for these two problems do yield a randomized polynomial time algorithm to determine whether a graph is globally rigid in $\mathbb{R}^d$ (see Exercise 15.15).

From these considerations, it follows that more combinatorial necessary and/or sufficient conditions for generic global rigidity are valuable. There is a nontrivial necessary condition for global rigidity [**Hendrickson 1992**], which reveals that generic global rigidity is related to generic (local) rigidity in interesting ways.

THEOREM 15.37. *If a graph G on at least $d+2$ nodes is generically globally rigid in $\mathbb{R}^d$, then it is $(d+1)$-connected and deleting any edge it remains generically rigid in $\mathbb{R}^d$.*

PROOF. The first condition follows from (the easy part of) Theorem 15.35.

To prove the second, suppose that $G' = G\setminus e$ is not rigid for some edge $e = ab \in E$. This means that for some (or any) generic labeling $\mathbf{u}$ in $\mathbb{R}^d$, the framework $(G', \mathbf{u})$ is not infinitesimally rigid. Since $(G, \mathbf{u})$ is infinitesimally rigid, it follows that $\mathrm{rk}(R_{G,\mathbf{u}}) = dn - \binom{d+1}{2}$ and $\mathrm{rk}(R_{G',\mathbf{u}}) = dn - \binom{d+1}{2} - 1$.

Every realization of the linkage $(G', \ell_{\mathbf{u}})$ spans $\mathbb{R}^d$. Indeed, let $(G'', \mathbf{u})$ be a maximal stress-free subgraph of $(G', \mathbf{u})$. Then, using that $n > d+1$,

$$|E(G'')| \geq \mathrm{rk}(R_{G'',\mathbf{u}}) = \mathrm{rk}(R_{G',\mathbf{u}}) = dn - \binom{d+1}{2} - 1 > (d-1)n - \binom{d}{2},$$

and hence every $(d-1)$-dimensional realization of $(G'', \ell_{\mathbf{u}})$ would have to carry a stress, contradicting Lemma 15.24.

By Corollary 15.28, the space of reduced realizations of the linkage $(G', \ell_{\mathbf{u}})$ is a compact manifold of dimension 1, i.e., it consists of one or more closed Jordan curves. Let J be the curve through the congruence class of $\mathbf{u}$.

The function $f(\mathbf{x}) = |\mathbf{x}_a - \mathbf{x}_b|^2$ is invariant under congruences, and hence it can be viewed as a (smooth) function on J. The vector labeling $\mathbf{u}$ cannot be a stationary point of f: this would give an algebraic equation between the coordinates of $\mathbf{u}$, which would be identically true, and hence f would be a constant, and moving along J we would get a flexing of the whole graph G (not only of G'). Hence $\mathbf{u}$ is neither the maximizing nor the minimizing argument of f, and hence there is another realization $\mathbf{v}$ along J such that $f(\mathbf{u}) = f(\mathbf{v})$. This means that $(G, \mathbf{v})$ is a noncongruent realization of the linkage $(G, \ell_{\mathbf{u}})$, and so $(G, \mathbf{u})$ is not globally rigid. □

The converse of Theorem 15.37 is not valid [**Connelly 2005**]: $K_{5,5}$ is not globally rigid in $\mathbb{R}^3$, but it meets the necessary conditions in Hendrickson's Theorem. On the other hand, these conditions are sufficient in the plane [**Jackson–Jordán 2005**]. In fact, Jackson and Jordán prove the following result, quite analogous to Theorem 15.11 (for the proof, the reader is referred to the original paper).

THEOREM 15.38. *For a graph G on at least 4 nodes, the following are equivalent:*

(a) *G is globally rigid in the plane;*

(b) *G is 3-connected and deleting any edge, it remains rigid in the plane;*

(c) *G can be obtained from K_4 by a succession of the second Henneberg constructions* (H2) *and edge additions.* □

EXERCISE 15.1. Prove that a framework $(G, \mathbf{u})$ is flexible on the line if and only if G is disconnected.

EXERCISE 15.2. Prove that in the plane $i \mapsto \mathbf{e}_1$, $i \mapsto \mathbf{e}_2$ and $i \mapsto R\mathbf{u}_i$ (where R denotes rotation by 90°) are infinitesimal congruences, which generate Rig. In 3-space, $i \mapsto \mathbf{e}_1$, $i \mapsto \mathbf{e}_2$, $i \mapsto \mathbf{e}_3$, $i \mapsto \mathbf{e}_1 \times \mathbf{u}_i$, $i \mapsto \mathbf{e}_2 \times \mathbf{u}_i$ and $i \mapsto \mathbf{e}_3 \times \mathbf{u}_i$ generate Rig. For which frameworks are these generators linearly independent?

EXERCISE 15.3. (a) Show that (15.5) defines an infinitesimal motion. (b) Let $U(t)$ ($t \in [0,1]$) be a smooth family of congruences in $\mathbb{R}^d$. Prove that $(d/dt)(0)\mathbf{u} = \mathbf{w} + A\mathbf{u}$, where $\mathbf{w} \in \mathbb{R}^d$ and A is a skew symmetric matrix. Conversely, every affine transformation $\mathbf{u} \mapsto \mathbf{w} + A\mathbf{u}$, where $\mathbf{w} \in \mathbb{R}^d$ and A is a skew symmetric matrix, arises from a smooth family of congruences.

EXERCISE 15.4. Let $\mathbf{u}$ be a representation of G in $\mathbb{R}^d$. Prove that a vector $\tau \in \mathbb{R}^E$ is in the orthogonal complement of $\mathsf{Str}(G, u)$ if and only if there are vectors $\mathbf{v}_i \in \mathbb{R}^d$ such that $\tau_{ij} = (\mathbf{v}_i - \mathbf{v}_j)^\mathsf{T}(\mathbf{u}_i - \mathbf{u}_j)$ for every edge ij.

EXERCISE 15.5. Let $(G, \mathbf{u})$ be a rigid framework that has at least one cable or strut. Prove that it has a nonzero stress.

EXERCISE 15.6. Let G be the octahedron graph, where $V = \{1, 2, 3, 4, 5, 6\}$ and $\overline{E} = \{12, 34, 56\}$. Consider the following vector labeling in $\mathbb{R}^3$: let $\mathbf{u}_1, \mathbf{u}_2, \mathbf{u}_3, \mathbf{u}_4$ be points of the unit circle (in this cyclic order) of the plane $z = 0$ so that $\measuredangle(\mathbf{u}_2, \mathbf{u}_3) = \measuredangle(\mathbf{u}_4, \mathbf{u}_1)$. Let $\mathbf{u}_5$ and $\mathbf{u}_6$ be arbitrary points on the z-axis. Prove that $(G, \mathbf{u})$ is flexible.

EXERCISE 15.7. Let $\mathbf{u}, \mathbf{v}^1, \mathbf{v}^2 \dots$, be noncongruent realizations of the same linkage (G, ℓ), and suppose that $\mathbf{v}_i^n \to \mathbf{u}_i$ for all nodes. Prove that $(G, \mathbf{u})$ has a nontrivial flexing.

EXERCISE 15.8. Let $(G, \mathbf{u})$ be a framework and $\mathbf{v}$, an infinitesimal motion. Prove that the frameworks $(G, \mathbf{u} + \mathbf{v})$ and $(G, \mathbf{u} - \mathbf{v})$ define the same linkage (have the same edge lengths).

EXERCISE 15.9. Let $(G, \mathbf{u})$ be a generic framework with a stress S. Prove that there is a nonzero polynomial F in m variables x_{ij} ($ij \in E$) such that $S_{ij} = F_{ij}(\dots, |\mathbf{u}_s - \mathbf{u}_t|^2, \dots)$.

EXERCISE 15.10. Is the configuration space of the Peaucellier–Lipkin linkage (Example 15.1) a closed manifold?

EXERCISE 15.11. Prove that if linkage with algebraically independent edge lengths has a realization, then it has a generic realization.

EXERCISE 15.12. . Prove that a graph G is generically globally rigid on the line if and only if it is 2-connected.

EXERCISE 15.13. For every $k \geq 1$, construct a k-connected framework on the line that is not globally rigid.

EXERCISE 15.14. Prove that if a framework $(G, \mathbf{u})$ is globally rigid up to an affine transformation, but not globally rigid, then there is a nonzero matrix $Y \in \mathbb{R}^{V \times V}$ such that $Y_{ij} = 0$ for $ij \in E \cup \triangle$, and $YS = 0$ for every stress on $(G, \mathbf{u})$.

EXERCISE 15.15. Prove that generic global rigidity of a graph in $\mathbb{R}^d$ can be decided in randomized polynomial time.

EXERCISE 15.16. Prove that the framework in Figure 15.12(d) is globally rigid in the plane.

CHAPTER 16

The Colin de Verdière Number

Let G be a connected simple graph. We know that (by the Perron–Frobenius Theorem) the largest eigenvalue of the adjacency matrix has multiplicity 1. What can we say about the multiplicity of the second largest?

The eigenvalues of most graphs are all different, so the multiplicity of any of them gives any information about the graph in rare cases only. Therefore, it makes sense to try to maximize the multiplicity of the second largest eigenvalue by weighting the edges by positive numbers. The diagonal entries of the adjacency matrix do not carry any information, so we allow putting there arbitrary real numbers. The off-diagonal matrix entries that correspond to nonadjacent pairs of nodes remain 0.

There is a technical restriction, transversality (Strong Arnold Property as defined in Section 10.5), to exclude choices of edgeweights and diagonal entries that are too degenerate.

Since we do not put any restriction on the diagonal entries, we may add a constant to the diagonal entries to shift the spectrum so that the second largest eigenvalue becomes 0, without changing its multiplicity. The multiplicity in question is then the corank of the matrix (the dimension of its nullspace).

Finally, we multiply the matrix by -1, to follow convention. Recall that we have met such matrices in Chapter 14: canonical braced stress matrices of 3-polytopes have all these properties (cf. Proposition 14.14).

This multiplicity, as a graph parameter, was introduced and first studied in [**Colin de Verdière 1986**] and [**Colin de Verdière 1991**]; for a survey, see [**van der Holst et al. 1999**]. Its most interesting feature is that it is related to topological properties of the graph: for example, its value is at most 3 if and only if the graph is planar. For this book, its connection to geometric representations makes it particularly interesting. For example, we are going to show that for 3-connected planar graphs, the optimizing matrices in the definition above correspond to Steinitz representations (representations as skeletons of convex 3-polytopes).

16.1. Basics

We continue with the formal definition. Recall that the a G-matrix A is *transversal*, or has the *Strong Arnold Property*, if $AX \neq 0$ for every nonzero $\overline{G}$-matrix X with zero diagonal.

A well-signed G-matrix with exactly one negative eigenvalue and with the Strong Arnold Property will be called a *Colin de Verdière matrix* of the graph G. The *Colin de Verdière number* $\mu(G)$ of the graph G is defined as the maximum corank of any Colin de Verdière matrix of G.

We can formulate this problem as a minimum dimension problem for orthogonal representations—not in Euclidean space, but in Minkowski space. To define this space, we equip the linear space $\mathbb{R}^n$ with an indefinite inner product

$$\langle \mathbf{x}, \mathbf{y} \rangle = x_1 y_1 + \cdots + x_{n-1} y_{n-1} - x_n y_n.$$

In other words, with the matrix

$$\Delta = \Delta_n = \begin{pmatrix} I_{n-1} & 0 \\ 0 & -1 \end{pmatrix},$$

we have $\langle \mathbf{x}, \mathbf{y} \rangle = \mathbf{x}^{\mathsf{T}} \Delta \mathbf{y}$. Using this formalism, every symmetric matrix $A \in \mathbb{R}$ with one negative eigenvalue can be written as a "Gram matrix": $A_{ij} = \langle \mathbf{u}_i, \mathbf{u}_j \rangle$ with appropriate vectors $\mathbf{u}_i \in \mathbb{R}^d$, where d is the rank. This explains the analogies in our treatments here and in Section 10.5; but the indefiniteness of this inner product causes difficulties to be taken into account.

EXAMPLE 16.1. It is clear that $\mu(K_1) = 0$. We have $\mu(G) > 0$ for every graph with at least two nodes. Indeed, we can put "generic" numbers in the diagonal of $-A_G$ to make all the eigenvalues different, and then we can subtract the second smallest eigenvalue from all diagonal entries to get one negative and one 0 eigenvalue. Transversality, as noted in Section 10.5.2, is automatic for matrices with corank 1.

This discussion implies the useful fact that if G has at least two nodes, then it is sufficient to require that M has *at most one* negative eigenvalue. Indeed, if M is positive semidefinite, then 0 is its smallest eigenvalue, and so the Perron–Frobenius Theorem implies that it has multiplicity 1. So the only case when dropping the condition that M has a negative eigenvalue would allow larger corank is when $\mu(G) = 0$, i.e., when $G = K_1$. ♦

EXAMPLE 16.2. For a complete graph K_n with $n > 1$ nodes, it is easy to guess $-J$ as a Colin de Verdière matrix. It is trivial that $-J$ is a well-signed G-matrix with one negative eigenvalue. Transversality is trivial again. The corank of $-J$ is $n-1$, and one cannot beat this, since at least one eigenvalue must be negative. Thus $\mu(K_n) = n-1$. ♦

EXAMPLE 16.3. Next, let us consider the graph $\overline{K}_n$ consisting of $n \geq 2$ isolated nodes. All entries of a Colin de Verdière matrix M, except for the entries in the diagonal, must be 0. We must have exactly one negative entry in the diagonal. Trying to minimize the rank, we would like to put 0's in the rest of the diagonal, getting corank $n-1$. But this is where the Strong Arnold Property enters: we can put at most one 0 in the diagonal! In fact, assuming that (say) $M_{1,1} = M_{2,2} = 0$, the matrix X with

$$X_{i,j} = \begin{cases} 1, & \text{if } \{i,j\} = \{1,2\}, \\ 0, & \text{otherwise} \end{cases}$$

violates the Strong Arnold Property. (We could have appealed to Lemma 10.26.) So we must put $n-2$ positive numbers in the diagonal, and we are left with a single 0. It is easy to check that this matrix will satisfy (M3), and hence $\mu(\overline{K}_n) = 1$. ♦

EXAMPLE 16.4. Consider the path P_n on $n \geq 2$ nodes, labeled $1, 2, \ldots, n$ in their order on the path. Consider any Colin de Verdière matrix M for P_n, and delete its first column and last row. The remaining matrix has negative numbers in the diagonal and 0's above the diagonal, and hence it is nonsingular. Thus the corank of M is at most 1. We have seen that a corank of 1 can always be achieved. Thus $\mu(P_n) = 1$. ♦

EXAMPLE 16.5. Consider a complete bipartite graph $K_{p,q}$, where $p \geq q \geq 2$. In analogy with K_n, one can try to guess a Colin de Verdière matrix with low rank. The natural guess is

$$M = \begin{pmatrix} 0 & -J \\ -J^{\mathsf{T}} & 0 \end{pmatrix},$$

where J is the $p \times q$ all-1 matrix. This clearly belongs to $\mathcal{M}^1_{K_{p,q}}$ and has rank 2, which seems to imply $p+q-2$ for $\mu(K_{p,q})$. But it turns out that this matrix does not have the Strong Arnold Property unless $p, q \leq 3$. In fact, consider a matrix

$$X = \begin{pmatrix} Y & 0 \\ 0 & Z \end{pmatrix},$$

where Y is a $p \times p$ symmetric matrix with 0's in its diagonal and Z is a $q \times q$ symmetric matrix with 0's in its diagonal. The condition $MX = 0$ is then equivalent to requiring that Y and Z have 0 row-sums. It is easy to see that this implies that $Y = Z = 0$ if $p, q \leq 3$, but not if $p \geq 4$.

This establishes that $\mu(K_{p,q}) \geq p+q-2$ if $p, q \leq 3$, and equality holds here (Exercise 16.2); but if $p \geq 4$, then our guess for the matrix M realizing $\mu(K_{p,q})$ does not work. We will see that in this case μ will be smaller (equal to $\min\{p, q\}+1$, in fact). ♦

REMARK 16.6. There is a quite surprising fact in this last Example, which also underscores some of the difficulties associated with the study of μ. The graph $K_{4,4}$ (say) has a node-transitive and edge-transitive automorphism group, and so one would expect that at least some optimizing matrix in the definition of μ will share these symmetries: it will have the same diagonal entries and the same nonzero off-diagonal entries. We have seen that an analogous statement is valid for orthogonal representations in the definition of the theta-function (Theorem 11.5). But here this is not the case: it is easy to see that this would force us to consider the matrix we discarded above. So the optimizing matrix must break the symmetry!

16.1.1. Properties.

Let us summarize some basic properties of the Colin de Verdière number. We start with developing the combinatorial condition in Lemma 10.26 further for Colin de Verdière matrices.

LEMMA 16.7. *Let M be a Colin de Verdière matrix of a graph G, let $S \subseteq V$, and let M_S be the submatrix of M formed by the rows and columns in S. If M_S is positive semidefinite, then* $\operatorname{cork}(M_S) \leq 3$.

PROOF. Since M_S is a well-signed positive semidefinite $G[S]$-matrix, $\operatorname{cork}(M_S)$ is equal to the number of connected components of $G[S]$ for which the corresponding diagonal block of M_S is singular. Let $G_1, \dots, G_r$ be these components of $G[S]$, and let $M_1, \dots, M_r$ be the corresponding diagonal blocks of M_S. By the Perron–Frobenius Theorem, $\operatorname{Ker}(M_i)$ is 1-dimensional and contains a vector $x_i > 0$. Let $z > 0$ be an eigenvector belonging to the negative eigenvalue of M. We extend x_i with 0's to a vector in $\mathbb{R}^V$. Then $z^{\mathsf{T}} x_i > 0$, and we may scale x_i so that $z^{\mathsf{T}} x_i = 1$. The difference $x_i - x_j$ satisfies $z^{\mathsf{T}}(x_i - x_j) = 0$, and hence it belongs to the subspace generated by the eigenvectors of M with nonnegative eigenvalues. Since trivially $(x_i - x_j)^{\mathsf{T}} M (x_i - x_j) = 0$, this implies that $M(x_i - x_j) = 0$.

Now if $r \geq 4$, then $x_1 - x_2$ and $x_3 - x_4$ are two vectors in $\operatorname{Ker}(M)$ with disjoint supports that are not connected by any edge, contradicting Lemma 10.26. □

It is easy to see that if G is a graph with at least one edge and with connected components $H_1, \dots, H_k$, then

$$\mu(G) = \max\{\mu(H_1), \dots, \mu(H_k)\} \tag{16.1}$$

(Exercise 16.3). It follows, in particular, that if G has at least one edge, and we add a new, isolated node, then $\mu(G)$ does not change. The behavior of μ under the opposite way of adding a new node is also easy to describe but a bit harder to prove.

LEMMA 16.8. *Let G be a graph with at least one edge, and let us add a new node v connected to all of the other nodes to get a graph G'. Then $\mu(G') = \mu(G)+1$.*

PROOF. Let M be an optimal Colin de Verdière matrix of G, and let x be a positive eigenvector belonging to the negative eigenvalue. We may assume that $Mx = -x$. Consider the matrix

$$M' = \begin{pmatrix} M & -x \\ -x^{\mathsf T} & x^{\mathsf T}x \end{pmatrix}$$

(where v corresponds to the last row). Clearly M' is a well-signed G'-matrix. It is easy to check that it has the Strong Arnold Property: We have to check matrices of the form $X' = \begin{pmatrix} X & 0 \\ 0 & 0 \end{pmatrix}$, where X is a nonzero $\overline{G}$-matrix with zero diagonal. For a matrix of this form $M'X' = 0$ implies that $MX = 0$, which implies that $X = 0$, since M has the Strong Arnold Property. It is easy to see that $\mathrm{rk}(M') = \mathrm{rk}(M)$, which shows that

$$\mu(G') \geq \mathrm{cork}(M') = \mathrm{cork}(M)+1 = \mu(G)+1.$$

To prove that equality holds here is a bit more cumbersome. We may assume that $\mu(G') \geq 3$, else the assertion is obvious. Let M' be an optimal Colin de Verdière matrix of G', then it is of the form

$$M' = \begin{pmatrix} M & \mathbf{u} \\ -\mathbf{u}^{\mathsf T} & t \end{pmatrix}$$

with some vector $\mathbf{u} < 0$ and real number t. It is clear that M is a well-signed G-matrix. By interlacing eigenvalues, M has at most one negative eigenvalue. It is easy to see that $\mathrm{cork}(M) \geq \mathrm{cork}(M')-1$. Assuming that M' has the Strong Arnold Property, the definition of μ and the remark in Example 16.1 imply that

$$\mu(G) \geq \mathrm{cork}(M) \geq \mathrm{cork}(M')-1 = \mu(G')-1.$$

So it suffices to prove that M has the Strong Arnold Property. Let X be a nonzero $\overline{G}$-matrix with zero diagonal, and suppose that $MX = 0$. Then $X' = \begin{pmatrix} X & 0 \\ 0 & 0 \end{pmatrix}$ is a nonzero $\overline{G'}$-matrix with zero diagonal. Computing the product

$$M'X' = \begin{pmatrix} MX & 0 \\ -\mathbf{u}^{\mathsf T}X & 0 \end{pmatrix} = \begin{pmatrix} 0 & 0 \\ -\mathbf{u}^{\mathsf T}X & 0 \end{pmatrix},$$

we see that to get a contradiction it suffices to show that $\mathbf{u}^{\mathsf T}X = 0$. This will follow if we show that $\mathbf{u}$ is in the column space of M, since then $\mathbf{u} = M\mathbf{v}$ with some vector $\mathbf{v}$, and so $\mathbf{u}^{\mathsf T}X = \mathbf{v}^{\mathsf T}MX = 0$.

Suppose that this is not the case, then adding the column $\mathbf{u}$ to M increases its rank, and adding the row $(\mathbf{u}^{\mathsf T}, t)$ increases it even further, so $\mathrm{rk}(M') \geq \mathrm{rk}(M)+2$, and so $\mathrm{cork}(M) \geq \mathrm{cork}(M')+1$. By Interlacing Eigenvalues, this can only happen if M is positive semidefinite. So $\mathrm{cork}(M)$ is the number of those components of G for which the corresponding submatrix of M is singular (i.e., not positive definite). By Lemma 16.7, there are at most 3 such components, so $\mathrm{cork}(M') \leq 2$, which contradicts the assumption that $\mu(G') \geq 3$. □

Probably the most important property of μ is minor-monotonicity. The proof follows the same line of thought as for the algebraic width (see the proof of Theorem 10.33), and is not reproduced here.

THEOREM 16.9. *If graph H is a minor of graph G, then $\mu(H) \leq \mu(G)$.* □

EXAMPLE 16.10. Let $K_{p,q}^+$ denote the graph obtained from the complete bipartite graph $K_{p,q}$ by adding all edges in the color class of size p. We are interested in the case $q \geq 3$. We can also think of $K_{p,q}^+$ as taking q cliques $G_1, \dots, G_k$ with $p+1$ nodes each, selecting a p-subset of nodes in each, and glue these together. For $p = 1$ the graph is just a star, which (as we have seen) has $\mu(K_{1,q}^+) = 2$. For $p > 1$, the graph G is obtained from a star by repeatedly adding a new node and connecting it to all previous nodes, so by lemma 16.8, $\mu(K_{p,q}^+) = p+1$. ♦

EXAMPLE 16.11. We can settle the important example of the complete bipartite graph $K_{p,q}$, where $p \leq q$. We have seen that $\mu(K_{1,1}) = 1$ and $\mu(K_{p,q}) = p+q-2$ if $2 \leq q \leq 3$. Suppose that $q \geq 4$. Since $K_{p,q} \subset K_{p,q}^+$, it follows that $\mu(K_{p,q}) \leq \mu(K_{p,q}^+) = p+1$. On the other hand, contracting $p-4$ disjoint edges of $K_{p,q}$ we get a graph H which can also be obtained from $K_{3,3}$ by adding $p-3$ nodes and connecting them to all other nodes. By Lemma 16.8, $\mu(H) = \mu(K_{3,3})+p-3 = 4+(p-3) = p+1$, and so by minor-monotonicity, $\mu(K_{p,q}) \geq p+1$. ♦

Next, we study what happens to the Colin the Verdière number if the graph is decomposed along a cutset that is a clique. In other words, the graph is obtained by gluing together $k \geq 2$ graphs $G_1, \dots, G_k$ along cliques of the same size. Can $\mu(G)$ be larger than the maximum of the $\mu(G_i)$? This can happen, as shown by Example 16.10. But in a sense this is the worst case.

Let us fix a simple graph G, let $S \subseteq V$ induce a clique, and let $H_1, ...H_k$ be the connected components of $G \setminus S$. Let $V_i = V(H_i)$, and let G_i be the subgraph of G induced by $S \cup V_i$. For a G-matrix M, let M_i be the restriction of M to the rows and columns of H_i. Gluing two or more graphs together along a complete subgraph of each is often called a *clique-sum*. So the graph G is the clique sum of the graphs H_i (along the specified cliques $H_i[S]$). The following lemma describes the behavior of the Colin de Verdière number under clique-sums.

LEMMA 16.12. *With the notation above,*

$$\max\{\mu(G_1), \dots, \mu(G_k)\} \leq \mu(G) \leq 1+\max\{\mu(G_1), \dots, \mu(G_k)\}$$

If the upper bound is attained, then $|S| = \mu(G)-2$, and we can contract two or three of the components and delete the rest to obtain the graph $K_{\mu(G)-2,3}^+$.

PROOF. Since every G_i is an induced subgraph of G, it follows by Lemma 16.9 that $\mu(G) \geq \max\{\mu(G_1), \dots, \mu(G_k)\}$.

To prove the upper bound, we may assume that $N(V_i) = S$ for all i, since otherwise we could replace S by $N(V_i)$, which leads to a tighter inequality (the discussion of the case of attaining the upper bound takes some work, which is left to the reader). The matrix M has the following form:

$$M = \begin{pmatrix} M_0 & N_1 & \dots & N_k \\ N_1^{\mathsf{T}} & M_1 & & 0 \\ \vdots & & \ddots & \\ N_k^{\mathsf{T}} & 0 & & M_k \end{pmatrix},$$

where the first block corresponds to S and the other blocks, to $V_1, \dots, V_k$.

We start similarly as in the proof of Lemma 16.7. Let z and x_i be eigenvectors belonging to the smallest eigenvalues of M and M_i, respectively, where we extend x_i with 0's to a vector in $\mathbb{R}^V$. By the Perron–Frobenius Theorem, we have $z > 0$ and $x_i \geq 0$ (in fact, $x_i|_{V_i} > 0$). This implies that $z^{\mathsf{T}}x_i > 0$, and we may scale x_i so that $z^{\mathsf{T}}x_i = 1$.

The difference $x_i - x_j$ satisfies $z^{\mathsf{T}}(x_i - x_j) = 0$, and hence it belongs to the subspace spanned by the eigenvectors of M with nonnegative eigenvalue. This implies that

$$(x_i - x_j)^{\mathsf{T}}M(x_i - x_j) \geq 0 \qquad (1 \leq i, j \leq k). \tag{16.2}$$

The matrix $M_1 \oplus \dots \oplus M_k$ is a symmetric submatrix of M, and hence (by Interlacing Eigenvalues) it has at most one negative eigenvalue. So the matrices M_i are positive semidefinite with at most one exception.

Case 1. One of the matrices M_i (say, M_1) has a negative eigenvalue. We claim that the other M_i are positive definite. Suppose not, and let (say) M_2 have a zero eigenvalue. Then $x_2|_{V_2} \in \mathrm{Ker}(M_2)$, and hence $\mathrm{supp}(Mx_2) \subseteq S$. Using this, we get that

$$(x_1 - x_2)^{\mathsf{T}}M(x_1 - x_2) = x_1^{\mathsf{T}}Mx_1 < 0,$$

which contradicts (16.2).

So $M_2, \dots, M_k$ are positive definite, and, in particular, nonsingular. We can use row operations and simultaneous column operations to annulate the submatrices N_i and N_i^{T}, to get the matrix

$$\widehat{M} = \begin{pmatrix} M_0' & N_1 & 0 & \dots & 0 \\ N_1^{\mathsf{T}} & M_1 & 0 & \dots & 0 \\ 0 & 0 & M_2 & \dots & 0 \\ & & & \ddots & \\ 0 & 0 & 0 & \dots & M_k \end{pmatrix}$$

with $M_0' = M_0 - \sum_{i\geq 2} N_i M_i^{-1} N_i^{\mathsf{T}}$. Consider the matrix

$$M' = \begin{pmatrix} M_0' & N_1 \\ N_1^{\mathsf{T}} & M_1 \end{pmatrix}$$

By the Inertia Theorem, the eigenvalues of M have the same signs as the eigenvalues of the matrix $\widehat{M}$, which are clearly the eigenvalues of M' together with the eigenvalues of $M_2, \dots, M_k$. Thus M' has one negative eigenvalue and $\mathrm{cork}(M)$ zero eigenvalues.

We claim that all off-diagonal entries of M' are nonpositive. Indeed, this holds for M_0 and M_1, and the matrices N_i and N_i^{T} are nonpositive, so it suffices to argue only that the off-diagonal entries of the submatrix $M_0' = M_0 - \sum_{i\geq 2} N_i M_i^{-1} N_i^{\mathsf{T}}$ are nonpositive; they are in fact negative. The matrix M_i^{-1} is nonnegative by Lemma 4.14 for $i \geq 2$. Hence all entries of $N_i M_i^{-1} N_i^{\mathsf{T}}$ are nonnegative, and subtracting from M_0, the off-diagonal entries remain negative.

It follows that M' is a well-signed G_1-matrix. We claim that it has the Strong Arnold property. Indeed, let X be a nonzero $\overline{G_1}$-matrix with zero diagonal, and suppose that $M'X = 0$. Since S induces a clique, we can write

$$X = \begin{pmatrix} 0 & Y \\ Y^{\mathsf{T}} & Z \end{pmatrix}.$$

For any set of $V_1 \times V_j$ matrices U_j ($j = 2, \dots, k$), the matrix

$$W = \begin{pmatrix} 0 & Y & 0 & \dots & 0 \\ Y^{\mathsf{T}} & Z & U_2 & \dots & U_k \\ 0 & U_2^{\mathsf{T}} & 0 & \dots & 0 \\ & \vdots & & & \\ 0 & U_k^{\mathsf{T}} & 0 & \dots & 0 \end{pmatrix}$$

is a nonzero $\overline{G}$-matrix with zero diagonal. We want to choose the matrices U_i so that $MW = 0$; a little playing with these matrices shows that $U_i = -Y^{\mathsf{T}} N_i M_i^{-1}$ is a right choice. This contradicts the Strong Arnold Property of M.

Thus we see that M' has the Strong Arnold Property, and so it is a valid Colin de Verdière matrix for G_1. Hence

$$\mu(G) = \operatorname{cork}(M) = \operatorname{cork}(M') \leq \mu(G_1) \leq \max\{\mu(G_1), \dots, \mu(G_k)\}.$$

This completes Case 1. Note that the upper bound in the lemma is never attained in this case.

Case 2. All matrices M_i are positive semidefinite. Since contracting V_1 we get a complete $(|S|+1)$-graph as a minor of G_1, we have $\mu(G_1) \geq |S|$. So we are done unless $|S| < \mu(G)$; let us assume that this is the case.

Let $D = \{x \in \operatorname{Ker}(M) : x|_S = 0\}$. We have $\dim(\operatorname{Ker}(M)) = \mu(G) > |S|$, which implies that $D \neq 0$. Since $D \subseteq \bigoplus_{i=1}^k \operatorname{Ker}(M_i)$, at least one of the matrices M_i must be singular. Let (say) $M_1, \dots, M_r$ be singular, then their least eigenvalues are 0, and the corresponding eigenvectors x_i have multiplicity 1. Since $z^{\mathsf{T}} x_1 = 1 \neq 0$, we have $x_1 \notin D$, and so

$$\dim(D) < \sum_{i=1}^k \dim(\operatorname{Ker}(M_i)) = r.$$

Not all vectors on S are restrictions of vectors in $\operatorname{Ker}(M)$; for example, $\mathbb{1}_S$ is not. Indeed, if $\mathbb{1}_S = x|_S$ for some $x \in \operatorname{Ker}(M)$, then

$$0 = x_1^{\mathsf{T}} M x = x_1^{\mathsf{T}} N_1 x|_S + x_1^{\mathsf{T}} M_1 x|_{V_1} = x_1^{\mathsf{T}} N_1 \mathbb{1}_S < 0,$$

since all entries of $x_1^{\mathsf{T}} N_1$ are negative. This contradiction implies that $\dim(D) > \mu(G) - |S| \geq 1$ and so $r \geq 1 + \dim(D) \geq 3$.

Since $r \leq 3$ follows by Lemma 16.7, we must have equality in the estimates above. In particular, $|S| = \mu(G) - 1$, which means that G_1 can be contracted to $K_{\mu(G)}$, and so $\mu(G_1) \geq \mu(G) - 1$.

Furthermore, $G \setminus S$ must have at least three components. Contracting three of these and deleting the others, we get the graph $K^+_{\mu(G)-1,3}$. □

COROLLARY 16.13. *If $\mu(G) > 2$, then $\mu(G)$ is invariant under subdivision.*

PROOF. Minor-monotonicity implies that if we subdivide an edge, then $\mu(G)$ cannot decrease. Lemma 16.12 implies that if we glue a triangle onto an edge of G, then $\mu(G)$ does not increase (as long as it is larger than 2). Deleting the common edge of the triangle does not increase μ (again by minor-monotonicity), which means that subdividing the edge does not increase $\mu(G)$. □

We quote without proof the following result obtained by [**Bacher–Colin de Verdière 1995**]; the proof proceeds along similar lines as the previous one, with more complications. A $\Delta - Y$ *transformation* of a graph G means creating a new node, selecting a triangle T in G, connecting the new node to the nodes of T, and deleting the edges of T. The reverse operation is called $Y - \Delta$: this means that we delete a node of degree 3, and connect each pair of its neighbors (this may create parallel edges).

THEOREM 16.14. *If $\mu(G) > 3$, then $\mu(G)$ is invariant under $\Delta - Y$ transformations.* □

16.1.2. Small and large values. Graphs with $\mu \leq 4$ have very interesting topological characterizations. Statements (a)–(c) of the following theorem were proved in [**Colin de Verdière 1990**]; an elementary proof of (c) is given in [**van der Holst 1995**]; (d) was proved in [**Lovász–Schrijver 1998**].

THEOREM 16.15. *Let G be a connected graph.*

(a) *$\mu(G) \leq 1$ if and only if G is a path;*

(b) *$\mu(G) \leq 2$ if and only if G is outerplanar;*

(c) *$\mu(G) \leq 3$ if and only if G is planar;*

(d) *$\mu(G) \leq 4$ if and only if G is linklessly embeddable in 3-space.*

Graphs with large Colin de Verdière number, $n-4$ and up, have been studied in [**Kotlov–Lovász–Vempala 1997**]. The results are in a sense analogous to those for small numbers, but they are stated for the complementary graph (this is natural, since these graphs are very dense), one has to exclude twin nodes (this is a minor technical complication), and they are less complete (this is unfortunate). We only state two cases.

THEOREM 16.16. *Let G be a simple graph without twins.*

(a) *If $\overline{G}$ is outerplanar, then $\mu(G) \geq n-4$; if $\overline{G}$ is not outerplanar, then $\mu(G) \leq n-4$;*

(b) *If $\overline{G}$ is planar, then $\mu(G) \geq n-5$; if $\overline{G}$ is not planar, then $\mu(G) \leq n-5$.*

So for graphs with $\mu(G) = n-5$, their complement may or may not be planar; we do know that their complement is not outerplanar.

The proof of these theorems will take up most of this chapter. Our main tools will be two vector-labelings discussed in the next two sections.

16.2. Small values and nullspace representation

16.2.1. Nullspace representations. Every G-matrix M defines a vector-labeling of the graph G in cork(M) dimensions as follows. We take a basis $\mathbf{v}_1, \dots, \mathbf{v}_d$ of the null space of M, write them down as row vectors, to get a matrix U. Each node i of the graph corresponds to a column $\mathbf{u}_i$ of the matrix obtained, which is a vector in $\mathbb{R}^d$, and so we get the *nullspace representation* of the graph. Saying this in a fancier way, each coordinate function is a linear functional on the nullspace, and so the nullspace representation is a representation in the dual space of the nullspace.

The nullspace representation satisfies the equation $UM = 0$; in other words, M defines a braced stress on $(G, \mathbf{u})$. The nullspace representation is uniquely determined up to a linear transformation of $\mathbb{R}^d$. Furthermore, if D is a diagonal matrix with nonzero diagonal, then $(D^{-1}\mathbf{u}_i : i \in V)$ is a nullspace representation derived from the scaled version DMD of M (which in many cases can replace M). If all the vectors $\mathbf{u}_i$ are nonzero, then a natural choice for D is $D_{ii} = |\mathbf{u}_i|$: this scales the vectors $\mathbf{u}_i$ to unit vectors, i.e., we get a mapping of V into the unit sphere. We will see that other choices of this scaling may give even nicer representations.

16.2.2. Colin de Verdière number and planarity. We start with proving assertions (a)–(c) of Theorem 16.15. Part (d) will need more preparation.

PROOF. (a) We have seen (Example 16.4) that every path P_n has $\mu(P_n) = 1$. Hence, so does every subgraph of a path with at least one edge (i.e., a graph whose connected components are paths). On the other hand, if a graph G is not a subgraph of a path, then either it has a node of degree at least 3, or it contains a cycle. This implies that either K_3 or $K_{3,1}$ is a minor of G. Since $\mu(K_3) = \mu(K_{1,3}) = 2$ (Examples 16.2 and 16.11), this implies that $\mu(G) \geq 2$ by minor-monotonicity.

(b) Suppose that G is outerplanar. To show that $\mu(G) \leq 2$, we may assume that G is maximal outerplanar, i.e., a triangulation of a polygon. In this case, G can be obtained from a triangle by repeatedly taking clique-sums with a triangle. By Lemma 16.12, this implies that $\mu(G) = 2$.

If G is not outerplanar, then it contains either K_4 or $K_{3,2}$ as a minor. These graphs have $\mu(K_4) = \mu(K_{3,2}) = 3$, and hence $\mu(G) \geq 3$ by minor-monotonicity.

(c) Suppose that G is planar. Since M is a well-signed G-matrix with exactly one negative eigenvalue, Theorem 14.12 implies that $\mathrm{cork}(M) \leq 3$, and hence $\mu(G) \leq 3$.

The converse follows by the same type of argument as in the previous cases: If G is not planar, then it contains either K_5 or $K_{3,3}$ as a minor. Since $\mu(K_5) = \mu(K_{3,3}) = 4$, this implies that $\mu(G) \geq 4$. □

16.2.3. From Colin de Verdière matrices to Steinitz representations. The theorem proved above implies that every planar graph has Colin de Verdière number 3. Our main tool was the Discrete Nodal Theorem 14.10 (through its corollaries). We show that we get much more out of this.

Consider a 3-connected planar graph G, a Colin de Verdière matrix M of G, and the nullspace representation $\mathbf{u}$ in $\mathbb{R}^3$ obtained from M. Recall that this satisfies

$$\sum_j M_{ij}\mathbf{u}_j = 0 \qquad (i \in V). \tag{16.3}$$

This is a vector labeling of G in $\mathbb{R}^3$. We have seen another important vector labeling of such graphs, namely their Steinitz representation by a convex polytope. It turns out that these two vector labelings are closely related [**Lovász–Schrijver 1999**, **Lovász 2001**]: The vectors $\mathbf{u}_i$ can be scaled to give a Steinitz representation, and conversely, every Steinitz representation yields a Colin de Verdière matrix in a natural way. We discuss these constructions in this and in the next section. (See Exercises 16.4 and 16.5 for scaling results in other, simpler cases.)

The main step in the construction is the nullspace representation obtained from the Colin de Verdière matrix, as defined in Section 14.2. As a special case of this construction, any Colin de Verdière matrix yields a representation of a graph G in $\mu(G)$-space. In this case we have $M_{ij} < 0$ for every edge ij, and so writing the definition as

$$\sum_{j\in N(i)} (-M_{ij})\mathbf{u}_j = M_{ii}\mathbf{u}_i \qquad \text{for all } i \in V,$$

we see that each vector $\mathbf{u}_i$ is in the cone spanned by its neighbors, or in the negative of this cone (depending on the sign of M_{ii}). It follows that if a hyperplane contains $\mathbf{u}_i$, then either it contains all neighbors i, or i has neighbors on both sides of the hyperplane.

For the rest of this section, let G be a 3-connected planar graph, let M be a Colin de Verdière matrix of G with corank 3, and let $(\mathbf{u}_i : i \in V)$ be the nullspace representation derived from M. Since G is 3-connected, its embedding in the sphere is essentially unique, and so we can talk about the countries of G.

We say that the Colin de Verdière matrix M of G is *polyhedral* if the vectors $\mathbf{u}_i$ are the vertices of a convex polytope P, and the map $i \mapsto \mathbf{u}_i$ is an isomorphism between G and the skeleton of P. Note that the vectors $\mathbf{u}_i$ are determined by M up to a linear transformation of $\mathbb{R}^3$, and the skeleton graph of this polytope does not depend on this linear transformation.

Recall that Corollary 14.11 implies that for every plane H through the origin, the induced subgraphs $G[V_H^+]$ and $G[V_H^-]$ are connected. Another property of the representation $\mathbf{u}$ we need is the following.

LEMMA 16.17. *If a, b, c are distinct nodes of the same country, then $\mathbf{u}_a$, $\mathbf{u}_b$ and $\mathbf{u}_c$ are linearly independent.*

In particular, no node is represented by the 0 vector, and any two adjacent nodes are represented by nonparallel vectors.

PROOF. Suppose not. Then there exists a plane H through the origin that contains $\mathbf{u}_a, \mathbf{u}_b$ and $\mathbf{u}_c$. Corollary 14.11 implies that the sets of nodes V_H^+ and V_H^- induce nonempty connected subgraphs of G. We know that $|V_H| \geq 3$; clearly either $V_H' = V_H$ or V_H' is a cutset, and hence $|V_H'| \geq 3$. Thus G has three pairwise node-disjoint paths, connecting a, b and c to distinct points in V_H'. We may assume that the internal nodes of these paths are not contained in V_H', and so they stay in V_H. Contracting every edge on these paths as well as the edges induced by V_H^+ and V_H^-, we obtain a planar graph in which a, b and c are still on one country, and they have two common neighbors, which is clearly impossible. □

This lemma implies that each $\mathbf{u}_i$ is nonzero, and hence we may scale them to unit vectors (through the appropriate scaling of the matrix M). We call such a nullspace representation *normalized*. The previous lemma also implies that $\mathbf{u}_i$ and $\mathbf{u}_j$ are nonparallel if i and j are adjacent, and so for adjacent nodes i and j there exists a unique shortest geodesic on S^2 connecting $\mathbf{u}_i$ and $\mathbf{u}_j$. This gives an extension of the mapping $i \to \mathbf{u}_i$ to a mapping $\psi : G \to S^2$. We will show that this is in fact an embedding.

The main result in this section is the following [**Lovász–Schrijver 1998, Lovász 2001**]:

THEOREM 16.18. *Every Colin de Verdière matrix of a 3-connected planar graph G can be scaled to be polyhedral.*

PROOF. The proof will consist of two parts: first we show that the nullspace representation with unit vectors gives an embedding of the graph in the sphere; then we show how to rescale these vectors to get a Steinitz representation.

Claim 1. *Let F_1 and F_2 be two countries of G sharing an edge ab. Let H be the plane through 0, $\mathbf{u}_a$ and $\mathbf{u}_b$. Then all the other nodes of F_1 are in V_H^+, and all the other nodes of F_2 are in V_H^- (or the other way around).*

Suppose not, then there is a node c of F_1 and a node d of F_2 such that $c, d \in V_H^+$ (say). (Note that by Lemma 16.17, no node of F_1 or F_2 belongs to V_H.) By Corollary 14.11, there is a path P connecting c and d with $V(P) \subseteq V_H^+$. Both a and b have neighbors in V_H^+, and as we have seen, this implies that they have neighbors a' and b' in V_H^- as well. By Corollary 14.11 again, there is a path connecting a' and b' contained in V_H^-. This gives us a path Q connecting a and b, with at least one internal node, contained in V_H^- (up to its endpoints).

What is important from this is that P and Q are disjoint. But trying to draw this in the planar embedding of G, we see that this is impossible (see Figure 16.1).

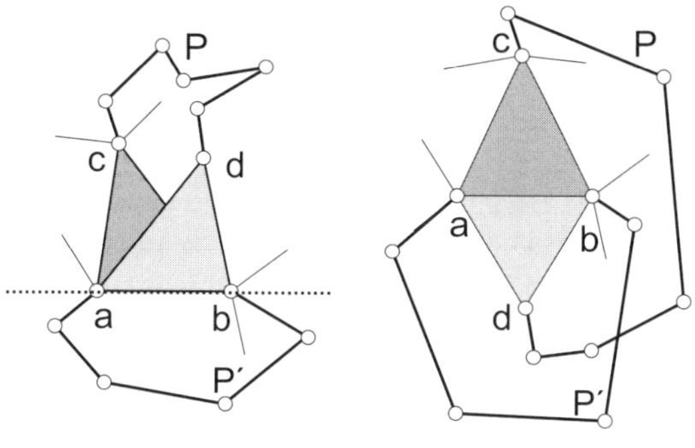

FIGURE 16.1. Why two countries do not fold.

Claim 1 implies the following.

Claim 2. *If $1, \dots, k$ are the nodes of a country, in this cyclic order, then $\mathbf{u}_1, \dots, \mathbf{u}_k$ are the extremal rays of the convex cone they generate in $\mathbb{R}^3$, in this cyclic order.*

Next we turn to the cyclic order around nodes.

Claim 3. *Let $a \in V$, and let (say) $1, \dots, k$ be the nodes adjacent to a, in this cyclic order in the planar embedding. Let H denote a plane through 0 and $\mathbf{u}_a$, and suppose that $\mathbf{u}_1 \in V_H^+$ and $\mathbf{u}_k \in V_H^-$. Then there is an h, $1 \le h \le k-1$ such that $\mathbf{u}_1, \dots, \mathbf{u}_{h-1} \in V_H^+$ and $\mathbf{u}_{h+1}, \dots, \mathbf{u}_k \in V_H^-$.*

If the Claim fails to hold, then there are two indices $1 < i < j < k$ such that $\mathbf{u}_i \in V_H^-$ and $\mathbf{u}_j \in V_H^+$. By Corollary 14.11, there is a path P connecting 1 and j in V_H^+, and a path P' connecting i and k in V_H^-. In particular, P and P' are disjoint. But this is clearly impossible in the planar drawing.

Claim 4. *With the notation of Claim 3, the geodesics on S^2 connecting $\mathbf{u}_a$ to $\mathbf{u}_1, \dots, \mathbf{u}_k$ issue from $\mathbf{u}_a$ in this cyclic order.*

Indeed, let us consider the halfplanes T_i bounded by the line ℓ through 0 and u_a, and containing $\mathbf{u}_i$ $(i = 1, \dots, k)$. Claim 1 implies that T_{i+1} is obtained from T_i by rotation about ℓ in the same direction. Furthermore, it cannot happen that these halfspaces T_i rotate about ℓ more than once for $i = 1, \dots, k, 1$: in such a case any plane through ℓ not containing any $\mathbf{u}_i$ would violate Claim 3.

Claim 5. *The normalized nullspace mapping is an embedding of G in the sphere. Furthermore, each country f of the embedding is the intersection of a convex polyhedral cone C_f with the sphere.*

To prove this, note that Claims 2 and 4 imply that we can extend the map $\psi: G \to S^2$ to a mapping $\overline{\psi}: S^2 \to S^2$ such that $\overline{\psi}$ is a local homeomorphism (i.e., every point $x \in S^2$ has a neighborhood that is mapped homeomorphically to a neighborhood of $\overline{\psi}(x)$.) This implies (in the case of the 2-sphere) that $\overline{\psi}$ is a homeomorphism (see Exercise 16.6).

As a second part of the proof of Theorem 16.18, we construct appropriate scaling of a nullspace representation. The construction will start with the polar polytope. Let $G^* = (V^*, E^*)$ denote the dual graph of G.

Claim 6. *We can assign a vector $\mathbf{w}_f$ to each $f \in V^*$ so that whenever $ij \in E$ and fg is corresponding edge of G^*, then*

$$\mathbf{w}_f - \mathbf{w}_g = M_{ij}(\mathbf{u}_i \times \mathbf{u}_j). \tag{16.4}$$

Let $\mathbf{v}_{fg} = M_{ij}(\mathbf{u}_i \times \mathbf{u}_j)$. It suffices to show that the vectors $\mathbf{v}_{fg}$ sum to 0 over the edges of any cycle in G^*; then we can invoke the potential argument. Since G^* is a planar map, it suffices to verify this for the cycles bounding the countries of G^*. Expressing this in terms of the edges of G, it suffices to show that

$$\sum_{j \in N(i)} M_{ij}(\mathbf{u}_i \times \mathbf{u}_j) = 0.$$

But this follows from the basic equation $\sum_j M_{ij}\mathbf{u}_j = 0$ upon multiplying by $\mathbf{u}_i$, taking into account that $\mathbf{u}_i \times \mathbf{u}_i = 0$ and $M_{ij} = 0$ for $j \notin N(i) \cup \{i\}$. This proves Claim 6.

An immediate property of the vectors $\mathbf{w}_f$ is that if f and g are two adjacent nodes of G^* such that the corresponding countries of G share the node $i \in V$, then $\mathbf{u}_i^\mathsf{T}\mathbf{w}_f = \mathbf{u}_i^\mathsf{T}\mathbf{w}_g$. This follows immediately from (16.4), upon scalar multiplication by $\mathbf{u}_i$. Hence the inner product $\mathbf{u}_i^\mathsf{T}\mathbf{w}_f$ is the same number α_i for every country f incident with $\mathbf{u}_i$. In other words, these vectors $\mathbf{w}_f$ all lie on the plane $\mathbf{u}_i^\mathsf{T}x = \alpha_i$.

Let P^* be the convex hull of the vectors $\mathbf{w}_f$.

Claim 7. *Let $f \in V^*$, and let $\mathbf{u} \in \mathbb{R}^3$, $\mathbf{u} \neq 0$. The vector $\mathbf{w}_f$ maximizes $\mathbf{u}^\mathsf{T}x$ over P^* if and only if $\mathbf{u} \in C_f$.*

First suppose that $\mathbf{w}_f$ maximizes $\mathbf{u}^\mathsf{T}x$ over P^*. Let $i_1, \dots, i_s$ be the nodes of the country of G corresponding to f (in this counterclockwise order). By Claim 2, the vectors $\mathbf{u}_{i_t}$ are precisely the extreme rays of the cone C_f.

Let fg_t be the edge of G^* corresponding to $i_t i_{t+1}$. Then $\mathbf{u}^\mathsf{T}(\mathbf{w}_f - \mathbf{w}_{g_t}) \geq 0$ for $t = 1, \dots, s$, and hence by (16.4),

$$\mathbf{u}^\mathsf{T} M_{i_t i_{t+1}}(\mathbf{u}_{i_{t+1}} \times \mathbf{u}_{i_t}) \geq 0,$$

or

$$\mathbf{u}^\mathsf{T}(\mathbf{u}_{i_t} \times \mathbf{u}_{i_{t+1}}) \geq 0.$$

This means that $\mathbf{u}$ is on the same side of the plane through $\mathbf{u}_{i_t}$ and $\mathbf{u}_{i_{t+1}}$ as C_f. Since this holds for every i, it follows that $\mathbf{u} \in C_f$.

Conversely, let $\mathbf{u} \in C_f$. Arbitrarily close to $\mathbf{u}$ we may find a vector $\mathbf{u}'$ in the interior of C_f. Let $\mathbf{w}_g$ be a vertex of P^* maximizing $(\mathbf{u}')^\mathsf{T} x$ over $x \in P^*$. Then by the first part of our Claim, $\mathbf{u} \in C_g$. Since the cones C_h, $h \in V^*$ are nonoverlapping, it follows that $f = g$. Thus $(\mathbf{u}')^\mathsf{T} x$ is optimized over $x \in P^*$ by $\mathbf{w}_f$. Since $\mathbf{u}'$ may be arbitrarily close to $\mathbf{u}$, it follows that $\mathbf{u}^\mathsf{T} x$ too is optimized by $\mathbf{w}_f$. This proves Claim 7.

This claim immediately implies that every vector $\mathbf{w}_f$ $(f \in V^*)$ is a vertex of P^*.

Claim 8. *The vertices $\mathbf{w}_f$ and $\mathbf{w}_g$ form an edge of P^* if and only if $fg \in E^*$.*

Let $fg \in E^*$, and let ij be the edge of G corresponding to $fg \in E^*$. Let $\mathbf{u}$ be a vector in the relative interior of the cone $C_i \cap C_j$. Then by Claim 7, $\mathbf{u}^\mathsf{T} x$ is maximized by both $\mathbf{w}_f$ and $\mathbf{w}_g$. No other vertex $\mathbf{w}_h$ of P^* maximizes $\mathbf{u}^\mathsf{T} x$, since then we would have $\mathbf{u} \in C_h$ by Claim 7, but this would contradict the assumption that $\mathbf{u}$ is in the relative interior of $C_f \cap C_g$. Thus $\mathbf{w}_f$ and $\mathbf{w}_g$ span an edge of P^*.

Conversely, assume that $\mathbf{w}_f$ and $\mathbf{w}_g$ span an edge of P^*. Then there is a nonzero vector $\mathbf{u}$ such that $\mathbf{u}^\mathsf{T} x$ is maximized by $\mathbf{w}_f$ and $\mathbf{w}_g$, but by no other vertex of P^*. By Claim 7, $\mathbf{u}$ belongs to $C_f \cap C_g$, but to no other cone C_h, and thus the cones C_f and C_g must share a facet, proving that $fg \in E^*$. This completes the proof of Claim 8. As an immediate corollary, we get the following.

Claim 9. *Every vector $\mathbf{w}_f$ is a vertex of P^*, and the map $f \mapsto \mathbf{w}_f$ is an isomorphism between G^* and the skeleton of P^*.*

To complete the proof, let us note that the vectors $\mathbf{w}_f$ are not uniquely determined by (16.4): we can add the same (but otherwise arbitrary) vector to each of them. Thus we may assume that the origin is in the interior of P^*. Then the polar P of P^* is well-defined, and its skeleton is isomorphic to G. Furthermore, the vertices of P are orthogonal to the corresponding facets of P^*, i.e., they are positive multiples of the vectors $\mathbf{u}_i$. Thus the vertex of P representing node i is $\mathbf{u}_i' = \alpha_i \mathbf{u}_i$, where $\alpha_i = \mathbf{u}_i^\mathsf{T} \mathbf{w}_f > 0$ for any country f incident with i. □

The scaling of M constructed in this proof is almost canonical. There were only two free steps: the choice of the basis in the nullspace of M, and the translation of P^* so that the origin be in its interior. The first corresponds to a linear transformation of P. It is not difficult to show that if this linear transformation is A, then the scaling constants are multiplied by $|\det(A)|$.

The second transformation is a bit more complicated. Translation of P^* corresponds to a projective transformation of P. Suppose that instead of P^*, we take the polyhedron $P^* + \mathbf{w}$, then the new scaling constant becomes $\alpha_i' = \mathbf{u}_i^\mathsf{T}(\mathbf{w}_f + \mathbf{w}) = \alpha_i + \mathbf{u}_i^\mathsf{T}\mathbf{w}$. Note that here $(\mathbf{u}_i^\mathsf{T}\mathbf{w} : i \in V)$ is just an arbitrary vector in the nullspace of M. Thus we get that *in the above construction, the scaling constants α_i are* *uniquely determined by M up to multiplication by the same positive number and addition of a vector in the nullspace of M.*

If you want a canonical choice, you can make it orthogonal to $\mathrm{Ker}(M)$. This can be expressed explicitly by solving for $\mathbf{w}$ the equation

$$\sum_i (\alpha_i + \mathbf{u}_i^\mathsf{T}\mathbf{w})\mathbf{u}_i = 0,$$

which gives

$$\mathbf{w} = -(UU^\mathsf{T})^{-1}U\alpha.$$

Of course, there might be many other scalings that yield polyhedra. For example, if P is simple (every facet is a triangle), then any scaling "close" to the one constructed above would also yield a Steinitz representation.

16.2.4. From Steinitz representations to Colin de Verdière matrices. Let P be any polytope in $\mathbb{R}^3$, containing the origin in its interior. Let G be the skeleton of P. Let P^* be its polar, and $G^* = (V^*, E^*)$, the skeleton of P^*.

Let $\mathbf{u}_i$ and $\mathbf{u}_j$ be two adjacent vertices of P, and let $\mathbf{w}_f$ and $\mathbf{w}_g$ be the endpoints of the corresponding edge of P^*. Then by the definition of polar, we have

$$\mathbf{u}_i^\mathsf{T}\mathbf{w}_f = \mathbf{u}_i^\mathsf{T}\mathbf{w}_g = \mathbf{u}_j^\mathsf{T}\mathbf{w}_f = \mathbf{u}_j^\mathsf{T}\mathbf{w}_g = 1.$$

This implies that the vector $\mathbf{w}_f - \mathbf{w}_g$ is orthogonal to both vectors $\mathbf{u}_i$ and $\mathbf{u}_j$, and hence it is parallel to the vector $\mathbf{u}_i \times \mathbf{u}_j$. Thus we can write

$$\mathbf{w}_f - \mathbf{w}_g = M_{ij}(\mathbf{u}_i \times \mathbf{u}_j),$$

where the labeling of $\mathbf{w}_f$ and $\mathbf{w}_g$ is chosen so that $M_{ij} < 0$; this means that $\mathbf{u}_i$, $\mathbf{u}_j$ and $\mathbf{w}_g$ form a right-handed basis.

Let $i \in V$, and consider the vector

$$\mathbf{u}_i' = \sum_{j \in N(i)} M_{ij}\mathbf{u}_j.$$

Then

$$\mathbf{u}_i \times \mathbf{u}_i' = \sum_{j \in N(i)} M_{ij}(\mathbf{u}_i \times \mathbf{u}_j) = \sum (\mathbf{w}_f - \mathbf{w}_g) = 0,$$

because the last sum extends over all edges fg of the facet of P^* corresponding to i, oriented counterclockwise, and hence it vanishes. Thus $\mathbf{u}_i$ and $\mathbf{u}_i'$ are parallel, and we can write

$$\mathbf{u}_i' = -M_{ii}\mathbf{u}_i \tag{16.5}$$

with some real M_{ii}. We complete the definition of a matrix $M = M(P)$ by setting $M_{ij} = 0$ whenever i and j are distinct nonadjacent nodes of G.

THEOREM 16.19. *The matrix $M(P)$ is a Colin de Verdière matrix for the graph G.*

PROOF. It is obvious by construction that M has the right pattern of 0's and negative elements. (16.5) can be written as

$$\sum_j M_{ij}\mathbf{u}_j = 0,$$

which means that each coordinate of the $\mathbf{u}_i$ defines a vector in the null space of M. Thus M has corank at least 3.

For an appropriately large constant $C > 0$, the matrix $CI - M$ is nonnegative and irreducible. Hence it follows from the Perron–Frobenius Theorem that the smallest eigenvalue of M has multiplicity 1. In particular, it cannot be 0 (which we know has multiplicity at least 3), and so it is negative. Thus M has at least one negative eigenvalue.

The most difficult part of the proof is to show that M has only one negative eigenvalue. This would follow from Theorem 14.15, but we give another proof. Observe that if we start with any true Colin de Verdière matrix $\overline{M}$ of the graph G, and we construct the polytope $\overline{P}$ from its null space representation as in Section 16.2.3, and then we construct a matrix $M(\overline{P})$ from $\overline{P}$, then we get back (up to positive scaling) the matrix $\overline{M}$. Thus there is at least one polytope $\overline{P}$ with the given skeleton for which $M(\overline{P})$ has exactly one negative eigenvalue.

Steinitz proved that any two 3-dimensional polytopes with isomorphic skeletons can be transformed into each other continuously through polytopes with the same skeleton (see [**Richter-Gebert 1996**]). Each vertex moves continuously, and hence so does their center of gravity. So we can translate each intermediate polytope so that the center of gravity of the vertices stays 0. Then the polar of each intermediate polytope is well-defined, and also transforms continuously. Furthermore, each vertex and each edge remains at a positive distance from the origin, and therefore the entries of $M(P)$ change continuously. It follows that the eigenvalues of $M(P)$ change continuously.

Assume that $M(P_0)$ has more than 1 negative eigenvalue. Let us transform P_0 into a polytope P_1 for which $M(P_1)$ has exactly one negative eigenvalue. There is a last polytope P_t with at least 5 nonpositive eigenvalues. By construction, each $M(P)$ has at least 3 eigenvalues that are 0, and we know that it has at least one that is negative. Hence it follows that $M(P_t)$ has exactly four 0 eigenvalues and one negative. But this contradicts Proposition 14.12.

Next we show transversality. Let X be a nonzero $\overline{G}$-matrix with zero diagonal, and assume that $MX = 0$. Every column of X is in the nullspace of M. We know that this nullspace is 3-dimensional, and hence it is spanned by the coordinate vectors of the $\mathbf{u}_i$. This means that there are vectors $\mathbf{h}_i \in \mathbb{R}^3$ $(i \in V)$ such that $X_{ij} = \mathbf{u}_i^\mathsf{T}\mathbf{h}_j$. We know that $X_{ij} = 0$ if $j = i$ or $j \in N(i)$, and so $\mathbf{h}_i$ must be orthogonal to $\mathbf{u}_i$ and all its neighbors. Since $\mathbf{u}_i$ and its neighbors obviously span $\mathbb{R}^3$, it follows that $X = 0$, a contradiction.

Finally, it is trivial by definition that M is polyhedral. □

16.2.5. Further geometric properties. Let M be a polyhedral Colin de Verdière matrix of G, and let P be the corresponding convex 3-polytope. Let P^* be the polar polytope with 1-skeleton $G^* = (V^*, E^*)$. The key relations between these objects are:

$$\sum_j M_{ij}\mathbf{u}_j = 0 \qquad \text{for every } i \in V, \tag{16.6}$$

and

$$\mathbf{w}_f - \mathbf{w}_g = M_{ij}(\mathbf{u}_i \times \mathbf{u}_j) \text{ for every } ij \in E \text{ and corresponding } fg \in E^*. \tag{16.7}$$

There are some consequences of these relations that are worth mentioning.

First, taking the vectorial product of (16.7) with $\mathbf{w}_g$, we get $(\mathbf{w}_f - \mathbf{w}_g)\times\mathbf{w}_g = \mathbf{w}_f\times\mathbf{w}_g$ on the left hand side and

$$M_{ij}\big((\mathbf{u}_i\times\mathbf{u}_j)\times\mathbf{w}_g\big) = M_{ij}\big((\mathbf{u}_i^\mathsf{T}\mathbf{w}_g)\mathbf{u}_j - (\mathbf{u}_j^\mathsf{T}\mathbf{w}_g)\mathbf{u}_i\big) = M_{ij}(\mathbf{u}_j - \mathbf{u}_i).$$

on the right. Thus

$$\mathbf{w}_f\times\mathbf{w}_g = M_{ij}(\mathbf{u}_j - \mathbf{u}_i). \tag{16.8}$$

Equation (16.8) gives the following relation between the Colin de Verdière matrices of a 3-connected planar graph and its dual. Let M be a polyhedral Colin de Verdière matrix for a planar graph G. Let P be a polytope defined by M. Let V_f denote the volume of the cone spanned by the origin and the facet f. If $ij\in E$ incident with facets f and g, then let $M^*_{fg} = 1/M_{ij}$. Let $M^*_{ff} = -V_f - \sum_{g\in N(f)} M^*_{fg}$; finally, let $M^*_{fg} = 0$ if f and g are distinct nonadjacent facets. These numbers form a symmetric matrix $M^*\in\mathbb{R}^{V^*\times V^*}$.

COROLLARY 16.20. *The matrix M^* is a polyhedral Colin de Verdière matrix of the dual graph G^*.* □

16.3. Linkless embeddings

In this section we return to part (d) of Theorem 16.15. The condition is more involved than the others, and needs an explanation, which we provide before turning to the proof.

16.3.1. Preliminaries: topology. Every (finite) graph can be embedded in 3-space, but there are ways to distinguish "simpler" embeddings. First, we have to define when two disjoint Jordan curves J_1 and J_2 in $\mathbb{R}^3$ are "linked". There are several nonequivalent ways to define this. Perhaps the most direct (and let this serve as our definition) is that two curves are *linked* if there is no topological 2-sphere in $\mathbb{R}^3$ separating them (so that one curve is in the interior of the sphere, while the other is in its exterior). This is equivalent to saying that we can continuously shrink either one of the curves to a single point without ever intersecting the other.

An alternative definition uses the notion of *linking number.* We consider an embedding of the 2-disc D in $\mathbb{R}^3$ with boundary J_1, which intersects J_2 a finite number of times, and always transversally. Walking along J_2, we count the number of times D is intersected from one side to the other, and subtract the number of times it is intersected in the other direction. If this linking number is 0, we say that the two curves are *homologically unlinked.* Reasonably elementary topological arguments show that the details of this definition are correct: such a 2-disc D exists, the linking number as defined is independent of the choice of D, and whether or not two curves are homologically linked does not depend on which of them we choose as J_1.

In a third version, we count all intersection points of J_2 with the disk, and we only care about the parity of this number. If this parity is even, we say that the two curves are *unlinked modulo* 2. If two curves are unlinked, then they are homologically unlinked, which in turn implies that they are unlinked modulo 2. (Neither one of these implications can be reversed.)

An embedding of a graph G in $\mathbb{R}^3$ is called *linkless* if any two disjoint cycles in G are unlinked. A graph G is *linklessly embeddable* if it has a linkless embedding in $\mathbb{R}^3$ (Figure 16.2).

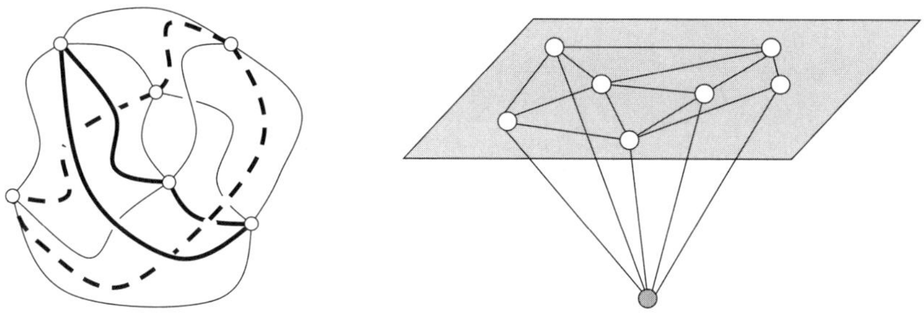

FIGURE 16.2. Left: A graph embedded in 3-space with two cycles linked. Right: An apex graph (a planar graph with an additional node). This embedding is linkless (even flat).

There is an even stronger definition of linkless embeddings: An embedding of G in $\mathbb{R}^3$ is called *flat*, if for each cycle C in G there is a disk D bounded by C (a 'panel') whose interior is disjoint from G.

Clearly, each flat embedding is linkless in any of the senses above, but the converse does not hold in general. (A trivial example: embed a single cycle in 3-space so that the image is a proper knot.) However, if G has an embedding that is linkless in the weakest sense (the linking number of any two disjoint cycles is even), then it has a flat embedding (linkless in the strongest sense) as well. So the class of linklessly embeddable graphs is very robust with respect to variations in the definition. This fact follows from [**Robertson–Seymour–Thomas 1995**], as a byproduct of Theorem 16.21 below.

Suppose that we have a graph G linklessly embedded in 3-space (in any particular sense). It is trivial that if we delete an edge from G, then the embedding remains linkless. It takes a bit more effort to verify that if we contract an edge e, then the resulting graph G/e is linklessly embeddable: informally, we can continuously deform the embedding to make the edge e shorter and shorter, until in the limit we get an embedding of G/e. Two linked cycles in G/e would yield two linked cycles in G. So linkless embeddability is a minor-closed property.

For either one of the versions of linkless embedding, if a graph is linklessly embedded, and we perform a $\Delta - Y$ or $Y - \Delta$ transformation, then there is a way to define an embedding of the new edges so that the resulting graph is linkless as well. This is quite easy to see for $Y - \Delta$, and a bit more difficult for $\Delta - Y$ (see Exercise 16.8).

A basic result in the graph minor theory of Robertson, Seymour and Thomas says that every minor-closed graph property can be characterized by a finite number of forbidden minors. To find this list explicitly is, in general, very difficult, and it is very remarkable that, verifying a conjecture of Sachs, a deep forbidden minor characterization of linklessly embeddable graphs was proved in [**Robertson–Seymour–Thomas 1995**].

THEOREM 16.21. *A graph G is flatly embeddable in 3-space if and only of it does not contain any of the graphs in Figure 16.3 as a minor.* □

We note that the graphs in the Petersen family (Figure 16.3) cannot be linklessly embedded even in the weakest sense: every embedding of them contains two disjoint cycles with an odd linking number. Exercise 16.9 gives a hint how to prove

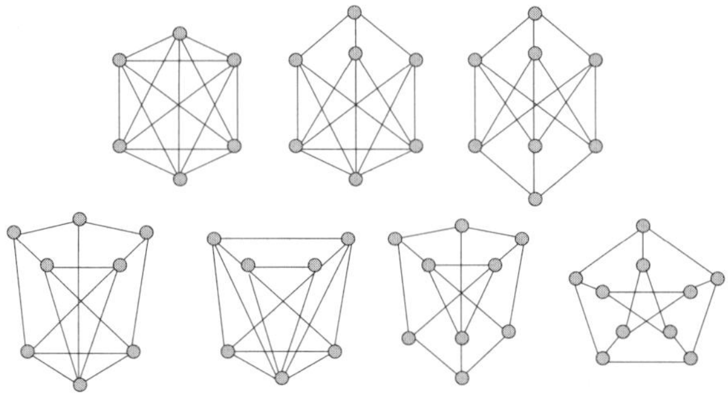

FIGURE 16.3. Excluded minors for linklessly embeddable graphs (the Petersen Family).

this in the case of K_6; the other graphs in the Petersen family arise from K_6 by $\Delta - Y$ and $Y - \Delta$ transformations, which preserve this property.

The rest of this section is devoted to the proof of part (d) of Theorem 16.15, giving a different characterization of linklessly embeddable graphs.

16.3.2. Preliminaries: simultaneous paneling. We do not claim that in a flat embedding, the panels belonging to different cycles are disjoint from each other. For example, let us take the graph $K_{2,2,2}$, embedded in 3-space as the skeleton of the regular octahedron. This embedding is flat (the skeleton of any convex 3-polytope is), but one cannot panel the facets and the three 4-cycles without creating some intersection between the interiors of the panels. If each member of a family of cycles can be paneled so that the interiors of the panels are disjoint, we say that this family can be *simultaneously paneled.* Reasonably large families of cycles can be paneled simultaneously, as the following result of [**Böhme 1990**] shows:

LEMMA 16.22 (Paneling Lemma). *Let $\mathcal{C}$ be a family of cycles in a flatly embedded graph G. Assume that the intersection of any two cycles in $\mathcal{C}$ is either a path or empty. Then the cycles in $\mathcal{C}$ can be simultaneously paneled.* □

We need the following special case of the paneling lemma. Let T be a spanning tree of a flatly embedded graph G. For each edge $e \in E \setminus E(T)$, let C_e be its fundamental cycle (the unique cycle in $T+e$). The cycles C_e satisfy the conditions of the Paneling Lemma, and so they can be simultaneously paneled. We call this paneling a *fundamental paneling* with respect to T.

Let Y be the union of the panels in a fundamental paneling. It is easy to see that Y, as a topological space, is contractible. This implies, in particular, that every cycle of G can be contracted to a single point inside Y, and also that the complement of Y is connected.

As an application of the Paneling Lemma, we prove a fact about extending linkless embeddings.

LEMMA 16.23. *Let G be a flatly embedded graph, and let $S \subseteq V$ induce a connected subgraph. Define $G_1 = G[S \cup N(S)]$ and $G_2 = G \setminus S$, and let $i, j \in V(G_2)$. Then the embedding can be extended to an embedding of $G+ij$ in which no cycle in G_2+ij is linked with any cycle in G_1.*

PROOF. Consider a spanning tree of $G[S]$, and extend it to a spanning tree T of G_1 so that all nodes in $N(S)$ are leaves of T. Consider a fundamental paneling of

G_1 with respect to T (as described after the Paneling Lemma 16.22), and let Y be the union of these panels. As remarked, the complement of Y is arcwise connected, and hence we can connect i and j by a Jordan arc J whose interior is disjoint from Y. We may perturb J a little to make its interior disjoint from the rest of G as well.

We claim that using J to embed the edge ij, we get an embedding of $G+ij$ that satisfies the conditions of the lemma. Indeed, let C_1 be a cycle in G_1 and C_2, a cycle in G_2+ij disjoint from C_1. We want to prove that C_1 and C_2 are not linked. We may assume that $ij \in E(C_2)$ (else, the assertion is trivial since G is flatly embedded).

If an edge $e \in E(G_1)$ is disjoint from C_2, then the whole cycle C_e is disjoint from C_2, since all the nodes of C_e other than the endpoints of e are contained in S. Hence the panel of the cycle C_e is disjoint from C_2 as well. Since C_1 can be contracted inside the union of these panels, it cannot be linked with C_2. □

16.3.3. Preliminaries: polytopes. We need a couple of results about polytopes. First, a technical lemma. Let us say that a segment $\mathbf{xy}$ *crosses* a hyperplane H, if $\mathbf{x}$ and $\mathbf{y}$ belong to different sides of H. In other words, the segment intersects H in a single point that is an interior point of it.

LEMMA 16.24. *Let $\mathcal{F}$ be a finite family of linear hyperplanes in $\mathbb{R}^d$. Then there is a centrally symmetric simplicial d-polytope P such that no edge of P crosses any hyperplane in $\mathcal{F}$.*

PROOF. If P is a convex polytope containing the origin in its interior, and $\mathbf{x}$ is a point on its boundary, then *pulling* $\mathbf{x}$ is the operation of replacing P with the convex hull P' of P and $(1+\varepsilon)\mathbf{x}$. Here we assume that $\varepsilon > 0$ is small enough so that for every facet not containing $\mathbf{x}$, the point $\mathbf{x}' = (1+\varepsilon)\mathbf{x}$ is on the same side of the facet as P. If P is simplicial, then pulling preserves this property. The new point $\mathbf{x}'$ is a vertex; every other vertex remains a vertex of P', except if $\mathbf{x}$ was a vertex of P. Every edge of P remains an edge, except if $\mathbf{x}$ was an interior point of an edge e of P; in this case, e ceases to be an edge. New edges connect $\mathbf{x}'$ to all vertices on faces containing $\mathbf{x}$ (Figure 16.4).

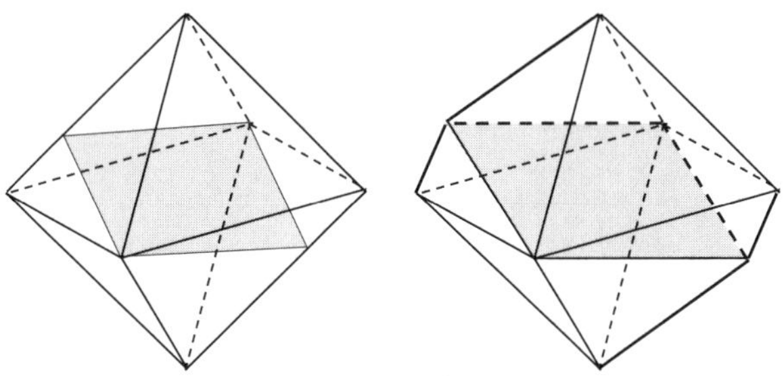

FIGURE 16.4. Left: an octahedron and a plane through its center intersecting two edges. Right: Pulling the intersection points of the edges and the plane to create new vertices, but eliminate the intersections. The illustration is in 3 dimensions, in the proof we need 5.

If P is centrally symmetric (with respect to the origin), then we can pull both $\mathbf{x}$ and $-\mathbf{x}$ by the same ε to keep central symmetry.

Starting with any simplicial centrally symmetric polytope in $\mathbb{R}^d$ (say, with the cross-polytope), we pick a hyperplane $H \in \mathcal{F}$, and pull all intersection points of H with the interiors of edges, maintaining central symmetry. Let us call this "cleaning H". We clean all hyperplanes in $\mathcal{F}$ one-by-one, to get a polytope P'.

We claim that P' satisfies the requirements of the lemma: cleaning a hyperplane H gets rid of all crossings between edges and H without creating new crossings with hyperplanes already cleaned. Indeed, when pulling an interior point $\mathbf{x}$ of an edge e, all the new edges are incident with $\mathbf{x}'$, and so they do not cross H. Let $H' \in \mathcal{F}$ be a hyperplane already cleaned. If H' contains $\mathbf{x}$, then it contains $\mathbf{x}'$, and so no new edge crosses it. If H' does not contain $\mathbf{x}$, then every face containing $\mathbf{x}$ is fully contained in the closed halfspace bounded by H' containing $\mathbf{x}$ (since H' is already cleaned). Since $\mathbf{x}'$ is contained in this closed halfspace as well, no new edge will cross H'. □

Next, we relate polytopes and the topological considerations in the previous section. We have seen that the skeleton of any 3-polytope, as a graph embedded in 3-space, is linkless (even flat). Among skeletons of 4-polytopes we find both linklessly embeddable ones (say, K_5), and nonembeddable ones (every complete n-graph, $n \geq 5$, is the skeleton of a 4-polytope). The following results [**Lovász–Schrijver 1998**] imply that the skeleton of a 5-polytope is never linklessly embeddable.

To state the result, we need a definition from convex geometry. Let P be a d-polytope. We say that two faces F_1 and F_2 of P are *antipodal*, if there are two parallel supporting hyperplanes H_1 and H_2 such that $F_1 \subseteq H_1$ and $F_2 \subseteq H_2$ (Figure 16.5). It is easy to see that every d-polytope has antipodal faces F_1 and F_2 with $\dim(F_1)+\dim(F_2) = d-1$: for example, we can take a facet and a vertex farthest from its hyperplane.

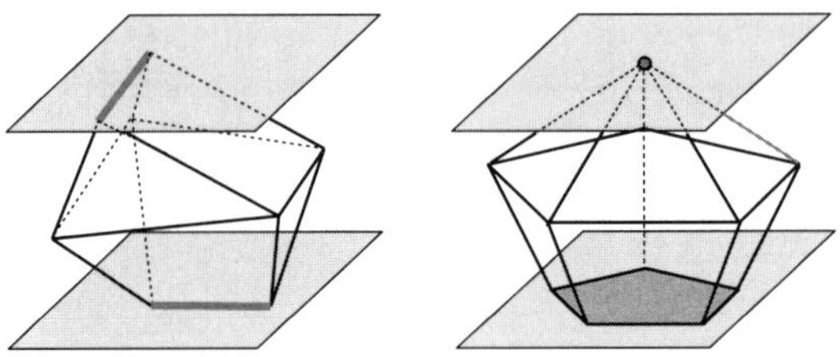

FIGURE 16.5. Antipodal faces of a polytope

We call two faces A_1 and A_2 *parallel* if their affine hulls are disjoint but these affine hulls contain parallel lines. Formally, $\mathrm{aff}(A_1) \cap \mathrm{aff}(A_2) = \emptyset$ but $\mathrm{lin}(A_1 - A_1) \cap \mathrm{lin}(A_2 - A_2) \neq 0$. Two antipodal faces with $\dim(F_1)+\dim(F_2) \geq d$ are always parallel.

It will be convenient to work with polytopes with no parallel faces. This can be achieved by taking a hyperplane H that it disjoint from P, does not contain any nonempty intersection $\mathrm{aff}(A_1) \cap \mathrm{aff}(A_2)$, and is not parallel to any nonnull linear subspace $\mathrm{lin}(A_1 - A_1) \cap \mathrm{lin}(A_2 - A_2)$. Moving H to infinity by a projective transformation maps P onto a polytope with the same lattice of faces but with no parallel faces.

One advantage of having no parallel faces is that for two antipodal faces F_1 and F_2, there are two parallel supporting hyperplanes H_1 and H_2 such that $F_i = H_i \cap P$ (Exercise 16.11). A further simple property we need is the following.

LEMMA 16.25. *Let P be a d-polytope having no parallel faces, and let A_1 and A_2 be antipodal facets with* $\dim(A_1)+\dim(A_2) = d-2$. *Then the number of pairs of antipodal facets $\{F_1, F_2\}$ with $A_1 \subseteq F_1$, $A_2 \subseteq F_2$ and* $\dim(F_1)+\dim(F_2) = d-1$ *is* 2.

PROOF. Let H_1 and H_2 be two parallel supporting hyperplanes such that $A_1 = H_1 \cap P$ and $A_2 = H_2 \cap P$. Let $L_i = \text{lin}(A_i - A_i)$, and let H be the hyperplane obtained by translating H_1 to the origin. Then $L_1 \cap L_2 = 0$ (since A_1 and A_2 are not parallel) and $L_1 + L_2 \subseteq H$. Since $\dim(L_1 + L_2) = d-2 = \dim(H)-1$, we can rotate H about $L_1 + L_2$ in two directions. This corresponds to simultaneous rotation of H_1 and H_2 keeping them parallel and preserving $A_i \subseteq H_i$. In both directions, we can rotate until one of the H_i hits a face F_i of P with $\dim(F_i) > \dim(A_i)$.

Now the hypothesis that P has no parallel faces has some useful consequences. It cannot happen that both H_1 and H_2 hit higher-dimensional faces F_1 and F_2 simultaneously, since these would be antipodal faces with $\dim(F_1)+\dim(F_2) \geq d$, and therefore parallel. By a similar argument, we must have $\dim(F_i) = \dim(A_i)+1$, and so $\{F_i, A_{3-i}\}$ is an antipodal pair with $\dim(F_i)+\dim(A_{3-i}) = d-1$. So rotation in either direction yields such a pair, two of them altogether. It is easy to see that no further such pair exists. □

A continuous map $\phi : \partial P \to \mathbb{R}^m$ will be called *generic*, if the images of the interiors of a k-face and an l-face do not intersect if $k+l < m$, these images have a finite number of intersection points if $k+l = m$, and at these points they intersect transversally.

Let $\check{F}$ denote the relative interior of the face F. For a vertex point $\mathbf{x}$, we define $\check{\mathbf{x}} = \mathbf{x}$.

LEMMA 16.26. *Let $P \subseteq \mathbb{R}^d$ be a convex polytope having no parallel faces and let $\phi : \partial P \to \mathbb{R}^{d-1}$ be a generic continuous map. Then $\sum |\phi(\check{F}_1) \cap \phi(\check{F}_2)|$ is odd, where the summation extends over all antipodal pairs $\{F_1, F_2\}$ of faces with* $\dim(F_1)+\dim(F_2) = n-1$.

In particular, there exists a pair of antipodal faces F_1 and F_2 with $\dim(F_1)+\dim(F_2) = d-1$ such that $|\phi(\check{F}_1) \cap \phi(\check{F}_2)|$ is odd. This extends a result of [**Bajmóczy–Bárány 1979**], which asserts the existence of such a pair of faces with $\phi_t(\check{F}_1) \cap \phi_t(\check{F}_2) \neq \emptyset$.

PROOF. There is at least one map for which this is true: the map ϕ_0 defined by projecting ∂P onto a facet F of P from a point outside P but very close to an interior point of F (this map was used to construct the Schlegel diagram). By the assumption that P has no parallel faces it follows that there is a unique vertex $\mathbf{v}$ antipodal to F. Then $\sum |\phi_0(\check{F}_1) \cap \phi_0(\check{F}_2)| = 1$, since the only nonzero term in the sum is obtained when $\{F_1, F_2\} = \{F, \mathbf{v}\}$.

Now we deform ϕ_0 through a family of maps $\phi_t : \partial P \to \mathbb{R}^{d-1}$ $(0 \leq t \leq 1)$ so that ϕ_1 is an arbitrary generic continuous map. We need that each ϕ_t has the following properties. Let A_i be a k_i-dimensional face of P $(i = 1, 2)$.

(i) If $k_1 + k_2 < d-2$, then $\phi_t(\check{A}_1)$ and $\phi_t(\check{A}_2)$ do not intersect.

(ii) If $k_1 + k_2 = d-2$, then $\phi_t(\check{A}_1)$ and $\phi_t(\check{A}_2)$ intersect for a finite number of values of t only, and then $|\phi_t(\check{A}_1) \cap \phi_t(\check{A}_2)| = 1$, and this happens only for one pair of faces at the same time.

(iii) If $k_1+k_2 = d-1$, then $\phi_t(\breve{A}_1)$ and $\phi_t(\breve{A}_2)$ have a finite number of intersection points, and at these points they intersect transversally.

To see that such a deformation exists takes some (routine) arguments in topology, which are omitted.

The number $|\phi_t(\breve{F}_1)\cap\phi_t(\breve{F}_2)|$, where $\dim(F_1)+\dim(F_2) = d-1$, does not change unless $\phi_t(\breve{F}_1)$ intersects $\phi_t(\breve{A}_2)$, where A_2 is a facet of F_2, or the other way around. So t must be one of the special values from (ii), and in fact the intersecting pair A_1 and A_2 must be antipodal. Fixing this A_1 and A_2, the target sum $\sum |\phi_t(\breve{F}_1)\cap\phi_t(\breve{F}_2)|$ changes by 1 for every pair $\{F_1, F_2\}$ of antipodal faces with $\dim(F_1)+\dim(F_2) = d-1$ such that $A_1 \subseteq F_1$ and $A_2 \subseteq F_2$. By Lemma 16.25, there are two such pairs, so the sum changes by an even number. □

Now we are able to state and prove our main topological tool, establishing the connection between linear algebra (convex polytopes) and linking cycles. The fact may be interesting in itself as a Borsuk-type theorem about linking. The theorem extends to polytopes in any odd dimension with essentially the same proof (see [**Lovász–Schrijver 1998**] for the details).

THEOREM 16.27. *Let P be a convex polytope in $\mathbb{R}^5$ and let ϕ be an embedding of its 1-skeleton into $\mathbb{R}^3$. Then there exist antipodal 2-faces F_1 and F_2 of P such that $\phi(\partial F_1)$ and $\phi(\partial F_2)$ have an odd linking number.*

PROOF. Let us extend ϕ to a continuous mapping $\phi' : \partial P \to \mathbb{R}^4$ (this is clearly possible) and define

$$\psi(\mathbf{x}) = \begin{pmatrix}\phi'(\mathbf{x})\\ f(\mathbf{x})\end{pmatrix},$$

where $f(\mathbf{x}) = 0$ if $\mathbf{x} \in P_1$ and $f(\mathbf{x}) > 0$ otherwise (for example, $f(\mathbf{x})$ can be the distance of $\mathbf{x}$ from P_1). We can perturb f a little to make ψ generic.

By Lemma 16.26, P has a pair of antipodal faces F_1 and F_2 such that $\dim(F_1)+\dim(F_2) = 4$ and $|\psi(\breve{F}_1)\cap\psi(\breve{F}_2)|$ is odd. Note that the intersection points of $\psi(\breve{F}_1)$ and $\psi(\breve{F}_2)$ cannot be between $\psi(P_1)$ and $\psi(\partial P\setminus P_1)$, since the former is contained in the hyperplane $x_4 = 0$, while the latter is contained in the halfspace $x_4 > 0$. So we must have $\dim(F_1) = \dim(F_2) = 2$.

We claim that $\psi(\partial F_1)$ and $\psi(\partial F_2)$ have an odd linking number. Let $\psi_1 : F_1 \to \mathbb{R}^3$ be an extension of $\psi|_{\partial F_1}$, such that $\psi_1(F_1)$ intersects G transversally a finite number of times. Let $\psi_2 : F_2 \to \mathbb{R}^4$ be an extension of $\psi|_{\partial F_2}$, such that the fourth coordinate of every point of $\psi_2(\breve{F}_2$ is negative. Then $\psi|_{\partial F_1}\cup\psi_1$ and $\psi|_{\partial F_2}\cup\psi_2$ can be thought of as two maps of the 2-sphere into $\mathbb{R}^4$. The common points of the images are the intersection points of $\psi(\breve{F}_1)$ and $\psi(\breve{F}_2)$, and the intersection points of $\psi_1(F_1)$ and $\psi(\partial F_2)$. The number of these intersections must be even. Since $|\psi(\breve{F}_1)\cap\psi(\breve{F}_2)|$ is odd, so is $|\psi_1(F_1)\cap\psi(\partial F_2)|$, which proves that $\phi(\partial F_1)$ and $\phi(\partial F_2)$ have an odd linking number. □

16.3.4. Completing the proof: Nullspace representation. We want to prove that a graph G is linklessly embeddable if and only if $\mu(G) \le 4$. The fact that this condition is sufficient was noted by Robertson, Seymour and Thomas. Indeed, if a graph is not linklessly embeddable, then by Theorem 16.21 it contains one of the graphs in the Petersen family (Figure 16.3) as a minor. All these graphs have $\mu(G) = 5$. This can be verified directly, or it can be justified by the following argument: $\mu(K_6) = 5$ by Example 16.2, and hence Theorem 16.14 implies that

$\mu(G) = 5$ for every graph that arises from K_6 by $\Delta - Y$ and $Y - \Delta$ transformations. It is not too tedious to check that these are exactly the graphs in the Petersen family.

So a graph G that is not linklessly embeddable has a minor with $\mu = 5$; by minor monotonicity, this implies that $\mu(G) \geq 5$.

To prove the converse, let G be a linklessly embeddable graph, and fix a flat embedding of G in $\mathbb{R}^3$. Suppose (by way of contradiction) that $\mu(G) \geq 5$. Let $\mathbf{u} : V \to \mathbb{R}^5$ be the nullspace representation obtained from a Colin de Verdière matrix M of G. By Lemma 16.24, there exists a centrally symmetric polytope $P \subseteq \mathbb{R}^5$ such that no edge of P crosses any of the hyperplanes $\mathbf{u}_i^\perp$, $(\mathbf{u}_i + \mathbf{u}_j)^\perp$, or $(\mathbf{u}_i - \mathbf{u}_j)^\perp$. In other words, for every proper face F of P, any of these hyperplanes either contains F or is disjoint from its interior.

We are going to work with three geometrically represented graphs. First, there is the linkless embedding of G in $\mathbb{R}^3$; second, there is the nullspace representation of M in $\mathbb{R}^5$ (it may be useful to also keep in mind the arrangement of the hyperplanes $\mathbf{u}_i^\perp$); and third, we will consider the polytope P and its 1-skeleton P_1, also in $\mathbb{R}^5$. Our goal is to map P_1 into $\mathbb{R}^3$ continuously. Roughly speaking, we want to map P_1 into the graph G (as embedded in $\mathbb{R}^3$). Theorem 16.27 will provide a pair of linked cycles in G.

For every vector $\mathbf{x} \in \mathbb{R}^5$, let $V_0(\mathbf{x}) = \{i \in V : \mathbf{x}^\mathsf{T}\mathbf{u}_i = 0\}$, $V_+(\mathbf{x}) = \{i \in V : \mathbf{x}^\mathsf{T}\mathbf{u}_i > 0\}$ and $V_-(\mathbf{x}) = \{i \in V : \mathbf{x}^\mathsf{T}\mathbf{u}_i < 0\}$. For every vertex $\mathbf{v}$ of P, let $\phi(\mathbf{v})$ be any node in $V_+(\mathbf{v})$ (which is nonempty by the Discrete Nodal Theorem 14.10).

To extend the map ϕ to the edges of P, we need a simple observation.

Claim 1. *Let F be a face of P, let $\mathbf{z}$ be an interior point of F, and let $i \in V$ be any node of G. If $\mathbf{z}^\mathsf{T}\mathbf{u}_i = 0$, then $\mathbf{v}^\mathsf{T}\mathbf{u}_i = 0$ for every vertex $\mathbf{v} \in V(F)$; if $\mathbf{z}^\mathsf{T}\mathbf{u}_i > 0$, then $\mathbf{v}^\mathsf{T}\mathbf{u}_i \geq 0$ for every vertex $\mathbf{v} \in V(F)$, and $\mathbf{v}^\mathsf{T}\mathbf{u}_i > 0$ for at least one vertex $\mathbf{v} \in V(F)$.*

This follows from the fact that the hyperplane $\mathbf{u}_i^\perp$ does not separate any two vertices of F.

Since $\mathbf{v}^\mathsf{T}\mathbf{u}_{\phi(\mathbf{v})} > 0$ by the definition of $\phi(\mathbf{v})$, it follows that $\mathbf{z}^\mathsf{T}\mathbf{u}_{\phi(\mathbf{v})} > 0$ for every vertex $\mathbf{v} \in V(F)$. In other words, $\phi(\mathbf{v}) \in V_+(\mathbf{z})$.

Now we describe how to extend the map ϕ to the edges of P. Let $\mathbf{vw}$ be an edge of P, let $i = \phi(\mathbf{v})$ and $j = \phi(\mathbf{w})$, and let $\mathbf{z}$ be an internal point of the edge $\mathbf{vw}$. Then $i, j \in V_+(\mathbf{z})$ by Claim 1. In the (easy) case when i and j belong to the same connected component of $G[V_+(\mathbf{z})]$, we can find a path of G in $V_+(\mathbf{z})$ connecting i and j, and map the edge $\mathbf{vw}$ onto this path (Figure 16.6).

The more difficult case is when i and j belong to different components of $G[V_+(\mathbf{z})]$. Such an edge of P is called *broken.* This in particular implies that $G[V_+(\mathbf{z})]$ is disconnected, so the exceptional case (ii) applies in the Discrete Nodal Theorem 14.10 (it is useful to look at the geometric description at the end of that section). Thus $G[V_0(\mathbf{z})]$ has three connected components H_1, H_2 and H_3, where $\mathbf{z}$ is positive on H_1 and H_2, and negative on H_3. We may assume that $i \in V(H_1)$ and $j \in V(H_2)$. By Lemma 16.23, we can extend the embedding of G to an embedding of $G+ij$ so that no cycle in $G\big[V(H_3) \cup N\big(V(H_3)\big)\big]$ is linked with any cycle in $(G+ij) \setminus V(H_3)$. We map the edge $\mathbf{vw}$ of P onto the edge ij of $G+ij$.

We carry out this construction for every broken edge of P_1; we can make sure that the new edges added are disjoint (except for their endpoints). Thus we have

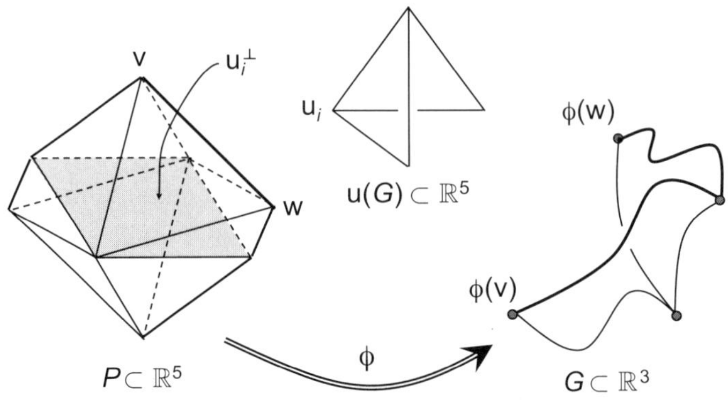

FIGURE 16.6. Illustration of constructing the map ϕ. Top middle: nullspace representation $\mathbf{u}$ in $\mathbb{R}^5$. Left: the polytope $P \subseteq \mathbb{R}^5$ with a hyperplane $\mathbf{u}_i^\perp$. Right: G embedded in $\mathbb{R}^3$.

extended G to a graph $\widehat{G}$ embedded in $\mathbb{R}^3$ (this embedding will not be linkless in general). This gives a continuous mapping $\phi: \ P_1 \to \mathbb{R}^3$. We can perturb this mapping a very little to get a continuous injective mapping ψ.

Claim 2. *Let* $\mathbf{vw}$ *be a broken edge of* P. *Then the* 3*-dimensional subspace* $\{\mathbf{v},\mathbf{w}\}^\perp$ *is generated by the vectors* $\mathbf{u}_i$ *contained in it.*

Indeed, by the Nodal Lemma, there is a 3-dimensional subspace $S \subseteq \mathbf{z}^\perp$ (where $\mathbf{z}$ is an internal point of the edge $\mathbf{vw}$), and this is generated by the vectors $\mathbf{u}_i$ contained in it. Since every hyperplane $\mathbf{u}_i^\perp$ containing $\mathbf{z}$ also contains $\mathbf{v}$ by Claim 1, it follows that $\mathbf{v} \in S^\perp$. Similarly $\mathbf{w} \in S^\perp$, and hence $S = \{\mathbf{v},\mathbf{w}\}^\perp$.

Claim 3. *For every face* F *of* P, *the faces* F *and* $-F$ *together have at most one broken edge.*

Suppose that $\mathbf{vw}$ and $\mathbf{v}'\mathbf{w}'$ are different broken edges of a face F. By Claim 2, $\{\mathbf{v},\mathbf{w}\}^\perp$ is generated by three vectors $\mathbf{u}_{i_1}$, $\mathbf{u}_{i_2}$ and $\mathbf{u}_{i_3}$, and similarly, $\{\mathbf{v}',\mathbf{w}'\}^\perp$ is generated by three vectors $\mathbf{u}_{j_1}$, $\mathbf{u}_{j_2}$ and $\mathbf{u}_{j_3}$. Since $\{\mathbf{v},\mathbf{w}\}^\perp \neq \{\mathbf{v}',\mathbf{w}'\}^\perp$, we may assume that $\mathbf{v}^\mathsf{T}\mathbf{u}_{j_1} \neq 0$ and $\mathbf{v}'^\mathsf{T}\mathbf{u}_{i_1} \neq 0$. But then the edge $\mathbf{vv}'$ of F crosses one of the hyperplanes $(\mathbf{u}_{i_1}+\mathbf{u}_{j_1})^\perp$ or $(\mathbf{u}_{i_1}-\mathbf{u}_{j_1})^\perp$, a contradiction.

A similar argument applies when $\mathbf{v}'\mathbf{w}'$ is an edge of $-F$ different from the antipode of $\mathbf{vw}$. The antipode of a broken edge is not broken, since $\phi(-vb), \phi(-wb) \in V_+(-\mathbf{z}) = V_-(\mathbf{z})$, and $V_(\mathbf{z})$ induces a connected subgraph by (ii) of the Discrete Nodal Theorem 14.10 again. This proves the claim.

The map $\phi: \ P_1 \to \mathbb{R}^3$ is not necessarily injective, but we can modify it by arbitrarily little to get an injective continuous map $\psi: \ P_1 \to \mathbb{R}^3$. By Theorem 16.27, P has two antipodal 2-dimensional faces F and $-F$ such that the cycles $\psi(C)$ and $\psi(-C)$ have an odd linking number, where $C = \partial(F)$.

Claim 4. *The subgraphs* $\phi(C)$ *and* $\phi(-C)$ *of* $\widehat{G}$ *are node-disjoint.*

Indeed, let $\mathbf{z}$ be any interior point of F, then $\mathbf{z}^\mathsf{T}\mathbf{u}_{\phi(\mathbf{v})} > 0$ for every vertex $\mathbf{v} \in V(F)$ by Claim 1, i.e., $\phi(\mathbf{v}) \in V_+(\mathbf{z})$. It follows by a similar argument that if f is an unbroken edge of F, then $V\big(\phi(f)\big) \subseteq V_+(\mathbf{z})$. Since C has at most one unbroken edge by Claim 3, whose image is a single edge of $\widehat{G}$, we have $V(\phi(C)) \subseteq V_+(\mathbf{z})$. Similarly $V(\phi(-C)) \subseteq V_-(\mathbf{z})$, hence $V(\phi(C)) \cap V(\phi(-C)) = \emptyset$ as claimed.

Now it is easy to complete the proof. Since the closed curves $\phi(C)$ and $\phi(-C)$ are disjoint, and they are very close to the closed curves $\psi(C)$ and $\psi(-C)$ with an odd linking number, it follows that they have an odd linking number. This implies that they contain cycles $A \subseteq \phi(C)$ and $B \subseteq \phi(-C)$ with an odd linking number.

If neither F nor $-F$ has a broken edge, then A and B are two linked cycles in G, which contradicts the assumption that G is flatly embedded.

By Claim 3, this leaves us with the alternative that F (say) contains a single broken edge $\mathbf{vw}$, and $-F$ contains none. Let $i = \phi(\mathbf{v})$ and $j = \phi(\mathbf{w})$, and let $\mathbf{z}$ be an interior point of the edge $\mathbf{vw}$. Then $V(A) \subseteq V(\phi(C)) \subseteq V_+(\mathbf{z})$, and similarly $V(B) \subseteq V_-(\mathbf{z})$. Since we are in the exceptional case (ii) of the Nodal Theorem, $S = V_-(\mathbf{z})$ induces a connected subgraph containing B, and A is a subgraph of $G[V \setminus S] + ij$. By the construction of $\phi(\mathbf{vw})$, these two cycles cannot be linked, a contradiction again.

16.4. Engaged representations

The Colin de Verdière number can be characterized in terms of a geometric representation that is similar to orthogonal representations [**Kotlov–Lovász–Vempala 1997**]. This will lead to the proof of Theorem 16.16.

A vector-labeling $\mathbf{u}: V \to \mathbb{R}^d$ of a graph G will be called 1-*engaged*, if

$$\mathbf{u}_i^\mathsf{T}\mathbf{u}_j \begin{cases} = 1 & \text{if } ij \in E, \\ < 1 & \text{if } ij \in \overline{E}. \end{cases} \tag{16.9}$$

We impose no condition for $i = j$; if we have, in addition, that $|\mathbf{u}_i| > 0$ for every node i, then we call the 1-engaged representation *strong*.

It is useful to compare 1-engaged representations with orthogonal representations. We can think of orthogonal representations as an attempt to write the adjacency matrix as a Gram matrix, sticking to the 0's in nonadjacent positions (but allowing arbitrary entries in the diagonal and in adjacent positions). The above definition can be thought of as trying to write the adjacency matrix as a Gram matrix, where the 1's in adjacent positions are required, the diagonal is arbitrary, and the 0's in nonadjacent positions are relaxed to be anything less than 1.

EXAMPLE 16.28 (Planar triangulations). Recall from Section 5.5.3 that every planar triangulation with no separating 3- and 4-cycles has a representation by orthogonal proper caps on the sphere. Each cap has a unique pole (the point in space from which the part of the sphere you see is exactly the given cap). An important observation is that *two circles are orthogonal if and only if the corresponding poles have inner product* 1 (Figure 16.7). This translates a representation with orthogonal circles into a strong 1-engaged representation. ♦

EXAMPLE 16.29 (Edge-transitive polytopes). Let P be a polytope in $\mathbb{R}^d$ whose congruence group is transitive on its edges. We may assume that 0 is its center of gravity. Then there is a real number c such that $\mathbf{u}^\mathsf{T}\mathbf{v} = c$ for the endpoints of every edge. We will assume that $c > 0$ (this excludes only very simple cases, see Exercise 10.16). Then we may also assume that $c = 1$.

We have to verify that $\mathbf{u}^\mathsf{T}\mathbf{v} < 1$ for any two nonadjacent vertices. Suppose not, then $1 \leq \mathbf{u}^\mathsf{T}\mathbf{v} < |\mathbf{u}|\,|\mathbf{v}|$, and so we may assume that $|\mathbf{u}| > 1$. Let $\mathbf{w}$ be the

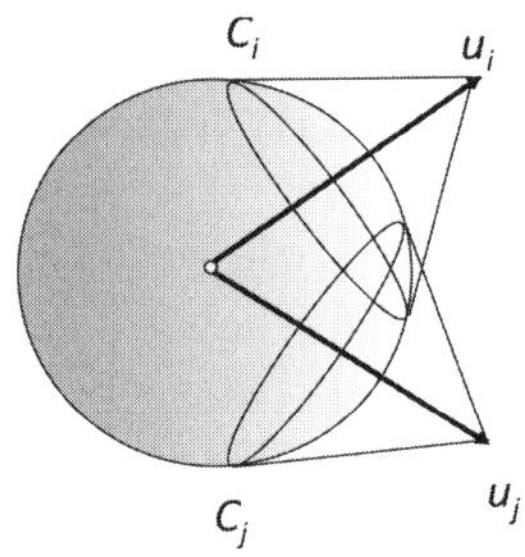

FIGURE 16.7. Poles of circles. If the two circles are orthogonal, then the pole of each circle is contained in the plane of the other, and the poles have inner product 1.

vertex of P maximizing the linear function $\mathbf{u}^\mathsf{T}\mathbf{x}$. If $\mathbf{w} \neq \mathbf{u}$, then there is a path on the skeleton of P from $\mathbf{u}$ to $\mathbf{w}$ along which the linear functional $\mathbf{u}^\mathsf{T}\mathbf{x}$ is monotone increasing. Let $\mathbf{z}$ be the first vertex on this path after $\mathbf{u}$, then $\mathbf{u}^\mathsf{T}\mathbf{z} > \mathbf{u}^\mathsf{T}\mathbf{u} > 1$, contradicting the fact that $\mathbf{uz}$ is an edge of P.

So we know that $\mathbf{u}^\mathsf{T}\mathbf{w} < \mathbf{u}^\mathsf{T}\mathbf{u}$ for every vertex $\mathbf{w}$. There is a path on the skeleton of P from $\mathbf{v}$ to $\mathbf{u}$ along which the linear functional $\mathbf{u}^\mathsf{T}\mathbf{x}$ is monotone increasing. Let $\mathbf{y}$ be the last vertex on this path before $\mathbf{u}$, then $\mathbf{u}^\mathsf{T}\mathbf{y} > \mathbf{u}^\mathsf{T}\mathbf{v} \geq 1$, a contradiction again since $\mathbf{uy}$ is an edge.

So we know that $\mathbf{u}^\mathsf{T}\mathbf{v} = 1$ for the endpoints of every edge, and $\mathbf{u}^\mathsf{T}\mathbf{v} < 1$ for two nonadjacent vertices. Thus the vertices of the polytope provide a strong 1-engaged representation of its skeleton. ♦

The next lemma implies that if the 1-engaged representation is strong, then the representing vectors are the vertices of a convex polytope, and the edges are edges of this polytope (perhaps not every edge). We only state this lemma for twin-free graphs; see [**Kotlov–Lovász–Vempala 1997**] for a version covering the general case.

LEMMA 16.30. *Let G be a twin-free graph without isolated nodes. Let $\mathbf{u} : V \to \mathbb{R}^d$ be a 1-engaged representation of G. Let P be the convex hull of the vectors $\mathbf{u}_i$ $(i \in V)$ and the origin 0, and let P' be the union of all faces of P not containing 0. For $i \in V$, define $Q_i = \{\mathbf{x} \in P : \mathbf{u}_i^\mathsf{T}\mathbf{x} = 1\}$.*

(a) *The vectors $\mathbf{u}_i$ are nonzero and different from each other.*

(b) *If $|\mathbf{u}_i| > 1$, then $\mathbf{u}_i$ is a vertex of P.*

(c) *If $ij \in E$, then either $\mathbf{u}_i$ or $\mathbf{u}_j$ is a vertex of P. If both of them are vertices of P, then the segment connecting $\mathbf{u}_i$ and $\mathbf{u}_j$ is an edge of P.*

(d) *If $\mathbf{u}_i$ is not a vertex of P, then Q_i is a face of P, whose vertices are exactly the vectors representing the neighbors of i.*

(e) *$Q_j \neq Q_i$ for $j \neq i$.*

PROOF. (a) The assumption that G has no isolated node implies that for every node i there is a node j with $\mathbf{u}_i^\mathsf{T}\mathbf{u}_j = 1$, and hence $\mathbf{u}_i \neq 0$. Since (16.9) determines the neighbors of a node i once the vector $\mathbf{u}_i$ is given, the assumption that G is twin-free implies that no two nodes are represented by the same vector.

(b) This follows from the observation that the plane defined by $\mathbf{u}_i^\mathsf{T}\mathbf{x} = 1$ separates $\mathbf{u}_i$ from every other $\mathbf{u}_j$ and also from 0.

(c) We have $\mathbf{u}_i^\mathsf{T}\mathbf{u}_j = 1$, which implies that one of $\mathbf{u}_i$ and $\mathbf{u}_j$ is longer than 1, and hence it is a vertex of P by (b). Let, say, $|\mathbf{u}_i| > 1$. If both $\mathbf{u}_i$ and $\mathbf{u}_j$ are vertices of P, but $[\mathbf{u}_i, \mathbf{u}_j]$ is not an edge of P, then some point $\alpha\mathbf{u}_i + (1-\alpha)\mathbf{u}_j$ $(0 < \alpha < 1)$ can be written as a convex combination of other vertices:

$$\alpha\mathbf{u}_i + (1-\alpha)\mathbf{u}_j = \sum_{k\neq i,j} \lambda_k\mathbf{u}_k,$$

where $\lambda_k \geq 0$ and $\sum_k \lambda_k = 1$. Multiplying by $\mathbf{u}_i$, we get

$$\alpha\mathbf{u}_i^\mathsf{T}\mathbf{u}_i + (1-\alpha)\mathbf{u}_i^\mathsf{T}\mathbf{u}_j = \sum_{k\neq i,j} \lambda_k\mathbf{u}_i^\mathsf{T}\mathbf{u}_k \leq \sum_{k\neq i,j} \lambda_k \leq 1.$$

On the other hand,

$$\alpha\mathbf{u}_i^\mathsf{T}\mathbf{u}_i + (1-\alpha)\mathbf{u}_i^\mathsf{T}\mathbf{u}_j = \alpha(|\mathbf{u}_i|^2 - 1) + 1 > 1,$$

a contradiction. This proves (c).

(d) By (a), we have $|\mathbf{u}_i| \leq 1$, and so $\mathbf{u}_i^\mathsf{T}\mathbf{u}_j \leq 1$ for every vertex of P. Hence the hyperplane $\{\mathbf{x} : \mathbf{u}_i^\mathsf{T}\mathbf{x} = 1\}$ supports P, and it touches P at a face Q_i. Condition (16.9) implies that among the nodes of G different from i, all neighbors of i, and only those, are contained in Q_i. It follows by (b) that every neighbor of i is represented by a vertex of P, and hence, by a vertex of Q_i. The vector $\mathbf{u}_i$ may or may not belong to Q_i, but it is not a vertex of it (since it is different from its neighbors). □

(e) If $Q_i = Q_j$ then i and j are twins.

Corollary 16.31. *If a twin-free graph G has a 1-engaged representation in $\mathbb{R}^d$, then it is isomorphic with a subgraph of the skeleton of a d-polytope. If the 1-engaged representation is strong, then the convex hull of the representing vectors can serve as this polytope.*

Proof. Let $\varepsilon > 0$ be a sufficiently small number. For every node i with $|\mathbf{u}_i| \leq 1$, let $\mathbf{u}_i'$ be a point outside P, at a distance ε from Q_i, and at distance ε^2 from P. For nodes i with $|\mathbf{u}_i| \leq 1$, let $\mathbf{u}_i' = \mathbf{u}_i$. Then every vector $\mathbf{u}_i'$ is a vertex of the polytope $Q = \text{conv}(\mathbf{u}'(V))$, and every edge of G corresponds to an edge of Q. □

The assertion of the theorem is not really interesting for $d > 3$, since every graph can be embedded into the 1-skeleton of a 4-polytope. However, the explicit construction of this embedding may be useful.

16.4.1. Transversality of 1-engaged representations. We need the condition of transversality for 1-engaged representations. Applying the general definition, a 1-engaged representation is *transversal*, if the hypersurfaces in $\mathbb{R}^{dn}$ defined by the equations in (16.9) (corresponding to the edges of G) intersect transversally at $\mathbf{u}$. This leads to a very similar condition as for orthogonal representations: the representation $\mathbf{u}$ is transversal if and only if $(G, \mathbf{u})$ carries no homogeneous stress, i.e., there is no nonzero G-matrix X with zero diagonal such that $\sum_j X_{ij}\mathbf{u}_j = 0$ for every node i. (In the case of orthogonal representations, the equations in the definition correspond to the edges, so there we excluded homogeneous stress on the complement of G.)

As before, it is useful to find some reformulations of transversality. Suppose that $(G, \mathbf{u})$ carries a homogeneous stress X. Multiplying the equation $\sum_j X_{ij}\mathbf{u}_j = 0$ by $\mathbf{u}_i$, we get

$$\sum_j X_{ij}\mathbf{u}_j^{\mathsf{T}}\mathbf{u}_i = \sum_j X_{ij} = 0,$$

and this in turn gives that $\sum_j X_{ij}(\mathbf{u}_j - \mathbf{u}_i) = 0$. So a homogeneous stress on a 1-engaged representation is automatically a stress.

The converse of this statement is almost, but not quite, true. If we have a stress X on $(G, \mathbf{u})$, then multiplying the equation $\sum_j X_{ij}(\mathbf{u}_j - \mathbf{u}_i) = 0$ by $\mathbf{u}_i$, we get

$$\sum_j X_{ij}(\mathbf{u}_j - \mathbf{u}_i)^{\mathsf{T}}\mathbf{u}_i = \Big(\sum_j X_{ij}\Big)(|\mathbf{u}_i|^2 - 1) = 0.$$

This means that $\sum_j X_{ij} = 0$ holds whenever $|\mathbf{u}_i| \neq 1$, and in this case it follows that $\sum_j X_{ij}\mathbf{u}_j = 0$ also holds.

These considerations show that every strong 1-engaged representation of a twin-free graph in dimension 3 is transversal: Lemma 16.30 implies that as a framework the graph is a subgraph of the skeleton of a 3-polytope, by Cauchy's Theorem 14.5, this representation has no stress, and this implies that it has no homogeneous stress.

16.4.2. The Colin de Verdière number and 1-engaged representations. We denote by $\gamma(G)$ the smallest dimension in which G has a transversal 1-engaged representation. This quantity is closely related to the Colin de Verdière number.

THEOREM 16.32. *For every graph G with at least 3 nodes, we have $\gamma(G) = n - 1 - \mu(\overline{G})$.*

While this theorem establishes a useful connection between a graph and its complement, it is not really deep (and not a real duality): the quantity $\mu(\overline{G})$ is defined as a maximum, $\gamma(G)$ is defined as a minimum. So the statement of the lemma has the form minimum = minimum.

PROOF. First, let $k = \mu(\overline{G})$, and let M be a Colin de Verdière matrix of $\overline{G}$. Then M is a well-signed $\overline{G}$-matrix with exactly one negative eigenvalue. Let $\lambda_1 < 0 = \lambda_2 = \cdots = \lambda_{k+1} < \lambda_{k+2} \leq \ldots$ be the eigenvalues of M. By scaling, we may assume that $\lambda_1 = -1$. Let $\mathbf{v}_1$ be the eigenvector of unit length belonging to λ_1. Since M is well-signed, we may assume that all entries of $\mathbf{v}_1$ are positive. We form the diagonal matrix $D = \operatorname{diag}(\mathbf{v}_1)$.

We start with getting rid of the negative eigenvalue, and consider the matrix

$$M' = D^{-1}\big(M + \mathbf{v}_1\mathbf{v}_1^{\mathsf{T}}\big)D^{-1} = D^{-1}MD^{-1} + J.$$

This matrix is symmetric and positive semidefinite, and so we can write it as a Gram matrix $M' = \mathsf{Gram}(\mathbf{u})$, where $\mathbf{u} : V \to \mathbb{R}^{n-r}$. From the fact that M is a well-signed $\overline{G}$-matrix, it follows that

$$\mathbf{u}_i^{\mathsf{T}}\mathbf{u}_j = M'_{ij}\begin{cases} = 1 & \text{for } ij \in E, \\ < 1 & \text{for } ij \in \overline{E}. \end{cases} \tag{16.10}$$

(The somewhat artificial scaling above has been introduced to guarantee this.) The diagonal entries of M' are positive, but they can be larger than, equal to, or smaller than 1.

Thus we get a vector-labeling $\mathbf{u}$ in $\mathbb{R}^{n-k-1}$ satisfying (16.10). We claim that $(\overline{G}, \mathbf{u})$ carries no homogeneous stress. Suppose that X is such a homogeneous stress. We have seen that this implies that $\sum_j X_{ij} = 0$. In other words, $JX = 0$. Let $Y = D^{-1}XD^{-1}$, then

$$MY = D(M' - J)DD^{-1}XD^{-1} = DM'XD^{-1} = 0.$$

Since $Y \neq 0$ and has the same 0 entries as X, this contradicts the assumption that M is transversal.

The converse construction is straightforward. Let $i \mapsto \mathbf{u}_i \in \mathbb{R}^d$ be a 1-engaged representation that carries no homogeneous $\overline{G}$-stress, let $A = \mathsf{Gram}(\mathbf{u})$, and let $M = AA^{\mathsf{T}} - J$. It is clear that M is a well-signed G-matrix with at most one negative eigenvalue, and has corank at least $n-d-1$.

The only nontrivial part of the proof is to show that M has the Strong Arnold Property. Suppose that $MX = 0$ for a nonzero $\overline{G}$-matrix X with zero diagonal. Then $AA^{\mathsf{T}}X = JX$, and hence $\operatorname{rk}(A^{\mathsf{T}}X) = \operatorname{rk}(X^{\mathsf{T}}AA^{\mathsf{T}}X) = \operatorname{rk}(X^{\mathsf{T}}JX) \leq \operatorname{rk}(J) = 1$. (We have used the fact that $\operatorname{rk}(B) = \operatorname{rk}(B^{\mathsf{T}}B)$ for every real matrix B.) If $\operatorname{rk}(A^{\mathsf{T}}X) = 0$, then $A^{\mathsf{T}}X = 0$, and so X is a homogeneous stress on $(\overline{G}, \mathbf{u})$, which contradicts the hypothesis. So we must have $\operatorname{rk}(A^{\mathsf{T}}X) = 1$, and we can write $A^{\mathsf{T}}X = \mathbf{u}\mathbf{v}^{\mathsf{T}}$, where $\mathbf{u} \in \mathbb{R}^d$ and $\mathbf{v} \in \mathbb{R}^V$ are nonzero vectors. Hence $A\mathbf{u}\mathbf{v}^{\mathsf{T}} = AA^{\mathsf{T}}X = JX = \mathbb{1}(X\mathbb{1})^{\mathsf{T}}$, which implies that (rescaling $\mathbf{u}$ and $\mathbf{v}$ if necessary) $A\mathbf{u} = \mathbb{1}$ and $\mathbf{v} = X\mathbb{1}$. Hence $\mathbf{v} = XA\mathbf{u} = \mathbf{v}\mathbf{u}^{\mathsf{T}}\mathbf{u}$, and so $|\mathbf{u}| = 1$. It follows that

$$(A - \mathbb{1}^{\mathsf{T}}\mathbf{u})(A - \mathbb{1}^{\mathsf{T}}\mathbf{u})^{\mathsf{T}} = AA^{\mathsf{T}} - J = M,$$

so M is positive semidefinite. But then its smallest eigenvalue is 0, which has multiplicity at least $n-d-1>1$, contradicting the Perron–Frobenius Theorem. □

16.4.3. 1-engaged representations in low dimension. The following theorem is a reformulation of Theorem 16.16.

THEOREM 16.33. *Let G be a simple twin-free graph.*

(a) *If G is outerplanar, then $\gamma(G) \leq 3$; if G is not outerplanar, then $\gamma(G) \geq 3$;*

(b) *If G is planar, then $\gamma(G) \leq 4$; if G is not planar, then $\gamma(G) \geq 4$.*

PROOF. We describe the proof of part (b); the proof of part (a) uses similar, but much simpler arguments, and is not given here. Corollary 16.31 implies the second assertion, so what we need to prove is that G is planar, then $\gamma(G) \leq 4$.

We may assume that G is 3-connected, since adding edges to G can only increase $\gamma(G)$ by minor-monotonicity. Let $\mathbf{v} : V \to \mathbb{R}^3$ be the cage representation in Theorem 5.10 of a planar graph G. For $ij \in E$, the segment connecting $\mathbf{v}_i$ and $\mathbf{v}_j$ touches the unit sphere at a point $\mathbf{w}$. Then $\mathbf{v}_i - \mathbf{w}$ and $\mathbf{v}_j - \mathbf{w}$ are parallel vectors, orthogonal to $\mathbf{w}$, and pointing in opposite directions. Their length is easily computed by Pythagoras' Theorem: $|\mathbf{v}_i - \mathbf{w}| = \sqrt{|\mathbf{v}_i|^2 - 1}$ and $|\mathbf{v}_j - \mathbf{w}| = \sqrt{|\mathbf{v}_j|^2 - 1}$. This gives

$$\mathbf{v}_i^{\mathsf{T}}\mathbf{v}_j = 1 + (\mathbf{v}_i - \mathbf{w})^{\mathsf{T}}(\mathbf{v}_j - \mathbf{w}) = 1 - \sqrt{|\mathbf{v}_i|^2 - 1}\sqrt{|\mathbf{v}_j|^2 - 1}.$$

Introducing the vectors

$$\mathbf{u}_i = \begin{pmatrix} \mathbf{v}_i \\ \sqrt{|\mathbf{v}_i|^2 - 1} \end{pmatrix} \in \mathbb{R}^4,$$

we get that $\mathbf{u}_i^\mathsf{T}\mathbf{u}_j = 1$. It follows similarly that if the segment connecting the two vectors intersects the sphere, then $\mathbf{u}_i^\mathsf{T}\mathbf{u}_j < 1$.

Thus we get a 1-engaged representation in $\mathbb{R}^4$. This representation is transversal: any stress on $(G, \mathbf{u})$ would be a stress on $(G, \mathbf{v})$ as well, which is impossible by Cauchy's Theorem 14.5. □

REMARK 16.34. Perhaps the most interesting special case is that of a planar but not outerplanar graph G. Theorem 16.33 implies that $\gamma(G) = 3$ or $\gamma(G) = 4$. In [**Kotlov–Lovász–Vempala 1997**] it is proved that a triangulation of the sphere (maximal planar map) with at least 8 nodes has $\gamma(G) = 3$ if and only if no 3- or 4-cycle separates it into two parts with at least two nodes. (The case of non-maximal planar graphs is open.)

16.5. Related representations

A related representation was studied in [**Kang et al. 2011**], where the following conditions were imposed:

$$\mathbf{u}_i^\mathsf{T}\mathbf{u}_j \begin{cases} \geq 1 & \text{if } ij \in E, \\ < 1 & \text{if } ij \in \overline{E}. \end{cases} \tag{16.11}$$

The results are similar to some of those described above: Every outerplanar graph has a representation in $\mathbb{R}^3$ satisfying (16.11), and every planar graph has such a representation in $\mathbb{R}^4$. There are outerplanar graphs that do not have such a representation in $\mathbb{R}^2$, and there are planar graphs that do not have such a representation in $\mathbb{R}^3$. Estimates on the minimal dimension in terms of other graph parameters were given in [**Reiterman–Rödl–Šinajová 1989, Reiterman–Rödl–Šinajová 1992, Brightwell–Scheinerman 1993**].

We have repeatedly come across representations in which the subgraph induced by the nodes placed in certain halfspaces are connected. A *halfspace* in the following discussion is the (open) set $\{\mathbf{x} \in \mathbb{R}^d : \mathbf{a}^\mathsf{T}\mathbf{x} > b\}$, where $\mathbf{a} \in \mathbb{R}^d$ is a nonzero vector. If $b = 0$, we call the halfspace *linear*. If we want to emphasize that this is not assumed, we call it *affine*. We say that a representation $\mathbf{v} : V \to \mathbb{R}^d$ of a graph G has the *connectivity property with respect to a family of subspaces*, if for every halfspace in the family, the set $\mathbf{v}^{-1}(H)$ induces a connected subgraph of G. We allow here that this subgraph is empty; if we insist that it is nonempty, we say that the representation has the *nonempty connectivity property*.

EXAMPLE 16.35. The skeleton of a convex polytope in $\mathbb{R}^d$ has the connectivity property with respect to all halfspaces (cf. Appendix C.1). ♦

EXAMPLE 16.36. The rubber band representation of a 3-connected planar graph, with the nodes of a country nailed to the vertices of a convex polygon, has the connectivity property with respect to all halfplanes (Claim 1 in the proof of Theorem 3.2). ♦

EXAMPLE 16.37. The nullspace representation defined by a well-signed G-matrix M with exactly one negative eigenvalue, has the nonempty connectivity property with respect to all linear halfspaces whose boundary hyperplane is spanned by representing vectors (Theorem 14.10), and also with respect to all linear halfspaces whose boundary contains no representing vectors.

If the graph is 3-connected planar, then the nullspace representation defined by a well-signed G-matrix has the nonempty connectivity property with respect to all linear halfspaces (Corollary 14.11). ♦

A general study of this property was initiated by Van der Holst, Laurent and Schrijver [**van der Holst et al. 1995a**]. In one version, the authors consider full-dimensional representations in $\mathbb{R}^d$ with the connectivity property with respect to all halfspaces. It is not hard to see that *not* having such a representation is a property closed under minors. They give the following (nice and easy) characterization:

THEOREM 16.38. *A graph has a full-dimensional representation in $\mathbb{R}^d$ with the connectivity property with respect to all halfspaces if and only if it is connected and has a K_{d+1} minor.* □

We note that this result does not supply a "good characterization" for graphs not having a K_{d+1} minor (both equivalent properties put this class into co-NP). We know from the Robertson–Seymour theory of graph minors [**Robertson–Seymour 2003**] (see [**Lovász 2006**] for a survey) that not having a K_{d+1} minor is a property in P, and explicit certificates (structure theorems) for this are known for $d \leq 5$. To describe an explicit certificate for this property for general d is an outstanding open problem.

To sketch the other version, let us consider full-dimensional representations in $\mathbb{R}^d$ with the nonempty connectivity property with respect to all linear halfspaces. No full characterization of graphs admitting such a representation is known, but the following is proved in [**van der Holst et al. 1995a**]:

THEOREM 16.39. *A graph has a full-dimensional representation in $\mathbb{R}^3$ with the nonempty connectivity property for all linear halfspaces if and only if it can be obtained from planar graphs by taking clique-sums and subgraphs.* □

EXERCISE 16.1. Let $f(G)$ denote the maximum multiplicity of the largest eigenvalue of any matrix obtained from the adjacency matrix of a graph G, where 1's are replaced by arbitrary positive numbers and the diagonal entries are changed arbitrarily. Prove that $f(G)$ is the number of connected components of the graph G.

EXERCISE 16.2. The complete graph K_n is the only graph on $n \geq 3$ nodes with $\mu = n - 1$.

EXERCISE 16.3. If G has at least one edge, then $\mu(G) = \max_H \mu(H)$, where H ranges over all connected components of G.

EXERCISE 16.4. Let M be a Colin de Verdière matrix of a path P_n, let π be an eigenvector of M belonging to the negative eigenvalue, and let $w : V(P_n) \to \mathbb{R}$ be a nullspace representation of M. Then the mapping $i \mapsto w_i/\pi_i$ gives an embedding of G in the line [**Lovász–Schrijver 1999**].

EXERCISE 16.5. Let M be a Colin de Verdière matrix of a 2-connected outerplanar graph G, and let $w : V \to \mathbb{R}^2$ be a nullspace representation of M. Then the mapping $i \mapsto \mathbf{v}_i = \mathbf{w}_i^0$, together with the segments connecting $\mathbf{v}_i$ and $\mathbf{v}_j$ for $ij \in E$, gives an embedding of G in the plane as a convex polygon with noncrossing diagonals.

EXERCISE 16.6. Let $\overline{\psi} : S^2 \to S^2$ be a local homeomorphism. (a) Prove that there is an integer $k \geq 1$ such that $|\psi^{-1}(x)| = k$ for each $x \in S^2$. (b) Let G be any 3-connected planar graph embedded in S^2, then $\psi^{-1}(G)$ is a graph embedded in S^2, with kn nodes, km edges, and $k|F(G)|$ faces. (c) Use Euler's Formula to prove that $k = 1$ and ψ is bijective.

EXERCISE 16.7. Let ij be a cut-edge of the graph G. Let A be a G-matrix with $A_{ij} \neq 0$, and let $\mathbf{u}$ be the nullspace representation belonging to A. Then $\mathbf{u}_i$ and $\mathbf{u}_j$ are parallel.

EXERCISE 16.8. Prove that if a graph is embedded in $\mathbb{R}^3$ in such a way that every pair of disjoint cycles have an even linking number, and we perform a $\Delta - Y$ or $Y - \Delta$ transformation, then the new edges can be embedded so that the resulting embedded graph has the same property.

EXERCISE 16.9. Prove that for every embedding of K_6 in $\mathbb{R}^3$, the sum of linking numbers of disjoint pairs of triangles (10 pairs) is odd.

EXERCISE 16.10. All triangles in a flatly embedded simple graph can be simultaneously paneled.

EXERCISE 16.11. Let P be a d-polytope with no parallel faces, and let F_1 and F_2 be antipodal faces. Prove that there are two parallel supporting hyperplanes H_1 and H_2 such that $F_i = H_i \cap P$. Show by an example that this does not hold if parallel faces are allowed.

EXERCISE 16.12. Prove that if G is the disjoint union of graphs G_1 and G_2, then $\gamma(G) \leq \min\{\gamma(G_1)+|V(G_2)|,\ \gamma(G_2)+|V(G_1)|\}$.

EXERCISE 16.13. A graph is planar if and only if its 1-subdivision has a 1-engaged representation in $\mathbb{R}^3$.

EXERCISE 16.14. Every graph has a subdivision with $\gamma \leq 4$.

EXERCISE 16.15. If G is a planar map of girth at least 5, then $\gamma(G) \leq 3$.

EXERCISE 16.16. Prove that a graph G has a full-dimensional representation in $\mathbb{R}^d$ with the connectivity property with respect to all linear halfspaces, such that the origin is not contained in the convex hull of representing vectors, if and only if it has a K_d minor.

EXERCISE 16.17. Prove that if a graph G has a full-dimensional representation in $\mathbb{R}^d$ in general position with the connectivity property with respect to all linear halfspaces, then G is d-connected.

EXERCISE 16.18. Prove the following characterizations of graphs having full-dimensional representations in $\mathbb{R}^d$ with the nonempty connectivity property with respect to all linear halfspaces, for small values of d: (a) for $d = 1$: forests; (b) for $d = 2$: series-parallel graphs (graphs with no K_4 minors).

CHAPTER 17

Metric Representations

Given a graph, we would like to embed it in a Euclidean space so that the length of shortest paths between nodes in the graph (the *graph metric*) should be the same as, or at least close to, the geometric distance of the representing vectors. Our goal is to illustrate some of the embedding techniques, and, mainly, the use of such embeddings to prove graph-theoretic theorems and to design algorithms.

We start with discussing a few embedding techniques that preserve distances. It is not hard to see, however, that in most nontrivial cases every embedding will necessarily have some distortion. For example, the graph metric on the "claw" $K_{1,3}$ cannot be embedded isometrically in a Euclidean space of any dimension. We get more general and useful results if we study embeddings where the distances may change, but in controlled manner. To emphasize the difference, we will distinguish *distance preserving* (isometric) and *distance respecting* representations.

The complete k-graph can be embedded isometrically in a Euclidean space with dimension $k-1$, but not in lower dimensions. There is often a trade-off between dimension and distortion. This motivates our concern with the dimension of the space in which we represent our graph.

Most of the methods that have been developed for such embeddings extend to more general (often all) finite metric spaces. Recall that a metric space is a set V endowed with a distance function $\mathsf{d}_2 : V \times V \to \mathbb{R}_+$ such that $\mathsf{d}_2(u,v) = 0$ if and only if $u = v$, $\mathsf{d}_2(v,u) = \mathsf{d}_2(u,v)$, and $\mathsf{d}_2(u,w) \le \mathsf{d}_2(u,v) + \mathsf{d}_2(v,w)$ for all $u, v, w \in V$. Sometimes it will be convenient to extend the representability problem to semimetric spaces (to spaces (V, d_2) where $\mathsf{d}_2 : V \times V$ is nonnegative, symmetric and satisfies the triangle inequality, but $\mathsf{d}_2(u,v) = 0$ is allowed for different points $u, v \in V$).

There is a large literature of embeddings of one metric space into another so that distances are preserved (isometric embeddings) or at least not distorted too much. These results are often very combinatorial, and have important applications to graph theory and combinatorial algorithms; see [**Indyk–Matoušek 2004**], [**Matoušek 2002**], [**Matoušek 2013**] for more.

Here we restrict our interest to embeddings of (finite) metric spaces into basic normed spaces. It will be often convenient to use the scaled version of the standard ℓ_p-norm, namely

$$\|x\|_p = \frac{1}{d^{1/p}} |x|_p = \Big(\frac{1}{d} \sum_{i=1}^{d} |x_i|^p \Big)^{1/p}$$

for $x \in \mathbb{R}^d$ and $p \ge 1$. In the case of $p = \infty$, we have $|x|_\infty = \|x\|_\infty = \max_i |x_i|$. This version of the norms can be considered as function norms over the (finite) probability space $\{1, \dots, d\}$ endowed with the uniform measure. With this scaling, we have $\|\mathbf{x}\|_p \le \|\mathbf{x}\|_q$ for $1 \le p < q \le \infty$.

An *ℓ_p-representation* of a finite metric space (V, d_2) is a mapping $\mathbf{u}: V \to \mathbb{R}^m$ such that $\mathsf{d}_2(i,j) = \|\mathbf{u}_i - \mathbf{u}_j\|_p$. We say that (V, d_2) is *ℓ_p-representable*, or shortly an *ℓ_p-metric*. Note that in the definition of this representability we could use the $|.|_p$-norm instead of the $\|.\|_p$-norm, by scaling the representing vectors by a factor of $d^{1/p}$.

In our limited treatment of metric embeddings, our main interest will lie in one embedding technique, the Bourgain random subset representation, which has a variety of graph theoretic applications. It is perhaps natural that we need to do some preliminary work discussing some examples of isometric embeddings which will be needed later. We also need a fundamental tool, the Johnson–Lindenstrauss Lemma, which helps to control the dimension of the embeddings.

After treating the random subset representation and its applications, we introduce the fascinating notion of volume-respecting embedding, which can be considered as a quantitative version of "general position". The random subset representation method can be extended to this case, but this will be somewhat hard going, requiring several facts from high dimensional geometry, which are collected in the Appendix.

17.1. Preserving distances

17.1.1. Supremum norm. The embedding problem is easy if the supremum norm is used in the target space.

THEOREM 17.1 (Fréchet). *Any finite metric space on n points can be embedded isometrically into $(\mathbb{R}^n, \|.\|_\infty)$.*

The theorem is valid for infinite metric spaces as well (when infinite dimensional L_∞ spaces must be allowed for the target space), with a slight modification of the proof. We describe the simple proof for the finite case, since its idea will be used later on.

PROOF. Let (V, d_2) be a finite metric space, and consider the vector labeling

$$v \mapsto \mathbf{u}_v = (\mathsf{d}_2(v,i): \ i \in V) \qquad (v \in V).$$

By the triangle inequality,

$$\|\mathbf{u}_u - \mathbf{u}_v\|_\infty = \max_i |\mathsf{d}_2(u,i) - \mathsf{d}_2(v,i)| \le \mathsf{d}_2(u,v).$$

On the other hand, the coordinate corresponding to $i = v$ gives

$$\|\mathbf{u}_u - \mathbf{u}_v\|_\infty \ge |\mathsf{d}_2(u,v) - \mathsf{d}_2(v,v)| = \mathsf{d}_2(u,v).$$ □

17.1.2. Manhattan distance. Another representation with special combinatorial significance is a representation in $\mathbb{R}^n$ with the ℓ_1-distance (often called Manhattan distance). For a fixed underlying set V, all ℓ_1-semimetrics form a closed convex cone: if $i \mapsto \mathbf{u}_i \in \mathbb{R}^m$ is an ℓ_1-representation of (V, d_2) and $i \mapsto \mathbf{u}'_i \in \mathbb{R}^{m'}$ is an ℓ_1-representation of (V, d'_2), then

$$i \mapsto (m+m') \begin{pmatrix} \frac{\alpha}{m}\mathbf{u} \\ \frac{\alpha'}{m'}\mathbf{u}' \end{pmatrix}$$

is an ℓ_1-representation of $\alpha \mathsf{d}_2 + \alpha' \mathsf{d}'_2$ for $\alpha, \alpha' \ge 0$.

A very useful way of representing ℓ_1-semimetrics is by *Hamming metrics.* We have a measurable space (X,μ), and a family of measurable subsets of finite measure $X_i \subseteq X$ $(i \in V)$. Define

$$\mathsf{d}_2(i,j) = \mu(X_i \triangle X_j).$$

It is trivial to check that this defines a semimetric. Since we are interested in finite metric spaces, we can always assume that the space X is finite: the sets X_i partition X into a finite number of parts, and we can aggregate each part to a single atom (with the same weight).

Hamming metrics on the same set V form a convex cone: clearly they can be scaled by any positive number by scaling the measure μ, and the sum of two Hamming metrics d_2 and d_2' is again a Hamming metric. To see this, let (V,d_2) be represented by subsets X_i in a measurable space (X,μ), and (V,d_2') be represented by subsets X_i' in a measurable space (X',μ'), where we may assume that X and X' are disjoint. Then $(V,\mathsf{d}_2+\mathsf{d}_2')$ is represented by the sets $X_i \cup X_i'$ in the measurable space $(X\cup X', \mu\cup\mu')$ (the measure $\mu\cup\mu'$ being defined by $(\mu\cup\mu')(S) = \mu(S\cap X) + \mu'(S\cap X')$).

This last remark is superfluous in the light of the following fact (but it is used in its proof):

PROPOSITION 17.2. *Hamming semimetrics on a finite set coincide with ℓ_1-semimetrics.*

PROOF. Let (V,d_2) be represented by subsets X_i in a finite measurable space (X,μ). Define a vector $\mathbf{u}_i \in \mathbb{R}^X$ by

$$(\mathbf{u}_i)_j = \begin{cases} \dfrac{\mu(j)}{|X|}, & \text{if } j \in X_i, \\ 0, & \text{otherwise.} \end{cases}$$

It is straightforward to check that this defines an ℓ_1-representation of (V,d_2).

Conversely, let $\mathbf{u}: V \to \mathbb{R}^d$ be an ℓ_1-representation of (V,d_2). If $k=1$, then $i \mapsto [0,\mathbf{u}_i]$ is a Hamming representation of (V,d_2) in the measurable space $(\mathbb{R},\lambda)$. If $k>1$, we construct a Hamming representation for each coordinate, and add them up as described above. □

Of special interest are the ℓ_1-semimetrics (V,d_S) defined by a subset $S\subseteq V$,

$$\mathsf{d}_S(i,j) = \begin{cases} 1, & \text{if } |\{i,j\}\cap S| = 1, \\ 0, & \text{otherwise,} \end{cases}$$

which we call a 2-*partition distance.* (These are often called cut-semimetrics, but I do not want to use this term since "cut norm" and "cut distance" are used in this book in a different sense.) 2-partition distances and their connection to Manhattan metrics are discussed in detail, with many applications from number theory to optimization, in the book [**Deza–Laurent 1997**].

PROPOSITION 17.3. *A finite semimetric space (V,d) is an ℓ_1-semimetric if and only if it can be written as a nonnegative linear combination of 2-partition distances.*

PROOF. Every 2-partition distance d_S can be represented in the 1-dimensional space by the map $\mathbb{1}_S$. It follows that every nonnegative linear combination of 2-partition distances on the same underlying set is an ℓ_1-semimetric. Conversely, every ℓ_1-semimetric is a sum of ℓ_1-semimetrics representable in $\mathbb{R}^1$ (just consider the coordinates of any ℓ_1-representation). So it suffices to consider ℓ_1-semimetrics (V,d) for which there is a representation $i \mapsto u_i \in \mathbb{R}$ such that $\mathsf{d}(i,j) = |u_i - u_j|$. We may assume that $V = [n]$ and $u_1 \le \cdots \le u_n$, then for $i < j$,

$$\mathsf{d}(i,j) = \sum_{k=1}^{n-1} (u_{k+1} - u_k)\mathsf{d}_{\{1,\dots,k\}}$$

expresses d as a nonnegative linear combination of 2-partition distances. □

There are a lot of results relating the ℓ_1-metric to other classical metrics on linear spaces, most notably to ℓ_p-metrics. We only need the following fact.

PROPOSITION 17.4. *If a finite metric space* (V,d) *is* ℓ_2*-representable, then it is* ℓ_1*-representable.*

PROOF. Let (V,d) be a finite metric space and let $\mathbf{u} : V \to \mathbb{R}^d$ be an ℓ_2-representation. We may assume (by trivial scaling) that $\mathbf{u}(V) \subseteq B^d$.

By Proposition 17.2, it suffices to represent (V,d) as a Hamming distance. Consider the measurable space (B^k, μ), where the measure is defined by

$$\mu(Y) = C \int_Y \frac{1}{|\mathbf{x}|^{1-d}}\, d\mathbf{x},$$

(the constant C will be determined later). It is easy to see that

$$\mu(Y) \sim \int_{S^{d-1}} \lambda([-\mathbf{x},\mathbf{x}] \cap Y)\, d\sigma(\mathbf{x}),$$

where λ is the 1-dimensional Lebesgue measure and σ is the $(d-1)$-dimensional Lebesgue measure on S^{d-1}.

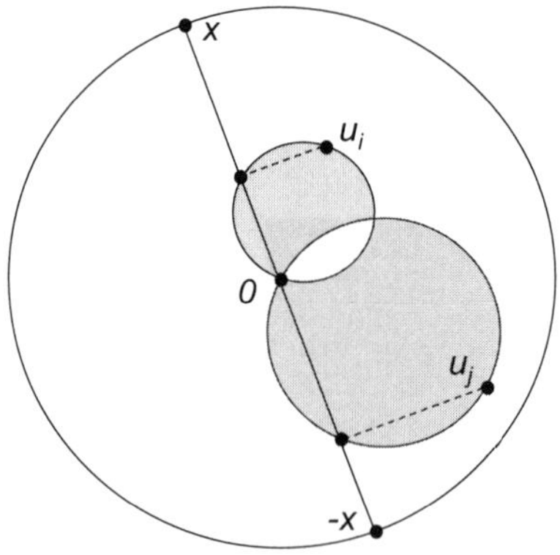

FIGURE 17.1. Representing an ℓ_2 metric in ℓ_1.

The sets X_i representing the points in V are easy to define: X_i is the "Thales ball" over the segment $[0, \mathbf{u}_i]$. In formula, this means that

$$X_i = \{\mathbf{x} \in \mathbb{R}^k : \mathbf{x}^\mathsf{T}(\mathbf{u}_i - \mathbf{x}) \ge 0\}.$$

To calculate the measure of $X_i \triangle X_j$, note that for any $\mathbf{x} \in S^{k-1}$, the intersection $[-\mathbf{x},\mathbf{x}] \cap (X_i \triangle X_j)$ is just the segment between the orthogonal projections of $\mathbf{u}_i$ and $\mathbf{u}_j$ onto the diameter $[-\mathbf{x},\mathbf{x}]$ (Figure 17.1). Hence

$$\lambda\big([-\mathbf{x},\mathbf{x}] \cap (X_i \triangle X_j)\big) = |\mathbf{x}^\mathsf{T}\mathbf{u}_i - \mathbf{x}^\mathsf{T}\mathbf{u}_j| = |\mathbf{x}^\mathsf{T}(\mathbf{u}_i - \mathbf{u}_j)| = |\mathbf{u}_i - \mathbf{u}_j|\,|\mathbf{x}^\mathsf{T}(\mathbf{u}_i - \mathbf{u}_j)^0|.$$

Integrating over $\mathbf{x} \in S^{k-1}$, the integral of $|\mathbf{x}^{\mathsf{T}}(\mathbf{u}_i - \mathbf{u}_j)^0|$ is independent of $(\mathbf{u}_i - \mathbf{u}_j)^0$ (which is just a unit vector). Hence

$$\int_{S^{k-1}} \lambda\big([-\mathbf{x},\mathbf{x}] \cap (X_i \triangle X_j)\big)\, d\sigma(\mathbf{x}) \sim |\mathbf{u}_i - \mathbf{u}_j|,$$

and the scaling factor C above has been chosen so that $\mu(X_i \triangle X_j) = |\mathbf{u}_i - \mathbf{u}_j|$. $\square$

The converse of the last proposition does not hold: there are ℓ_1-representable finite metric spaces that are not ℓ_2-representable, as shown e.g. by the graph distance in the 4-cycle.

17.2. Respecting distances

As we have seen, the possibility of a geometric representation that exactly reflects the graph distance is limited, and therefore we allow distortion. Let $F : V_1 \to V_2$ be a mapping of the metric space (V_1, d_1) into the metric space (V_2, d_2). We define the *distortion* of F as

$$\sup_{\substack{u,v \in V_1 \\ u \neq v}} \frac{\mathsf{d}_2\big(F(u),F(v)\big)}{\mathsf{d}_1(u,v)} \Big/ \inf_{\substack{u,v \in V_1 \\ u \neq v}} \frac{\mathsf{d}_2\big(F(u),F(v)\big)}{\mathsf{d}_1(u,v)}.$$

Note that to have finite distortion, the map F must be injective. The distortion does not change if all distances in one of the metric spaces are scaled by the same factor. So if we are looking for embeddings in Euclidean spaces (more generally, in normed linear spaces), then we may scale the vectors, and restrict our attention to embeddings that are *contractive*, that is, satisfying $\mathsf{d}_2\big(F(u),F(v)\big) \le \mathsf{d}_1(u,v)$ for all $u, v \in V_1$.

We present some basic theorems about embeddings, and give an important application in flow theory.

17.2.1. Dimension reduction. In many cases, we can control the dimension of the ambient space based on general principles, and we start with a discussion of these. The dimension problem is often easy to handle, due to a fundamental lemma [**Johnson–Lindenstrauss 1984**].

LEMMA 17.5 (Johnson–Lindenstrauss). *For every $0 < \varepsilon < 1$, every n-point set $S \subset \mathbb{R}^n$ can be mapped into $\mathbb{R}^d$ with $d < (80 \ln n)/\varepsilon^2$ with distortion at most ε in the Euclidean distance.*

PROOF. Orthogonal projection onto a random d-dimensional subspace L does the job. First, let us see what happens to the distance of a fixed pair of points $\mathbf{x}, \mathbf{y} \in S$. Instead of projecting the fixed segment of length $|\mathbf{x}-\mathbf{y}|$ on a random subspace, we can project a random vector of the same length on a fixed subspace. Then (C.17) in the Appendix applies, and tells us that with probability at least $1-4e^{-\varepsilon^2 d/4}$, the projections $\mathbf{x}'$ and $\mathbf{y}'$ of every fixed pair of points $\mathbf{x}$ and $\mathbf{x}'$ satisfy

$$\frac{1}{1+\varepsilon}\sqrt{\frac{d}{n}} \le \frac{|\mathbf{x}'-\mathbf{y}'|}{|\mathbf{x}-\mathbf{y}|} \le (1+\varepsilon)\sqrt{\frac{d}{n}}.$$

It follows that with probability at least $1-\binom{n}{2}4e^{-\varepsilon^2 d/4}$, this inequality holds simultaneously for all pairs $x, y \in S$, and then the distortion of the projection is at most $(1+\varepsilon)^2 < 1+3\varepsilon$. Replacing ε by $\varepsilon/3$ and choosing d as in the Theorem, this probability is positive, and the lemma follows. $\square$

REMARK 17.6. More careful computation gives a constant of 8 instead of 80. It will also be important later that the mapping constructed in the proof is linear, and it can be assumed that it is nonsingular when restricted to any d-dimensional affine subspace determined by S.

The Johnson–Lindenstrauss Lemma does not extend to all values of p. To be more precise, there are examples of n-point subsets S of $\mathbb{R}^n$ for which every mapping $\phi: S \to \mathbb{R}^d$ $(d < n^c)$ has a distortion of at least $\Omega(c^2)$, if the ℓ_1-distance is used in both S and $\phi(S)$ [**Brinkman–Charikar 2005**]. See [**Matoušek 2008**] for a survey of the many extensions and variants of the Johnson–Lindenstrauss Lemma.

17.2.2. Embedding with small distortion. We describe an important construction for the embedding of a general metric space into a Euclidean (ℓ_2) space [**Bourgain 1985**], which has low distortion (and other interesting and useful properties, as we will see later).

THEOREM 17.7. *Every metric space with n points can be embedded in the Euclidean space $\mathbb{R}^d$ with distortion $O(\log n)$ and dimension $d = O(\log n)$.*

To motivate the construction, let us recall Fréchet's Theorem 17.1: we embed a general metric space (V, d) isometrically into ℓ_∞ by assigning a coordinate to each point $w \in V$, and considering the representation

$$\mathbf{x}: \ i \mapsto (\mathsf{d}(w,i): \ w \in V).$$

If we want to represent the metric space by Euclidean metric with a reasonably small distortion, this construction will not work, since it may happen that all points except u and v are at the same distance from u and v, and once we take a sum of coordinates instead of the maximum to compute the norm, the contribution from u and v will become negligible. The remedy will be to take distances from sets rather than from points; it turns out that we need sets with sizes of all orders of magnitude, and this is where the logarithmic factor is lost.

We start with describing a simpler version, which is only good "on the average". For a while, it will be more convenient to work with ℓ_1 distances instead of Euclidean distances. Both of these deviations from our goal will be easy to fix.

Let (V, d) be a metric space on n elements. We may assume that the diameter of the space is at most 1. Let $m = \lceil \log n \rceil$, and for every $1 \le r \le m$, choose a random subset $\mathbf{A}_r \subseteq V$, putting every element $v \in V$ into $\mathbf{A}_r$ with probability 2^{-r}. Let $\mathsf{d}(v, \mathbf{A}_i)$ denote the distance of node v from the set $\mathbf{A}_i$ (if $\mathbf{A}_i = \emptyset$, we set $\mathsf{d}(v, \mathbf{A}_i) = 1$). Consider the vector labeling $\mathbf{x}: \ V \to \mathbb{R}^m$ defined by

$$\mathbf{x}_u = \big(\mathsf{d}(u, \mathbf{A}_1), \dots, \mathsf{d}(u, \mathbf{A}_m)\big)^{\mathsf{T}}. \tag{17.1}$$

We call this the *random subset representation* of (V, d) in ℓ_p.

LEMMA 17.8. *The random subset representation is contractive with respect to any ℓ_p-norm.*

PROOF. For $u, v \in V$, by the triangle inequality,

$$|\mathsf{d}(u, \mathbf{A}_i) - \mathsf{d}(v, \mathbf{A}_i)| \le \mathsf{d}(u, v),$$

and hence

$$\|\mathbf{x}_u - \mathbf{x}_v\|_p \le \|\mathbf{x}_u - \mathbf{x}_v\|_\infty \le \mathsf{d}(u, v).$$

□

To prove deeper properties of this representation, we have to use that it depends on the random choice of the sets A_i, and so we have to make probabilistic statements about it. It will be convenient to introduce, for any real-valued random variable X, the *median* $\mathsf{median}(X)$, defined as a real number such that $\mathsf{P}\big(X < \mathsf{median}(X)\big) \le 1/2$ and $\mathsf{P}\big(X > \mathsf{median}(X)\big) \le 1/2$. There may be a whole interval of such values, in which case we denote by $\mathsf{median}(X)$ the midpoint of this interval (just to make it well defined).

It is easy to see that every median of X minimizes $\mathsf{E}(|X-y|)$ over $y \in \mathbb{R}$ (assuming the expectation is finite). In other words, $\mathsf{E}(|X-\mathsf{median}(X)|) \le \mathsf{E}(|X-y|)$ for every $y \in \mathbb{R}$. For a random vector $\mathbf{X} = (X_1, \dots, X_m)$, we define $\mathsf{median}(\mathbf{X}) = \big(\mathsf{median}(X_1), \dots, \mathsf{median}(X_m)\big)$.

Our next lemma says that for a fixed node u, the vector $\mathbf{x}_u$ cannot be too concentrated.

LEMMA 17.9. *Let $u \in V$ and $0 < \delta \le \operatorname{diam}(V, \mathsf{d})$. Define $X_r = \min(d(u, \mathbf{A}_r), \delta)$ and $\mathbf{X} = (X_1, \dots, X_m)$. Then*

$$\mathsf{E}\big(|\mathbf{X} - \mathsf{median}(\mathbf{X})|_1\big) \ge \frac{\delta}{4}.$$

PROOF. For any integer $r \in [m]$ and real number $t \in (0, \delta)$, the inequality $X_r > t$ is equivalent to $d(u, A_r) > t$, which in turn means that $A_r \cap B(u,t) = \emptyset$. Hence

$$\mathsf{P}(d(k, A_r) > x) = (1-p_r)^{|B(k,x)|} = (2k)^{-s^{-r}|B(k,x)|}. \tag{17.2}$$

Define $r(x) = \lfloor a \log |B(k,x)| \rfloor$, then clearly $1 \le r(x) \le m$. Formula (17.2) implies that

$$\frac{1}{8k} < \mathsf{P}(d(k, A_{r(x)}) > x) \le \frac{1}{2k}. \tag{17.3}$$

Hence

$$\frac{1}{4} \le \mathsf{P}(X_{r(t)} > t) \le \frac{3}{4} \tag{17.4}$$

for every t.

Whatever the value of $\mathsf{median}(X_{r(t)})$ is, $X_{r(t)}$ is on the other side of t with probability at least $1/4$ by (17.4). Hence $\mathsf{P}\big(t \in [X_{r(t)}, \mathsf{median}(X_{r(t)})]\big) \ge 1/4$. The coordinate $r(t)$ depends on t, but we get rid of this dependence if we sum over r:

$$\sum_{r=1}^{m} \mathsf{P}\big(t \in [X_r, \mathsf{median}(X_r)]\big) \ge \frac{1}{4}.$$

Now let $T \in [0, \delta]$ be a uniform random number, then for every choice of x and y, $\mathsf{P}\big(T \in [x,y]\big) \le |x-y|/\delta$, and so

$$\begin{aligned} \mathsf{E}\big(|\mathbf{X} - \mathsf{median}(\mathbf{X})|_1\big) &= \sum_{r=1}^{m} \mathsf{E}\big(|X_r - \mathsf{median}(X_r)|\big) \\ &\ge \sum_{r=1}^{m} \delta \mathsf{P}(T \in [X_r, \mathsf{median}(X_r)]) \ge \frac{\delta}{4}. \end{aligned}$$ □

As an easy application of this lemma, we bound the distance of two representative points, at least in expectation.

LEMMA 17.10. *The random subspace representation* (17.1) *satisfies*

$$\mathsf{E}\big(\|\mathbf{x}_u - \mathbf{x}_v\|_p\big) \geq \frac{d(u,v)}{8m}.$$

for every pair of points $u, v \in V$.

PROOF. By the inequality relating different ℓ_p-norms, it suffices to prove this for $p = 1$. Define $\delta = d(u,v)/2$, $X_r = \min(d(u, \mathbf{A}_r), \delta)$, $Y_r = \min(d(v, \mathbf{A}_r), \delta)$, $\mathbf{X} = (X_1, \dots, X_m)^\mathsf{T}$ and $\mathbf{Y} = (Y_1, \dots, Y_m)^\mathsf{T}$. Since $\mathbf{X}$ and $\mathbf{Y}$ depend only on how $\mathbf{A}_r$ behaves in the disjoint open neighborhoods of radius δ of u and v, respectively, these quantities are independent random variables. Fix $\mathbf{Y}$ and invoke Lemma 17.9:

$$\mathsf{E}\big(\|\mathbf{X} - \mathbf{Y}\|_1\big) \geq \frac{\delta}{4m}.$$

It is easy to check that

$$|d(u, \mathbf{A}_i) - d(v, \mathbf{A}_i)| \geq |X_i - Y_i|,$$

and so

$$\mathsf{E}\big(\|\mathbf{x}_u - \mathbf{x}_v\|_1\big) \geq \mathsf{E}\big(\|\mathbf{X} - \mathbf{Y}\|_1\big) \geq \frac{\delta}{4m} = \frac{d(u,v)}{8m}. \qquad \square$$

Now we are ready to prove the theorem.

Proof of Theorem 17.7. Lemma 17.8 says that the random subset embedding is contractive, so it suffices to prove that it does not contract any distance by more than a factor of $m = O(\log n)$, with large probability. The previous lemma only bounds the expectation of these distances. The Law of Large Numbers suggests the remedy: the distance $\|\mathbf{x}_u - \mathbf{x}_v\|_1$ is the average of m independent bounded random variables, and hence it will be close to its expectation with high probability, if m is large enough. We can multiply m basically for free. To be more precise, let us generate N independent copies $\mathbf{x}^1, \dots, \mathbf{x}^N$ of the representation (17.1), and consider the mapping $\mathbf{y} : V \to \mathbb{R}^{Nm}$, defined by

$$\mathbf{y}_u = \begin{pmatrix} \mathbf{x}_u^1 \\ \vdots \\ \mathbf{x}_u^N \end{pmatrix} \tag{17.5}$$

Then

$$\|\mathbf{y}_u - \mathbf{y}_v\|_1 \leq d(u,v) \quad \text{and} \quad \mathsf{E}(\|\mathbf{y}_u - \mathbf{y}_v\|_1) = \mathsf{E}(\|\mathbf{x}_u - \mathbf{x}_v\|_1) \geq \frac{1}{8m} d(u,v).$$

For every u and v, $\|\mathbf{y}_u - \mathbf{y}_v\|_1 \to \mathsf{E}(\|\mathbf{x}_u - \mathbf{x}_v\|_1) \geq d(u,v)/(8m)$ almost surely by the Law of Large Numbers, and hence if N is large enough, then with high probability

$$\|\mathbf{y}_u - \mathbf{y}_v\|_1 \geq \frac{d(u,v)}{9m} \qquad \text{for all } u, v \in V. \tag{17.6}$$

So $\mathbf{y}$ is a representation in ℓ_1 with distortion at most $9m$.

The assertions above extend to every ℓ_p-distance instead of the ℓ_1-distance (in particular, to the Euclidean distance): first, $\|\mathbf{y}_u - \mathbf{y}_v\|_p \leq \|\mathbf{y}_u - \mathbf{y}_v\|_\infty \leq d(u,v)$ just like above, and second, $\|\mathbf{y}_u - \mathbf{y}_v\|_p \geq \|\mathbf{y}_u - \mathbf{y}_v\|_1 \geq d(u,v)/(9m)$ with high probability for all $u, v \in V$.

We are not done yet, since choosing a large N will result in a representation in a very high dimension. For the Euclidean distance ($p = 2$), we can apply the Johnson–Lindenstrauss Lemma 17.5 (in other words, we apply a random projection), to reduce the dimension as stated in the Theorem. □

REMARK 17.11. The theorem remains valid for all $p \geq 1$, but the proof needs additional work, since the Johnson–Lindenstrauss Lemma does not hold for all p. One can estimate the concentration of $\|\mathbf{x}_u - \mathbf{x}_v\|_1$ using the Chernoff-Hoeffding Inequality; the details of this computation are omitted.

Deep connections of the random subset representation with graph theory were described by [**Linial–London–Rabinovich 1995**]. They showed how to construct an embedding satisfying the conditions in Theorem 17.7 by a deterministic algorithm, and gave the important application to flow theory to be described in Section 17.2.3. They also showed that Bourgain's embedding is essentially optimal, in the sense that the $O(\log n)$ factor cannot be improved (see also [**Matoušek 1997**]):

PROPOSITION 17.12. *Let G be an a-regular graph ($a \geq 3$), which is a b-edge-expander ($b > 0$), and let (V, d) be the graph metric of G, then every embedding of (V, d_2) in a Euclidean space has distortion at least $b \log(n/2)/\log a$.*

PROOF. Since the graph is a-regular, the number of nodes accessible from a given node in at most t steps ($t \geq 1$) is bounded by $1 + a + a(a-1) + \cdots + a(a-1)^{t-1} < a^t$. So in fewer than $\log(n/2)/\log a$ steps fewer than half of the nodes can be reached, which implies that

$$\sum_{u,v \in V} d(u,v) \geq \frac{\log(n/2)}{\log a} \frac{n^2}{2}.$$

Since the number of edges is $an/2$, it follows that the graph distance satisfies the inequality

$$\sum_{u,v \in V} d(u,v) \geq \frac{n}{a} \frac{\log(n/2)}{\log a} \sum_{uv \in E} d(u,v). \tag{17.7}$$

On the other hand, for the 2-partition distance d_S determined by a set $S \subseteq V$, we have

$$\sum_{u,v \in V} d_S(u,v) = |S|\,|V \setminus S|,$$

and by the hypothesis about the expansion of the graph,

$$\sum_{uv \in E} d_S(u,v) = e_G(S, V \setminus S) \geq \frac{ab}{n} |S|\,|V \setminus S|,$$

whence

$$\sum_{u,v \in V} d_S(u,v) \leq \frac{n}{ab} \sum_{uv \in E} d_S(u,v).$$

Since every ℓ_1-semimetric is a nonnegative linear combination of 2-partition distances by Proposition 17.3, it follows that every ℓ_1-semimetric d_1 satisfies

$$\sum_{u,v \in V} d_1(u,v) \leq \frac{n}{ab} \sum_{uv \in E} d_1(u,v). \tag{17.8}$$

Comparing (17.7) and (17.8), we see that every ℓ_1-metric approximating the graph distance must have distortion at least $\big(\log(n/2)\big)/\log a$. By Proposition 17.4, the same holds for every ℓ_2-metric. □

17.2.3. Multicommodity flows and approximate Max-Flow-Min-Cut. The following fundamental result in the theory of multicommodity flows was proved in [**Leighton–Rao 1998**]. Stated in a larger generality, as proved in [**Linial–London–Rabinovich 1995**], it says the following. Suppose that we have a multicommodity flow problem on a graph G. This means that we are given k pairs of nodes (s_i, t_i) $(i = 1, \dots, k)$, and for each such pair, we are given a *demand* $d_i \ge 0$. Every edge e of the graph has a *capacity* $c_e \ge 0$. We would like to design a flow f^i from s_i to t_i of value d_i for every $1 \le i \le k$, so that for every edge the capacity constraint is satisfied.

We state the problem more precisely. We need a reference orientation of G; let $\overrightarrow{E}$ denote the set of these oriented edges. We want an (s_i, t_i)-flow f^i for every $1 \le i \le k$, with prescribed flow values d_i, so that the total flow through an edge does not exceed the capacity of the edge:

$$\sum_{i=1}^{k} |f^i(e)| \le c_e \quad \text{for all } e \in \overrightarrow{E}(G) \tag{17.9}$$

(this clearly does not depend on which orientation of the edge we consider).

Let us hasten to point out that the solvability of a multicommodity flow problem is just the feasibility of a linear program (we treat the values $f(e)$ as variables). So the multicommodity flow problem is polynomial time solvable. We can also apply the Farkas Lemma, and derive a necessary and sufficient condition for solvability. If you work out the dual, quite likely you will get a condition that is not transparent at all; however, there is a very nice form [**Iri 1967**, **Shahroki–Matula 1990**], which fits particularly well into the topic of this book.

Consider a semimetric D on V. Let us describe an informal (physical) derivation of the conditions. Think of an edge $e = uv$ as a pipe with cross section c_e and length $D(e) = D(u, v)$. Then the total volume of the system is $\sum_e c_e D(e)$. If the multicommodity problem is feasible, then every flow f^i occupies a volume of at least $d_i D(s_i, t_i)$, and hence

$$\sum_i d_i D(s_i, t_i) \le \sum_e c_e D(e). \tag{17.10}$$

We call this inequality the *volume condition*. This condition, when required for every semimetric, is also sufficient:

THEOREM 17.13. *Let G be a graph, let $s_i, t_i \in V$, $d_i \in \mathbb{R}_+$ $(i = 1, \dots, k)$, and $c_e \in \mathbb{R}_+$ $(e \in E)$. Then there exist (s_i, t_i)-flows $(f^i : i \in [k])$ satisfying the demand conditions* $\mathrm{val}(f^i) = d_i$ *and the capacity constraints* (17.9) *if and only if the volume condition* (17.10) *holds for every semimetric D in V.*

We leave the exact derivation of the necessity of the condition, as well as the proof of the converse based on linear programming duality, to the reader.

The obvious *cut conditions* obtained by specializing (17.10) for 2-partition metrics, provide a system of necessary conditions for the problem to be feasible: If the multicommodity flows exist, then for every $S \subseteq V$, we must have

$$\sum_{\mathsf{tl}(e)\in S,\mathsf{hd}(e)\notin S} c_e \geq \sum_{i:\ S\cap\{s_i,t_i\}=1} d_i. \tag{17.11}$$

In the case of one or two commodities ($k \leq 2$), these conditions are necessary and sufficient; this is the content of the Max-Flow-Min-Cut Theorem of Ford and Fulkerson ($k = 1$) and the Gomory–Hu Theorem ($k = 2$). However, for $k \geq 3$, the cut conditions are not sufficient any more for the existence of multicommodity flows. The theorem of Leighton and Rao asserts that *if the cut-conditions are satisfied, then relaxing the capacities by a factor of $O(\log n)$, the problem becomes feasible.* The relaxation factor was improved by Linial, London and Rabinovich to $O(\log k)$; we state the result in this tighter form, but for simplicity of presentation prove the original (weaker) form.

THEOREM 17.14. *Suppose that for a multicommodity flow problem, the cut conditions* (17.11) *are satisfied. Then replacing every edge capacity c_e by $20c_e \log k$, the problem becomes feasible.*

PROOF. Using Theorem 17.13, it suffices to prove that for every semimetric D on V,

$$\sum_i d_{i=1}^k D(s_i,t_i) \leq 20(\log n) \sum_{e\in E} c_d D(e). \tag{17.12}$$

The cut conditions are equivalent to the volume conditions (17.10) for 2-partition distances, which in turn implies their validity for semimetrics that are nonnegative linear combinations of 2-partition distances, which, as we have seen, are exactly the ℓ_1-semimetrics. By Bourgain's Theorem 17.7, there is an ℓ_1-metric D' such that

$$D'(u,v) \leq D(u,v) \leq 20(\log n)D'(u,v).$$

We know that D' satisfies the semimetric condition (17.10), and hence D satisfies the relaxed semimetric conditions (17.12). □

17.3. Respecting the volume

A very interesting extension of the notion of small distortion was formulated in [**Feige 1998**]. We want an embedding of a metric space in a Euclidean space such that is contractive, and at the same time volume respecting up to a given size s, which means that every set of at most s nodes spans a simplex whose volume is almost as large as possible. Obviously, the last condition above needs an explanation, and for this, we will have to define a certain "volume" of an arbitrary metric space (a fascinating notion in itself). Respecting the volume can also be viewed as a quantitative version of "general position". We describe an algorithmic application of these results to the approximate computation of the bandwidth of a graph.

17.3.1. Dimension reduction for volume. Before defining the volume in general metric spaces, we stay for a while with the traditional notion of volume, and prove a generalization of the Johnson–Lindenstrauss Lemma 17.5 [**Magen 2007**]. To simplify notation, for a finite set $S \subseteq \mathbb{R}^d$ with k elements, we define $\mathsf{vol}(S)$ as the $(k-1)$-dimensional volume of $\mathrm{conv}(S)$. For a vector labeling $\mathbf{x} : S \to \mathbb{R}^d$ of a finite set S with k elements, we denote by $\mathsf{vol}_{\mathbf{x}}(S)$ the $(k-1)$-dimensional volume of $\mathrm{conv}(\mathbf{x}(S))$. If $\mathbf{x}$ is injective, then this is the same as $\mathsf{vol}(\mathbf{x}(S))$. Note that $\mathsf{vol}(S)$ can be zero if S is not affine independent, and similar warning holds for $\mathsf{vol}_{\mathbf{x}}(S)$.

LEMMA 17.15. *For every real number $\varepsilon > 0$ and integer $k \geq 2$, and for every n-point set $V \subset \mathbb{R}^n$, there is a linear map $\phi : \mathbb{R}^n \to \mathbb{R}^d$ with $d = O(\varepsilon^{-2} k \log n)$ such that for every $S \subseteq V$ with $|S| = r \leq k$, we have the bounds*

$$(1-\varepsilon)^{r-1}\mathsf{vol}(S) \leq \mathsf{vol}(\phi(S)) \leq \mathsf{vol}(S).$$

Note that such a mapping is necessarily contractive with respect to the Euclidean distance and bijective on V.

PROOF. This lemma can be proved extending the proof of Lemma 17.5, but a more elementary proof can be obtained by reducing it to the Johnson–Lindenstrauss Lemma. Recall that this lemma provides a *linear* map $\phi : \mathbb{R}^n \to \mathbb{R}^d$ that is *contractive* and *almost isometric* on V.

The property that the map is contractive does not imply that (say) areas of triangles are not increased, and the property that distances are not decreased by much does not imply that areas of triangles are not decreased by much. The idea is to embed V into a larger finite set W, so that approximately preserving distances between points in W will imply approximate preservation of volumes of r-dimensional simplices with vertices in V for all $r \leq k$.

To show how this idea works, let us start with proving the upper bound. For every point $\mathbf{v} \in S$ and every affine space $A \subseteq \mathbb{R}^n$, let $\mathbf{v}_A$ denote the orthogonal projection of $\mathbf{v}$ onto A. Let U be the set of all points $\mathbf{v}_{\mathrm{aff}(X)}$, where $X \subseteq V$, $|X| < k$. (Note that $V \subseteq U$.) Let $\phi : \mathbb{R}^n \to \mathbb{R}^{d_1}$ be a linear map into some dimension d_1 that is contractive on U.

Consider a particular r-dimensional simplex $\mathrm{conv}(Y)$, where $r \leq k$ and $Y \in \binom{V}{r+1}$. Choose any $\mathbf{v} \in Y$, and let $X = Y \setminus \mathbf{v}$ and $A = \mathrm{aff}(X)$. We may assume (for notational convenience) that $\mathbf{v}_A = 0$, then $\phi(\mathbf{v}_A) = 0$, A is a linear subspace, and $\mathbf{v} \perp A$. Let $\mathbf{y}$ denote the orthogonal projection of $\phi(\mathbf{v})$ onto $\phi(A)$ (Figure 17.2), and let $h = |\mathbf{v}|$ and $h' = |\phi(\mathbf{v}) - \mathbf{y}|$ denote the distance of $\mathbf{v}$ from A and of $\phi(\mathbf{v})$ from $\phi(A)$, respectively.

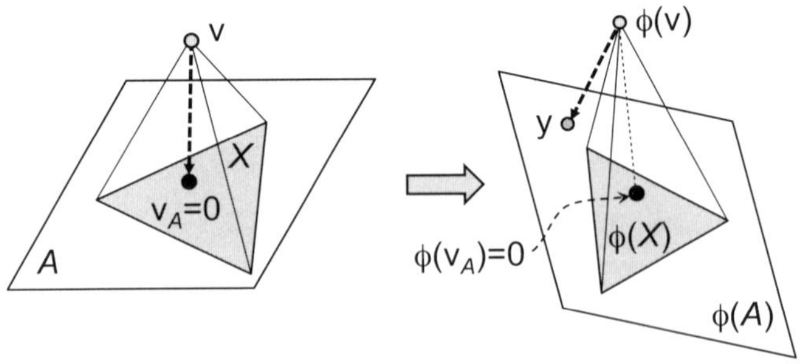

FIGURE 17.2. Adding projections to V, and the image of a simplex under the Johnson–Lindenstrauss mapping.

By induction on r, we may assume that

$$\mathsf{vol}(\phi(X)) \leq \mathsf{vol}(X). \tag{17.13}$$

Linearity of ϕ implies that $\text{lin}(\phi(X)) = \phi(A)$, and since ϕ is contractive, $h' = |\phi(\mathbf{v})-\mathbf{y}| \le |\phi(\mathbf{v})| \le |\mathbf{v}| = h$. Using (17.13), we get

$$\text{vol}(\phi(Y)) = \frac{h'}{r}\text{vol}(\phi(X)) \le \frac{h}{r}\text{vol}(X) = \text{vol}(Y).$$

This proves the upper bound in the lemma.

Unfortunately, this simple way of extending V is not sufficient to guarantee that volumes do not shrink much. We extend U further by considering, for every $\mathbf{u} \in V$ and every affine subspace A spanned by points in V with $\dim(A) = r < k$ and $\mathbf{u} \notin A$, a cubic lattice $L_{\mathbf{u},A}$ in A of edge-length $|\mathbf{u}-\mathbf{u}_A|/\sqrt{r}$, containing $\mathbf{u}_A$, and its finite subset $W_{\mathbf{u},A} = \{\mathbf{z} \in L_{\mathbf{u},A} : |\mathbf{z}| \le 3|\mathbf{u}-\mathbf{u}_A|\}$. We add to V all of the sets $W_{\mathbf{u},A}$ to get a set W.

We have to construct the contractive map more carefully as above: We apply the Johnson–Lindenstrauss Lemma 17.5 to the set W with error bound $\varepsilon_1 = \varepsilon/22$ to get a linear map $\phi: \mathbb{R}^n \to \mathbb{R}^{d_2}$ for the appropriate dimension d_2.

To show that this mapping does not shrink the volumes of simplices of dimension at most k too much, consider $Y, X, A, \mathbf{v}$ and $\mathbf{y}$ as above. The argument for the upper bound in the lemma remains valid, but now we can do a similar computation for the lower bound as well. Induction on r gives

$$\text{vol}(\phi(X)) \ge (1-\varepsilon)^{r-1}\text{vol}(X). \tag{17.14}$$

Linearity of ϕ implies that $\phi(L_{\mathbf{v},A})$ is a lattice in $\phi(A)$; not cubic, but the contractive property of ϕ implies that the images of lattice cubes, which are the basic parallelopipeds of $\phi(L_{\mathbf{v},A})$, have diameter at most h. Let C be the lattice cube in $L_{\mathbf{v},A}$ such that $\mathbf{y} \in \phi(C)$ (Figure 17.3).

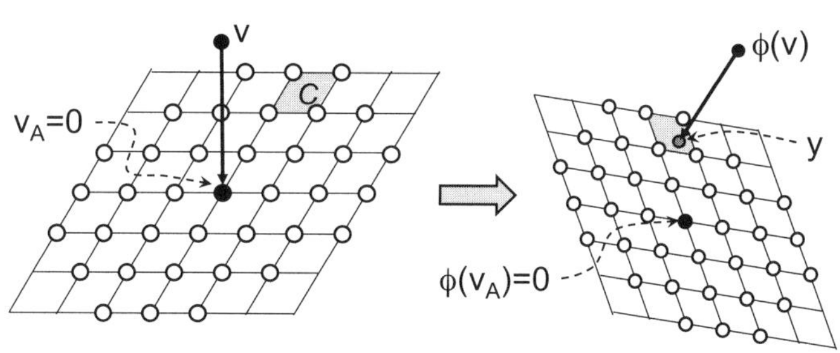

FIGURE 17.3. The lattice added to V, and its image under the Johnson–Lindenstrauss mapping.

Since $|\mathbf{y}| \le |\phi(\mathbf{v})| \le |\mathbf{v}|$, it follows that for every vertex $\mathbf{z}$ of C, we have $|\phi(\mathbf{z})| \le |\mathbf{y}|+|\phi(\mathbf{z})-\mathbf{y}| \le 2h$. Hence $|\mathbf{z}| \le 3h$, and so $\mathbf{z} \in W_{\mathbf{v},A}$. By the distance-respecting properties of ϕ,

$$|\phi(\mathbf{v})-\phi(\mathbf{z})|^2 \ge (1-\varepsilon_1)^2|\mathbf{z}-\mathbf{v}|^2 = (1-\varepsilon_1)^2(|\mathbf{z}|^2+h^2),$$

and (considering the point $-\mathbf{z}$)

$$|\phi(\mathbf{v})+\phi(\mathbf{z})|^2 \le |\mathbf{v}+\mathbf{z}|^2 = |\mathbf{z}|^2+h^2.$$

Expanding and subtracting, we get

$$4\phi(\mathbf{v})^{\mathsf{T}}\phi(\mathbf{z}) \le (2\varepsilon_1-\varepsilon_1^2)(|\mathbf{z}|^2+h^2) < 2\varepsilon_1(|\mathbf{z}|^2+h^2) \le 20\varepsilon_1 h^2. \tag{17.15}$$

This linear inequality holds for every vertex $\phi(\mathbf{z})$ of the parallelotope $\phi(C)$, so it must hold for $\mathbf{y} \in \phi(C)$ as well. Hence

$$|\mathbf{y}|^2 = \phi(\mathbf{v})^{\mathsf{T}}\mathbf{y} < 20\varepsilon_1 h^2.$$

Thus

$$|\phi(\mathbf{v}) - \mathbf{y}|^2 = |\phi(\mathbf{v})|^2 - |\mathbf{y}|^2 \geq (1-\varepsilon_1)^2 h^2 - 20\varepsilon_1 h^2 \geq (1-22\varepsilon_1)^2 h^2 = (1-\varepsilon)^2 h^2,$$

and

$$\mathsf{vol}\big(\phi(Y)\big) = \frac{h'}{r}\mathsf{vol}\big(\phi(X)\big) \geq (1-\varepsilon)\frac{h}{r}(1-\varepsilon)^{r-1}\mathsf{vol}(X) = (1-\varepsilon)^r \mathsf{vol}(Y).$$

To bound the dimension in the application of Lemma 17.5, we have to estimate the size of W. Quite elementary computations yield that $|W| < (20n)^k$. Thus

$$d_2 \leq 80\frac{1}{\varepsilon_1^2}\ln|W| < 10^5\frac{k}{\varepsilon^2}\ln n. \qquad \square$$

17.3.2. Volume in metric spaces. Let (V, d) be a finite metric space with $|V| = n$. For a tree T on $V(T) = V$, we define $\Pi_T = \prod_{uv\in E(T)} d(u,v)$, and we define the *tree-volume* of (V, d) by

$$\mathsf{vol}_{\mathrm{tr}}(V, \mathsf{d}) = \frac{1}{(n-1)!}\min_T \Pi_T.$$

Note that the tree T that gives here the minimum is also minimizing the total length of edges (this follows from the fact that it can be found by the Greedy Algorithm).

Since we only consider one metric, denoted by d, in this section, we will not indicate it in $\mathsf{vol}_{\mathrm{tr}}(V, \mathsf{d})$ and similar quantities to be defined below. We will, however, change the underlying set, so we keep V in the notation. The following lemma relates the volume of a simplex in an embedding with the tree-volume.

LEMMA 17.16. *For every contractive map* $\mathbf{x} : V \to \mathbb{R}^N$ *of a metric space* (V, d), $\mathsf{vol}_{\mathbf{x}}(V) \leq \mathsf{vol}_{\mathrm{tr}}(V)$.

PROOF. We prove by induction on n that for any tree T on V,

$$\mathsf{vol}_{\mathbf{x}}(V) \leq \frac{1}{(n-1)!}\Pi_T.$$

For $n = 2$ this is obvious. Let $n > 2$, and let $i \in V$ be a node of degree 1 in T, let j be the neighbor of i in T, and let h be the distance of i from the affine hull of $\mathbf{x}(V \setminus i)$. Then by induction,

$$\begin{aligned}\mathsf{vol}_{\mathbf{x}}(V) &= \frac{h}{n-1}\mathsf{vol}_{\mathbf{x}}(V \setminus i) \leq \frac{d(i,j)}{n-1}\mathsf{vol}_{\mathbf{x}}(V \setminus i) \\ &\leq \frac{d(i,j)}{n-1}\frac{1}{(n-2)!}\Pi_{T-i} = \frac{1}{(n-1)!}\Pi_T. \qquad \square\end{aligned}$$

The main result about tree-volume is that the upper bound given in Lemma 17.16 can be attained, up to a polylogarithmic factor, for all small subsets simultaneously.

THEOREM 17.17. *Every finite metric space* (V, d) *with* n *points has a contractive map* $\mathbf{x} : V \to \mathbb{R}^d$ *(with respect to the Euclidean metric) with* $d = O\big((\log n)^4 \log\log n\big)$*, such that for every* $2 \le k \le \log n$ *and every* $S \in \binom{V}{k}$*,*

$$\mathsf{vol}_{\mathbf{x}}(S)^{\frac{1}{k-1}} \ge \frac{1}{O\big(\log n\sqrt{\log k}\big)} \mathsf{vol}_{\mathrm{tr}}(S)^{\frac{1}{k-1}}.$$

PROOF. The proof of this theorem is built on the proof of Bourgain's Theorem 17.7, but it is of course more involved. It will be convenient to assume that all distances in (V, d) are distinct; this can be achieved by an arbitrarily small perturbation.

Instead of powers of 1/2, we define a finer sequence of probabilities for the densities of the random sets. We use the following parameters:

$$s = 1 + \frac{1}{\log\log n}, \qquad m = \lceil (\log\log n)\log n \rceil, \qquad p_r = 1 - \frac{1}{(2m)^{s^{-r}}}. \tag{17.16}$$

It is clear that $0 < p_r < 1$, and so we can construct a random subset $A_r \subseteq V$, where every element $v \in V$ is put into A_r with probability p_r, independently of each other ($1 \le r \le m$). For notational convenience, we assume that $V = [n]$. For a while, we focus on a single set S, say $S = [k]$.

Consider the balls $B_j = B\big(j, \frac{1}{2}d(j, S \setminus j)\big)$ for $j \in S$. It is easy to see that these sets are disjoint. We may assume that $|B_k| = \min_{j \in S} |B_j|$.

Just as in the proof of Theorem 17.7, we need to show that for an appropriate r, the distance $d(k, A_r)$ is not too concentrated as a random variable; we need, however, a more complicated formulation of this fact than before. We let $\delta = d(k, S \setminus k)/2$.

Claim 1. *There are real numbers* $x_1, x_2 \in [0, \delta]$ *with* $x_2 \ge x_1 + \frac{\delta}{2m}$*, and an integer* $r \in [m]$ *such that the following inequalities hold:*

$$\mathsf{P}\big(d(k, A_r) \le x_1\big) \ge \frac{1}{2}, \quad \mathsf{P}\big(d(k, A_r) \ge x_2\big) \ge \frac{1}{4m}, \quad \mathsf{P}\big(d(k, A_r) \ge \delta\big) < \frac{1}{2m}.$$

For any integer $r \in [m]$ and real number $0 \le x \le \delta$, the inequality $d(k, A_r) > x$ is equivalent to $A_r \cap B(k, x) = \emptyset$. Hence

$$\mathsf{P}\big(d(k, A_r) > x\big) = (1 - p_r)^{|B(k,x)|} = \frac{1}{(2m)^{|B(k,x)|/s^r}}. \tag{17.17}$$

Define $r(x) = \lfloor (\log\log n)\log |B(k, x)| \rfloor$, then clearly $1 \le r(x) \le m$. Formula (17.17) implies that

$$\frac{1}{4m} < \mathsf{P}\big(d(k, A_{r(x)}) > x\big) \le \frac{1}{2m}. \tag{17.18}$$

There is a specific value of r that belongs to at least a $1/m$ fraction of all x values, and so there are two values $0 \le x_1 < x_2 \le \delta$ with $r(x_1) = r(x_2) = r$ and $x_2 \ge x_1 + \delta/m$. By (17.18), we have

$$\mathsf{P}\big(d(k, A_r) \ge x_2\big) \ge \frac{1}{4m},$$

and

$$\mathsf{P}\big(d(k, A_r) \ge \delta\big) \le \mathsf{P}\big(d(k, A_r) \ge x_2\big) \le \frac{1}{2m}.$$

The following claim is the main step in the proof.

Claim 2. *There are integers $r_1, \dots, r_{k-1} \in [m]$ such that the matrix*

$$M = \begin{pmatrix} 1 & d(1, A_{r_1}) & \dots & d(1, A_{r_{k-1}}) \\ 1 & d(2, A_{r_1}) & \dots & d(2, A_{r_{k-1}}) \\ \vdots & & & \vdots \\ 1 & d(k, A_{r_1}) & \dots & d(k, A_{r_{k-1}}) \end{pmatrix} \tag{17.19}$$

satisfies

$$\mathsf{E}\big(|\det(M)|\big) \ge \frac{\mathsf{vol}_{\mathrm{tr}}(S)}{(16m)^k}.$$

To prove this claim, we use induction on k. For $k = 1$ the assertion is trivial. Let $k > 1$ and $S' = S \setminus k$. By induction, we have a choice of $r_1, \dots, r_{k-2} \in [m]$ so that

$$\mathsf{E}\big(|\det(M')|\big) \ge \frac{\mathsf{vol}_{\mathrm{tr}}(S')}{(16m)^{k-1}}, \tag{17.20}$$

where M' is the matrix constructed by (17.19) for S'. We may assume (rearranging the rows if necessary) that $\det(M') > 0$.

Consider the numbers x_1, x_2 and r from Claim 1, and define $r_{k-1} = r$. Expanding $\det(M)$ by its last column, we get

$$\det(M) = \sum_{j=1}^{k} (-1)^{k-j} d(u_j, A_r) \det(M^{jk}),$$

where M^{jk} is the submatrix of M obtained by deleting row j and the last column k. We can write this as

$$\det(M) = d(k, A_r) \det(M') + Y,$$

where Y is the sum of the first $k-1$ terms above. Here $\det(M')$ and $d(k, A_r)$ are independent random variables, but Y is not independent of either one of them.

Recall the proof of Theorem 17.7: there we forced independence by cutting off the outliers: we replaced the numbers $d(j, A_r)$ by $\min\{\delta, d(j, A_r)\}$. Here we want to do something similar, but just cutting off the tail would not work: while this did not increase the distances in the case $k = 2$, the volumes (determinants) could increase or decrease for $k \ge 3$. So we have to use a more tricky way to force some degree of independence of the distances from various given points.

Let $X = \min\{\delta, d(k, A_r)\}$, and let W be the event that $B_j \cap A_r \neq \emptyset$ for all $j \le k-1$. This event is independent of X, since X depends only on the choice of A_r inside B_k, but the event W depends only on how A_r is chosen outside B_k, since $B_j \cap B_k = \emptyset$. Furthermore, if we condition on W, then Y will be independent of X, because under this condition $d(j, A_r)$ depends only on the choice of A_r outside B_k for $j < k$, and the random variable $\det(M^{jk})$ is independent of the choice of A_r.

Next we show that, conditioning on W, with probability at least $1/(4m)$, we have

$$|\det(M)| \ge \frac{\delta}{2m} \det(M'). \tag{17.21}$$

First, let us condition on W and also on the event $Z = (Y \ge \det(M')(x_1 + x_2)/2)$. We know by Claim 1 that $d(k, A_r) \ge x_2$ with probability at least $1/(4m)$; in this case, $X \geq x_2$ also holds. The random variable X is independent of events W and Z, so even under these conditions, we have

$$d(k, A_r) \geq X \geq x_2 \geq \frac{x_1+x_2}{2} + \frac{\delta}{2m}$$

with probability at least $1/(4m)$. If this occurs, then

$$\begin{aligned}\det(M) &= d(k, A_r)\det(M')+Y \geq d(k, A_r)\det(M') - \frac{x_1+x_2}{2}\det(M')\\ &\geq \Big(x_2 - \frac{x_1+x_2}{2}\Big)\det(M') \geq \frac{\delta}{2m}|\det(M')|.\end{aligned}$$

Conditioning on W and on $\neg Z = (Y < \det(M')(x_1+x_2)/2)$, we conclude similarly: with probability at least $1/(4m)$,

$$\begin{aligned}\det(M) &= d(k, A_r)\det(M')+Y \leq d(k, A_r)\det(M') - \frac{x_1+x_2}{2}\det(M')\\ &\leq \Big(x_1 - \frac{x_1+x_2}{2}\Big)\det(M') \leq -\frac{\delta}{2m}|\det(M')|.\end{aligned}$$

This proves (17.21). Hence

$$\begin{aligned}\mathsf{E}\big(|\det(M)| \mid W\big) &\geq \frac{1}{4m}\frac{\delta}{2m}|\det(M')| = \frac{d(k, S')}{8m^2}\frac{\mathsf{vol}_{\mathrm{tr}}(S')}{(16m^2)^{k-1}}\\ &\geq 2\frac{\mathsf{vol}_{\mathrm{tr}}(S)}{(16m^2)^k}.\end{aligned}$$

To conclude, we use that

$$\begin{aligned}\mathsf{E}\big(|\det(M)|\big) &= \mathsf{E}\big(|\det(M)| \mid W\big)P(W) + \mathsf{E}\big(|\det(M)| \mid \neg W\big)\big(1-P(W)\big)\\ &\geq \mathsf{E}\big(|\det(M)| \mid W\big)P(W) \geq 2\frac{\mathsf{vol}_{\mathrm{tr}}(S)}{(16m)^k}P(W).\end{aligned}$$

To estimate the probability of W, we use that $|B_j| \geq |B_k|$, and so

$$\mathsf{P}\big(d(j, A_r) > \delta\big) = \mathsf{P}(B_j \cap A_r = \emptyset) \leq \mathsf{P}(B_k \cap A_r = \emptyset) = \mathsf{P}(d(k, A_r) > \delta) < \frac{1}{2m}.$$

Thus

$$\mathsf{P}(W) \geq 1 - \sum_{j=1}^{k}\mathsf{P}\big(d(j, A_r) > \delta\big) \geq \frac{1}{2}.$$

This completes the proof of Claim 2.

Now we are ready to complete the proof. Let $A_r^1, \dots, A_r^N$ be random subsets with density p_r. We assume that the sets A_r^j are independently generated for $j = 1, \dots, N$, $r = 1, \dots, m$. To each point $u \in V$, we associate the row vector

$$\mathbf{y}_u = \frac{1}{\sqrt{mN}}\big(1, d(u, A_r^j) : j = 1, \dots, N,\ r = 1, \dots, m\big).$$

Let $S \subseteq V$, $|S| = k$, and assume again that $S = [k]$. Consider the $k \times (m+1)$ matrix Y with rows $\mathbf{y}_1, \dots, \mathbf{y}_k$. Let $\mathcal{A}$ be the family of $k \times k$ submatrices of Y containing the first column. By the Binet–Cauchy Formula (A.6) and elementary determinant transformations, we get

$$\mathsf{vol}_{\mathbf{y}}(S)^2 = \frac{1}{(k-1)!}\sum_{U \in \mathcal{A}}\det(U)^2. \tag{17.22}$$

By Claim 2, there is a subset U such that

$$\mathsf{E}(|\det(U)|) \geq \frac{1}{(mN)^{k/2}} \frac{\mathsf{vol}_{\mathrm{tr}}(S)}{(16m)^k}.$$

There are, in fact, not only one, but at least $N(N-1)\cdots(N-k+1)$ such sets. (The reason why not quite N^k is because the integers $r_1, \dots, r_k$ in Claim 2 may not be different, but we need to pick different columns to insure probabilistic independence.) Hence

$$\mathsf{E}(\mathsf{vol}_{\mathbf{y}}(S)^2) \geq \frac{N(N-1)\cdots(N-k+1)}{N^k} \frac{\mathsf{vol}_{\mathrm{tr}}(S)^2}{256m^{6k}}. \tag{17.23}$$

As $N \to \infty$, the random variable $\mathsf{vol}_{\mathbf{y}}(S)^2$ will get more and more concentrated around its expectation, by (17.22) and the Law of Large Numbers. (The terms in (17.22) are not all independent, but can be partitioned into substantially large independent subfamilies; we leave verification of the details to the reader). The first factor in (17.23) tends to 1. So if N is large enough, then

$$|\mathsf{vol}_{\mathbf{y}}(S)| \geq \frac{\mathsf{vol}_{\mathrm{tr}}(S)}{20m^{3k}}.$$

with probability arbitrarily close to 1. It follows that this inequality holds for all sets $S \in \binom{V}{k}$ with high probability.

The last step needed is to reduce the dimension, using Lemma 17.15. $\square$

17.4. Bandwidth

The theory of volume-respecting representations described above was developed in [**Feige 1998**] in order to design an efficient approximation algorithm for the bandwidth of a graph. This application is described in this section. I believe that this deep theory will have further important applications.

17.4.1. Bandwidth and density. Let G be a simple graph, and let d denote the graph-distance in G. The *bandwidth* $\mathsf{w}_{\mathrm{bd}}(G)$ of a graph G is defined as the smallest integer b such that the nodes of G can be labeled by $1, \dots, n$ so that $|i-j| \leq b$ for every edge ij. The name refers to the fact that for this labeling of the nodes, all 1's in the adjacency matrix will be contained in a band of width $2b+1$ around the main diagonal.

The bandwidth of a graph is NP-hard to compute, or even to approximate within a constant factor [**Dubey–Feige–Unger 2010**]. As an application of volume-respecting embeddings, we describe an algorithm that finds a polylogarithmic approximation of the bandwidth in polynomial time [**Feige 1998**]. The ordering of the nodes which approximates the bandwidth will be obtained by projecting the representation onto a random line, in a way reminiscent of the Goemans–Williamson algorithm in Section 13.3.

We need some preliminary observations about the bandwidth. It is clear that if there is a node of degree D, then $\mathsf{w}_{\mathrm{bd}}(G) \geq D/2$. More generally, if there are k nodes of the graph at graph-distance at most t from a node v (not counting v), then $\mathsf{w}_{\mathrm{bd}}(G) \geq k/(2t)$. Let $B(v,t)$ denote the set of elements at distance at most t from v, and define the *local density* of the graph G by

$$\mathsf{d}_{\mathrm{loc}} = \mathsf{d}_{\mathrm{loc}}(G) = \max_{v,t} \frac{|B(v,t)|-1}{t}.$$

Then

$$\mathsf{w}_{\mathrm{bd}} \geq \frac{1}{2}\mathsf{d}_{\mathrm{loc}}. \tag{17.24}$$

It is not hard to see that equality does not hold here, and the ratio $\mathsf{w}_{\mathrm{bd}}/\mathsf{d}_{\mathrm{loc}}$ can be as large as $\log n$ (see Example 17.18 below).

We need the following related quantity, which we call the *harmonic density*:

$$\mathsf{d}_{\mathrm{har}} = \mathsf{d}_{\mathrm{har}}(G) = \max_{v\in V} \sum_{u\in V\setminus v} \frac{1}{d(u,v)}. \tag{17.25}$$

The following inequality establishes a relationship between these two density notions.

$$\mathsf{d}_{\mathrm{loc}} \leq \mathsf{d}_{\mathrm{har}} \leq (1+\ln n)\mathsf{d}_{\mathrm{loc}}. \tag{17.26}$$

Indeed, for an appropriate $t > 0$ and $v \in V$,

$$\mathsf{d}_{\mathrm{loc}} = \frac{|B(v,t)|-1}{t} \leq \sum_{u\in B(v,t)\setminus v} \frac{1}{d(u,v)} \leq \sum_{u\in V\setminus v} \frac{1}{d(u,v)} \leq \mathsf{d}_{\mathrm{har}}.$$

On the other hand, $\mathsf{d}_{\mathrm{har}}$ is the sum of $n-1$ positive numbers, among which the k-th largest is at most $\mathsf{d}_{\mathrm{loc}}/k$ by the definition of the local density. Hence

$$\mathsf{d}_{\mathrm{har}} \leq \sum_{k=1}^{n-1} \frac{\mathsf{d}_{\mathrm{loc}}}{k} \leq \Bigl(\sum_{k=1}^{n-1} \frac{1}{k}\Bigr)\mathsf{d}_{\mathrm{loc}},$$

which proves the second inequality in (17.26).

The following examples show that the above bounds cannot be improved (except for the constants), but also that the bandwidth can be a quite tricky quantity.

EXAMPLE 17.18. Consider the path P_n on n nodes. Clearly $\mathsf{w}_{\mathrm{bd}}(P_n) = 1$. Since $|B(v,t)| \leq 2t+1$, and equality holds for $t < n/2$, we have $\mathsf{d}_{\mathrm{loc}}(P_n) = 2$. Thus inequality (17.24) is sharp for paths. Furthermore, for odd n and the midpoint v of the path, we have

$$\mathsf{d}_{\mathrm{har}} = \sum_{u\in V\setminus v} \frac{1}{d(u,v)} = 2 \sum_{k=1}^{(n-1)/2} \frac{1}{k} \sim 2\ln n,$$

showing that the upper bound in (17.26) is sharp up to a constant factor. ♦

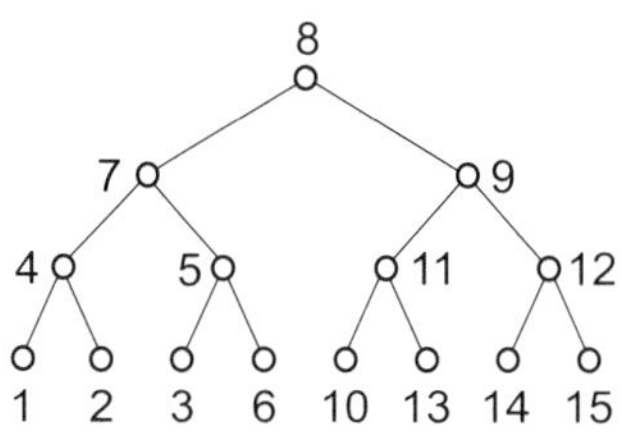

$$\begin{pmatrix}
0&0&0&1&0&0&0&0&0&0&0&0&0&0&0\\
0&0&0&1&0&0&0&0&0&0&0&0&0&0&0\\
0&0&0&0&1&0&0&0&0&0&0&0&0&0&0\\
1&1&0&0&0&0&1&0&0&0&0&0&0&0&0\\
0&0&1&0&0&1&1&0&0&0&0&0&0&0&0\\
0&0&0&0&1&0&0&0&0&0&0&0&0&0&0\\
0&0&0&1&1&0&0&1&0&0&0&0&0&0&0\\
0&0&0&0&0&0&1&0&1&0&0&0&0&0&0\\
0&0&0&0&0&0&0&1&0&0&1&1&0&0&0\\
0&0&0&0&0&0&0&0&0&0&1&0&0&0&0\\
0&0&0&0&0&0&0&0&1&1&0&0&1&0&0\\
0&0&0&0&0&0&0&0&1&0&0&0&0&1&1\\
0&0&0&0&0&0&0&0&0&0&1&0&0&0&0\\
0&0&0&0&0&0&0&0&0&0&0&1&0&0&0\\
0&0&0&0&0&0&0&0&0&0&0&1&0&0&0
\end{pmatrix}$$

FIGURE 17.4. Optimal labeling of a binary tree for bandwidth, specified as a labeling and as a matrix.

EXAMPLE 17.19. Let T_r denote the full binary tree of depth r. This tree has $n = 2^{r+1} - 1$ nodes, and if v denotes the root, then $|B(v,r)| = n$; it is easy to see that this is the choice of v and r defining the local density, so

$$\mathsf{d}_{\text{loc}} = \frac{n-1}{r} = \frac{2^{r+1}-2}{r} \qquad \text{and} \qquad \mathsf{w}_{\text{bd}} \geq \frac{\mathsf{d}_{\text{loc}}}{2} = \frac{n-1}{2r} = \frac{2^r-1}{r}.$$

The bandwidth of $\lceil (n-1)(2r) \rceil$ can indeed be realized [**Chung 1997**] (see Figure 17.4, and also Exercise 17.8). The harmonic density is given by

$$\mathsf{d}_{\text{har}} = \sum_{j=1}^{r} 2^j \frac{1}{j} \sim \frac{2^{r+1}}{r}.$$

This shows that the lower bound in (17.26) is asymptotically sharp. ♦

We conclude this preparation for the main result with a couple of inequalities relating the harmonic density to the tree-volume. Specializing from the previous section, we recall the notation $\Pi_T = \prod_{uv \in E(T)} d(u,v)$ for a tree T on some nodes of the graph G (not necessarily a subgraph of G).

LEMMA 17.20. *Let G be a graph on n nodes and T, a tree on k nodes. Then*

$$\sum_{\phi\colon V(T) \hookrightarrow V} \frac{1}{\Pi_{\phi(T)}} \leq n \mathsf{d}_{\text{har}}(G)^{k-1}.$$

PROOF. This is clear for $k = 1$. Let $k > 1$. Select any leaf v of T, denote its neighbor by u, and set $T' = T \setminus v$ and $\phi' = \phi|_{V(T')}$. Then

$$\sum_{\phi} \frac{1}{\Pi_{\phi(T)}} = \sum_{\phi'} \sum_{x \in V \setminus \phi(T')} \frac{1}{\Pi_{\phi(T')} d(x, \phi'(u))} \leq \sum_{\phi'} \frac{1}{\Pi_{\phi(T')}} \mathsf{d}_{\text{har}}.$$

From this the lemma follows by induction. □

As an application of this bound, we prove:

LEMMA 17.21. *For every graph G on n points,*

$$\sum_{S \in \binom{V}{k}} \frac{1}{\mathsf{vol}_{\text{tr}}(S)} \leq (k-1)!\, n\, (4\mathsf{d}_{\text{har}})^{k-1}.$$

PROOF. We use the elementary fact that the number of isomorphism types of trees on k nodes is bounded by 4^{k-1} (see e.g. [**Lovász 1991**], Problem 4.18). So summing the bound in Lemma 17.20 over all isomorphism types of trees, we get

$$\sum_{T} \sum_{\phi\colon V(T) \hookrightarrow V} \frac{1}{\Pi_{\phi(T)}} \leq n(4\mathsf{d}_{\text{har}})^{k-1},$$

and hence

$$\begin{aligned}\sum_{S \in \binom{V}{k}} \frac{1}{\mathsf{vol}_{\text{tr}}(S)} &\leq \sum_{S \in \binom{V}{k}} \sum_{\substack{T \text{ tree} \\ V(T)=S}} \frac{(k-1)!}{\Pi_T} = (k-1)! \sum_{T} \sum_{\phi\colon V(T) \hookrightarrow V} \frac{1}{\Pi_{\phi(T)}} \\ &\leq (k-1)!\, n\, (4\mathsf{d}_{\text{har}})^{k-1}.\end{aligned}$$

□

17.4.2. Approximating the bandwidth. Now we come to the main theorem describing the algorithm to get an approximation of the bandwidth of a graph.

THEOREM 17.22. *Let G be a graph on n nodes, let $\eta \ge 1$, let $k = \lceil \ln n \rceil$, and let $\mathbf{x} : V \to \mathbb{R}^m$ be a contractive vector labeling such that $\mathsf{vol}_{\mathbf{x}}(S) \ge \eta^{1-k}\mathsf{vol}_{\mathrm{tr}}(S)$ for every k-set $S \subseteq V$. Project the representing points onto a random line L through the origin, and label the nodes by $1, \dots, n$ according to the order of their projections on the line. Then with high probability, $|i-j| \le 480\eta \mathsf{d}_{\mathrm{har}} \log n$ for every edge ij.*

PROOF. Let $\mathbf{y}_i$ be the projection of $\mathbf{x}_i$ on L. Let ij be an edge; since $\mathbf{x}$ is contractive we have $|\mathbf{x}_i - \mathbf{x}_j| \le 1$, and hence by (C.17) it follows that with high probability, $|\mathbf{y}_i - \mathbf{y}_j| \le 2|\mathbf{x}_i - \mathbf{x}_j|/\sqrt{m} \le 2/\sqrt{m}$ for all edges ij. We will assume below that this is indeed the case.

We call a set $S \subseteq V$ *squeezed*, if $\operatorname{diam}\big(\mathbf{y}(S)\big) \le 2/\sqrt{m}$. So every edge is squeezed; but we are interested in squeezed k-sets, which we count in two different ways. Let X be the number of squeezed k-sets.

By (C.22) (with m in place of n, $k-1$ in place of d, and $2/\sqrt{m}$ in place of S), we have for every k-set S,

$$\mathsf{P}(S \text{ is squeezed}) \le \frac{(m-k+1)\pi_{k-1}^2 \pi_{m-k+1}}{m^{(k+1)/2}\pi_m 2^{k-1}\mathsf{vol}_{\mathbf{x}}(S)} \le *** \tag{17.27}$$

Using Lemma 17.21, we can bound the expected number of squeezed k-sets. Since $n \le 3^k$ by the choice of k, we get

$$\begin{aligned}\mathsf{E}(X) &\le \Big(\frac{40\eta}{k-1}\Big)^{k-1} k \sum_{S \in \binom{V}{k}} \frac{1}{\mathsf{vol}_{\mathrm{tr}}(S)} \\ &\le \Big(\frac{40\eta}{k-1}\Big)^{k-1} (k-1)! n (4\mathsf{d}_{\mathrm{har}})^{k-1} \le (240\eta \mathsf{d}_{\mathrm{har}})^{k-1}.\end{aligned}$$

On the other hand, let $ij \in E$ $(i < j)$. Every k-subset of $\{i, i+1, \dots, j\}$ is squeezed, and so

$$X \ge \binom{j-i+1}{k} > \Big(\frac{j-i}{k}\Big)^{k-1}$$

(the last step is valid for $j - i \ge k-1$, which we may assume), and comparing with the upper bound, we get

$$\mathsf{E}\Big(\Big(\frac{j-i}{k}\Big)^{k-1}\Big) \le (240\eta \mathsf{d}_{\mathrm{har}})^{k-1}.$$

Using Markov's Inequality and taking $(k-1)$-st roots, we get with high probability that

$$j - i \le 480\eta \mathsf{d}_{\mathrm{har}} k.$$

□

As a consequence, we get that $\mathsf{d}_{\mathrm{loc}}$ approximates w_{bd} up to a polylogarithmic factor:

COROLLARY 17.23. *For every graph with n points, its bandwidth and local density satisfy*

$$\mathsf{d}_{\mathrm{loc}} \le \mathsf{w}_{\mathrm{bd}} \le O\big((\log n)^3 \sqrt{\log \log n}\big)\mathsf{d}_{\mathrm{loc}}.$$

REMARK 17.24. We can generalize the notion of bandwidth to any finite metric space (V, d): the parameter $\mathsf{w}_{\mathrm{bd}} = \mathsf{w}_{\mathrm{bd}}(V, \mathsf{d})$ is the smallest real number a such that there is a bijection $f : V \to [n]$ such that $|f(i) - f(j)| \le a\, d(i,j)$ for all $i, j \in V$.

It takes a minute to realize that this notion does generalize graph bandwidth: the definition of bandwidth requires this condition only for the case when ij is an edge, but this already implies that it holds for every pair of nodes due to the definition of graph distance. (We do loose the motivation for the word "bandwidth".)

Most of what we proved in this section extends to the more general setting of finite metric spaces; see Exercises 17.11–17.13.

EXERCISE 17.1. Let (V, d) be a metric space. Prove that $(V, \sqrt{d})$ is a metric space.

EXERCISE 17.2. The graph metric C_4 is ℓ_1-embeddable, but not ℓ_2-embeddable.

EXERCISE 17.3. The graph $K_{2,3}$ is not ℓ_1-embeddable.

EXERCISE 17.4. A *tree-metric* is a graph metric defined by an edge-weighted tree. Prove that every tree-metric is ℓ_1-embeddable.

EXERCISE 17.5. A metric space (V, d) is called *hypermetric*, if for every $k \geq 1$, every $2k+1$ points $p_1, \dots, p_{k+1}, q_1, \dots, q_k \in V$ satisfy

$$\sum_{1 \leq i < j \leq k+1} d(p_i, p_j) + \sum_{1 \leq i < j \leq k} d(q_i, q_j) \leq \sum_{\substack{1 \leq i \leq k+1 \\ 1 \leq j \leq k}} d(p_i, q_j).$$

Prove that every ℓ_1-embeddable metric is a hypermetric.

EXERCISE 17.6. Prove that every finite metric space (V, d) with n points has a contractive vector-labeling $\mathbf{x}$ such that $\mathsf{vol}_{\mathbf{x}}(V) \geq 2^{1-n/2}\mathsf{vol}_{\mathrm{tr}}(V)$.

EXERCISE 17.7. Let (V, d) be a finite metric space and $v \in V$. Prove that (V, d) is isometrically embeddable in a Euclidean space if and only if the $(V \setminus v) \times (V \setminus v)$ matrix A defined by $A_{ij} = d(v,i)^2 + d(v,j)^2 - d(i,j)^2$ is positive semidefinite.

EXERCISE 17.8. Prove that the bandwidth of the complete k-ary tree of depth r is $\lceil (k^{r+1} - k)/(2r(k-1)) \rceil$.

EXERCISE 17.9. Let T be a spanning tree on a matrix space (V, d), and let $i, j \in V$, and let $T_1, \dots, T_k$ be the spanning trees that can be obtained from T by adding the edge ij and deleting one of the other edges of the arising cycle. Prove that

$$\frac{1}{\Pi(T)} \leq \frac{1}{\pi(T_1)} + \dots + \frac{1}{\pi(T_k)}.$$

EXERCISE 17.10. For every spanning tree T on a matrix space (V, d), we have

$$\frac{1}{\Pi(T)} \leq \sum_P \frac{1}{\Pi(P)},$$

where P ranges over all spanning paths on V.

EXERCISE 17.11. Show how to construct for every finite metric space (V, d) another finite metric space (V, d') such that $1 \leq d'(i,j) \leq n$ and $\mathsf{w}_{\mathrm{bd}}(V, d) = \mathsf{w}_{\mathrm{bd}}(V, d')$.

EXERCISE 17.12. Define the notions of local and harmonic densities for every finite metric space, and prove that inequalities (17.24) and (17.26), and Lemmas 17.20 and 17.21 remain valid.

EXERCISE 17.13. Generalize Theorem 17.22 to finite metric spaces.

CHAPTER 18

Matching and Covering in Frameworks

In this chapter we study vector labelings from their purely linear algebraic perspective—where only linear dependence and independence of the underlying vectors play a role. Replacing "cardinality" by "rank" in an appropriate way, we get generalizations of ordinary graph-theoretic parameters and properties that are often quite meaningful and relevant. Furthermore, these results about frameworks can be applied to purely graph-theoretic problems, where the vector-labeling is used to encode information about other parts, or additional structure, of the graph.

We restrict our attention to the study of two basic notions in graph theory: matchings (sets of pairwise disjoint edges) and node-covers (sets of nodes that meet all edges). For a graph G, we denote by $\nu(G)$ its *matching number*, the maximum cardinality of a matching, and by $\tau(G)$ the *node-cover number*, the minimum cardinality of a node-cover. It is clear that to cover the edges of a matching we need to use different nodes, and hence

$$\nu(G) \le \tau(G). \tag{18.1}$$

While these problems are in a sense dual to each other, their complexity is quite different even for ordinary graphs: the matching problem is polynomially solvable, while the node-cover problem is NP-hard.

The situation is similar for frameworks $(G, \mathbf{v})$ where $\mathbf{v}: V \to \mathbb{R}^d$. To avoid some trivial complications, let us assume that the graph has no isolated nodes and the node positions (vector labels) generate the whole space $\mathbb{R}^d$. For $S \subseteq V$, we define $\mathbf{v}(S) = \{\mathbf{v}_i : i \in S\}$ and (whenever there is only one vector labeling around) $\operatorname{rk}(S) = \operatorname{rk}\big(\mathbf{v}(S)\big)$. The *node-cover number* $\tau(G, \mathbf{v})$ of a framework is the minimum rank of a node-cover. A *matching* in $(G, \mathbf{v})$ is a set of node-disjoint edges such that all their endnodes are labeled by linearly independent vectors, i.e., $\operatorname{rk}(V(M)) = 2|M|$. We denote by $\nu(G, \mathbf{v})$ the maximum cardinality of a matching.

Ordinary graphs can be considered as the special case when nodes are labeled by linearly independent vectors. We denote such a vector labeling by $\mathbf{u}_{\text{free}}$ (since we are only concerned with the linear independence structure of the vector labels, it does not matter which linearly independent set of vectors we use). It is easy to check that $\tau(G, \mathbf{u}_{\text{free}}) = \tau(G)$ and $\nu(G, \mathbf{u}_{\text{free}}) = \nu(G)$.

The extension of inequality (18.1) to frameworks remains valid:

$$\nu(G, \mathbf{v}) \le \tau(G, \mathbf{v}). \tag{18.2}$$

In this chapter, we formulate extensions of basic characterization results like Kőnig's and Tutte's matching theorems to frameworks. Several of these results are known from matroid theory, and have many applications in graph theory and other combinatorial problems. The node-cover number τ is NP-hard even in the case of ordinary graphs, so no similarly complete results can be expected; however, the geometric methods providing partial results will be important in the proof of a classification theorem of cover-critical ordinary graphs (whose statement does not involve vector labelings).

While the results on matchings are more complete and perhaps more important, we start with a study of covering, since some results developed for covering will be useful in the study of matchings.

Since only the linear structure is used, most of the considerations in this chapter could be formulated over any field (or at least over any sufficiently large field). Indeed, most elementary arguments would work over any matroid; however, two of the main results (Theorems 18.8(b) and 18.15) do make use of the additional linear structure provided by the underlying field.

Let us describe an example of a combinatorially interesting matching in frameworks.

EXAMPLE 18.1 (Branchings). Let G be a directed graph; we define an (unoriented) framework $(H, \mathbf{v})$ as follows. For every edge $e = ij \in E$, we create two nodes h_e and t_e of H, and connected them by an (unoriented) edge $\widehat{e}$. We label h_e by the unit vector $\mathbf{v}(h_e) = \mathbf{e}_j$, and t_e by the difference $\mathbf{v}(t_e) = \mathbf{e}_{j+n} - \mathbf{e}_{i+n}$.

In the framework $(H, \mathbf{v})$, edges $\widehat{e}_1, \dots, \widehat{e}_k$ form a matching if and only if no two of the edges $e_1, \dots, e_k$ enters the same node of G (equivalently, the vectors $\mathbf{v}(h_e)$ are linearly independent) and they form a forest in G in the unoriented sense (equivalently, the vectors $\mathbf{v}(t_e)$ are linearly independent). Such a set of edges is called a *branching*. It follows that $\nu(H, \mathbf{v}) \le n-1$; if equality holds for some set of edges, then they form a *spanning branching* or *arborescence*: a spanning tree (in the undirected sense) which has a root r so the branching contains a directed path from the root to every other node. So *every matching with $n-1$ edges in $(H, \mathbf{v})$ corresponds to a spanning branching in G, and vice versa.* ◆

18.1. Cover-critical ordinary graphs

We start with recalling some basic facts about ordinary graphs and their node-cover numbers. A graph G is called *cover-critical*, if deleting any edge from it, $\tau(G)$ decreases. This is equivalent to saying that deleting any edge, $\alpha(G)$ increases, and for this reason these graphs are often called α-critical. Since isolated nodes play no role here, we will assume that cover-critical graphs have no isolated nodes. These graphs have an interesting theory, initiated in [**Erdős–Gallai 1961**]. See also [**Berge 1973**], [**Lovász 1991**], [**Lovász–Plummer 1986**].

There are some easy but very useful observations we can make. In a cover-critical graph G every node v is contained in a minimum node-cover. Indeed, if we delete v, the rest of the graph can be covered by $\tau(G) - 1$ nodes; adding v to this set, we get a minimum node-cover containing v. It is just slightly more complicated to see that every node is missed by some minimum node-cover.

For a cover-critical graph G, the following quantity, called its *covering defect*, will play a crucial role: $\delta(G) = 2\tau(G) - n = n - 2\alpha(G)$. The next theorem summarizes some of the basic properties of cover-critical graphs.

THEOREM 18.2. *For every cover-critical graph G,*

(a) [**Erdős–Gallai 1961**] $\delta(G) \ge 0$; *in other words, $n \le 2\tau(G)$.*

(b) [**Erdős–Hajnal–Moon 1964**] $m \le \binom{\tau(G)+1}{2}$.

(c) [**Hajnal 1965**] *Every node has degree at most $\delta(G) + 1$.*

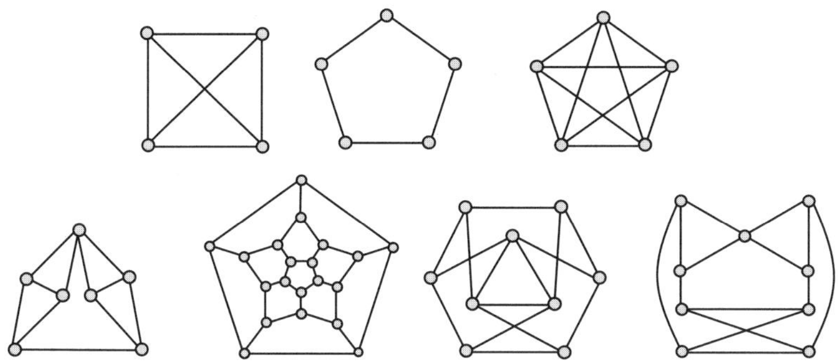

FIGURE 18.1. Some cover-critical graphs

These results will follow easily from more general considerations concerning cover-critical frameworks.

Cover-critical graphs can be classified by their defect. Since a graph is cover-critical if and only if each of its connected components are cover-critical, we may restrict our attention to connected graphs. The only connected cover-critical graph with $\delta = 0$ is K_2 [**Erdős–Gallai 1961**], and the only connected cover-critical graphs with $\delta = 1$ are odd cycles (this follows from Hajnal's Theorem 18.2(c)).

The fact that all odd cycles are cover-critical with the same defect motivates the following simple observation: *if G' is obtained from G by subdividing an edge by two nodes, then G' is cover-critical if and only if G is, and $\delta(G) = \delta(G')$.* By this fact, for $\delta \geq 2$ it suffices to describe those cover-critical graphs that do not have two adjacent nodes of degree 2.

There is a slightly less trivial version of this observation: Let us split a node of a graph G into two (both having positive degree), and connect both of them to a new node. The resulting graph G' is cover-critical if and only if G is, and $\delta(G) = \delta(G')$. Together with Exercise 18.1, this implies that it is enough to describe connected cover-critical graphs with minimum degree at least 3. Let us call these graphs *basic cover-critical.*

There is only one basic cover-critical graph with $\delta = 2$, namely K_4 [**Andrásfai 1967**], and four basic cover-critical graphs with $\delta = 3$, the last four graphs in Figure 18.1 [**Surányi 1973**]. This motivates the following theorem [**Lovász 1978**]:

THEOREM 18.3. *For every $\delta \geq 2$, the number of basic cover-critical graphs with $\delta(G) = \delta$ is finite.*

The proof will show that every basic cover-critical graph has at most $2^{3\delta^2}$ nodes. (This bound is probably very rough.) The proof uses a generalization of the notion of cover-critical graphs to frameworks, and is postponed to section 18.2.4.

18.2. Covering in frameworks

18.2.1. Extending the rank function. Our goal is to generalize the notions of covers and criticality to frameworks [**Lovász 1977**], [**Lovász 1978**]. We start with extending the notion of rank from sets to vectors.

Let $\mathbf{u} : V \to \mathbb{R}^d$ be a vector labeling of a finite set V. We can think of the rank function $\mathrm{rk}(S) = \dim\big(\mathbf{u}(S)\big)$ as a function defined on the subsets of V, but also as a function defined on the set of all 0-1 vectors in $\mathbb{R}^V$. We extend the definition to all vectors $\mathbf{w} \in \mathbb{R}_+^V$ as follows. There is a unique decomposition

$$\mathbf{w} = \sum_{i=1}^{t} \alpha_i \mathbb{1}_{S_i}, \quad \alpha_i > 0, \quad S_1 \subset S_2 \subset \cdots \subset S_t. \tag{18.3}$$

(Here $t = 0$ if $\mathbf{w} = 0$.) To construct this decomposition, we let S_1 be the set of nodes with maximal w_i, and let α_1 be the difference of this maximal w_i and the next largest entry of $\mathbf{w}$. Then $\mathbf{w} - \alpha_1 \mathbb{1}_{S_1}$ has more maximal entries. We repeat this until only zero entries are left. In other words, we choose the level-sets of $\mathbf{w}$ for the sets S_i.

Using this decomposition, we define

$$\mathrm{rk}(\mathbf{w}) = \sum_{i=1}^{t} \alpha_i \,\mathrm{rk}(S_i). \tag{18.4}$$

If $\mathbf{w}$ is 0-1 valued, then this is just the rank of its support.

The decomposition (18.3) is unique, but it does not remain unique if we drop the condition that the sets S_i form a chain. In other words, we have many decompositions of the form

$$\mathbf{w} = \sum_{i=1}^{t} \alpha_i \mathbb{1}_{T_i}, \quad \alpha_i > 0. \tag{18.5}$$

For such decompositions, we can still assert an inequality:

LEMMA 18.4. *For any decomposition of a vector* $\mathbf{w} \in \mathbb{R}_+^V$ *of the form* (18.5), *we have* $\mathrm{rk}(\mathbf{w}) \le \sum_i \alpha_i \,\mathrm{rk}(T_i)$.

In other words, among all representations of $\mathbf{w}$ in the form (18.5), the value $\sum_i \alpha_i \,\mathrm{rk}(T_i)$ is minimized when the sets T_i form a chain.

PROOF. The proof is a straightforward application of a technique called "uncrossing". Suppose that there are two incomparable sets T_i and T_j. Let, say, $\alpha_i \le \alpha_j$. Let us remove T_i from the family, but add $T_i \cap T_j$ and $T_i \cup T_j$ with coefficient α_i, while keeping T_j with coefficient $\alpha_j - \alpha_i$. (If either one of $T_i \cap T_j$ and $T_i \cup T_j$ was already used in the representation, we add α_i to their coefficient. If $\alpha_i = \alpha_j$, we remove T_j as well.) Since

$$\mathbb{1}_{T_i} + \mathbb{1}_{T_j} = \mathbb{1}_{T_i \cap T_j} + \mathbb{1}_{T_i \cup T_j},$$

this way we get a expression for the same vector $\mathbf{w}$, and since

$$\mathrm{rk}(T_i) + \mathrm{rk}(T_j) \ge \mathrm{rk}(T_i \cap T_j) + \mathrm{rk}(T_i \cup T_j),$$

this representation is not worse than the one we started with.

To conclude the proof, it suffices to argue that repeatedly performing this uncrossing operation, we end up with a representation where the sets T_i form a chain in a finite number of steps. This follows from the observation that the quantity $\sum_i \alpha_i |T_i|^2$ is strictly increasing at each uncrossing step. □

COROLLARY 18.5. *The extended rank function* $\mathrm{rk} : [0,1]^V \to \mathbb{R}$ *is homogeneous, subadditive and convex.*

This corollary can be easily derived from the results of [**Edmonds 1970**]; this form was stated in [**Lovász 1983a**].

PROOF. It is clear that rk is homogeneous. To prove that it is subadditive, we prove that for $\mathbf{w}, \mathbf{w}_1, \mathbf{w}_2 : [0,1]^V \to \mathbb{R}$ and $\lambda > 0$,

$$\mathrm{rk}(\lambda\mathbf{w}) = \lambda\,\mathrm{rk}(\mathbf{w}) \quad (\lambda \geq 0), \qquad \mathrm{rk}(\mathbf{w}_1 + \mathbf{w}_2) \leq \mathrm{rk}(\mathbf{w}_1) + \mathrm{rk}(\mathbf{w}_2).$$

The first equation is trivial, since the same sets S_i occur in the decomposition (18.3) of $\mathbf{w}$ and $\lambda\mathbf{w}$. If we substitute for $\mathbf{w}_1$ and $\mathbf{w}_2$ their representation (18.3), we get a decomposition of $\mathbf{w}_1 + \mathbf{w}_2$ of the form (18.5), and so the inequality follows by Lemma 18.4.

Finally, convexity is a consequence of homogeneity and subadditivity. □

18.2.2. Multiple covers. A *k-cover* in an ordinary graph is a weighting $\mathbf{w}$ of nodes by nonnegative integers such that for every edge, the sum of weights of the two endpoints is at least k. For a node-cover T, the indicator function $\mathbb{1}_T$ is a 1-cover (and every minimal 1-cover arises this way).

It is a bit more complicated to describe 2-covers. For every set $T \subseteq V$, define the weighting

$$\mathbb{2}_T = \mathbb{1}_T + \mathbb{1}_{N(V \setminus T)}.$$

(Recall that $N(V \setminus T)$ is the set of all nodes having a neighbor in $V \setminus T$, so it may contain nodes of T as well as nodes of $V \setminus T$). We claim that $\mathbb{2}_T$ is a 2-cover of G for every set $T \subseteq V$ (Figure 18.2). If both endpoints of an edge are covered by T, then this edge is covered twice. If one of them is in T, but the other is one not, then the first endpoint is in $N(V \setminus T)$, and so it has weight 2 in $\mathbb{2}_T$. If neither endpoint is contained in T, then both of them are covered by $N(V \setminus T)$.

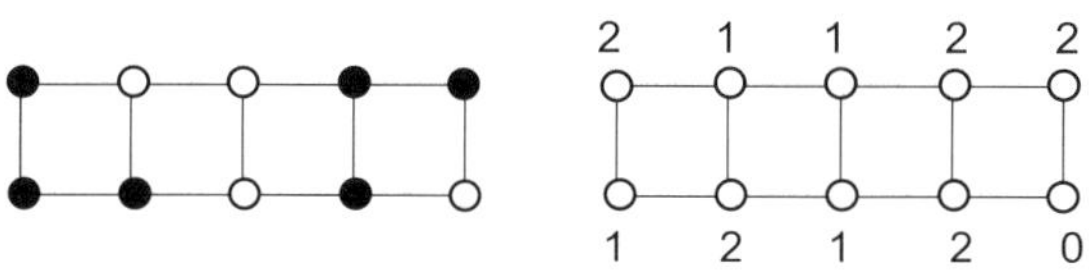

FIGURE 18.2. A set T of nodes and the 2-cover $\mathbb{2}_T$ constructed from it.

If $T = V$, then $\mathbb{2}_T = \mathbb{1}_V$. The same holds of $T = \emptyset$, provided G has no isolated nodes. This illustrates the important fact that $\mathbb{2}_T$ does *not* depend on T in any monotone way. If T is a node-cover, then we have $N(V \setminus T) \subseteq T$, and so the support of $\mathbb{2}_T$ is T. If T is a minimal node-cover (with respect to inclusion), then $N(V \setminus T) = T$, and so $\mathbb{2}_T = 2\mathbb{1}_T$.

If we allow the trivial operation of increasing some of the weights, then this construction will produce all 2-covers. Indeed, for a 2-cover $\mathbf{w}$, let T be its support. Then T must be a node-cover, and all nodes in $N(V \setminus T)$ must have weight 2, so $\mathbf{w} \geq \mathbb{2}_T$. This implies that $\mathbf{w}$ is a 2-cover if and only if there is a set T such that $\mathbf{w} \geq \mathbb{2}_T$. If the 2-cover $\mathbf{w}$ is minimal (i.e., no node-weight can be decreased and still have a 2-cover), then $\mathbf{w} = \mathbb{2}_T$ for some set T. This set T is by no means unique; for example, it can always be chosen from among node-covers.

One would expect that to describe k-covers for larger values of k becomes increasingly complicated, perhaps impossible. But in fact knowing 1-covers and 2-covers gives us all.

The sum of a k-cover and an l-cover is, trivially, a $(k+l)$-cover. A 2-cover is not necessarily the sum of two 1-covers (for example, the all-1 weighting of the nodes of a triangle). But every 3-cover is the sum of a 2-cover and a 1-cover; more generally, for $k \ge 2$ every k-cover $\mathbf{w} = (w_i : i \in V)$ can be decomposed into the sum of a 2-cover $\mathbf{w}'$ and a $(k-2)$-cover $\mathbf{w}''$ by the following simple formulas:

$$w'_i = \begin{cases} 2, & \text{if } w_i \ge k, \\ 1, & \text{if } 1 \le w_i \le k-1, \\ 0, & \text{if } w_i = 0, \end{cases} \qquad w''_i = w_i - w'_i. \tag{18.6}$$

It is easy to check that $\mathbf{w}'$ is a 2-cover and $\mathbf{w}''$ is a $(k-2)$-cover. Repeating this argument, we can decompose every k-cover as a sum of $\lfloor k/2 \rfloor$ 2-covers and at most one 1-cover (Figure 18.3).

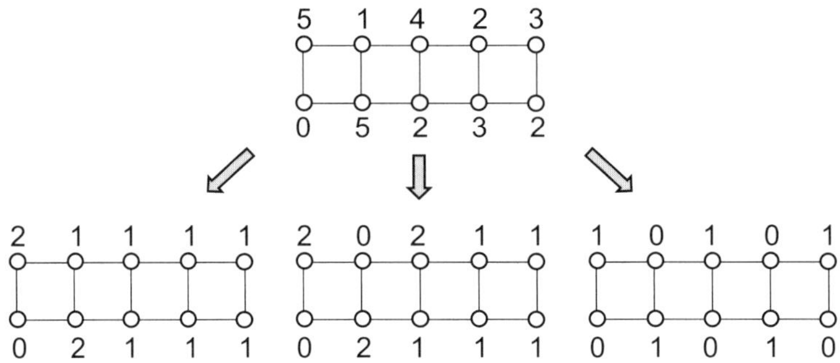

FIGURE 18.3. Decomposing a 5-cover into two 2-covers and a 1-cover.

The notion of k-covers is independent of the vector labeling, but the geometry enters when we replace "size" by "rank". Recalling the definition (18.4), we see that the rank of the 2-cover defined by a node-cover T is

$$\operatorname{rk}(2_T) = \operatorname{rk}(T) + \operatorname{rk}\big(N(V \setminus T)\big). \tag{18.7}$$

In the decomposition (18.6) of a k-cover, every level-set of $\mathbf{w}'$ or $\mathbf{w}''$ is a level-set of $\mathbf{w}$, and hence we can use the level sets of $\mathbf{w}$ to define the ranks of $\mathbf{w}'$ and $\mathbf{w}''$ (some of them may have coefficient 0). This implies that

$$\operatorname{rk}(\mathbf{w}') + \operatorname{rk}(\mathbf{w}'') = \operatorname{rk}(\mathbf{w}). \tag{18.8}$$

The *k-cover number* $\tau_k(G, \mathbf{v})$ for a framework $(G, \mathbf{v})$ is defined as the minimum rank of any k-cover. For the (most important) case of $k = 1$, we set $\tau(G, \mathbf{v}) = \tau_1(G, \mathbf{v})$. In other words, $\tau(G, \mathbf{v})$ is the minimum dimension of a linear subspace $L \subseteq \mathbb{R}^d$ such that at least one endpoint of every edge of G is labeled by a vector in L.

Since the sum of a k-cover and an m-cover is a $(k+m)$-cover, we have $\tau_{k+m}(G, \mathbf{v}) \le \tau_k(G, \mathbf{v}) + \tau_m(G, \mathbf{v})$; in particular

$$\tau_k(G, \mathbf{v}) \le k\tau(G, \mathbf{v}). \tag{18.9}$$

This implies that the difference

$$\delta(G, \mathbf{v}) = 2\tau(G, \mathbf{v}) - \tau_2(G, \mathbf{v})$$

is nonnegative. We call this quantity the *covering defect* of the framework $(G, \mathbf{v})$. (This quantity could be described as an "integrality gap" in integer programming; cf. Exercise 18.4.) The covering defect is defined for all frameworks, in particular for ordinary graphs, and it is not hard to see that $\delta(G, \mathbf{u}_{\text{free}}) = \delta(G)$ for all cover-critical graphs G.

The following simple lemma is the key to the study of cover-critical graphs and frameworks.

LEMMA 18.6. *Let S be any node-cover in a framework $(G, \mathbf{v})$, and let T be a node-cover with minimum rank. Then* $\text{rk}(\mathbb{2}_{S\cup T}) \le \text{rk}(\mathbb{2}_S)$.

PROOF. If we decompose the 3-cover $\mathbf{w} = \mathbb{1}_T + \mathbb{2}_S$ by (18.6), we get $\mathbf{w} = \mathbf{w}' + \mathbf{w}''$, where $\mathbf{w}' = \mathbb{2}_{S\cup T}$ is a 2-cover and $\mathbf{w}''$ is a 1-cover. Using the convexity of the rank function and (18.8), we get

$$\text{rk}(T) + \text{rk}(\mathbb{2}_S) \ge \text{rk}(\mathbf{w}) = \text{rk}(\mathbb{2}_{S\cup T}) + \text{rk}(\mathbf{w}'').$$

Since $\text{rk}(T) \le \text{rk}(\mathbf{w}'')$ by the minimality of $\text{rk}(T)$, this implies the lemma. □

Applying the lemma repeatedly, we get that if S is a node-cover, and $T_1, \dots, T_k$ are node-covers with minimum rank, then

$$\text{rk}(\mathbb{2}_{S\cup T_1\cup\dots\cup T_k}) \le \text{rk}(\mathbb{2}_S). \tag{18.10}$$

This inequality will be most useful when there are many node-covers with minimum rank, which will happen in the "cover-critical" case, to be discussed next.

18.2.3. Cover-critical frameworks. A framework is *cover-critical*, if deleting any node or edge, its node-cover number decreases. This condition implies, in particular, that the graph has no isolated nodes, and no node is labeled by the zero vector.

If $(G, \mathbf{v})$ is cover-critical, and we project the representing vectors onto a generic $\tau(G)$-dimensional space, then ranks up to $\tau(G)$ don't change, and hence the resulting framework is still cover-critical. Hence we could always assume that the representation is in $\tau(G)$-dimensional space. Informally, there is no hyperplane meeting every edge, but deleting any edge, there will be. (This reduction is, however, not always convenient.)

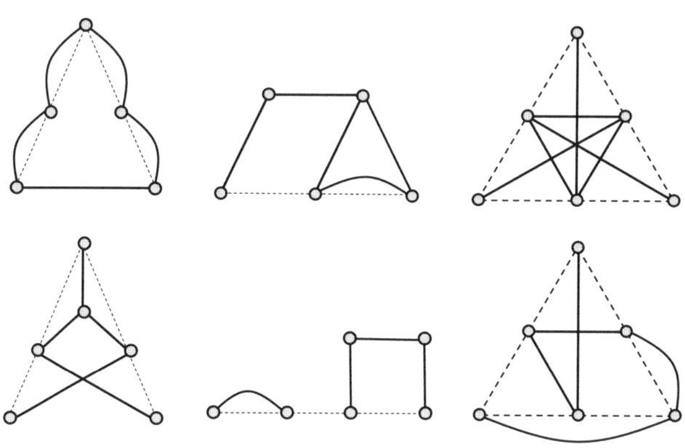

FIGURE 18.4. Some cover-critical frameworks. The figures are drawn in projective plane, so that 1-dimensional linear subspaces are depicted as points, 2-dimensional linear subspaces are depicted as lines etc. The reader is recommended to use this way of drawing frameworks to follow the arguments in this chapter.

A framework $(G, \mathbf{u}_{\text{free}})$ is cover-critical if and only if G is cover-critical as an ordinary graph. This way the theory of cover-critical frameworks generalizes the theory of cover-critical graphs.

Some basic results concerning cover-critical graphs can be generalized to the vector-labeled case without difficulty. For example, every node is contained in a minimum rank node-cover, and missed by another minimum rank node-cover.

There are properties that are trivial in the unlabeled case, but need a proof in our case.

LEMMA 18.7. *In a cover-critical framework $(G, \mathbf{v})$, the neighbors of any node are represented by linearly independent vectors.*

PROOF. Let $b \in N(a)$. By criticality, $\tau(G \setminus ab, \mathbf{v}) < \tau(G, \mathbf{v})$, and hence there is a $(\tau(G,\mathbf{v})-1)$-dimensional subspace L that contains one of the endnode positions of every edge except ab. Clearly, $\mathbf{v}_a, \mathbf{v}_b \notin L$ (otherwise L would be a node-cover of G). Hence L must contain $\mathbf{v}_j$ for every $j \in N(a)$, $j \neq b$. It follows that $\mathbf{v}_b \notin \text{lin}\{\mathbf{v}_j : j \in N(a) \setminus \{b\}\}$. This holds for every $b \in N(a)$, showing that the vectors $\mathbf{v}_b$ $(b \in N(a))$ are linearly independent. $\square$

The last lemma implies that the degree of each node is at most $\text{rk}(V)$. We are going to prove a much sharper bound on the degrees, as part of the following generalization of Theorem 18.2 to frameworks. For $a \in V$, set $\varepsilon_a = \text{rk}(V) - \text{rk}(V \setminus \{a\})$. In other words, $\varepsilon_a = 1$ if $\mathbf{v}_a$ is linearly independent of the vectors $\{\mathbf{v}_j : j \neq a\}$, and $\varepsilon_a = 0$ otherwise.

THEOREM 18.8. *Let $(G, \mathbf{v})$ be a cover-critical framework and $a \in V$. Then*

(a) $\text{rk}(V) = \tau_2(G, \mathbf{v})$.

(b) $m \leq \binom{\tau(G,\mathbf{v})+1}{2}$.

(c) $\deg(a) \leq \delta(G, \mathbf{v}) + \varepsilon_a$.

PROOF. (a) follows easily from our considerations about multiple covers. Since V is a 2-cover, we have $\tau_2(G, \mathbf{v}) \leq \text{rk}(V)$. On the other hand, let $T_1, \dots, T_k$ be all minimum node-covers, and S, a further node-cover. Then, as remarked, $T_1 \cup \dots \cup T_k = V$, and so by (18.10), we have

$$\text{rk}(2_S) \geq \text{rk}(2_{S \cup T_1 \cup \dots \cup T_k}) = \text{rk}(2_V) = \text{rk}(V).$$

This proves that $\text{rk}(V) = \tau_2(G, \mathbf{v})$.

(b) Let $(G, \mathbf{v})$ be a cover-critical framework in dimension d. We may assume that $d = \tau(G, \mathbf{v})$. Consider the matrices $M_{ij} = \mathbf{v}_i \mathbf{v}_j^\mathsf{T} + \mathbf{v}_j \mathbf{v}_i^\mathsf{T}$ $(ij \in E)$. We claim that these matrices are linearly independent.

For any edge ab, the graph $G \setminus ab$ has a node-cover T that does not span the space $\mathbb{R}^d$; let $\mathbf{u} \in \mathbb{R}^d$ $(\mathbf{u} \neq 0)$ be orthogonal to the span of T. Clearly $\mathbf{u}^\mathsf{T}\mathbf{v}_a \neq 0$ and $\mathbf{u}^\mathsf{T}\mathbf{v}_b \neq 0$, else $\{j \in V : \mathbf{u}^\mathsf{T}\mathbf{v}_j\} = 0$ would be a set covering all edges of G of rank $d-1 < \tau(G)$. Then

$$\mathbf{u}^\mathsf{T} M_{ij} \mathbf{u} = \mathbf{u}^\mathsf{T}\mathbf{v}_i \mathbf{v}_j^\mathsf{T}\mathbf{u} + \mathbf{u}^\mathsf{T}\mathbf{v}_j \mathbf{v}_i^\mathsf{T}\mathbf{u} = 2(\mathbf{u}^\mathsf{T}\mathbf{v}_i)(\mathbf{u}^\mathsf{T}\mathbf{v}_j) = 0$$

for every edge $ij \neq ab$ (since one of i and j must be in T), but

$$\mathbf{u}^\mathsf{T} M_{ab} \mathbf{u} = 2(\mathbf{u}^\mathsf{T}\mathbf{v}_a)(\mathbf{u}^\mathsf{T}\mathbf{v}_b) \neq 0.$$

This shows that M_{ab} is linearly independent of the matrices M_{ij}, $ij \neq ab$. Since ab was an arbitrary edge, this implies that the matrices M_{ij} are linearly independent. Hence their number cannot be larger than the dimension of the space of symmetric $d\times d$ matrices, which is $\binom{d+1}{2}$.

(c) Let $T \subseteq V\setminus a$ be a node-cover. We claim that

$$\deg(a) \leq \mathrm{rk}(2_T) - \mathrm{rk}(V) + \varepsilon_a. \tag{18.11}$$

Applying this to an optimal node-cover not containing a, and using (a), assertion (c) of the theorem will follow. We prove this inequality by induction on $|V\setminus T|$.

Let $S_1,\dots,S_k$ be all node-covers of minimum rank that do not contain a, and let $T' = T\cup S_1\cup\dots\cup S_k$. By Lemma 18.6, we have $\mathrm{rk}(2_{T'}) \leq \mathrm{rk}(2_T)$.

If $T' = V\setminus a$, then $2_{T'} = \mathbb{1}_{T'} + \mathbb{1}_{N(a)}$, and so by Lemma 18.7,

$$\deg(a) = \mathrm{rk}\big(N(a)\big) = \mathrm{rk}(2_{T'}) - \mathrm{rk}(T') \leq \mathrm{rk}(2_T) - \mathrm{rk}(V) + \varepsilon_a,$$

and we are done. So suppose that $A = V\setminus T' \neq \{a\}$, and so $B = A\setminus a$ is nonempty. Note that A and B are stable sets of nodes.

Claim. Let $i\in N(a)$. Then $i\notin N(B)$, and $\mathbf{v}_i$ is linearly independent of the set $\mathbf{v}(N(A)\setminus i)$.

Indeed, by the cover-critical property of $(G,\mathbf{v})$, the framework $(G\setminus ai,\mathbf{v})$ has a node-cover Z with rank $\tau(G,\mathbf{v})-1$. Clearly $\mathbf{v}_a$ and $\mathbf{v}_i$ are linearly independent of $\mathbf{v}(Z)$. The set $Z\cup\{i\}$ is a node-cover of G with minimum rank, and hence it is one of the sets S_j. By the definition of A, it follows that $Z\cap A = \emptyset$. Since Z covers all edges except ai, there is no edge connecting i to B, and $N(A)\setminus i \subseteq Z$. So $\mathbf{v}_i$ is linearly independent of $\mathbf{v}(N(A)\setminus i)$. This proves the Claim.

The Claim implies that $\mathrm{rk}\big(N(A)\big) = \mathrm{rk}\big(N(B)\big) + \deg(a)$. Let $b\in B$. Applying (18.11) with $V\setminus B$ in place of T (which we can do by the induction hypothesis), we get

$$\deg(b) \leq \mathrm{rk}(2_{V\setminus B}) - \mathrm{rk}(V) + \varepsilon_b = \mathrm{rk}(V\setminus B) + \mathrm{rk}\big(N(B)\big) - \mathrm{rk}(V) + \varepsilon_b.$$

Since $\deg(b)\geq 1$ and $\mathrm{rk}(V\setminus B)\leq \mathrm{rk}(V\setminus A)+1$, this implies that $\mathrm{rk}\big(N(B)\big)\geq \mathrm{rk}(V) - \mathrm{rk}(V\setminus A) - \varepsilon_b$, and hence

$$\begin{aligned}\deg(a) &= \mathrm{rk}\big(N(A)\big) - \mathrm{rk}\big(N(B)\big) \leq \mathrm{rk}\big(N(A)\big) + \mathrm{rk}(V\setminus A) - \mathrm{rk}(V) + \varepsilon_b\\ &= \mathrm{rk}(2_{T'}) - \mathrm{rk}(V) + \varepsilon_b \leq \mathrm{rk}(2_T) - \mathrm{rk}(V) + \varepsilon_b.\end{aligned}$$

We are done, unless $\varepsilon_a = 0$ and $\varepsilon_b = 1$ for all $b\in B$. In this latter case $\mathbf{v}_a$ depends linearly on $\{\mathbf{v}_j : j\in V\setminus a\}$, but in this dependence the elements of B cannot be involved. So $\mathbf{v}_a$ depends linearly on $\{\mathbf{v}_j : j\in V\setminus A\}$. This shows that $\mathrm{rk}(V\setminus B) = \mathrm{rk}(V\setminus A)$, and we gain 1 in the above computation, proving (18.11). □

Assertion (a) implies that the set V is an optimal 2-cover, and

$$\mathrm{rk}(V) \leq 2\tau(G,\mathbf{v}). \tag{18.12}$$

If equality holds in (18.12), then part (c) of the theorem implies that $\deg(a) = 1$ and $\varepsilon_a = 1$ for every node a, and so G is a matching. This characterizes cover-critical frameworks with $\delta = 0$.

Theorem 18.8(b) generalizes Theorem 18.2(b), giving a good upper bound on the number of edges. What about the number of nodes? Since we do not allow isolated nodes, a cover-critical framework in $\mathbb{R}^d$ can have at most $d(d+1)$ nodes by Theorem 18.8(b). One cannot say more in general, since from every cover-critical framework we can construct another one by splitting every node to nodes of degree 1, labeled by the same vector. From the complete $(d+1)$-graph with the free labeling, we get a framework $(G, \mathbf{v})$ with $d(d+1)$ nodes and $\tau(G, \mathbf{v}) = d$.

The next step should be to generalize Theorem 18.3 to frameworks. However, it seems that the structure of cover-critical frameworks can be much more complicated than the structure of cover-critical ordinary graphs, and a classification theorem seems to be out of reach. But, somewhat surprisingly, Theorem 18.3 and some other results about "ordinary" cover-critical graphs will be proved using vector-labelings.

Before returning to ordinary graphs, we need a couple of technical lemmas, nothing deep, but involving some case-distinction. The first one analyzes how changing the label of a single node in certain ways influences the cover-critical property. Given a linear subspace $L \subseteq \mathbb{R}^d$ and a finite set $S \subset \mathbb{R}^d$, we say that a point $\mathbf{x} \in L$ is *in general position on L relative to S*, if $\mathbf{x} \in \text{lin}(S')$ for some $S' \subseteq S$ only if $L \subseteq \text{lin}(S')$. It is clear every subspace contains a point in general position relative to any finite set.

LEMMA 18.9. *Let $(G, \mathbf{v})$ be a framework in $\mathbb{R}^d$, let $j \in V$, and construct a new labeling $\bar{\mathbf{v}}$ by replacing $\mathbf{v}_j$ by a vector $\bar{\mathbf{v}}_j$ in general position on* $\text{lin}\big(\mathbf{v}(N(j))\big)$ *relative to $\mathbf{v}(V\setminus j)$ (we keep every other position). Then $\tau(G, \bar{\mathbf{v}}) \ge \tau(G, \mathbf{v})$. If, in addition, the original label $\mathbf{v}_j$ is in general position in $\mathbb{R}^d$ relative to $\mathbf{v}(V\setminus j)$, then $\tau(G, \bar{\mathbf{v}}) = \tau(G, \mathbf{v})$. If, in addition, $(G, \mathbf{v})$ is cover-critical, then $(G, \bar{\mathbf{v}})$ is cover-critical as well.*

PROOF. Let T be a node-cover of G for which $\text{rk}\,\big(\bar{\mathbf{v}}(T)\big)$ is minimal. If $\text{rk}\,\big(\mathbf{v}(T)\big) \le \text{rk}\,\big(\bar{\mathbf{v}}(T)\big)$, then $\tau(G, \mathbf{v}) = \text{rk}\,\big(\mathbf{v}(T)\big) \le \text{rk}\,\big(\bar{\mathbf{v}}(T)\big) \le \tau(G, \bar{\mathbf{v}})$, and we are done. So suppose that $\text{rk}\,\big(\bar{\mathbf{v}}(T)\big) < \text{rk}\,\big(\mathbf{v}(T)\big)$, then $j \in T$ and $\bar{\mathbf{v}}_j \in \text{lin}\big(\mathbf{v}(T\setminus j)\big)$. By the assumption that $\bar{\mathbf{v}}_j$ is in general position on $\text{lin}\big(\mathbf{v}(N(j))\big)$, it follows that $\text{lin}\big(\mathbf{v}(N(j))\big) \subseteq \text{lin}\big(\mathbf{v}(T\setminus j)\big)$. But then $T' = T\setminus j \cup N(j)$ is a node-cover of G with $\text{rk}\,\big(\mathbf{v}(T')\big) = \text{rk}\,\big(\mathbf{v}(T\setminus j)\big) \le \text{rk}\,\big(\bar{\mathbf{v}}(T)\big) = \tau(G, \bar{\mathbf{v}})$, a contradiction.

Now suppose that $\mathbf{v}_j$ is in general position relative to $\mathbf{v}(V\setminus j)$, and let T be a node-cover of G for which $\text{rk}\,\big(\mathbf{v}(T)\big)$ is minimal. If $j \notin T$, then $\tau(G, \bar{\mathbf{v}}) \le \text{rk}\,\big(\bar{\mathbf{v}}(T)\big) = \text{rk}\,\big(\mathbf{v}(T)\big) = \tau(G, \mathbf{v})$. Suppose that $j \in T$. By the assumption about $\mathbf{v}_j$, we have either $\mathbf{v}_j \notin \text{lin}\big(\mathbf{v}(T\setminus j)\big)$, or $\text{lin}\big(\mathbf{v}(T\setminus j)\big) = \text{lin}\big(\mathbf{v}(V\setminus j)\big)$. In the first case,

$$\tau(G, \bar{\mathbf{v}}) \le \text{rk}\,\big(\bar{\mathbf{v}}(T)\big) \le \text{rk}\,\big(\mathbf{v}(T\setminus j)\big) + 1 = \text{rk}\,\big(\mathbf{v}(T)\big) = \tau(G, \mathbf{v}).$$

In the second,

$$\tau(G, \bar{\mathbf{v}}) \le \text{rk}\,\big(\mathbf{v}(V\setminus j)\big) = \text{rk}\,\big(\mathbf{v}(T\setminus j)\big) \le \text{rk}\,\big(\mathbf{v}(T)\big) = \tau(G, \mathbf{v}).$$

Finally, suppose that G is cover-critical, and let $e \in E$. If e is not incident with j, then the same argument as above gives that $\tau(G\setminus e, \bar{\mathbf{v}}) = \tau(G\setminus e, \mathbf{v}) < \tau(G, \mathbf{v}) = \tau(G, \bar{\mathbf{v}})$. If e is incident with j, then a node-cover T of $G\setminus e$ with $\text{rk}\,\big(\mathbf{v}(T)\big) < \tau(G, \mathbf{v})$ cannot contain j, and hence

$$\tau(G\setminus e, \bar{\mathbf{v}}) \le \text{rk}\,\big(\bar{\mathbf{v}}(T)\big) = \text{rk}\,\big(\mathbf{v}(T)\big) < \tau(G, \mathbf{v}) = \tau(G, \bar{\mathbf{v}}). \qquad \square$$

The second technical lemma deals with conditions under which deleting a node preserves the cover-critical property.

LEMMA 18.10. *Let $(G, \mathbf{v})$ be a framework, and let j be a node such that $\mathbf{v}_j \in \mathrm{lin}\big(\mathbf{v}(N(j))\big)$. Let $\mathbf{v}'$ be the orthogonal projection of $\mathbf{v}$ onto the hyperplane $\mathbf{v}_j^{\perp}$. Then $\tau(G\setminus j, \mathbf{v}') = \tau(G, \mathbf{v}) - 1$. If, in addition, $(G, \mathbf{v})$ is cover-critical, then so is $(G\setminus j, \mathbf{v}')$.*

PROOF. Let T be a node-cover of $G\setminus j$ with $\mathrm{rk}\,\big(\mathbf{v}'(T)\big) = \tau(G\setminus j, \mathbf{v}')$. Then $T \cup \{j\}$ is a node-cover of G with $\mathrm{rk}\,\big(\mathbf{v}(T)\big) \le \mathrm{rk}\,\big(\mathbf{v}'(T)\big) + 1$, showing that $\tau(G, \mathbf{v}) \le \tau(G\setminus j, \mathbf{v}') + 1$.

Conversely, let T be a node-cover of G with $\mathrm{rk}\,\big(\mathbf{v}(T)\big) = \tau(G, \mathbf{v})$. The node-cover T contains either j or $N(j)$. In both cases, $\mathbf{v}_j \in \mathrm{lin}\big(\mathbf{v}(T)\big)$. Since $T\setminus j$ is a node-cover in $G\setminus j$, this implies that

$$\tau(G\setminus j, \mathbf{v}') \le \mathrm{rk}\,\big(\mathbf{v}'(T\setminus j)\big) = \mathrm{rk}\,\big(\mathbf{v}'(T)\big) < \mathrm{rk}\,\big(\mathbf{v}(T)\big) = \tau(G, \mathbf{v}).$$

Finally, assume that $(G, \mathbf{v})$ is cover-critical, and let $e \in E(G\setminus j)$. There exists a node-cover T of $G\setminus e$ with $\mathrm{rk}\,\big(\mathbf{v}(T)\big) = \tau(G, \mathbf{v}) - 1$. It follows as above that $\mathrm{rk}\,\big(\mathbf{v}'(T)\big) = \tau(G, \mathbf{v}) - 2 = \tau(G\setminus j, \mathbf{v}') - 1$. This shows that $(G\setminus j, \mathbf{v}')$ is cover-critical. □

18.2.4. Classifying cover-critical graphs. Combining the previous lemmas with Theorem 18.8, we get an important corollary for ordinary graphs, saying that a large cover-critical graph with small cover defect is almost bipartite: a minimal cover T spans a small number of edges only, while its complement is, of course, a stable set.

COROLLARY 18.11. *Let G be a cover-critical graph, and let $T \subseteq V$ be a node-cover. Then*

$$|E(G[T])| \le \binom{\tau(G) + |T| - n + 1}{2}.$$

In particular, if T is a minimum node-cover, then

$$|E(G[T])| \le \binom{\delta(G) + 1}{2}.$$

PROOF. Let $A = V\setminus T$. Starting with the framework $(G, \mathbf{u}_{\mathrm{free}})$, we apply the constructions in Lemmas 18.9 and 18.10: we replace $\mathbf{v}_i$ $(i \in A)$ by a point $\bar{\mathbf{v}}_i \in \mathbf{v}\big(N(i)\big)$ in general position, and then delete i and project the remaining labels onto $\bar{\mathbf{v}}_i^{\perp}$. Repeating this with all nodes of A, we are left with a cover-critical framework $(G[T], \mathbf{u})$ with $\tau(G[T], \mathbf{u}) = \tau(G) - |A|$. An application of Theorem 18.8(b) completes the proof. □

We need one further lemma, which is "pure" graph theory.

LEMMA 18.12. *Let G be a bipartite graph, and let $S_1, \dots, S_k \subseteq V$ be such that $G\setminus S_i$ has a perfect matching for every i. Assume, furthermore, that deleting any edge of G this property does not hold any more. Let $s = \max_i |S_i|$. Then the number of nodes of G with degree at least 3 is at most $2s^3\binom{k}{3}$.*

PROOF. Suppose that G has more than $2s^3\binom{k}{3}$ nodes of degree at least 3. Let us label every edge e with an index i for which $G\setminus e\setminus S_i$ has no perfect matching, and let M_i denote a perfect matching in $G\setminus S_i$. Clearly all edges labeled i belong to M_i.

Let us label every node with degree at least 3 by a triple of the edge-labels incident with it. There will be more than $2s^3$ nodes labeled by the same triple, say by $\{1,2,3\}$.

The set $M_1 \cup M_2$ consists of the common edges, of cycles alternating with respect to both matchings, and of alternating paths ending at $S_1 \cup S_2$. A node labeled $\{1,2,3\}$ cannot belong to a common edge of M_1 and M_2 (trivially), and it cannot belong to an alternating cycle, since then we could replace the edges of M_1 on this cycle by the other edges of the cycle, to get a perfect matching in $G \setminus S_i$ that misses an edge labeled i, contrary to the definition of the label. So it must lie on one of the alternating paths. The same argument holds for the sets $M_1 \cup M_3$ and $M_2 \cup M_3$. The number of alternating M_i-M_j paths is at most s, so there are three nodes labeled $\{1,2,3\}$ that lie on the same M_i-M_j paths for all three choices of $\{i,j\} \subseteq \{1,2,3\}$. Two of these nodes, say u and v, belong to the same color class of G.

We need one more combinatorial preparation: the alternating M_i-M_j path through u and v has a subpath Q_{ij} between u and v, which has even length, and therefore starts with an edge labeled i and ends with an edge labeled j, or the other way around. It is easy to see that (permuting the indices $1,2,3$ if necessary) we may assume that Q_{12} starts at u with label 1, and Q_{23} starts at u with label 2.

Now traverse Q_{23} starting at u until it first hits Q_{12} (this happens at v at the latest), and return to u on Q_{12}. Since G is bipartite, this cycle is even, and hence alternates with respect to M_2. But it contains an edge labeled 2, which leads to a contradiction just like above. □

Now we are able to prove the finite basis theorem for cover-critical graphs.

Proof of Theorem 18.3. Let G be a basic cover-critical graph with $\delta(G) = \delta$, let T be a minimum node-cover in G, let F be the set of edges induced by T, and let G' be obtained by deleting the edges of F from G. By Corollary 18.11, $|F| \le \binom{\delta+1}{2}$.

Let $R_1, \dots, R_k$ be the minimal sets of nodes that cover all edges of F. Trivially $|R_i| \le |F|$ and $k \le 2^{|F|}$. Let $\tau_i = \tau(G' \setminus R_i)$; clearly $\tau_i \ge \tau(G) - |R_i|$. Since $G' \setminus R_i$ is bipartite, it has a matching M_i of size τ_i. Let $S_i = V \setminus V(M_i)$. (We can think of the matching M_i as a "certificate" that R_i cannot be extended to a node-cover of size less than $\tau(G)$.)

By this construction, the graphs $G' - S_i$ have perfect matchings. We claim that deleting any edge from G' this does not remain true. Indeed, let $e \in E(G')$, then $G \setminus e$ has a node-cover T with $|T| < \tau(G)$. Since T covers all edges of F, it contains one of the sets R_i. Then $T - R_i$ covers all edges of $G' - R_i$ except e, and so $\tau(G' - R_i - e) \le |T| - |R_i| < \tau(G) - |R_i| \le \tau_i$. Thus $G' - R_i - e$ cannot contain a matching of size τ_i, and hence $G' - S_i - e$ cannot contain a perfect matching.

We can estimate the size of the sets S_i as follows:

$$|S_i| = n - 2\tau_i \le n - 2(\tau(G) - |R_i|) = 2|R_i| - \delta \le 2|F| - \delta \le \delta^2.$$

We can invoke Lemma 18.12, and conclude that G' has at most $2\binom{k}{3}\delta^6$ nodes of degree at least 3. Putting back the edges of F can increase this number by at most $2|F|$, and hence

$$n \le 2\binom{k}{3}\delta^6 + 2\delta(\delta+1) < 2^{3\delta^2}.$$

□

18.3. Matchings

18.3.1. Gallai's Identity. Let G be a graph without isolated nodes, and let $\rho(G)$ denote the minimum number of edges covering all nodes. A simple but important fact in graph theory is Gallai's Identity:

$$\rho(G)+\nu(G)=n. \tag{18.13}$$

As a warm-up to the topic of matchings in frameworks, we prove a generalization of this identity. Let $(G,\mathbf{v})$ be a framework in $\mathbb{R}^d$, where we assume that the graph G has no isolated nodes and the vector labels span the whole space. Recall that $\nu(G,\mathbf{v})$ denotes the maximum cardinality of a matching. Let us call a set of edges an *edge-cover*, if their endpoint labels span the space $\mathbb{R}^d$, and let $\rho(G,\mathbf{v})$ denote the minimum cardinality of an edge-cover. We claim that

$$\rho(G,\mathbf{v})+\nu(G,\mathbf{v})=d. \tag{18.14}$$

To prove this, first let $M\subseteq E$ be a matching of size $\nu(G,\mathbf{v})$. The vector labels of edges in M span a subspace L of dimension $2\nu(G,\mathbf{v})$. If this is not the whole space, there must be an edge f with at least one endpoint label not in L, so adding this edge to M we increase the dimension of L. Repeating this at most $d-2\nu(G,\mathbf{v})$ times, we get an edge-cover. Thus

$$\rho(G,\mathbf{v})\le\nu(G,\mathbf{v})+\big(d-2\nu(G,\mathbf{v})\big)=d-\nu(G,\mathbf{v}).$$

Conversely, let S be an edge-cover of minimum size, and let $M\subseteq S$ be a matching of maximum size among subsets of S. For every edge $e\in S\setminus M$, the set $M\cup\{e\}$ is not a matching, and hence $\mathrm{rk}\big(V(M)\cup V(e)\big)\le 2|M|+1$. Thus

$$\begin{aligned} d=\mathrm{rk}\big(V(S)\big)&=\mathrm{rk}\big(V(M)\cup V(S\setminus M)\big)\le 2|M|+|S\setminus M|\\ &=|S|+|M|\le\rho(G,\mathbf{v})+\nu(G,\mathbf{v}). \end{aligned}$$

This completes the proof of (18.14).

EXAMPLE 18.13 (Pinning a framework). Let $(G,\mathbf{u})$ be a bar-and-joint framework in the plane. Our goal is to pin down some of the nodes so that we prevent all infinitesimal motions. Recall the basic equation system (15.4):

$$(\mathbf{u}_i-\mathbf{u}_j)^{\mathsf{T}}(\mathbf{v}_i-\mathbf{v}_j)=0\quad(ij\in E), \tag{18.15}$$

where the unknowns are the $2n$ coordinates of the velocities $\mathbf{v}$. A pin at node i adds the equations $v_{i,1}=0$ and $v_{i,2}=0$. Note that neither one of these equations is automatically fulfilled, since a rigid translation in a non-coordinate direction does not satisfy them.

Recall that $\mathsf{Inf}\subseteq\mathbb{R}^{2\times V}$ denotes the linear space of infinitesimal motions. A condition $v_{i\nu}=0$ defines a hyperplane in Inf; let $f_{\nu i}$ be its normal vector in Inf. Pinning a set $S\subseteq V$ eliminates all infinitesimal motions if and only if the vectors $f_{\nu i}$, $(\nu=1,2,\ i\in S)$ generate the whole space Inf.

Define a graph H on the set $W=\{1,2\}\times V$ by connecting $(1,i)$ and $(2,i)$ for every $i\in V$, and label node (ν,i) by $f_{\nu i}$. Without the vector labels this is a trivial graph, but the vector labels make it interesting. By our discussion above, the minimum number of nodes of G whose pinning fixes the framework is equal to $\rho(H,f)$. By Gallai's Identity (18.14), this is equal to $\dim(\mathsf{Inf})-\nu(H,f)$. To determine the dimension of Inf is a straightforward computation in linear algebra. We'll see that $\nu(H,f)$ is computable in polynomial time, so the pinning number is computable in polynomial time. ♦

18.3.2. Bipartite frameworks. We want to extend the fundamental Marriage Theorem, which states that for bipartite graphs, equality holds in (18.1). Let us define a framework to be *bipartite* if its node set has a bipartition $V = V_1 \cup V_2$ such that every edge connects V_1 to V_2, and $\mathrm{rk}(V) = \mathrm{rk}(V_1) + \mathrm{rk}(V_2)$. In other words, the space has a decomposition $\mathbb{R}^d = L_1 \oplus L_2$ such that every edge has an endpoint labeled by a vector in L_1 and an endpoint labeled by a vector in L_2.

THEOREM 18.14. *If $(G, \mathbf{v})$ is a bipartite framework, then $\nu(G, \mathbf{v}) = \tau(G, \mathbf{v})$.*

A proof follows almost immediately from the results of the previous section (this proof would extend to the non-representable case easily).

PROOF. We have seen that $\nu(G, \mathbf{v}) \le \tau(G, \mathbf{v})$, so it suffices to prove the reverse inequality. We may assume that $(G, \mathbf{v})$ is cover-critical (just delete edges and/or nodes as long as $\tau(G, \mathbf{v})$ remains unchanged; it is enough to find a matching of size $\tau(G, \mathbf{v})$ in this reduced graph). Let $V = V_1 \cup V_2$ be the bipartition of $(G, \mathbf{v})$, then both V_1 and V_2 are node-covers, and hence

$$\tau(G, \mathbf{v}) \le \min\{\mathrm{rk}(V_1), \mathrm{rk}(V_2)\} \le \frac{\mathrm{rk}(V_1) + \mathrm{rk}(V_2)}{2} = \frac{\mathrm{rk}(V)}{2}.$$

By (18.12) and the remark after it, this implies that $(G, \mathbf{v})$ is a matching, and so $\nu(G, \mathbf{v}) = m \ge \tau(G, \mathbf{v})$. □

18.3.3. General frameworks. Tutte's Theorem characterizing the existence of a perfect matching and Berge's Formula describing the maximum size of matchings extend to frameworks as well [**Lovász 1980**]. Various forms of this theorem are called the "Matroid Parity Theorem" or "Matroid Matching Theorem" or "Polymatroid Matching Theorem". In spite of these names, this result does not belong to matroid theory; it uses more from the structure of linear spaces than what matroids can capture.

In a framework, we can consider the 2-dimensional subspaces determined by the pairs of vectors assigned to the endpoints of an edge. The matching problem can be formulated in terms of these subspaces, and in fact this provides useful geometric insight.

Recall an elementary exercise from geometry: if a family of lines in projective d-space has the property that every pair intersects, then either all lines lie in the same plane, or they all go through one and the same point. In the language introduced in Appendix A.4, *a set of lines contains no two independent lines if and only if either they are coplanar or concurrent.*

What happens if we exclude three independent lines? Figure 18.5 shows four configurations with this property. Theorem 18.15 below will imply that these are all.

We can rephrase this as a question about 2-dimensional linear subspaces of real linear space instead of lines in a projective space. Our starting observation is then that if a family of 2-dimensional linear subspaces in a real linear space contains no two linearly independent subspaces, then either they are all contained in a 3-dimensional subspace, or they all contain a common 1-dimensional subspace. The general question is: what is the maximum number of linearly independent subspaces in a family of 2-dimensional subspaces?

To show how this question relates to the matching number of frameworks, select a finite set V of vectors so that each of the given subspaces contains at least two

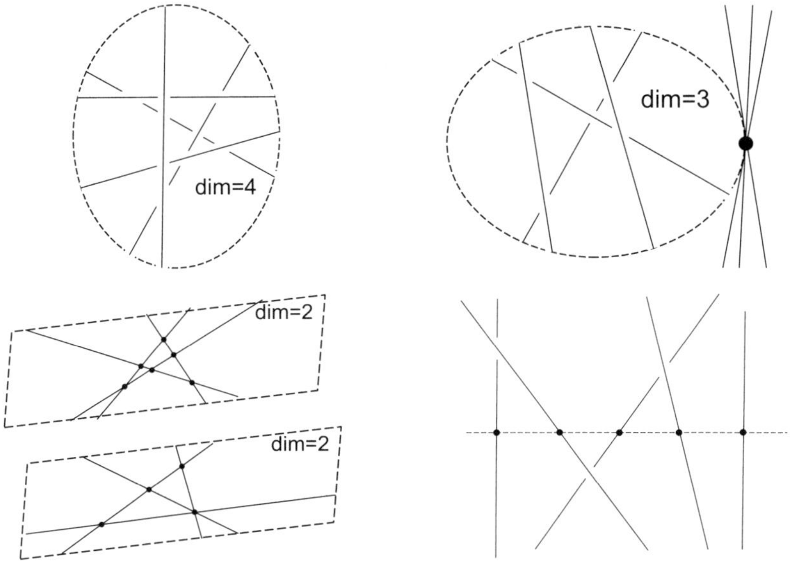

FIGURE 18.5. Four configurations of lines in a projective space (of arbitrary dimension) not containing three independent lines: (a) lines in a 4-dimensional subspace; (b) lines in a 3-dimensional subspace, together with lines going through a particular point of this subspace; (c) lines in two 2-dimensional subspaces; (d) lines intersecting a given line.

of them, and for each of the given subspaces, join two of the points of V in it by an edge. The condition on the subspaces says that the matching number of the resulting framework is less than k.

Conversely, consider a framework. If the vectors representing the endpoints of an edge are not linearly independent, then this edge will not be a member of any matching, and we can forget about it. So we may assume that the vectors representing the endpoints of any edge generate 2-dimensional subspace. A set of edges forms a matching if and only if the corresponding 2-dimensional subspaces are linearly independent.

There is a trivial bound on the matching number of a framework in $\mathbb{R}^d$:

$$\nu(G,\mathbf{v}) \le \left\lfloor \frac{d}{2} \right\rfloor. \tag{18.16}$$

Combining the ideas behind the bounds (18.2) and (18.16), we can formulate an upper bound that, for an optimal choice, gives equality [**Lovász 1980**].

THEOREM 18.15. *Let $(G,\mathbf{u})$ be a framework in $\mathbb{R}^d$, and let $\mathcal{P} = \{E_1,\dots,E_k\}$ range over all partitions of its edge set and let L range over linear subspaces of $\mathbb{R}^d$. Let $L_i = \mathrm{lin}\big(\mathbf{u}(V(E_i))\big)$. Then*

$$\nu(G,\mathbf{v}) = \min_{\mathcal{P},L}\Big\{\dim(L) + \sum_{i=1}^{k}\Big\lfloor \frac{1}{2}\dim(L_i/L)\Big\rfloor\Big\}.$$

The bound (18.16) corresponds to the choice $k=1$, $L=0$, $L_1=\mathbb{R}^d$. To get the bound (18.2), let L be the linear span of a node-cover T, and let $V\setminus T=\{i_1,\dots,i_k\}$, and let E_j consist of all edges contained in the linear span of L and $\mathbf{u}_{i_j}$.

The proof of the $\le$ direction in Theorem 18.15 is easy: for any subspace L and any partition $\mathcal{P}$, we have

$$\nu(G,\mathbf{v})\le \dim(L)+\nu(G,\mathbf{v}/L)$$

and for any matching M in $(G,\mathbf{v}/L)$, we have

$$|M\cap E_i|\le \left\lfloor\frac{\dim(L_i/L)}{2}\right\rfloor,$$

and hence

$$|M|=\sum_{i=1}^k|M\cap E_i|\le\sum_{i=1}^k\left\lfloor\frac{\dim(L_i/L)}{2}\right\rfloor.$$

The proof of the $\ge$ direction is quite lengthy, and has been reproduced in another book [**Lovász–Plummer 1986**], so we do not present it here.

EXAMPLE 18.16 (Graph matching). Let us apply the above result to "ordinary" matchings. If the nodes of the graph G are labeled by the standard basis $(\mathbf{e}_i:\ i\in V)$ in $\mathbb{R}^V$, then $\nu(G,\mathbf{e})=\nu(G)$. So the graph G has a perfect matching if and only if $\nu(G,\mathbf{e})=n/2$. By Theorem 18.15, this holds if and only if

$$\dim(L)+\sum_{i=1}^k\left\lfloor\frac{1}{2}\dim\mathbf{v}(E_i)\right\rfloor\ge\frac{n}{2}\tag{18.17}$$

for every partition $\mathcal{P}=\{E_1,\dots,E_k\}$ of the edge set and every linear subspace L, where $\mathbf{v}_i=\mathbf{e}_i/L$. If L is spanned by some of the vectors $\mathbf{e}_i$, then we can directly see that this condition is equivalent to Tutte's. In the general case, we need a little induction argument.

If any $\mathbf{v}_i=0$, then we can delete this node and proceed by induction. If the graph is disconnected, we can use induction on the connected components.

So suppose that G is connected and the vectors $\mathbf{v}_i$ are nonzero. If $k>1$, then, since the graph is connected, there are two partition classes, say E_1 and E_2, with $V(E_1)\cap V(E_2)\ne 0$. Then

$$\dim\mathbf{v}(V(E_1\cup E_2))\le\dim\mathbf{v}(V(E_1))+\dim\mathbf{v}(V(E_2))-1,$$

which implies that

$$\left\lfloor\frac{1}{2}\dim\mathbf{v}(V(E_1\cup E_2))\right\rfloor\le\left\lfloor\frac{1}{2}\dim\mathbf{v}(V(E_1))\right\rfloor+\left\lfloor\frac{1}{2}\dim\mathbf{v}(V(E_1))\right\rfloor,$$

so we can merge E_1 and E_2 to get a partition for which the inequality (18.17) is tighter. Going on similarly, we get that it suffices to check (18.17) for the partition of the edge set into a single class. Then we need

$$\dim(L)+\left\lfloor\frac{1}{2}\dim\mathbf{v}(E)\right\rfloor\ge\frac{n}{2}$$

Since $\dim\mathbf{v}(E)=n-\dim(L)$, this holds unless $\dim(L)=0$ and n is odd, so Tutte's condition is violated. ♦

18.3.4. Matching and multilinear algebra. Let $(G, \mathbf{u})$ be a framework, where $\mathbf{u} : V \to \mathbb{R}^d$. To every edge $e = ij$, we assign the antisymmetric tensor $A_e = \mathbf{u}_i \wedge \mathbf{u}_j$. (If you prefer a more elementary language, we can instead consider the skew symmetric $d \times d$ matrix $A_e = \mathbf{u}_i \mathbf{u}_j^\mathsf{T} - \mathbf{u}_j \mathbf{u}_i^\mathsf{T}$.) The sign of A_e depends on the orientation of the edge e, and we fix an orientation for each edge (this will not play an important role).

By basic exterior algebra, it is clear that the edges $e_1, \dots, e_m$ form a matching if and only if $A_{e_1} \wedge \dots \wedge A_{e_m} \neq 0$. It would be nice to have a proof of Theorem 18.15 based on exterior algebra, but this seems to be elusive. However, the following theorem (see [**Lovász–Plummer 1986**]) generalizes a characterization of graphs with perfect matchings [**Tutte 1947**], and it does provide a useful tool for randomized matching algorithms.

THEOREM 18.17. *For every framework* $(G, \mathbf{u})$,

$$\nu(G, \mathbf{u}) = \frac{1}{2} \max_{x_e \in \mathbb{R}} \operatorname{rk}\Big(\sum_{e \in E} x_e A_e\Big).$$

PROOF. First, we prove the $\le$ direction. Let $k = \nu(G, \mathbf{u})$, and let M be a maximum matching. The assertion is independent of the basis in $\mathbb{R}^d$, so we may assume that the endpoints of the edges in M are labeled by $\mathbf{e}_1, \dots, \mathbf{e}_{2k}$. Choosing $x = \mathbb{1}_{V(M)}$, the matrix $\sum_e x_e A_e$ will consist of $2k$ 1's in different rows and columns and zeros everywhere else, so its rank will be $2k$.

Second, let $2r$ be the rank of the matrix $A = \sum_e x_e A_e$ for an arbitrary choice of the numbers x_i (the rank of a skew symmetric matrix is always even). Then, by elementary multilinear algebra,

$$\underbrace{A \wedge \dots \wedge A}_{r} \neq 0$$

Expanding this exterior product, there must be a nonzero term, i.e., there must be edges $e_1, \dots, e_r$ such that $A_{e_1} \wedge \dots \wedge A_{e_r} \neq 0$. Let $e_k = i_k j_k$, then

$$A_{e_1} \wedge \dots \wedge A_{e_r} = \mathbf{u}_{i_1} \wedge \mathbf{u}_{j_1} \wedge \dots \wedge \mathbf{u}_{i_r} \wedge \mathbf{u}_{j_r},$$

and if this is nonzero, then $\mathbf{u}_{i_1}, \mathbf{u}_{j_1}, \dots, \mathbf{u}_{i_r}, \mathbf{u}_{j_r}$ are linearly independent, and so $\{e_1, \dots, e_r\}$ is a matching. This proves that $\nu(G, \mathbf{u}) \ge r$. □

Theorem 18.17 is of the form max = max, and so it does not provide a good characterization of the matching number. But the maximum on the right side is clearly attained for algebraically independent values of the x_i, and so the maximum is attained for almost all random choices, say from the uniform distribution in $[0, 1]$. By the Schwartz–Zippel Lemma 3.12, the maximum is attained with high probability if the coefficients x_i are chosen randomly from (say) $\{0, 1, \dots, n^2\}$. So the theorem provides an efficient randomized algorithm for computing $\nu(G, \mathbf{u})$.

EXAMPLE 18.18 (Gyrators). A *gyrator* g is an ideal device in an electrical network: it has a real parameter R_g (its *resistance*), and it forces the currents and voltage differences of two given directed edges ij and kl as

$$U_{ij} = R_g I_{kl} \quad U_{kl} = -R_g I_{ij}. \tag{18.18}$$

Note that R_g depends on the order as well as on the orientation of the two edges: if we reverse ij, then R_g changes sign, and so it does when the edges ij and kl are interchanged. We denote this gyrator by $g(ij,kl)$ if the participating edges need emphasis.

Perhaps it helps understand this if we compare it with an (ideal) transformer, which also establishes a relationship between two edges ij and kl:

$$U_{ij} = aU_{kl} \quad I_{kl} = aI_{ij}.$$

The behavior of a gyrator is more complicated, mostly because of the negative sign in the second equation.

Consider an electrical network composed of (ohmic) resistances, voltage and current sources, and gyrators. Determining the currents and voltages on the edges means the solution of a system of linear equations in the variables U_{ij} and I_{ij}. This system may be overdetermined or underdetermined, depending on the graph and on the numerical values of the resistances.

But let us assume that the parameters (resistances) of these devices are algebraically independent: the network is "generic". (This is not such an unreasonable assumption: in the case of real physical devices, their resistances are only approximately specified by the production process, and so no two will be exactly the same, and more generally, no exact algebraic relation will hold between them.) In this case, the combinatorial information (the graph and the pairs of edges that are related by gyrators) determines whether the network is over- or underdetermined. The following theorem was proved by [**Milić 1974**]:

THEOREM 18.19. *Let G be a network consisting of voltage and current sources, resistances, and gyrators. Suppose that the resistances and the parameters of the gyrators are algebraically independent over the rationals. Then there is a unique assignment of currents and voltage differences to the edges if and only if G has a spanning tree which contains all voltage sources, no current source, and either both or none of the edges of each gyrator.*

PROOF. We work with the electric potential function $p: V \to \mathbb{R}$. For the sake of simplicity, suppose that the network contains only gyrators. Then setting all currents and voltages to zero, we get a solution; we want to know whether this is unique. The voltage of an edge ij can be expressed as $U_{ij} = p_j - p_i$, and the currents are determined by

$$I_{ij} = -\frac{1}{R_g}(p_l - p_k), \quad I_{kl} = \frac{1}{R_g}(p_j - p_i) \tag{18.19}$$

for the two edges ij and kl forming a gyrator g. The only condition that is left is Kirchhoff's Current Law:

$$\sum_{j\in N(i)} I_{ij} = 0. \tag{18.20}$$

So we have n equations in n unknowns p_i. These equations are not independent, since their sum is trivially zero; but this is fine, since we need only the potential differences $p_j - p_i$. For the equations to determine the differences $p_j - p_i$ uniquely, the rank of the system of these equations must be $n-1$.

Substituting the expressions (18.19) in (18.20), we get after some transformations

$$\Big(\sum_{g=g(ij,kl)} \frac{1}{R_g}((\mathbf{e}_i - \mathbf{e}_j)\wedge(\mathbf{e}_k - \mathbf{e}_l))\Big)p = 0,$$

where the summation extends over all gyrators. By Theorem 18.17, the rank of the matrix of this system is twice the matching number of the framework $(H, \mathbf{v})$ on node set E, in which two edges are adjacent if and only if they form a gyrator, and an edge $e = ij$ (as a node of H) is labeled with the vector $\mathbf{v}_{ij} = \mathbf{e}_j - \mathbf{e}_i$. It is easy to check that $(n-1)/2$ edges of H form a matching in $(H, \mathbf{v})$ if and only if the $n-1$ edges of G corresponding to them form a spanning tree in G. □

So the gyrator problem can be reduced to the matching problem of frameworks. Theorem 18.15 gives a good characterization (NP∩co-NP). ♦

EXERCISE 18.1. Let G be a cover-critical graph with a node u of degree two, different from the triangle. Prove that the neighbors of u ar nonadjacent, and have no neighbor in common other than u.

EXERCISE 18.2. Prove that the nodes of a cover-critical graph can be covered by node-disjoint edges and odd cycles of the graph.

EXERCISE 18.3. Let $(G, \mathbf{v})$ be a cover-critical framework. Prove that for every node there is an optimal node-cover that contains it and also one that avoids it.

EXERCISE 18.4. A *fractional cover* of a graph G is a map $\omega : V \to \mathbb{R}_+$ such that $\omega(i) + \omega(j) \geq 1$ for every edge ij. Let $\tau^*(G) = \min \sum_i \omega(i)$, where the minimum ranges over all fractional covers of G. Prove that $\tau^*(G) = \tau_2(G)/2$.

EXERCISE 18.5. Let V be a finite set and let $f : 2^V \to \mathbb{R}$ be a setfunction. Extend f to a function $\widehat{f} : \mathbb{R}_+^V \to \mathbb{R}$ by $\widehat{f}(\mathbf{w}) = \sum_{i=1}^t \alpha_i f(S_i)$, where the α_i and S_i are defined as in (18.3). Prove that $\widehat{f}$ is convex if and only of f is submodular.

EXERCISE 18.6. Prove that for any framework, $\tau_{2k}(G, \mathbf{v}) = k\tau_2(G, \mathbf{v})$ and $\tau_{2k+1}(G, \mathbf{v}) = k\tau_2(G, \mathbf{v}) + \tau_1(G, \mathbf{v})$.

EXERCISE 18.7. For every representation $\mathbf{v}$ of G, the setfunction $\mathrm{rk}(\mathbb{2}_X)$ is submodular.

EXERCISE 18.8. Let i be a node of a framework $(G, \mathbf{v})$. Let us delete all edges incident with i, create a new node j, connect it to i, and represent it by a vector $\mathbf{v}_j$ that is in general position in the space $\mathrm{lin}(\mathbf{v}(N(i)))$. Let G' be the graph obtained. Prove that $\tau(G, \mathbf{v}) = \tau(G', \mathbf{v})$, and if $(G, \mathbf{v})$ is cover-critical, then so is $(G', \mathbf{v})$.

EXERCISE 18.9. For a framework $(G, \mathbf{v})$, split every node i into $\deg(i)$ nodes of degree 1, and represent each of these by the vector $\mathbf{v}_i$. Let $(G', \mathbf{v}')$ be the framework obtained. Prove that $\tau(G, \mathbf{v}) = \tau(G', \mathbf{v})$, and $(G, \mathbf{v})$ is cover-critical if and only if $(G', \mathbf{v}')$ is cover-critical.

EXERCISE 18.10. Prove that if a family of lines in projective d-space contains no three independent lines, then it is one of the configurations in Figure 18.5.

EXERCISE 18.11. Let G be a simple graph and $\mathbf{u} : V \to \mathbb{R}^d$, a full-dimensional vector labeling. We say that $\mathbf{u}$ is *independence preserving*, if every independent (stable) set of nodes is labeled by linearly independent vectors. Prove that $\mathbf{u}$ is independence preserving if and only if its Gale dual $\mathbf{v} : V \to \mathbb{R}^{n-d}$ satisfies $\tau(G, \mathbf{v}) = \tau(G)$.

EXERCISE 18.12. Let B be a skew symmetric matrix of dimension $p = 2n$. For each partition $P = \{\{i_1, j_1\}, \ldots, \{i_n, j_n\}\}$ of the set $\{1, \ldots, 2n\}$ into pairs, let π_P denote the permutation $2k-1 \mapsto i_k$, $2k \mapsto j_k$ of $\{1, \ldots, 2n\}$, and let $b_P = \operatorname{sgn}(\pi_P) b_{i_1 j_1} \cdots b_{i_n j_n}$ (where sgn denotes the sign of the permutation). Define the *Pfaffian* of B by Pfaff $B = \sum_P b_P$.

(a) Prove that b_P depends neither on the order in which the classes of the partition are listed nor on the order of the two elements of a class. (So Pfaff B is well defined.)

(b) Prove that $\det B = (\text{Pfaff } B)^2$.

(c) Let B be obtained from the adjacency matrix of G by replacing each pair of 1's corresponding to the same edge by x_e and $-x_e$, where the numbers x_e are algebraically independent transcendentals. Prove that G has a perfect matching if and only if $\det(B) \neq 0$.

CHAPTER 19

Combinatorics of Subspaces

In this chapter we have a look at several results from previous chapters, from a more general point of view. In fact, we do so in two stages.

First, we discuss some results about combinatorial properties of subspaces, which generalize results concerning stresses and rigidity, covering and matching in frameworks, and more. Collections of subspaces (points, lines, planes) in Euclidean, affine, or projective spaces have many interesting and important combinatorial properties, which are treated in detail in books about combinatorial geometry, and could themselves fill one or more monographs. There are deep studies of the numbers of cells of different dimensions they form, and of the topological and algebraic geometric properties of their union or its complement, just to name a few. Some of their properties are generalizing graph-theoretic results treated in this book, and we restrict our attention to results of this nature.

The second level of more abstract and more general results uses matroid theory. This theory has been the topic of several monographs, and it is difficult to fit it in a book whose aim is to establish more direct connections between graph theory and geometry. On the other hand, matroid theory provides deeper insight into several of the results presented in this book. In the last section of this chapter, a minimalist introduction to matroids is followed by a survey of how some of the key constructions and key results can be put in the framework of matroid theory.

19.1. Covering of hypergraphs and subspaces

Several questions about covering the edges of graphs extend to covering the edges of hypergraphs, and even to covering the members of a family of subspaces. [**Bollobás 1965**] proved the following generalization of Theorem 18.2(b):

THEOREM 19.1. *Let $A_1, \dots, A_n$ be p-element sets and $B_1, \dots, B_n$, q-element sets. Suppose that $A_i \cap B_i = 0$ for all i, but $A_i \cap B_j \neq 0$ for $i \neq j$. Then $n \leq \binom{p+q}{p}$.*

To obtain Theorem 18.2, we take $p = 2$ and $q = \tau(G) - 1$, let the sets A_i be the edges of G, and let B_i be a minimal node-cover of $G \setminus \{A_i\}$. Recall that we proved and used a generalization (Theorem 18.8(b)), where B_i was a subspace of dimension $\tau(G, \mathbf{u}) - 1$ covering all edges of the framework $(G, \mathbf{u}) \setminus \{A_i\}$. An analogous generalization of Theorem 19.1 can also be proved (see Exercise 19.5). One can even replace both families of subsets by families of subspaces [**Lovász 1979a**]:

THEOREM 19.2. *Let $A_1, \dots, A_n$ be p-dimensional subspaces and $B_1, \dots, B_n$, q-dimensional subspaces of $\mathbb{R}^d$. Suppose that $A_i \cap B_i = 0$ for all i, but $A_i \cap B_j \neq 0$ for $i \neq j$. Then $n \leq \binom{p+q}{p}$.*

PROOF. We may assume that $d = p+q$. Indeed, trivially $d \geq p+q$. If $d > p+q$, then the orthogonal projections A_i' and B_i' of A_i and B_i onto a generic $(p+q)$-dimensional space satisfy the same conditions as the original subspaces A_i and B_i, so it suffices to prove the lemma for these projected subspaces of a $(p+q)$-dimensional space.

As described in Appendix A.5.2, we can assign an antisymmetric r-form Ψ_A. to every linear subspace A so that $\Psi_A \wedge \Psi_B = 0$ if and only if the subspaces A and B have a nonzero intersection. In particular, $\Psi_{A_i} \wedge \Psi_{B_i} \neq 0$, but $\Psi_{A_i} \wedge \Psi_{B_j} = 0$ for $i \neq$.

We claim that the tensors $\Psi_{A_1}, \dots, \Psi_{A_n}$ are linearly independent. Suppose (by way of contradiction) that there is a linear dependence $\sum_i \alpha_i \Psi_{A_i} = 0$, and let j be any subscript with $\alpha_j \neq 0$. Then

$$0 = \Big(\sum_i \alpha_i \Psi_{A_i}\Big) \wedge \Psi_{B_j} = \sum_i \alpha_i (\Psi_{A_i} \wedge \Psi_{B_j}) = \alpha_j (\Psi_{A_j} \wedge \Psi_{B_j}) \neq 0,$$

a contradiction.

It follows that n is not larger than the dimension of the space of antisymmetric p-tensors in a $(p+q)$-dimensional space, which is $\binom{p+q}{p}$. □

REMARK 19.3. The proof shows that the condition that $A_i \cap B_j \neq 0$ for $i \neq j$ could be weakened: it would suffice to assume that $A_i \cap B_j \neq 0$ for $i < j$. Theorem 19.1 can be proved by a similar linear algebra argument, replacing exterior product of vectors by their symmetric product, or (essentially equivalently) by the product of variables.

These theorems can be applied to bound the size of critical families of sets. A *hypergraph* is a finite family of finite sets. The *node-cover number* $\tau(\mathcal{H})$ of a hypergraph $\mathcal{H}$ is the minimum cardinality of a set meeting every member of $\mathcal{H}$. The hypergraph is called *cover-critical*, if $\tau(\mathcal{H}') < \tau(\mathcal{H})$ for every $\mathcal{H}' \subset \mathcal{H}$.

For a family $\mathcal{F}$ of subspaces, we define its *linear covering number* $\tau_{\text{lin}}(\mathcal{F})$ as the minimum dimension of a linear subspace intersecting every subspace in $\mathcal{F}$ in a nonzero subspace. The family $\mathcal{F}$ is called *cover-critical*, if $\tau_{\text{lin}}(\mathcal{F}') < \tau_{\text{lin}}(\mathcal{F})$ for every $\mathcal{F}' \subset \mathcal{F}$. The first assertion of the next theorem was proved in [**Bollobás 1965**]; the second, in [**Lovász 1977**].

THEOREM 19.4.
(a) *If $\mathcal{H}$ is a cover-critical family of k-sets, then* $|\mathcal{H}| \leq \binom{\tau(\mathcal{H})+k-1}{k}$.

(b) *If $\mathcal{F}$ is a cover-critical family of k-dimensional subspaces, then* $|\mathcal{F}| \leq \binom{\tau_{\text{lin}}(\mathcal{F})+k-1}{k}$.

PROOF. We prove the second assertion; the proof of the first one is similar. Let $\mathcal{F} = \{A_1, \dots, A_n\}$. For each $1 \leq i \leq n$, the family $\mathcal{F} \setminus \{A_i\}$ can be covered by a subspace B_i of dimension $l-1$. This means that $A_j \cap B_i \neq 0$ for $j \neq i$. On the other hand, B_i cannot cover $\mathcal{F}$, and hence $A_i \cap B_i = 0$. By Lemma 19.2, it follows that $n \leq \binom{l+k-1}{k}$. □

19.2. Transcendence degree and transcendence rank

Notions like "general position" and "generic" have played an important role in several chapters of this book. In this section we put these notions in a more general context, leading to a couple of combinatorial results in the next section that, as we shall see, generalize several previously presented results.

We fix a real closed field $\mathbb{K} \subset \mathbb{R}$ (that is, the square root of every positive number in $\mathbb{K}$ is in $\mathbb{K}$, and every polynomial of odd degree with coefficients in $\mathbb{K}$ has a root in $\mathbb{K}$). Most of the time, you could think of this as the field of real algebraic numbers without losing any of the essential ideas (many of the arguments would even work over the rational field), but for a few arguments we need this generality. It is well known that every real number that is algebraic over $\mathbb{K}$ belongs to $\mathbb{K}$. We need, in addition, that $\mathbb{K}$ is not too large: we assume that $\mathbb{R}$ has infinite transcendence degree over $\mathbb{K}$ (so that we can find arbitrarily many real numbers that are transcendental over $\mathbb{K}$). For a set A of real numbers, we denote by $\mathbb{K}(A)$ the real closed field generated by $\mathbb{K}$ and the elements of A. For a finite set A of vectors and/or matrices, we define $\mathbb{K}(A) = K(\widehat{A})$, where $\widehat{A}$ is the set of all entries of all elements in A.

Let L be a linear subspace of $\mathbb{R}^d$. It is clear that if $L \neq 0$, then $\mathbb{K}(L) = \mathbb{R}$ is uninteresting. To measure how "transcendent" L is, we consider "simple" vectors generating L. Let $k = \dim(L)$. There is at least one k-tuple of coordinates such that the projection of L onto these coordinates is k-dimensional; let us call such a k-tuple of coordinates *basic*. Unless said otherwise, we fix a basic k-tuple, and (for notational convenience) reorder the coordinates so that this basic k-tuple consists of the first k coordinates. Let $\mathbb{R}^{d,k}$ denote the linear space of all vectors in $\mathbb{R}^d$ supported on the first k coordinates. The spaces $\mathbb{R}^{d,k}$ and $\mathbb{R}^k$ are isomorphic in a trivial way: the map $\Pi_k : \mathbb{R}^d \to \mathbb{R}^k$ deleting the last $d-k$ coordinates provides this isomorphism; but it will be useful to distinguish them.

There is a unique basis in L whose projection onto the first k coordinates is the standard basis in $\mathbb{R}^k$; we call this the *standard basis in* L (with respect to the basic set $\{1,\dots,k\}$). The matrix whose columns are the vectors in the standard basis can be written as $\binom{I_k}{X}$, where $X \in \mathbb{R}^{(d-k)\times k}$. Any other basis of L can be obtained as the columns of a $d\times k$ matrix $\binom{B}{XB}$ with some nonsingular $k\times k$ matrix B. We'll abuse terminology and say that a $d\times k$ matrix A is a basis of L if its columns form a basis of L.

We say that the subspace is *algebraic over* $\mathbb{K}$, if it has a basis consisting of vectors with all entries in $\mathbb{K}$. We are going to omit "over $\mathbb{K}$" if no confusion can arise.

LEMMA 19.5. *For every linear subspace $L \subseteq \mathbb{R}^d$, there is a unique smallest real closed field over which it is algebraic. This field is generated (as a real closed field) by the entries of any standard basis.*

PROOF. Let $\binom{I_k}{X}$ be a standard basis of L. Consider any other basis $A = \binom{B}{XB}$. Here B is a submatrix of A, and so the entries of B^{-1} belong to the field $\mathbb{K}(A)$. Hence

$$\mathbb{K}(A) \supseteq \mathbb{K}(B^{-1}A) = \mathbb{K}\begin{pmatrix} I_k \\ X \end{pmatrix} = \mathbb{K}(X).$$

This shows that $\mathbb{K}(X)$ is the smallest field generated by any basis of L. □

We denote the field in the lemma by $\mathbb{K}_L$. The proof above shows that L is algebraic over a field $\mathbb{K}$ if and only if the standard basis (with respect to one, or all, basic k-tuples of coordinates) has all of its entries in $\mathbb{K}$. Another consequence of Lemma 19.5 is that

$$\mathbb{K}_{L^\perp} = \mathbb{K}_L, \tag{19.1}$$

since the space $L^\perp$ has a standard basis $\binom{-X^\mathsf{T}}{I_{d-k}}$, and so $\mathbb{K}_{L^\perp} = \mathbb{K}(-X^\mathsf{T}) = \mathbb{K}(X) = \mathbb{K}_L$.

For a field $\mathbb{K}$ and a set of real numbers X, we define the *transcendence degree* $\deg_{\rm tr}(X|\mathbb{K})$ of X over $\mathbb{K}$ as the maximum number of elements of X that are algebraically independent over $\mathbb{K}$. In the case of the field denoted by $\mathbb{K}$, we often use the simplified notation $\deg_{\rm tr}(X)$, where "over $\mathbb{K}$" is understood. For a linear subspace $L \subseteq \mathbb{R}^d$, we define its *transcendence rank* as the transcendence degree of $\mathbb{K}_L$ over $\mathbb{K}$:

$$\theta(L) = \theta(L|\mathbb{K}) = \deg_{\rm tr}(\mathbb{K}_L) = \min\{\deg_{\rm tr}(A|\mathbb{K}) : \ A \text{ is a basis of } L\}. \tag{19.2}$$

We need some simple properties of the transcendence rank. From the proof of Lemma 19.5 we see that for any k-dimensional subspace $L \subseteq \mathbb{R}^d$, we have

$$\theta(L) \le k(d-k). \tag{19.3}$$

Equation (19.1) implies that

$$\theta(L^\perp) = \theta(L). \tag{19.4}$$

It is easy to see that

$$\theta(L_1+L_2) \le \theta(L_1)+\theta(L_2). \tag{19.5}$$

We have noted that "bad" bases of a subspace L can have arbitrary transcendental entries. But no basis can be "too transcendental".

LEMMA 19.6. *For every k-dimensional subspace $L \subseteq \mathbb{R}^d$, we have*

$$\max\{\deg_{\rm tr}(A) : \ A \text{ is a basis of } L\} = \theta(L)+k^2.$$

PROOF. Let $\binom{I_k}{X}$ be a standard basis of L, then $\deg_{\rm tr}(X) = \theta(L)$. Let $\binom{B}{XB}$ be any other basis, where B is a nonsingular $k\times k$ real matrix. Since the entries of $\binom{B}{XB}$ can be computed from the entries of $\binom{B}{X}$ algebraically, and vice versa, we have

$$\deg_{\rm tr}\begin{pmatrix} B \\ XB \end{pmatrix} = \deg_{\rm tr}\begin{pmatrix} B \\ X \end{pmatrix} = \deg_{\rm tr}(X)+\deg_{\rm tr}(B|\mathbb{K}(X)) = \theta(L)+\deg_{\rm tr}(B|\mathbb{K}(X)).$$

The last term is maximized when all entries of B are algebraically independent over $\mathbb{K}(X)$. □

19.2.1. Transcendental subspaces. A subspace is "generic", if it does not satisfy any special condition it does not absolutely have to... This is not a definition. One way of making it precise is the following.

We say that a set of vectors in $\mathbb{R}^d$ is *fully transcendental* (over $\mathbb{K}$) if all their entries are algebraically independent over $\mathbb{K}$. It is easy to see that a fully transcendental set of k vectors is linearly independent as long as $k \le d$.

A k-dimensional subspace $L \subseteq \mathbb{R}^d$ is *transcendental* (over $\mathbb{K}$), if it has a fully transcendental basis. By Lemma 19.6, this is equivalent to saying that

$$\theta(L) = k(d-k). \tag{19.6}$$

By (19.4), this implies that a subspace $L \subseteq \mathbb{R}^d$ is transcendental if and only if $L^\perp$ is.

Applying a nonsingular linear transformation with entries in $\mathbb{K}$, every algebraic subspace is mapped onto an algebraic subspace, and every transcendental subspace is mapped onto a transcendental subspace. If $\mathbb{K} \subseteq \mathbb{K}' \subseteq \mathbb{R}$ are fields, then any subspace algebraic over $\mathbb{K}$ is algebraic over $\mathbb{K}'$, and any subspace transcendental over $\mathbb{K}'$ is transcendental over $\mathbb{K}$. If J is a transcendental subspace of $\mathbb{R}^d$, then any $k = \dim(J)$ coordinates are basic, since the projection of a fully transcendental basis of J to $\mathbb{R}^k$ is a fully transcendental basis of $\mathbb{R}^k$.

Let $L \subseteq \mathbb{R}^d$ by any subspace of dimension k. We say that a subspace of L is *transcendental within* L, if it is the intersection of L with a subspace that is transcendental over $\mathbb{K}_L$. Note that 0 and L are both transcendental subspaces of L.

In the special case when $L = R^d$, to be transcendental within L is equivalent to being a transcendental subspace of $\mathbb{R}^d$. More generally, we can state the following fact, which is absolutely natural, but it takes some work to prove it (Exercise 19.2).

LEMMA 19.7. *A subspace $J \subseteq \mathbb{R}^{d,k}$ is transcendental within $\mathbb{R}^{d,k}$ if and only if its projection $\Pi_k J$ is a transcendental subspace of $\mathbb{R}^k$.* □

If L is any algebraic subspace of dimension k, then we can apply a linear transformation with entries in $\mathbb{K}$ that maps L onto $\mathbb{R}^{d,k}$, and this preserves transcendentality of subspaces of L. Sometimes we will need this linear transformation to be orthogonal. For this, we need to compute an orthonormal basis B in L, to be mapped onto the standard basis of $\mathbb{R}^{d,k}$; in general, this basis cannot be expressed rationally by the entries of a standard basis of L, but the entries of B are in $\mathbb{K}$ nevertheless (here we use that $\mathbb{K}$ is real closed). We will refer to this procedure as "normalization".

Using normalization, we can extend several simple previous observations about transcendental subspaces to subspaces that are transcendental within a subspace. For example, we have

$$\theta_{\mathbb{K}_J}(H) \le \dim(H)\big(\dim(J) - \dim(H)\big) \tag{19.7}$$

for any two subspaces $H \subseteq J$, with equality if and only if H is transcendental within J. As another consequence, we have seen that the orthogonal complement of a transcendental subspace of $\mathbb{R}^d$ is transcendental. This can be extended to transcendental subspaces of any subspace: *If H is transcendental within J, then so is $J \cap H^\perp$.*

LEMMA 19.8. *Let $H \subseteq J$ be subspaces of $\mathbb{R}^d$ with $\dim(H) = q$ and $\dim(J) = r$. Then*

$$\theta(J) - (r-q)(d-r) \le \theta(H) \le \theta(J) + (r-q)q$$

If equality holds in the first inequality, then $J \cap H^\perp$ is transcendental within $H^\perp$. If equality holds in the second, then H is transcendental within J.

PROOF. Since H and $J \cap H^\perp$ generate J, we have

$$\theta(J) \le \theta(H) + \theta_{\mathbb{K}_H}(J \cap H^\perp).$$

Here by (19.1) and (19.7),

$$\theta_{\mathbb{K}_H}(J \cap H^\perp) \le \dim(J \cap H^\perp)\big(\dim(H^\perp) - \dim(J \cap H^\perp)\big) = (k-r)(d-k). \tag{19.8}$$

This proves the first inequality. If equality holds, then we have equality in (19.8), which means that $J \cap H^\perp$ is transcendental within $H^\perp$.

The second inequality follows from the first by replacing H and J by $J^\perp$ and $H^\perp$, using the relations $\theta(J^\perp) = \theta(J)$ and $\theta(H^\perp) = \theta(H)$. In the case of equality we know that $H^\perp \cap J$ is transcendental within J. Then so is its orthogonal complement within J, which is just H. □

19.2.2. The intersection of algebraic and transcendental subspaces. The following lemma summarizes a number of useful relations between an algebraic subspace and a transcendental subspace.

LEMMA 19.9. *Let L be a algebraic subspace and J, a transcendental subspace of $\mathbb{R}^d$. Let $k = \dim(L)$ and $r = \dim(J)$. Then*

(a) *either $L+J = \mathbb{R}^d$ or $L \cap J = \emptyset$;*

(b) *if $L+J \neq \mathbb{R}^d$, then $\theta(L+J) = r(d-k-r)$;*

(c) *if $L \cap J \neq 0$, then $\theta(L \cap J) = (k+r-d)(d-r)$;*

(d) *J is transcendental within $L+J$, and $L \cap J$ is transcendental within L.*

PROOF. We may assume by normalization that $L = \mathbb{R}^{d,k}$.

(a) First, assume that $k+r = d$. Let B be a fully transcendental basis of J. The submatrix of B formed by the last r rows is nonsingular, and hence no vector in J is supported on the first $k = d-r$ rows. Thus $L \cap J = R^{d,k} \cap J = 0$. This also implies that $L+J = \mathbb{R}^d$.

Next, if $k+r < d$, then we extend L arbitrarily to a $(d-r)$-dimensional algebraic subspace L', and by the above, $L' \cap J = 0$ and so $L \cap J = 0$. It follows by a similar argument that if $k+r > d$, then $L+J = \mathbb{R}^d$.

(b) We have $\dim(L+J) = k+r < d$ by (a). We may assume that the last r coordinates are basic in J, so J has a basis of the form $\binom{B}{I_r}$, where $B \in \mathbb{R}^{(d-r)\times r}$. Let B' be the $(d-r-k) \times r$ matrix obtained from B by deleting its first k rows. Then the columns of the matrix

$$A = \begin{pmatrix} I_k & 0 \\ 0 & B' \\ 0 & I_r \end{pmatrix}$$

form a basis of $L+J$, showing that $\theta(L+J) \le \deg_{\rm tr}(A) = \deg_{\rm tr}(B') \le (d-k-r)r$. On the other hand, J is transcendental in $\mathbb{R}^d$, so $\theta(J) = r(d-r)$, Lemma 19.8 implies that $\theta(L+J) \ge \theta(J) - kr = r(d-k-r)$.

(c) This follows by applying (b) to the algebraic subspace $L^\perp$ and transcendental subspace $J^\perp$.

(d) Both assertions follow from Lemma 19.8, since the pairs $J \subseteq L+J$ and $L \cap J \subseteq L$ satisfy the inequality in the lemma with equality. □

Here is a heuristic idea; the rest of this section will be devoted to make it precise. Let us choose a "generic" subspace J and a subspace $H \subseteq J$. We expect that $J \cap H^\perp$ is a "generic" subspace in $H^\perp$. Otherwise, making a "generic" choice U among the subspaces of $H^\perp$ of the same dimension as $J \cap H^\perp$, the subspace $H+U$ would be even "more generic" than J.

This conclusion is, however, false in general:

EXAMPLE 19.10. Let a, b and c be algebraically independent over $\mathbb{K}$, and consider the 2-dimensional subspace $J \subseteq \mathbb{R}^3$ defined by the equation $ax+by+cz = 0$. Then $J^\perp$ is generated by the fully transcendental vector (a,b,c), and hence J is a transcendental subspace of $\mathbb{R}^3$. Let H be the 1-dimensional subspace generated by the vector $(1/a, 1/b, -2/c)^\mathsf{T}$. Clearly $H \subset J$, and $\mathbb{K}_H = \mathbb{K}(a/c, b/c) = \mathbb{K}_J$. A vector spanning the subspace $J \cap H^\perp$ can be computed in $\mathbb{K}_H = \mathbb{K}_J$, and so it is not a transcendental subspace of $H^\perp$. ♦

Lemma 19.8 gives a sufficient condition on J and H for the above conclusion to hold. A simple special case when this applies is if $H = J \cap L$ for some algebraic subspace L. More generally,

LEMMA 19.11. *Let J be a transcendental subspace of $\mathbb{R}^d$, and let $H = (L_1 \cap J) \oplus \cdots \oplus (L_m \cap J)$, where $L_1, \dots, L_m$ are algebraic. Then $J \cap H^\perp$ is transcendental within $H^\perp$.*

Note the "$\oplus$" operation sign, which implies that the subspaces $L_1 \cap J, \dots, L_m \cap J$ are linearly independent.

PROOF. We may assume that the intersections $L_i \cap J$ are nonzero subspaces. Let $r = \dim(J)$ and $k_i = \dim(L_i)$, then $\dim(L_i \cap J) = k_i + r - d$ by Lemma 19.9(a), and so $\dim(H) = \sum_{i=1}^m (k_i + r - d)$. By Lemma 19.9(d), the subspace $L_i \cap J$ is transcendental within L_i, and hence $\theta(L_i \cap J) = (k_i + r - d)(d - r)$. Thus

$$\begin{aligned}\theta(H) &\le \sum_{i=1}^m \theta(L_i \cap J) = \sum_{i=1}^m (k_i + r - d)(d - r) = \dim(H)(d - r)\\ &= \theta(J) - (r - \dim(H))(d - r).\end{aligned}$$

By Lemma 19.8, this implies that $J \cap H^\perp$ is transcendental within $H^\perp$. □

19.3. Generic points

19.3.1. The rank of generic points. After these rather technical lemmas, we come to the first main result in this topic. We say that a point $\mathbf{a}$ is a *generic point* of a rational subspace L, if $\text{lin}(\mathbf{a})$ is a transcendental subspace of L over $\mathcal{K}$. The definition guarantees that a generic point of a rational subspace L does not have any property expressible by algebraic relations (with coefficients in $\mathbb{K}$) between the entries of the point, unless all points of L have this property. For example, if $\mathbf{a}$ is a generic point of a rational subspace L, and $\mathbf{a} \in J$ for another rational subspace J, then $L \subseteq J$.

Informally, if we introduce in some arguments below a "generic point" $\mathbf{a}$, we mean that it is fully transcendental over the field generated by all coordinates of all vectors introduced in the argument previously (including appropriate bases of all the subspaces we considered).

LEMMA 19.12 (Generic Point Lemma). *Let $\mathcal{F}$ be a family of subspaces of $\mathbb{R}^d$. For every $A \in \mathcal{F}$, select a generic point $\mathbf{x}_A \in A$. Then*

$$\text{rk}\{\mathbf{x}_A : A \in \mathcal{F}\} = \min_{\mathcal{G} \subseteq \mathcal{F}} \big(|\mathcal{F} \setminus \mathcal{G}| + \text{rk}(\mathcal{G})\big).$$

By the remark above, we should select the points $\mathbf{x}_A$ one by one so that they are generic taking into account the previously selected points. This property does not depend on the order in which these points are selected.

PROOF. For every $\mathcal{G} \subseteq \mathcal{F}$, we have $\mathrm{rk}\{\mathbf{x}_A : A \in \mathcal{G}\} \le |\mathcal{G}|$ and $\mathrm{rk}\{\mathbf{x}_A : A \in \mathcal{G}\} \le \mathrm{rk}(\mathcal{G})$, whence

$$\begin{aligned}\mathrm{rk}\{\mathbf{x}_A : A \in \mathcal{F}\} &\le \mathrm{rk}\{\mathbf{x}_A : A \in \mathcal{F}\setminus\mathcal{G}\} + \mathrm{rk}\{\mathbf{x}_A : A \in \mathcal{G}\} \\ &\le |\mathcal{F}\setminus\mathcal{G}| + \mathrm{rk}(\mathcal{G}).\end{aligned} \tag{19.9}$$

(This inequality holds, in fact, for every choice of representatives $\mathbf{x}_A \in A$, not only for a generic choice.)

To prove that equality holds for a generic choice of the representatives and for a suitable $\mathcal{G}$, let us consider any subspace $A \in \mathcal{F}$, and its representative $\mathbf{x}_A$. If $\mathbf{x}_A \notin \mathrm{lin}\{\mathbf{x}_B : B \in \mathcal{F}\setminus\{A\}\}$, then we can delete A from $\mathcal{F}$ and apply induction:

$$\mathrm{rk}\{\mathbf{x}_B : B \in \mathcal{F}\} = 1 + \mathrm{rk}\{\mathbf{x}_B : B \in \mathcal{F}\setminus\{A\}\} = 1 + |\mathcal{F}\setminus\{A\}\setminus\mathcal{G}| + \mathrm{rk}\,\mathcal{G}$$

for a suitable $\mathcal{G} \subseteq \mathcal{F}\setminus\{A\}$. This shows that this set $\mathcal{G}$ gives equality in (19.9) as well.

So we may assume that $\mathbf{x}_A \in \mathrm{lin}\{\mathbf{x}_B : B \in \mathcal{F}\setminus\{A\}\}$. Using the assumption that $\mathbf{x}_A$ is a generic point of A, it follows that $A \subseteq \mathrm{lin}\{\mathbf{x}_B : B \in \mathcal{F}\setminus\{A\}\}$, and so, even more, $A \subseteq \mathrm{lin}\{\mathbf{x}_B : B \in \mathcal{F}\}$. This holds for every subspace $A \in \mathcal{F}$, and hence $\cup\mathcal{F} \subseteq \mathrm{lin}\{\mathbf{x}_B : B \in \mathcal{F}\}$. Clearly equality must hold, whence

$$\mathrm{rk}\{\mathbf{x}_A : A \in \mathcal{F}\} = \mathrm{rk}(\mathcal{F}),$$

showing that $\mathcal{G} = \mathcal{F}$ gives equality in (19.9). □

The following version of the Generic Point Lemma does not involve generic points.

COROLLARY 19.13. *Let $A_1, \dots, A_n$ be finite subsets of a linear space $\mathbb{R}^d$. Then*

$$\max_{\mathbf{a}_i \in A_i} \mathrm{rk}\{\mathbf{a}_1, \dots, \mathbf{a}_n\} = \min_{S \subseteq [n]} \left(\mathrm{rk}\Big(\bigcup_{i \in S} A_i\Big) + n - |S| \right).$$

We could replace the expression on the left by

$$\max\{|X| : X \subseteq \mathbb{R}^d,\ X \text{ linearly independent},\ |X \cap A_i| \le 1\}.$$

PROOF. The max $\le$ min direction is clear by the same simple argument as above. To prove the reverse inequality, let $L_i = \mathrm{lin}(A_i)$, and let $\mathbf{x}_i$ be a generic point in L_i. Then

$$\mathrm{rk}\{\mathbf{x}_1, \dots, \mathbf{x}_n\} = \min_{S \subseteq [n]} \left(\mathrm{rk}\Big(\bigcup_{i \in S} A_i\Big) + n - |S| \right)$$

by the Generic Point Lemma 19.12. We want to replace the points $\mathbf{x}_i$ by appropriate points $\mathbf{a}_i \in A_i$ so that the dimension of the set does not decrease. This can be done one-by-one: if $\mathbf{x}_1$ is linearly dependent on $\mathbf{x}_2, \dots, \mathbf{x}_n$, then we can replace it by any element of A_1. If it is linearly independent from $\mathbf{x}_2, \dots, \mathbf{x}_n$, then $L_1 \not\subseteq \mathrm{lin}\{\mathbf{x}_2, \dots, \mathbf{x}_n\}$, and hence A_1 has an element $\mathbf{a}_1 \notin \mathrm{lin}\{\mathbf{x}_2, \dots, \mathbf{x}_n\}$, and we can replace $\mathbf{x}_1$ by $\mathbf{a}_1$. Repeating this procedure for every $\mathbf{x}_i$, the corollary follows. □

We can also formulate a probabilistic version of the theorem.

COROLLARY 19.14. *Let $\mathcal{F}$ be a family of nonnull subspaces of $\mathbb{R}^d$. For every $A \in \mathcal{F}$, select a random point $\mathbf{x}_A \in A$ uniformly from the unit sphere in A. Then almost surely,*

$$\operatorname{rk}\{\mathbf{x}_A : A \in \mathcal{F}\} = \min_{\mathcal{G} \subseteq \mathcal{F}} \big(|\mathcal{F} \setminus \mathcal{G}| + \operatorname{rk} \mathcal{G}\big).$$

We could replace the uniform distribution on the sphere by any probability distribution on A that is not concentrated on a finite number of hyperplanes in A.

19.3.2. Applications of generic points. As an application of the Generic Point Lemma, let us prove the Matroid Sum Theorem, at least in the linear case.

THEOREM 19.15. *Let us consider k vector labelings $\mathbf{v}^j : V \to \mathbb{R}^{d_j}$ $(j = 1, \dots, k)$ of the same set $V = \{1, \dots, n\}$. Let $Y_i \subseteq V$ $(i = 1, \dots, k)$ range over sets such that $\mathbf{v}^i(Y_i)$ is a linearly independent set of vectors. Then*

$$\max |Y_1 \cup \dots \cup Y_k| = \min_{X \subseteq V} |V \setminus X| + \sum_{i=1}^{k} \operatorname{rk}\big(\mathbf{v}^i(X)\big).$$

Clearly we could restrict the sets Y_i to be disjoint, or, going in the other direction, we could require that Y_i is a basis for $\mathbf{v}^i$.

PROOF. The max $\le$ min direction is easy to check. To prove the converse, consider the set of vectors

$$A_j = \left\{ \begin{pmatrix} \mathbf{v}_j^1 \\ 0 \\ \vdots \\ 0 \end{pmatrix}, \begin{pmatrix} 0 \\ \mathbf{v}_j^2 \\ \vdots \\ 0 \end{pmatrix}, \begin{pmatrix} 0 \\ 0 \\ \vdots \\ \mathbf{v}_j^k \end{pmatrix} \right\} \qquad (j = 1, \dots, k) \tag{19.10}$$

in $\mathbb{R}^{d_1} \oplus \dots \oplus \mathbb{R}^{d_k}$.

Let $Y_1, \dots, Y_k \subseteq V$ be disjoint sets such that $\mathbf{v}^i(Y_i)$ is a linearly independent set. For each $j \in Y_i$, choose $\mathbf{a}_j \in A_j$ so that the i-th block of $\mathbf{a}_j$ is nonzero. For the indices in $V \setminus (Y_1 \cup \dots \cup Y_k)$, choose $\mathbf{a}_j \in A_j$ arbitrarily. Then the set $\mathbf{a}(Y_1 \cup \dots \cup Y_k)$ is linearly independent, and so

$$\operatorname{rk}(\mathbf{a}(V)) \ge |Y_1 \cup \dots \cup Y_k|.$$

It is easy to reverse this argument, and show that for every choice of vectors $\mathbf{a}_j \in A_j$ we can find disjoint sets $Y_i \subseteq V$ such that $\mathbf{v}^i(Y_i)$ is linearly independent and

$$\operatorname{rk}(\mathbf{a}(V)) = |Y_1 \cup \dots \cup Y_k|.$$

So

$$\max_{\mathbf{a}} \operatorname{rk}(\mathbf{a}(V)) = \max_{Y_1, \dots, Y_k} |Y_1 \cup \dots \cup Y_k|. \tag{19.11}$$

To find the maximum on the left, we use Corollary 19.13, and get the theorem. □

As a special case, we describe the answer to the following question: *Given a finite set of vectors $S \subseteq \mathbb{R}^d$, when can we partition it into k linearly independent subsets?* A trivial necessary condition is that $|S| \le dk$. This condition, when stipulated for every subset of S, is also sufficient.

Let $S = \{\mathbf{u}_1, \dots, \mathbf{u}_n\}$, and invoke Theorem 19.15 with $\mathbf{v}^1 = \dots = \mathbf{v}^k = \mathbf{u} : [n] \to \mathbb{R}^d$. The set of vectors S can be partitioned into k linearly independent subsets if and only if the maximum on the left side of the min-max formula in

Theorem 19.15 is n; this means that the right side must be at least n for every $X \subseteq S$: $k\,\mathrm{rk}(X) + |S \setminus X| \ge n$. Rearranging, we get the following corollary:

COROLLARY 19.16. *A finite set $S \subseteq \mathbb{R}^d$ can be partitioned into k linearly independent subsets if and only if*

$$k\,\mathrm{rk}(X) \ge |X|$$

for every $X \subseteq S$.

From this Corollary, it is not hard to derive the theorems of Tutte and Nash-Williams on covering a graph by forests and on disjoint spanning trees (Theorem 15.16; see Exercise 19.9).

19.3.3. Bipartite matching revisited. As an application of Theorem 19.15, let us give a new proof of the Marriage Theorem for frameworks (Theorem 18.14), or, more precisely, in the following form (which is just the Matroid Intersection Theorem [**Edmonds 1970**], for representable matroids; see Section 19.5).

THEOREM 19.17. *Let $\mathbf{u} : V \to \mathbb{R}^d$ and $\mathbf{v} : V \to \mathbb{R}^e$ be two vector labelings of the same finite set. Then*

$$\max\{|X| : X \subseteq V,\ \mathbf{u}(X) \textit{ and } \mathbf{v}(X) \textit{ linearly independent}\}$$
$$= \min\{\mathrm{rk}(\mathbf{u}(Y)) + \mathrm{rk}(\mathbf{v}(V \setminus Y)) :\ Y \subseteq V\}.$$

To see the connection with the Marriage Theorem, first note that Theorem 19.17 is a special case of Theorem 18.14, when the bipartite framework consists of disjoint edges. Conversely, starting with an arbitrary bipartite framework, we can delete isolated nodes and split each node into nodes of degree 1 with the same vector label. This does not change the values of τ and ν, and so it reduces the general case to the case when each node has degree 1.

PROOF. With the notation of the theorem, we may assume that $\mathbf{u}(V)$ spans $\mathbb{R}^d$ and $\mathbf{v}(V)$ spans $\mathbb{R}^e$. For a set X to be considered on the left side, $\mathbf{u}(X)$ must be linearly independent, so X extends to a basis for $\mathbf{u}$, and similarly, X extends to a basis Z for $\mathbf{v}$. So the left side can be formulated as follows:

$$\max\{|Y \cap Z| :\ \mathbf{u}(Y) \text{ is a basis of } \mathbb{R}^d, \mathbf{v}(Z) \text{ is a basis of } \mathbb{R}^e\}.$$

Let $\mathbf{w} :\ V \to \mathbb{R}^{n-e}$ be a Gale dual of $\mathbf{v}$, then Z is a basis of $\mathbf{v}$ if and only if $Z' = V \setminus Z$ is a basis of $\mathbf{w}$. Since

$$|Y \cap Z| = |Y \cup Z'| - |Z'| = |Y \cup Z'| - (n - e),$$

finding the maximum of $|Y \cap Z|$ is equivalent to finding the maximum of $|Y \cup Z'|$, where Y is a basis for $\mathbf{u}$ and Z' is a basis for $\mathbf{w}$. But this is answered by the special case $k = 2$ of Theorem 19.15:

$$\max |Y \cup Z'| = \min_{X \subseteq V} \Big\{ |V \setminus X| + \mathrm{rk}\big(\mathbf{u}(X)\big) + \mathrm{rk}\big(\mathbf{w}(X)\big) \Big\}$$
$$= \min_{X \subseteq V} \Big\{ n - |X| + \mathrm{rk}\big(\mathbf{u}(X)\big) + \mathrm{rk}(\mathbf{v}(V \setminus X)) + |X| - e \Big\}.$$

and so

$$\max |Y \cap Z| = \max |Y \cup Z'| - (n - e) = \min \mathrm{rk}\big(\mathbf{u}(X)\big) + \mathrm{rk}(\mathbf{v}(V \setminus X))$$

as claimed. $\square$

19.4. Generic hyperplanes

19.4.1. The rank of generic intersections. The following theorem was proved in [**Lovász 1977**] (with a somewhat sketchy proof; see also Section 19.5 below for its matroid-theoretical predecessors).

LEMMA 19.18 (Generic Hyperplane Lemma). *Let $\mathcal{F}$ be a family of nonzero subspaces in $\mathbb{R}^d$, and let H be a linear hyperplane that is transcendental over the field generated by the fields $\mathbb{K}_L$ ($L \in \mathcal{F}$). Let $\mathcal{F}' = \{H \cap A : A \in \mathcal{F}\}$. Then*

$$\mathrm{rk}(\mathcal{F}') = \min\{\mathrm{rk}(\mathcal{G}_1)+\dots+\mathrm{rk}(\mathcal{G}_k)-k\},$$

where $\{\mathcal{G}_1,\dots,\mathcal{G}_k\}$ ranges over all partitions of the set $\mathcal{F}$.

PROOF. Define $A' = H \cap A$ for subspaces $A \subseteq \mathbb{R}^d$, and let $\mathcal{G}' = \{A' : A \in \mathcal{G}\}$ for $\mathcal{G} \subseteq \mathcal{F}$. Note that $\mathrm{lin}(\mathcal{G}') \subseteq \mathrm{lin}(\mathcal{G})'$ (but equality does not always hold). Hence $\mathrm{rk}(\mathcal{G}') \le \mathrm{rk}(\mathcal{G})-1$ if $\mathcal{G}' \ne \emptyset$ (here we use already that H is transcendental), and so for any partition $\{\mathcal{G}_1,\dots,\mathcal{G}_k\}$ of $\mathcal{F}$,

$$\mathrm{rk}(\mathcal{F}') \le \sum_{i=1}^{k} \mathrm{rk}(\mathcal{G}_i') \le \sum_{i=1}^{k} \big(\mathrm{rk}(\mathcal{G}_i)-1\big).$$

We want to prove that equality holds here for a suitable partition of $\mathcal{F}$.

Let $Q = \mathrm{lin}(\mathcal{F}')$. Clearly $Q \subseteq H$. For every $A \in \mathcal{F}$, we have $\dim(A') = \dim(A)-1$ (since H is transcendental). Since $A' \subseteq Q$ (by the definition of Q), but $A \not\subseteq Q$ (as $A \not\subseteq H$), this implies that $Q \cap A = A'$. Hence $\dim(Q \cup A) = \dim(Q)+\dim(A)-\dim(A') = \dim(Q)+1$.

Let $N_1,\dots,N_k$ be all the different subspaces of the from $Q+A$ ($A \in \mathcal{F}$). It follows from the above that $\dim(N_i) = \dim(Q)+1$ and $N_i \cap N_j = Q$ for $i \ne j$. Let $\mathcal{G}_i = \{A \in \mathcal{F} : Q+A = N_i\}$. Clearly $\mathcal{G}_1,\dots,\mathcal{G}_k$ form a partition of $\mathcal{F}$. Set $M_i = \mathrm{lin}(\mathcal{G}_i)$ and $L_i = \mathrm{lin}(\mathcal{G}_i')$. So $L_i \subseteq M_i' \subset M_i \subseteq N_i$ and $L_1+\dots+L_k = Q$ (Figure 19.1).

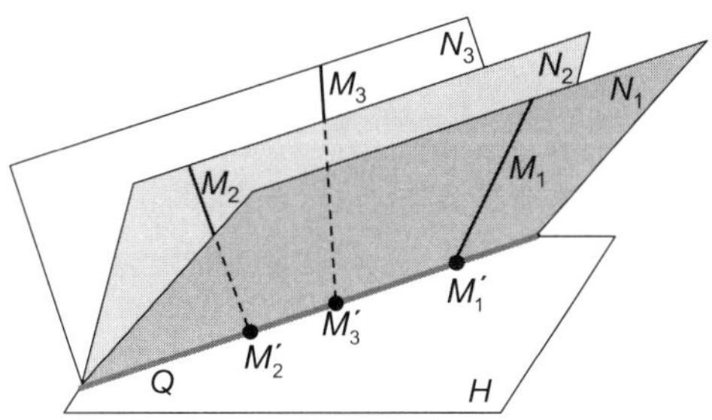

FIGURE 19.1. Intersecting a family of subspaces $\{M_1, M_2, M_3\}$ by a generic hyperplane H (the picture uses the representation in projective space). Here $k = 3$, and each $\mathcal{G}_i$ consists of a single 2-dimensional subspace. If the (1-dimensional) subspaces M_1', M_2' and M_3' are linearly dependent, then projecting from M_1', the image of H will go through the common image of M_2' and M_3', so H cannot be generic.

The key step in the proof is the following claim.

Claim 1. *The subspaces M_i' are linearly independent.*

Suppose this is not the case, and let $M'_1,\dots,M'_r$ be a minimal family of these subspaces that are not linearly independent, so there are nonzero vectors $\mathbf{x}_i \in M'_i$ such that $\mathbf{x}_1+\dots+\mathbf{x}_r = 0$. Let $X = M'_1+\dots+M'_{r-2}$, then $\mathbf{x}_{r-1}/X \neq 0$ (else, $M'_1\dots,M'_{r-1}$ would be dependent), and $\mathbf{x}_{r-1}/X+\mathbf{x}_r/X = 0$. Hence $M_{r-1}/X \cap M_r/X \neq 0$.

On the other hand, $(M_{r-1}+X)\cap(M_r+X) \subseteq N_{r-1}\cap N_r = Q$, and hence $M_{r-1}/X\cap M_r/X = Q/X \subseteq H/X$. Notice that $M'_1,\dots,M'_{r-2}$ are linearly independent by the minimality of r, and so $X = M'_1\oplus\dots\oplus M'_{r-2}$, where every M'_i is the intersection of a rational subspace G'_i with H. So Lemma 19.11 can be applied, and we obtain that $H/X = H\cap X^\perp$ is transcendental within $X^\perp$. It is easy to see that the subspaces M_{r-1}/X and M_r/X are algebraic within $X^\perp$, and so H/X cannot contain $M_{r-1}/X\cap M_r/X \neq 0$, a contradiction.

We can strengthen this Claim a bit:

Claim 2. *The subspaces $M_1, M'_2,\dots,M'_k$ are linearly independent.*

Indeed, if $\mathbf{x}_1 \in M_1$, $\mathbf{x}_i \in M'_2$ for $i = 2,\dots,k$, and $\mathbf{x}_1+\dots+\mathbf{x}_k = 0$, then $\mathbf{x}_1 \in H$ (since all the other $\mathbf{x}_i$ belong to H), and so $\mathbf{x}_i \in M'_i$ for all i. By the Claim, this implies that every $\mathbf{x}_i = 0$.

Now it is easy to complete the proof. Claim 1 implies that $L_1,\dots,L_k$ are linearly independent:

$$\dim(L_1)+\dots+\dim(L_k) = \dim(L_1+\dots+L_k) = \dim(Q). \tag{19.12}$$

Claim 2 implies that

$$\dim(M_1)+\dim(L_2)+\dots+\dim(L_k) = \dim(M_1+Q) = \dim(N_1) = \dim(Q)+1.$$

It follows that

$$\dim(L_1) = \dim(M_1)-1. \tag{19.13}$$

Applying (19.13) to the other subspaces M_i as well, and using (19.12), we get

$$\mathrm{rk}(\mathcal{F}') = \dim(Q) = \sum_{i=1}^k \dim(L_i) = \sum_{i=1}^k (\dim(M_i)-1) = \sum_{i=1}^k (\mathrm{rk}(\mathcal{G}_i)-1). \qquad \square$$

19.4.2. Generic hyperplanes and rigidity. As an application of the Generic Hyperplane Lemma, we give a description of generically rigid graphs and new proofs of Theorem 15.11 (the equivalence of (a) and (b)) and Theorem 15.14.

Consider a framework $(G,\mathbf{u})$ in the plane. As shown in Section 15.2, the framework is stress-free if and only if the matrices $U\mathbf{b}_e\mathbf{b}_e^\mathsf{T}$ are linearly independent, where U is the $2\times V$ matrix whose columns are the vectors $\mathbf{u}_i$, and $\mathbf{b}_e = \mathbf{e}_j-\mathbf{e}_i$ for $e = ij \in E$. These matrices have two properties: first, they are of the form $\mathbf{x}\mathbf{b}^\mathsf{T}$, where $\mathbf{x}_e \in \mathbb{R}^2$; second, they satisfy the equation $(RU)\cdot X = 0$, where R is the counterclockwise rotation by $90°$ in the plane. This equality is easy to verify by direct computation, or through the argument

$$(RU)\cdot(U\mathbf{b}_e\mathbf{b}_e^\mathsf{T}) = \mathrm{tr}(RU\mathbf{b}_e^\mathsf{T}\mathbf{b}_e U^\mathsf{T}) = R\cdot(U\mathbf{b}_e^\mathsf{T}\mathbf{b}_e U^\mathsf{T}) = 0,$$

since R is a skew symmetric 2×2 matrix, while $U\mathbf{b}_e^\mathsf{T}\mathbf{b}_e U^\mathsf{T}$ is a symmetric 2×2 matrix.

So for every edge e we have a 2-dimensional linear subspace L_e of $\mathbb{R}^{2\times V}$, consisting of the vectors $\mathbf{x}\mathbf{b}_e^\mathsf{T}$. Furthermore, we have a hyperplane H defined by

$(RU)\cdot X = 0$. The matrix $U\mathbf{b}_e\mathbf{b}_e^{\mathsf{T}}$ is a (typically nonzero) vector in the intersection of this line with this hyperplane.

Now assume that U is generic; then the hyperplane is generic (with respect to the subspaces L_e) $U\mathbf{b}_e\mathbf{b}_e^{\mathsf{T}} \neq 0$, and the intersections $H\cap L_e$ are 1-dimensional, i.e., they consist of scalar multiples of $U\mathbf{b}_e\mathbf{b}_e^{\mathsf{T}}$. The Generic Hyperplane Lemma 19.18 can be applied, and gives that

$$\operatorname{rk}\{U\mathbf{b}_e\mathbf{b}_e^{\mathsf{T}}:\ e\in E\} = \min_{E_1,\dots,E_k}\sum_{i=1}^{k}(\operatorname{rk}\{L_e:\ e\in E_i\}-1), \tag{19.14}$$

where $\{E_1,\dots,E_k\}$ ranges over all partitions of E.

It is not hard to figure out the rank of $\{L_e:\ e\in E_i\}$. If the edges in E_i form a connected graph G_i, then the vectors $\mathbf{b}_e$ ($e\in E_i$) generate all vectors supported on $V_i = V(G_i)$ whose entries sum to 0. The dimension of this space is $|V_i|-1$, and so $\operatorname{rk}(\{L_e:\ e\in E_i\}) = 2(|V_i|-1)$. If G_i is disconnected, then we have to sum this over all components, but in this case we improve the value on the right side of (19.14) if we split E_i into its connected components. So we get that

$$\operatorname{rk}\{U\mathbf{b}_e\mathbf{b}_e^{\mathsf{T}}:\ e\in E\} = \min_{E_1,\dots,E_k}\sum_{i=1}^{k}(2|V_i|-3). \tag{19.15}$$

To get a rigid graph, this value must be $2n-3$ (the dimension of Mat in this case), which is just the partition condition in Theorem 15.14. To get a stress-free graph, we need that this dimension is at least m, in other words, $\sum_i(2|V_i|-3)\geq m$ for every partition $\{E_1,\dots,E_k\}$ of E. We can write this condition as

$$\sum_{i=1}^{k}(2|V_i|-3-|E_i|)\geq 0. \tag{19.16}$$

If Laman's condition holds, then all terms here are nonnegative, and so (19.16) is valid, and the graph is stress-free. Together with the trivial implication (a)$\Rightarrow(b)$, this proves the equivalence of conditions (a) and (b) in Theorem 15.11. (This approach does not yield the equivalence of Henneberg's condition 15.11(c).)

19.5. Matroids

Matroids are common generalizations of several "independence" notions (linear independence, algebraic independence, circuit-freeness in a graph). This concept was introduced by [**Whitney 1935**]; independently, a couple of years later [**van der Waerden 1937**] observed that linear and algebraic independence satisfy the same basic properties. Here we recall some very basic facts only, which are necessary for describing the matroid theoretic aspects of our discussions in this book. There are several excellent monographs on the subject [**Recski 1989**], [**Oxley 1992**], covering much more. In particular, the reader is referred to these monographs or to any of the numerous survey articles for the proofs of the results quoted below.

19.5.1. Basics. A matroid is a pair $(E,\mathcal{M})$, where E is a finite set and $\mathcal{M}\subseteq 2^E$ is a family of its subsets called *independent sets*, satisfying the following axioms:

(M1) $\emptyset\in\mathcal{M}$;

(M2) If $X\in\mathcal{M}$ and $Y\subseteq X$, then $Y\in\mathcal{M}$;

(M3) If $X,Y \in \mathcal{M}$ and $|X| > |Y|$, then there is an element $x \in X \setminus Y$ such that $Y \cup \{x\} \in \mathcal{M}$.

These axioms impose a lot of structure on the matroid. It follows that all maximal (by inclusion) independent subsets of a set $S \subseteq E$ have the same cardinality, which we call the *rank* of S, and denote by $r(S)$. Maximal independent subsets of E are called *bases*; they all have the same cardinality $r(E)$, which we call the *rank of the matroid*, and denote it simply by r.

Maximal subsets of S with a given rank are called *flats*. Flats of rank $r-1$ are called *hyperplanes*. Minimal dependent sets are called *circuits*. There are many important properties of the rank function, of flats and of circuits; we need some basic properties of the rank function only.

It will be useful to consider a larger class of setfunctions. Let $f : 2^E \to \mathbb{R}$ be a setfunction. We say that f is *monotone*, if $Y \subseteq X \subseteq E$ implies that $f(Y) \le f(X)$; it is called *submodular*, if

$$f(X \cap Y) + f(X \cup Y) \le f(X) + f(Y)$$

for every two sets $X,Y \subseteq E$. In this brief section we need integer-valued setfunctions only, but there are interesting examples of submodular setfunctions that are not integer-valued (see Exercises 19.11-19.13). An integer valued, monotone, submodular setfunction f with $f(\emptyset) = 0$ is called a *polymatroid*. Restricting a polymatroid to the set of elements of rank at most 1 we get a matroid.

The rank function of a matroid is integer valued, monotone and submodular, and in addition it satisfies the trivial conditions that $r(\emptyset) = 0$ and $r(\{x\}) \le 1$ for every $x \in E$. Conversely, these conditions characterize rank functions of matroids.

The terminology above already suggests that matroids are motivated by two basic examples.

Example 19.19 (Linear matroids). In this most basic example, the underlying set of the matroid is a finite set $E \subseteq \mathbb{K}^d$ (where $\mathbb{K}$ is any field), and independence means linear independence. The rank of a set is the dimension of the subspace it generates, and flats are intersections of E with linear subspaces. We also say that E *represents* the matroid over the field $\mathbb{K}$.

Which matroids are representable over which fields? This question leads to a deep and rich theory, which is, alas, beyond the scope of this book.

Many properties of sets of vectors used in this book are "matroidal" (depend only on the matroid induced by these vectors). For example, E is in general position if and only if it induces the matroid consisting of all sets of cardinality at most d. This matroid is called the *uniform matroid of rank d*. As a special case, we get the *free matroid*, in which every subset of E is independent. ♦

Example 19.20 (Graphic matroids). Let $G = (V,E)$ be a multigraph, and assume for convenience that it is connected. Let $\mathcal{M}$ consist of those subsets of E that contain no cycles; in other words, they are edge-sets of those subgraphs of G that are forests. Then $(E,\mathcal{M})$ is a matroid, called the *graphic matroid induced by G*. Then rank of $(E,\mathcal{M})$ is $n-1$, the bases of $(E,\mathcal{M})$ are the spanning trees of G, and the circuits of $(E,\mathcal{M})$ are the cycles in G.

Graphic matroids are all linear; in fact, they can be represented by vectors over an arbitrary field K. ♦

Let us describe a third class of matroids, which is not as well studied as the previous two, but is tied to the contents of this book, in particular to Section 19.2.

EXAMPLE 19.21 (Algebraic matroids). Let KK be a subfield of a field $\mathcal{K}_1$, and let $E \subseteq \mathcal{K}_1$ be a finite set. Let $\mathcal{M}$ consist of those set $X \subset E$ that are algebraically independent over $\mathcal{K}$. The $(E, \mathcal{M})$ is a matroid, which we call an *algebraic matroid* over the field $\mathcal{K}$. The transcendence degree of a set of numbers, used in Section 19.2, is just the rank in this matroid. Some basic matroidal properties of algebraic independence have been used implicitly in that section. ♦

Let us describe an example of very unstructured matroids, which can serve as counterexamples of many conjectures motivated by the more structured examples above.

EXAMPLE 19.22 (Paving matroids). For a given rank $r > 0$ and underlying set E, start with a family $\mathcal{S}$ of subsets of E such that the intersection of any two of them has at most $r-2$ elements. Let $\mathcal{M}$ consist of all subsets $X \subseteq E$ such that $|X| \le r$, and $|X \cap A| < r$ for every $A \in \mathcal{S}$. (In particular, every set of size $r-1$ or less belongs to $\mathcal{M}$.) Then $(E, \mathcal{M})$ is a matroid, called the *paving matroid* induced by $\mathcal{S}$.

The paving matroid of rank 4 on the set $\{1, \dots, 8\}$ induced by the family of 5 sets $\big\{\{1,2,3,4\}, \{1,2,5,6\}, \{1,2,7,8\}, \{3,4,5,6\}, \{3,4,7,8\}\big\}$, called the *Vámos matroid* is an example of a matroid that is neither linear nor algebraic. ♦

We also need the following very simple example of a submodular setfunction.

EXAMPLE 19.23. Let $\mathcal{H}$ be a hypergraph. We define a setfunction on the subsets of $E = E(\mathcal{H})$ by

$$f(\mathcal{S}) = |\cup \mathcal{S}| = \Big| \bigcup_{A \in \mathcal{S}} A \Big|.$$

More generally, we can consider a family $\mathcal{F}$ of flats in a matroid (as the most important special case, a family of finite dimensional linear subspaces of a linear space), and define setfunction on the subsets of $\mathcal{F}$ by

$$f(\mathcal{S}) = r(\cup \mathcal{S}) \qquad (\mathcal{S} \subseteq \mathcal{F}).$$

This setfunction is trivially integer valued, monotone, and satisfies $f(\emptyset) = 0$. It is not hard to see that it is submodular, so it is a polymatroid. A polymatroid obtained from a family of linear subspaces is called *linear*. ♦

19.5.2. Duality. For a matroid $(E, \mathcal{M})$, we construct its *dual matroid* $(E, \mathcal{M}^*)$, in which a set $X \subseteq E$ is independent if and only if $E \setminus X$ spans the whole space in $(E, \mathcal{M})$ (in other words, it contains a basis). It is not hard to verify that this is indeed a matroid. It also follows that $(\mathcal{M}^*)^* = \mathcal{M}$. The bases of $\mathcal{M}^*$ are the complements of the bases of $\mathcal{M}$. The rank function r^* in the dual matroid can be expressed by the formula

$$r^*(X) = r(E \setminus X) + |X| - r(E).$$

The construction of the Gale dual in Section C.3.4 implies that if a matroid $(E, \mathcal{M})$ is representable over a given field, then so is $(E, \mathcal{M}^*)$. Note that while the Gale dual is not uniquely determined, its underlying matroid is.

The dual of a graphic matroid is not graphic in general; in fact, it is graphic if and only if the graph is planar. If the graph is a planar map, then the dual of its matroid is the matroid of the dual map.

19.5.3. Generic points and hyperplanes. The Generic Point Lemma is closely related to a purely matroid-theoretical result [**Edmonds 1970**]. One way to introduce this is to construct an extension of a polymatroid (E, f) with a new element a, of which we think of (for the sake of a geometric motivation) as a generic k-dimensional "subspace" of an element $b \in E$ in the polymatroid. Here the word "generic" should mean that a "sticks out" of the span of any subset $X \subseteq E$ as much as submodularity allows. In formula:

$$f(X \cup \{a\}) = \min\{f(X)+k, f(X \cup \{b\}). \tag{19.17}$$

It is not hard to verify that this extended function is still a polymatroid on the subsets of $E \cup \{a\}$. In particular, we can extend (E, f) by a point (element of rank 1). Repeating this with all elements of E, and keeping track of the rank function, we get the following theorem.

THEOREM 19.24. *If $f : 2^V \to \mathbb{Z}$ is a polymatroid, then*

$$\mathcal{M}_f = \{X \subseteq V : \ |X \cap S| \le f(S) \quad \textit{for all } S \subseteq V\}$$

is a matroid, whose rank function is given by

$$r(X) = \min_{Y \subseteq X} \big(f(Y) + |X \setminus Y|\big).$$

Lemma 19.12 concerns the special case when the polymatroid is linear. In this setting, the content of the theorem is that a generic choice of representative points provides a linear representation of the matroid M_f. So Lemma 19.12 says less than Theorem 19.24 in the sense that it concerns only linearly representable submodular functions, but more in the sense that it gives a linear representation of the induced matroid. This geometric view of matroid-theoretical results was worked out by [**Mason 1977**].

Similarly, the Generic Hyperplane Theorem is closely related to a result about submodular functions [**Edmonds 1970**]:

THEOREM 19.25. *If $f : 2^V \to \mathbb{Z}$ is a submodular setfunction that is nonnegative on nonempty sets, then*

$$f'(X) = \min\big(f(X_1) + \cdots + f(X_k)\big)$$

(where $\{X_1, \ldots, X_k\}$ ranges over all partitions of X into nonempty sets) defines a submodular setfunction. □

Clearly $f'(\emptyset = 0$. If $f(\emptyset) \ge 0$, then $f'(X) = f(X)$ for every nonempty set; so the only interesting case is when $f(\emptyset) < 0$. As a result on submodular setfunctions, the last theorem says that f' "fixes" this problem of f: it is the unique submodular setfunction that is nonnegative, $f'(\emptyset) = 0$, $f'(X) \le f(X)$ for $X \ne \emptyset$, and which majorizes every other submodular setfunction with these properties.

What Lemma 19.18 adds to this result is that if g is a linear polymatroid defined by nonnull subspaces, and we define $f = g - 1$, then f' is a polymatroid realizable by the family of intersections of these subspaces with a transcendental hyperplane.

This result can be combined with Theorem 19.24:

COROLLARY 19.26. *Let $f : 2^V \to \mathbb{Z}$ be a submodular setfunction that is positive on nonempty sets. Then*

$$\mathcal{M}'_f = \{A \subseteq V : \ |A \cap X| \le f(X) - 1 \quad \textit{for all } X \subseteq V, \ X \ne \emptyset\}$$

is a matroid, whose rank function is given by

$$r(X) = \min_{Y_0, Y_1, \dots, Y_k} |Y_0| + (f(Y_1) - 1) + \cdots + (f(Y_k) - 1),$$

where $\{Y_0, \dots, Y_k\}$ *ranges over all partitions of* X. □

Again, if f is linearly representable, then so is the matroid $\mathcal{M}'_f$. Starting with the free matroid on a set V, we get the graphic matroid induced by the complete graph on V.

19.5.4. Matroid sum. We have come to a point where a matroid theoretical formulation of the results is not only be more general, but allows for a more elegant, more conceptual treatment. Given two matroids $(E, \mathcal{M}_1)$ and $(E, \mathcal{M}_2)$ on a common underlying set, we can construct the set system $\mathcal{M}_1 \vee \mathcal{M}_2 = \{X_1 \cup X_2 : X_i \in \mathcal{M}_i\}$, which we call the *sum* of the matroids $(E, \mathcal{M}_1)$ and $(E, \mathcal{M}_2)$.

THEOREM 19.27. *The set-system* $(E, \mathcal{M}_1 \vee \mathcal{M}_2)$ *is a matroid. The rank function* $\widehat{r}$ *of* $(E, \mathcal{M}_1 \vee \mathcal{M}_2)$ *can be expressed in terms of the rank functions* r_i *of* $(E, \mathcal{M}_i)$ *as follows:*

$$\widehat{r}(X) = \max\nolimits_{Y \subseteq X}(r_1(Y) + r_2(X \setminus Y)) = \min\nolimits_{Y \subseteq X}\{r_1(Y) + r_2(Y) + |X \setminus Y|\}.$$ □

Note that this gives a min-max (NP∩co-NP) characterization of this rank function.

It is straightforward to define the sum $(E, \mathcal{M}_1 \vee \cdots \vee \mathcal{M}_k)$ of k matroids $(E, \mathcal{M}_j)$ on the same underlying set and generalize the above expressions for its rank function. Corollary 19.16 is the special case when we add up the same linear matroid k times, and it generalizes to matroids in a straightforward manner (Exercise 19.14).

19.5.5. Matroids and rigidity. Consider a vector labeling $\mathbf{u}$ of the complete graph K_n in $\mathbb{R}^d$, where we assume that $n \geq d+1$ to avoid some trivial complications. Stress-free sets of edges of K_n define a matroid on $E(K_n)$, which we call the d-dimensional *rigidity matroid* of $(K_n, \mathbf{u})$. For an explicitly known representation $\mathbf{u}$, we can do computations in this matroid using elementary linear algebra.

It is not hard to see that for a full-dimensional vector labeling $\mathbf{u}$, the rank of this matroid is $dn - \binom{d+1}{2}$ (this specializes to our favorite value of $2n-3$ for $d = 2$). A graph on $V(K_n)$ is rigid in this realization if and only if its edges form a basis of this matroid.

Now assume that $\mathbf{u}$ is generic; trivially, this leads to the richest matroid on $E(K_n)$ (with the largest family of independent sets), which we call the *generic rigidity matroid on n elements.*

In the case $d = 2$, we can invoke matroid theoretical results through the following argument. The setfunction $|V(E)|$ $(E \subseteq E(K_n))$ is submodular (a special case of Example 19.23), and hence so is the setfunction $f(E) = 2|V(E)| - 3$. This setfunction is nonnegative on every nonempty set of edges, but negative for $E = \emptyset$. By Theorem 19.25, we can define a nonnegative setfunction f' by the formula

$$f'(X) = \min \sum_{i=1}^{k} f(X_i),$$

where $\{X_1, \dots, X_k\}$ ranges over all partitions of X into nonempty sets. We see that $f'(\{x\}) = 1$ for every singleton set $\{x\}$, and so f' is the rank function of a matroid, which is just the rigidity matroid. The graph G is generically rigid if and only if $f'(E) = 2n-3$, and hence we get the partition condition for generic rigidity.

For higher d, we can also consider the submodular setfunction $f(X) = d|V(X)| - \binom{d+1}{2}$, and "fix" it similarly as for $d = 2$. This is more complicated than for $d = 2$, but the main problem is that, as we have seen, it does not capture the generic rigidity matroid; it is a freer matroid for $d > 2$ and large enough n (having more independent sets).

19.5.6. Graphs on matroids. The problems treated in Chapter 18 offer themselves to a generalization to matroids. We can label the nodes of a graph G by the elements of a matroid; it will be convenient to identify the nodes with their labels, so we consider a graph whose nodes are the elements of a matroid. So we consider pairs $(G, \mathcal{M})$, where G is a finite graph and $(V(G), \mathcal{M})$ is a matroid.

All the quantities introduced in Chapter 18 can be defined. We define $\tau(G, \mathcal{M})$ as the minimum rank of a node-cover. We define a *matching* as a family of disjoint edges whose union is independent in the matroid, and $\nu(G, \mathcal{M})$ as the maximum cardinality of a matching.

Most of the results about covering and cover-critical graphs in Sections 18.2.1, 18.2.2 and 18.2.3 extend to this more general setting with essentially the same proofs. The most notable exception is Theorem 18.8(b), whose proof uses more of the algebra of linear spaces than just dependence-independence. For cover-critical graphs on general matroids, only a weaker bound

$$m \le 2^{\tau(G,\mathcal{M})-1} \tag{19.18}$$

seems to be known.

For matchings, the situation is worse: The problem of computing $\nu(G, \mathcal{M})$ contains NP-hard problems for explicitly described matroids. To show a simple example, let G be a graph and $k \ge \omega(G)$. We want to decide whether $\omega(G) = k$ or $\omega(G) < k$. It is clear that this is an NP-hard task.

Consider the graph $H = G \boxtimes K_2$, and let $\mathcal{F}$ be the family of $(2k)$-cliques in H (this may be an empty family). We define an almost trivial graph on $U = V(H)$ by connecting nodes $(i,1)$ and $(i,2)$ for every $i \in V(G)$. All the information we need about G goes into the matroid $(U, \mathcal{M})$, which is the paving matroid of rank $2k$ defined by the family $\mathcal{F}$. (Every $(2k)$-clique in H is of the form $B \times V(K_2)$, where B is a k-clique in H, so any two sets in $\mathcal{F}$ have at most $2k-2$ elements in common.) Now $\nu(G, \mathcal{M}) = k$ if and only if $\omega(G) = k$. So to determine $\nu(G, \mathcal{M})$ for this graph on this matroid is NP-hard.

The matching problem for graphs on matroids can be solved in polynomial time when the matroid is linear, but not in general. One can go beyond the class of linear matroids: for example, it can be solved for algebraic matroids [**Dress–Lovász 1987**]. To illustrate how the same geometric argument can lead to very different results when applied in different matroid models, let us state one version of this result.

THEOREM 19.28. *Let $K_1 \subset K_2$ be two fields, and let $u_1, \dots, u_n, v_1, \dots, v_n \in K_2$. Then the maximum cardinality of a subset $\{i_1, \dots, i_k\} \subseteq \{1, \dots, n\}$ such that $u_{i_1}, v_{i_1}, \dots, u_{i_k}, v_{i_k}$ are algebraically independent over K_1 is given by the formula*

$$\min\Big\{\deg_{\rm tr}(K|K_1) + \sum_{j=1}^{r} \Big\lfloor \frac{1}{2}\deg_{\rm tr}\big(\{u_i, v_i : i \in S_j\} \,|\, K\big)\Big\rfloor,$$

where the minimum is taken over all field extensions K of K_1, and over all partitions $\{S_1, \ldots, S_r\}$ of $\{1, \ldots, n\}$. □

A larger class of matroids to which the min-max theorem about matchings can be extended was studied in [**Björner–Lovász 1987**]. However, the complexity of the matroid matching problem is still not completely understood.

EXERCISE 19.1. Let $\nu(\mathcal{F})$ denote the maximum number of elements in a linearly independent family $\mathcal{F}$, and let $\tau_{\text{lin}}(\mathcal{F})$ be the minimum dimension of a subspace intersecting every element of $\mathcal{F}$ in a nonzero vector. Prove that $\nu(\mathcal{F}) \le \tau_{\text{lin}}(\mathcal{F})$, and show that equality does not hold in general.

EXERCISE 19.2. Prove Lemma 19.2.

EXERCISE 19.3. Construct a vector-labeling $\mathbf{u}$ of K_4 such that $\tau(K_4, \mathbf{u}) = 3$ but $\tau_{\text{lin}}(\mathcal{F}_{K_4,\mathbf{u}}) = 2$.

EXERCISE 19.4. For each edge ij of a framework $(G, \mathbf{v})$, let $L_{ij} = \text{lin}\{\mathbf{v}_i, \mathbf{v}_j\}$, and let $\mathcal{F}_{G,\mathbf{v}}$ be the family of these subspaces. (a) Prove that $\nu(\mathcal{F}_{G,\mathbf{v}}) = \nu(G, \mathbf{v})$. (b) Prove that if the vector labels are linearly independent, then $\tau_{\text{lin}}(\mathcal{F}_{G,\mathbf{v}}) = \tau(G, \mathbf{v})$. (c) Prove that $\tau_{\text{lin}}(\mathcal{F}_{G,\mathbf{v}}) \le \tau(G, \mathbf{v})$ for all vector labels.

EXERCISE 19.5. Let $A_1, \ldots, A_n$ be p-sets of points in $\mathbb{R}^n$, and let $B_1, \ldots, B_n$ be q-dimensional subspaces of $\mathbb{R}^d$.. Suppose that $A_i \cap B_i = 0$ for all i, but $A_i \cap B_j \ne 0$ for $i < j$ (no condition on $i > j$). Prove that $n \le \binom{p+q}{p}$.

EXERCISE 19.6. Let $\mathcal{H}$ be a family of 2-dimensional subspaces of a linear space, such that $\text{rk}(\mathcal{H}) = 2|\mathcal{H}| - 2$ and $\text{rk}(\mathcal{H} \setminus \{A\}) = 2|\mathcal{H}| - 2$ for every $A \in \mathcal{H}$.
(a) Prove that there is a unique partition $\mathcal{H} = \{\mathcal{H}_1, \ldots, \mathcal{H}_k\}$ such that

$$\text{rk}(\mathcal{F}) = \begin{cases} 2|\mathcal{F}| - 2 & \text{if } \mathcal{F} = \mathcal{H}, \\ 2|\mathcal{F}| - 1 & \text{if } \mathcal{F} \text{ contains exactly } k-1 \text{ of the classes } \mathcal{H}_i, \\ 2|\mathcal{F}|, & \text{otherwise.} \end{cases}$$

(b) Prove that if $k \ge 3$, then $\text{lin}(\mathcal{H} \setminus \mathcal{H}_1) \cap \cdots \cap \text{lin}(\mathcal{H} \setminus \mathcal{H}_k) \ne 0$.

EXERCISE 19.7. Let L be a algebraic subspace of $\mathbb{R}^d$ and J, any other subspace. (a) Prove that $\mathbb{K}_{L+J} \subseteq \mathbb{K}_J$ and $\mathbb{K}_{L \cap J} \subseteq \mathbb{K}_J$. (b) Prove that L/J is algebraic within $J^\perp$.

EXERCISE 19.8. Let $L \subseteq \mathbb{R}^d$ and $J \subseteq L$ be subspaces. Prove that J is transcendental within L if and only if J has a basis obtained by taking r linear combinations of a standard basis of L with coefficients algebraically independent over $\mathbb{K}_L$.

EXERCISE 19.9. Derive Theorem 15.16 from Corollary 19.16.

EXERCISE 19.10. Prove that every polymatroid can be represented by a family of flats of a matroid as in Example 19.23.

EXERCISE 19.11. Let $A_1, \ldots, A_n$ be random events, and for $X \subseteq [n]$, define

$$f(X) = \mathsf{P}\Big(\bigwedge_{i \in X} A_i\Big).$$

Prove that f is a submodular setfunction.

EXERCISE 19.12. Let $\mathbf{a}_1, \ldots, \mathbf{a}_n$ be linearly independent vectors in $\mathbb{R}^n$. For $X = \{i_1, \ldots, i_k\} \subseteq [n]$, let

$$f(X) = \ln \text{vol}_k(P(\mathbf{a}_{i_1}, \ldots, \mathbf{a}_{i_k})),$$

where $P(\mathbf{a}_{i_1}, \ldots, \mathbf{a}_{i_k})$ denotes the (k-dimensional) parallelotope spanned by the vectors $\mathbf{a}_{i_1}, \ldots, \mathbf{a}_{i_k}$. Prove that f is a submodular setfunction.

EXERCISE 19.13. Let A be an $n \times n$ positive definite matrix, and let A_X denote the submatrix formed by its rows and columns in X ($X \subseteq [n]$). Prove that $\ln \det(A_X)$ is a submodular setfunction.

EXERCISE 19.14. Let $(E, \mathcal{M})$ be a matroid. Prove that E can be partitioned into k independent subsets if and only if

$$k\, r(X) \geq |X|$$

for every $X \subseteq E$.

EXERCISE 19.15. Let $(V, \mathcal{M})$ be a matroid and $\mathcal{H} \subseteq 2^V$, a family of r-sets. Define $\tau_{\mathcal{M}}(\mathcal{H})$ as the minimum rank of a subset $X \subseteq V$ meeting every set in $\mathcal{H}$. Suppose that $\mathcal{H}$ is critical in the sense that $\tau_{\mathcal{M}}(H \setminus \{A\}) < \tau_{\mathcal{M}}(\mathcal{H})$ for every $A \in \mathcal{H}$. Prove that

$$|\mathcal{H}| \leq r^{\tau_{\mathcal{M}}(\mathcal{H})}.$$

CHAPTER 20

Concluding Thoughts

We have seen a fairly large variety of geometric representations. Is there a way to fit these into a single theory? Perhaps not, considering the abundance of possibilities how the graph structure can be reflected in geometry. But while writing this book, more and more common ideas between this variety of representations have emerged. Let us describe some of these (admitting that they form a germ of a coherent theory at most).

20.1. Non-degeneracy

A common theme in various geometric representations is that imposing non-degeneracy conditions on the representation often makes it easier to analyze and therefore more useful (most often, by eliminating the possibility of numerical coincidence). There are at least four types of non-degeneracy conditions that play a role in several types of geometric representations (and so they pop up in several chapters of this book). It is interesting that different non-degeneracy conditions can be imposed on one and the same representation model, leading to different combinatorial conditions; we have seen this in connection with orthogonal representations as well as with rigidity and motions of frameworks.

Faithfulness. This is perhaps the most natural non-degeneracy condition. As a simple example, we want to represent a graph so that adjacent nodes are at unit distance (recall Section 13.2.1). Such a representation is faithful, if nonadjacent nodes have distance different from 1. We have seen in Section 10.4 that faithful orthogonal representations behave quite differently from arbitrary orthogonal representations. This is a very natural but often rather weak condition.

General position. Recall that a vector representation in $\mathbb{R}^d$ is in *general position* if no $d+1$ of the points are contained in a hyperplane. This property of quite different geometric representations is often connected to the connectivity of the graph. In Section 3.3, we characterized graph connectivity by the general position of rubber band representations; in Section 10.3, by the existence of orthogonal representations in general position; and in Section 15.5, by the existence of globally rigid bar-and-joint frameworks in general position. Orthogonal representations in general position were the key to several properties of the variety of orthogonal representations. A quantitative form of general position, "respecting the volume", lead us to an efficient approximation algorithm for the bandwidth of a graph.

Genericity. The most powerful nondegeneracy condition is that the representation is *generic*, which means that the coordinates of node positions (sometimes other numerical data) are algebraically independent real numbers. This cannot happen, of course, for a unit distance representation of any graph with at least one edge; but if we allow arbitrary prescribed distances instead of unit distances, then generic representations may exist, playing a central role in the study of rigidity properties and flexing of bar-and-joint frameworks (Sections 15.3 and 15.5).

Using generic points and projections was a basic tool in the study of matching and covering in frameworks (Chapter 18). We did a general study of "genericity" in subspaces, and applied it in the combinatorics of subspaces (Sections 19.2 and 19.4).

Genericity certainly avoids unwanted numerical coincidences between the data, but it may seem quite far from anything practical; transcendental numbers are almost impossible to explicitly describe and do computations with. But there is a very simple way out: choose the coordinates randomly, independently and uniformly from $[0, 1]$. With probability 1, these coordinates will be algebraically independent. This seems to be of no help though, since to describe a random point of $[0, 1]$, we need infinitely many digits. Here is where general lemmas (see the Schwartz–Zippel Lemma 3.12) come to help, showing that if any substitution for the variables eliminates unwanted algebraic relations, then random choice from any sufficiently large finite set does so.

Transversality. To be generic is a condition that is often too strong; this leads us to what is (I feel) the deepest nondegeneracy notion, *transversality* (closely related to the Strong Arnold Property of matrices). This condition was introduced in a general context in Section 10.5; recall that it means that, writing the constraints on the representation as algebraic equations, the corresponding hypersurfaces intersect transversally. Thus the existence and properties of the representation (the intersection point of these hypersurfaces) are not just accidental, but forced by some more fundamental structure. If a solution of the system is transversal, and we perturb each equation by a small amount, then we get another solvable system with a solution nearby.

In graph theory, this condition was first used by Colin de Verdière, in his beautiful work on the "Colin de Verdiére number" (Chapter 16) and on a version of tree-width (algebraic width, see Section 10.5) But in a sense, in the study of rigidity of bar-and-joint frameworks, the transversality condition played a central role ever since the study of such frameworks began in the 19th century: for such a framework, it corresponds to stress-freeness, a central notion in Chapters 14 and 15.

Can we tell what kind of combinatorial properties are related to these different kinds of nondegeneracy conditions? Probably not in general, but a couple of points are worth mentioning.

As we have seen, representations in general position are often related to connectivity properties of the graph. Connectivity properties of graphs have interesting and somewhat surprising connections with geometric representations, and sometimes, through those, with topological properties. Various versions of the connectivity property with respect to halfspaces (Section 16.5) are a good example.

Transversality was instrumental in proving minor-monotonicity of the Colin de Verdière number and of algebraic width. Minor monotonicity is a basic feature of a graph property or a graph parameter, and its connection with transversality is, I feel, worth of further exploration.

Genericity was used most substantially in Chapters 15 and 19, and has strong matroid-theoretic connections. The interplay between the two matroids involved (the linear matroid of the vector representation and the algebraic matroid of the coordinates) is, I feel, not completely understood.

20.2. Duality

Many related but different forms of duality have played a role in this book.

• Gale duality, introduced in Section 19.4, is perhaps the simplest form of duality. This construction is known under many aliases (dual chain group in matroid theory, dual code in coding theory, Gale diagram in the theory of hyperplane arrangements). We can carry out this construction for any vector representation of a graph G, to get a *dual vector representation*. In some cases, this gives interesting new constructions; for example, from cover preserving representations we get independence preserving representations (recall Exercise 18.11).

• Matroid duality generalizes Gale duality, as well as the duality of planar maps.

• Linear programming duality is a basic construction in linear programming and through this, in much of applied mathematics.

• Semidefinite duality is a generalization of linear programming duality, treated in detail in Chapter 13.

• Polarity is a duality construction in convex geometry, which is closely related to linear programming duality. Related duality notions are the blocking and antiblocking relations for polyhedra in the nonnegative orthant (Section C.3). Polarity of 3-polytopes is a metric form of planar graph duality. A duality between resistance and energy (as can be observed in Section 4.2.1) is related to blocking duality.

• The Weak Perfect Graph Theorem, asserting that the complement of a perfect graph is perfect, expresses a certain duality. This can be extended to a full-blown duality (anti-blocking relation) between the TSTAB bodies of a graph and its complement, as described in Section 11.4. This duality goes back, ultimately, to semidefinite duality, so it is quite interesting that it connects complementation of graphs as well as to antiblocking of polytopes.

• Self-duality in various situations is also worth mentioning. We have met self-polar polytopes (Example 11.16) and self-complementary graphs (Corollary 11.7). Self-duality of nonnegative orthant and of the positive semidefinite cone are the key ingredients behind the duality theory of linear and semidefinite programming, respectively. Discrete holomorphic forms are the self-dual subspace of the space of circulations on a map (those circulations which define a circulation on the dual graph).

Not all duality-type phenomena are fully understood. The Gale dual is only determined up to an affine transformation; for geometric representations with metric properties (which is the majority), this dualization does not seem to be useful. Yet in some cases more than the basic linear structure is dualized. We don't have a general theory for this, but let us briefly describe two examples.

We have mentioned a complete duality of the TSTAB bodies of a graph and its complement. This is a duality between the *profiles* of the orthogonal representations. However, there is a duality-type relationship between the optimal orthogonal representations themselves. Consider an orthogonal representation $(\mathbf{u}, \mathbf{c})$ of a graph G that is optimal with respect to a weighting w, and construct from it the following matrix, adding the handle as a last column:

$$U = ((\mathbf{c}^\mathsf{T}\mathbf{u}_1)\mathbf{u}_1, \dots, (\mathbf{c}^\mathsf{T}\mathbf{u}_1)\mathbf{u}_n, \mathbf{c})$$

Also construct a similar matrix from an optimal dual representation (with a little twist in the sign of the last column):

$$W = ((\mathbf{d}^\mathsf{T}\mathbf{v}_1)\mathbf{v}_1, \dots, (\mathbf{d}^\mathsf{T}\mathbf{v}_1)\mathbf{v}_n, -\mathbf{d}).$$

Then (11.48) expresses that every row of U is orthogonal to every row of W. If, in addition, $\mathbf{u}$ is d-dimensional and $\mathbf{v}$ is r-dimensional with $d+r = n+1$, then W is a Gale dual of U. These matrices have the additional property that (disregarding the last column) any two columns are orthogonal either in U or in W. Is there a deeper duality principle behind this observation?

Turning to our second example, Theorems 16.15 and 16.16 seem to indicate a duality between the Colin de Verdère numbers of outerplanar/planar graphs and their complements. This is thanks to two almost-dual geometric representations, the nullspace representation of the graph and the 1-engaged representation of its complement. But while the 1-engaged representation has strong metric properties, it is unclear how to impose those on the nullspace representation, and the theory is incomplete.

20.3. Very large graphs and their limits

It is more and more apparent that the notion of a *network* is fundamental for many branches of science. The networks one studies are often very large, even huge with billions of nodes, and investigating such large networks raises novel problems and takes novel tools. For example, the graph may not be fully known or only be inspectable by sampling (see the Introduction of [**Lovász 2012**] for a detailed description).

Based on the applications of geometric representations in this book, one would like to see how these geometric tools have to be modified to be applicable to very large graphs. As an illustration, we may want to compute a useful representation, say an optimal orthogonal representation in the sense of Section 11.1. Everything we proved about such representations remains valid independently of how large the graph is, but to compute the representation we need to know the whole graph, and even then, its size may be prohibitive. What can be computed from the limited information available (for example, from sampling)?

Very large graphs are often better understandable if we introduce limit objects for sequences of graphs whose size goes to infinity, while becoming more and more similar in an appropriate sense. Random graphs with a fixed edge density are simple examples. We want to generalize geometric representation notions and results to these limit objects. The expectation is that if we succeed in generalizing a graph parameter or other graph-theoretic concept to the limiting case, then this indicates its tractability for very large finite graphs.

There are several limit notions for graphs, and the possibilities of extending geometric representation theory is quite different for different types of graphs and limit notions. We discuss three of these. I tried to collect the rather sporadic results about the types of representations considered in this book, and also add some speculation about the possibilities for some of our main representation concepts.

Dense limit. Convergence of a dense graph sequences was introduced in [**Borgs et al. 2008**]. The limit object for a sequence of dense graphs has several avatars. Geometric representations may mean different things even for equivalent models.

• A *graphon* is defined as a symmetric, measurable function $[0,1]^2 \to [0,1]$ [**Lovász–Szegedy 2006**]. This is the most commonly used description of the limit object. Various basic notions of graph theory have been extended to graphons: subgraph densities, connectivity, spectrum, matchings, coloring etc. (See [**Lovász 2012**] and also [**Hladký–Rocha 2017**] and the references there.)

Some geometric representations have nice extensions to graphons; for example, the A-squared representation introduced in Section 9.3 can be defined for graphons, as a vector-labeling from a Hilbert space (or from some other Lebesgue space). In particular, the 2-neighborhood distance can be defined, and regularity partitions can be generalized. Perhaps most importantly, several crucial compactness properties of graphons as well as of the space of graphons can be proved with the help of this representation.

On the other hand, it seems to be difficult to generalize orthogonal representations to graphons. Can we generalize at least the graph parameter $\vartheta(G)$? One has to be careful about the normalization, to guarantee that the values converge for an appropriately convergent graph sequence. Proposition 11.4 and inequality (11.35) suggest that $\vartheta(G)/\sqrt{n}$ may be an interesting normalization. Let me raise a specific question: For every graphon W and every n, we can sample from the graphon and obtain a "W-random graph" $G(n,W)$. Is it true that for every graphon W, $\vartheta(G(n,W))/\sqrt{n}$ converges to a limit almost surely as $n \to \infty$ (the limit may be 0 or ∞)?

• Another equivalent (cryptomorphic) form of limits of dense graph sequences is a *local random graph model* on a countable set, which is a probability distribution on (infinite) graphs on node set $\mathbb{N}$, which is invariant under permutations of the nodes, and has the "locality" property that observing two disjoint finite subsets of the nodes (two "windows"), the subgraphs induced by them are independent (in the probabilistic sense). This version has the advantage that the node set is countable, so (for example) we can generate independent random weights for the nodes, which (among others) allows us to define a random ordering of the nodes. Unfortunately, I don't know of any work exploiting this equivalent but different model.

• A third equivalent description of a limit object is a *graph parameter* expressing the limiting densities of finite graphs in a convergent sequence of dense graphs. Such parameters can be characterized by linear inequalities and semidefiniteness conditions. This formulation is analogous (in a sense generalizes) the theory of moments of random variables (see [**Borgs–Chayes–Lovász 2010**] and [**Lovász–Szegedy 2010**] for details). Since moments of random variables have many geometric connections, it is likely that such connections will be discovered for these graph parameters as well.

Scaling limit. Turning to the case of sparse graphs (which are much more relevant for applications), the first thing to note is that these graphs may be interesting on various scales. The simplest way of defining these scales is the graph distance. Since for sparse graphs the diameter may be very large, we may want to scale the distances so that we keep the diameter of the graph fixed, or we may want to keep the edge lengths fixed (and not worry about the diameter). A limit object in the first sense is called the *scaling limit*; this preserves global properties of the graphs. A limit object in the second sense is called the *local limit*; this preserves local properties of the graphs. As a very simple but useful example, the limit of an $n \times n$ grid, as $n \to \infty$, may be a full square in the plane, but also an infinite grid—depending on what properties we are interested in.

Scaling limit is a rich and extensively studied limit notion for graphs, and we encountered this notion several times, mostly in Chapter 8, but also (not under this name) in Section 5.3, where conformal mappings were obtained as scaling limits of coin representations.

Generally, we talk about scaling limits when the graph already has a geometric representation. This is often rather simple (say, a lattice domain), and it is some other feature whose limiting behavior is interesting. To define and describe scaling limits without a predefined geometry (say, for bounded degree graphs which are not planar or almost planar) would be very interesting.

Local limit. Local limit theory has been developed in the extreme case, when the degrees are universally bounded by a number D (some extensions to the case when the degrees are "almost bounded" are also available). There are again different ways of describing limit objects; these are not completely equivalent, but closely related.

• *Involution-invariant random rooted graphs* (also called *unimodular networks*) is the original limit concept introduced in [**Benjamini–Schramm 2001**]. This is a probability distribution on bounded-degree connected graphs with a specified root, under a measure-preservation condition.

In this latter setting one can define random countable planar graphs, and use an appropriate generalization of the coin representation (Theorem 5.1) to prove, among others, that the random walk on a random countable planar graph is recurrent with probability one. Appropriate generalizations of the square tiling theorems (Chapter 6) have applications in a similar fashion [**Benjamini–Schramm 1996a**].

Dual orthogonal representations in a finite dimension $d > D$ exist in this setting, and $\overline{G}$-orthogonalization makes sense. Results in Section 10.3 remain valid *mutatis mutandis*, but how many new problems and phenomena the infinite setting provides remains an unexplored question.

• A *graphing* is a bounded-degree graph on $[0, 1]$, whose edge-set is a Borel set in the unit square, and a certain measure-preserving condition is satisfied. An easy example is a circle, where diagonals of a fixed length are the edges.

The underlying space of a graphing is often a manifold, and in this case a dual orthogonal representation means some sort of a vector field. Dual orthogonal representations in dimension $d > D$ exist on every connected component, but it is unclear whether they can be chosen in a measurable (let alone continuous or smooth) way. Would such a representation tell us anything interesting about the combinatorial structure of the graphing?

20.4. Graphs and geometry

The interplay between graph problems and geometry has many forms. Often the graph we want to study comes together with a predefined geometric representation, and the issue is to test certain mathematical and physical properties of the resulting structure, or compute some parameters that connect combinatorial and geometric structure. Typical of this problem setting is the rigidity of bar-and-joint frameworks. Some basic graph theoretic questions have interesting and natural extensions to this situation, as in Chapter 18, where instead of asking for the minimum number of nodes covering all edges, we asked for the minimum *rank* of such a set. Besides raising interesting questions, this extension of the problem was essential for the solution of a basic problem of classifying cover-critical graphs in the original (geometry-free) setup.

This leads us to the really interesting combination of geometric and combinatorial ideas: we start with just a graph and a purely combinatorial problem. We try to find a way to represent a graph in a "good" form, helping to find the answer to the combinatorial question. Visualizing the structure of the graph this way may in itself be very useful, but the most interesting cases are when a proof uses this geometric structure in an essential way. Determining the Shannon capacity of the pentagon (as described in the Introduction) is a good example of such a proof.

As it is to be expected in mathematics, the two different starting aspects above cannot be strictly separated. For example, Tutte's rubber band method produces a nice planar drawing for planar graphs, but then it can be applied in connectivity algorithms, and the basic physical notion it leads to, namely the notion of a stress, is fundamental in rigidity theory.

These connections with physics are classical, but new connections seem to emerge, as discussed Chapters 8 and 12. The connection with quantum physics is particularly tantalizing. After all, in quantum physics one assigns vectors in a linear space to the states of a physical system, even if these states form a discrete set; and the mathematical methods discussed in this book do the same. I think that the material in Chapter 12 is only the tip of the iceberg.

The boundaries of geometry cannot be sharply drawn, and further connections between graph theory and other areas of geometric nature could have been included in our book:

• It is difficult to draw a border between geometry and linear algebra, and there are many applications of linear algebra in graph theory and other areas of combinatorics, which can now be considered classical, and are not treated in this book. The lecture notes [**Babai–Frankl 1992**] provide an excellent introduction to this method.

• One highly developed interface between geometry and combinatorial questions (mostly combinatorial optimization) is *polyhedral combinatorics*. Since thousands of pages have been written about this topic, I decided not to treat it in this book as a topic on its own right. (Of course, some of its basic results I had to use in the presentation of other topics.) I refer to [**Schrijver 2003**] for a comprehensive treatment of polyhedral combinatorics.

• The applications of combinatorial and algebraic topology in the solution of combinatorial problems also has its successes, but it is a method with a quite different flavor, and I have not treated it in this book. [**Matoušek 2003**] gives a nice introduction to this area.

• In this book, we considered geometry over the real field (with some excursions to the complex numbers and a brief excursion to finite fields in Section 3.3). One could expect that geometry over finite fields, while perhaps somewhat less expressive visually, is even more closely tied to combinatorial problems. There are indeed many connections, from error-correcting codes to expander graphs to quasirandom graphs to Ramsey theory.

I am certain that many new results of this nature will be obtained in the future (or are already in the literature, sometimes in a quite different disguise). Whether these will be collected and combined in another monograph, or integrated into science through some other platform provided by the fast changing technology of communication, I cannot predict. But the beauty of nontrivial connections between combinatorics, geometry, algebra and physics will remain here to inspire research.

APPENDIX A

Linear Algebra

A.1. Vectors, matrices and vector labelings

Usually we define vectors as elements of $\mathbb{R}^d$ for some nonnegative integer d, and we think of them as column vectors (unless said otherwise). But it will be useful to consider, for a finite set V, every map $V \to \mathbb{R}$ as a vector. These vectors form the linear space $\mathbb{R}^V$. If we label the elements of V by $1, \dots, n$, then $\mathbb{R}^V$ is isomorphic $\mathbb{R}^n$ in a trivial way. Using the notation $[n] = \{1, \dots, n\}$, the spaces $\mathbb{R}^{[n]}$ and $\mathbb{R}^n$ are identical.

Similarly, matrices can mean any map $A : U \times V \to \mathbb{R}$, where U and V are finite sets; we call this a $U \times V$ matrix, and the space of all such matrices will be denoted by $\mathbb{R}^{U\times V}$. We also say that the rows of A are indexed by the elements of U, and the columns, by elements of V.

A $U \times V$ matrix A can also be thought of as a map $\mathbf{a} : V \to \mathbb{R}^U$. In this way, it corresponds to a *vector labeling* of the set V, assigning to $i \in V$ the vector $\mathbf{a}_i \in \mathbb{R}^U$. We also write $(\mathbf{a}_i : i \in V)$ for such a matrix, and define the *rank* of the set of labeling vectors as $\text{rk}\{\mathbf{a}_i : i \in V\} = \text{rk}(A)$.

A *basis* of a set of vectors $S \subseteq \mathbb{R}^V$ is a subset of S that is linearly independent and spans the same linear subspace as S. A *column basis* of a matrix $A \in \mathbb{R}^{U\times V}$ is a basis of the set of its columns. If we think of A as a vector labeling $\mathbf{a} : V \to \mathbb{R}^U$, then it will be convenient to call a subset $X \subset V$ a *basis for* $\mathbf{a}$ if $\{\mathbf{a}_i : i \in X\}$ is a column basis of A.

The all-1 vector in dimension d is denoted by $\mathbb{1}_d$. The $d \times d$ identity matrix is denoted by I_d, the all-one matrix of the same dimensions, by J_d. The vector $\mathbb{1}_S \in \mathbb{R}^V$ and the matrices $I_S, J_S \in \mathbb{R}^{S\times S}$ are defined similarly for every finite set S. In the case when the coordinates are indexed by the set V (usually the node set of the graph G) we omit the subscript: $\mathbb{1} \in \mathbb{R}^V$ and $I, J \in \mathbb{R}^{V\times V}$. If A is some logical assertion, then $\mathbb{1}(A) = 1$ if A is true and $\mathbb{1}(A) = 0$ otherwise. The vector $\mathbb{1}_S$ is the incidence vector of the set S: $(\mathbb{1}_S)_i = \mathbb{1}(i \in S)$ (where the universe is understood). The vectors $\mathbf{e}_i \in \mathbb{R}^d$ having 1 in the i-th position and 0 elsewhere form the *standard basis* of $\mathbb{R}^d$. The standard basis of $\mathbb{R}^V$ can be defined analogously.

For two subsets $A, B \subseteq \mathbb{R}^d$, we define $A+B = \{a+b : a \in A, b \in B\}$. The set $A-B$ is defined analogously. For a subset $A \subseteq \mathbb{R}^d$, we denote by $A^\perp$ the set of vectors orthogonal to all vectors in A. We need, for linear subspaces A and B, the elementary relations $(A^\perp)^\perp = A$ and $(A+B)^\perp = A^\perp \cap B^\perp$.

For $\mathbf{v} \in \mathbb{R}^d$, we denote by $|\mathbf{v}|$ its Euclidean norm, and by $\mathbf{v}^0 = (1/|\mathbf{v}|)\mathbf{v}$, the unit vector in the direction of $\mathbf{v}$ (this is only defined for $\mathbf{v} \neq 0$).

A.2. Eigenvalues

Let A be an $n\times n$ real matrix. An *eigenvector* of A is a nonzero vector such that Ax is parallel to x; in other words, $Ax = \lambda x$ for some real or complex number λ. This number λ is called the *eigenvalue* of A belonging to eigenvector v. Clearly λ is an eigenvalue if and only if the matrix $A-\lambda I$ is singular, equivalently, if and only if $\det(A-\lambda I) = 0$. This is an algebraic equation of degree n for λ, and hence has n roots (with multiplicity).

The *trace* of the square matrix $A = (A_{ij})$ is defined as

$$\mathrm{tr}(A) = \sum_{i=1}^{n} A_{ii}.$$

The trace of A is equal to the sum of the eigenvalues of A, each taken with the same multiplicity as it occurs among the roots of the equation $\det(A-\lambda I) = 0$.

We can think of $n\times m$ matrices as vectors with nm coordinates. For two matrices of the same shape, $A, B \in \mathbb{R}^{n\times m}$, we define their *inner product* by

$$A\cdot B = \mathrm{tr}(A^{\mathsf{T}}B) = \sum_{i=1}^{n}\sum_{j=1}^{m} A_{ij}B_{ij}.$$

This should not be confused with the matrix product, but it is related:

$$A\cdot B = \mathrm{tr}(A^{\mathsf{T}}B).$$

If the matrix A is symmetric, then its eigenvalues and eigenvectors are particularly well behaved. All the eigenvalues are real, and there is an orthonormal basis $v_1, \dots, v_n$ of the space consisting of eigenvectors of A. Let $\lambda_1 \geq \dots \geq \lambda_n$ be the corresponding eigenvalues, then they are precisely the roots of $\det(A-\lambda I) = 0$. The matrix A can be written as

$$A = \sum_{i=1}^{n} \lambda_i v_i v_i^{\mathsf{T}}.$$

Another way of saying this is that every symmetric matrix can be written as $U^{\mathsf{T}}DU$, where U is an orthogonal matrix and D is a diagonal matrix. The eigenvalues of A are just the diagonal entries of D.

The largest eigenvalue of a symmetric matrix can be expressed by the Rayleigh quotient:

$$\lambda_1 = \max_{x\in\mathbb{R}^n\setminus 0} \frac{x^{\mathsf{T}}Ax}{|x|^2},$$

and every maximizer is an eigenvector belonging to λ_1. This formula generalizes: if we know the first $i-1$ eigenvectors $v_1, \dots, v_{i-1}$, then the i-th largest eigenvalue can be expressed by the i-th Rayleigh quotient:

$$\lambda_i = \max \frac{x^{\mathsf{T}}Ax}{|x|^2},$$

where the maximization is over all vectors $x \in \mathbb{R}^n$ such that $x \neq 0$ but $v_j^{\mathsf{T}}x = 0$ for $j = 1, \dots, i-1$. Again, every maximizer is a corresponding eigenvector.

A *principal minor* of A is a submatrix B obtained by deleting some rows and the corresponding columns. The following important theorem was proved by Cauchy.

THEOREM A.1 (Interlacing eigenvalues). *Let A be an $n\times n$ symmetric matrix with eigenvalues $\lambda_1 \geq \cdots \geq \lambda_n$. Let B be an $(n-k)\times(n-k)$ principal minor of A with eigenvalues $\mu_1 \geq \cdots \geq \mu_{n-k}$. Then*

$$\lambda_i \geq \mu_i \geq \lambda_{i+k}.$$

We conclude this little overview with a further basic fact about eigenvectors of nonnegative matrices. The following classical theorem concerns the largest eigenvalues and the eigenvectors corresponding to it.

THEOREM A.2 (Perron-Frobenius). *If an $n\times n$ matrix has nonnegative entries, then it has a nonnegative real eigenvalue λ which has maximum absolute value among all eigenvalues. This eigenvalue λ has a nonnegative real eigenvector. If, in addition, the matrix has no block-triangular decomposition (i.e., it does not contain a $k\times(n-k)$ block of 0-s disjoint from the diagonal), then λ has multiplicity 1 and the corresponding eigenvector is positive.*

For us, the case of symmetric matrices is important; in this case the fact that the eigenvalue λ is real is trivial, but the information about its multiplicity and the signature of the corresponding eigenvector is used throughout the book.

A.3. Semidefinite matrices

A symmetric $n\times n$ matrix A is called *positive semidefinite*, if all of its eigenvalues are nonnegative. This property is denoted by $A \succeq 0$. The matrix is *positive definite*, if all of its eigenvalues are positive.

There are many equivalent ways of defining positive semidefinite matrices, some of which are summarized in the Proposition below. We need a definition: the *Gram matrix* of n vectors $\mathbf{u}_1, ..., \mathbf{u}_n \in \mathbb{R}^d$ is the matrix $\mathsf{Gram}(\mathbf{u}) = \mathsf{Gram}(\mathbf{u}_1,\dots,\mathbf{u}_n) = (a_{ij})_{i,j=1}^n$, where $a_{ij} = \mathbf{u}_i^\mathsf{T}\mathbf{u}_j$. If U is the matrix with columns $\mathbf{u}_1,\dots,\mathbf{u}_n$, then $A = U^\mathsf{T}U$. The matrix A does not determine the vectors $\mathbf{u}_i$ uniquely, but almost: since the inner products of the vectors $\mathbf{u}_i$ are determined, so are their lengths and mutual distances, and hence two representations of the same matrix differ only in an isometric linear transformation of the space.

PROPOSITION A.3. *For a real symmetric $n\times n$ matrix A, the following are equivalent:*

(i) *A is positive semidefinite;*

(ii) *the quadratic form x^TAx is nonnegative for every $x \in \mathbb{R}^n$;*

(iii) *A can be written as the Gram matrix of vectors $\mathbf{u}_1,\dots,\mathbf{u}_n \in \mathbb{R}^d$ for some d. The smallest dimension d from where the vectors $\mathbf{u}_i$ can be chosen is the rank of A.*

(iv) *A is a nonnegative linear combination of matrices of the type xx^T;*

(v) *The determinant of every principal minor of A is nonnegative.* □

It follows that the diagonal entries of any positive semidefinite matrix are nonnegative, and it is not hard to work out the case of equality: all entries in a row or column with a 0 diagonal entry are 0 as well. In particular, the trace of a positive semidefinite matrix A is nonnegative, and $\mathrm{tr}(A) = 0$ if and only if $A = 0$.

Even though above we have described a large collection of equivalent necessary and sufficient conditions for semidefiniteness, simpler sufficient conditions are useful. One of these is being *diagonally dominant*: If a real symmetric matrix $A \in \mathbb{R}^{n\times n}$ satisfies $a_{ii} \geq \sum_{j\neq i} |a_{ij}|$ for every row, then A is positive semidefinite.

The sum of two positive semidefinite matrices is again positive semidefinite (this follows e.g. from (ii) again). The simplest positive semidefinite matrices are of the form aa^{T} for some vector a. These matrices are precisely the positive semidefinite matrices of rank 1. Property (iv) above shows that every positive semidefinite matrix can be written as a sum of rank-1 positive semidefinite matrices.

The product of two positive semidefinite matrices A and B is not even symmetric in general (and so it is not positive semidefinite); but the following can still be claimed about the product:

If A and B are positive semidefinite matrices, then $A\cdot B = \operatorname{tr}(AB) \geq 0$, and equality holds if and only if $AB = 0$. A matrix A is positive semidefinite if and only if $A\cdot B \geq 0$ for every positive semidefinite matrix B.

Property (v) above provides a way to check whether a given matrix is positive semidefinite. This works well for small matrices, but it becomes inefficient very soon, since there are many principal minors to check. An efficient method to test if a symmetric matrix A is positive semidefinite is the following algorithm:

Carry out Gaussian elimination on A, pivoting always on diagonal entries. If you ever find a negative diagonal entry, or a zero diagonal entry whose row contains a nonzero, stop: the matrix is not positive semidefinite. If you obtain an all-zero matrix (or eliminate the whole matrix), stop: the matrix is positive semidefinite.

If this simple algorithm finds that A is not positive semidefinite, it also provides a certificate in the form of a vector v with $v^{\mathsf{T}}Av < 0$.

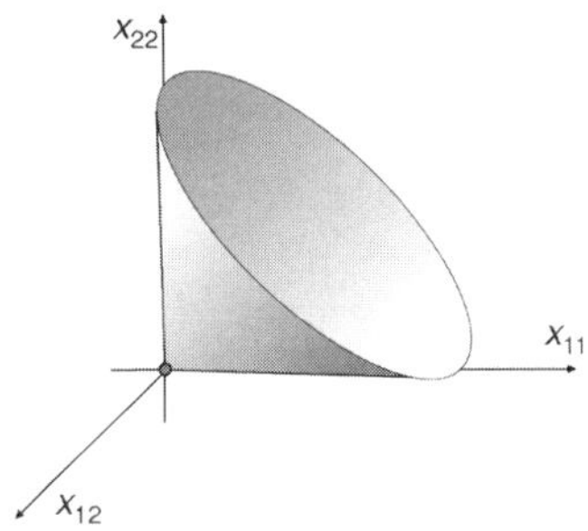

FIGURE A.1. The semidefinite cone for dimension 2.

Positive semidefinite matrices have some important properties in terms of the geometry of the space $\mathbb{R}^{n\times n}$. The fact that the sum of two positive semidefinite matrices is again positive semidefinite (together with the trivial fact that every positive scalar multiple of a positive semidefinite matrix is positive semidefinite), translates into the geometric statement that *the set of all positive semidefinite matrices forms a convex closed pointed cone $\mathcal{P}$ in $\mathbb{R}^{n\times n}$ with vertex* 0.

This cone $\mathcal{P}$ is important, but its structure is quite nontrivial. In particular, it is not polyhedral for $n \geq 2$; for $n = 2$ it is a nice rotational cone (Figure A.1; the fourth coordinate x_{21}, which is always equal to x_{12} by symmetry, is suppressed). For $n \geq 3$ the situation becomes more complicated, because $\mathcal{P}$ is neither polyhedral nor smooth: any matrix of rank less than $n-1$ is on the boundary, but the boundary is not differentiable at such points.

A.4. Geometric spaces

A.4.1. Linear, affine and projective spaces. While from an algebraic point of view linear spaces are most basic, in a geometric setting it is often more convenient to use (real) affine and projective spaces. An affine space is simply a linear space where the location of the origin is forgotten. To be precise, all we need to do is to define an *affine subspace* A of a linear space $\mathbb{R}^d$ as a translation of a linear subspace $L \subseteq \mathbb{R}^d$ by a vector a: $A = L + a$. Two affine subspaces of the same dimension are *parallel*, if they are translates of the same linear subspace.

Projective spaces can be defined in two ways. The shortest way to construct a projective space P^{d-1} from a linear space $\mathbb{R}^d$ is by deleting the origin and identifying parallel vectors. Projective subspaces are the images of linear subspaces under this identification. An important point to remember is that the dimension of P^{d-1} is one less than the dimension of $\mathbb{R}^d$. In particular, a projective plane is obtained from a linear 3-space this way.

If we consider coordinates, then a point in the projective $(d-1)$-space can be represented by a nonzero vector in $\mathbb{R}^d$, whose entries are called its *homogeneous coordinates*. We have to keep in mind that two vectors which are nonzero scalar multiples of each other represent the same projective point. Each projective $(d-2)$-dimensional hyperplane corresponds to a $(d-1)$-dimensional linear subspace. This can be described by an equation $\mathbf{a}^\mathsf{T}\mathbf{x} = 0$. The entries of the vector $\mathbf{a}$ are the *homogeneous coordinates* of the hyperplane. From this description it is clear that we can interchange the roles of points and hyperplanes in projective spaces.

Another way of constructing a projective space is to start with an affine space, and extend it with *ideal points*, where an ideal point is created for each maximal family of parallel lines. The ideal point becomes an element of each line in this parallel family. The set of all ideal points is the *ideal hyperplane*. Every affine subspace is extended to a projective subspace by including the ideal points of lines contained in it; projective subspaces constructed this way are called *ordinary*. Additional *ideal* projective subspaces are formed by the ideal points added to an affine subspace; in other words, ideal subspaces are the intersections of ordinary subspaces with the idea hyperplane. Extension with ideal points does not change the dimension of an affine space.

In the first construction of projective spaces, it is clear that all subspaces of the same dimension are "alike", i.e., they can be transformed into each other by linear transformations of V. So even if the second construction seems to distinguish between ordinary and ideal subspaces, this is just an artifact of the construction method.

For this book, the significance of these constructions is that they allow us to represent configurations of linear subspaces of a d-dimensional linear space as configurations of subspaces of a $(d-1)$-dimensional projective space (Figure A.2, which we can visualize in a $(d-1)$-dimensional affine space (where we either don't need to think about ideal points, or visualize them as "infinitely distant").

A.4.2. Circle geometry. Linear, affine and projective geometries are built on (straight) lines, planes and flats. There is another beautiful piece of geometry

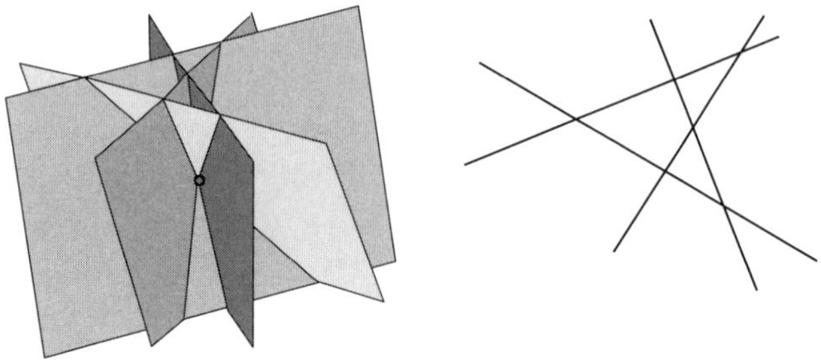

FIGURE A.2. A family of 2-dimensional linear subspaces in 3-space, and their picture as lines in the projective plane.

built on circles and spheres. We are going to need some of its basic facts, which are summarized here.

For a radius $r > 0$, a ball is the set of points at a distance at most r from a given point $\mathbf{x}$ called its *center*. Its boundary is a $(d-1)$-*sphere*. The *unit ball* in $\mathbb{R}^d$ is the set $B^d = \{\mathbf{x} \in \mathbb{R}^d : |\mathbf{x}| \le 1\}$, and the *unit sphere* is the set $S^{d-1} = \{\mathbf{x} \in \mathbb{R}^d : |\mathbf{x}| = 1\}$. We can talk about a lower dimensional ball or sphere in $\mathbb{R}^d$: this is a ball or sphere in an affine subspace. The intersection of a sphere with an affine subspace or with another sphere is either a sphere, or a point, or empty.

The intersection of a sphere S with a closed affine halfspace H, assuming it is not empty, not the whole sphere, and not a single point, is called a *cap*. The cap is *proper* if H does not contain the center of the sphere.

For a cap C on the sphere S, there are three possible points that can be thought of as its "center". Let us assume, for simplicity, that $S = S^{d-1}$, and we are working in $\mathbb{R}^d$. The line ℓ through the center of S orthogonal to the halfspace H intersects the sphere at two points, one of which belongs to the cap; we call this the *center of the cap*. The line ℓ intersects the boundary of H at a point, which we call the *inner center* of the cap. Finally, assuming that the cap is proper, ℓ contains a point from which the visible part of the sphere is exactly C; we call this point the *pole* of the cap.

Inversion. Also called "reflection in a sphere", inversion is an involution of the d-space. To be more precise, inversion with respect to the a $(d-1)$-dimensional sphere $S \subset \mathbb{R}^d$ centered at the origin and with radius r is defined as the map

$$\mathbf{x} \mapsto \frac{r^2}{|\mathbf{x}|^2}\mathbf{x}. \tag{A.1}$$

Inversion with respect to a sphere with another center can be defined by shifting the coordinate system accordingly. This definition is valid in every dimension, but we will use it mostly in dimension 2.

The points of the sphere are fixed by this transformation, and interior points are mapped onto exterior points and vice versa. For the origin, inversion is not defined by (A.1). We can say that this involution is a transformation of $\mathbb{R}^d \setminus \{0\}$, or—often better—we can add a "point at infinity", denoted by ∞, and define 0 and ∞ to be images of each other. The inner center and pole of a proper cap are inverses of each other (Figure A.3).

Perhaps the most important property of involution is that it is sphere-preserving. More precisely, inversion preserves the set of $(d-1)$-dimensional spheres

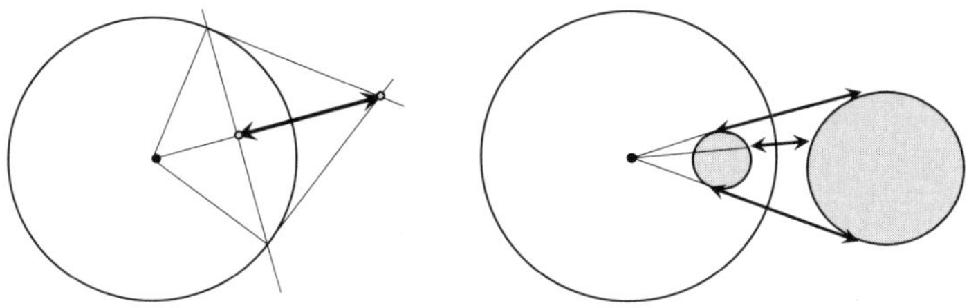

FIGURE A.3. Left: A pair of inverse points as the center and pole of a cap on a 1-dimensional sphere. Right: The inverse of a circle.

and hyperplanes (we can consider hyperplanes as degenerate spheres). Hyperplanes through the origin are mapped onto themselves. Hyperplanes not going through the origin are mapped onto spheres going through the origin, and vice versa. Spheres not going through the origin are mapped onto spheres not going through the origin.

Inversion also preserves angles between two smooth curves intersecting each other, in particular it preserves angles between intersecting circles and/or lines.

Stereographic projection. Let S be a sphere in the d-dimensional space, and let $\mathbf{a} \in S$ be any point on the sphere, and let $\mathbf{b}$ be its antipode. Let H be a hyperplane orthogonal to the diagonal $\mathbf{a}-\mathbf{b}$, not going through ab. Projecting the sphere from ab onto H is called *stereographic projection*. While this definition is important in every dimension, we will use it mostly in dimension 3. Sometimes we call $\mathbf{a}$ the "North pole", and its antipode, the "South pole" (Figure A.4).

The choice of H is not very important: replacing it by another, parallel hyperplane results only in a homothetical transformation in the plane. The plane H is sometimes chosen to be tangent to S at the South pole, sometimes passing through the center of the sphere; the former choice has the advantage that it is almost an isometry near the South pole; the latter, that it leaves the equator invariant. The *inverse stereographic projection*, projecting the plane onto the sphere from $\mathbf{a}$ to S, is also used here and there in this book.

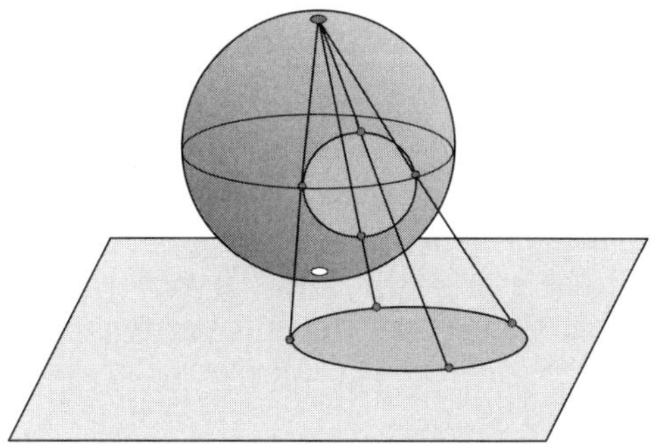

FIGURE A.4. Stereographic projection of a circle.

A little computation shows that (assuming H is the hyperplane tangent to S at the South pole) stereographic projection can be thought of as inversion with respect to the sphere centered at the North pole and going through the South pole. This implies that stereographic projection and its inverse map circles and lines onto circles and/or lines, and preserve the angle between them.

Möbius transformations. Let us restrict our attention to a 2-sphere S. We consider bijective homeomorphisms of S onto itself that map circles onto circles; we call such a map *circle-preserving*. A trivial example is the reflection of the sphere in a plane through its center; this reverses the orientation of circles. Circle-preserving maps that also preserve the orientation are called *Möbius transformations*. It is easy to see that every circle-preserving orientation-reversing map can be obtained as a Möbius transformation followed by a reflection.

These transformations of a sphere S can be described in various ways. Möbius transformations are exactly those maps of the form

$$z \mapsto \frac{az+b}{cz+d}, \qquad \frac{-c}{d} \mapsto \infty, \qquad \infty \mapsto \frac{a}{c}.$$

(where a, b, c, d are complex numbers such that $ad-bc \neq 0$, and the sphere S is identified with $\mathbb{C}\cup\{\infty\}$). Circle-preserving transformations of S can be obtained from projective transformations of the projective 3-space keeping S invariant, by restricting them to S. We can also apply a stereographic projection of S to a plane not necessarily tangent to S (keeping in mind that the North pole is mapped onto an additional point ∞), move the plane to some other position in space, and project it back. Inversion with respect to another sphere S' orthogonal to S, restricted to S, is a circle preserving, but orientation reversing transformation of S, and every circle preserving transformation of S can be decomposed as a product of such inversions.

Circle-preserving transformations of S^2 form a group $\overline{\mathcal{M}}$, of which Möbius transformations form a subgroup $\mathcal{M}$ of index 2. Another subgroup is the group $\mathcal{O}(3)$ of orthogonal transformations (isometries of the sphere). We will need the fact that every finite subgroup of $\overline{\mathcal{M}}$ is conjugate to a subgroup of $\mathcal{O}(3)$.

A.4.3. Cross ratio. The *cross ratio* $(\mathbf{v}_1 : \mathbf{v}_2 : \mathbf{v}_3 : \mathbf{v}_4)$ of four vectors $\mathbf{v}_1, \mathbf{v}_2, \mathbf{v}_3, \mathbf{v}_4$ in a 2-dimensional subspace of a linear space, no two of which are parallel, is defined as follows: we can write $\mathbf{v}_3 = \lambda_1\mathbf{v}_1 + \lambda_2\mathbf{v}_2$ and $\mathbf{v}_4 = \mu_1\mathbf{v}_1 + \mu_2\mathbf{v}_2$, and define

$$(\mathbf{v}_1 : \mathbf{v}_2 : \mathbf{v}_3 : \mathbf{v}_4) = (\lambda_1\mu_2)/(\lambda_2\mu_1).$$

This number is invariant under scaling of the vectors $\mathbf{v}_i$ by nonzero real numbers, which implies that the cross-ratio can be defined for any four distinct collinear points in a projective space. Under duality, cross-ratio can be defined for any four distinct hyperplanes sharing a hyperline.

A much more significant property of the cross-ratio of vectors is that it is invariant under linear transformations of the linear space (equivalently, under projective transformations of the projective space). In particular, projecting a line ℓ in the projective space onto another line ℓ' from any point not on either one of them, the cross ratio of any four points is preserved.

Cross ratios are also important in circle geometry. For four points u, v, w, z in the complex plane, we can define the *complex cross ratio*

$$(u : v : w : z) = \frac{(w-u)(z-v)}{(w-v)(z-u)}.$$

This quantity is real if and only of the four points are on a circle or on a line. It is invariant under Möbius transformations of the complex plane (see Section A.4.2).

A.4.4. Rank, dimension and hull. One potentially confusing thing about these different spaces is that notions like "hull" and "rank" mean different constructions. In each case, the "hull" of a set is the smallest subspace containing it, a set of points is "independent", if no element of it is contained in the hull of the others, and the "rank" of a set is the maximum size of an independent subset of it. Specifically, we define its *linear hull* of a subset $S \subseteq \mathbb{R}^d$ as

$$\mathrm{lin}(S) = \Big\{\sum_{i=1}^{k} x_i \mathbf{a}_i : \mathbf{a}_i \in S,\, x_i \in \mathbb{R},\, k \in \mathbb{N}\Big\},$$

and its *affine hull* as

$$\mathrm{aff}(S) = \Big\{\sum_{i=1}^{k} x_i \mathbf{a}_i : \mathbf{a}_i \in S,\, x_i \in \mathbb{R},\, \sum_i x_i = 1, k \in \mathbb{N}\Big\}.$$

The *rank* $\mathrm{rk}(S) = \dim(\mathrm{lin}(S))$ is just the rank of the matrix A whose columns are the vectors in S. We define the *affine rank* $\mathrm{rk}_{\mathrm{aff}}(S)$ is the rank of the matrix $\binom{\mathbb{1}}{A}$.

We define the *convex hull*: of $S \subseteq \mathbb{R}^d$ by

$$\mathrm{conv}(S) = \Big\{\sum_{i=1}^{k} x_i \mathbf{a}_i : \mathbf{a}_i \in S,\, x_i \in \mathbb{R},\, x_i \geq 0,\, \sum_i x_i = 1, k \in \mathbb{N}\Big\}.$$

There are some relations between these notions, like

$$\mathrm{conv}(S) \subseteq \mathrm{aff}(S) \subseteq \mathrm{lin}(S), \quad \text{and} \quad \mathrm{aff}(S) = S + \mathrm{lin}(S - S).$$

For a family $\mathcal{F}$ of subsets of $\mathbb{R}^d$, we use the notation

$$\mathrm{rk}(\mathcal{F}) = \mathrm{rk}(\cup\mathcal{F}) = \mathrm{rk}\Big(\bigcup_{L\in\mathcal{F}} L\Big).$$

If $J \subseteq \mathbb{R}^d$ is a subspace, then for every vector $\mathbf{a}$, we denote by $\mathbf{a}/J$ its orthogonal projection onto the orthogonal complement $J^\perp$ of J. For any subset $X \subseteq \mathbb{R}^d$, the set X/J is defined analogously.

Generalizing linear independence of vectors, we will call a family $\{L_1, \dots L_m\}$ of non-null linear subspaces of $\mathbb{R}^d$ *linearly independent*, if every set of nonzero vectors $x_1, \dots, x_n$, where $x_i \in L_i$, is linearly independent. Two subspaces are linearly independent if any only if their intersection is the zero subspace. There are equivalent definitions, whose equivalence is easy to prove, but it is useful to collect them in a Proposition.

PROPOSITION A.4. *For a family $\mathcal{L} = \{L_1, \dots L_m\}$ of non-null linear subspaces of $\mathbb{R}^d$, the following are equivalent:*

(a) *$\{L_1, \dots L_m\}$ are linearly independent.*

(b) *Every L_i intersects* $\mathrm{lin}(L_1 \cup \dots \cup L_{i-1})$ *in the zero subspace ($i = 2, \dots, m$).*

(c) *Whenever $x_i \in L_i$ and $x_1 + \dots x_m = 0$, then $x_1 = \dots = x_m = 0$.*

(d) $\dim(L_1 + \dots + L_m) = \dim(L_1) + \dots + \dim(L_m)$.

(e) *For any two indices $1 \leq i < j \leq m$, we have*

$$\mathrm{lin}(\mathcal{L} \setminus \{L_i\}) \cap \mathrm{lin}(\mathcal{L} \setminus \{L_j\}) = \mathrm{lin}(\mathcal{L} \setminus \{L_i, L_j\})$$

In the language of projective spaces, we say that a set of projective subspaces is *independent*, if the corresponding linear subspaces are linearly independent. Proposition A.4 can be reformulated: for example, a family of subspaces in a projective space is independent if and only if each of them is disjoint from the (projective) subspace spanned by the others. If the family consists of k lines, then they are independent if and only if they span a $(2k-1)$-dimensional subspace.

A.5. Exterior algebra

A.5.1. Cross product. The following construction is probably familiar from physics to most readers. For $\mathbf{a}, \mathbf{b} \in \mathbb{R}^3$, we define their *cross product* as the vector

$$\mathbf{a}\times\mathbf{b} = |\mathbf{a}|\,|\mathbf{b}|\,\sin\phi\,\mathbf{u}, \tag{A.2}$$

where ϕ is the angle between $\mathbf{a}$ and $\mathbf{b}$ $(0 \le \phi \le \pi)$, and $\mathbf{u}$ is a unit vector in $\mathbb{R}^3$ orthogonal to the plane of $\mathbf{a}$ and $\mathbf{b}$, so that the triple $(\mathbf{a}, \mathbf{b}, \mathbf{u})$ is right-handed (positively oriented). The definition of $\mathbf{u}$ is ambiguous if $\mathbf{a}$ and $\mathbf{b}$ are parallel, but then $\sin\phi = 0$, so the cross product is 0 anyway. The length of the cross product gives the area of the parallelogram spanned by $\mathbf{a}$ and $\mathbf{b}$.

The cross product is distributive with respect to linear combination of vectors, it is anticommutative: $\mathbf{a}\times\mathbf{b} = -\mathbf{b}\times\mathbf{a}$, and $\mathbf{a}\times\mathbf{b} = 0$ if and only if $\mathbf{a}$ and $\mathbf{b}$ are parallel. The cross product is not associative; instead, it satisfies the *Expansion Identity*

$$(\mathbf{a}\times\mathbf{b})\times\mathbf{c} = (\mathbf{a}^{\mathsf{T}}\mathbf{c})\mathbf{b} - (\mathbf{b}^{\mathsf{T}}\mathbf{c})\mathbf{a}, \tag{A.3}$$

which implies the *Jacobi Identity*

$$(\mathbf{a}\times\mathbf{b})\times\mathbf{c} + (\mathbf{b}\times\mathbf{c})\times\mathbf{a} + (\mathbf{c}\times\mathbf{a})\times\mathbf{b} = 0. \tag{A.4}$$

Another useful replacement for the associativity is the following.

$$(\mathbf{a}\times\mathbf{b})^{\mathsf{T}}\mathbf{c} = \mathbf{a}^{\mathsf{T}}(\mathbf{b}\times\mathbf{c}) = \det(\mathbf{a}, \mathbf{b}, \mathbf{c}) \tag{A.5}$$

(here $(\mathbf{a}, \mathbf{b}, \mathbf{c})$ is the 3×3 matrix with columns $\mathbf{a}$, $\mathbf{b}$ and $\mathbf{c}$).

Sometimes we use the cross product in the special case when the vectors lie in a fixed plane Π. Let $\mathbf{k}$ be a unit vector normal to Π, then $\mathbf{a}\times\mathbf{b}$ is $A\mathbf{k}$, where A is the signed area of the parallelogram spanned by $\mathbf{a}$ and $\mathbf{b}$ (this means that A is positive if and only if a positive rotation takes the direction of $\mathbf{a}$ to the direction of $\mathbf{b}$, when viewed from the direction of $\mathbf{k}$). In this case all the information about $\mathbf{a}\times\mathbf{b}$ is contained in this scalar A, which in tensor algebra will be denoted by $\mathbf{a}\wedge\mathbf{b}$ (see below).

A.5.2. Exterior product. A couple of times we need the more general construction of exterior product. We define a *k-tensor* over the vector space $V = \mathbb{R}^d$ as a map $T : V^k \to \mathbb{R}$ that is linear in each variable. This is more restrictive than the usual definition, but it will suffice for our purposes. The simplification means that we identify V with the space V^* of linear functionals on V: a vector $\mathbf{y} \in \mathbb{R}^d$ acts on $\mathbb{R}^d$ by $\mathbf{x} \mapsto \mathbf{y}^{\mathsf{T}}\mathbf{x}$. All k-tensors form a linear space of dimension d^k.

A tensor is *antisymmetric*, if interchanging two variables multiplies the function value by -1. Perhaps the most important example is the determinant $\det(\mathbf{v}_1, \dots, \mathbf{v}_d)$ $(\mathbf{v}_1, \dots, \mathbf{v}_d \in \mathbb{R}^d)$. All antisymmetric k-tensors form a linear space $\wedge^k V$ of dimension $\binom{d}{k}$. In particular, there are no nontrivial antisymmetric k-tensors for $k > d$, and antisymmetric d-tensors are scalar valued and just constant multiples of the determinant.

Every k-tensor T gives rise to an antisymmetric one by *antisymmetrization*:

$$\mathcal{A}T(\mathbf{v}_1, \dots, \mathbf{v}_k) = \frac{1}{k!} \sum_{\pi \in S_k} \operatorname{sgn}(\pi) T(\mathbf{v}_{\pi(1)}, \dots, \mathbf{v}_{\pi(k)}).$$

The *tensor product* of two tensors T and S is defined simply as the product of functions, with disjoint sets of variables:

$$(T \circ S)(\mathbf{v}_1, \dots, \mathbf{v}_d, w_1, \dots, w_r) = T(\mathbf{v}_1, \dots, \mathbf{v}_d) S(w_1, \dots, w_d).$$

For two antisymmetric tensors, this product is not antisymmetric, but it can be made antisymmetric; this way we define their *exterior product*

$$T \wedge S = \mathcal{A}(T \circ S).$$

Every vector $\mathbf{u} \in \mathbb{R}^d$ can be considered as 1-tensor, since it defines a linear map $\mathbf{x} \mapsto \mathbf{u}^\mathsf{T}\mathbf{x}$, which is trivially antisymmetric. We can define the exterior product of two vectors:

$$\mathbf{u} \wedge \mathbf{v} = \frac{1}{2}(\mathbf{u} \circ \mathbf{v} - \mathbf{v} \circ \mathbf{u}),$$

or indeed, any number of vectors:

$$\mathbf{u}_1 \wedge \dots \wedge \mathbf{u}_k = \mathcal{A}(\mathbf{u}_1 \circ \dots \circ \mathbf{u}_k).$$

Antisymmetric k-tensors of the form $\mathbf{u}_1 \wedge \dots \wedge \mathbf{u}_k$ generate the space $\wedge^k V$; in fact, it suffices to restrict $\mathbf{u}_1, \dots, \mathbf{u}_k$ to elements of any fixed basis. We consider the tensors $\mathbf{e}_{i_1} \wedge \dots \wedge \mathbf{e}_{i_k}$ $(1 \le i_1 < i_2 < \dots < i_k \le d)$ as the standard basis in $\wedge^k V$. It is not hard to see that $\mathbf{u}_1 \wedge \dots \wedge \mathbf{u}_k \neq 0$ if and only if $\mathbf{u}_1, \dots, \mathbf{u}_k$ are linearly independent.

From the dimension formula it follows that $\wedge^k V \cong \wedge^{d-k} V$. In fact, we can consider a tensor $\mathbf{v}_1 \wedge \dots \wedge \mathbf{v}_{d-k}$ as a linear functional on $\wedge^k V$ acting as

$$(\mathbf{v}_1 \wedge \dots \wedge \mathbf{v}_{d-k})[\mathbf{u}_1 \wedge \dots \wedge \mathbf{u}_k] = \det(\mathbf{u}_1, \dots, \mathbf{u}_k, \mathbf{v}_1, \dots, \mathbf{v}_{d-k})$$

(extending linearly to $\wedge^k V$ and $\wedge^{d-k} V$). This gives an explicit identification between $\wedge^k V$ and $\wedge^{d-k} V$. If $\mathbf{u}, \mathbf{v} \in \mathbb{R}^3$, then $\mathbf{u} \wedge \mathbf{v} \in \wedge^2 \mathbb{R}^3 \cong \mathbb{R}^3$, and so $\mathbf{u} \wedge \mathbf{v}$ can be considered as a vector in $\mathbb{R}^3$. It is easy to check that this vector is just the cross product $\mathbf{u} \times \mathbf{v}$.

If $\mathbf{u}_1, \dots, \mathbf{u}_k$ and $\mathbf{v}_1, \dots, \mathbf{v}_k$ are two sets of linearly independent vectors, then $\mathbf{u}_1 \wedge \dots \wedge \mathbf{u}_k$ and $\mathbf{v}_1 \wedge \dots \wedge \mathbf{v}_k$ are nonzero scalar multiples of each other if and only if $\operatorname{lin}(\mathbf{u}_1, \dots, \mathbf{u}_k) = \operatorname{lin}(\mathbf{v}_1, \dots, \mathbf{v}_k)$. So if U is a linear subspace of $\mathbb{R}^d$, then the exterior product of a basis of U is an encoding of the subspace that is (almost, up to a nonzero scalar) independent of the basis selected, and we will denote it by Ψ_U. If U and W are two linear subspaces, then $\Psi_U \wedge \Psi_V \neq 0$ if and only if $U \cap V = 0$; in this case, $\Psi_U \wedge \Psi_V = \Psi_{U+V}$.

The following identity about determinants, called the Binet–Cauchy Formula, will be useful. For a matrix A and subset S of its rows, we denote by A_S the submatrix consisting of the rows in S. Let $A, B \in \mathbb{R}^{n \times m}$, where $m \le n$. Then

$$\det(A^\mathsf{T} B) = \sum_{S \in \binom{[n]}{m}} \det(A_S^\mathsf{T} B_S). \tag{A.6}$$

APPENDIX B

Graphs

B.1. Basics

A *graph*, unless otherwise emphasized, is a simple graph, i.e., it has no loops (edges connecting a node to itself) or parallel edges. We use the word *node* for the node of a graph, the word *vertex* for the vertex of a polytope, and the word *point* for points in the plane or in other spaces. If we need multiple edges, we use the word *multigraph*. If we need loops, we say this explicitly.

For the graph denoted by G, we denote its node set by V and its edge set by E, and the numbers of nodes and edges of G by n and m, respectively. Unless we state otherwise, we tacitly assume that $V = [n] = \{1, \dots, n\}$. For $S \subseteq V$, we denote by $G[S] = (S, E[S])$ the subgraph induced by S. For a simple graph G, we denote by $\overline{G} = (V, \overline{E})$ the complementary graph, often called the *complement* of G. The number of connected components of a graph G will be denoted by $c(G)$.

A set of mutually adjacent nodes is called a *clique*, and a set of mutually non-adjacent nodes, a *stable set*. Nonadjacent nodes are often called "independent", and so a stable set of nodes is called an "independent set"; but this word is overloaded even in the context of this book, since we have to talk about "linearly independent vectors", "algebraically independent numbers", and also about "independent events".

We denote the neighborhood of a node v in a G (the set of nodes adjacent to it) by $N_G(v)$, or in the case when the graph is understood, by $N(v)$. The *degree* of the node v, the number of edges with an endpoint in v, will be denoted by $\deg(v)$. For a simple graph, $\deg(v) = |N(v)|$. Two nodes of a graph are *twins*, if they have the same neighborhoods. We say that a node w *distinguishes* u and v, if it is connected to one but not to the other. So two nodes are twins if and only if they are not distinguished by any node. If no loops are allowed (most of the time in this book), then twin nodes must be nonadjacent.

The *linegraph* $L(G)$ of G is defined on $V\big(L(G)\big) = E$, where to edges of G are adjacent in $L(G)$ if and only if they have a node in common.

We have to introduce several basic graph parameters that will be used throughout this book.

$\omega(G)$: clique number, the number of nodes of the largest clique;

$\alpha(G)$: stability number, the number of nodes of the largest stable set;

$\chi(G)$: chromatic number, the minimum number of stable sets into which the node set can be partitioned.

Clearly the stability number of a graph is the clique number of its complement; we do not introduce a separate expression for the chromatic number $\chi(\overline{G})$ of the complement. We define a metric d_G on V, where $d_G(u,v)$ is the minimum length of path in G connecting u and v.

For $X \subseteq V$, we denote by $N(X)$ the set of neighbors of X, i.e., the set of nodes $j \in V$ for which there is a node $i \in V$ such that $ij \in E$. Warning: $N(X)$ may intersect X (if X is not a stable set). We denote by $\overline{N}$ the neighborhood in the complementary graph: $\overline{N}(i) = V \setminus N(i) \setminus i$.

Let $G = (V, E)$ be a directed graph. For an edge (arc) e, we denote by $\mathsf{hd}(e)$ and $\mathsf{tl}(e)$ the head and the tail of e.

B.2. Flows and cuts

Let G be a directed graph and $s, t \in V$. A vector $z \in \mathbb{R}_+^E$ is called an *s-t flow*, if the flow into any node other than s or t is equal to the flow out of this node:

$$\sum_{e:\,\mathsf{hd}(e)=v} z_e = \sum_{e:\,\mathsf{tl}(e)=v} z_e \qquad (v \in V \setminus \{s,t\}).$$

It is easy to check that

$$\sum_{e:\,\mathsf{hd}(e)=t} z_e - \sum_{e:\,\mathsf{tl}(e)=t} z_e \sum_{e:\,\mathsf{tl}(e)=s} z_e - \sum_{e:\,\mathsf{hd}(e)=s} z_e;$$

the common value ω of the two sides is the *value* of the flow.

From the rich theory of flows, we only recall its central result, the Max-Flow-Min-Cut Theorem [**Ford–Fulkerson 1962**] (see e.g. [**Schrijver 2003**] for much more). An s-t cut ($s, t \in V$) is defined by a set of nodes U with $s \in U$ but $t \notin U$, as the set of edges $e \in E$ with $\mathsf{tl}(e) \in U$ but $\mathsf{hd}(e) \notin U$. Every edge e of the graph has a *capacity* $c_e \geq 0$ assigned to it. A flow z is *feasible*, if $0 \leq z_e \leq c_e$ for every edge e. The *capacity* of a cut C is the sum of capacities of its edges.

THEOREM B.1 (Max-Flow-Min-Cut Theorem). *The maximum value of a feasible s-t flow is equal to the minimum capacity of an s-t cut.* □

A fundamental theorem in graph theory, Menger's Theorem is closely related to flow theory; it can be derived from the Max-Flow-Min-Cut Theorem. For $X, Y \subseteq V$, we denote by $\kappa(X,Y)$ *connectivity* between X and Y, defined as the minimum number of nodes that cover all X-Y paths in G (paths connecting a node in X to a node in Y).

THEOREM B.2 (Menger's Theorem I). $\kappa(X,Y)$ *the maximum number of node disjoint* X-Y *paths.*

(If X and Y are not disjoint, some of these paths may consist of a single node.)

A graph G is k-connected if and only if $n > k$ and $\kappa(X,Y) = k$ for any two k-subsets X and Y. The largest k for which this holds is the *node-connectivity* of G, denoted $\kappa(G)$. The complete graph K_n is $(n-1)$-connected but not n-connected. A version of Menger's Theorem characterizes node connectivity:

THEOREM B.3 (Menger's Theorem II). *A simple graph G is k-connected if and only if $n > k$ and any two nodes are connected by k paths that are disjoint except for their endpoints.*

Further versions of Menger's Theorem concern edge-disjoint paths and separating sets by edges; for these, we refer to any basic graph theory book.

B.3. Matrices associated with graphs

Let G be a (finite, undirected, simple) graph with node set $V(G) = [n]$. The *adjacency matrix* of G is be defined as the $n \times n$ matrix $A_G = A = (A_{ij})$ in which

$$A_{ij} = \begin{cases} 1, & \text{if } i \text{ and } j \text{ are adjacent,} \\ 0, & \text{otherwise.} \end{cases}$$

(We suppress the subscript G if the graph is understood.)

We can extend this definition to the case when G has multiple edges: we just let A_{ij} be the number of edges connecting i and j. We can also have weights on the edges, in which case we let A_{ij} be the weight of the edge. We could also allow loops and include this information in the diagonal.

The *Laplacian* of the graph is defined as the $V \times V$ matrix $L_G = L = (L_{ij})$ in which

$$L_{ij} = \begin{cases} \deg(i), & \text{if } i = j, \\ -A_{ij}, & \text{if } i \neq j. \end{cases}$$

In the case of edge-weighted graphs, we define $\deg(i) = \sum_j A_{ij}$. So $L = D - A$, where $D = D_G$ is the diagonal matrix of the degrees of G. Clearly $L\mathbb{1} = 0$. The *normalized Laplacian* is defined as $\widehat{L} = D^{-1/2} L D^{-1/2}$.

The (generally nonsquare) *incidence matrix* of G comes in two flavors. For an undirected graph G, we define its node-edge incidence matrix $M = M_G$ as the $E \times V$ matrix for which

$$M_{ev} = \begin{cases} 1 & \text{if } v \text{ is an endpoint of } e, \\ 0 & \text{otherwise.} \end{cases}$$

Often, however, the following matrix is more useful. We consider a directed graph G, and let $B = B_G$ denote the $E \times V$ matrix for which

$$B_{ev} = \begin{cases} 1 & \text{if } v = \mathsf{hd}(e), \\ -1 & \text{if } v = \mathsf{tl}(e), \\ 0 & \text{otherwise.} \end{cases}$$

While B_G depends on the orientation of G, changing the orientation only means scaling some rows by -1, which often does not matter much, and we can use B for any undirected graph, giving it an arbitrary reference orientation. Column $b_v = B_G^{\mathsf{T}} \mathbb{1}_v$ of B, corresponding to node v, satisfies $|b_v|^2 = \deg(v)$. For an edge $e = uv$, the corresponding column of B^{T} is $\mathbb{1}_v - \mathbb{1}_u$. In topological terms, B is the (0-dimensional) *coboundary operator*. Its transpose B^{T} is the (1-dimensional) *boundary operator*.

It is easy to check that, independently of the orientation, $L = B^{\mathsf{T}} B$. This implies that the Laplacian L is positive semidefinite. It is worth while to express this equation in terms of quadratic forms:

$$\text{(B.1)} \qquad \mathbf{x}^{\mathsf{T}} L \mathbf{x} = (B\mathbf{x})^{\mathsf{T}} (B\mathbf{x}) = \sum_{ij \in E}^{n} (x_i - x_j)^2.$$

We need some general terminology. Given a graph G, a *G-matrix* is a symmetric matrix $M \in \mathbb{R}^{V \times V}$ such that $M_{ij} = 0$ for every edge $ij \in \overline{E}$. A G-matrix is *well-signed*, if $M_{ij} < 0$ for all $ij \in E$. (Note that we have not imposed any condition on the diagonal entries.) So the adjacency matrix A is a G-matrix, and the Laplacian L is a well-signed G-matrix. We denote the linear space of G-matrices by $\mathcal{M}_G$.

B.3.1. Nullspaces and ranges. Let us fix a connected graph G. Every vector x in the (right) nullspace of B satisfies $x_i = x_j$ for every edge ij. So x is scalar multiple of $\mathbb{1}$. The row space of B is the orthogonal complement of this space, i.e., the set of all vectors in $\mathbb{R}^V$ with coordinates summing to 0. (B.1)) implies that the nullspace of L is the same as the nullspace of B, and similar conclusion holds for the row space of L. It follows that $\text{rk}(L) = \text{rk}(B) = n-1$.

To determine the column space of B (equivalently, the range of B^T), notice that $(B\mathbf{x})_e = x_{\mathsf{hd}(e)} - x_{\mathsf{tl}(e)}$ for $\mathbf{x} \in \mathbb{R}^V$. A vector on the (oriented) edges arising as differences of a vector on the nodes is called a *potential*. Potentials form a linear space $\mathcal{A} = \mathcal{A}_G = \text{Rng}(B)$, with $\dim(\mathcal{A}) = \text{rk}(B) = n-1$. (See Section 4.1 for more on potentials.)

The nullspace of B^T is more interesting. A vector $z \in \mathbb{R}^E$ is called a *circulation* if $B^\mathsf{T} z = 0$. Explicitly, this means that z satisfies the *flow condition* (conservation condition) at each node:

$$(B^\mathsf{T} z)_v = \sum_{e:\,\mathsf{hd}(e)=v} z_e - \sum_{e:\,\mathsf{tl}(e)=v} z_e = 0.$$

Every cycle C in G defines two circulations, obtained by sending one unit of flow around the cycle, in either direction. In other words, the circulation f_C defined by a cycle C with an orientation (not necessarily an oriented cycle!) is defined by

$$(f_C)_e = \begin{cases} 1, & \text{if } e \text{ is a forward edge of } C, \\ -1, & \text{if } e \text{ is a backward edge of } C, \\ 0, & \text{otherwise.} \end{cases} \tag{B.2}$$

Circulations form a linear space $\mathcal{A}^\perp \subseteq \mathbb{R}^E$, which is generated by the special circulations f_C. It is easy to see that $\dim(\mathcal{A}^\perp) = m-n+1$.

Reversing the orientation of an edge corresponds to multiplying the corresponding coordinate by -1, both for $\mathcal{A}$ and $\mathcal{A}^\perp$. Reversing all edges does not change these linear spaces.

The rank, nullspace and range of the adjacency matrix A do not have such a nice and easy descriptions (but see Section 9.1).

B.4. Spectra of graphs

Eigenvalues of matrices associated with graphs are often closely related to their geometric representations, but of course they represent a topic about which several monographs have been written [**Chung 1997**, **Brouwer–Haemers 2011**, **Colin de Verdière 1998b**]. Here we give a very incomplete survey of some of the properties that play a role in our treatment, or at least motivate some of our considerations.

To fix notation, let $\lambda_1 \ge \lambda_2 \ge \cdots \ge \lambda_n$ be the eigenvalues of the adjacency matrix A of a graph G, let $\mu_1 \le \mu_2 \le \cdots \le \mu_n$ be the eigenvalues of the Laplacian L, and let $\nu_1 \le \nu_2 \le \cdots \le \nu_n$ be the eigenvalues of the normalized Laplacian $\widehat{L} = D^{-1/2} L D^{-1/2}$. For a d-regular graph, we have $\mu_i = d - \lambda_i$ and $\nu_i = \mu_i / d$.

B.4.1. The leading eigenvalue. The Perron–Frobenius Theorem implies immediately that if G is connected, then λ_1 has multiplicity 1. This eigenvalue is relatively uninteresting, it is a kind of "average degree". More precisely, let $\overline{d}$, $d_{\max}$ and $d_{\min}$ denote the average, maximum and minimum degree, respectively. Then the following inequalities hold:

$$\max\{\overline{d}, \sqrt{d_{\max}}\} \le \lambda_1 \le d_{\max}. \tag{B.3}$$

In particular, $\lambda_1 = d$ for every d-regular graph. Inequalities (B.3) imply that $d_{\min} \le \lambda_1$, which in turn implies by induction the following bound on the chromatic number:

$$\chi(G) \le \lambda_1 + 1. \tag{B.4}$$

Hence

$$\omega(G) \le \lambda_1 + 1, \tag{B.5}$$

which is of course weaker than (B.4), but it is useful as a starting point of generalizations in Chapter 11.

The smallest eigenvalue of the Laplacian L_G is really uninteresting, since $\mu_1 = 0$. It follows from (B.1) that the Laplacian L_G is singular and positive semidefinite. Furthermore, the multiplicity of 0 as an eigenvalue is equal to the number of connected components of G. Similar statement is not true for the adjacency matrix (if the largest eigenvalues of the connected components of G are different, then the largest eigenvalue of the whole graph has multiplicity 1). This illustrates the phenomenon that the Laplacian is often better behaved algebraically than the adjacency matrix.

B.4.2. The smallest eigenvalue. It is a nice elementary fact that *a graph is bipartite if and only if its spectrum is symmetric about the origin.* For connected graphs, it suffices to look at the largest and smallest eigenvalues: *A connected graph G is bipartite if and only if $\lambda_n(G) = -\lambda_1(G)$.*

For chromatic number larger than 2, no similar full characterization in terms of eigenvalues can be given; but at least the necessity part of the result above can be generalized [**Hoffman 1970**]. The ratio between the largest and smallest eigenvalues can be used to give a lower bound on the chromatic number in terms of the spectrum (recall that (B.4) provides an upper bound):

$$\chi(G) \ge 1 - \frac{\lambda_1}{\lambda_n}. \tag{B.6}$$

The smallest eigenvalue is closely related to the characterization of linegraphs. It was noted in [**Hoffman 1977**] that the smallest eigenvalue of the linegraph of a simple graph is at least -2. This property "almost" characterizes linegraphs. It was proved in [**Cameron et al. 1976**] that *if G is a simple graph on at least 37 nodes, then $\lambda_n(G) \ge -2$ if and only if G is obtained from the line-graph $L(H)$ by deleting a set of disjoint edges from each clique of $L(H)$ corresponding to the edges of H connecting a node to nodes of degree 1.*

B.4.3. The eigenvalue gap. The difference between the second and the first eigenvalues (the "eigenvalue gap") is an extremely important parameter in many branches of mathematics, and graph theory is no exception. We start with discussing the Laplacian L of a connected graph G. The smallest eigenvalue of the Laplacian, which is zero, has multiplicity 1. It is often important to bound the next smallest eigenvalue λ_2 from below (and sometimes, from above). This is often called the *Fiedler value* of the graph. There are many good upper and lower bounds for λ_2 in terms of various graph parameters; for us, it will suffice to know that $\lambda_2 \ge n^{-2}$ (see [**Zhang 2007**] for a survey). In Section 5.4, we illustrate the use of geometric methods by proving an upper bound on the Fiedler value of planar graphs.

Since the smallest and the next smallest eigenvalues of L are equal if and only if the graph is disconnected, we can expect that the gap between them is some kind of connectivity measure of the graph. Indeed, fundamental results relate the eigenvalue gap to expansion (isoperimetric) properties of graphs. These results (which have many related, but not equivalent, versions) can be considered as discrete analogues of Cheeger's inequality in differential geometry.

We illustrate this connection by two spectral estimates of the node-expansion and edge-expansion of a graph.

Expanders. This highly nontrivial class of graphs (more precisely, class of graph sequences) is the key element in many graph-theoretic constructions. An *expander* is a regular graph with small degree in which the number of neighbors of any set containing at most half of the nodes is at least a constant factor of its size. To be precise, the *node-expansion* $\mathrm{xp}(G)$ of a graph G is defined as the largest ε such that $|N(S)\setminus S| \ge \varepsilon|S|$ for every set $S \subset V$ with $|S| \le n/2$. An (a, b)-*expander* is an a-regular graph with $\mathrm{xp}(G) \ge b$.

Expanders play an important role in many applications of graph theory, in particular in computer science. The most important expanders are d-regular expanders, where $d \ge 3$ is a small constant. Such graphs are not easy to construct. One method is to do a random construction: for example, we can pick d random perfect matchings on $2n$ nodes (independently, uniformly over all perfect matchings), and let G be the union of them. Deterministic constructions are much more difficult to obtain, and are typically based on deep algebraic facts. The first construction was given in [**Margulis 1973**]; see [**Lubotzky–Phillips–Sarnak**] for a construction that is optimal in some sense.

The following spectral bounds on the node-expansion of a d-regular graph [**Alon–Milman 1985, Alon 1986**], play an important role in analyzing the above mentioned algebraic constructions:

$$\frac{1}{2d}\mu_2 \le \mathrm{xp}(G) \le 3\sqrt{\mu_2}. \tag{B.7}$$

Since we are considering only regular graphs, these bounds are easily translated to the spectrum of the adjacency matrix or of the transition matrix.

Edge expansion. There is an equally important connection between the eigenvalue gap of the normalized Laplacian $\widehat{L}$ and a quantity that can be viewed as an edge-counting version of the expansion. For every subset $S \subseteq V$, let $\deg(S) = \sum_{i\in S} \deg(i)$, and define

$$\Phi(G) = \min_{\emptyset\subset S\subset V} \frac{2\,m\,e_G(S, V\setminus S)}{\deg(S)\,\deg(V\setminus S)}.$$

For a d-regular graph, this can be written as

$$\Phi(G) = \min_{\emptyset\subset S\subset V} \frac{n}{d}\frac{e_G(S, V\setminus S)}{|S|\,|V\setminus S|}.$$

The following basic inequality [**Jerrum–Sinclair 1989**] is a discrete analogue of Cheeger's Inequality in differential geometry:

$$\frac{\Phi(G)^2}{16} \le 1 - \nu_2 \le \Phi(G). \tag{B.8}$$

This inequality plays a central role in the theory of rapidly mixing Markov chains, which is however not the subject of this book.

B.4.4. Multiplicity of eigenvalues. Multiplicity of eigenvalues usually corresponds to symmetries in the graph (although the correspondence is not exact). It was proved in [**Mowshowitz 1971**] and [**Petersdorf–Sachs 1970**] that *if all eigenvalues of A are different, then every automorphism of A has order* 1 *or* 2.

There is also an interesting graph-theoretic property related to the other extreme case, when the graph has very few distinct eigenvalues. A graph G is called *strongly regular*, if it is regular, and there are two nonnegative integers a and b such that for every pair i, j of nodes the number of common neighbors of i and j is

$$\begin{cases} a, & \text{if } i \text{ and } j \text{ are adjacent,} \\ b, & \text{if } i \text{ and } j \text{ are nonadjacent.} \end{cases}$$

Many interesting, highly symmetrical graphs are strongly regular. These graphs can be characterized in terms of their spectra: *A connected graph G is strongly regular if and only if it is regular and A_G has at most* 3 *different eigenvalues.*

The multiplicity of the second largest eigenvalue is also important; it is discussed in Chapter 16 in connection with the Colin de Verdière number.

APPENDIX C

Convex Bodies

C.1. Polytopes and polyhedra

A *convex body* in $\mathbb{R}^d$ is a closed, bounded convex set that has an interior point. The convex hull of a finite set of points in $\mathbb{R}^d$ is called a (convex) *polytope*. The intersection of a finite number of halfspaces in $\mathbb{R}^d$ is called a (convex) *polyhedron*. (We'll drop the adjective "convex", because we never need to talk about nonconvex polyhedra.) Every polytope is a polyhedron. A polyhedron is a polytope if and only if it is bounded.

For every polyhedron, there is a unique smallest affine subspace that contains it, its affine hull. The *dimension* of a polyhedron is the dimension of its affine hull. A polyhedron [polytope] in $\mathbb{R}^d$ that has dimension d (equivalently, that has an interior point) is called a *d-polyhedron* [*d-polytope*].

A hyperplane H is said to *support* the polyhedron if it has a point in common with the polyhedron and the polyhedron is contained in one of the closed halfspaces with boundary H. A *face* of a polyhedron is its intersection with a supporting hyperplane. A face of a polyhedron that has dimension one less than the dimension of the polyhedron is called a *facet*. A face of dimension 0 (i.e., a single point) is called a *vertex*, a face of dimension 1 is called an *edge*. The vertices and edges of a polytope P form a simple graph G_P, which we call the *skeleton* of the polytope.

Every face of a polyhedron [polytope] is a polyhedron [polytope]. Every vertex of a face is a vertex of the polyhedron. Every polytope has a finite number of faces. Every polytope is the convex hull of its vertices. The set of vertices is the unique minimal finite set of points whose convex hull is the polytope.

Every facet of a d-polyhedron P spans a (unique) hyperplane, and this hyperplane is the boundary of a uniquely determined halfspace that contains the polyhedron. The polyhedron is the intersection of the halfspaces determined by its facets this way.

A *convex cone* in $\mathbb{R}^n$ is a set of vectors is closed under sum and multiplication by positive scalars. Note that according to this definition, the set $\mathbb{R}^n$ is a convex cone. We call the (topologically) closed cone *pointed*, if the origin is a vertex of it; equivalently, if it does not contain a line. Any system of homogeneous linear inequalities

$$a_1^{\mathsf{T}} x \geq 0, \; \ldots, \; a_m^{\mathsf{T}} x \geq 0$$

defines a convex cone; convex cones defined by such (finite) systems are called *polyhedral*.

Let P be a polytope in $\mathbb{R}^d$, let $\mathbf{a} \in \mathbb{R}^d$, and let $\mathbf{u}$ be a vertex of P. Suppose that P has a vertex $\mathbf{v}$ such that $\mathbf{a}^{\mathsf{T}}\mathbf{u} < \mathbf{a}^{\mathsf{T}}\mathbf{v}$. Then P has a vertex $\mathbf{w}$ such that $\mathbf{uw}$ is an edge of P, and $\mathbf{a}^{\mathsf{T}}\mathbf{u} < \mathbf{a}^{\mathsf{T}}\mathbf{w}$.

Another way of formulating this is that if we consider the linear objective function $\mathbf{a}^\mathsf{T}\mathbf{x}$ on a polytope P, then from any vertex we can walk on the skeleton to a vertex that maximizes the objective function so that the value of the objective function increases at every step. This important fact is the basis for the *Simplex Method*. For our purposes, the following corollary this fact will be important as well: If G is the skeleton of a d-polytope, H is an (open or closed) halfspace containing an interior point of the polytope, then the subgraph of G induced by those vertices of P contained in this halfspace is connected.

C.2. Polyhedral combinatorics

As mentioned in the Preface, many applications of geometry in combinatorics make use of polyhedra defined by combinatorial structures. We cannot develop this large theory here, but we need some of these polyhedra.

The Stable Set Polytope of a graph is an important example studied in Section 11.4.2. Here we describe a couple of other families related to flows and cuts only. We formulate these examples for directed graphs, but we can get the corresponding definitions and facts for undirected graphs by replacing every undirected edge by two, oppositely directed edges.

A few special properties of polytopes (more generally, of convex sets) are often encountered in combinatorial situations. A convex set $K \subseteq \mathbb{R}^d$ is called *ascending*, if $\mathbf{x} \in K$, $\mathbf{y} \in \mathbb{R}^d$ and $\mathbf{y} \geq \mathbf{x}$ implies that $\mathbf{y} \in K$. For any convex body K, we define its *dominant* as

$$\widehat{K} = \{y \in \mathbb{R}^d : \exists x \in K,\ x \leq y\}.$$

This dominant is an ascending convex set. A convex set $K \subseteq \mathbb{R}^d$ is called a *convex corner*, if $K \subseteq \mathbb{R}^d_+$, and $\mathbf{x} \in K$, $\mathbf{y} \in \mathbb{R}^d_+$ and $\mathbf{y} \leq \mathbf{x}$ implies that $\mathbf{y} \in K$. A polytope which is a convex corner will be called a *corner polytope*.

C.2.1. Flows and cuts.

In the following examples of particularly important combinatorial polyhedra, we fix a directed graph G and two nodes $s, t \in V$. While there will be perhaps more versions than one would like to memorize, these are often related to each other in simple ways (containment, domination). We consider directed graphs, but corresponding notions and facts for undirected graphs can be obtained by replacing each edge by a pair of oppositely oriented edges.

We start with the *path polytope*, which is a polytope in $\mathbb{R}^d_+$, defined as the convex hull of indicator vectors of edge-sets of directed s-t paths. Every vector in the path polytope is an s-t flow of value 1. If we add circulations to these flows, we get all s-t flows of value 1. Such flows form the *1-valued flow polyhedron* (this is not a polytope, since circulations can be unbounded).

The *flow polyhedron* is the set of all s-t-flows. The 1-valued flow polyhedron is the intersection of it with a hyperplane.

Often one considers the *feasible flow polytope*: every edge e has a capacity $c_e \geq 0$, and we impose the conditions $x_e \leq c_e$ on the flow values x_e. Those capacity assignments that allow a feasible flow of value 1 also form a convex polyhedron; by the Max-Flow-Min-Cut Theorem, this is described by the inequalities

$$\begin{aligned} c_e &\geq 0, \\ \sum_{e \in C} c_e &\geq 1 \qquad \text{(for every } s\text{-}t \text{ cut } C). \end{aligned}$$

This polyhedron is just the dominant of the path polytope, and we call it the *path polyhedron*.

We can also work on the node set of G. The *node-path polytope* is the convex hull of indicator vectors of interior nodes of directed s-t paths. This is a polytope in $\mathbb{R}^{V\setminus\{s,t\}}$. Its dominant, the *node-path polyhedron* can be described by the inequalities

$$x_v \geq 0,$$
$$\sum_{v\in S} x_v \geq 1 \qquad (S \subseteq V \text{ separates } s \text{ and } t).$$

Closely related to the above constructions is another class of polyhedra. The *cut polytope* is the convex hull of indicator vectors of directed s-t cuts. Its dominant, the *cut polyhedron* consists of all assignments of nonnegative "lengths" x_e to the edges for which the length of every s-t path is at least 1; formally,

$$x_e \geq 0,$$
$$\sum_{e\in E(P)} x_e \geq 1 \qquad (\text{for every } s\text{-}t \text{ path } P).$$

Again, we can consider the version for nodes. The *node-cut polytope*, defined in $\mathbb{R}^{V\setminus\{s,t\}}$, is the convex hull of indicator vectors of s-t node-cuts (minimal subsets of $V\setminus\{s,t\}$ intersecting every s-t path). The *node-cut polyhedron* is the dominant of the node-cut polytope. The node-cut polyhedron can be described by the inequalities

$$x_v \geq 0,$$
$$\sum_{v\in S} x_v \geq 1 \qquad (S \subseteq V \text{ separates } s \text{ and } t).$$

C.3. Polarity and variations

C.3.1. Polarity. Let K be a convex body containing the origin as an interior point. The *polar* of K is defined by

$$K^* = \{\mathbf{x} \in \mathbb{R}^d : \mathbf{x}^\mathsf{T}\mathbf{y} \leq 1 \ \forall \mathbf{y} \in K\}.$$

It is clear that K^* is a convex body as well, containing the origin in its interior. For every convex body K we have $(K^*)^* = K$.

The polar of a polytope (containing the origin) is a polytope. Platonic bodies in the 3-dimensional space come in polar pairs, except that the regular tetrahedron is congruent to its own polar (Figure C.1).

For every vertex $\mathbf{v}$ of a polytope P, the inequality $\mathbf{v}^\mathsf{T}\mathbf{x} \leq 1$ defines a facet of P^*, and vice versa. The vector $\mathbf{v}$ is a normal vector of the facet $\mathbf{v}^\mathsf{T}\mathbf{x} \leq 1$. More generally, if $\mathbf{v}_0, \dots, \mathbf{v}_m$ are the vertices of a k-dimensional face F of P, then

$$F^\perp = \{\mathbf{x} \in P^* : \mathbf{v}_0^\mathsf{T}\mathbf{x} = 1, \dots, \mathbf{v}_m^\mathsf{T}\mathbf{x} = 1\}$$

defines a $d-k-1$-dimensional face of P^*. Furthermore, $(F^\perp)^\perp = F$ (Figure C.2).

We can, at least formally, define K^* for more general convex sets. If K is convex cone with its apex at the origin, then the condition $\mathbf{x}^\mathsf{T}\mathbf{y} \leq 1$ for all $\mathbf{y} \in K$ implies that $\mathbf{x}^\mathsf{T}\mathbf{y} \leq 0$. So the polar of a convex cone is a convex cone. It is often more convenient to work with the negative of this, called the *dual cone*, defined as

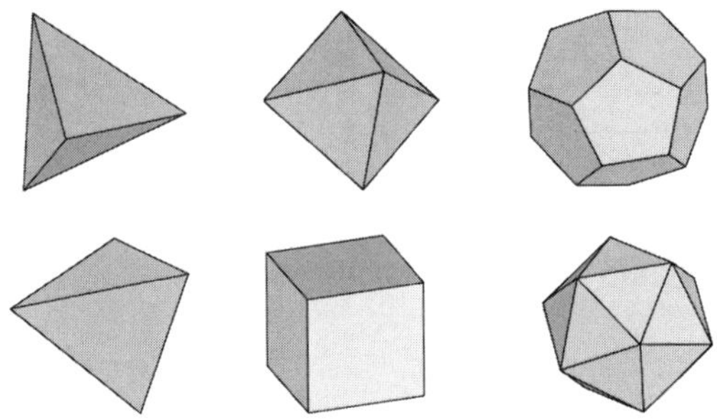

FIGURE C.1. The 5 platonic bodies, in polar pairs (the tetrahedron is congruent to its own polar).

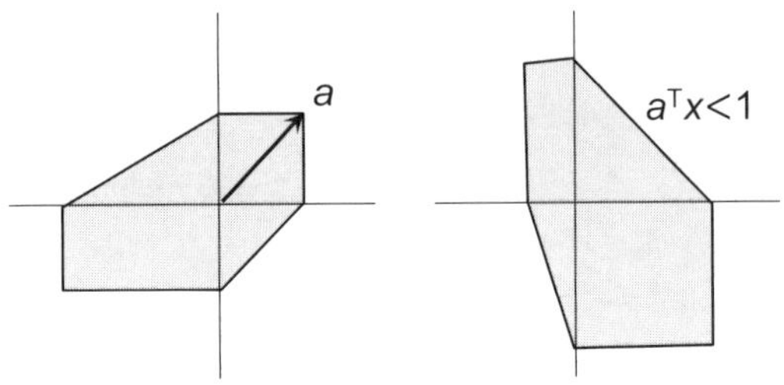

FIGURE C.2. A 2-dimensional polytope and its polar.

$\{x \in \mathbb{R}^n : x^\mathsf{T} y \geq 0 \ \forall y \in K\}$. This is again a convex cone. If K is closed (in the topological sense), then applying either duality or polarity twice, we get K back.

The nonnegative orthant is self-dual, and so is the cone of positive semidefinite matrices. If K is a linear subspace, then its polar (or dual) is the orthogonal subspace $K^\perp$.

C.3.2. Blocking. There are two constructions similar to polarity that concern convex bodies not containing the origin in their interior; rather, contained in the nonnegative orthant.

The *blocker* of an ascending convex set $K \subseteq \mathbb{R}^d_+$ is defined by

$$K^{\mathrm{bl}} = \{\mathbf{x} \in \mathbb{R}^d_+ : \ \mathbf{x}^\mathsf{T}\mathbf{y} \geq 1 \forall \mathbf{y} \in K\}$$

(Figure C.3).

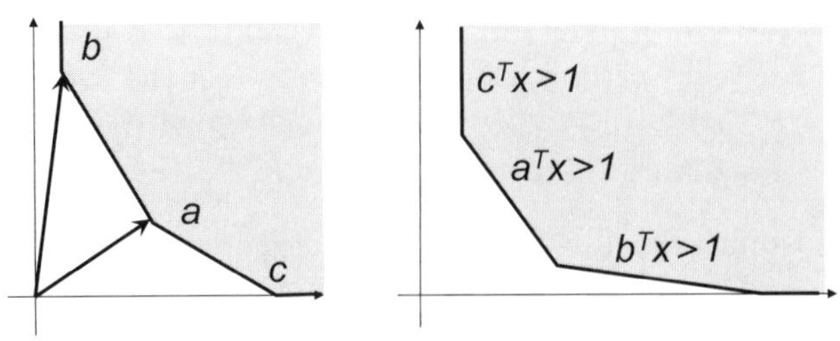

FIGURE C.3. A 2-dimensional ascending polytope and its blocker.

The blocker of an ascending convex set is a closed ascending convex set. The blocker of an ascending polyhedron is an ascending polyhedron. For every ascending closed convex set K, we have $(K^{\mathrm{bl}})^{\mathrm{bl}} = K$.

The correspondence between faces of a polyhedron P and those of P^{bl} is a bit more complicated than for polarity, and we describe the relationship between vertices and facets only. Every vertex $\mathbf{v}$ of P gives rise to a facet of P^{bl}, determined by the supporting hyperplane $\mathbf{v}^{\mathsf{T}}\mathbf{x} \geq 1$. This construction gives all the facets of P^{bl}, except possibly those corresponding to the nonnegativity constraints $x_i \geq 0$, which may or may not define facets.

EXAMPLE C.1. Fixing two nodes s and t of a graph G, the blocker of the path polyhedron is the cut polyhedron. The blocker of the node-path polyhedron is the node-cut polyhedron. ♦

It is easy to see that every ascending closed convex set K has a unique point $\mathbf{x}$ which is closest to the origin. (This point has combinatorial significance in several cases.) It is not hard to prove that this point is closely related to the analogous point of the blocker.

PROPOSITION C.2. *Let $K \subset \mathbb{R}_+^d$ be an ascending closed convex set, and let $\mathbf{x} \in K$ minimize the objective function $|\mathbf{x}|^2$. Then $\mathbf{y} = (1/|\mathbf{x}|^2)\mathbf{x}$ is in the blocker K^{bl}, and it minimizes the objective function $|\mathbf{y}|^2$ over K^{bl}. It follows in particular that $d(0,P)\,d(0,P^{\mathrm{bl}}) = 1$.* □

C.3.3. Antiblocking. The *antiblocker* of a convex corner K is defined by

$$K^{\mathrm{abl}} = \{\mathbf{x} \in \mathbb{R}_+^d : \mathbf{x}^{\mathsf{T}}\mathbf{y} \leq 1 \forall \mathbf{y} \in K\}$$

(see Figure C.4).

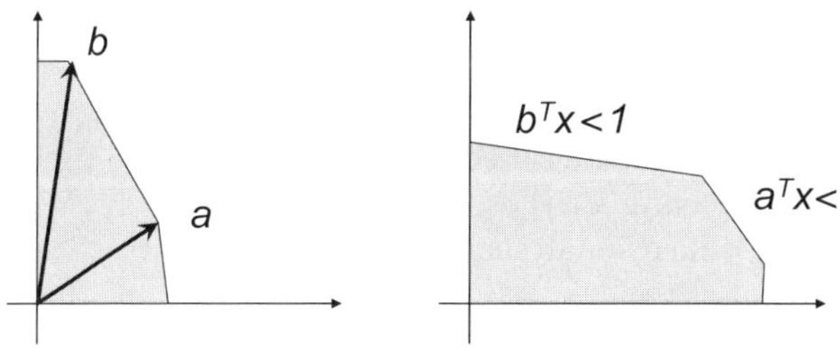

FIGURE C.4. A 2-dimensional convex corner and its antiblocker

The antiblocker of a convex corner is a convex corner. For every convex corner K, we have $(K^{\mathrm{abl}})^{\mathrm{abl}} = K$.

If the convex corner P is a polytope, then so is its antiblocker P^{abl}. To describe the correspondence between the faces of P and those of P^{abl} is again a more complicated than for the polars. The nonnegativity constraints $x_i \geq 0$ always define facets, and they do not correspond to vertices in the antiblocker. All other facets of P correspond to vertices of P^{abl}. Not every vertex of P defines a facet in P^{abl}. The origin is a trivial exceptional vertex, but there may be further exceptional vertices. A vertex of P defines a facet of P^* if and only if there is no other vertex $\mathbf{w}$ such that $\mathbf{v} \leq \mathbf{w}$.

C.3.4. Gale duality. Let V be a finite set of n elements, and let $\mathbf{u}: V \to \mathbb{R}^d$ be a vector labeling of V, where we assume that $\mathbf{u}(V)$ spans $\mathbb{R}^d$. Consider $\mathbf{u}$ as a $d \times V$ matrix A, then $\mathrm{rk}(A) = d$, and the row space L of A d-dimensional subspace L of $\mathbb{R}^V$. Let $b_1, \dots, b_{n-d}$ be a basis of $L^\perp$, which we consider as the rows of an $(n-d) \times V$ matrix B. Let $\{\mathbf{v}_j : j \in V\}$ be the columns of the matrix B. The vector labeling $\mathbf{v}$ of V is called a *Gale dual* of $\mathbf{u}$.

Being Gale duals is a symmetric relation between vector labelings of the same set. The Gale dual is not uniquely determined: choosing a different basis in $L^\perp$ results in a different Gale dual. But all Gale duals are images of each other under linear transformations of $\mathbb{R}^{n-d}$.

Let $\mathbf{v}: V \to \mathbb{R}^{n-d}$ be a Gale dual of the vector labeling $\mathbf{u}: V \to \mathbb{R}^d$. Let $X \subseteq V$, $|X| = d$ be a basis for $\mathbf{u}$. Then $V \setminus X$ is a basis for $\mathbf{v}$. More generally, we can express the rank of a subset of V in the labeling $\mathbf{v}$ in terms of the rank of its complement in the labeling $\mathbf{u}$:

$$\mathrm{rk}(\mathbf{v}(X)) = \mathrm{rk}(\mathbf{u}(V \setminus X)) + |X| - d. \tag{C.1}$$

C.4. Balls, spheres and caps

C.4.1. Volume. We denote by π_d the volume of the d-dimensional unit ball B^d. It is known that

$$\pi_d = \frac{\pi^{d/2}}{\Gamma(1+\frac{d}{2})} = \begin{cases} \dfrac{\pi^k}{k!} & \text{if } d = 2k, \\[2ex] \dfrac{2^{k+1}\pi^k}{(2k+1)!!} & \text{if } d = 2k+1. \end{cases} \tag{C.2}$$

For both the even and odd case, Stirling's Formula gives the asymptotics

$$\pi_d \sim \frac{1}{\sqrt{\pi d}} \Big(\frac{2e\pi}{d}\Big)^{d/2}. \tag{C.3}$$

(It is funny that both classical transcendental numbers e and π occur together here.) The surface area of this ball (the $(d-1)$-dimensional measure of the sphere S^{d-1}) is $d\pi_d$.

A cap on this ball can be described as the set $C_{\mathbf{u},s} = \{x \in S^{d-1} : \mathbf{u}^T x \ge s\}$, where $\mathbf{u} \in S^{d-1}$ is the center of the cap, and $-1 \le s \le 1$. We will need good bounds on the $(d-1)$-dimensional surface area of this cap; it suffices to restrict our attention to the case when $s > 0$. In dimension $d = 3$, the area of this cap is $2\pi(1-s)$. If the spherical radius of this cap is ρ, then this area can also be expressed as $4\pi \sin^2(\rho/2)$.

In higher dimension, the surface area cannot be expressed so explicitly. It is easy to derive the integral formula:

$$\mathrm{vol}_{n-1}(C_{\mathbf{u},s}) = (n-1)\pi_{n-1} \int_s^1 (1-x^2)^{(n-3)/2}\, dx. \tag{C.4}$$

but it is hard to bring this into a more manageable form. Instead of explicit formulas, the following estimates will be more useful for us:

$$\begin{aligned}\Big(1 - \frac{2}{s\sqrt{n-1}}\Big)\frac{\pi_{n-1}}{s}(1-s^2)^{(n-1)/2} &\le \mathrm{vol}_{n-1}(C_{\mathbf{u},s}) \\ &\le \frac{\pi_{n-1}}{s}(1-s^2)^{(n-1)/2}.\end{aligned} \tag{C.5}$$

(The lower bound is meaningful for $s \geq 2/\sqrt{n-1}$.) The upper bound follows from the estimate

$$\int\limits_s^1 (1-x^2)^{(n-3)/2}\,dx \leq \frac{1}{s}\int\limits_s^1 x(1-x^2)^{(n-3)/2}\,dx = \frac{1}{(n-1)s}(1-s^2)^{(n-1)/2}.$$

The lower bound uses a similar argument with a twist. Let $s \leq t \leq 1$, then

$$\int\limits_s^1 (1-x^2)^{(n-3)/2}\,dx \geq \int\limits_s^t (1-x^2)^{(n-3)/2}\,dx \geq \frac{1}{t}\int\limits_s^t x(1-x^2)^{(n-3)/2}\,dx$$
$$= \frac{1}{(n-1)t}\Big((1-s^2)^{(n-1)/2} - (1-t^2)^{(n-1)/2}\Big)$$

Optimizing for t, we get the estimate with a somewhat tedious but routine computation.

C.5. Volume of convex bodies

C.5.1. Classical inequalities. We start with a simple formula. Let K be a convex body in $\mathbb{R}^n$ containing the origin in its interior, and for every unit vector $\mathbf{v}$, let $h_K(\mathbf{v})$ denote the length of the segment in K of the semiline determined by $\mathbf{v}$. Choosing $\mathbf{v}$ randomly and uniformly from S^{n-1}, we have

(C.6) $$\mathrm{vol}(K) = \pi_n \mathsf{E}(h_K(\mathbf{v})^n).$$

Let $K_1, \dots, K_d$ be convex bodies in $\mathbb{R}^d$. It is not hard to see that the function $f(x_1, \dots, x_d) = \mathrm{vol}(x_1K_1 + \dots + x_dK_d)$ is a homogeneous polynomial of degree d in the variables $x_1, \dots, x_r \geq 0$. The coefficient of $x_1 \dots x_d$ in this polynomial, divided by $d!$, is called the *mixed volume* of $K_1, \dots, K_d$, denoted by $V(K_1, \dots, K_d)$.

This quantity has many deep and useful properties. It is easy to see that $V(K, \dots, K) = \mathrm{vol}(K)$. It is less trivial that the mixed volume is nonnegative. A deep inequality concerning mixed volume is the Alexandrov–Fenchel Inequality:

(C.7) $$V(K_1, K_1, K_3, \dots, K_d)V(K_2, K_2, K_3, \dots, K_d) \leq V(K_1, K_2, K_3, \dots, K_d)^2.$$

One can derive the Brunn–Minkowski Theorem from the Alexandrov–Fenchel Inequality (but it is easier to prove):

For any two convex bodies $A, B \subset \mathbb{R}^d$, the value $\mathrm{vol}(tA + (1-t)B)^{1/d}$ *is a concave function of* $0 \leq t \leq 1$.

We need a consequence of this theorem. For a convex body K, we define its *difference body* $K - K = \{\mathbf{x} - \mathbf{y} : \mathbf{x}, \mathbf{y} \in K\}$. It is trivial that $\mathrm{vol}(K - K) \geq \mathrm{vol}(K)$ (since $K - K$ contains a translated copy of K). The Brunn–Minkowski Theorem implies a much sharper bound:

(C.8) $$\mathrm{vol}(K - K) = 2^d\,\mathrm{vol}\Big(\frac{1}{2}K + \frac{1}{2}(-K)\Big) \geq 2^d\,\mathrm{vol}(K).$$

The volumes of a convex body and its polar are closely related. For every 0-symmetric convex body $K \subseteq \mathbb{R}^d$,

(C.9) $$\frac{\pi^d}{d!} \leq \mathrm{vol}(K)\,\mathrm{vol}(K^*) \leq \pi_d^2.$$

Using (C.2), it is easy to see that the upper bound is less than $(2\pi)^d/d!$, which is attained when both K and K^* are balls; this was proved for $d \leq 3$ by

[**Blaschke 1917**] and for all dimensions by [**Santaló 1949**]. The lower bound (with a smaller constant instead of π) was proved by [**Bourgain–Milman 1987**]. The currently known best bound above was proved by [**Kuperberg 2008**]. Mahler's unsettled conjecture states that the minimum of this product over n-dimensional 0-symmetric convex bodies is attained when K is a cube and K^* is a cross-polytope (or the other way around), which would give a lower bound of $4^d/d!$.

C.5.2. Projection. We conclude with some inequalities concerning the length of projections in high dimensions; while these are well known and often used, it does not seem easy to find their proofs, and so I include proofs for several of them.

For a set $K \subseteq \mathbb{R}^n$ and linear subspace $L \subseteq \mathbb{R}^n$, let K_L denote the orthogonal projection of K onto L. In the case when L is a one-dimensional subspace generated by a vector $\mathbf{v}$, then we also use the notation $K_\mathbf{v} = K_L$.

The question about the area of caps can be generalized as follows: given a number $0 < c < 1$, what is the surface area of S^{n-1} whose projection onto the subspace formed by the first d coordinates has length at most c? If $d = n-1$, then this set consists of two congruent caps, but if $d < n$, then the set is more complicated. The cases $c \le \sqrt{d/n}$ and $c > \sqrt{d/n}$ are quite different. We only need the first one in this book.

Let $\mathbf{x}'$ and $\mathbf{x}''$ denote the projections of a vector $\mathbf{x}$ onto the first d coordinates and onto the last $n-d$ coordinates, respectively. It is easy to see that if $\mathbf{u}$ is chosen randomly, uniformly from S^{n-1}, then

$$\mathsf{E}(|\mathbf{u}'|^2) = \frac{d}{n}. \tag{C.10}$$

Let us assume that $d \le n-2$ and (for the time being) $c < 1$. We want to find good estimates for the $(n-1)$-dimensional measure of the set

$$A = \{\mathbf{x} \in S^{n-1} : |\mathbf{x}'| \le s\}.$$

Define $\phi(\mathbf{x}) = \mathbf{x}' + (\mathbf{x}'')^0$, then ϕ maps A onto $A' = (sB^d) \times S^{n-d-1}$ bijectively. It is not hard to compute that the Jacobian of ϕ^{-1} (as a map $A' \to A$) is $(1-|\mathbf{x}'|^2)^{(n-d-2)/2} \le 1$, and hence

$$\mathrm{vol}_{n-1}(A) \le \mathrm{vol}_{n-1}(A') = s^d(n-d)\pi_d\pi_{n-d}. \tag{C.11}$$

(For $n = d+2$ the Jacobian is identically 1, whence $\mathrm{vol}_{n-1}(A) = \mathrm{vol}_{n-1}(A')$, which implies, for $n = 3$, the well-known formula for the surface area of caps mentioned before.)

Another way of formulating this inequality is that choosing a random vector $\mathbf{u}$ uniformly from S^{n-1}, it bounds the probability that its orthogonal projection to a d-dimensional subspace is small:

$$\mathsf{P}(|\mathbf{u}'| \le s) \le \frac{(n-d)\pi_d\pi_{n-d}}{n\pi_n} s^d. \tag{C.12}$$

Using the asymptotics for factorials, it is easy to derive the slightly weaker but more explicit bound

$$\mathsf{P}(|\mathbf{u}'| \le s) < \Big(2s\sqrt{\frac{n}{d}}\Big)^d. \tag{C.13}$$

This estimate is interesting when s is smaller than the expectation of $|\mathbf{u}'|$, say $s = \varepsilon\sqrt{d/n}$ for some $0 < \varepsilon < 1$. Then we get

$$\mathsf{P}\Big(|\mathbf{u}'| \le \varepsilon\sqrt{\frac{d}{n}}\Big) < (2\varepsilon)^d. \tag{C.14}$$

To prove tighter bounds on the concentration of $|\mathbf{u}'|$ around its expectation, we can use an explicit description of generating a uniform random element of S^{n-1}. We generate n independent standard Gaussian variables $X_1, \dots, X_n$, and normalize:

$$\mathbf{u} = \frac{1}{\sqrt{X_1^2 + \dots + X_n^2}}(X_1, \dots, X_n). \tag{C.15}$$

Here the random variable $X_1^2 + \dots + X_n^2$ is from a chi-squared distribution with parameter n, and many estimates are known for its distribution. We need the following estimate (see e.g. [**Massart 2003**])

$$\mathsf{P}(|X_1^2 + \dots + X_n^2 - n| > \varepsilon n) \le 2e^{-\varepsilon^2 n/8}. \tag{C.16}$$

If we generate a random unit vector $\mathbf{u}$ as described above, then

$$|\mathbf{u}'|^2 = \frac{X_1^2 + \dots + X_d^2}{X_1^2 + \dots + X_n^2}.$$

By (C.16), the numerator is concentrated around d, and the denominator is concentrated around n. Hence the length of $\mathbf{u}'$ is concentrated around $\sqrt{d/n}$. To be more precise, it follows that for every $\varepsilon > 0$,

$$\frac{1}{1+\varepsilon}\sqrt{\frac{d}{n}} \le |\mathbf{u}'| \le (1+\varepsilon)\sqrt{\frac{d}{n}} \tag{C.17}$$

with probability at least $1 - 4e^{-\varepsilon^2 d/4}$.

Next, we turn to projections of convex bodies. Let $K \subset \mathbb{R}^n$ be a convex body, let $\mathbf{v} \in S^{n-1}$ be chosen randomly from the uniform distribution, and let $K_\mathbf{v}$ be the projection of K onto the line containing $\mathbf{v}$. Let $|K_\mathbf{v}|$ denote the length of this projection, which is the width of K in direction $\mathbf{v}$. It is easy to check that

$$|K_\mathbf{v}| = \frac{1}{h_{(K-K)^*}(\mathbf{v})}. \tag{C.18}$$

Combining with (C.6), we can express the volume of $(K-K)^*$:

$$\mathrm{vol}\big((K-K)^*\big) = \pi_n \mathsf{E}\big(h_{(K-K)^*}(\mathbf{v})^n\big) = \pi_n \mathsf{E}\big(|K_\mathbf{v}|^{-n}\big). \tag{C.19}$$

By the Blaschke–Santaló Inequality (C.9) and (C.8), this implies

$$\mathsf{E}\big(|K_\mathbf{v}|^{-n}\big) = \frac{\mathrm{vol}\big((K-K)^*\big)}{\pi_n} \le \frac{\pi_n}{\mathrm{vol}(K-K)} \le \frac{\pi_n}{2^n\,\mathrm{vol}(K)}. \tag{C.20}$$

Markov's Inequality implies that for every $s > 0$,

$$\mathsf{P}(|K_\mathbf{v}| \le s) = \mathsf{P}(|K_\mathbf{v}|^{-n} s^n \ge 1) \le \frac{\pi_n s^n}{2^n\,\mathrm{vol}(K)}. \tag{C.21}$$

This inequality can be extended to the case when K is not full-dimensional, but d-dimensional ($d < n$), and so we have to consider a lower dimensional measure of its volume. Let $\mathbf{u}$ be the projection of $\mathbf{v}$ onto $\mathbb{R}^d$. The length $|\mathbf{u}|$ and the direction $\mathbf{u}^0$ are independent random variables. Clearly $|K_{\mathbf{v}}| = |\mathbf{u}|\,|K_{\mathbf{u}}|$, where $|\mathbf{u}|$ and $K_{\mathbf{u}} = K_{\mathbf{u}^0}$ are independent as random variables. Then, using (C.12),

$$\begin{aligned} \mathsf{P}(|K_{\mathbf{v}}| \le s) &= \mathsf{P}\Big(|\mathbf{u}| \le \frac{s}{|K_{\mathbf{u}}|}\Big) \le \mathsf{E}_{\mathbf{u}^0}\Big(\frac{(n-d)\pi_d\pi_{n-d}}{n\pi_n}\Big(\frac{s}{|K_{\mathbf{u}^0}|}\Big)^d\Big) \\ &= \frac{(n-d)\pi_d\pi_{n-d}}{n\pi_n} s^d \mathsf{E}_{\mathbf{u}^0}(|K_{\mathbf{u}^0}|^{-d}), \end{aligned}$$

and using (C.20) (noting that the projection $K \to K_{\mathbf{u}^0}$ happens in dimension d) we get

$$\mathsf{P}(|K_{\mathbf{v}}| \le s) \le \frac{(n-d)\pi_d^2\pi_{n-d}}{n\pi_n 2^d \operatorname{vol}(K)} s^d. \tag{C.22}$$

Applying the asymptotic formula for π_n, we get the more explicit bound

$$\mathsf{P}(|K_{\mathbf{v}}| \le s) \le \Big(\frac{4s\sqrt{n}}{d}\Big)^d \frac{1}{\operatorname{vol}(K)}. \tag{C.23}$$

Bibliography

M. Ajtai, J. Komlós, and E. Szemerédi, *A note on Ramsey numbers*, J. Combin. Theory Ser. A **29** (1980), no. 3, 354–360, DOI 10.1016/0097-3165(80)90030-8. MR600598

A. Y. Alfakih, *Graph connectivity and universal rigidity of bar frameworks*. part 3, Discrete Appl. Math. **217** (2017), no. part 3, 707–710, DOI 10.1016/j.dam.2016.10.008. MR3579945

A. Y. Alfakih and Y. Ye, *On affine motions and bar frameworks in general position*, Linear Algebra Appl. **438** (2013), no. 1, 31–36, DOI 10.1016/j.laa.2012.08.031. MR2993361

F. Alizadeh: Combinatorial optimization with semi-definite matrices, in: *Integer Programming and Combinatorial Optimization* (Proceedings of IPCO '92), (eds. E. Balas, G. Cornuéjols and R. Kannan), Carnegie Mellon University Printing (1992), 385–405.

F. Alizadeh, *Interior point methods in semidefinite programming with applications to combinatorial optimization*, SIAM J. Optim. **5** (1995), no. 1, 13–51, DOI 10.1137/0805002. MR1315703

N. Alon, *Eigenvalues and expanders*, Combinatorica **6** (1986), no. 2, 83–96, DOI 10.1007/BF02579166. Theory of computing (Singer Island, Fla., 1984). MR875835

N. Alon, *The Shannon capacity of a union*, Combinatorica **18** (1998), no. 3, 301–310, DOI 10.1007/PL00009824. MR1721946

N. Alon, *Explicit Ramsey graphs and orthonormal labelings*, Electron. J. Combin. **1** (1994), Research Paper 12, approx. 8. MR1302331

N. Alon: Lovász, vectors, graphs and codes,
`https://www.tau.ac.il/~nogaa/PDFS/l170.pdf`

N. Alon, I. Balla, L. Gishboliner, A. Mond and F. Mousset: The Minrank of Random Graphs over Arbitrary Fields,
`https://arxiv.org/abs/1809.01873`

N. Alon, E. Fischer, M. Krivelevich, and M. Szegedy, *Efficient testing of large graphs*, Combinatorica **20** (2000), no. 4, 451–476, DOI 10.1007/s004930070001. MR1804820

N. Alon and N. Kahale, *Approximating the independence number via the θ-function*, Math. Programming **80** (1998), no. 3, Ser. A, 253–264, DOI 10.1007/BF01581168. MR1603356

N. Alon and L. Lovász, *Unextendible product bases*, J. Combin. Theory Ser. A **95** (2001), no. 1, 169–179, DOI 10.1006/jcta.2000.3122. MR1840483

N. Alon and E. Lubetzky, *The Shannon capacity of a graph and the independence numbers of its powers*, IEEE Trans. Inform. Theory **52** (2006), no. 5, 2172–2176, DOI 10.1109/TIT.2006.872856. MR2234473

N. Alon and V. D. Milman, *λ_1, isoperimetric inequalities for graphs, and superconcentrators*, J. Combin. Theory Ser. B **38** (1985), no. 1, 73–88, DOI 10.1016/0095-8956(85)90092-9. MR782626

N. Alon and P. D. Seymour, *A counterexample to the rank-coloring conjecture*, J. Graph Theory **13** (1989), no. 4, 523–525, DOI 10.1002/jgt.3190130413. MR1010585

N. Alon, P. Seymour, and R. Thomas, *Planar separators*, SIAM J. Discrete Math. **7** (1994), no. 2, 184–193, DOI 10.1137/S0895480191198768. MR1271990

N. Alon and J. H. Spencer, *The probabilistic method*, Wiley-Interscience Series in Discrete Mathematics and Optimization, John Wiley & Sons, Inc., New York, 1992. With an appendix by Paul Erdős; A Wiley-Interscience Publication. MR1140703

N. Alon and M. Szegedy, *Large sets of nearly orthogonal vectors*, Graphs Combin. **15** (1999), no. 1, 1–4. MR1684496

A. Ambainis, J. Kempe, and O. Sattath, *A quantum Lovász local lemma*, J. ACM **59** (2012), no. 5, Art. 24, 24, DOI 10.1145/2371656.2371659. MR2995823
Rosenstiehl, Gordon and Breach, New York (1967), 9–19.

B. Andrásfai, *On critical graphs* (English, with French summary), Theory of Graphs (Internat. Sympos., Rome, 1966), gordon and Breach, New York; Dunod, Paris, 1967, pp. 9–19. MR0221964

E. M. Andreev, *Convex polyhedra in Lobačevskiĭ spaces* (Russian), Mat. Sb. (N.S.) **81 (123)** (1970), 445–478. MR0259734

E. M. Andreev, *Convex polyhedra of finite volume in Lobačevskiĭ space* (Russian), Mat. Sb. (N.S.) **83 (125)** (1970), 256–260. MR0273510

O. Angel and O. Schramm, *Uniform infinite planar triangulations*, Comm. Math. Phys. **241** (2003), no. 2-3, 191–213, DOI 10.1007/978-1-4419-9675-6_16. MR2013797

L. Asimow and B. Roth, *The rigidity of graphs*, Trans. Amer. Math. Soc. **245** (1978), 279–289, DOI 10.2307/1998867. MR511410

L. Babai and P. Frankl: *Linear Algebra Methods in Combinatorics, with Applications to Geometry and Computer Science*, Univ. Chicago Dept. Computer Sci. 1992 (unpublished preliminary version 2, 216 pages).

R. Bacher and Y. Colin de Verdière, *Multiplicités des valeurs propres et transformations étoile-triangle des graphes* (French, with English and French summaries), Bull. Soc. Math. France **123** (1995), no. 4, 517–533. MR1373946

E. G. Bajmóczy and I. Bárány, *On a common generalization of Borsuk's and Radon's theorem*, Acta Math. Acad. Sci. Hungar. **34** (1979), no. 3-4, 347–350 (1980), DOI 10.1007/BF01896131. MR565677

S. Beigi: Entanglement-assisted zero-error capacity is upper bounded by the Lovasz theta function, *Phys. Rev. A* **82** (2010), 010303(R).

J. S. Bell, *On the Einstein Podolsky Rosen paradox*, Phys. Phys. Fiz. **1** (1964), no. 3, 195–200, DOI 10.1103/PhysicsPhysiqueFizika.1.195. MR3790629

I. Benjamini and L. Lovász: Global Information from Local Observation, *Proc. 43rd Ann. Symp. on Found. of Comp. Sci.* (2002) 701–710.

I. Benjamini and L. Lovász, *Harmonic and analytic functions on graphs*, J. Geom. **76** (2003), no. 1-2, 3–15, DOI 10.1007/s00022-033-1697-8. Combinatorics, 2002 (Maratea). MR2005527

I. Benjamini and O. Schramm, *Harmonic functions on planar and almost planar graphs and manifolds, via circle packings*, Invent. Math. **126** (1996), no. 3, 565–587, DOI 10.1007/s002220050109. MR1419007

I. Benjamini and O. Schramm, *Random walks and harmonic functions on infinite planar graphs using square tilings*, Ann. Probab. **24** (1996), no. 3, 1219–1238, DOI 10.1214/aop/1065725179. MR1411492

I. Benjamini and O. Schramm, *Recurrence of distributional limits of finite planar graphs*, Electron. J. Probab. **6** (2001), no. 23, 13, DOI 10.1214/EJP.v6-96. MR1873300

C. H. Bennett, D. P. DiVincenzo, T. Mor, P. W. Shor, J. A. Smolin, and B. M. Terhal, *Unextendible product bases and bound entanglement*, Phys. Rev. Lett. **82** (1999), no. 26, 5385–5388, DOI 10.1103/PhysRevLett.82.5385. MR1789694

C. Berge, *Graphs and hypergraphs*, North-Holland Publishing Co., Amsterdam-London; American Elsevier Publishing Co., Inc., New York, 1973. Translated from the French by Edward Minieka; North-Holland Mathematical Library, Vol. 6. MR0357172

A. Björner and L. Lovász, *Pseudomodular lattices and continuous matroids*, Acta Sci. Math. (Szeged) **51** (1987), no. 3-4, 295–308. MR940934

W. Blaschke: Uber affine Geometrie VII: Neue Extremeingenschaften von Ellipse und Ellipsoid, *Ber. Verh. Sächs. Akad. Wiss. Math. Phys. Klass* **69** (1917), 412–420.

A. Blum and D. Karger, *An $\tilde{O}(n^{3/14})$-coloring algorithm for 3-colorable graphs*, Inform. Process. Lett. **61** (1997), no. 1, 49–53, DOI 10.1016/S0020-0190(96)00190-1. MR1439867

B. Bollobás, *On generalized graphs* (English, with Russian summary), Acta Math. Acad. Sci. Hungar. **16** (1965), 447–452, DOI 10.1007/BF01904851. MR0183653

A. I. Bobenko, C. Mercat, and Y. B. Suris, *Linear and nonlinear theories of discrete analytic functions. Integrable structure and isomonodromic Green's function*, J. Reine Angew. Math. **583** (2005), 117–161, DOI 10.1515/crll.2005.2005.583.117. MR2146854

A. I. Bobenko and B. A. Springborn, *Variational principles for circle patterns and Koebe's theorem*, Trans. Amer. Math. Soc. **356** (2004), no. 2, 659–689, DOI 10.1090/S0002-9947-03-03239-2. MR2022715

T. Böhme, *On spatial representations of graphs*, Contemporary methods in graph theory, Bibliographisches Inst., Mannheim, 1990, pp. 151–167. MR1126225

C. Borgs, J. Chayes, and L. Lovász, *Moments of two-variable functions and the uniqueness of graph limits*, Geom. Funct. Anal. **19** (2010), no. 6, 1597–1619, DOI 10.1007/s00039-010-0044-0. MR2594615

C. Borgs, J. T. Chayes, L. Lovász, V. T. Sós, and K. Vesztergombi, *Convergent sequences of dense graphs. I. Subgraph frequencies, metric properties and testing*, Adv. Math. **219** (2008), no. 6, 1801–1851, DOI 10.1016/j.aim.2008.07.008. MR2455626

J. Bourgain, *On Lipschitz embedding of finite metric spaces in Hilbert space*, Israel J. Math. **52** (1985), no. 1-2, 46–52, DOI 10.1007/BF02776078. MR815600

J. Bourgain and V. D. Milman, *New volume ratio properties for convex symmetric bodies in $\mathbf{R}^n$*, Invent. Math. **88** (1987), no. 2, 319–340, DOI 10.1007/BF01388911. MR880954

R. Bricard: Mémoire sur la théorie de l'octaedre articulé, *J. Math. Pures Appl.* **5** (1897), 113–148.

G. R. Brightwell and E. R. Scheinerman, *Representations of planar graphs*, SIAM J. Discrete Math. **6** (1993), no. 2, 214–229, DOI 10.1137/0406017. MR1215229

B. Brinkman and M. Charikar, *On the impossibility of dimension reduction in l_1*, J. ACM **52** (2005), no. 5, 766–788, DOI 10.1145/1089023.1089026. MR2176562

R. L. Brooks, C. A. B. Smith, A. H. Stone, and W. T. Tutte, *The dissection of rectangles into squares*, Duke Math. J. **7** (1940), 312–340. MR0003040

A. E. Brouwer and W. H. Haemers, *Spectra of graphs*, Universitext, Springer, New York, 2012. MR2882891

A. Cabello, S. Severini, A. Winter: Graph-Theoretic Approach to Quantum Correlations, *Phys. Rev. Lett.* **112** (2014), 040401.

P. J. Cameron, J.-M. Goethals, J. J. Seidel, and E. E. Shult, *Line graphs, root systems, and elliptic geometry*, J. Algebra **43** (1976), no. 1, 305–327, DOI 10.1016/0021-8693(76)90162-9. MR0441787

The electrical resistance of a graph captures its commute and cover times, *Proc. 21th ACM Symp. on Theory of Comput.* (1989), 574–586.

J. L. Cardy, *Critical percolation in finite geometries*, J. Phys. A **25** (1992), no. 4, L201–L206. MR1151081

D. Chelkak and S. Smirnov, *Discrete complex analysis on isoradial graphs*, Adv. Math. **228** (2011), no. 3, 1590–1630, DOI 10.1016/j.aim.2011.06.025. MR2824564

D. Chelkak and S. Smirnov, *Universality in the 2D Ising model and conformal invariance of fermionic observables*, Invent. Math. **189** (2012), no. 3, 515–580, DOI 10.1007/s00222-011-0371-2. MR2957303

J. Chen and N. Johnston, *The minimum size of unextendible product bases in the bipartite case (and some multipartite cases)*, Comm. Math. Phys. **333** (2015), no. 1, 351–365, DOI 10.1007/s00220-014-2186-7. MR3294952

J. Cheriyan and J. H. Reif, *Directed s-t numberings, rubber bands, and testing digraph k-vertex connectivity*, Proceedings of the Third Annual ACM-SIAM Symposium on Discrete Algorithms (Orlando, FL, 1992), ACM, New York, 1992, pp. 335–344. MR1173905

M. Chrobak and S.-i. Nakano, *Minimum-width grid drawings of plane graphs*, Comput. Geom. **11** (1998), no. 1, 29–54, DOI 10.1016/S0925-7721(98)00016-9. MR1633561

M. Chudnovsky, N. Robertson, P. Seymour, and R. Thomas, *The strong perfect graph theorem*, Ann. of Math. (2) **164** (2006), no. 1, 51–229, DOI 10.4007/annals.2006.164.51. MR2233847

F. R. K. Chung, *Labelings of graphs*, Selected topics in graph theory, 3, Academic Press, San Diego, CA, 1988, pp. 151–168. MR1205400
Series in Math. **92**, Amer. Math. Soc., Providence, RI (1997).

F. R. K. Chung, *Spectral graph theory*, CBMS Regional Conference Series in Mathematics, vol. 92, Published for the Conference Board of the Mathematical Sciences, Washington, DC; by the American Mathematical Society, Providence, RI, 1997. MR1421568

J. Chuzhoy, *Excluded grid theorem: improved and simplified*, STOC'15—Proceedings of the 2015 ACM Symposium on Theory of Computing, ACM, New York, 2015, pp. 645–654. MR3388244

V. Chvátal, *On certain polytopes associated with graphs*, J. Combinatorial Theory Ser. B **18** (1975), 138–154, DOI 10.1016/0095-8956(75)90041-6. MR0371732

V. Chvátal and J. Komlós, *Some combinatorial theorems on monotonicity*, Canad. Math. Bull. **14** (1971), 151–157, DOI 10.4153/CMB-1971-028-8. MR0337676

A. Coja-Oghlan, *The Lovász number of random graphs*, Approximation, randomization, and combinatorial optimization, Lecture Notes in Comput. Sci., vol. 2764, Springer, Berlin, 2003, pp. 228–239, DOI 10.1007/978-3-540-45198-3_20. MR2080795

A. Coja-Oghlan and A. Taraz, *Exact and approximative algorithms for coloring $G(n,p)$*, Random Structures Algorithms **24** (2004), no. 3, 259–278, DOI 10.1002/rsa.20007. MR2068869

Y. Colin de Verdière, *Sur la multiplicité de la première valeur propre non nulle du laplacien* (French), Comment. Math. Helv. **61** (1986), no. 2, 254–270, DOI 10.1007/BF02621914. MR856089

Y. Colin de Verdière, *Sur un nouvel invariant des graphes et un critère de planarité* (French), J. Combin. Theory Ser. B **50** (1990), no. 1, 11–21, DOI 10.1016/0095-8956(90)90093-F. MR1070462

Y. Colin de Verdière, *Un principe variationnel pour les empilements de cercles* (French), Invent. Math. **104** (1991), no. 3, 655–669, DOI 10.1007/BF01245096. MR1106755

Y. Colin de Verdière, *Multiplicities of eigenvalues and tree-width of graphs*, J. Combin. Theory Ser. B **74** (1998), no. 2, 121–146, DOI 10.1006/jctb.1998.1834. MR1654157

Y. Colin de Verdière, *Spectres de graphes* (French, with English and French summaries), Cours Spécialisés [Specialized Courses], vol. 4, Société Mathématique de France, Paris, 1998. MR1652692

D. Conlon and J. Fox, *Bounds for graph regularity and removal lemmas*, Geom. Funct. Anal. **22** (2012), no. 5, 1191–1256, DOI 10.1007/s00039-012-0171-x. MR2989432

R. Connelly, *A counterexample to the rigidity conjecture for polyhedra*, Inst. Hautes Études Sci. Publ. Math. **47** (1977), 333–338. MR0488071

R. Connelly, *Rigidity and energy*, Invent. Math. **66** (1982), no. 1, 11–33, DOI 10.1007/BF01404753. MR652643

R. Connelly, *Rigidity*, Handbook of convex geometry, Vol. A, B, North-Holland, Amsterdam, 1993, pp. 223–271. MR1242981

R. Connelly, *Generic global rigidity*, Discrete Comput. Geom. **33** (2005), no. 4, 549–563, DOI 10.1007/s00454-004-1124-4. MR2132290

R. Connelly, S.J. Gortler and L. Theran: Generic global and universal rigidity, `https://arxiv.org/abs/1604.07475v1`

R. Connelly, S. J. Gortler, and L. Theran, *Affine rigidity and conics at infinity*, Int. Math. Res. Not. IMRN **13** (2018), 4084–4102, DOI 10.1093/imrn/rnx014. MR3829178

R. Courant and D. Hilbert, *Methods of mathematical physics. Vol. I*, Interscience Publishers, Inc., New York, N.Y., 1953. MR0065391

T.S. Cubitt, D. Leung, W. Matthews, A. Winter: Improving zero-error classical communication with entanglement, *Phys. Rev. Lett.* **104** (2010), 230503.

T. S. Cubitt, D. Leung, W. Matthews, and A. Winter, *Zero-error channel capacity and simulation assisted by non-local correlations*, IEEE Trans. Inform. Theory **57** (2011), no. 8, 5509–5523, DOI 10.1109/TIT.2011.2159047. MR2849372

E. B. Davies, G. M. L. Gladwell, J. Leydold, and P. F. Stadler, *Discrete nodal domain theorems*, Linear Algebra Appl. **336** (2001), 51–60, DOI 10.1016/S0024-3795(01)00313-5. MR1855391

C. Delorme and S. Poljak, *Laplacian eigenvalues and the maximum cut problem*, Math. Programming **62** (1993), no. 3, Ser. A, 557–574, DOI 10.1007/BF01585184. MR1251892

C. Delorme and S. Poljak, *Combinatorial properties and the complexity of a max-cut approximation*, European J. Combin. **14** (1993), no. 4, 313–333, DOI 10.1006/eujc.1993.1035. MR1226579

M. M. Deza and M. Laurent, *Geometry of cuts and metrics*, Algorithms and Combinatorics, vol. 15, Springer-Verlag, Berlin, 1997. MR1460488

A. Dress and L. Lovász, *On some combinatorial properties of algebraic matroids*, Combinatorica **7** (1987), no. 1, 39–48, DOI 10.1007/BF02579199. MR905150

R. Duan, S. Severini, and A. Winter, *Zero-error communication via quantum channels, noncommutative graphs, and a quantum Lovász number*, IEEE Trans. Inform. Theory **59** (2013), no. 2, 1164–1174, DOI 10.1109/TIT.2012.2221677. MR3015725

C. Dubey, U. Feige, and W. Unger, *Hardness results for approximating the bandwidth*, J. Comput. System Sci. **77** (2011), no. 1, 62–90, DOI 10.1016/j.jcss.2010.06.006. MR2767125

R. J. Duffin, *Basic properties of discrete analytic functions*, Duke Math. J. **23** (1956), 335–363. MR0078441

R. J. Duffin, *Potential theory on a rhombic lattice*, J. Combinatorial Theory **5** (1968), 258–272. MR0232005

R. J. Duffin and E. L. Peterson, *The discrete analogue of a class of entire functions*, J. Math. Anal. Appl. **21** (1968), 619–642, DOI 10.1016/0022-247X(68)90268-0. MR0224840

A. M. Duval and V. Reiner, *Perron-Frobenius type results and discrete versions of nodal domain theorems*, Linear Algebra Appl. **294** (1999), no. 1-3, 259–268, DOI 10.1016/S0024-3795(99)00090-7. MR1693975

I. A. Dynnikov and S. P. Novikov, *Geometry of the triangle equation on two-manifolds* (English, with English and Russian summaries), Mosc. Math. J. **3** (2003), no. 2, 419–438, 742, DOI 10.17323/1609-4514-2003-3-2-419-438. Dedicated to Vladimir I. Arnold on the occasion of his 65th birthday. MR2025267

J. Edmonds, *Submodular functions, matroids, and certain polyhedra*, Combinatorial Structures and their Applications (Proc. Calgary Internat. Conf., Calgary, Alta., 1969), Gordon and Breach, New York, 1970, pp. 69–87. MR0270945

P. Erdős, *On even subgraphs of graphs* (Hungarian, with English summary), Mat. Lapok **18** (1967), 283–288. MR0238731

N. G. de Bruijn and P. Erdös, *A colour problem for infinite graphs and a problem in the theory of relations*, Nederl. Akad. Wetensch. Proc. Ser. A. **54** = Indagationes Math. **13** (1951), 369–373. MR0046630

P. Erdős and T. Gallai, *On the minimal number of vertices representing the edges of a graph.* (English, with Russian summary), Magyar Tud. Akad. Mat. Kutató Int. Közl. **6** (1961), 181–203. MR0144332

P. Erdős, F. Harary, and W. T. Tutte, *On the dimension of a graph*, Mathematika **12** (1965), 118–122, DOI 10.1112/S0025579300005222. MR0188096

P. Erdős, A. Hajnal, and J. W. Moon, *A problem in graph theory*, Amer. Math. Monthly **71** (1964), 1107–1110, DOI 10.2307/2311408. MR0170339

P. Erdős and L. Lovász, *Problems and results on 3-chromatic hypergraphs and some related questions*, Infinite and finite sets (Colloq., Keszthely, 1973; dedicated to P. Erdős on his 60th birthday), Vol. II, North-Holland, Amsterdam, 1975, pp. 609–627. Colloq. Math. Soc. János Bolyai, Vol. 10. MR0382050

I. Fáry, *On straight line representation of planar graphs*, Acta Univ. Szeged. Sect. Sci. Math. **11** (1948), 229–233. MR0026311

U. Feige, *Randomized graph products, chromatic numbers, and the Lovász ϑ-function*, Combinatorica **17** (1997), no. 1, 79–90, DOI 10.1007/BF01196133. MR1466577

U. Feige, *Approximating the bandwidth via volume respecting embeddings*, J. Comput. System Sci. **60** (2000), no. 3, 510–539, DOI 10.1006/jcss.1999.1682. 30th Annual ACM Symposium on Theory of Computing (Dallas, TX, 1998). MR1770195

U. Feige and M. Langberg, *The* RPR^2 *rounding technique for semidefinite programs*, J. Algorithms **60** (2006), no. 1, 1–23, DOI 10.1016/j.jalgor.2004.11.003. MR2228942

J. Ferrand, *Fonctions préharmoniques et fonctions préholomorphes* (French), Bull. Sci. Math. (2) **68** (1944), 152–180. MR0013411

C. M. Fiduccia, E. R. Scheinerman, A. Trenk, and J. S. Zito, *Dot product representations of graphs*, Discrete Math. **181** (1998), no. 1-3, 113–138, DOI 10.1016/S0012-365X(97)00049-6. MR1600755

L. R. Ford Jr. and D. R. Fulkerson, *Flows in networks*, Princeton University Press, Princeton, N.J., 1962. MR0159700

J. Fox and L. M. Lovász, *A tight lower bound for Szemerédi's regularity lemma*, Combinatorica **37** (2017), no. 5, 911–951, DOI 10.1007/s00493-016-3274-4. MR3737374

A. Föppl: *Theorie des Fachwerks*, Verlag Arthur Felix, Leipzig, 1880.

J.-C. Fournier, *Pavage des figures planes sans trous par des dominos: fondement graphique de l'algorithme de Thurston et parallélisation* (French, with English and French summaries), C. R. Acad. Sci. Paris Sér. I Math. **320** (1995), no. 1, 107–112, DOI 10.1016/0304-3975(95)00204-9. MR1320841

P. Frankl and H. Maehara, *The Johnson-Lindenstrauss lemma and the sphericity of some graphs*, J. Combin. Theory Ser. B **44** (1988), no. 3, 355–362, DOI 10.1016/0095-8956(88)90043-3. MR941443

P. Frankl and H. Maehara, *On the contact dimensions of graphs*, Discrete Comput. Geom. **3** (1988), no. 1, 89–96, DOI 10.1007/BF02187899. MR918182

P. Frankl and R. M. Wilson, *Intersection theorems with geometric consequences*, Combinatorica **1** (1981), no. 4, 357–368, DOI 10.1007/BF02579457. MR647986

H. de Fraysseix, P. O. de Mendez, and P. Rosenstiehl, *On triangle contact graphs*, Combin. Probab. Comput. **3** (1994), no. 2, 233–246, DOI 10.1017/S0963548300001139. MR1288442

H. de Fraysseix, J. Pach, and R. Pollack, *How to draw a planar graph on a grid*, Combinatorica **10** (1990), no. 1, 41–51, DOI 10.1007/BF02122694. MR1075065

A. Frieze and R. Kannan, *Quick approximation to matrices and applications*, Combinatorica **19** (1999), no. 2, 175–220, DOI 10.1007/s004930050052. MR1723039

Z. Füredi and J. Komlós, *The eigenvalues of random symmetric matrices*, Combinatorica **1** (1981), no. 3, 233–241, DOI 10.1007/BF02579329. MR637828

Z. Füredi and R. Stanley, *Sets of vectors with many orthogonal pairs*, Graphs Combin. **8** (1992), no. 4, 391–394, DOI 10.1007/BF02351595. MR1554361

M. R. Garey and D. S. Johnson, *The complexity of near-optimal graph coloring*, J. Assoc. Comp. Mach. **23** (1976), no. 1, 43–49, DOI 10.1145/321921.321926. MR0426496

J. Geelen and S. Iwata, *Matroid matching via mixed skew-symmetric matrices*, Combinatorica **25** (2005), no. 2, 187–215, DOI 10.1007/s00493-005-0013-7. MR2127610

M. Giustina, M.A.M. Versteegh, S. Wengerowsky, J. Handsteiner, A. Hochrainer, K. Phelan, F. Steinlechner, J. Kofler, J.-A. Larsson,C. Abellan, W. Amaya, V. Pruneri, M.W. Mitchell, J. Beyer, T. Gerrits, A.E. Lita, L.K. Shalm, S.W. Nam, T. Scheidl, R. Ursin, B. Wittmann, A. Zeilinger: A significant-loophole-free test of Bell's theorem with entangled photons, *Phys. Rev. Lett.* **115** (2015), 250401.

H. Gluck, *Almost all simply connected closed surfaces are rigid*, Geometric topology (Proc. Conf., Park City, Utah, 1974), Springer, Berlin, 1975, pp. 225–239. Lecture Notes in Math., Vol. 438. MR0400239

M. X. Goemans and D. P. Williamson, *Improved approximation algorithms for maximum cut and satisfiability problems using semidefinite programming*, J. Assoc. Comput. Mach. **42** (1995), no. 6, 1115–1145, DOI 10.1145/227683.227684. MR1412228

S. J. Gortler and D. P. Thurston, *Characterizing the universal rigidity of generic frameworks*, Discrete Comput. Geom. **51** (2014), no. 4, 1017–1036, DOI 10.1007/s00454-014-9590-9. MR3216675

S. J. Gortler, A. D. Healy, and D. P. Thurston, *Characterizing generic global rigidity*, Amer. J. Math. **132** (2010), no. 4, 897–939, DOI 10.1353/ajm.0.0132. MR2663644

J. Graver, B. Servatius, and H. Servatius, *Combinatorial rigidity*, Graduate Studies in Mathematics, vol. 2, American Mathematical Society, Providence, RI, 1993. MR1251062

M. Grötschel, L. Lovász, and A. Schrijver, *The ellipsoid method and its consequences in combinatorial optimization*, Combinatorica **1** (1981), no. 2, 169–197, DOI 10.1007/BF02579273. MR625550

M. Grötschel, L. Lovász, and A. Schrijver, *Polynomial algorithms for perfect graphs*, Topics on perfect graphs, North-Holland Math. Stud., vol. 88, North-Holland, Amsterdam, 1984, pp. 325–356, DOI 10.1016/S0304-0208(08)72943-8. MR778770

M. Grötschel, L. Lovász, and A. Schrijver, *Relaxations of vertex packing*, J. Combin. Theory Ser. B **40** (1986), no. 3, 330–343, DOI 10.1016/0095-8956(86)90087-0. MR842998

M. Grötschel, L. Lovász, A. Schrijver: Geometric Algorithms and Combinatorial Optimization, Springer (1988).

B. Grünbaum and T. S. Motzkin, *On polyhedral graphs*, Proc. Sympos. Pure Math., Vol. VII, Amer. Math. Soc., Providence, R.I., 1963, pp. 285–290. MR0153005

V. Guruswami, *Maximum cut on line and total graphs*, Discrete Appl. Math. **92** (1999), no. 2-3, 217–221, DOI 10.1016/S0166-218X(99)00056-6. MR1697554

W. Haemers, *On some problems of Lovász concerning the Shannon capacity of a graph*, IEEE Trans. Inform. Theory **25** (1979), no. 2, 231–232, DOI 10.1109/TIT.1979.1056027. MR521317

A. Hajnal, *A theorem on k-saturated graphs*, Canad. J. Math. **17** (1965), 720–724, DOI 10.4153/CJM-1965-072-1. MR0179103

J. Håstad, *Some optimal inapproximability results*, STOC '97 (El Paso, TX), ACM, New York, 1999, pp. 1–10. MR1715618

J. Håstad, *Clique is hard to approximate within* $n^{1-\epsilon}$, Acta Math. **182** (1999), no. 1, 105–142, DOI 10.1007/BF02392825. MR1687331

Z.-X. He and O. Schramm, *Hyperbolic and parabolic packings*, Discrete Comput. Geom. **14** (1995), no. 2, 123–149, DOI 10.1007/BF02570699. MR1331923

Z.-X. He and O. Schramm, *On the convergence of circle packings to the Riemann map*, Invent. Math. **125** (1996), no. 2, 285–305, DOI 10.1007/s002220050076. MR1395721

B. Hendrickson, *Conditions for unique graph realizations*, SIAM J. Comput. **21** (1992), no. 1, 65–84, DOI 10.1137/0221008. MR1148818

L. Henneberg: *Die graphische Statik der starren Systeme*, Leipzig, 1911.

B. Hensen, H. Bernien, A.E. Dréau, A. Reiserer, N. Kalb, M.S. Blok, J. Ruitenberg, R.F.L. Vermeulen, R.N. Schouten, C. Abellán, W. Amaya, V. Pruneri, M.W. Mitchell, M. Markham, D.J. Twitchen, D. Elkouss, S. Wehner, T.H. Taminiau and R. Hanson: Loophole-free Bell inequality violation using electron spins separated by 1.3 kilometres, *Nature* **526** (2015), 682–686.

J. Herzog, A. Macchia, S. Saeedi Madani, and V. Welker, *On the ideal of orthogonal representations of a graph in* $\mathbb{R}^2$, Adv. in Appl. Math. **71** (2015), 146–173, DOI 10.1016/j.aam.2015.09.009. MR3406962

J. Hladký, I. Rocha: Independent sets, cliques, and colorings in graphons, `https://arxiv.org/abs/1712.07367`

A. J. Hoffman, *Some recent results on spectral properties of graphs*, Beiträge zur Graphentheorie (Kolloquium, Manebach, 1967), Teubner, Leipzig, 1968, pp. 75–80. MR0245462

A. J. Hoffman, *On graphs whose least eigenvalue exceeds* $-1-\sqrt{2}$, Linear Algebra and Appl. **16** (1977), no. 2, 153–165, DOI 10.1016/0024-3795(77)90027-1. MR0469826

A. J. Hoffman, *On eigenvalues and colorings of graphs*, Graph Theory and its Applications (Proc. Advanced Sem., Math. Research Center, Univ. of Wisconsin, Madison, Wis., 1969), Academic Press, New York, 1970, pp. 79–91. MR0284373

H. van der Holst, *A short proof of the planarity characterization of Colin de Verdière*, J. Combin. Theory Ser. B **65** (1995), no. 2, 269–272, DOI 10.1006/jctb.1995.1054. MR1358989

H. van der Holst: *Topological and Spectral Graph Characterizations*, Ph.D. Thesis, University of Amsterdam, Amsterdam, 1996.

H. van der Holst, *Two tree-width-like graph invariants*, Combinatorica **23** (2003), no. 4, 633–651, DOI 10.1007/s00493-003-0038-8. MR2046827

H. van der Holst, M. Laurent, and A. Schrijver, *On a minor-monotone graph invariant*, J. Combin. Theory Ser. B **65** (1995), no. 2, 291–304, DOI 10.1006/jctb.1995.1056. MR1358991

H. van der Holst, L. Lovász, and A. Schrijver, *On the invariance of Colin de Verdière's graph parameter under clique sums*, Linear Algebra Appl. **226/228** (1995), 509–517, DOI 10.1016/0024-3795(95)00160-S. MR1344582

H. van der Holst, L. Lovász, and A. Schrijver, *The Colin de Verdière graph parameter*, Graph theory and combinatorial biology (Balatonlelle, 1996), Bolyai Soc. Math. Stud., vol. 7, János Bolyai Math. Soc., Budapest, 1999, pp. 29–85. MR1673503

M. Howard, J. Wallman, V. Veitch and J. Emerson: Contextuality supplies the 'magic' for quantum computation, *Nature* **510** (2014), 351–355.

J. E. Goodman and J. O'Rourke (eds.), *Handbook of discrete and computational geometry*, 2nd ed., Discrete Mathematics and its Applications (Boca Raton), Chapman & Hall/CRC, Boca Raton, FL, 2004. MR2082993

M. Iri, *On an extension of the maximum-flow minimum-cut theorem to multicommodity flows*, J. Operations Res. Soc. Japan **13** (1970/71), 129–135. MR0292487

R. Isaacs, Monodiffric functions, *Natl. Bureau Standards App. Math. Series* **18** (1952), 257–266.

B. Jackson and T. Jordán, *Connected rigidity matroids and unique realizations of graphs*, J. Combin. Theory Ser. B **94** (2005), no. 1, 1–29, DOI 10.1016/j.jctb.2004.11.002. MR2130278

B. Jackson and J. C. Owen, *A characterisation of the generic rigidity of 2-dimensional point-line frameworks*, J. Combin. Theory Ser. B **119** (2016), 96–121, DOI 10.1016/j.jctb.2015.12.007. MR3486339

B. Jackson and P. Keevash, *Necessary conditions for the global rigidity of direction-length frameworks*, Discrete Comput. Geom. **46** (2011), no. 1, 72–85, DOI 10.1007/s00454-011-9326-z. MR2794358

W. B. Johnson and J. Lindenstrauss, *Extensions of Lipschitz mappings into a Hilbert space*, Conference in modern analysis and probability (New Haven, Conn., 1982), Contemp. Math., vol. 26, Amer. Math. Soc., Providence, RI, 1984, pp. 189–206, DOI 10.1090/conm/026/737400. MR737400

M. Jerrum and A. Sinclair, *Approximating the permanent*, SIAM J. Comput. **18** (1989), no. 6, 1149–1178, DOI 10.1137/0218077. MR1025467

J. Jonasson and O. Schramm, *On the cover time of planar graphs*, Electron. Comm. Probab. **5** (2000), 85–90, DOI 10.1214/ECP.v5-1022. MR1781842

F. Juhász, *The asymptotic behaviour of Lovász' θ function for random graphs*, Combinatorica **2** (1982), no. 2, 153–155, DOI 10.1007/BF02579314. MR685042

G. A. Kabatjanskiĭ and V. I. Levenšteĭn, *Bounds for packings on the sphere and in space* (Russian), Problemy Peredači Informacii **14** (1978), no. 1, 3–25. MR0514023

R. J. Kang, L. Lovász, T. Müller, and E. R. Scheinerman, *Dot product representations of planar graphs*, Electron. J. Combin. **18** (2011), no. 1, Paper 216, 14. MR2853073

D. Karger, R. Motwani and M. Sudan: Approximate graph coloring by semidefinite programming, *Proc. 35th Ann. IEEE Symp. on Found. Comp. Science* (1994), 2–13.

H. Karloff and U. Zwick: A 7/8-Approximation Algorithm for MAX 3SAT? *Proc. 38th Ann. IEEE Symp. on Found. Comp. Science* (1997), 406–415.

B. S. Kashin and S. V. Konyagin, *Systems of vectors in Hilbert space* (Russian), Trudy Mat. Inst. Steklov. **157** (1981), 64–67, 235. Number theory, mathematical analysis and their applications. MR651759

G. Katona and G. Korvin, *Functions defined on a directed graph*, Theory of Graphs (Proc. Colloq., Tihany, 1966), Academic Press, New York, 1968, pp. 209–213. MR0233736

K.-I. Kawarabayashi and M. Thorup, *Coloring 3-colorable graphs with less than $n^{1/5}$ colors*, J. ACM **64** (2017), no. 1, Art. 4, 23, DOI 10.1145/3001582. MR3634492

R. Kenyon, *The Laplacian and Dirac operators on critical planar graphs*, Invent. Math. **150** (2002), no. 2, 409–439, DOI 10.1007/s00222-002-0249-4. MR1933589

R. Kenyon, *Conformal invariance of domino tiling*, Ann. Probab. **28** (2000), no. 2, 759–795, DOI 10.1214/aop/1019160260. MR1782431

R. Kenyon, *Lectures on dimers*, Statistical mechanics, IAS/Park City Math. Ser., vol. 16, Amer. Math. Soc., Providence, RI, 2009, pp. 191–230. MR2523460

R. Kenyon and J.-M. Schlenker, *Rhombic embeddings of planar quad-graphs* (English, with English and French summaries), Trans. Amer. Math. Soc. **357** (2005), no. 9, 3443–3458, DOI 10.1090/S0002-9947-04-03545-7. MR2146632

S. Khanna, N. Linial, and S. Safra, *On the hardness of approximating the chromatic number*, Combinatorica **20** (2000), no. 3, 393–415, DOI 10.1007/s004930070013. MR1774844

S. Khot, G. Kindler, E. Mossel, and R. O'Donnell, *Optimal inapproximability results for MAX-CUT and other 2-variable CSPs?*, SIAM J. Comput. **37** (2007), no. 1, 319–357, DOI 10.1137/S0097539705447372. MR2306295

C. O. Kiselman, *Functions on discrete sets holomorphic in the sense of Isaacs, or monodiffric functions of the first kind*, Sci. China Ser. A **48** (2005), no. suppl., 86–96, DOI 10.1007/BF02884698. MR2156493

A. A. Klyachko, M. A. Can, S. Binicioğlu, and A. S. Shumovsky, *Simple test for hidden variables in spin-1 systems*, Phys. Rev. Lett. **101** (2008), no. 2, 020403, 4, DOI 10.1103/PhysRevLett.101.020403. MR2430247

D. E. Knuth, *The sandwich theorem*, Electron. J. Combin. **1** (1994), Article 1, approx. 48. MR1269161

P. Koebe: Kontaktprobleme der konformen Abbildung, *Berichte über die Verhandlungen Sächs. Akad. Wiss.*, Math.-Phys. Klasse, **88** (1936) 141–164.

S. V. Konjagin, *Systems of vectors in Euclidean space and an extremal problem for polynomials* (Russian), Mat. Zametki **29** (1981), no. 1, 63–74, 155. MR604151

B. Korte and J. Vygen, *Combinatorial optimization*, 4th ed., Algorithms and Combinatorics, vol. 21, Springer-Verlag, Berlin, 2008. Theory and algorithms. MR2369759

A. Kotlov, *Rank and chromatic number of a graph*, J. Graph Theory **26** (1997), no. 1, 1–8, DOI 10.1002/(SICI)1097-0118(199709)26:1⟨1::AID-JGT1⟩3.3.CO;2-U. MR1464336

A. Kotlov, *Spectral characterization of tree-width-two graphs*, Combinatorica **20** (2000), no. 1, 147–152, DOI 10.1007/s004930070038. MR1770529

A. Kotlov, *Minors and strong products*, European J. Combin. **22** (2001), no. 4, 511–512, DOI 10.1006/eujc.2000.0428. MR1829745

A. Kotlov and L. Lovász, *The rank and size of graphs*, J. Graph Theory **23** (1996), no. 2, 185–189, DOI 10.1002/(SICI)1097-0118(199610)23:2⟨185::AID-JGT9⟩3.3.CO;2-X. MR1408346

A. Kotlov, L. Lovász, and S. Vempala, *The Colin de Verdière number and sphere representations of a graph*, Combinatorica **17** (1997), no. 4, 483–521, DOI 10.1007/BF01195002. MR1645686

G. Kuperberg, *From the Mahler conjecture to Gauss linking integrals*, Geom. Funct. Anal. **18** (2008), no. 3, 870–892, DOI 10.1007/s00039-008-0669-4. MR2438998

G. Laman, *On graphs and rigidity of plane skeletal structures*, J. Engrg. Math. **4** (1970), 331–340, DOI 10.1007/BF01534980. MR0269535

D. G. Larman and C. A. Rogers, *The realization of distances within sets in Euclidean space*, Mathematika **19** (1972), 1–24, DOI 10.1112/S0025579300004903. MR0319055

M. Laurent and S. Poljak, *On the facial structure of the set of correlation matrices*, SIAM J. Matrix Anal. Appl. **17** (1996), no. 3, 530–547, DOI 10.1137/0617031. MR1397243

M. Laurent and F. Rendl: Semidefinite Programming and Integer Programming, in: *Discrete Optimization*, Handbooks in Operations Research and Management Science **12** (2005), 393–514.

M. Laurent and A. Varvitsiotis, *Positive semidefinite matrix completion, universal rigidity and the strong Arnold property*, Linear Algebra Appl. **452** (2014), 292–317, DOI 10.1016/j.laa.2014.03.015. MR3201103

F.T. Leighton and S. Rao, An approximate max-flow min-cut theorem for uniform multicommodity flow problems with applications to approximation algorithms, *Proc. 29th Ann. IEEE Symp. on Found. Comp. Science* (1988), 422-431.

Y. Lin, G. Lippner, D. Mangoubi, and S.-T. Yau, *Nodal geometry of graphs on surfaces*, Discrete Contin. Dyn. Syst. **28** (2010), no. 3, 1291–1298, DOI 10.3934/dcds.2010.28.1291. MR2644790

N. Linial, E. London, and Y. Rabinovich, *The geometry of graphs and some of its algorithmic applications*, Combinatorica **15** (1995), no. 2, 215–245, DOI 10.1007/BF01200757. MR1337355

N. Linial, L. Lovász, and A. Wigderson, *Rubber bands, convex embeddings and graph connectivity*, Combinatorica **8** (1988), no. 1, 91–102, DOI 10.1007/BF02122557. MR951998

R. J. Lipton and R. E. Tarjan, *A separator theorem for planar graphs*, SIAM J. Appl. Math. **36** (1979), no. 2, 177–189, DOI 10.1137/0136016. MR524495

L. Lovász, *Normal hypergraphs and the perfect graph conjecture*, Discrete Math. **2** (1972), no. 3, 253–267, DOI 10.1016/0012-365X(72)90006-4. MR0302480

L. Lovász, *On minimax theorems of combinatorics* (Hungarian), Mat. Lapok **26** (1975), no. 3-4, 209–264 (1978). MR510823

L. Lovász, *Flats in matroids and geometric graphs*, Combinatorial surveys (Proc. Sixth British Combinatorial Conf., Royal Holloway Coll., Egham, 1977), Academic Press, London, 1977, pp. 45–86. MR0480111

L. Lovász, *Some finite basis theorems on graph theory*, Combinatorics (Proc. Fifth Hungarian Colloq., Keszthely, 1976), Colloq. Math. Soc. János Bolyai, vol. 18, North-Holland, Amsterdam-New York, 1978, pp. 717–729. MR519305

L. Lovász, *Topological and algebraic methods in graph theory*, Graph theory and related topics (Proc. Conf., Univ. Waterloo, Waterloo, Ont., 1977), Academic Press, New York-London, 1979, pp. 1–14. MR538032

L. Lovász, *On the Shannon capacity of a graph*, IEEE Trans. Inform. Theory **25** (1979), no. 1, 1–7, DOI 10.1109/TIT.1979.1055985. MR514926

L. Lovász, *Selecting independent lines from a family of lines in a space*, Acta Sci. Math. (Szeged) **42** (1980), no. 1-2, 121–131. MR576945

L. Lovász, *Submodular functions and convexity*, Mathematical programming: the state of the art (Bonn, 1982), Springer, Berlin, 1983, pp. 235–257. MR717403

L. Lovász, *Self-dual polytopes and the chromatic number of distance graphs on the sphere*, Acta Sci. Math. (Szeged) **45** (1983), no. 1-4, 317–323. MR708798

L. Lovász, *Singular spaces of matrices and their application in combinatorics*, Bol. Soc. Brasil. Mat. (N.S.) **20** (1989), no. 1, 87–99, DOI 10.1007/BF02585470. MR1129080

L. Lovász, *Combinatorial problems and exercises*, 2nd ed., North-Holland Publishing Co., Amsterdam, 1993. MR1265492

L. Lovász, *Random walks on graphs: a survey*, Combinatorics, Paul Erdős is eighty, Vol. 2 (Keszthely, 1993), Bolyai Soc. Math. Stud., vol. 2, János Bolyai Math. Soc., Budapest, 1996, pp. 353–397. MR1395866

L. Lovász, *Steinitz representations of polyhedra and the Colin de Verdière number*, J. Combin. Theory Ser. B **82** (2001), no. 2, 223–236, DOI 10.1006/jctb.2000.2027. MR1842113

L. Lovász, *Semidefinite programs and combinatorial optimization*, Recent advances in algorithms and combinatorics, CMS Books Math./Ouvrages Math. SMC, vol. 11, Springer, New York, 2003, pp. 137–194, DOI 10.1007/0-387-22444-0_6. MR1952986

L. Lovász, *Graph minor theory*, Bull. Amer. Math. Soc. (N.S.) **43** (2006), no. 1, 75–86, DOI 10.1090/S0273-0979-05-01088-8. MR2188176

L. Lovász, *Large networks and graph limits*, American Mathematical Society Colloquium Publications, vol. 60, American Mathematical Society, Providence, RI, 2012. MR3012035

L. Lovász and M. D. Plummer, *Matching theory*, North-Holland Mathematics Studies, vol. 121, North-Holland Publishing Co., Amsterdam; North-Holland Publishing Co., Amsterdam, 1986. Annals of Discrete Mathematics, 29. MR859549

L. Lovász, M. Saks, and A. Schrijver, *Orthogonal representations and connectivity of graphs*, Linear Algebra Appl. **114/115** (1989), 439–454, DOI 10.1016/0024-3795(89)90475-8. MR986889

L. Lovász and A. Schrijver, *A Borsuk theorem for antipodal links and a spectral characterization of linklessly embeddable graphs*, Proc. Amer. Math. Soc. **126** (1998), no. 5, 1275–1285, DOI 10.1090/S0002-9939-98-04244-0. MR1443840

L. Lovász and A. Schrijver, *On the null space of a Colin de Verdière matrix* (English, with English and French summaries), Ann. Inst. Fourier (Grenoble) **49** (1999), no. 3, 1017–1026. Symposium à la Mémoire de François Jaeger (Grenoble, 1998). MR1703435

L. Lovász and B. Szegedy, *Limits of dense graph sequences*, J. Combin. Theory Ser. B **96** (2006), no. 6, 933–957, DOI 10.1016/j.jctb.2006.05.002. MR2274085

L. Lovász and B. Szegedy, *Szemerédi's lemma for the analyst*, Geom. Funct. Anal. **17** (2007), no. 1, 252–270, DOI 10.1007/s00039-007-0599-6. MR2306658

L. Lovász and B. Szegedy: The graph theoretic moment problem, `https://arxiv.org/abs/1010.5159`

L. Lovász and K. Vesztergombi, *Geometric representations of graphs*, Paul Erdős and his mathematics, II (Budapest, 1999), Bolyai Soc. Math. Stud., vol. 11, János Bolyai Math. Soc., Budapest, 2002, pp. 471–498. MR1954739

L. Lovász and Y. Yemini, *On generic rigidity in the plane*, SIAM J. Algebraic Discrete Methods **3** (1982), no. 1, 91–98, DOI 10.1137/0603009. MR644960

A. Lubotzky, R. Phillips, and P. Sarnak, *Ramanujan graphs*, Combinatorica **8** (1988), no. 3, 261–277, DOI 10.1007/BF02126799. MR963118

H. Maehara and V. Rödl, *On the dimension to represent a graph by a unit distance graph*, Graphs Combin. **6** (1990), no. 4, 365–367, DOI 10.1007/BF01787703. MR1092585

A. Magen, *Dimensionality reductions in l_2 that preserve volumes and distance to affine spaces*, Discrete Comput. Geom. **38** (2007), no. 1, 139–153, DOI 10.1007/s00454-007-1329-4. MR2322120

P. Mani, *Automorphismen von polyedrischen Graphen* (German), Math. Ann. **192** (1971), 279–303, DOI 10.1007/BF02075357. MR0296808

G. A. Margulis, *Explicit constructions of expanders* (Russian), Problemy Peredači Informacii **9** (1973), no. 4, 71–80. MR0484767

J. H. Mason, *Matroids as the study of geometrical configurations*, Higher combinatorics (Proc. NATO Advanced Study Inst., Berlin, 1976), NATO Adv. Study Inst. Ser., Ser. C: Math. Phys. Sci., vol. 31, Reidel, Dordrecht-Boston, Mass., 1977, pp. 133–176. MR519783

P. Massart, *Concentration inequalities and model selection*, Lecture Notes in Mathematics, vol. 1896, Springer, Berlin, 2007. Lectures from the 33rd Summer School on Probability Theory held in Saint-Flour, July 6–23, 2003; With a foreword by Jean Picard. MR2319879

J. Matoušek, *On embedding expanders into l_p spaces*, Israel J. Math. **102** (1997), 189–197, DOI 10.1007/BF02773799. MR1489105

J. Matoušek, *Lectures on discrete geometry*, Graduate Texts in Mathematics, vol. 212, Springer-Verlag, New York, 2002. MR1899299

J. Matoušek, *Using the Borsuk-Ulam theorem*, Universitext, Springer-Verlag, Berlin, 2003. Lectures on topological methods in combinatorics and geometry; Written in cooperation with Anders Björner and Günter M. Ziegler. MR1988723

J. Matoušek, *On variants of the Johnson-Lindenstrauss lemma*, Random Structures Algorithms **33** (2008), no. 2, 142–156, DOI 10.1002/rsa.20218. MR2436844

J. Matoušek: *Lecture notes on metric embeddings*, `http://kam.mff.cuni.cz/~matousek/ba-a4.pdf`

J. Matoušek and A. Naor: Open problems on embeddings of finite metric spaces `http://kam.mff.cuni.cz/~matousek/metrop.ps`

R. J. McEliece, E. R. Rodemich, and H. C. Rumsey Jr., *The Lovász bound and some generalizations*, J. Combin. Inform. System Sci. **3** (1978), no. 3, 134–152. MR505587

C. Mercat, *Discrete Riemann surfaces and the Ising model*, Comm. Math. Phys. **218** (2001), no. 1, 177–216, DOI 10.1007/s002200000348. MR1824204

C. Mercat: Discrete polynomials and discrete holomorphic approximation (2002) `http://arxiv.org/abs/math-ph/0206041`

C. Mercat, *Exponentials form a basis of discrete holomorphic functions on a compact* (English, with English and French summaries), Bull. Soc. Math. France **132** (2004), no. 2, 305–326, DOI 10.24033/bsmf.2467. MR2076370

M. M. Milić, *General passive networks–solvability, degeneracies, and order of complexity*, IEEE Trans. Circuits and Systems **CAS-21** (1974), 177–183, DOI 10.1109/TCS.1974.1083845. MR0444328

G. L. Miller, S.-H. Teng, W. Thurston, and S. A. Vavasis, *Separators for sphere-packings and nearest neighbor graphs*, J. ACM **44** (1997), no. 1, 1–29, DOI 10.1145/256292.256294. MR1438463

B. Mohar and S. Poljak, *Eigenvalues and max-cut problem*, Czechoslovak Math. J. **40(115)** (1990), no. 2, 343–352. MR1046300

S. Morey and R. H. Villarreal, *Edge ideals: algebraic and combinatorial properties*, Progress in commutative algebra 1, de Gruyter, Berlin, 2012, pp. 85–126. MR2932582

T. S. Motzkin and E. G. Straus, *Maxima for graphs and a new proof of a theorem of Turán*, Canad. J. Math. **17** (1965), 533–540, DOI 10.4153/CJM-1965-053-6. MR0175813

A. Mowshowitz, *Graphs, groups and matrices*, Proceedings of the Twenty-Fifth Summer Meeting of the Canadian Mathematical Congress (Lakehead Univ., Thunder Bay, Ont., 1971), Lakehead Univ., Thunder Bay, Ont., 1971, pp. 509–522. MR0337682

C. St. J. A. Nash-Williams, *Random walk and electric currents in networks*, Proc. Cambridge Philos. Soc. **55** (1959), 181–194, DOI 10.1017/s0305004100033879. MR0124932

C. St. J. A. Nash-Williams, *Decomposition of finite graphs into forests*, J. London Math. Soc. **39** (1964), 12, DOI 10.1112/jlms/s1-39.1.12. MR0161333

Y. Nesterov and A. Nemirovskii, *Interior-point polynomial algorithms in convex programming*, SIAM Studies in Applied Mathematics, vol. 13, Society for Industrial and Applied Mathematics (SIAM), Philadelphia, PA, 1994. MR1258086

N. Nisan and A. Wigderson, *On rank vs. communication complexity*, 35th Annual Symposium on Foundations of Computer Science (Santa Fe, NM, 1994), IEEE Comput. Soc. Press, Los Alamitos, CA, 1994, pp. 831–836, DOI 10.1109/SFCS.1994.365711. MR1489253

C. v. Nuffelen, *Research Problems: A Bound for the Chromatic Number of a Graph*, Amer. Math. Monthly **83** (1976), no. 4, 265–266, DOI 10.2307/2318218. MR1538027

M. L. Overton, *On minimizing the maximum eigenvalue of a symmetric matrix*, SIAM J. Matrix Anal. Appl. **9** (1988), no. 2, 256–268, DOI 10.1137/0609021. SIAM Conference on Linear Algebra in Signals, Systems, and Control (Boston, Mass., 1986). MR938560

J. G. Oxley, *Matroid theory*, Oxford Science Publications, The Clarendon Press, Oxford University Press, New York, 1992. MR1207587

J. Pach and P. K. Agarwal, *Combinatorial geometry*, Wiley-Interscience Series in Discrete Mathematics and Optimization, John Wiley & Sons, Inc., New York, 1995. A Wiley-Interscience Publication. MR1354145

M. Petersdorf and H. Sachs, *Spektrum und Automorphismengruppe eines Graphen* (German), Combinatorial Theory and its Applications, III (Proc. Colloq., Balatonfüred, 1969), North-Holland, Amsterdam, 1970, pp. 891–907. MR0309792

A. Peres, *Two simple proofs of the Kochen-Specker theorem*, J. Phys. A **24** (1991), no. 4, L175–L178. MR1104166

S. Poljak and F. Rendl, *Nonpolyhedral relaxations of graph-bisection problems*, SIAM J. Optim. **5** (1995), no. 3, 467–487, DOI 10.1137/0805024. MR1344666

H. Pollaczek-Geiringer: Über die Gliederung ebener Fachwerke, *Z. Angew. Math. und Mech.* **7** (1927), 58–72.

G. Pólya, *Über eine Aufgabe der Wahrscheinlichkeitsrechnung betreffend die Irrfahrt im Straßennetz* (German), Math. Ann. **84** (1921), no. 1-2, 149–160, DOI 10.1007/BF01458701. MR1512028

L. Porkolab and L. Khachiyan, *On the complexity of semidefinite programs*, J. Global Optim. **10** (1997), no. 4, 351–365, DOI 10.1023/A:1008203903341. MR1457182

A. M. Raigorodskii, *On the chromatic numbers of spheres in* $\mathbb{R}^n$, Combinatorica **32** (2012), no. 1, 111–123, DOI 10.1007/s00493-012-2709-9. MR2927634

M. V. Ramana, *An exact duality theory for semidefinite programming and its complexity implications*, Math. Programming **77** (1997), no. 2, Ser. B, 129–162, DOI 10.1016/S0025-5610(96)00082-2. Semidefinite programming. MR1461379

A. Recski, *Matroid theory and its applications in electric network theory and in statics*, Algorithms and Combinatorics, vol. 6, Springer-Verlag, Berlin; Akadémiai Kiadó (Publishing House of the Hungarian Academy of Sciences), Budapest, 1989. MR1027839

J. Reiterman, V. Rödl, and E. Šiňajová, *Embeddings of graphs in Euclidean spaces*, Discrete Comput. Geom. **4** (1989), no. 4, 349–364, DOI 10.1007/BF02187736. MR996768

J. Reiterman, V. Rödl, and E. Šiňajová, *On embedding of graphs into Euclidean spaces of small dimension*, J. Combin. Theory Ser. B **56** (1992), no. 1, 1–8, DOI 10.1016/0095-8956(92)90002-F. MR1182453

J. Richter-Gebert, *Realization spaces of polytopes*, Lecture Notes in Mathematics, vol. 1643, Springer-Verlag, Berlin, 1996. MR1482230

I. Rivin, *A characterization of ideal polyhedra in hyperbolic* 3*-space*, Ann. of Math. (2) **143** (1996), no. 1, 51–70, DOI 10.2307/2118652. MR1370757

I. Rivin, *Combinatorial optimization in geometry*, Adv. in Appl. Math. **31** (2003), no. 1, 242–271, DOI 10.1016/S0196-8858(03)00093-9. MR1985831

N. Robertson and P. D. Seymour, *Graph minors. III. Planar tree-width*, J. Combin. Theory Ser. B **36** (1984), no. 1, 49–64, DOI 10.1016/0095-8956(84)90013-3. MR742386

N. Robertson and P. D. Seymour, *Graph minors. V. Excluding a planar graph*, J. Combin. Theory Ser. B **41** (1986), no. 1, 92–114, DOI 10.1016/0095-8956(86)90030-4. MR854606

N. Robertson and P. D. Seymour, *Graph minors. XVI. Excluding a non-planar graph*, J. Combin. Theory Ser. B **89** (2003), no. 1, 43–76, DOI 10.1016/S0095-8956(03)00042-X. MR1999736

N. Robertson, P. Seymour, and R. Thomas, *Sachs' linkless embedding conjecture*, J. Combin. Theory Ser. B **64** (1995), no. 2, 185–227, DOI 10.1006/jctb.1995.1032. MR1339849

B. Rodin and D. Sullivan, *The convergence of circle packings to the Riemann mapping*, J. Differential Geom. **26** (1987), no. 2, 349–360. MR906396

M. Rosenfeld, *Almost orthogonal lines in* E^d, Applied geometry and discrete mathematics, DIMACS Ser. Discrete Math. Theoret. Comput. Sci., vol. 4, Amer. Math. Soc., Providence, RI, 1991, pp. 489–492. MR1116372

H. Sachs, *Coin graphs, polyhedra, and conformal mapping*, Discrete Math. **134** (1994), no. 1-3, 133–138, DOI 10.1016/0012-365X(93)E0068-F. Algebraic and topological methods in graph theory (Lake Bled, 1991). MR1303402

L. A. Santaló, *An affine invariant for convex bodies of* n*-dimensional space* (Spanish), Portugaliae Math. **8** (1949), 155–161. MR0039293

A. Sard, *The measure of the critical values of differentiable maps*, Bull. Amer. Math. Soc. **48** (1942), 883–890, DOI 10.1090/S0002-9904-1942-07811-6. MR0007523

R. Schneider, *Convex bodies: the Brunn-Minkowski theory*, Encyclopedia of Mathematics and its Applications, vol. 44, Cambridge University Press, Cambridge, 1993. MR1216521

W. Schnyder: Embedding planar graphs on the grid, *Proc. First Annual ACM-SIAM Symp. on Discrete Algorithms*, SIAM, Philadelphia (1990), 138–148;

O. Schramm, *Existence and uniqueness of packings with specified combinatorics*, Israel J. Math. **73** (1991), no. 3, 321–341, DOI 10.1007/BF02773845. MR1135221

O. Schramm, *How to cage an egg*, Invent. Math. **107** (1992), no. 3, 543–560, DOI 10.1007/BF01231901. MR1150601

O. Schramm, *Square tilings with prescribed combinatorics*, Israel J. Math. **84** (1993), no. 1-2, 97–118, DOI 10.1007/BF02761693. MR1244661

O. Schramm, *Scaling limits of loop-erased random walks and uniform spanning trees*, Israel J. Math. **118** (2000), 221–288, DOI 10.1007/BF02803524. MR1776084

O. Schramm, *Conformally invariant scaling limits: an overview and a collection of problems*, International Congress of Mathematicians. Vol. I, Eur. Math. Soc., Zürich, 2007, pp. 513–543, DOI 10.4171/022-1/20. MR2334202

A. Schrijver, *A comparison of the Delsarte and Lovász bounds*, IEEE Trans. Inform. Theory **25** (1979), no. 4, 425–429, DOI 10.1109/TIT.1979.1056072. MR536232

A. Schrijver, *Combinatorial optimization. Polyhedra and efficiency. Vol. B*, Algorithms and Combinatorics, vol. 24, Springer-Verlag, Berlin, 2003. Matroids, trees, stable sets; Chapters 39–69. MR1956925

K. Schütte and B. L. van der Waerden, *Das Problem der dreizehn Kugeln* (German), Math. Ann. **125** (1953), 325–334, DOI 10.1007/BF01343127. MR0053537

J. T. Schwartz, *Fast probabilistic algorithms for verification of polynomial identities*, J. Assoc. Comput. Mach. **27** (1980), no. 4, 701–717, DOI 10.1145/322217.322225. MR594695

F. Shahrokhi and D. W. Matula, *The maximum concurrent flow problem*, J. Assoc. Comput. Mach. **37** (1990), no. 2, 318–334, DOI 10.1145/77600.77620. MR1072261

L.K Shalm, E. Meyer-Scott, B.G. Christensen, P. Bierhorst, M.A. Wayne, M.J. Stevens, T. Gerrits, S. Glancy, D.R. Hamel, M.S. Allman, K.J. Coakley, S.D. Dyer, C. Hodge, A.E. Lita, V.B. Verma, C. Lambrocco, E. Tortorici, A.L. Migdall, Y. Zhang, D.R. Kumor, W.H. Farr, F. Marsili, M.D. Shaw, J.A. Stern, C. Abellán, W. Amaya, V. Pruneri, T. Jennewein, M.W. Mitchell, P.G. Kwiat, J.C. Bienfang, R.P. Mirin, E. Knill, S.W. Nam: Strong loophole-free test of local realism, *Phy. Rev. Lett.* **115** (2015), 250402.

S. Smirnov, *Critical percolation in the plane: conformal invariance, Cardy's formula, scaling limits* (English, with English and French summaries), C. R. Acad. Sci. Paris Sér. I Math. **333** (2001), no. 3, 239–244, DOI 10.1016/S0764-4442(01)01991-7. MR1851632

S. Smirnov, *Conformal invariance in random cluster models. I. Holomorphic fermions in the Ising model*, Ann. of Math. (2) **172** (2010), no. 2, 1435–1467, DOI 10.4007/annals.2010.172.1441. MR2680496

S. Smirnov, *Discrete complex analysis and probability*, Proceedings of the International Congress of Mathematicians. Volume I, Hindustan Book Agency, New Delhi, 2010, pp. 595–621. MR2827906

D. A. Spielman, *Algorithms, graph theory, and linear equations in Laplacian matrices*, Proceedings of the International Congress of Mathematicians. Volume IV, Hindustan Book Agency, New Delhi, 2010, pp. 2698–2722. MR2827990

D.A. Spielman and S.-H. Teng: Disk Packings and Planar Separators, *12th Annual ACM Symposium on Computational Geometry* (1996a), 349–358.

D. A. Spielman and S.-H. Teng, *Spectral partitioning works: planar graphs and finite element meshes*, 37th Annual Symposium on Foundations of Computer Science (Burlington, VT, 1996), IEEE Comput. Soc. Press, Los Alamitos, CA, 1996, pp. 96–105, DOI 10.1109/SFCS.1996.548468. MR1450607

D. A. Spielman and S.-H. Teng, *Nearly linear time algorithms for preconditioning and solving symmetric, diagonally dominant linear systems*, SIAM J. Matrix Anal. Appl. **35** (2014), no. 3, 835–885, DOI 10.1137/090771430. MR3228466

E. Steinitz: Polyeder und Raumabteilungen, in: *Encyclopädie der Math. Wissenschaften* **3** (Geometrie) 3AB12, 1–139, 1922.

K. Stephenson, *Introduction to circle packing*, Cambridge University Press, Cambridge, 2005. The theory of discrete analytic functions. MR2131318

L. Surányi, *On line critical graphs*, Infinite and finite sets (Colloq., Keszthely, 1973; dedicated to P. Erdős on his 60th birthday), Vol. III, North-Holland, Amsterdam, 1975, pp. 1411–1444. Colloq. Math. Soc. Jańos Bolyai, Vol. 10. MR0401554

M. Szegedy: A note on the θ number of Lovász and the generalized Delsarte bound, *Proc. 35th Ann. IEEE Symp. on Found. Comp. Science* (1994), 36–39.

E. Szemerédi, *On sets of integers containing no k elements in arithmetic progression*, Acta Arith. **27** (1975), 199–245, DOI 10.4064/aa-27-1-199-245. Collection of articles in memory of Juriĭ Vladimirovič Linnik. MR0369312

E. Szemerédi, *Regular partitions of graphs* (English, with French summary), Problèmes combinatoires et théorie des graphes (Colloq. Internat. CNRS, Univ. Orsay, Orsay, 1976), Colloq. Internat. CNRS, vol. 260, CNRS, Paris, 1978, pp. 399–401. MR540024

R. Tamassia, *On embedding a graph in the grid with the minimum number of bends*, SIAM J. Comput. **16** (1987), no. 3, 421–444, DOI 10.1137/0216030. MR889400

R. Tamassia, *Graph drawing*, Handbook of discrete and computational geometry, CRC Press Ser. Discrete Math. Appl., CRC, Boca Raton, FL, 1997, pp. 815–832. MR1730200

R. Tamassia and I. G. Tollis, *Planar grid embedding in linear time*, IEEE Trans. Circuits and Systems **36** (1989), no. 9, 1230–1234, DOI 10.1109/31.34669. MR1009276

T. Tao, *Szemerédi's regularity lemma revisited*, Contrib. Discrete Math. **1** (2006), no. 1, 8–28. MR2212136

W.P. Thurston: *The finite Riemann Mapping Theorem,* lecture at the International Symposium at Purdue University on the occasion of the proof of the Bieberbach Conjecture (1985).

W. Thurston: Groups, tilings and finite state automata, AMS colloquim lectures, Summer 1989.

W. P. Thurston, *Three-dimensional geometry and topology. Vol. 1*, Princeton Mathematical Series, vol. 35, Princeton University Press, Princeton, NJ, 1997. Edited by Silvio Levy. MR1435975

W. T. Tutte, *The factorization of linear graphs*, J. London Math. Soc. **22** (1947), 107–111, DOI 10.1112/jlms/s1-22.2.107. MR0023048

W. T. Tutte, *On the problem of decomposing a graph into n connected factors*, J. London Math. Soc. **36** (1961), 221–230, DOI 10.1112/jlms/s1-36.1.221. MR0140438

W. T. Tutte, *How to draw a graph*, Proc. London Math. Soc. (3) **13** (1963), 743–767, DOI 10.1112/plms/s3-13.1.743. MR0158387

R. J. Vanderbei, *Linear programming*, 2nd ed., International Series in Operations Research & Management Science, vol. 37, Kluwer Academic Publishers, Boston, MA, 2001. Foundations and extensions. MR1845638

V. V. Williams, *Multiplying matrices faster than Coppersmith-Winograd [extended abstract]*, STOC'12—Proceedings of the 2012 ACM Symposium on Theory of Computing, ACM, New York, 2012, pp. 887–898, DOI 10.1145/2213977.2214056. MR2961552

V. V. Vazirani, *Approximation algorithms*, Springer-Verlag, Berlin, 2001. MR1851303

L. Vandenberghe and S. Boyd, *Semidefinite programming*, SIAM Rev. **38** (1996), no. 1, 49–95, DOI 10.1137/1038003. MR1379041

B. L. van der Waerden, *Algebra. Teil I* (German), Sechste Auflage der modernen Algebra. Die Grundlehren der Mathematischen Wissenschaften, Band 33, Springer-Verlag, Berlin, 1964. MR0177027

K. Wagner: Bemerkungen zum Vierfarbenproblem, *Jahresber. der Deutschen Math.-Ver.* **46** (1936), 26–32.

W. Whiteley, *Infinitesimally rigid polyhedra. I. Statics of frameworks*, Trans. Amer. Math. Soc. **285** (1984), no. 2, 431–465, DOI 10.2307/1999446. MR752486

H. Whitney, *On the Abstract Properties of Linear Dependence*, Amer. J. Math. **57** (1935), no. 3, 509–533, DOI 10.2307/2371182. MR1507091

H. Wolkowicz, *Some applications of optimization in matrix theory*, Linear Algebra Appl. **40** (1981), 101–118, DOI 10.1016/0024-3795(81)90143-9. MR629610

H. Wolkowicz, R. Saigal, and L. Vandenberghe (eds.), *Handbook of semidefinite programming*, International Series in Operations Research & Management Science, vol. 27, Kluwer Academic Publishers, Boston, MA, 2000. Theory, algorithms, and applications. MR1778223

D. Zeilberger, *A new approach to the theory of discrete analytic functions*, J. Math. Anal. Appl. **57** (1977), no. 2, 350–367, DOI 10.1016/0022-247X(77)90265-7. MR0450576

D. Zeilberger and H. Dym, *Further properties of discrete analytic functions*, J. Math. Anal. Appl. **58** (1977), no. 2, 405–418, DOI 10.1016/0022-247X(77)90218-9. MR0435417

S.-b. Hu, *The Laplacian eigenvalue of graphs* (English, with English and Chinese summaries), J. Math. (Wuhan) **27** (2007), no. 6, 661–663. MR2367618

R. Zippel, *Probabilistic algorithms for sparse polynomials*, Symbolic and algebraic computation (EUROSAM '79, Internat. Sympos., Marseille, 1979), Lecture Notes in Comput. Sci., vol. 72, Springer, Berlin-New York, 1979, pp. 216–226. MR575692

U. Zwick: Computer assisted proof of optimal approximability results, *Proc. 13th Ann. ACM-SIAM Symp. on Discrete Algorithms* (2002), 496–505.

Author Index

Subject Index

SELECTED PUBLISHED TITLES IN THIS SERIES

65 **László Lovász,** Graphs and Geometry, 2019
64 **Kathrin Bringmann, Amanda Folsom, Ken Ono, and Larry Rolen,** Harmonic Maass Forms and Mock Modular Forms: Theory and Applications, 2017
63 **Cornelia Druţu and Michael Kapovich,** Geometric Group Theory, 2018
62 **Martin Olsson,** Algebraic Spaces and Stacks, 2016
61 **James Arthur,** The Endoscopic Classification of Representations, 2013
60 **László Lovász,** Large Networks and Graph Limits, 2012
59 **Kai Cieliebak and Yakov Eliashberg,** From Stein to Weinstein and Back, 2012
58 **Freydoon Shahidi,** Eisenstein Series and Automorphic L-Functions, 2010
57 **John Friedlander and Henryk Iwaniec,** Opera de Cribro, 2010
56 **Richard Elman, Nikita Karpenko, and Alexander Merkurjev,** The Algebraic and Geometric Theory of Quadratic Forms, 2008
55 **Alain Connes and Matilde Marcolli,** Noncommutative Geometry, Quantum Fields and Motives, 2008
54 **Barry Simon,** Orthogonal Polynomials on the Unit Circle: Part 1: Classical Theory; Part 2: Spectral Theory, 2005
53 **Henryk Iwaniec and Emmanuel Kowalski,** Analytic Number Theory, 2004
52 **Dusa McDuff and Dietmar Salamon,** J-holomorphic Curves and Symplectic Topology, Second Edition, 2012
51 **Alexander Beilinson and Vladimir Drinfeld,** Chiral Algebras, 2004
50 **E. B. Dynkin,** Diffusions, Superdiffusions and Partial Differential Equations, 2002
49 **Vladimir V. Chepyzhov and Mark I. Vishik,** Attractors for Equations of Mathematical Physics, 2002
48 **Yoav Benyamini and Joram Lindenstrauss,** Geometric Nonlinear Functional Analysis, 2000
47 **Yuri I. Manin,** Frobenius Manifolds, Quantum Cohomology, and Moduli Spaces, 1999
46 **J. Bourgain,** Global Solutions of Nonlinear Schrödinger Equations, 1999
45 **Nicholas M. Katz and Peter Sarnak,** Random Matrices, Frobenius Eigenvalues, and Monodromy, 1999
44 **Max-Albert Knus, Alexander Merkurjev, Markus Rost, and Jean-Pierre Tignol,** The Book of Involutions, 1998
43 **Luis A. Caffarelli and Xavier Cabré,** Fully Nonlinear Elliptic Equations, 1995
42 **Victor W. Guillemin and Shlomo Sternberg,** Variations on a Theme by Kepler, 1991
40 **R. H. Bing,** The Geometric Topology of 3-Manifolds, 1983
38 **O. Ore,** Theory of Graphs, 1962
37 **N. Jacobson,** Structure of Rings, 1956
36 **Walter Helbig Gottschalk and Gustav Arnold Hedlund,** Topological Dynamics, 1955
32 **R. L. Wilder,** Topology of Manifolds, 1949
31 **E. Hille and R. S. Phillips,** Functional Analysis and Semi-groups, 1996
30 **Tibor Radó,** Length and Area, 1948
29 **A. Weil,** Foundations of Algebraic Geometry, 1946
28 **G. T. Whyburn,** Analytic Topology, 1942
27 **S. Lefschetz,** Algebraic Topology, 1942
26 **N. Levinson,** Gap and Density Theorems, 1940
25 **Garrett Birkhoff,** Lattice Theory, 1940
24 **A. A. Albert,** Structure of Algebras, 1939

For a complete list of titles in this series, visit the
AMS Bookstore at **www.ams.org/bookstore/collseries/**.